T0226128

Materials for Electrochemical Energy Storage and Conversion II—Batteries, Capacitors and Fuel Cells

MATERIALS RESEARCH SOCIETY
SYMPOSIUM PROCEEDINGS VOLUME 496

Materials for Electrochemical Energy Storage and Conversion II—Batteries, Capacitors and Fuel Cells

Symposium held December 1–5, 1997, Boston, Massachusetts, U.S.A.

EDITORS:

David S. Ginley
National Renewable Energy Laboratory
Golden, Colorado, U.S.A.

Daniel H. Doughty
Sandia National Laboratories
Albuquerque, New Mexico, U.S.A.

Bruno Scrosati
University "La Sapienza" Rome
Rome, Italy

Tsutomu Takamura
Petoca, Ltd.
Tokyo, Japan.

Zhengming (John) Zhang
Celgard LLC, Hoecht Group
Charlotte, North Carolina, U.S.A.

Materials Research Society
Warrendale, Pennsylvania

CAMBRIDGE UNIVERSITY PRESS
Cambridge, New York, Melbourne, Madrid, Cape Town,
Singapore, São Paulo, Delhi, Mexico City

Cambridge University Press
32 Avenue of the Americas, New York NY 10013-2473, USA

Published in the United States of America by Cambridge University Press, New York

www.cambridge.org
Information on this title: www.cambridge.org/9781107413511

Materials Research Society
506 Keystone Drive, Warrendale, PA 15086
http://www.mrs.org

First published 1998
First paperback edition 2013

Single article reprints from this publication are available through
University Microfilms Inc., 300 North Zeeb Road, Ann Arbor, MI 48106

CODEN: MRSPDH

ISBN 978-1-107-41351-1 Paperback

CONTENTS

*Invited Paper

v

*Invited Paper

*Invited Paper

*Invited Paper

*Invited Paper

*Invited Paper

PREFACE

Our energy-hungry world is increasingly relying on new methods to store and convert energy for portable electronics, as well as new, environmentally friendly modes of transportation and electrical energy generation. The availability of advanced materials is linked to the commercial success of improved power sources such as batteries, fuel cells and capacitors with higher specific energy and power, longer cycle life and rapid charge/discharge rates.

The symposium 'Materials for Electrochemical Energy Storage and Conversion II—Batteries, Capacitors and Fuel Cells' was held December 1-5, during the 1997 MRS Fall Meeting in Boston, Massachusetts. The papers were heavily weighted toward lithium batteries, with half of the 140 papers presented discussing various aspects of this rapidly advancing technology. In particular, cathode materials were discussed from the aspect of modeling, new synthesis techniques, and processing technology. *Ab initio* modeling methods, as well as empirical techniques, were applied to the family of transition metal oxides and sulfides that gave increased understanding of phase stability, lithium diffusion and intercalation mechanism. The effect of dopants on improved capacity retention was explained, and degradation mechanism of capacity loss at elevated temperature was elucidated.

Lithium ion rechargeable battery anode materials were also discussed, emphasizing carbon and the new family of tin oxides with very high capacity—up to 1000 mAh/g. Intercalation mechanism and Li bonding sites were the topic of several papers, with new ^{7}Li-NMR techniques identifying both reversible and irreversible lithium binding. Coupled with the semi-empirical modeling calculations on carbon, a fuller understanding of carbon anode materials emerged.

Closely related to Li battery materials, the area of supercapacitors has made significant progress in the development of improved electrodes and electrolytes for the rapid charge/discharge cycles required for these devices. Papers on the use of ruthenate perovskites, metal carbide and metal nitrides illustrate some of the novel materials being investigated. Papers describe innovative approaches leading to successful devices.

Finally, new colloidal deposition techniques and sol-gel processing procedures are described that enable the fabrication of thin solid-oxide fuel cells. These developments decrease the operating temperature, size and weight of fuel cells, making them more cost competitive.

In all, the results presented at this symposium gave a deeper understanding of the relationship between synthesis, properties and performance of power source materials. Substantial progress is evident since the last symposium on this topic (MRS Proceedings Volume 393), which gives us an encouraging outlook that further increases in performance are coming in the future.

This proceedings volume is organized into six sections highlighting: general papers on a wide variety of rechargeable battery technologies; new approaches to modeling of Li batteries; advances in fuel-cell technology; new work on Li battery cathodes; anodes and electrolytes; and work on super-capacitors. We think the volume is an excellent snapshot of the current state of the art in energy storage and conversion technologies, many of which will make a significant impact on society.

David S. Ginley
Daniel H. Doughty
Bruno Scrosati
Tsutomu Takamura
Zhengming (John) Zhang

February, 1998

ACKNOWLEDGMENTS

We wish to sincerely thank all of the speakers, authors, and referees for their contributions to the success of the symposium and of these proceedings.

It is our pleasure to acknowledge with gratitude the financial support provided for the symposium by: National Renewal Energy Laboratory, Sandia National Laboratories and the Materials Research Society. Without this help, we could not have successfully organized the symposium or produced the proceeding volume.

We gratefully acknowledge the invaluable assistance of Mrs. Carole Allman of the National Renewal Energy Laboratory in Golden, Colorado, for her indispensable assistance with the symposium and her diligence and care in assembling this proceedings volume.

Zhengming (John) Zhang acknowledges the great support from SKC America, Inc.

MATERIALS RESEARCH SOCIETY SYMPOSIUM PROCEEDINGS

MATERIALS RESEARCH SOCIETY SYMPOSIUM PROCEEDINGS

Part I

General Section: Rechargeable Batteries

Advances in Battery Technologies and Markets:
Material Science Aspects

A.J. Salkind
College of Engineering, Rutgers, The State University of New Jersey, Piscataway, NJ,
and UMDNJ-Robert Wood Johnson Medical School, 675 Hoes Ln, Piscataway, NJ 08854

Abstract

Advances in materials innovation, availability, processing, and volumetric and gravimetric utilization, have lead to dramatic changes in the technologies and markets for batteries, and other electrochemical storage devices, in recent years. The increased ability to cycle of the new materials and preparations has also resulted in a more dominant proportion of secondary batteries in the market. The growth of the various technology sectors of the battery and energy storage device industry will be compared to the relevant improvements in materials science. The developments needed in materials, composites, and processing techniques is reviewed, in order to match the ongoing changes in electronic circuits, power devices, transportation, and the growth of portable appliances. An analysis of the technical/market segments of the battery industry is presented.

INTRODUCTION

The dramatic increase in many technology segments and the overall battery market, over the last decade, is illustrated by the annual manufacturing data summary shown in Table I. Overall, the international level of production reported at manufacturing companies pricing for 1996 was approximately $32 billion . [However, the data is based on reports from companies that represent less than ½ the population of the world. A realistic estimate of overall production levels would be approximately $40 billion, translated into US currency]. The $33 billion can be compared to estimates of annual world production of $19 billion for 1990 and $30 for 1994. In overall, the battery business has more than doubled over the last decade.

In many cases, this increase in use and production resulted from improvements in materials science which allowed various battery systems to meet more stringent technical requirements of energy density, power density, shelf life, availability in special sizes and configurations, sealed or maintenance free construction, and ability to perform in any gravitational orientation.

The application growth areas have been mainly in the following six segments;
1. CCC [communications, cellular telephones, camcorder]
2. Notebook and palmtop computers
3. Primary cells, alkaline, hearing aid, non-US markets
4. Instruments and tools
5. Small and off-road traction; AGV, electric bikes (esp. China)
6. Traction and EV - non-US

Mat. Res. Soc. Symp. Proc. Vol. 496 © 1998 Materials Research Society

The systems and technologies to meet these growing and new markets fall into four main product types;

Small Rechargeables for Consumer Electronics [Ni-MH, Lithium ion, Rechargeable MnO_2]

Advanced Energy Storage Systems: Advanced Lead-Acid, Ni-MH, Rechargeable Lithium

New Manufacturing Techniques: continuous, high speed, environmentally benign, low labor content

Consumer primary cells, alkaline MnO2 worldwide, zinc-air hearing aid cells

Battery Materials Background

The materials of construction and assembly of batteries encompass a great variety of chemical and physical shapes. They include elements, inorganic and organic compounds, polymers, ceramics, aqueous and non-aqueous solutions, and films, fibers, and mats. There are six main function categories for materials in batteries; negative active material, positive active material, substrates and conductivity enhancers, electrolytes; separator systems, and packaging materials. These are illustrated in Table II. In several cases, use in batteries constitutes the major application for a particular material e.g. lead, cadmium, electrolytic MnO2. In other cases, batteries constitute a very small fraction of the world use of a material and it is difficult to obtain special material, e.g. cellophane, absorber fabrics. The general requirements for each of these classes of components are shown in Table III. In Table IV, the current status of common consumer size secondary cells is compared. In order to appreciate the role of material science in capacity and performance improvements in cells, the example is shown in Fig. 1 of the increase in capacity in AA size Nickel-Cadmium and Nickel-MH cells in the period of 1968-1995. In the same volume, the capacity of Ni-Cd cells improved by a factor of 2.5. This was mainly attributable to new substrate materials and in improvements in the packing density of $Ni(OH)_2$.

The characteristics and improvements in Nickel Hydroxide are a good example of the semiconductor nature of battery materials and pore structure can be modified to provide much improved energy density and performance. Table V illustrates the nickel hydroxide structure and compares the properties of conventional and "spherical" material. The latter is a name coined to reflect the appearance of the high density powder.[1]. Although the chemical structure is the same, the surface areas and pore structures are very different. In essence, the preparation of the spherical material eliminates large, and unnecessary, pores. Further improvements are under study. Reisner et.al. [2] reported that the preparation of nickel hydroxide by nanophase techniques lead to a more catalytic material which can have a higher packing density. Other improvements in nickel electrode active material included the recent development of a uniform foam [3] made by coating polyurethane foam by the carbonyl nickel process. In addition, a variety of nickel powders of high porosity have been developed. These can be used as additives to nickel active material or in some lithium cell systems. Very significant improvements in the performance of Ni-MH cells were recently announced by several groups. In consumer size cells, a new "C" size is reported to be capable of 40 ampere pulses [4]. Electric Vehicle size Ni-MH batteries supplied by at least two manufacturers [5] are now under

Table I - Technology and Market Segments

			values in US Dollars - Billions			
			World	World	World	World
			1990	1992	1994	1996
1.Conventional Primary						
Leclanche' & HD			4.40	4.40	4.40	4.30
Alkaline MnO$_2$			4.10	4.40	4.80	5.20
Lithium			0.50	1.00	1.10	1.13
Metal - Air			0.40	0.60	0.65	0.68
Silver/Other			0.60	0.60	0.50	0.48
Subtotal			~10.0	~11	~11.5	11.80
2.Conventional Secondary						
Lead - Acid SLI			7.70	8.00	8.20	8.50
Lead - Acid Ind			1.80	2.00	2.40	2.70
Lead - Acid Consumer			0.15	0.20	0.40	0.60
Ni-Cd Consumer			1.60	2.00	2.20	2.00
Ni-MH Consumer			0.00	0.20	0.80	1.50
Ni-Cd Industrial			0.40	0.50	0.50	0.60
MnO$_2$ - Zinc			0.00	0.00	0.10	0.20
Other			0.25	0.25	0.25	0.25
Lithium - Ion			0.00		0.18	0.90
Subtotal			11.70	~13	~15	17.30
3.Special Batteries						
Reserve/			0.10	0.15	0.15	0.15
Medical			0.12	0.15	0.15	0.20
Military/Aerospace			0.50	0.50	0.50	0.60
Fuel Cells			0.07	0.10	0.15	0.20
Other			0.30	0.30	0.30	0.30
Subtotal			~1.1	~1	1.10	1.50
OVERALL TOTALS (estimate)			~23	~26	~30	~33
Ratio Secondary/Primary			1.17	1.19	1.3	1.46

Table II- Categories of Battery Materials

a) Negative Active Materials
[anode/fuels]
Hydrogen
Metal Hydrides
Zinc
Cadmium
Iron
Lithium, Lithium ion
Lead
Aluminum
Magnesium
Sodium
Calcium

b) Positive Active Materials
[cathode/oxidants]

Oxygen, Air	$LiMn_2O_4$	SO_2
Iodine	$LiCoO_2$	$SOCl_2$
Bromine	$LiNiO_2$	V_2O_5
Sulfur	Organo-Sulfides	
FeS, FeS_2	MnO_2	TiS_2
CuS	Silver Vanadium Oxide	
PbO_2	$NiCl_2$	CuO
AgCl	$NiOOH/ Ni(OH)_2$	
AgO, Ag_2O	$(CFx)n$	
CuCl	m-DNB	
WO_3	$K_2Cr_2O_7$	

c) Substrates and Conductivty Enhancers
Carbon, Graphite, (powder, fiber)
Nickel: powder, foam, fiber, sinter
    screen, foil
Lead/Lead Alloy: Grid, sheet
Aluminum, Copper, and Silver;
    screen, foil, wire,
Stainless, Special Alloys
Tin, tin oxide

d) Electrolytes

Aqueous	*Non-Aqueous*	*Solid*
Sulfuric	Solvent	LiI
KOH	Propylene Carbonate	β Alumina
NaOH	DiMethyl Carbonate	
NH_4Cl	DiEthyl Carbonate, DiOxolane	
$ZnCl_2$	Glymes	
K_2CO_3(molten)		

Solutes
$LiClO_4, LiAsF_6$
$LiBF_4, LiPF_6$

e) Separators, Absorbers, Membranes
cellophane, grafted polyethylene
microporous PVC, Polypropylene
rubber
fabrics, non-woven and woven
cotton, nylon, polypropylene
paper; wood stock, glass fiber stock
glass mat, ceramics

f) Seals and Packaging Materials
rubber , glass, ceramic
Steel and Stainless
plastic cases; PVC, ABS,
            polysulfone, methacrylate
Seals; glass-metal
        ceramic-metal
        rubber, polymer, asphalt

TABLE III - GENERAL REQUIREMENTS OF BATTERY SYSTEMS COMPONENTS

1) ELECTRODES
Conductivity
Voltage (ΔG) at a reasonable level
high surface morphology
high density (true, bulk, and packed)
low cost, available locally worldwide
Stable in electrolyte with appropriate i_o
Safe and readily disposable

examples: Negatives: Li, Al, Cd, Zn, Pb, Mg, i.e. electron donors
Positives: PbO_2, MnO_2, NiOOH, Bromine, Chlorine, Iodine, HgO, CuCl, AgCl, AgO, Ag_2O, doped PAN, $SOCl_2$, BBFOS, O_2, S, CuS, SO_2, i.e. electron acceptors

2) ELECTROLYTES
Conductivity
interaction with electrodes; stability, i_o, passive film formation
safety, flammability,
cost, and availability
density

examples: H_2SO_4, KOH, NH_4Cl, $ZnCl_2$, Propylene Carbonate with dissolving Li Salt, Conducting solids e.g. LiI

3) ELECTRODE SUPPORTS AND CONDUCTIVITY ENHANCERS
Shape: powder, cast grid, expanded grid, woven grid, sintered powder structures, metallic foams, fibers, fiber mats

Materials: Graphite powders, rods, fibers, metallic plated graphite fibers and fiber mats, Metals; e.g. Nickel, Lead, Stainless Steel, Iron

4) OTHER COMPONENTS
Cases
Separators and spacers
Additives for throwing power (Tafel Curve modifiers), expanders
Connectors and leads
Vents
Gaskets and Seals
Catalytic Surface providers

Table IV - Current State of Performance of Consumer Secondary Cells

	Ni-Cd	Ni-MH	Li-ion
Voltage (V), discharge	1.2-1.0	1.2-1.0	3.7-3.0
Specific Energy (Wh/kg)	60-85	80-100	90-120
Energy Density (Wh/l)	140-180	240-300	300-380
Cost ($/Wh), (OEM)	0.3	0.4	0.6
Life cost ($/cycle)	0.06	0.1	0.08
Cycle life	1000 (with conditioning)	500+	500-1200
Cold performance (at -20 C, % rated)	50	50	40
Self Discharge (%/month)	5-10	10-20	5-10
Fast charge (C rate) Power density (W/l)	4 1000	3 800	2 500
Temperature Range, operational	-40 to +45	-40 to +45	-20 to +50
Service Life (years)	4-8	2-5	??
Charge control	voltage limit, *1.4V or thermal	voltage roll-over or voltage limit, 1.4V	voltage limit, 4.2V single cell control
Discharge control	typically 0.6V, but not critical	typically 0.8V, but not critical	typically 3.0V, is critical

*% of rated

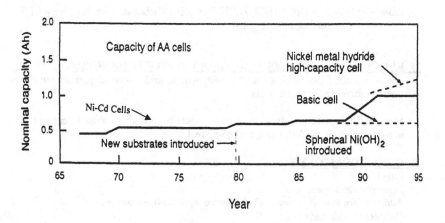

FIGURE 1 - HISTORICAL IMPROVEMENT IN CAPACITY OF "AA" SIZE NICKEL BATTERY SYSTEMS

TABLE V - TYPES AND CHARACTERISTICS OF NICKEL HYDROXIDES

The nickel hydroxide structure:
- is valence 2 in the fully discharged state
- is valence 3 (approximately) in the charge state
- is a layered lattice in which planes of nickel atoms are sandwiched between planes of hydroxyl atoms
- has only a 15% density change between fully charges and fully discharged

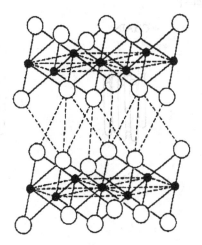

Comparison of Hg porosimetry results and BET surface areas for two types of nickel hydroxide

	Conventional	Spherical
Median pore diameter (volume) (μm)	5.07	4.71
Median pore diameter (area) (μm)	2.11	4.16
Bulk (geometric) density (g ml^{-1})	1.24	1.73
Apparent (skeletal) density (g ml^{-1})	1.69	2.69
Specific surface area (m^{2}g^{-1})	0.24	0.21
BET specific surface area (m^{2}g^{-1})	45.7	17.9

test. Lastly, an autometed new plant for the production of Ni-Cd and Nickel-MH EV batteries has been completed in France.

In rechargeable lithium cell systems the presently utilized lithium insertion structures are shown in Figure 2.

The need for new materials and systems can be illustrated by Figure 4, which shows the current state of implanted biomedical devices. Figure 3, illustrates the power needs of the various battery powered implants. Significant research is being supported in order to achieve high energy and high power needs of the heart assist program. Among many projects, bipolar Ni-MH, and Ni-Zn projects are being supported by the NIH.

Battery manufacturing requires skills in machine development and material handling. This is illustrated in Figure 5, a typical fabrication facility layout for the manufacture of lithium-ion cells.[6]

LAYER   CHAIN   FRAMEWORK

            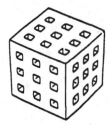

GRAPHITE $NbSe_3$ MnO_2
$LiCoO_2$ WO_3
MoS_2 V_2O_5
FeS_2 V_6O_{13}
                                                                 MoO_2
                                                                 Mo_6S_8
                                                                 $LiMn_2O_4$
                                                                 $LiNiO_2$

FIGURE 2 - LITHIUM INSERTION STRUCTURES

ARTIFICIAL HEART
DEFIBRILLATOR
CARDIOVERTER
NEURO-STIMULATOR   **FIGURE 3 - POWER**
DRUG PUMP   **REQUIREMENTS OF**
TACHYARRYTHMIA CONTROL   **POWERED IMPLANTS**
DUAL CHAMBER PACEMAKER
SINGLE CHAMBER PACEMAKER

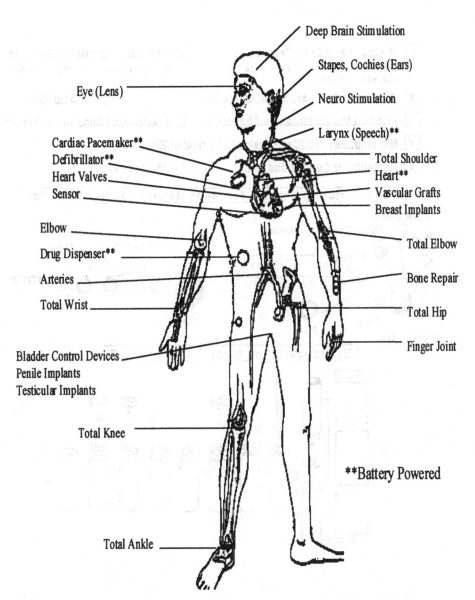

Deep Brain Stimulation

Stapes, Cochles (Ears)

Eye (Lens)

Neuro Stimulation

Larynx (Speech)**

Cardiac Pacemaker**
Defibrillator**
Heart Valves
Sensor

Total Shoulder
Heart**
Vascular Grafts
Breast Implants

Elbow
Drug Dispenser**
Arteries
Total Wrist

Total Elbow

Bone Repair

Total Hip

Finger Joint

Bladder Control Devices
Penile Implants
Testicular Implants

Total Knee

**Battery Powered

Total Ankle

FIGURE 4 - IMPLANTED BIOMEDICAL DEVICES

1 **Mixing of Anode or Cathode**
Example of composition and how to do with what kind of machine.

2 **Coating the Paste on Cu or Al Foil**
Example of preparation of good electrode can be produced.

3 **Roll-Press Process, Slitter Operation**

4 **Welding Tabs**

5 **Winding the Electrode and Separator**

6 **Electrolyte Filling under Vacuum**

7 **Beading, Sealing (Cell Closing) and related Process**

8 **Rectangular Cell Production**

9 **Safety Test**

Manufacturing Process for LI-Ion Battery Positive Electrode

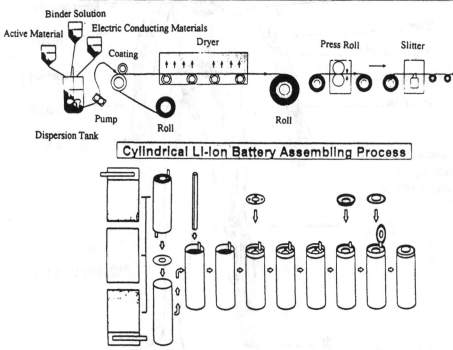

Cylindrical LI-Ion Battery Assembling Process

FIGURE 5 - TYPICAL LITHIUM-ION BATTERY PRODUCTION PROCESS
(Ref. Hohsen Corp. Advertisement)

TABLE VI-MOTIVATIONS FOR NEW BATTERY MATERIALS/ SYSTEMS

ENVIRONNMENT

Electric Vehicles, Toxic Materials Disposal, Renewal energy sources, Batteries for storage and leveling of variable/intermittent power generation, New manufacturing processes and technologies, portable sensors, etc.

ELECTRONIC DEVICE DEVELOPMENTS

Trend to 3V or lower semiconductor logic, fewer cells needed, Flash memories and other novel memory devices, nanodevices involving fewer electrons.

PORTABLE PRODUCTS PROLIFERATION

Laptops, beepers, cellular telephones, power tools, smart cards, recreational items.

MEDICAL NEEDS

Defibrillators, ventricular asist devices, drug delivery systems, etc.

MILITARY AND SPACE NEEDS

Space station, planetary probe, robot, soldier of the future

TABLE VII- TYPICAL EXPERIMENTS IN PROGRESS AND NEW DEVELOPMENTS

COMMERCIAL

- "C" size Ni-MH Cell with brief 40 amp pulse capabilty
- High rate alkaline Zn- MnO_2 primary cell
- High volumetric density Li-ion and Ni-MH cells
- Production plant for EV batteries (France)
- Fuel Cell direct conversion of Natural Gas to Electricity
- High power lead-acid cells
- Rechargeable Zinc-Air Batteries in EV application
- Uniform nickel foam by vapor deposition process

RESEARCH

- Nanostructured Battery active materials [MnO_2 and $Ni(OH)_2$]
- Research pilot line for cellulosic film separators (England)
- Advanced carbons for alkaline and lithium cells
- VRLA improvement studies
- Bipolar nickel-MH batteries
- Advanced Lead-Acid designs with horizontal electrodes.
- MH alloys with lower self-discharge and higher energy density
- Advanced rapid charging techniques
- New electrolytes for lithium cells
- Nickel-Zn batteries reach a 600 cycle to 80% capacity point.
- Bi doped MnO_2 under test for rechargeable alkaline zinc cells.

SUMMARY AND CONCLUSIONS:

1- THE BATTERY BUSINESS IS A LARGE BUT NICHE ORIENTED INDUSTRY.
ABOUT 60% OF THE BATTERIES PRODUCED IN THE INDUSTRIALIZED
AREAS OF THE WORLD ARE FABRICATED BY THE 20 LARGEST
MANUFACTURERS.

2- THE OVERALL GROWTH RATE OF BATTERY PRODUCTION VALUE IS
ABOUT 8%
IT IS MUCH HIGHER IN SPECIAL SEGMENTS SUCH AS
CONSUMER ELECTRONICS, TELEPHONES, COMPUTERS, PORTABLE
TOOLS, SMALL TRACTION VEHICLES.

3- THERE ARE VERY SIGNIFICANT IMPROVEMENTS NEEDED IN MATERIALS
INCLUDING:
Metal hydrides with lower Vapor Pressure and Higher hydrogen storage content
Polymer gel electrolytes (for lithium)
Higher rate inexpensive primary cells
Better separators, absorbers , and membranes
Lithium cell electrolytes stable at higher voltages
Higher valence change nickel hydroxide(greater than 1 valence change)
A safety shuttle mechanism for overcharge of lithium cells
Bipolar packaging materials for lead-acid, Ni-MH, and other systems
Improved VRLA materials and assembly techniques

REFERENCES:

1. M.Wataba, M. Ohnishi, Y. Harada, and M. Oshitani, " Development of a High Density
Pasted Nickel Electrode", Proc. 34th Int. Power Sources Symp. Cherry Hill, NJ (1990)

2. D. Reisner, A. Salkind, P. Strutt, and T.D. Xiao, " Nickel Hydroxide and Other
Nanophase Cathode Materials for Rechargeable Batteries", Proc. 20th Int. Power
Sources Symposium, Brighton, England, (April 1997)

3. V. Ettel (private communication) INCO company, Toronto, Canada (1997)

4. Energizer Powers Systems , Gainsville, Florida, Technical Literature for the
EMP- 2000Cs Cell (1997)

5. Commercial Literature, SAFT Battery Co.,Bordeaux, France; and the Ovonic Battery
Company , Michigan, USA.(1997)

6. Hohsen Corp. Commercial Literature. 10-20-904 Minami Senba, 4-Chome,
Chuo-ku, Osaka 542 , Japan (1997)

POLYACENE (PAS) BATTERIES

SHIZUKUNI YATA*, KAZUYOSHI TANAKA**, AND TOKIO YAMABE**
*Battery Business Promotion, Kanebo, Ltd., 5-90, 1-chome, Tomobuchi-cho, Miyakojima-ku, Osaka 534, Japan
**Department of Molecular Engineering, Graduate School of Engineering, Kyoto University, Yoshida-honmachi, Sakyo-ku, Kyoto 606-01, Japan

Introduction

Two crucial characteristics are expected for batteries. One is large capacity (large energy density) and the other high power. In the last decade, the batteries with large capacity and energy density have been strongly required in the field of portable electronics and communication equipments and, hence, the new batteries such as Ni-MH and Li-ion ones have been developed and commercialized. At present, Li-ion battery has the largest energy density of all the available batteries and its energy density has come up to 300Wh/l, twice as large as that of Ni-Cd one. In general, electric power of batteries signifies the amount of energy put out at unit time. Thus the output power discharged from a battery, storing a certain amount of capacity, in 1hour is ten times as much as that discharged in 10 hours. It is indeed an excellent ability to go on walking all the way with a heavy knapsack on one's back but so is it to output a tremendous power at a moment just as in the weightlifting. It is well known that Ni-Cd battery can output the highest power of all the batteries. It can be discharged at about 10C, namely in about 6 minutes. Large energy density and high output power seem to be similar, but they are essentially different.

There have been known two kinds of systems for storing electricity since early days. One is a condenser (capacitor) and the other a battery. The former can physically store free electrons on the surface of electrodes and the latter can accumulate electrons into the active materials in terms of chemical reaction. The amount of stored electricity is rather small in a condenser using the surface of materials. But it can be charged and discharged at an extremely high rate, as free electrons can move much rapidly. On the other hand, in a battery, electron transfer has to be converted into ion transfer through electrolyte. Therefore, the response becomes slower, and the charging and discharging rate is limited. But it can accumulate much electricity because charges are accommodated into the active materials in the molecular level. Comparing these two systems, it is understood that the common point is merely storing electricity. In fact, they have been used and played an important role in the each field until today.

In 1980s, however, since the research of "synthetic metals" whose flame is composed of organic substances commenced, the concept of "electric conductor" which had been applied only to metal became much more enlarged. Moreover, the role of electrons different from that of electrons used in the chemical bonding and that of the free electrons was clearly recognized with the discovery of conductive polymers. As

15

Mat. Res. Soc. Symp. Proc. Vol. 496 © 1998 Materials Research Society

the result, the new system to store electricity lying between condensers (capacitors) and batteries came out.

Polyacenic materials

Let us use the simplest chemical formulae to define a material. For example, polyethylene and polyacetylene are denoted as, respectively, $(CH_2)_x$ and $(CH)_x$ (Fig. 1). In this style, polyacene and polyacenacene are represented by $(C_2H)_x$ and $(C_3H)_x$, respectively (Fig. 2) [1]. Moreover, as illustrated in Fig. 3, $(C_{12}H)_x$ is a polymer with a wide hexagonal plane of carbon atoms like carbon materials. Polyacene and polyacenacene lying between the polymers and the carbon materials (in Figs. 1 and 3) are generally called one-dimensional graphite or polyacenic materials. The intercalation into carbon materials has been extensively studied. The interlayer distance of graphite, which is a crystalline form of carbon allotrope, is 3.35 Å. It has been known that graphite can be doped with Li up to the C_6Li stage (372mAh/g) with the expansion of the interlayer distance to 3.70 Å (Fig. 4).

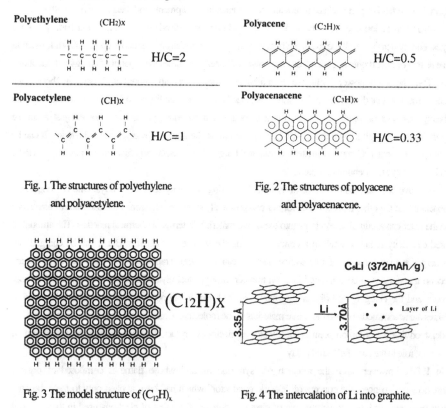

Polyethylene (CH₂)x H/C=2

Polyacene (C₂H)x H/C=0.5

Polyacetylene (CH)x H/C=1

Polyacenacene (C₃H)x H/C=0.33

Fig. 1 The structures of polyethylene and polyacetylene.

Fig. 2 The structures of polyacene and polyacenacene.

$(C_{12}H)x$

C₆Li (372mAh/g)

Li

Layer of Li

Fig. 3 The model structure of $(C_{12}H)_x$

Fig. 4 The intercalation of Li into graphite.

Polyacenic semiconductor (PAS) which we have originally developed, is a kind of polyacenic materials. It is prepared from pyrolytic treatment of phenol-formaldehyde resin at 500-700°C [2,3]. PAS consists of carbon and hydrogen, H/C (the molar ratio of hydrogen to carbon) of typical PAS being 0.2-0.4 (Table I) [3]. X-ray structural analyses have shown that the interlayer distance of PAS is more than 4 Å on average (Fig. 5) [4]. A model structure of molecular-order PAS is shown in Fig 6 [5]. Note that PAS has a rather loose structure that can store a lot of ions smoothly.

The electrical conductivity of the Li-doped PAS becomes larger as a matter of course. In fact, the conductivity of PAS doped to the C_3Li stage is larger than that of the pristine by about five orders of magnitude (Fig. 7) [6]. In the X-ray diffraction measurement, a significant shift in the main peak is not observed even after the 100th doping and undoping cycles (Fig. 8) [6]. This indicates that

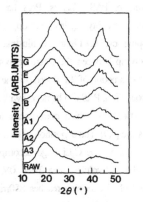

Fig. 5 XRD patterns of PAS.

Table I The elemental analysis and electric conductivity of PAS at room temperature.

Sample	Heat Reaction temperature (°C)	Elemental analysis (in wt. %)		H/C molar ratio	σ (Ω⁻¹cm⁻¹)
A	590	C:93.3	H:2.76	0.36	2×10^{-9}
B	610	C:93.5	H:2.55	0.33	2×10^{-7}
C	670	C:94.7	H:1.80	0.23	2×10^{-2}
D	740	C:94.7	H:1.42	0.18	2×10^{0}

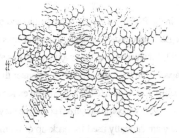

Fig. 6 A model structure of PAS.

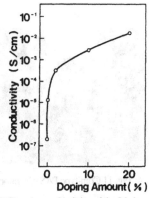

Fig. 7 Electric conductivity of the Li-doped PAS.

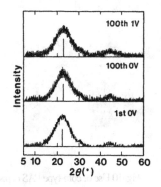

Fig. 8 XRD patterns of the Li-doped PAS.

17

Li doped to PAS causes no degradation of the structure of PAS. That is, PAS has large structural stability against the doping and undoping cycles. Figure 9 illustrates the relationship between the electrode potential and the amount of Li doped to the PAS electrode [7,8]. PAS can be doped with Li up to 1,100mAh/g, which corresponds to the C_2Li stage [6,9]. The deposition of Li metal is observed over the C_2Li stage accompanied with the flat-voltage area near 0V. The amount of Li-undoping reaches 850mAh/g, which exceeds the value of the theoretical capacity for graphite (C_6Li) [7,8].

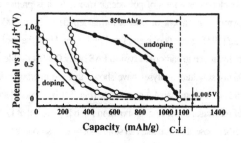

Fig. 9 The relationship between the electrode potential and the Li-doped amount.

PAS capacitors

In the PAS capacitors, anions go into and out from the positive electrode, and cations into and out from the negative electrode. Although the small ions are employed for cations, the bulky ions are such as BF_4^- and ClO_4^- for anions. Thus the amount of anions stored in the PAS positive electrode limits the capacity of the PAS capacitor. In this case, anions do not enter the internal structure of PAS but stored to some surface part. This enables ions to go into and out with an extremely rapid response.

In 1989, we commercialized coin-type capacitors employing PAS for both positive and negative electrodes. They are currently used for back-up of memory in the small-size electronic devices such as pagers, cellular phones, IC memory cards, and so on (Figs. 10 and 11). More recently, we have developed cylindrical-type

Fig. 10 Use of coin-type PAS capacitors.

Fig. 11 Coin-type PAS capacitors.

PAS capacitors and commercialized them in 1997 summer (Fig. 12) [10,11]. Using the PAS material for the both electrodes, they can be power sources with extremely high reliability. For instance, they are operable in environments of –20 to 60℃. They last for 10,000 cycles or more and can be used semi-permanently. The capacity reaches 140F (farad) in 18650-size, being about three times as much as those of other capacitors with carbon materials in the same size. The most attractive characteristics is rapid charging and discharging such that they can be fully charged and discharged in 30 seconds (Fig. 13). In 18650-size, they can be charged and discharged at the current

Fig. 12 Cylindrical PAS capacitors.

over 10A, corresponding to the rate over 100C. This indicates that they can output the highest power among all the available batteries and capacitors at the present time. These PAS capacitors are safe and harmless to environment, because they contain neither inflammable materials such as lithium itself nor heavy metals such as cadmium and mercury and, moreover, can be used semi-permanently.

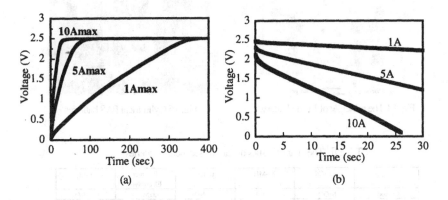

Fig. 13 The characteristics of rapid (a) charging and (b) discharging of cylindrical PAS capacitors (18650-size).

PAS batteries

Energy density

Lithium ion batteries typically necessitate a lithium transition metal oxide, like $LiCoO_2$ or $LiMn_2O_4$, as a positive electrode, carbon materials as a negative electrode, and solutions composed of organic solvents and salts as an electrolyte (Fig. 14). As mentioned above, carbon materials represented by graphite can be doped with Li up to the C_6Li stage. In this state, the amount of Li accumulated per volume is the one third of that in Li metal.

On the other hand, as described in the former section, the interlayer distance of PAS is over 4 Å on average. PAS has more space for Li to be inserted into and can accumulate Li three times as much as graphite, corresponding to the C_2Li stage [6,9]. The PAS in the C_2Li stage is storing as many Li per volume as Li metal but is safer, because the Li doped into PAS is not metallic but rather ionic. We have attempted the fabrication of the cylindrical PAS batteries (18650-size) using $LiCoO_2$ as the positive electrode and PAS as the negative electrode and investigated their fundamental characteristics (Fig. 15). They have an excellent capacity of 2300mAh and their volumetric energy density has reached 450Wh/l (Table II) [8, 12]. In these PAS batteries,

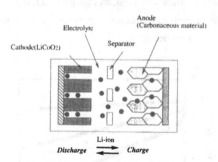

Fig. 14 The principle of Li-ion battery.

Fig. 15 Cylindrical PAS batteries.

Table II The characteristics of various secondary batteries.

	Ni-Cd	Ni-MH	Li-ion	Polymer Electrolyte	PAS
Cathode	NiOOH	NiOOH	LixMyOz	LixMyOz	LixMyOz
Electrolyte	KOH (aq.)	KOH (aq.)	Non-aqueous	Solid or Gel	Non-aqueous
Anode	Cd	Hydrogen Storage Alloy	Carbonaceous Materials	Carbonaceous Materials	Polyacene Semiconductor (PAS)
Average Voltage (V)	1.2	1.2	~3.6	~3.6	~3.2
Energy Density (Wh/l)	~150	~200	250~300	200~300	450
Note				Under Development	Under Development

20

the Li corresponding to 300mAh/g is pre-doped to the PAS anodes before the fabrication. The irreversible Li (Fig. 9) found in the first doping and undoping cycle can be compensated with the pre-doped Li into the PAS negative electrodes, which can decrease loss of the Li amount of the positive electrode.

Safety

The cells used for portable equipments at present are small in size and relatively easy to handle. In the near future, however, the scaling up of the cells will be unavoidable taking account of the use for household electrical products such as a vacuum cleaner and for electric vehicles. When the diameter of a cell is doubled, the calorific value of the cell should be reduced to the one fourth to ensure the equivalent safety.

The capacity of cylindrical PAS batteries (18650-size) is 2300mAh and they store the energy of about 7.4Wh per cell, corresponding to about 6,500 cal. Supposing that the calorific capacity of the cell is roughly 10 cal/℃ and that the stored electric energy is completely converted into heat and stored within the cell, the temperature increase of the cell is estimated to be about 650℃. This is extremely dangerous but, in fact, the short-circuit current is limited because the cell has the internal resistance. Even if the internal short-circuit occurs and the cell generates heat locally, the cell can be protected from the thermal runaway by controlling the internal resistance of the cell and increasing the heat dissipation rate compared with the heat generation rate. Thus the construction of the cell is extremely important to control the short-circuit current.

The most essential factor in the safety of a cell is a self-heat-generation, caused by (i) the reaction between the electrolyte and the negative electrode, (ii) thermal decomposition of the electrolyte, (iii) the reaction between the electrolyte and the positive electrode, and so on [13].

The calorific value accompanied with the reaction between the anode materials and the electrolyte is shown in Fig. 16. Two kinds of graphite (SFG-6, Lonza and MCMB 6-28, Osaka Gas) are doped with Li to 300mAh/g, and PAS is doped to 900mAh/g. In spite of containing more Li, the calorific value of the PAS anode is the smallest, being less than 10 cal/g. This may be caused by that PAS can retain Li more strongly than other carbon materials, although the calorific value also depends on the particle size and shape of the anode materials. Figure 17 shows the calorimetric result of $LiCoO_2$ positive electrode. As more lithium is extracted, the calorific value becomes larger and the safety decreases. As mentioned above, pre-doping of Li has effect of reducing the amount of Li extracted from $LiCoO_2$ and plays an important role in the safety of a cell.

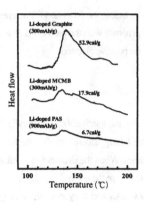

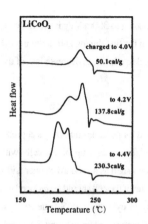

Fig. 16 DSC profiles of the Li-doped
PAS and other carbon materials.

Fig. 17 DSC profiles of LiCoO₂.

Summary

"Ecology" will be one of the key words in the coming next generation. For instance, the present fuel of a car is petroleum but, considering an environmental problem such as noise and exhaust fumes, an electric vehicle that is operated by electricity is eagerly expected. Batteries with larger energy density and higher power are obviously required for the electric vehicle to run at the sufficient speed and for a long way, although lead-acid and Ni-Cd batteries have been investigated for the use of electric vehicles. PAS batteries have the practical energy density suitable for this use. Moreover, they can be expected to guarantee the safety when scaled up.

So far, metal has played the principal role in the cell, whereas carbon and polymer materials have played merely a supporting role. But we have come up to the very starting point of the he new batteries' era utilizing the ability, especially the safe and large Li-storing potential, of PAS.

References

1. T. Yamabe, K. Tanaka, K. Ohzeki, and S. Yata, Solid State Commun., **44**, 823, (1982).

2. S. Yata, U.S. Patent #4,601,849

3. K. Tanaka, K. Ohzeki, T. Yamabe, and S. Yata, Synth. Met., **9**, 41, (1984).

4. K. Tanaka, M. Ueda, T. Koike, T. Yamabe, and S. Yata, Synth. Met., **25**, 265, (1988).

5. S. Yata, Y. Hato, S. Nagura, K. Tanaka, T. Yamabe, Polymer Preprints, Jpn., **44**, 381, (1995) (in Japanese).

6. S. Yata, H. Kinoshita, M. Komori, N. Ando, T. Kashiwamura, T. Harada, K. Tanaka, and T. Yamabe, Synth. Met., **62**, 153, (1994).

7. S. Yata, Y. Hato, H. Kinoshita, N. Ando, A. Anekawa, and T. Hashimoto, K. Tanaka, and T. Yamabe, Abs. of 34th Battery Symp. Jpn., (1994), p. 59-60 (in Japanese).

8. S. Yata, Y. Hato, H. Kinoshita, N. Ando, A. Anekawa, T. Hashimoto, M. Yamaguchi, K. Tanaka, and T. Yamabe, Synth. Met., **73**, 273, (1995).

9. S. Yata, H. Kinoshita, M. Komori, N. Ando, T. Kashiwamura, T. Harada, K. Tanaka, and T. Yamabe, Proceeding of the Symposium on New Sealed Rechargeable Batteries and Supercapacitors, edited by B.M. Barnett, E. Dowgiallo, G. Halpert, Y. Matsuda, and Z. Takehara, (Electrochem. Soc., Inc., Pennington, NJ, 1993), p. 502-512.

10. S. Yata, E. Okamoto, H. Satake, H. Kubota, M. Fujii, T. Taguchi, and H. Kinoshita, J. Power Sources, **60**, 207, (1996).

11. S. Yata, E. Okamoto, H. Satake, H. Kubota, M. Fujii, T. Taguchi, and H. Kinoshita, in Proceedings of the Symp. on Electrochemical Capatitors II, (The Electrochem. Soc., Inc., Pennington, NJ, 1996), p. 208-219.

12. S. Yata, Y. Hato, H. Kinoshita, N. Ando, and M. Yamaguchi, Abs. of 34th Battery Symp. Jpn., (1994), p. 61-62 (in Japanese).

13. J. Yamaki, in The Latest Technologies of The New Secondary Battery Materials, (CMC, Japan, 1997), p. 229-248 (in Japanese).

MECHANISMS CAUSING CAPACITY LOSS ON LONG TERM STORAGE IN NiMH SYSTEM

D. Singh, T. Wu, M. Wendling, P. Bendale, J. Ware , D. Ritter, L. Zhang

Energizer Power Systems, P.O Box 147114, Gainesville , Florida-32614-7114

ABSTRACT

Capacity recovery after long term storage and loaded storage is a critical issue with the NiMH system since its inception. A measurable loss in capacity is observed when cells are stored for long periods of time or discharged deeply to zero volts. The different mechanisms that are known to cause self discharge and capacity loss after storage and loaded storage will be the focus of this paper. Capacity loss after long term storage involves two main events. One is self discharge which causes the open circuit voltage(OCV) of the cell to drop. Self discharge is caused by decomposition of NiOOH, migration of metal ions and possible degradation of separator. Self discharge can be prevented by using separators which are stable at high temperatures and pH and have good ion trapping capability. Various separator types and treatments can play an important role in inhibiting metal ions from migrating thus reducing self discharge. Self discharge during storage causes a severe suppression in the voltage of the foam positive electrode. This drop in voltage causes a breakdown of the cobalt conductive network in the nickel positive electrode. Reduction of high valence cobalt(III) which forms the electrode's conductive network takes place at these low voltages. A permanent breakdown in the conductive network results in low efficiency of the cell on consecutive charge and discharge cycles. In addition, the cobalt in its lower valence states can migrate away from the electrode into the separator causing shorts. These events effect the charge and discharge efficiency of these cells thereby resulting in capacity loss. Various mechanisms causing self discharge which affect capacity recovery after long term storage and loaded storage are discussed in this paper.

INTRODUCTION

A loss in capacity , approximately 15% , is observed in NiMH cell with foam positive electrodes when they are stored for long periods of time in factory warehouses, store shelves or with customers in applications. This effect is more evident at higher storage temperatures. A similar loss in capacity is also observed when cells are discharged to very low voltages. For example, if a consumer forgets to shut off his cellular phone for days, the cells are deeply discharged to 0 volts. When these cells are charged and discharged again a measurable capacity loss is observed.

Figure 1 represents the cell capacity recovery as a function of the storage time at 40° C ambient temperature for foam positive/ MH negative cells. Cell open circuit voltage is also shown in the figure. The decline of cell open circuit voltage due to normal self discharge is one of the indicators that predicts a loss in capacity recovery depending on the electrode formulation. Even though it is an indicator it is not always accurate in predicting a loss in

Mat. Res. Soc. Symp. Proc. Vol. 496 © 1998 Materials Research Society

capacity recovery. The drop in open circuit voltage is greater at higher storage temperatures and time

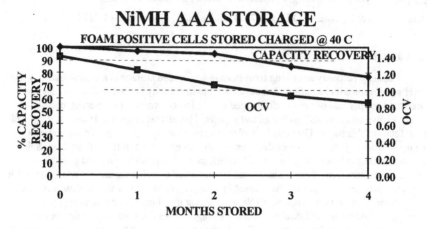

Figure 1: Capacity recovery after cells are stored at 40 deg C for 4 months.

EXPERIMENTAL

NiMH cells were made using the cylindrical jelly roll construction with foam nickel hydroxide positives and metal hydride negatives. Two sets of experiments were conducted. One, long term storage test , in which these cells were cycled ten times at C- rate charge and discharge to obtain an average capacity before storage. The cells were then stored in an oven at 40 deg C and the open circuit voltage and capacities were monitored every two weeks. These cells were also taken out at intervals of 1 month and analyzed for ion migration , separator degradation and loss in conductivity using techniques such as SEM, XRD, ICP and electrochemical testing. The capacities before and after storage were compared to determine capacity loss.

Secondly, these cells were tested for tolerance to loaded storage. In this test the cells were cycled ten times using C rate charge and discharge and then connected to a 5 ohm resistor for 1 wk. The cell voltage was forcibly driven down to 0 volts by using this procedure. After one week these cells were cycled ten more times using C rate charge and discharge and the capacities were compared before and after short down. Electrochemical tests were conducted on these cells to determine the cause for loss in capacity.

Cells with additional conductor, namely Ni210 were made and tested using the long term storage and the loaded storage regime. This was done to confirm our hypotheses and find solutions to reducing capacity loss after long term and loaded storage.

RESULTS

Causes For Capacity Loss

Loss in capacity after long term storage involves two main events. One is the self discharge mechanism which causes the open circuit voltage and charge retention of cells to drop and the second is the effect of low voltage on cell components which cause a loss in capacity. Individual voltages of the cell, positive and negative electrode are represented in figure 2. The voltage of the negative electrode is reversed for illustration and comparison with the positive electrode voltage. Higher voltage loss was observed in the positive electrode than the negative electrode after one month storage.

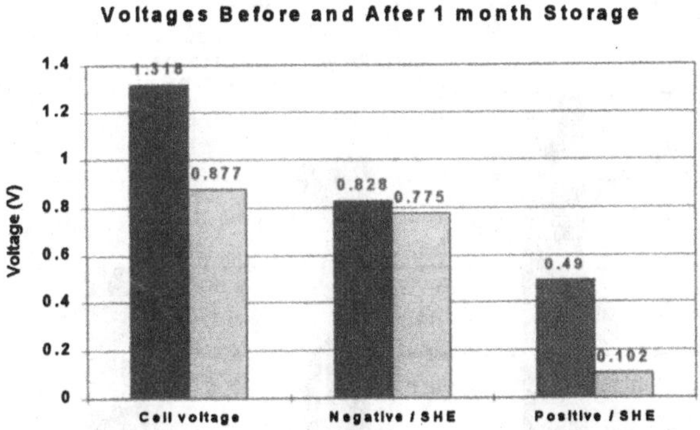

Voltages Before and After 1 month Storage

Figure 2: Voltage of the cell, positive and negative before and after 1 month storage.

Mechanism I : Self Discharge

Three main mechanisms cause self discharge in NiMH cells. First is the reduction of the active material in the positive electrode ,i.e., NiOOH. Second is the migration of metal ions from the positive and negative electrode such as Co, Mn, Al etc. which can cause shuttles between the two electrodes thus causing soft shorts and self discharge. Third is the degradation of the separartor at high temperatures and pH. The degradation products can cause shuttles and thus result in higher rates of self discharge.

1. NiOOH Reduction At the Positive Electrode

During self discharge NiOOH is discharged to $Ni(OH)_2$. The charged nickel electrode is thermodynamically unstable, and thus will slowly decompose giving off oxygen gas as the material reverts back to $Ni(OH)_2$. This reaction is shown in the equation below. Rapid self discharge is observed when the state of charge of the electrode is high . This is illustrated in figure 3.

$$2 NiOOH + H_2O \implies 2Ni(OH)_2 + 1/2\ O_2 \tag{1}$$

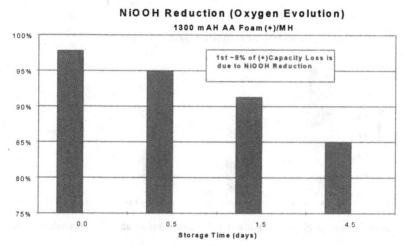

Figure 3: % Capacity loss due to NiOOH reduction.

Other reactions leading to reduction of NiOOH are the direct reaction of NiOOH with hydrogen and electrocatalytic hydrogen oxidation. These reactions are discussed below.

Direct Reaction of NiOOH With H_2:

Hydrogen from the metal hydride negative is known to diffuse and discharge the positive electrode as shown in the equation below. This mechanism is limited by the rate of hydrogen diffusion.

$$NiOOH + 1/2\ H_2 \implies Ni(OH)_2 \tag{2}$$

Electrocatalytic H_2 oxidation: The oxidation of dissolved hydrogen on the nickel electrode surface or the nickel metal substrate will simultaneously reduce NiOOH to $Ni(OH)_2$. This is illustrated in equations (3) to (5).

$$H_2(g) ==> H_2(M) ==> 2H\bullet(M) \tag{3}$$

$$2\ H\bullet\ (M) =====> 2H^+ + 2e^- \tag{4}$$

$$2\ NiOOH + 2H^+ + 2e^- =====> 2Ni(OH)_2 \tag{5}$$

2. Self Discharge Due To Migration Of Metal Ions

Migration Of Metal Ions From the Positive:

Capacity loss is also known to occur due to migration of cobalt from the positive electrode. Cobalt metal or its oxides are added to the $Ni(OH)_2$ foam electrodes to impart conductivity. Cobalt and its oxides are added in the (+2) oxidation state which is readily soluble in alkaline KOH which is used as an electrolyte in these cells. A formation process[2] involving electrochemical charge and discharge is performed on these cells to convert the cobalt(II) state to CoOOH (III) which forms the highly conductive network in the positive electrode. CoOOH is a very stable compound and is insoluble in KOH. Due to the solubility of the cobalt(II) compounds in KOH, any $HCoO_2^-$ ions which escape positive electrode can migrate through the separator to the negative electrode causing soft shorts. Soft shorts can severely supresses the cell voltage.

Metal Ions From The Negative:

Metal ions such as Co, La, Mn, Al have been also observed in the separator due to corrosion of the negative electrode. They can cause soft shorts or reach the positive electrode resulting in reduction of NiOOH. Etching of the electrode to remove the soluble ions on the surface has shown to improve the charge retention of these cells. This is illustrated in the figure 4 below.

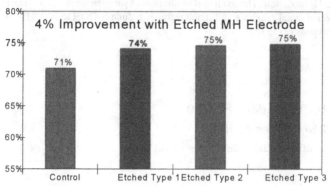

Figure 4: The amount of charge retained (%) as a function of MH electrode treatment. A 4% improvement in charge retention was observed by etching the negative electrode.

3. Self Discharge Due To Degradation Of Separator

Effect Of Storage Time And Temperature :

The stability of separator at high temperatures in the chemical environment has a significant effect on charge retention. The self discharge of the foam positive/MH cells was studied at different temperatures and times. Figures 5 and 6 below shows the charge retention for cells built with polyamide and polypropylene separator at three different temperatures, 25, 40 & 60 deg C. As seen from theses curves, the charge retention decreases with increase of the storage time and gets worse when cells are stored at elevated temperature. The use of polypropylene separator improves the charge retention significantly.

It is apparent that of polypropylene separator significantly improves the charge retention. The improvement is more significant at elevated temperature. This is believed to be due to the instability of the polyamide separator especially at elevated temperature in a caustic environment. Polyamide separator decomposes to form nitrate ions which are responsible for a shuttle reaction between the positive and negative. The following reaction illustrates the oxidation and hydrolysis of the polyamide separator in alkaline solution at high temperatures:

$$NH_3 + 6\ NiOOH + H_2O + OH^- ====> 6Ni(OH)_2 + NO_2^- \qquad (6)$$

$$NO_2^- + MH_x =========> MH_{x-6} + NH_3 + H_2O + OH^- \qquad (7)$$

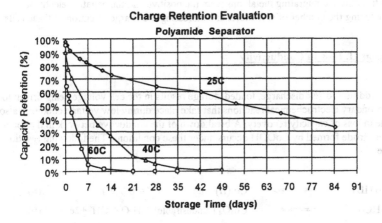

Figure 5 : Charge Retention for foam(+)/MH (-) with polyamide separator cells at 25, 40 & 60 C

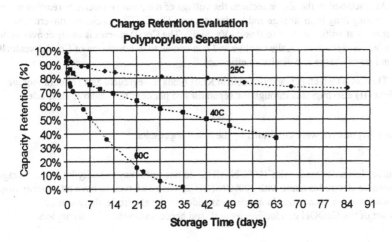

Figure 6 : Charge Retention for foam(+)/MH (-) with polypropylene separator cells at 25, 40 & 60 C

Surface modified polyolefin separators can be used to mitigate these reactions. Both acrylic (-COOH) and sulfonated (-SO₃H) groups exhibit high ion exchange capacity. This results in trapping the migrating metal ions from the positive and the negative electrode thereby reducing the number of soft shorts and increasing the charge retention of these cells.

Mechanisms II. Loss In Conductivity

In addition to self discharge, loaded storage results in the cell voltage being driven to 0 V. The secondary electrochemical reactions that take place during low voltage condition also play a role in loss of capacity recovery. A breakdown of the conductive network in the foam positive electrode formed by CoOOH occurs under these conditions. The following reactions take place:

$$Co\ (III)\ =====>\ Co\ (II)\ @\ \ \ (<=0.0\ V\ ) \tag{8}$$

$$3\ HCoO_2^-\ =====>\ Co_3O_4\ (\ undesirable)\ +\ H_2O\ +\ OH^-\ +2e^- \tag{9}$$

$$3\ HCoO_2^-\ +\ 1/2\ O_2\ ====>\ Co_3O_4\ +\ 3OH^- \tag{10}$$

As discussed in the above sections the voltage of the positive electrode remains at low voltages during long term storage and loaded storage. The Co(III) species is converted to Co(II) species at voltages close to 0 volts Vs. SHE. The Co(II) species is easily converted to Co_3O_4 which is electrochemically inactive and non conductive. Formation of Co_3O_4 eventually results in loss of charge and discharge efficiency.

The Co(II) which is readily soluble in KOH tends to segregate as a result of storage. The cobalt (II) also migrates through the separator and plates at the negative electrode.

The above hypotheses were confirmed by the following results:

1. Positive electrodes made with 100% Ni(OH)2 do not suffer any capacity loss after storage: Foam positive electrodes made with 100% Ni(OH)2) were observed to have no capacity loss after loaded storage. Since no Co species was added to the electrode, there was no breakdown of the CoOOH conductive network and hence, exhibited no capacity loss.

2. Cells become more rate dependent due to loss in conductivity after storage: Due to loss of conductivity in the positive electrode the capacity obtained at higher rates, such as C-rate, was much lower than before storage. Some portion of the lost capacity after storage was recovered at lower discharge rates(C/5). Table I below shows the capacity change before and after storage at C and C/5 discharge rates. As shown, 164 mAH of the C-rate capacity is lost after storage. However, about 40% of the lost capacity can be recovered at C/5 rate.

Table I: Rate Dependence Due To Loss in Conductivity In The Foam Positive

	C-rate Capacity (Ah)	C/5-rate Capacity (Ah)
Before Storage	1.192	0.038
After Storage	1.028	0.100

Possible Solutions To Reduce Capacity Loss

1. Increase conductivity by using high density foam substrate: Higher density (110 PPI) foam substrate was used to compare with a lower density (80 PPI) foam. The capacity recovery after storage was improved from 85% to 89% as shown in table II. The higher density foam is more conductive than the 80 PPI foam material.

Table II: Improve Conductivity By Using High Density Ni Foam Substrate

Foam Substrate	% Capacity Recovery
80 PPI	85
110 PPI	89

2. Addition Of Stable Conductive Powders Improves Capacity Recovery : Ni 210 which is has a very fine morphology and high surface area was added to the positive electrode at various levels, in addition to cobalt, in an attempt to improve conductivity. Ni 210 is very stable under high pH and storage environment which results in breakdown of the cobalt conductive network. The capacity recovery obtained after a loaded storage test is summarized in the figure 7 below. Cells were deeply discharged for 1 week using a 5 ohm resistor to 0 volts. The capacity of these cells were measured before and after loaded storage. A significant improvement was observed when additional conductors were added to the positive electrode formulation.

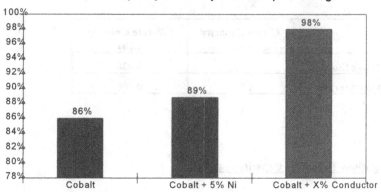

Figure 7 : Addition of stable conductors such as Ni210 can improve capacity recovery after long term storage and loaded storage

Effect Of Loss Of Conductivity On Cell Performance

It was determined that loss of the positive electrode's conductivity effects the capacity recovery in two ways. One is loss in charge acceptance and the other is loss in discharge efficiency. Figures 8 and 9 below show the charge acceptance and discharge efficiency for foam positive electrode without any additional conductors before and after loaded storage. The cells were charged at a charge rate of C/5 and discharged at C rate. As shown in the figures, a 7% reduction of the charge acceptance was observed after loaded storage. A 5% discharge efficiency loss was observed at C-rate. A similar test was performed on cells with additional conductors such as Ni 210. No loss in charge acceptance was observed and only a 2% loss in discharge efficiency was observed.

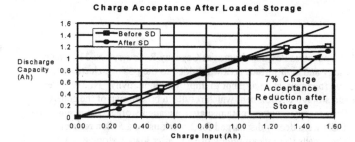

Figure 8 : Charge acceptance before and after loaded storage. A 7% loss is observed in charge acceptance after loaded storage.

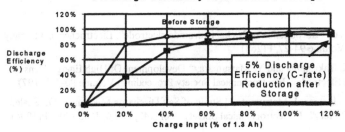

Discharge Efficiency After Loaded Storage

Figure 9: Discharge efficiency at C-Rate before and after loaded storage. A 5% loss is observed in discharge efficiency after loaded storage.

CONCLUSIONS

Capacity loss after long term storage is caused due to two mechanisms. First, normal self discharge results in cell voltage suppression. The foam positive electrode experiences a greater loss in voltage than the MH negative electrode. Secondly, at these low voltages a break down of the cobalt conductive network occurs in the foam positive electrode. This loss in conductivity affects the cell performance by reducing the charge and discharge efficiency of the positive electrode. Loss in conductivity is also the primary mechanism for capacity loss after loaded storage.

Capacity loss can be minimized by two ways. First, by reducing the rate of self discharge and secondly, by adding conductors that are stable to the positive electrode. Self discharge can be reduced by lowering migration of metal ions such as cobalt, aluminum etc. from both the positive and negative electrodes. In addition, separator materials with high temperature stability should be used . By doing this one can minimize harmful contaminants which come from degradation of the separator. Further, separators with good ion trapping ability should be used to prevent ions from migrating to either electrode. This will not only help in reducing soft shorts but also decreases the rate of self discharge.

Conductivity of the foam positive should be improved to obtained better capacity recovery after long term and loaded storage. Several ways of improving conductivity were examined. It was found that by increasing the density of the Ni foam substrate and by adding conductors which are stable at low voltages and high pH the capacity recovery can be significantly improved.

REFERENCES

1. A.H Zimmerman, "Introduction to Nickel hydroxide electrode", Electrochemical Society Proceedings, Volume 94-27, P(268).

2. M.Oshitani, K. Takashima and Y. Matsumara, " Development Of High Energy Density Pasted Nickel Electrode", Electrochemical Society Proceedings Volume 90-4, P(197)

3. M. Ikoma, Y. Hoshina, I. Matsumoto, " Study of Self Discharge Mechanism Of Sealed Type NiMH Battery", J. Electrochemical Society, Vol. 143, No. 6, June 1996, P(1904).

BENEFITS OF RAPID SOLIDIFICATION PROCESSING OF MODIFIED LaNi5 ALLOYS BY HIGH PRESSURE GAS ATOMIZATION FOR BATTERY APPLICATIONS

I. E. ANDERSON*, V. K. PECHARSKY*, J. TING*, C. WITHAM**, and R. C. BOWMAN**
*Ames Laboratory (USDOE), Ames, IA 50011-3020, andersoni@ameslab.gov
**California Institute of Technology, Pasadena, CA 91109

ABSTRACT

A high pressure gas atomization approach to rapid solidification has been employed to investigate simplified processing of Sn modified LaNi5 powders that can be used for advanced Ni/metal hydride (Ni/MH) batteries. The current industrial practice involves casting large ingots followed by annealing and grinding and utilizes a complex and costly alloy design. This investigation is an attempt to produce powders for battery cathode fabrication that can be used in an as-atomized condition without annealing or grinding. Both Ar and He atomization gas were tried to investigate rapid solidification effects. Sn alloy additions were tested to promote subambient pressure absorption/desorption of hydrogen at ambient temperature. The resulting fine, spherical powders were subject to microstructural analysis, hydrogen gas cycling, and annealing experiments to evaluate suitability for Ni/MH battery applications. The results demonstrate that a brief anneal is required to homogenize the as-solidified microstructure of both Ar and He atomized powders and to achieve a suitable hydrogen absorption behavior. The Sn addition also appears to suppress cracking during hydrogen gas phase cycling in particles smaller than about 25μm. These results suggest that direct powder processing of a $LaNi_{5-x}Sn_x$ alloy has potential application in rechargeable Ni/MH batteries.

INTRODUCTION

Nickel/metal hydride (Ni/MH) rechargeable batteries, especially small, sealed battery packs for consumer electronics, electrical power tools and appliances are steadily replacing Ni/Cd batteries because of their extended cycling life, absence of memory effects, enhanced energy storage capacity, and reduced environmental hazard. Recent research has focused also on replacement of Pb/acid automotive batteries with large, sealed cell Ni/MH batteries for all of the above reasons, as well as a weight reduction of at least 25% for double the storage capacity [1]. However, rapid growth in Ni/MH battery applications is hampered by the need for significant improvements in battery electrode material design, processing, and cell fabrication technologies to boost Ni/MH cell performance at extended cycling lifetimes and to reduce cell manufacturing cost, which depends critically on consistent high quality metal hydride particulate.

Negative electrodes in Ni/MH rechargeable batteries could be improved significantly if fabricated from metal alloy particulate, which conforms to a following requirements: 1) large hydrogen storage capacity; 2) full capacity hydrogen absorption-desorption between −25°C and +50°C below ambient pressure; 3) large initial surface area; 4) fracture resistance during battery cycling; 5) chemical stability in basic solutions; 6) suitability for high volume production; and 7) low cost. The scientific and patent literature contains an extensive number of reports regarding different aspects of the crystal chemistry and hydrogen storage properties of numerous materials, concluding that rare earth intermetallic alloys [2] are preferred for cathode fabrication. However, Ni/MH battery technology is still lacking basic knowledge of how to enhance electrode performance and lower processing complexity utilizing advanced processing methods with the existing hydrogen storage alloys. Perhaps more importantly, basic knowledge is lacking on design of improved hydrogen storage alloys with optimum performance and extended cycling

37

stability, based on new, less complicated and more affordable compositions that are adaptable to improvements in processing and high volume manufacturing.

Both AB_5 (LaNi$_5$ prototype) and AB_2 (TiNi$_2$ prototype) alloy compounds have been investigated for Ni/MH battery cathode material applications. Both alloys have about the same average hydrogen storage capacity of about 1.5% by weight, with the AB_2 alloys having a slightly larger theoretical maximum hydrogen storage capacity, 2.0 wt.%, compared to 1.6 wt.% for AB_5 alloys. However, to exploit the greater capacity (by weight) of the AB_2 alloys in an actual application would require a Ni/MH battery of a larger physical size, not desirable, for example, in a compact electric vehicle. This paradox arises because of the higher density for AB_5 alloys, about 8-8.5g/cm^3, compared to alloys of the AB_2 alloy type, about 5-7g/cm^3. Thus, a similar size Ni/MH battery of either AB_5 or AB_2 would have a similar storage capacity.

With such similar potential for practical battery application, the decision as to which compound type to pursue in this study for alloy simplification and reduction in processing complexity was determined from physical metallurgy principles. Commercial AB_5 and AB_2 alloys are quite complex, typically containing 5 to 9 elements, and both utilize some relatively expensive elements, especially Co. However, from a processing perspective, the microstructure that has been established as optimum for AB_5 cathode material performance is a homogenous, equilibrium phase [2]; compared to the optimum microstructure for AB_2 which is a highly disordered multiphase mixture of amorphous and crystalline regions [3]. For this study AB_5 alloys were chosen, because of the potential for development of a simplified processing approach without need for precise control of a metastable microstructure decomposition reaction.

The standard route for manufacturing the AB_5 battery materials includes the following initial steps: 1) melting and chill casting of large ingots; 2) extensive heat treatment of the ingots to eliminate microscopic compositional inhomogeneities, and; 3) grinding of the annealed ingots into fine powders, which may include hydriding and dehydriding to fracture large ingots into smaller pieces. Rapid solidification processing (RSP) of LaNi$_5$ powders by high pressure gas atomization (HPGA) has been suggested [4] as a means to suppress the Ni segregation effect observed during conventional chill casting of LaNi$_5$. If fully successful, HPGA could produce useful AB_5 powders directly, eliminating the homogenization anneal and the grinding steps. In other words, one of the key barriers to widespread commercialization of Ni/MH batteries for vehicle applications [1], processing cost, could be attacked by the ability to combine the above mentioned 3 steps of cathode particulate fabrication into one step by the HPGA approach.

This paper will focus on the results of our HPGA processing research on a simple AB_5 alloy, LaNi$_{5-x}$Sn$_x$, where Sn can substitutionally alloy for the Ni [5]. From previous work [5] on annealed ingot samples, the Sn substitutional addition is known to retain much of the excellent capacity for hydrogen storage of LaNi$_5$ (6 H-atoms per LaNi$_5$ formula unit) while lowering the H$_2$ absorption/desorption equilibrium pressure below atmospheric, practical for vehicle battery applications. Two levels of the Sn addition were tried in this study to reduce the H$_2$ absorption pressure and two atomization gases, Ar and He, were tried to test the influence of the external cooling rate on the rapid solidification effect in these alloys.

EXPERIMENT

The two alloys, LaNi$_{4.75}$Sn$_{0.25}$ and LaNi$_{4.85}$Sn$_{0.15}$, were selected based on their H$_2$ absorption pressure isotherms in fully homogenized alloy particles [5]. The latter alloy produced a slightly subambient absorption isotherm at ambient temperature of about 0.9 atm and the higher Sn level suppressed the isotherm to about 0.5 atm [5]. Alloy components had purity levels of at least 99.9% for La, 99.994% for Ni, and 99.99% for Sn. The alloys were prepared as a bottom poured chill cast ingots using a high purity Al$_2$O$_3$ crucible and a pouring temperature of 1550°C.

A high pressure gas atomization (HPGA) process [4] was used to produce powders of the cast ingot of each alloy. Argon gas (99.98% purity) was supplied at a pressure of 7.6Mpa (1100 psi) to the HPGA nozzle to atomize the LaNi$_{4.75}$Sn$_{0.25}$ melt, poured at a temperature of 1630°C from an Al_2O_3 crucible. Helium gas (99.98% purity) was supplied at a pressure of 5.5Mpa (800 psi) to the HPGA nozzle to atomize the $LaNi_{4.85}Sn_{0.15}$ melt, poured at a temperature of 1750°C from an Al_2O_3 crucible. A passivating surface film [4] was formed on the particles by injection of reaction gases at an intermediate point during atomized spray freefall within the spray chamber. Each powder yield was screened in ambient atmosphere with wire mesh sieves into size classes of dia. < 25 μm, 25 μm to 38 μm, and 38 μm to 55 μm for further analysis. Some of the dia. < 25 μm powder from each powder batch was sonic-sifted with an electro-formed Ni foil sieve in an inert atmosphere to produce a dia. < 10 μm sample. The particle size distribution of each powder batch was analyzed with an automated laser light scattering instrument.

Chemical composition of cast ingot and powder samples were analyzed for La, Ni, and Sn by inductively coupled plasma-atomic emission spectroscopy and by inert gas fusion for C, O, and N. X-ray diffraction (XRD) measurements with Cu-K_α radiation were performed in as-atomized and annealed powder samples. Following metallographic preparation of powder cross-section samples, SEM was used to examine the solidification microstructure of as-atomized powders as a function of size class. Electron microprobe analysis provided semi-quantitative analysis of solidification segregation in powder cross-section samples following calibration with a well annealed cast sample of the initial alloy. The H_2 absorption-desorption data were obtained on as-atomized and annealed powders with an all-metal Sieverts' gas-volumetric apparatus.

RESULTS

Several previous publications of our investigations [4,6,7] have reported the essential characteristics of $LaNi_5$-type powders produced by HPGA. The powders are generally spherical in shape [4,6], have a low oxygen content [7], e.g. 570 ppmw for dia. < 10um powders of $LaNi_{4.75}Sn_{0.25}$, and have a fine cellular solidification microstructure [6,7] that results from rapid solidification. The cellular solidification microstructure of as-atomized $LaNi_{4.75}Sn_{0.25}$ powders consists of a cell interior phase that is depleted in Sn and a cell boundary phase that is enriched in Sn [7], compared to the nominal composition. Because both cell interior and cell boundary phases are different compositions of the same crystal structure, we refer to this phenomenon as "phase" segregation. In addition, we found [4] that unannealed, stoichiometric $LaNi_5$ powders produced by HPGA have an x-ray diffraction pattern indicating a single phase $CaCu_5$-type structure. Their hydrogen absorption/desorption behavior and total hydrogen capacity in gas phase cycling experiments are identical to well-annealed $LaNi_5$ powders produced from a crushed ingot. These desirable hydrogen storage characteristics of HPGA powders probably are linked most closely to structural and chemical purity and minimal surface oxidation and to suppression of Ni segregation in the as-atomized powders. Our studies [6] also demonstrated that hydrogen gas phase cycling induced cracking of the unannealed, stoichiometric $LaNi_5$ powders in all but the finest sizes of the HPGA powders, less than about 5um. Initial experiments on the Sn addition [6] have demonstrated a reduction in particle cracking from dilation stresses [8] during hydrogen cycling and a depression of the hydrogen absorption equilibrium pressure at ambient temperature to a suitable subambient pressure. As a direct result of this work [6] and other previous studies [5], the current investigation further explores substitutional addition of Sn to stoichiometric $LaNi_5$, producing $LaNi_{5-x}Sn_x$ alloys.

The reason for choosing He as an alternative to Ar for the atomization gas is related to the well-known increase in the melt cooling rate during He gas atomization. In other words, an enhanced solidification rate due to the increased quenching capability of He atomization gas may suppress phase segregation in the HPGA powders of $LaNi_{5-x}Sn_x$ alloys. However, the first benefit that was noted from production of $LaNi_{4.85}Sn_{0.15}$ alloy powder by use of He atomization gas was an increased yield of extremely fine powders, compared to the size distribution results from Ar-atomized $LaNi_{4.75}Sn_{0.25}$ powder. In fact, 84 wt.% of the He-atomized powder was less than 20 μm compared to the d_{84} value of 43 μm for the Ar-atomized powder. Such a high yield of fine powder is highly desirable from a rapid solidification perspective regardless of the atomization gas choice because a high ratio of atomized droplet surface area/volume would enhance the gas quenching effect, important for RSP. Certainly, extremely fine powders also are preferred for battery electrodes because of enhanced surface area.

Figures 1a and b show typical powder particle cross-section microstructures from dia. <10μm samples of Ar-atomized and He-atomized powders, respectively, to exhibit the as-solidified morphology. The backscattered electron images reveal the phase segregation pattern in both powders, although the pattern is less distinct in the He-atomized powder microstructure. The cell spacing is less than 1 μm in both powders.

Electron microprobe studies of the particle microstructure cross-sections provided a semi-quantitative characterization of the effect of annealing on the phase segregation effect. The results in Figures 2a on an unannealed Ar-atomized $LaNi_{4.75}Sn_{0.25}$ particle show that the La content remained close to constant while the Sn and Ni content varied in an opposite manner. Microprobe results on an annealed particle of this alloy, given in Figure 2b, indicate that high temperature diffusion produced a homogeneous composition across the full particle diameter.

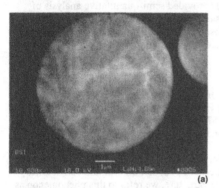 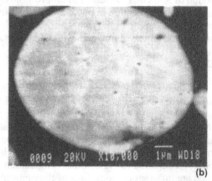

(a) (b)

Figure 1. SEM micrographs using backscattered electron imaging of unetched cross-sections of a) Ar-atomized $LaNi_{4.75}Sn_{0.25}$ and b) He-atomized $LaNi_{4.85}Sn_{0.15}$ showing the Sn-enriched cell boundary phase in light contrast.

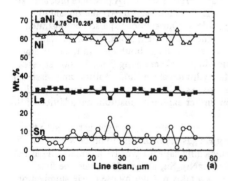 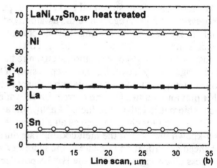

Figure 2. Comparison of electron microprobe data for a representative incremental trace across a random particle cross-section, analyzing for La, Ni, and Sn for; a) an as-atomized $LaNi_{4.75}Sn_{0.25}$ particle and b) a $LaNi_{4.75}Sn_{0.25}$ particle that was annealed at 950°C for 4 hours in vacuum.

Structural characterization of the powder by XRD was utilized to determine a minimum annealing time required for particle homogenization at an annealing temperatures greater than 900°C, as shown in Figure 3. Figure 3b reveals that a suitable compositional uniformity of the hexagonal phase was achieved in only 5 min of annealing time for the He-atomized LaNi$_{4.85}$Sn$_{0.15}$ powder. The same symmetric Bragg peaks shape was achieved in 4 hours of annealing for Ar-atomized LaNi$_{4.75}$Sn$_{0.25}$ powder, as Figure 3a indicates.

A characterization of the H$_2$ gas absorption and desorption behavior of the HPGA powder is shown in Figure 4 for both of the as-atomized and annealed powder samples. The as-atomized absorption and desorption data of a 25 μm to 38 μm LaNi$_{4.75}$Sn$_{0.25}$ powder sample shows a sub-ambient pressure trend in Figure 4a, as desired, but has an enhanced slope compared to the ideal plateau [5] of the powder annealed for 4 hours. This suggests that the phase segregation in the solidification microstructure of the as-atomized powder provides more resistance to the transport and exchange of hydrogen, as expected. Figure 4b shows the results for He-atomized powders and illustrates that an ideal plateau-like trend was achieved by only 5 min of annealing at 900°C, consistent with the XRD data of Figure 3b that indicates phase homogenization was achieved by this rapid anneal.

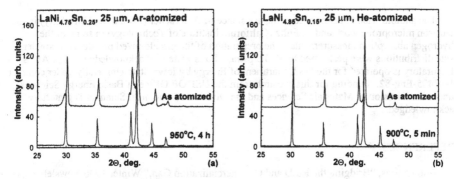

Figure 3. Comparison of XRD patterns of as-atomized alloy powder to annealed powder from HPGA powders with dia. < 25um of a) Ar-atomized LaNi$_{4.75}$Sn$_{0.25}$ and b) He-atomized LaNi$_{4.85}$Sn$_{0.15}$.

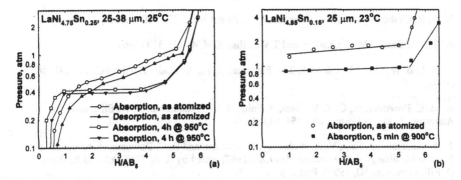

Figure 4. Summary of the hydrogen exchange behavior at ambient temperatures in as-atomized and annealed powder samples of a) Ar-atomized LaNi$_{4.75}$Sn$_{0.25}$ and b) He-atomized LaNi$_{4.85}$Sn$_{0.15}$.

CONCLUSIONS

The results of this investigation build on the previous studies [4,6,7] indicate that a high pressure gas atomization approach can produce high quality AB_5 powders for Ni/MH battery applications. A change from Ar to He atomization gas resulted in a very desirable increase in yield of ultrafine (dia. < 20 μm) spherical powders, although complete suppression of phase segregation in these $LaNi_{5-x}Sn_x$ alloy powders was not achieved in spite of the enhanced heat transfer coefficient of the He. The penalty in hydrogen absorption behavior of the phase segregated microstructure in the as-atomized powders was completely erased by a very brief anneal at high temperature. A good possibility exists that such an anneal could be incorporated in the atomization process sequence before removal of the powder from the unit to maintain process simplicity. Future work will involve some additional alloy composition adjustments and further modification of the HPGA process, including design of an in-situ passivation treatment for the gas atomized powder that would both minimize corrosion and enhance activation, guided by extensive electrochemical cycling experiments.

ACKNOWLEDGMENTS

The authors wish to acknowledge the assistance of A. Kracher (Iowa State University) in the electron microprobe work and B. Fultz (California Institute of Technology) in some of the hydrogen absorption measurements. The preparation of the gas atomized powder and powder size distributions were performed by R. Terpstra, who is gratefully acknowledged. The Ames Laboratory is operated for the US Department of Energy by Iowa State University under contract W-7405-Eng-82. Funding for this research from the USDOE Office of Basic Energy Sciences, under the Division of Materials Sciences and the Division of Chemical Sciences is gratefully acknowledged.

REFERENCES

1. Anonymous, "Bridging the R&D and Commercialization Gap," Winter 1996 Newsletter, United States Automotive Battery Consortium (USABC), Troy, MI.

2. T. Sakai, M. Matsuoka, and C. Iwakura, in *Handbook on the Physics and Chemistry of Rare Earths*, edited by K.A. Gschneider, Jr. and L. Eyring (Elsevier Science B.V., 1995) v.21, p.133.

3. S. R. Ovshinsky and M.A. Fetcenko, U. S. Patent No. 5 277 999 (January 11, 1994).

4. I.E. Anderson, M.G. Osborne and T.W. Ellis, JOM, **48** (3), 38 (1996).

5. S. Luo, W. Luo, J.D. Clewley, T.B. Flanagan, and L.A. Wade, J. Alloys Comp. **231**, 467 (1995).

6. R. C. Bowman, Jr., C. K. Witham, B. Fultz, B. V. Ratnakumar, T. W. Ellis, and I. E. Anderson, J. Alloys Comp., **253-254**, 613 (1997).

7. E. Anderson, J. Ting, V. K. Pecharsky, C. Witham, and R. C. Bowman, in *Advances in Powder Metallurgy & Particulate Materials-1997*, edited by R. A. McKotch and R. Webb (MPIF, Princeton, NJ, 1997), Part 5, p.31.

8. S. B. Biner, J. Mater. Sci. (to be published).

INFLUENCE OF CARBON STRUCTURE AND PHYSICAL PROPERTIES ON THE CORROSION BEHAVIOR IN CARBON BASED AIR ELECTRODES FOR ZINC AIR BATTERIES

M. NEAL GOLOVIN, IRENA KUZNETSOV, IYI ATIJOSAN, LAWRENCE A. TINKER, CHRISTOPHER S. PEDICINI
AER Energy Resources, Inc., 1500 Wilson Way, Suite 250, Smyrna, GA 30082

Introduction

Rechargeable zinc-air batteries are a long run time solution for portable electronics. Cells with a nominal voltage of 1.05 V have a specific energy of 169 Wh kg^{-1} and aenergy density of 219 Wh L^{-1}. This cell is capable of delivering 16 – 18 Ah of capacity at 2 A (16.7 mA cm^{-2}) discharge rate. However, the cycle life of these cells is artificially shortened because of the generation of carbon dioxide in the alkaline electrolyte during the cycling of the cell. The carbon dioxide has the effect of removing the OH$^-$ (equation 1) needed by the anode

$$CO_2 + 2OH^- \rightarrow CO_3^{-2} + H_2O \qquad (1)$$

$$Zn + 4OH^- \rightarrow Zn(OH)_4^{-2} + 2 e^- \qquad (2)$$

$$Zn(OH)_4^{-2} \rightarrow ZnO + 2 OH^- + H_2O \qquad (3)$$

half reaction (equations 2 and 3). The need for hydroxide is demonstrated in equation 2. There is literature evidence [1,2] that the CO_2 was the result of corrosion of the carbon matrix of the air electrode. We report here the results of our investigation of this corrosion reaction and what properties of the carbon contribute to the corrosion reaction rate.

Experimental

All bifunctional air electrodes were made according to the standard AER Energy formulation [3], using different carbons in the active layer. The carbons tested were Shawinigan Acetylene Black (AB50; Chevron Chemical), Vulcan XC-72 (Cabot), Ketjen EC-600J (KB, Akzo-Nobel). and two proprietary carbon black samples, labeled Carbon A and Carbon B. One of the proprietary (Carbon B) samples was graphitized by the manufacturer at 2700 °C. Air electrodes made with AB50 are also referred to as A100 electrodes. Samples of Vulcan and KB were graphitized at 2700 °C (Fiber Materials, Biddeford, ME). Carbons treated with Co tetra(4-methoxyphenyl)porphyrin (CoTMPP; Aldrich) were prepared by mixing the appropriate carbon and CoTMPP in a V-blender, followed by heat treatment at 400 °C for 1 hour under N_2.

All corrosion experiments were carried out using the AER Energy 13220 prismatic cell [5], without an anode. 45% KOH was used as the cell electrolyte. The two cathodes were electronically isolated from each other and one was discharged against the other at 1 A for 200 hours. The electrolyte was then analyzed, using standard acid/base titration techniques, for KOH and K_2CO_3. The corrosion rate was calculated as a corrosion current based on the amount of OH$^-$ consumed and only oxidation of carbon to CO_2 (*i.e.* 4 e$^-$ oxidation) is considered in the rate calculation. Relative corrosion rates are the corrosion current normalized to the rate for A100 electrodes. Table I shows the data obtained from this test procedure. Both corrosion cell cycling and standard cell cycling studies were conducted using a MACCOR automated cell cycling system.

Carbon	I_{corr} $\mu A\ cm^{-2}$	Rel I_{corr}
AB50	61.6	1.0
AB50/2% CoTMPP	226.1	3.7
AB50/3% CoTMPP	178	2.9
Graphitized Vulcan (GV)	38.4	0.6
GV/2% CoTMPP	104.9	1.3
Carbon A	57.7	0.9
Carbon A/2% CoTMPP	119.8	1.9
70% AB50/30% KB	198.4	3.2
Carbon B	41.4	0.7
Carbon B/2% CoTMPP	69.8	1.1
70/30 AB50/GK	61.6	1.0
AB50 Acetone washed air dry	175.6	2.1
AB50 Acetone washed oven dry	143.5	1.7

Table I: Corrosion data for air electrodes with active layers produced with different carbons. Current density is based on geometric area.

Double layer capacitance measurements were made by dc measurements at constant current of 0.2 A during the discharge of the 13220 cell, measuring the slope of the voltage *vs.* time curve between 1.55 V and 1.28 V and using equation 4,

$$I = C_{dl} \frac{dV}{dt} \tag{4}$$

where I is the current in A and the term dV/dt is the slope of the discharge profile in V s^{-1}, and C_{dl} is the double layer capacitance in F. Data were collected for at least three cells of each chemistry tested, since the C_{dl} changes with cycling. Data were recorded for the third cycle for all chemistries.

Thermogravimetric Analysis (TGA) data were obtained with a Thermal Analysis model 951 TGA. The data were collected under a dried compressed air stream flowing at 100 cc/min. Data were collected at a scan rate of 10 °C/min between ambient and 900 °C. Data were recorded as weight loss versus temperature and as the derivative of the weight loss with respect to time. The temperature at which the peak in the derivative data was observed was recorded.

Gas chromatography (GC) data were obtained using a Carle model 111 analytical GC using a thermal conductivity detector. The data were obtained using a He carrier gas (at 30 cc/min) with a 6' molecular sieve (5A) column operated isothermally at 65 °C. Corrosion cells were operated in an enclosed container of fixed volume (175 mL) fitted with a silicone septum. 500 μL samples were withdrawn at specified times. The peak at a retention time of 2.5 min was integrated and the area was compared with a known calibration curve to determine the number of moles of CO observed.

Other physical measurements were made off sight. IC Laboratories of Katonah, NY obtained x-ray diffraction data. Data were collected using CuKα radiation. The 002 peak at 2θ ~ 25° was fitted to determine the full width at half max for Scherrer analysis. Values for the d_{002} spacing were calculated as well. Nitrogen adsorption surface area data were obtained by PMI of Ithaca, NY. Multiple point data were obtained and the surface area was calculated using the BET adsorption isotherm (BET surface area).

Results and Discussion

The hypothesis that a corrosion reaction was affecting the life of rechargeable zinc-air cells was made based on the observation that electrolytes of cells cycled to end of life (defined as 50% of rated capacity) contained extremely high amounts of K_2CO_3 (>10% by weight). One important objection to the corrosion hypothesis was that the cell, by its very nature, is open to the air. Carbonation of the electrolyte (45% KOH solution) could result from influx of CO_2 in the ambient air. We tested this with cells run under ambient air and CO_2 scrubbed air. Figure 1 shows the capacity fade for these two cells. There was little

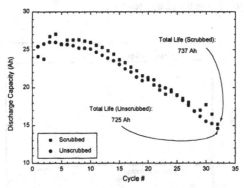

Figure 1: Capacity fade data for a standard 13220 cell in ambient air and in a CO_2 scrubbed air stream. Discharge at 2 A to a 0.8 V end. Charge at 2 A to 2.02 V limit tapering the current until 100% of discharge capacity was recharged (AP=1).

difference in the capacity fade for the two cells. The scrubbed cell was titrated for hydroxide and carbonate content. It was observed that at the end of life, the scrubbed cell's electrolyte contained 15% (w/w) K_2CO_3 despite there being no observable CO_2 in the air stream. The carbon dioxide had to come from somewhere, and the only source of carbon is the electrode material. Indeed, it was found that the total life was related to the observed corrosion rates. For the three best studied air electrode chemistries, the data in Table II demonstrate the relation between corrosion and life. An important point to note is that the addition of CoTMPP to the carbon increases its corrosion rate. CoTMPP was added as a discharge catalyst [5] for the 4-electron reduction of O_2 to OH^-.

Cathode Base	I_{corr} (μA cm^{-2})	Rel. I_{corr}	Cycle Life (Ah)
A100	61.6	1.0	650 - 750
AB50/3%CoTMPP	178	2.9	350 - 500
GV	38.4	0.6	900 - 1000

Table II: Comparison of corrosion and cycle life data for different air electrode chemistries.

Ross [2] showed that the oxidation of carbon in alkaline solution occurs at potentials above 450 mV vs. Hg/HgO (~1.85 V vs. Zn). Figure 2 shows the typical voltage profile of the 13220 cell. The charge voltage is well within this region. Through simple substitution experiments, we have found that the "charge package" catalysts ($FeWO_4$, WC/12%Co, NiS in a 1:1:1 weight ratio) are critical. In particular, through substitution of K_2WO_4 for $FeWO_4$, we have found that the iron is critical for the charge reaction. Cells with K_2WO_4, tungsten carbide/12%Co, and NiS in 1:1:1 weight ratio will not accept charge at 2.02 V even after 300 hours. Thus, we believe that the charge reaction occurs on the iron atoms of the iron tungstate. We are unsure of the roles of the other constituents, though the cells work better with than without them. Thus we hypothesize that the corrosion of carbon is a chemical redox process, rather than an electrochemical process. The oxidant is either O_2 or some long lived intermediate generated during the charge reaction.

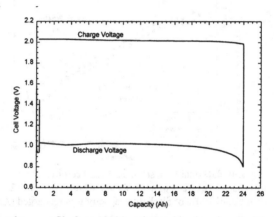

Figure 2: Typical voltage profile for a 13220 rechargeable zinc air cell. Discharge at 4 A for 5 min, then 3 A to a 0.8 V End. Recharge at 2 A to a 2.02 limit tapering the current to an endpoint of AP=1.

The corrosion rate data were calculated with the assumption that the only oxidation product from the carbon corrosion is CO_2. Ross and Sokol found that CO was also an important corrosion product. Figure 3 shows the generation of CO in a catalyzed and an uncatalysed AB50 cell. The data show that the CO generation is small (~400ppm max for AB50/3% CoTMPP). The data also show that the amount of CO in the enclosed container did not increase after roughly 2 hours. Thus, we feel that the generation of CO, at the potential where the cell is charging, is unimportant relative to CO_2 production

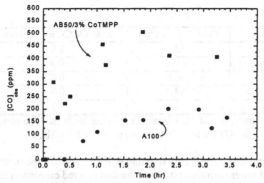

Figure 3: Production of CO as a function of time for corrosion cells in an enclosed chamber.

Since the charge reaction is the electrochemical oxidation of OH⁻, it seems reasonable that an important consideration in choosing a carbon is the surface area. The question becomes what is the proper measure of the surface area. One of the most common methods of surface area measurement is the BET surface area. This method has the drawback that it is a gas phase measurement. One should be careful when comparing solution phase phenomena with gas phase measurement. The corrosion rates are a good example of this. Table III shows comparison of several carbons with similar BET areas. A good measure of the electrochemically available surface area is the double layer capacitance [6], which represents the electrode/electrolyte interfacial region. The double

Carbon	BET Area $m^2\ gram^{-1}$	C_{dl} F	Relative I_{corr}
AB50	88	34.5	1.0
AB50/3% CoTMPP	88	120.7	2.9
Carbon A	83	26.5	0.9
GV/2% CoTMPP	89	61.0	1.3
70/30 AB50/GK	145	33.9	1.0

Table III: Comparison of BET surface area and double layer capacitance data as it relates to carbon corrosion. GK is graphitized Ketjen EC-600J (heated to 2700 °C).

layer capacitance is, in effect, a measure of the surface area of the electrode in contact with the electrolyte. The data in Table III show that a series of carbons with similar BET areas can have different corrosion rates. The corrosion reaction is dependent on the wettability of the carbon base of the electrode. It should be noted that the CoTMPP seems to enhance the wetting phenomenon without any change to the BET surface area of the carbon. The BET area of GV is 92 $m^2\ gram^{-1}$. The reactivity of the carbon was also related to its chemical oxidation sensitivity. This sensitivity was measured by TGA. The carbon samples were scanned between 20 °C and 900 °C. In all cases, the carbons completely burned away (*i.e.* 100% weight loss) by 825 °C. The inflection temperature of the weight loss curve was obtained from the peak in the *d(weight loss)/dT* curves. The data in Table IV show that the carbons that are more sensitive to chemical oxidation are also more sensitive to corrosion.

Carbon	TGA Peak T (°C)	Relative Corrosion Rate
AB50	732.6	1.0
Carbon A	761.2	0.9
GV	766.0	0.8
Carbon B	788.2	0.8
KB	684.0	3.2

Table IV: Data for corrosion and TGA in air for different carbons. The corrosion data listed for KB are for a 30% (w/w) mix of KB in AB50

There was also a relation between the structure of the carbon and the corrosion rate. More graphitic carbons had lower corrosion rates than did less ordered carbons. Table V shows x-ray diffraction and corrosion data for several carbons. Recall Carbon A is a

Carbon	L_C (Å)	d_{002} (Å)	Rel. I_{corr}
AB50	14	3.55	1.0
GV	46	3.49	0.8
Carbon A	29	3.53	0.9
Carbon B	40	3.49	0.8

Table V: Data comparing diffraction results and corrosion testing for graphitized and untreated carbons

Carbon black and Carbon B is a graphitized black. The carbons with more crystallinity had lower corrosion rates. This is in concert with the observations of Ross et al.[7], where they looked at the electrochemistry of graphitic versus non-graphitic carbons. The higher degree of crystallinity imparts some resistance to corrosion.

Conclusions

We have studied the phenomenon of the carbonation of the 45% KOH electrolyte of AER Energy's rechargeable zinc-air cells. The goal was to characterize the important physical and chemical parameters that are important to the carbonation process. Towards this end, we have found that the carbonation of the electrolyte is the result of the oxidation of the carbon matrix of the air electrode. This carbonation is noticed after hundreds of hours in service, and probably occurs during the charge half cycle. The wetted electrode surface area (as measured by the double layer capacitance) is where the reaction occurs and the reactivity of a given carbon is related to the wetted area. The reactivity of a given carbon is also related to its resistance to air oxidation at elevated temperatures. X-ray diffraction shows that the least reactive carbons are the most crystalline. These data, combined, lead us to conclude that a long life cell may be made from a graphitic carbon. While power considerations have not been considered, it must be kept in mind when considering a carbon for use in a cell. Thus, the appropriate choice of wetting characteristics and crystallinity are important for making long life cells with acceptable power performance.

Acknowledgements

We would like to thank Dr. V. R. Shepard for running the corrosion cell tests. We would like to thank D. J. Brose and Dr. T. A. Reynolds of Chemica Technologies, Inc. for assistance in running the GC experiments.

References

1. K. Kinoshita Carbon: Electrochemical and Physicochemical Properties (John Wiley and Sons, New York, 1988), pp. 293-387

2. P. N. Ross, H. Sokol, J. Electrochem. Soc., **131,** 1742 (1984)

3. V. R. Shepard, Y. G. Smalley, R. D. Betz, U. S. Patent No. 5,306,579

4. M. N. Golovin, L. A. Tinker, E. S. Buzzelli, Proc. Electrochem. Soc. **95-14,** 147 (1995)

5. E. Yeager, Electrocatalysts for Oxygen Reduction, U. S. Government Report #LBL-24788 (NTIS Order # DE88008397) (1988)

6. A. J. Bard and L. R. Faulkner, Electrochemical Methods: Fundamentals and Applications, (John Wiley and Sons, New York, 1980), pp. 488 – 552

7. P. N. Ross and M. Sattler, J. Electrochem. Soc., **135,** 1464 (1988)

VARIATION OF STRESSES AHEAD OF THE INTERNAL CRACKS IN ReNi₅ POWDERS DURING HYDROGEN CHARGING AND DISCHARGING CYCLES

S.B. Biner
Ames Laboratory, Iowa State University, Ames IA 50011 U.S.A

ABSTRACT

In this study, the evolution of the stress-states ahead of the penny shaped internal cracks in both spherical and disk shaped ReNi₅ particles during hydrogen charging and discharging cycles were investigated using coupled diffusion/deformation FEM analyses. The results indicate that large tensile stresses, on the order of 20-50% of the modulus of elasticity, develop in the particles. The disk shaped particles, in addition to having faster charging/discharging cycles, may offer better resistance to fracture than the spherical particles.

INTRODUCTION

The application of hydride-forming ReNi₅ compounds, where Re denotes the rare earths La, Ce, and Misch-metals, in re-chargeable nickel metal hyride batteries is well known[1-4]. In the case of LaNi₅, a large amount of hydrogen can be absorbed to form LaNi₅H₆.₇ at nearly room temperature[1]. The calculated density of the absorbed hydrogen is about a factor of two higher than the density of liquid hydrogen[5]. The lattice dimensions of unsaturated LaNi₅ are about a=5.017A° and c=3.982A° , and after saturation with hydrogen, the lattice parameters increase to about a=5.440A° and c=4.310A° which represents a volume expansion of over 25% [3]. Associated with this large volume expansion, the hydrogenation cycles produce very fine powders through a cleavage fracture of initial particles leading to reductions in battery performance.

In this study, to elucidate the stress-states in the particles during the hydrogen charging and discharging cycles a set of coupled diffusion/deformation finite element analyses are performed. The role of the particle shape on the evolution of the stress-states ahead of the penny shaped internal cracks is investigated.

DETAILS OF THE FEM ANALYSES

The hydrogen diffusion into ReNi₅ is assumed to be a bulk diffusion process driven by chemical potential [5]which can be described by the general behavior:

$$J = -sD\left[\kappa_s \frac{\partial}{\partial x}(\ln \tilde{\theta}) + \frac{\partial \phi}{\partial x}\right] \quad (1)$$

where J is the flux concentration of the diffusing phase, $D(c,\tilde{\theta})$ is the diffusivity, $s(\tilde{\theta})$ is the solubility, $\kappa_s(c,\tilde{\theta})$ is the "Soret effect" factor providing diffusion because of a possible temperature gradient, $\tilde{\theta}$ is the absolute temperature, and ϕ is the normalized concentration (often also referred to as the "activity" of the diffusing material), $\phi = c / s$ in which c is the mass concentration of the diffusing material in the base material. During the hydrogen charging and discharging cycles, the stress-strain behavior of the powders is assumed to be elastic:

51

$$\sigma = \mathbf{D}\big(\varepsilon_D + \varepsilon_V(\phi)\big) \tag{2}$$

where **D** is the elasticity matrix, ε_D is the deviatoric component of the total strain and $\varepsilon_V(\phi)$ is the volumetric strain which is a function of the normalized hydrogen concentration. During the analyses, it was assumed that both the diffusion parameters and the elastic properties remain constant in spite of possible phase changes; also, temperature effects were neglected. Throughout the study, the time values were normalized with the coefficient of the diffusion, all the stress values were normalized with the value of the Young's modulus and the concentration values were normalized with the solubility value, thus yielding results that are independent of the material parameters.

RESULTS

In the next set of simulations the charging and discharging behavior of a spherical particle containing a penny shaped internal crack were studied. To model the crack tip singularity, collapsed elements having shifted mid-side nodes[6] were used at immediate crack tip region in these analyses. Penetration of the crack faces resulting from the volume expansion was prevented by using a series of non-dimensional interface elements[7] along the crack surfaces. No hydrogen was present in the particles at the beginning of the solution. The evolution of the stress state in the particle during the hydrogen charging cycle is summarized in Fig. 1 (left). In this figure, the crack tip is located at $x/r = 0.4$; x/r values smaller than this value represent the crack wake region and larger values are for the regions ahead of the crack tip. Because of the embedded nature of the crack, the development of the hydrogen concentration profiles was identical to that seen for a spherical particle without an internal crack. As can be seen from the figure, large normal stress values develop ahead of the crack tip even at very early stages of the charging. This increase continuous and reaches a value as high as half of the modulus of elasticity. The attainment of this peak value occurred shortly before reaching a fully charged stage, then a slight reduction took place. Nevertheless, the stress level remained significantly high at the fully charged stage. Fig. 1 (right) shows the stress-states in the same particle during the discharging cycle. The zero value of the normalized time corresponds the fully charged state in the figure. With the loss of the hydrogen in the regions near to the particle surface, a stress redistribution takes place and the crack tip stresses relax significantly at very early stages of the discharging cycle. As can be seen from the figure, the stress distribution at the interior regions of the particle becomes quite uniform, as if there is no crack, resulting from the closure of the crack faces due to overall contraction.

In next set of simulations the behavior of a disk shaped particle containing a penny shaped internal crack were investigated. The volume of this disk shaped particle and also the size of the penny shaped crack were identical to that of spherical particle. The development of the stress-states in the disk shaped particle containing the penny shaped crack during the hydrogen charging cycle is shown in Fig. 2 (left). The normalized time of 0.0518 corresponds to fully charged state in the figure. When a comparison is made with the previous spherical particle case, a significant reduction in the charging time can be observed. This reduction arises from the larger surface-area/volume ratio of the disk shaped particle. As can be seen from the figure, in spite of the presence of an internal crack, the stress distribution at fully charged state was very uniform which is opposite to that seen for the spherical particle containing a penny shaped crack (Fig.1). The shape changes of this disk shaped particle during the same time intervals given in

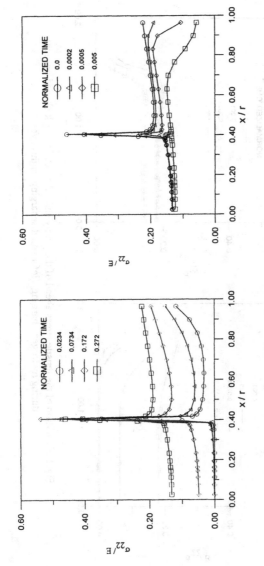

Fig.1 Evolution of the tensile stress ahead of the penny shaped crack in a shperical partical during hydrogen charging (left) and discharging (right) cycles.

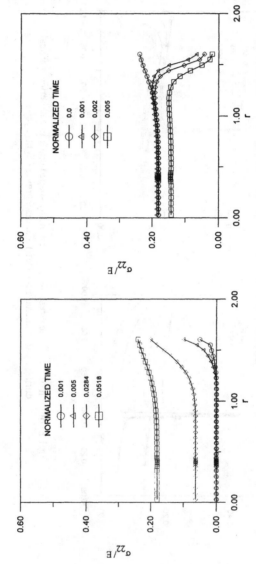

Fig.2 Evolution of the tensile stress ahead of the penny shaped crack in a disk shaped partical during hydrogen charging (left) and discharging (right) cycles.

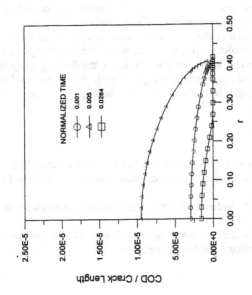

Fig.4 Crack opening profiles of the penny shaped internal crack in the disk shaped particle during hydrogen charging cycle.

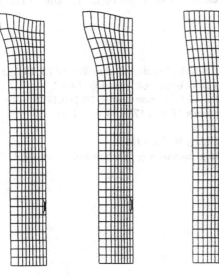

Fig.3 The shape changes in the disk shaped particle with a penny shaped crack during hydrogen charging cycle. The normalized times are: top, 0.001, middle, 0.0284 and bottom, 0.0518.

Fig. 2 (left) are shown in Fig. 3. As can be seen from the figure, resulting from the fast diffusion of the hydrogen from both top and side surfaces, a large expansion occurs near the corner regions. This expansion introduces a compression in the central regions leading to the stress distribution seen in Fig. 2 (left). This can also be substantiated from the crack opening behavior during the charging cycle as seen in Fig. 4. Shortly after opening of the crack, the crack tip region closes again due to the evolution of the compression in the central regions of the particle as described earlier. The variations in the stress values during the discharging cycle of the same particle with the internal crack are summarized in Fig. 2 (right). As can be seen from the figure, no stress elevation also takes place during the discharging cycle associated with the aforementioned shape changes.

CONCLUSIONS

In this study, the development of the stress-states in ReNi$_5$ particles during hydrogen charging and discharging cycles was investigated by using coupled diffusion/deformation FEM analyses. The results indicate that:
1. Large tensile stresses, on the order of 20-30% of the modulus of elasticity, develop in the particles.
2. The disk shaped particles, in addition to having faster charging/discharging cycles, may offer better resistance to fracture than the spherical particles

ACKNOWLEDGMENT

This work was performed for the United States Department of Energy by Iowa State University under contract W-7405-82. This research was supported by the Director of Energy Research, Office of Basic Sciences.

REFERANCES

1. J.H.N vanVucht, F.A. Kuijpers and H.C.A.M Bruing, *Philips Res. RPTS*, **25**, 133, 1970
2. M. H. Mintz and Y. Zeiri, *J. Alloys and Compounds*, **216**, 159, 1994.
3. K.H.J. Buschow and H.H. vanMal, *J. Less Common Metals*, **29**, 203,1972.
4. Z. Zuchner and T. Rauf, *J. Less Common Metals*, **172-174**, 611, 1991
5. R. Wang, *Mat. Res. Bull.* **11**, 281, 1976.
6. R.S. Barsoum, *Int. J. Num. Meth. Engng*, **10**, 25, 1976
7. S.B. Biner, O. Buck and W.A. Spitzig, *Engineering Fracture Mechanics*, **47**, 1, 1994.

THE CORROSION PHENOMENA IN THE COIN CELL BR2325 OF THE "SUPERSTOICHIOMETRIC FLUOROCARBON - LITHIUM" SYSTEM

Valentin N. Mitkin, Peter S. Galkin, Tatjana N. Denisova, Ol'ga V. Koreneva,
Sergey V. Filatov, Eugene A. Shinelev
3, Lavrentjeva ave., Institute of Inorganic Chemistry SB RAS, Novosibirsk, 630090, RUSSIA
Alexander B. Alexandrov, Vladimir L. Afanasiev, Alexander A. Enin,
Viktor V. Moukhin, Vladimir V. Rozhkov, Vladlen V. Telezhkin
94, B. Khmelnitsky str., JS Novosibirsk Chemical Concentrates Plant, Novosibirsk, 630110, RUSSIA,

Introduction

It was noted at the earlier study and at the longer observations of the novel various types of superstoichiometric fluorocarbon materials CF_{1+x}, where x = 0.1-0.33 (FCM) and their behavior, that despite of their known hygroscopity during a storage of samples in laboratory and technological utensils nevertheless occurs an appreciable sorption of atmospheric moisture. The color of samples does not change but sometimes there is appears a smell of hydrogen fluoride and even corrosion of glasswares at a long storage. On the basis of these facts was assumed that at a long storage the slow reactions of HF producing with a sorption moisture can proceed. This phenomena is necessary to take into account for successful manufacturing of long life lithium cells based on superstoichiometric fluorocarbon composite cathodes (FCC).

The chemistry of such slow hydrolytic process and especially of processes which can proceed at manufacturing of FCC earlier was not investigated also of any data in the literature in this occasion is not present. Just for this reason we undertook a study of the corrosion phenomena which can proceed in industrial sources of a current at a long storage under influence of slow hydrolysis of C-F bonds by moisture. The goal of our study was to search long term damages in the "slightly wet" FCM and based on these materials cathodic composites for fluorocarbon-lithium cells. As a model for corrosion process investigation we have chosen a standard coin lithium battery of a type BR2325.

Experimental

For this work there were assembled more than 2000 pilot BR2325 coin cells. During their manufacturing the microimpurity of a moisture at a level of 0.0N mass. % were introduced into the fluorocarbon cathodic composites (FCC). Thus we meant a study not only the electromigration behavior the potential hydrolysis of C-F bond's product - HF but also the "destiny" of various constructional units of battery-anode, separator, cathode and body of a cell. The samples were stored at room conditions during 12 months and then the research of the reasons and consequences of corrosion processes was begun.

Before researches near to 1300 BR2325 pilot samples assembled in the 1996 spring of an experimental batch of these elements were classified by these ranges of "short currents circuit" (SC) and open circuit of voltage (OCV) and divided to a separate classes "OCV - SC". At it was received "spatial" histogram of their distribution on classes. These data are given in a fig. 1A. The character of histogram shows that in 12 month's storage all experienced cells of type BR2325 can conditionally be shared into two main groups, which are allocated and separated from each other on the diagram. In the first group ("A"- group) there were cells with OCV range from 2.85 up to 3.00 V and "short" currents from 0 up to 5 mA and them has appeared about 80% in the total volume of series. In the second group "B" there were cells with OCV range 3.25-3.50 V and with SC level of 10-40 mA. In one half-year for a second independent series ca. 300 stored series of BR2325 coin cells has been received the 2nd distribution histogram on "OCV - SC" classes which is given in a fig. 1B.

More than 320 BR2325 pilot cells from all "OCV - SC" classes and different terms of storage were subjected to discharge tests for loading R=5.6 kOhm at room temperature. The results of the certain discharge capacities for all "OCV - SC" classes are shown in table 1 and are also submitted as histogram in a fig. 1C. Besides them the results of the various "OCV - SC" class's discharge tests as the data on average electrical capacities for ten chosen classes from groups "A" and "B" were collected at the diagrams. These data are given separately for both groups (dependence on SC) in a fig. 2. The typical discharge curves are shown in a fig. 3.

The characterization of chemical and electrochemical corrosion phenomena was carried out by methods of the chemical analysis of products of corrosion in the cathode, anode, separator and cathodic cover and also by digit tests of BR2325 cells of the storage's different terms.

The analyses were carried out on lithium distribution by the àcidymetry with HCl titration after immediately water treatment of the "opened" cells and also by the spectrophotometry with "thorone" color reaction (for the additional control of Li-contents). Iron, chromium and nickel as elements of cell's material (stainless steel 12X18N9T) were determined by flamed variant of atomic absorption after the special procedure of chemical pretreatment in various details of a cell construction of a current. The acidity and free-fluoride contents in FCC's materials of different storage terms of BR2325 series were also determined. At these

57

A. Distribution of Classes "OCV - Short Current" in corroded series of BR2325 (N = 1420, May 1997

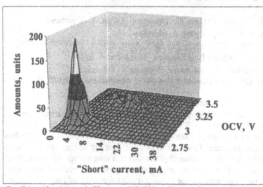

"Short" current, mA

B. Distribution of Classes "OCV - Short Current" in corroded series of BR2325 (N = 336, October 1997)

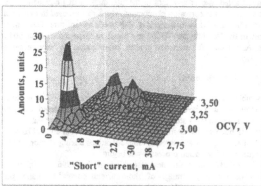

"Short" current, mA

C. Distribution of discharge capacities in Classes "OCV - Short Current" of BR2325 corroded series. Tests of May-October 1997; room temperature; R = 5.6 kOhm.

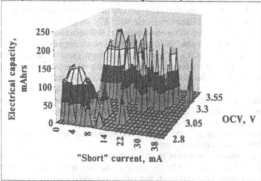

"Short" current, mA

Fig 1. Distribution diagrams of corroded BR2325 experimental series - 12 and 18 months storage.

FCC cathode's analysis were used a two variants of immediately treatment of the cathode from the opened cell both by a mix of i-PrOH with a pure water and by specially dried i-PrOH. The acidity of the i-PrOH solutions was determined by potentiometrical titration with 0.008M NaOH. The free-fluoride was determined also by potentiomentical titration with solution of lanthanum nitrate. Data of the chemical analysis on the Li contents, acidities, and also Fe, Cr and Ni contents in some parts of BR2325 cell's are given in table 1-2. It is well visible a lawful decreasing of all corrosion damages in group "B". The comparison of histogram's character from fig. 1A and 1B shows that during 1-1.5 years of BR2325 series storage the character of self-distribution between groups "A" and "B" has stayed almost the same however the relative amounts of samples in a group "A" a little has increased and has made already more than 90%. Thus also there was also shifted a top of most intensive "mountain" from "OCV-SC" area of 3.00 V-3.0 mA into the area of 2.95 V -2.0 mA.

The analysis of BR2325 cell's cathodic composites on acidity has shown that in 12 months of a storage more than 90 % of cathodes from a series "A" has appeared to be acidic and in a series "B" in the same time a large part of cathodes (more than 80 %) was practically neutral. However in 18 months of a storage the rather significant part of cathodes from group "A" has got a visible weak acidity while in group "B" was begun to be displayed the tendency to reduction of cathode's acidity.

There was found out that used for acidity determination our standard treatment of i-PrOH or the alcohol-water mixtures without long boiling under reflux leads to find out only the acidity from the "open" pores of not unloaded cathodes. After discharge tests from different "OCV - SC" classes in all cases it is found out the more large or less additional acidity that earlier haven't been found in the fluorocarbon cathodes. This additional acidity caused by "opening" of blocked pores in the cathodic composite during or after discharge process in BR2325 cell..

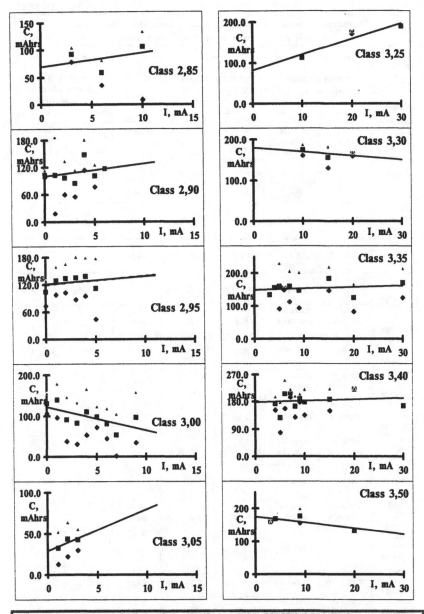

Fig. 2. Discharge test's data of electrical capacities determinations in Low- and High
- "OCV - Short Current" corroded BR2325 classes at room temperature. R = 5.6
kOhm. Squares - average values. Rhomboids and triangles - levels of variation

Table 1.

Corrosion phenomena distribution in experimental series of BR 2325 till and after discharge tests							
Classes of OCV - Short current, V - mA	**Lithium content, mg-atom**		**The total sum of Fe,Cr,Ni, mg**		**Acidity of cathode**	**Electrical discharge capacity at room temperature**	
	In anode and separator as Metal	In cathode, Lithium Salts	In anode and separator as Metal complexes	In cathode and cover as Metal complexes	Estimation as HF content, mg-mol	Till 20 mV 5.6 kOhm mAhrs	Till 2 V, 5.6 kOhm mAhrs
2.85-5 till discharge	3.7	6.2	7.3	16.7	0.15 June		
2.85-6 after discharge	1.2-2.0	7.7-10.7	5.1 ± 2.5	14.3 ± 5.2	0.12 June	46 ± 1	19 ± 13
2.90-3 till discharge	1.74-2.75	3.7-7.1	6.8 ± 1.4	16.8 ± 6.0	1.1 Sept.		
2.90-3 after discharge	3.41-8.11	3.7-8.2	9.3 ± 1.9	20.6 ±6.5	0.3 June	94 ± 16	57 ± 12
2.90-4 till discharge	Data in process	Data in process	8.2	2.2	6.0 Sept.		
2.90-4 after discharge	1.50	10.0	8.0	19.1	0.3 June	148 ± 34	55 ± 19
2.95-2 till discharge	Data in process	Data in process	No metals June	5.2 ± 2.0 June	1.5 Sept.		
2.95-3 after discharge	Data in process	Data in process	11.0 ± 1.7	5.5 ± 2.3	Data in process	159 ± 24	74 ± 21
2.95-4 till discharge	Data in process	Data in process	16.2 ± 2.0	9.2 ± 3.6	1.0 Sept		
2.95-4 after discharge	1.54	8.47	13.3 ± 1.2	16.5 ± 4.9	0.17 June	156 ± 17	97 ± 8
3.00-2 till discharge	Data in process	Data in process	2.2 ± 0.5	27.4 ± 1.4	0.18 Sept.		
3.00-2 after discharge	1.7	8.49	12.6	21.9	0.06 June	137 ± 18	90 ± 16
3.00-3 till discharge	8.15-8.80	1.1-1.6	9.5 ± 2.5	19.3 ± 6.9	0.18 Sept.		
3.00-3 after discharge	Data in process	Data in process	8.8	17.5	0.12 June	129 ± 24	78 ± 8

Table 2.

Iron, chromium and nickel distribution in details of construction BR 2325 till and after discharge tests (mg)

Classes of OCV -Short current,	Metal content in anode and separator			Metal content in cathodic mass			Metal content in positive cover			Total metal content		
V - mA	Fe	Cr	Ni	Fe	Cr	Ni	Fe	Cr	Ni	Fe	Cr	Ni
2.85-5 till discharge	5.5	0.7	1.1	6.3	2.6	0.7	5.5	1.0	0.7	17.2	4.3	2.5
2.85-6 after discharge, n=2, June	3.9 ±2	0.4 ±0.2	0.8 ±0.3	5.4 ±0.2	2.0 ±0.5	0.6 ±0.3	3.2 ±1.3	0.7 ±0.2	0.5 ±0.2	14.1 ±3.5	3.1 ±0.9	2.2 ±0.8
Classes OCV = 2.900 - 2.950 V, short currents - 2; 3; 4; 5 mA												
2.90-3 till discharge, n=4, Sept.	5.1 ±1.1	0.4 ±0.1	1.3 ±0.2	6.8 ±1.8	2.3 ±0.9	0.9 ±0.1	3.4 ±1.3	0.9 ±0.3	0.5 ±0.2	17.3 ±4.2	3.6 ±1.3	2.7 ±0.5
2.90-3 after discharge n=3, June	7.2 ±1.7	0.6 ±0.1	1.5 ±0.1	9.4 ±2	3.5 ±0.6	1.1 ±0.3	4.9 ±0.4	1.0 ±0.1	0.7 ±0.1	21.5 ±4.1	5.1 ±0.8	3.3 ±0.6
2.90-4 after discharge n=1, July	6.5	0.3	1.2	14.6	3.3	1.4	-	-	-	21.1	3.4	2.6
2.90-4 till discharge n=1, Sept.				8.1	0.4	1.9	-	-	-	8.1	0.4	1.9
2.90-5 till discharge n=1,2 June	7.3 ±0.1	1.3 ±0.5	0.9 ±0.3	7.6	2.4	0.9	3.3	0.9	0.5	12.8 ±5.5	2.9 ±1.1	2.3 ±0.3
Classes OCV = 2.950 - 3.000 V, short currents - 2; 3; 4 mA												
2.95-2 till discharge n=2, June				3.9 ±1.4	0.14 ±0.1	1.2 ±0.5				3.9 ±1.4	0.14 ±0.1	1.2 ±0.5
2.95-3 after discharge n=2, June	8.7 ±1.2	1.9 ±0.4	0.4 ±0.1	3.6 ±0.6	0.2 ±0.0	0.9 ±0.3				12.3 ±1.7	2.0 ±0.3	1.3 ±0.3
2.95-4 till discharge n=2, July	11.6 ±1.6	4.3 ±0.3	0.35 ±0.1	6.9 ±1.9	0.5 ±0.1	1.7 ±0.6				18.5 ±2.9	4.9 ±0.5	1.8 ±0.2
2.95-4 after discharge, n=2, Sept.	11.6 ±0.9	0.4 ±0.1	1.3 ±0.2	6.8 ±1.8	2.3 ±0.9	0.9 ±0.1	4.0 ±0.5	0.9 ±0.3	0.6 ±0.1	22.4 ±3.2	3.6 ±1.3	2.8 ±0.4
Classes OCV = 3.000 - 3.050 V, short currents - 1; 2; 3 mA												
3.00-2 till discharge n=2, Sept.	1.8 ±0.3	0.1 ±0.1	0.25 ±0.1	13.9 ±0.5	5.6 ±1.2	0.9 ±0.1	4.7 ±1.5	1.6 ±0.7	0.7 ±0.1	20.4 ±0.4	7.3 ±0.7	1.9 ±0.3
3.00-2 after discharge n=1 June	8.7	1.4	2.5	12.8	3.1	0.5	2.9	1.0	0.6	24.4	5.5	3.6
Classes OCV = 3.350 - 3.450 V, short currents - 7-10; 10-15; 15-20; 30 mA Experimental data only for tests till discharge												
3.35-15 n=2, June	No corrosion traces			0.4 ±.02	0.01 ±.01	0.04 ±.01	No corrosion traces			0.4 ±.02	0.01 ±.01	0.04 ±.01
3.40-(7-10) n=4, June	No corrosion traces			0.6 ±0.2	0.04 ±.02	0.03 ±.01	No corrosion traces			0.6 ±0.2	0.04 ±.02	0.03 ±.01
3.40-(10-15) n=2, June	No corrosion traces			0.5 ±.02	0.02 ±.00	0.02 ±.01	No corrosion traces			0.5 ±.02	0.02 ±.00	0.02 ±.01
3.45-7 n=1, June	1.48	0.20	0.19	0.45	0.04	0.03	No corrosion traces			1.93	0.24	0.21

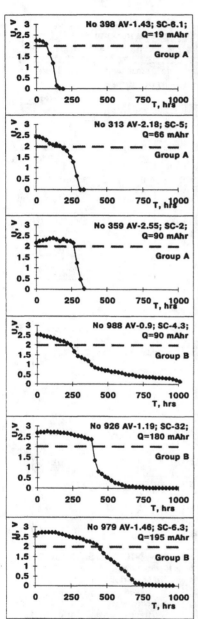

Fig. 3. Typical discharge curves for various types of corrosion phenomena in BR2325 series at R=5.6 kOhm, room temperature.

Simultaneously with the increase of the degradation phenomena of stored BR2325 cells is accompanying also corrosion of the anode and stainless steel cathodic subcover contacted with the FCC's material and also in separator and at surface of the anode begins to collect iron, chromium and nickel compounds - as corrosion products. Thus the tendency of dependence between lithium distribution in the details of BR2325 construction and the electrochemical and chemical or migration of Fe, Cr and Ni to the same construction parts of cell for different "OCV - SC" classes are displayed in tables 1 and 2. In these tables also was shown the decrease of discharge capacity and influence of cathodic acidity onto these processes. For an explanation of the observed phenomena we consider necessary to carry out rather large additional researches including the study of these process's kinetics. Nowadays it is possible to formulate briefly the only primary assumptions of the character and reasons of the observable corrosion phenomena in the "slightly wet" fluorocarbon-lithium coin cells BR2325.

In particular we consider that the general circuit of chemical and electrochemical corrosion in the pilot BR2325 cells with cathodes containing even insignificant amounts of a moisture is connected first of all due to formation of hydrogen fluoride in the real cell internal units by the slow hydrolysis of C-F bonds by moisture traces.

Conclusion

1. The main reason of the corrosion phenomena in the CF_{1+x}- Li cells is the generation of HF small amounts in system due to stayed and not removed moisture presence even at a level of tens PPM.

2. The role of HF microamounts generated in a fluorocarbon-lithium cells under long storage conditions at the reason at a slow hydrolysis by moisture traces can be functional not only for a direct reaction with lithium metal and constructive materials of the cell but also in secondary participation namely, in our assumption, by the autocatalytic transferring of metals from materials of the cover's and subcover's body as a various kinds of fluorocomplexes or mixed fluoro- and solvocomplexes onto separator and lithium anode with the subsequent reduction of these complexes by metal lithium and other corrosion reactions. These corrosion phenomena decreases the storage life of CF_{1+x}- Li cells and accelerates self-discharge processes.

3. The establishment of the pathways and mechanisms of such reactions with participation of microimpurities of water has the importance for understanding of happening processes, which are rather essential to an explanation of the corrosion phenomena and also for prediction of such fluorocarbon cathode's behavior under storage especially for long life and also for the development of principle and chemical methods of the effective suppression in system of the corrosion able microagents and creation of the new "know-how" lithium cells with 15-20 year's life.

Part II

Lithium Ion Rechargeable Batteries – Modeling

APPLICATION OF AB INITIO METHODS TO SECONDARY LITHIUM BATTERIES

M. K. AYDINOL, A. VAN DER VEN, G. CEDER
Massachusetts Institute of Technology, Department of Materials Science and Engineering,
Cambridge, MA 02139

ABSTRACT

Ab initio methods have started to be widely used in materials science for the prediction of properties of metals, alloys and compounds. These methods basically require only the atomic numbers of the constituent species. Such methods not only provide us with predictions of some of the properties of the material (even before synthesizing it) but also help us in understanding the phenomena that control those properties.

The use of ab initio methods in the field of electrochemistry is, however, quite recent and rare [1-4]. In this study, we demonstrate how ab initio methods can be used to investigate the properties of secondary lithium batteries. Particular examples will be given in predicting average insertion voltages in spinel Li-Mn and Li-Co oxides and in layered $LiMO_2$ (M = Ti, V, Mn, Fe, Co and Al) compounds. Additionally, the stability of these compounds to metal reduction and structural stability of $LiCoO_2$ upon lithium removal is investigated. We find that the oxygen anion plays an active role in the electrochemical intercalation of lithium. The amount of electron transfer to oxygen occurring upon lithium intercalation correlates strongly with the cell voltages. The more electron transfer to oxygen occurs, the higher lithium intercalation potential is obtained.

INTRODUCTION

Today's electronic devices are highly dependent on the properties of their power source. The properties of a good power source should include: light weight, cost effectiveness, high energy density, rechargeability and safety. Designing a power source to meet all these requirements is challenging. Lithium batteries are one of the best candidates to meet these requirements due to their potential for very high density and low environmental impact.

In this work we demonstrate, how computational methods can help us to design new materials for rechargeable lithium batteries. We first develop a thermodynamic model to calculate the average output voltage of a hypothetical battery with a transition metal oxide cathode of any choice of transition metal ion or crystal structure. We then use this to perform a systematical study of the effect of structure and chemistry on the average cell voltage. We then point out the structural stability issues in these oxides and investigate the possible transition metal-vacancy exchange at the cationic sites in the fully delithiated MO_2 compounds. This mechanism is one of the major reasons of capacity decay observed in several host structures, especially in $LiNiO_2$ [5,6], upon successive cycling. In the search and understanding of failure mechanisms, we investigate another possible mechanism which is the metal reduction instead of lithium during charging of the battery. We finally give an example from our efforts to find novel cathode materials.

METHODLOGY

Figure 1 shows schematically the operation of a rechargeable lithium battery. Upon discharge, Li^+ ions from the anode (high lithium chemical potential side) flow through an ionically conducting (but electronically insulating) electrolyte and insert into the host cathode (low lithium

65

chemical potential side). Meanwhile, to compensate charge neutrality in the host cathode, electrons flow through the external circuit. Charging of the battery is the reverse of this process. Rechargeability of the battery requires that this cycle can be performed several hundred times without loosing its capacity. This requires a cathode material where Li ions can easily be inserted and removed. In addition, the material must retain its structural integrity during this process and chemical reactions between electrode and electrolyte need to be prevented.

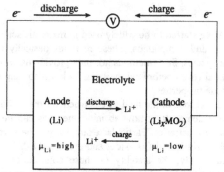

Figure 1. Schematics of a rechargeable lithium battery.

Li is accommodated in the cathode by intercalation into transition metal oxides such as $LiCoO_2$, $LiNiO_2$ and $LiMn_2O_4$. $LiCoO_2$ and $LiNiO_2$ adapt the layered α-$NaFeO_2$ structure whereas $LiMn_2O_4$ has a spinel structure. However, both $LiNiO_2$ and $LiMn_2O_4$ suffer severe capacity loss during repeated recharging. For $LiMn_2O_4$, cycling performance can be enhanced at the expense of capacity by substituting excess Li for Mn [7].

The open circuit voltage obtained from a lithium insertion reaction, is directly related to the difference in chemical potential of lithium in the anode and in the cathode [8];

$$E = -\frac{\mu_{Li}^{cathode} - \mu_{Li}^{anode}}{ze} \qquad (1)$$

where z is the number of electrons transferred and e is the electronic charge. If the anode is composed of pure metallic lithium, the anode chemical potential is constant and the variation of open cell voltage during the insertion process can be associated with changes of the lithium chemical potential in the cathodic host material.

Although it is difficult to compute the lithium chemical potential in the cathode as a function of lithium content [9-11], the *average* potential can be determined more easily. By integrating Equation 1 between the end-of-charge composition (x_1) and end-of-discharge composition (x_2), the average open cell voltage can be found as,

$$\overline{E} = -\frac{\Delta G_r}{(x_2 - x_1)zF} \qquad (2)$$

where F is the Faraday constant and ΔG_r is the Gibbs free energy change (in Joules) in the reaction,

$$\left(x_2 - x_1\right) \text{Li (anode)} + \text{Li}_{x_1} \text{Host (cathode)} \xrightarrow{\text{discharge}} \text{Li}_{x_2} \text{Host (cathode)} \qquad (3)$$

Assuming that the entropy and volume effects are negligible, ΔG_r ($\equiv \Delta E_r + P\Delta V_r - T\Delta S_r$) can be approximated by only the change in the internal energy (ΔE_r) at 0 °K. ΔE_r is approximately 3 to 4 eV, the term $P\Delta V_r$ is of the order of 10^{-5} eV and the entropy term is of the order of thermal energy, $k_B T$ (about 25×10^{-3} eV at room temperature). Therefore the above approximation is quite valid. Computing the variations in potential during the intercalation process can be done in a similar way. However, since ordered arrangements of Li and vacancies can occur during lithium removal [9-12], computing the variation in potential necessitates the information on structural changes in the host structure as the lithium content varies. This generally involves computing the energy of large supercells which can be tedious with accurate quantum mechanical methods [13,14]. Therefore, we restrict ourselves to the computation of the average insertion potential between the simple cathode composition limits. More elaborate analysis on phase stability and transformations, and potential changes during lithium removal from Li_xCoO_2 can be found elsewhere [9,10].

To calculate the total energies of the compounds investigated in this study, we used the *ab initio* pseudopotential method. This method has been known to be one of the most accurate quantum mechanical methods and its predictive power on the properties of the oxides has been well demonstrated [15]. In this method the Shrödinger equation is solved within the Local Density Approximation of the Density Functional Theory for the valence electrons only. The effect of the core electrons and the nucleus is represented by a pseudopotential [16]. The wave functions of the valence electrons are expanded in plane waves. We used the ultrasoft pseudopotential technique as implemented in the VASP code [17,18]. Ultrasoft pseudopotentials allow the use of moderately low energy cutoffs for the expansion of the plane waves. In this study, the energy cutoff for plane waves was set to 600 eV for the metal oxides and to 400 eV for the metallic elements. The reciprocal space sampling was done with 116 *k*-points for the oxides and 256 *k*-points for the metals in the irreducible Brillouin zone. All degrees of freedom in the structures were fully optimized so as to obtain the minimal ground state energy in the calculations.

INTERCALATION VOLTAGE AND EFFECT OF STRUCTURE AND CHEMISTRY

To assess the accuracy of our approach we have shown in Table I, the calculated and experimentally measured average voltages for three known metal oxide cathodes. $LiCoO_2$ and $LiVO_2$ are in the layered α-$NaFeO_2$ structure and $Li_xMn_2O_4$ is spinel. In $Li_xMn_2O_4$, there are two experimentally observed voltage plateaus [19,20]. These plateaus correspond to the intercalation of lithium between $0<x<1$ and $1<x<2$ in $Li_xMn_2O_4$. They are generally referred to as the 4 V and 3 V plateaus, respectively. Therefore, in Table I these plateau values for $Li_xMn_2O_4$ are given separately. The 1 V drop between the 4 V and 3 V plateaus has been attributed to the observed Jahn-Teller distortion of the Mn octahedron, when the average Mn valence state reaches 3.5^+ [20]. However recent ab initio calculations revealed that Jahn-Teller distortions can not be responsible for this drop [21]. It is seen from Table I that, agreement with experiments is quite reasonable, with the calculated value consistently lower than the measured.

Given the predictive character of our method, we investigated the effect of chemistry and structure on lithium intercalation potential in transition metal oxides. The effect of chemistry was systematically studied by changing the transition metal in $LiMO_2$ (M = Ti, V, Mn, Fe, Co, Zn and Al), keeping the system in the α-$NaFeO_2$ structure. In addition, the effect of structure was studied by calculating the insertion potential for $LiCoO_2$ in six different structures. The structures and the anion coordination around the metal ions are presented in Table II. The intercalation potential as a function of the host structure and metal ion chemistry is shown in Figure 2. In Figure 2 the spinel results are given over the full intercalation cycle between MO_2 and $Li_2M_2O_4$.

Table I. Calculated and measured cell voltages in known cathodes. Spinel $Li_xMn_2O_4$ values are given separately for the two voltage plateaus observed in the system.

	Calculated	Measured
Li_xCoO_2	3.7	4.0
$Li_xMn_2O_4$	3.9 and 2.6	4.1 and 2.7
Li_xVO_2	3.0	3.1

Table II. Crystal structures and their oxygen coordination of the metal ions.

	Structure	O coordination of M
γ-LiFeO$_2$	Body centered tetragonal	octahedral
MoS$_2$	Hexagonal	trigonal prismatic
CuFeS$_2$	Body centered tetragonal	tetrahedral
CdI$_2$	Hexagonal	octahedral
α-NaFeO$_2$	Trigonal	octahedral
Al$_2$MgO$_4$	Face centered cubic	octahedral

It is clear that both chemistry and structure have an effect on the battery voltage, but that of chemistry is considerably larger. For LiCoO$_2$, the difference between the structure with lowest and highest potential is only 0.5 V. Substitutions of the metal ion, on the other hand, lead to larger variations. A hypothetical LiAlO$_2$ cathode, for example, has an intercalation potential around 5.4 V.

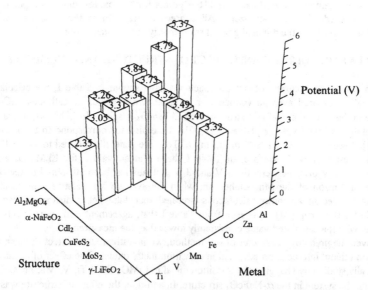

Figure 2. Calculated average cell potentials in LiMO$_2$ oxides having different structures

To understand why different transition metals yield different average voltages, it is important to know what electronic changes are occurring in the cathode during lithium intercalation. For this purpose we plot in Figure 3 the difference in the valence electron density between $LiCoO_2$ and CoO_2. This figure shows how incoming electrons are distributed in the crystal structure of the host oxide upon Li intercalation. Surprisingly, electrons are transferred not only to the Co ion but also to the oxygen. Table III gives the electron transfer to each oxygen and metal in $LiMO_2$ compounds. These numbers are obtained by integrating the difference of the electron density in $LiMO_2$ and MO_2 (for this purpose in identical geometry) in a sphere of radii 0.98 Å and 1.15 Å respectively for the metal and the oxygen ions. Since there are two oxygen ions per metal ion, the total electron transfer to oxygen is actually higher than to the transition metal. This result may force us to rethink the role played by the transition metal, which is commonly believed to be the electron acceptor in intercalation compounds. Clearly, Figure 3 shows that the oxygen anions participate in the electron reduction process. The electrochemical activity of the oxygen increases as one goes to the right in the 3d transition metal series as seen in Table III. Comparing Table III with Figure 2, shows a direct correlation between electron transfer to oxygen and high intercalation voltage. The very high intercalation voltage for the $LiAlO_2$ can therefore be almost considered as an upper bound on the average voltage in layered $LiMO_2$ compounds. Due to the strongly fixed valence of Al, electron exchange in Li_xAlO_2 is solely with the oxygen bands, as can be seen from Figure 4.

Table III. Fractional charge accepted by metal and each oxygen ion upon Li intercalation.

	Ti	V	Mn	Co	Zn	Al
Metal	0.262	0.253	0.289	0.276	0.067	0.030
Oxygen	0.216	0.245	0.254	0.25	0.273	0.322

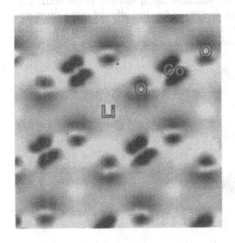

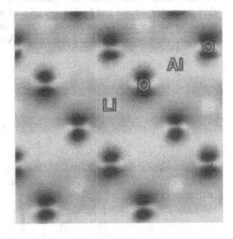

Figure 3. Positive part of the difference in electron density in $LiCoO_2$ in the (11•0) plane of the α-NaFeO$_2$ structure. Darker indicates higher electron density.

Figure 4. Positive part of the difference in electron density in $LiAlO_2$ in the (11•0) plane of the α-NaFeO$_2$ structure. Darker indicates higher electron density,

STRUCTURAL STABILITY OF THE HOST AND CATION-VACANCY EXCHANGE

The capacity of the battery depends directly on the amount of lithium that can be inserted and removed reversibly from the lithium metal oxide cathode. Many current battery designs suffer from a reduction of capacity as the battery is repeatedly charged and discharged. A complete picture of all the mechanisms that contribute to battery degradation for secondary lithium batteries does not yet exist. However, retaining the structural stability of the cathode host, during repeated lithium insertion and removal is one of the required properties for rechargeability. Phase transformations occurring in the cathode host, during lithium removal and insertion, can degrade the capacity of the battery by restricting Li accommodation in the host. In addition, volume changes accompanying these transformations can break the integrity of the battery assembly causing a mechanical failure. Especially, in deeply charged layered $LiNiO_2$ cathodes, it has been observed that [5,22], Ni ions migrate to the vacant Li planes in the structure, occupying the available sites for the guest lithium ions. Therefore limiting the amount of lithium ions that can be inserted [5,6].

We investigated this mechanism of cation-vacancy exchange by ab initio methods in fully delithiated layered CoO_2 and AlO_2. Although it is very unlikely that the AlO_2 limit can ever be obtained we use it here to compare with CoO_2. For this purpose we set up a supercell composed of nine primitive cells of the layered α-$NaFeO_2$ structure and computed the energies of the system at five distinct geometry along a most possible path of metal migration to a nearest neighbor lithium site. The geometry is shown in Figure 5. Along the migration path there are three intermediate geometry of interest. They are shown as solid triangles and a square in Figure 5. First, the metal ion should migrate through three oxygen ions having a triangular coordination. Then it reaches the tetrahedral site and another triangular site. Finally the metal ion reaches the octahedrally coordinated lithium site which is shown in Figure 5. We have performed total energy calculations at these five distinct geometry using the pseudopotential method. In the initial and final geometry, M site and Li site, we relaxed all the degrees of freedom allowed in the structure, including the volume, to get the equilibrium energies. For the intermediate geometry, we relaxed only the ionic positions within a sphere of 3 Å radii located at the corresponding site.

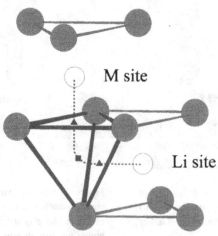

Figure 5. The path for M ion to migrate to Li plane in layered α-$NaFeO_2$ structure

The results are shown in Figure 6. As can be seen, for CoO_2, the energy of the system when Co is migrated to the Li site is above the one when Co remains at its M site. This indicates that even when there are no lithium ions present in the lithium plane, cobalt ions will not migrate to these unoccupied sites. In experimental literature there is no evidence for Co ions migrating to these empty Li sites. This has been attributed to the instability of Co^{3+} ions at the tetrahedral site along the path of migration [23]. However, our calculations clearly show that, there is no even driving force for this migration to occur in the Co system. The situation is, however, different for the delithiated Al system. Although there is a certain activation barrier (about 300 meV), Al ions may migrate to the empty lithium planes to reach to a lower energy state. The occupation of the available sites for Li intercalation, by the aluminum ions can degrade the capacity of the battery. However the results shown in Figure 6 are valid only when there are no lithium ions present at the Li sites. Presence of some lithium at these sites, on the other hand, may prevent some of the Al migration.

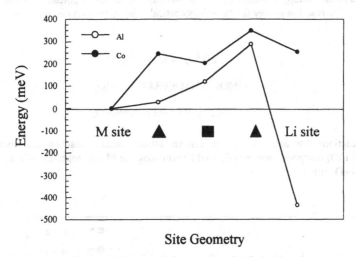

Figure 6. Energy of site geometry in Figure 5.

STABILITY AGAINST METAL REDUCTION

In this section, we investigate another instability, namely, the possible reduction of the metal ion instead of the lithium ion upon charging of the battery, which also can cause a capacity loss. In principle, a high enough charging potential will ultimately reduce all cations by driving them to the anode. Our objective is to investigate whether the equilibrium potential at which the transition metal is reduced from Li_xMO_2 in the α-$NaFeO_2$ structure is below the reduction potential for lithium.

Metal extraction and reduction may be given by the general reaction:

$$LiMO_2 \text{ (cathode)} \rightarrow xM \text{ (anode)} + \{Li, (1-x)M, 2O\} \text{ (cathode)} \qquad (4)$$

{Li, (1 - x)M, 2O} could be any product (single or multiphase) with the right stoichiometric composition. It is difficult to estimate the products of such reactions, but any reasonable product can serve to provide an *upper bound* for the reduction potential of the M ions.

Using the same thermodynamics, as in equations (1-3), the reduction potential for the metal can be found as:

$$V_M^{reduction} = \frac{G_{\{Li, (1-x)M, 2O\}} + x G_M - G_{LiMO_2}}{zF} \qquad (5)$$

The most stable {Li, (1 - x)M, 2O} product will result in the lowest reaction free energy and therefore give the lowest reduction potential. Any other product (real or hypothetical) will give a higher value (hence an upper bound) for the thermodynamic reduction potential. In this study we will calculate the reaction energy for three hypothetical solid state reactions which include metal reduction.

$$LiMO_2 \rightarrow M + LiO_2 \qquad (6.a)$$

$$LiMO_2 \rightarrow \tfrac{1}{4}M + \tfrac{1}{2}Li_2O + \tfrac{3}{4}MO_2 \qquad (6.b)$$

$$LiMO_2 \rightarrow \tfrac{1}{2}M + \tfrac{1}{2}Li_2O_2 + \tfrac{1}{2}MO_2 \qquad (6.c)$$

Whereas reactions 6.b and 6.c contain known lithium oxides, reaction 6.a contains the hypothetical LiO_2 compound which is derived by removing the M ions topotactically from $LiMO_2$ in the α-NaFeO$_2$ structure.

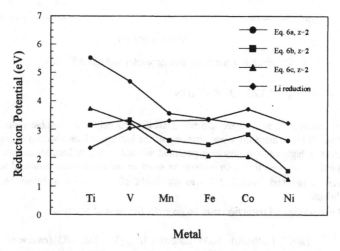

Figure 7. Metal and lithium reduction potentials upon charging of the battery.

In Figure 7, we present the results for $z = 2$ in Equation 5. This choice is somewhat arbitrary and will depend on the choice of the electrolyte. In most organic electrolytes high valence states are unlikely. In batteries with a $Li_xMn_2O_4$ cathode, Mn^{+2} has been detected in the electrolyte [24]. It can be seen from Figure 7 that equation 6.c is the always thermodynamically favored reaction, except for Ti, where the formation of Li_2O is more favorable than the other products. More importantly, the reduction potentials of the transition metals Mn, Fe, Co and Ni fall below that of the lithium reduction curve. This indicates that, thermodynamically these transition metals can be reduced from their states in $LiMO_2$. Whether such reduction actually takes place, will be determined by kinetic factors such as diffusion of the metal ion in the solid state and transport through the electrolyte. Recently evidence was presented for metal reduction in $LiCoO_2$ cells [25]. They found that significant amounts of Co deposited on the anode after charging the cell to 4.5V. Similar evidence for Mn deposits at the anode of cells $LiMn_2O_4$ cathodes has also been reported [24,26] and related to the capacity loss observed in this system.

DESIGNING A NEW CATHODE

In the above sections we showed how ab initio methods can be used to understand materials' properties of the cathodes used in rechargeable lithium batteries. In this section, we give an example how the understanding obtained from ab initio methods can be used to design new materials for battery applications.

The previous results, pointing at the role of the oxygen in electron transfer process during lithium intercalation, indicate that at least in principle, non-transition metals can also be used in lithium-metal-oxides for lithium insertion. Since most non-transition metals have fixed valence states, they will shift the electron transfer process more towards oxygen, thereby raising the lithium potential. One element of particular interest due to its abundance, low density and cost is Al. We found an average intercalation voltage of 5.4 V for Li_xAlO_2 in the layered α-$NaFeO_2$ structure, which is actually known as α-$LiAlO_2$ in this structure [27]. Pure $LiAlO_2$ is a wide band gap insulator, but in solid solution with lithium-transition metal oxides it may become conductive enough for use as a cathode. From computing the enthalpies of mixing at 0 °K, we found that lithium-transition metal oxides do not mix well in the solid solution with $LiAlO_2$. The mixing enthalpy in the $LiCoO_2$-$LiAlO_2$ system, however, is small (max. 30 meV), so that miscibility is expected at elevated temperatures. Table IV shows the average lithium insertion voltage for $Li_x(Al_yCo_{1-y})O_2$ mixtures. Clearly, Al increases the lithium intercalation voltage, in accordance with our expectations based on electrochemical activity of the oxygen anion. While a high insertion voltage can lead to a cathode with high energy density it is not always desirable as many electrolytes are only stable up to about 4.5 V. New gel-based electrolytes [28,29] however claim stability up to 5 V which may make these high voltage cathodes practical. In that case, Al has significant potential for decreasing cost and weight of the active cathode material.

Table IV. Effect of adding Al into $LiCoO_2$ on the average voltage.

%Al	0	25	33.33	50	66.67	75	100
Ave. V	3.7	4.1	4.2	4.4	4.6	4.8	5.4

CONCLUSIONS

In this study, we outlined how ab initio methods can be used in the understanding of the structure-property relationships of the oxide cathode materials used in rechargeable lithium batteries.

These methods allow us to predict the average voltages that can be obtained from any cathode material before actually synthesizing it. They can also provide us with the information of the working principles of the battery as well as potential failure mechanisms. This knowledge will ultimately help us designing new battery materials or improving the properties of the existing ones.

ACKNOWLEDGEMENTS

We gratefully thank the Pittsburgh Supercomputing Center, San Diego Supercomputing Center and the Center for Theoretical and Computational Materials Science of the National Institute of Technology for providing us with computing resources. Support from Idaho National Engineering Laboratory is gratefully acknowledged. Earlier parts of this study were supported by Furukawa Electric.

REFERENCES

1. J. N. Reimers and J. R. Dahn, Phys. Rev. B **47**, 2995 (1993).

2. K. Miura, A. Yamada and M. Tanaka, Electrochim. Acta **41**, 249 (1996).

3. M. K. Aydinol, A. F. Kohan, G. Ceder, K. Cho and J. Joannopoulos, Phys. Rev. B **56**, 1354 (1997).

4. G. Ceder, M. K. Aydinol and A. F. Kohan, Comput. Mat. Sci. **8**, 161 (1997).

5. T. Ohzuku, A. Ueda and M. Nagayama, J. Electrochem. Soc. **140**, 1862 (1993).

6. J. R. Dahn, U. von Sacken, M. W. Juskow and J. Al-Janabi, J. Electrochem. Soc. **138**, 2207 (1991).

7. R. J. Gummow, A. de Kock and M. M. Thackeray, Solid State Ionics **69**, 59 (1994).

8. W. R. McKinnon, in Solid State Electrochemistry, edited by P. G. Bruce (Cambridge University Press 1995) p. 163.

9. A. Van der Ven, M. K. Aydinol and G. Ceder, J. Electrochem. Soc., submitted for publication (1998).

10. A. Van der Ven, M. K. Aydinol and G. Ceder, These proceedings, 1997 MRS Fall Meeting, Boston, MA, 1997.

11. C. Wolverton and A. Zunger, These proceedings, 1997 MRS Fall Meeting, Boston, MA, 1997.

12. J. N. Reimers, J. R. Dahn and U. von Sachen, J. Electrochem. Soc. **140**, 2752 (1993).

13. G. Ceder, Comput. Mat. Sci. **1**, 144 (1993).

14. D. de Fontaine, in Solid State Physics, edited by H. Ehrenreich, D. Turnbull (Academic Press 1994) vol. 47, p. 33.

15. A. F. Kohan and G. Ceder, Comput. Mat. Sci. **8**, 142 (1997).

16. M. C. Payne, M. P. Teter, D. C. Allan, T. A. Arias and J. D. Joannopoulos, Rev. Mod. Phys. **64**, 1045 (1992).

17. G. Kresse and J. Furthmuller, Comput. Mat. Sci. **6**, 15 (1996).

18. G. Kresse and J. Furthmuller, Phys. Rev. B **54**, 11169 (1996).

19. J. M. Tarascon, E. Wang, F. K. Shokoohi, W. R. McKinnon and S. Colson, J. Electrochem. Soc. **138**, 2859 (1991).

20. T. Ohzuku, M. Kitagawa and T. Hirai, J. Electrochem. Soc. **137**, 769 (1990).

21. M. K. Aydinol and G. Ceder, J. Electrochem. Soc. **144**, 3832 (1997).

22. A. Rougier, P. Graverau and C. Delmas, J. Electrochem. Soc. **143**, 1168 (1996).

23. G. G. Amatucci, J. M. Tarascon and L. C. Klein, J. Electrochem. Soc. **143**, 1114 (1996).

24. D. H. Jang, Y. J. Shin and S. M. Oh, J. Electrochem. Soc. **143**, 2204 (1996).

25. G. G. Amatucci, J. M. Tarascon and L. C. Klein, Solid State Ionics **83**, 167 (1996).

26. J. M. Tarascon, W. R. McKinnon, F. Coowar, T. N. Bowmer, G. Amatucci and D. Guyomard, J. Electrochem. Soc. **141**, 1421 (1994).

27. T. Ohzuku, A. Ueda and M. Kouguchi, J. Electrochem. Soc. **142**, 4033 (1995).

28. D. Guyomard and J. M. Tarascon, J. Power Sources **54**, 92 (1995).

29. J. M. Tarascon and D. Guyomard, Solid State Ionics **69**, 293 (1994).

FIRST-PRINCIPLES THEORY OF CATION AND INTERCALATION ORDERING IN Li_xCoO_2

C. WOLVERTON and ALEX ZUNGER
National Renewable Energy Laboratory, Golden, CO 80401

ABSTRACT

Several types of cation- and vacancy-ordering exist in the Li_xCoO_2 battery material. The ordering patterns are of interest due to the fact that they can control the voltage in rechargeable Li batteries. We present a first-principles total energy theory which can predict both cation- and vacancy-ordering patterns at both zero and finite temperatures. Also, by calculating the energetics of the Li intercalation reaction, this theory can provide first-principles predictions of battery voltages of Li_xCoO_2/Li cells. Our calculations allow us to search the entire configurational space to predict the lowest-energy ground state structures, search for large voltage cathodes, explore metastable low-energy states, and extend our calculations to finite temperatures, thereby searching for order-disorder transitions and states of partial disorder.

INTRODUCTION: TYPES OF ORDERING IN Li_xCoO_2

The $LiMO_2$ oxides form a series of structures based on an octahedrally-coordinated network with anions (O) on one fcc sublattice and cations (Li and M) on the other. [1] In this paper, we examine the energetics and thermodynamics of ordering tendencies in the Li_xCoO_2 oxide. The $LiCoO_2$ compound is used as a cathode material in rechargeable Li batteries. [2] When Li is de-intercalated from the compound, it creates a vacancy (denoted □) that can be positioned in different lattice locations. Hence, we will examine three types of ordering problems: (i) Li/Co ordering in $LiCoO_2$ ($x=1$) leads to ordered $R\bar{3}m$ at low temperature and disordered Li/Co (rock-salt) at high temperature. (ii) Similarly, □/Co ordering in the completely deintercalated $□CoO_2$ ($x=0$) is also of interest. (iii) The vacancies left behind by Li extraction can form ordered vacancy compounds in partially deintercalated Li_xCoO_2, leading to a □/Li ordering problem for intermediate compositions $0 \leq x \leq 1$. We refer to the first two of these ordering problems as *cation ordering* since they occur at fixed composition (either fully intercalated or deintercalated) and deal with the positions of the cations. The third type of ordering we call *intercalation ordering* since it deals with a fixed CoO_2 framework and the ordering may occur upon intercalation or deintercalation of Li as a function of composition.

METHOD

We use a three-step first-principles approach to study ordering. The salient points are described here, and more details of the method may be found in [3]:

(1) *Total energy calculations:* We calculate the $T=0$ total energy of a set of (not necessarily stable) ordered structures via the full potential, all-electron linearized augmented plane wave method (LAPW) [4] with all atomic positions fully relaxed via quantum mechanical forces. We then map those energies onto a

(2) *Cluster expansion (CE)*. The cluster expansion (CE) technique (see, e.g. Refs. [5, 6]) consists of an Ising-like expression in which each substitutional unit is associated with the site of an ideal lattice, and the "spin variable" S_i is given the value $+1(-1)$ if an $A(B)$ atom is assigned to site i. (Note that in the three ordering problems studied, A/B = Li/Co, □/Co, and □/Li.)

Mat. Res. Soc. Symp. Proc. Vol. 496 © 1998 Materials Research Society

	Disordered	CuPt-type (CP)	D4	CuAu-type (CA)	Chalcopyrite (CH)	Y2	W2	V2	Z2
Octahedral (Rocksalt)	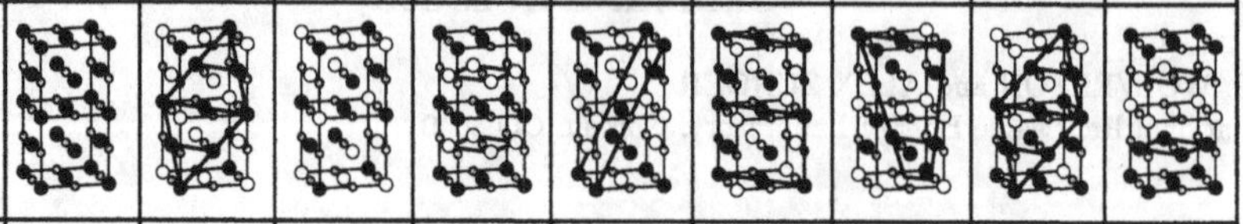								
Cation Superlattice	—	(1,1) along [111]	—	(1,1) along [001]	(2,2) along [201]	(2,2) along [110]	(2,2) along [311]	(2,2) along [111]	(2,2) along [100]

Figure 1: Cation arrangements in octahedral oxide networks. The black and white atoms represent the cations, while the grey atoms are the anions.

Within this description, the energy of *any* configuration σ can be written as:

$$E_{\mathrm{CE}}(\sigma) = \sum_f D_f J_f \overline{\Pi}_f(\sigma), \tag{1}$$

where f is a figure comprised of several lattice sites (pairs, triplets, etc.), D_f is the number of figures per lattice site, J_f is the Ising-like interaction for the figure f, and $\overline{\Pi}_f$ is a function defined as a product over the figure f of the variables S_i, averaged over all symmetry equivalent figures of lattice sites. We determine $\{J_f\}$ by fitting $E_{\mathrm{CE}}(\sigma)$ of N_σ structures to LDA total energies $E_{\mathrm{LDA}}(\sigma)$, given the matrices $\{\overline{\Pi}_f(\sigma)\}$ for these structures. Once the coefficients of the expansion are known, the Ising-like expression may be easily evaluated for any substitutional configuration. Thus, one can calculate (via first-principles) the total energy of *a few* cation arrangements, but then effectively search the space of 2^N configurations. Having obtained such a general and computationally simple parameterization of the configuration energy, we subject it to

(3) *Monte Carlo simulated annealing.* The third step of our approach involves subjecting the cluster expansion to Monte Carlo simulations so as to investigate the thermodynamic properties of the cation or intercalation ordering. System sizes of 8^3-32^3=512-32,768 sites were used with between 200-200,000 spin flips per site. For $T \neq 0$, both canonical (fixed composition) and grand canonical (fixed chemical potential) simulations have been performed. Phase boundaries are located by monitoring the composition and energy, as well as their second- and fourth-order cumulants. For first-order transitions, strong hysteresis necessitates the use of thermodynamic integration (which has been successfully applied to Monte Carlo studies of the Ising fcc and triangular antiferromagnets. [7]) For second-order (or weakly first-order) transitions, simulations were performed for a variety of cell sizes in order to investigate finite-size effects. For $T = 0$, we use a simulated annealing algorithm to predict the lowest energy states among the astronomical (2^N) number of possibilities.

CATION (Li/Co) ORDERING: x=0 AND 1

Here, we construct three separate cluster expansions to describe three different types of structural energetics:

(a) <u>Li/Co ordering</u>: Formation enthalpies for different Li/Co arrangements σ of LiCoO$_2$ (x=1) on the fcc lattice:

$$\Delta H_f(\sigma, \underline{\mathrm{LiCoO_2}}) = E_{\mathrm{tot}}(\sigma, \underline{\mathrm{LiCoO_2}}) - E_{\mathrm{tot}}(\mathrm{LiO}, B1) - E_{\mathrm{tot}}(\mathrm{CoO}, B1) \tag{2}$$

where the last two terms refer to LiO and CoO in the NaCl ($B1$) structure with lattice constants obtained by minimizing the respective total energies with respect to hydrostatic deformation.

(b) □/Co ordering: Formation enthalpies for different □/Co arrangements σ of $\square CoO_2$ (x=0) on the fcc lattice:

$$\Delta H_f(\sigma, \underline{\square CoO_2}) = E_{tot}(\sigma, \underline{\square CoO_2}) + E_{tot}(Li, bcc) - E_{tot}(LiO, B1) - E_{tot}(CoO, B1) \quad (3)$$

where $E_{tot}(Li, bcc)$ is the total energy of Li in the bcc structure with lattice constant obtained from total energy minimization.

(c) Effect of cation ordering on average voltage: For a $Li_x CoO_2/Li$ cell, the voltage $V(x)$ as a function of Li composition is given by [8, 9] the Li chemical potential difference between cathode ($Li_x CoO_2$) and anode (Li metal):

$$-eV(x) = \mu_{Li}(Li_x CoO_2) - \mu_{Li}(Li \text{ metal}) \quad (4)$$

From this expression, it is straightforward to show the following (see, e.g., Refs. [10] or [11]):

$$\Delta H_{react}^{\sigma}(x_2, x_1) = -F \int_{x_1}^{x_2} dx V(x) = -(x_2 - x_1) F \overline{V}(\sigma, x_2, x_1) \quad (5)$$

where F is the Faraday constant and the reaction energy of Li intercalation between two Li compositions x_1 and x_2 is given by

$$\Delta H_{react}^{\sigma}(x_2, x_1) = E_{tot}(Li_{x_2} CoO_2, \sigma) - E_{tot}(Li_{x_1} CoO_2, \sigma) - (x_2 - x_1) E_{tot}(Li, bcc) \quad (6)$$

E_{tot} is the total energy of a system of electrons in the Coulomb potential due to the nuclei. $\Delta H_{react}^{\sigma}$ is the energy gained upon de-intercalation of Li from $LiCoO_2$, relative to Li metal. Therefore, by computing the energetics of Li intercalation in Eq. (6), we ascertain the average battery voltage.

Energetics of Li/Co ordering in $LiCoO_2$

The formation energies [Eq. (2)] of $LiCoO_2$ in various cation arrangements are given in the second column of Table I. We note that the "cubic" D4 structure is only slightly higher in energy than the "layered" CuPt structure, the latter being the observed phase of $LiCoO_2$ grown at high temperature. This competition is interesting because $LiCoO_2$ has been synthesized in the D4 structure by solution growth at low temperature (see, e.g., [12, 13]).

Energetics of □/Co ordering in $\square CoO_2$

The formation energies [Eq. (3)] of $\square CoO_2$ in various □/Co arrangements are given in the third column of Table I. These configurations correspond to various arrangements of Co and □. We note the following:

(1) The relative order of energetics is similar in $\square CoO_2$ as in $LiCoO_2$. There is only one qualitative difference: CH drops in energy significantly upon extraction of Li, and is lower in energy than the Y2 structure, whereas the reverse is true for $LiCoO_2$.

(2) The separation in energy between "layered" CuPt and "cubic" D4 increases in $\square CoO_2$ compared to $LiCoO_2$, due to the symmetry of the phases: Upon extraction of Li in the rhombohedral CuPt structure, the c/a ratio decreases significantly, providing a significant source of energy lowering for $\square CoO_2$ - CuPt. D4, on the other hand, is not a layered superlattice in any direction and has cubic symmetry. Hence, the cell parameters of $\square CoO_2$ (D4) cannot distort in any preferred direction, and consequently, $\square CoO_2$ (D4) does not relax as much as CuPt.

(3) The CuPt structure of $\square CoO_2$ has an ABC... stacking of the cation planes. However, recent electrochemical measurements of Amatucci et al. [14] have succeeded in completely de-intercalating Li from $LiCoO_2$, forming a $\square CoO_2$ structure with the stacking of planes in an AAA... arrangement which we call "CuPt (AAA)". We have performed total energy calculations

Table I: FLAPW calculated formation energies (eV/formula unit) of various cation arrangements in LiCoO$_2$ and $\square$CoO$_2$+Li(bcc): $\Delta H_f(\sigma, \text{LiCoO}_2)$, $\Delta H_f(\sigma, \square\text{CoO}_2)$, (formation energies of σ) and $\Delta H_{\text{react}}(\sigma)$ (average intercalation voltage of LiCoO$_2$ relative to Li) are defined in Eqs. (2), (3), and (6) respectively.

Cation Structure	LiCoO$_2$ $\Delta H_f(\sigma)$ Eq. (2)	$\square$CoO$_2$+Li(bcc) $\Delta H_f(\sigma)$ Eq. (3)	ΔH_{react} Eq. (6)
CuPt	-3.38	+0.40	-3.78
D4	-3.37	+0.54	-3.91
Y2	-3.07	+0.80	-3.87
CH	-2.84	+0.64	-3.48
W2	-2.82	+0.94	-3.76
CuAu	-2.23	+1.65	-3.88
V2	-2.02	+2.88	-4.90
Z2	-2.13	+2.38	-4.51
Random(η=0,SRO=0)	-2.68	+1.31	-3.99
Disordered(η=0,SRO$\neq$0)	-2.95	+0.91	-3.86
CuPt(η=0.88,SRO=0)	-3.22	+0.60	-3.82
D4(η=0.88,SRO=0)	-3.21	+0.71	-3.92

of $\square$CoO$_2$ in both the CuPt and CuPt (AAA) structures. Consistent with the observations of Amatucci *et al.* [14], we find that the $\square$CoO$_2$ in the AAA stacking is lower in energy than the CuPt structure by $\sim$0.05 eV/formula unit.

(4) We find that LiCoO$_2$ in the CuPt (AAA) structure is higher in energy than the CuPt structure (with ABC stacking) by $\sim$0.15 eV/formula unit, in agreement with the fact that the observed CuPt ground state in LiCoO$_2$ has ABC stacking.

Effect of cation arrangement on average voltages

Table I gives the calculated reaction energies given in Eq. (6) for each of the cation arrangements σ studied here. The average voltages for all cation arrangements considered are in the $\sim$4 V range. In particular, the average voltage for LiCoO$_2$ in the CuPt structure (3.78 V) is in reasonable agreement with measured values (4.0-4.2 V). [2]

Ground States of Li/Co Ordering and $\square$/Co Ordering

The simulated annealing algorithm finds the "layered" CuPt structure as the predicted thermodynamic low temperature state. This is interesting because the CuPt cation structure is observed when LiCoO$_2$ is grown at high temperatures, but when grown at low temperatures, [12, 13] the cubic D4 phase results. Our calculations indicate that the layered CuPt phase is the thermodynamic ground state of LiCoO$_2$, and the growth of the D4 structure must therefore be due to kinetic limitations of low-temperature growth. In Table I, it was simply noted that the CuPt structure was the lowest in energy of the eight structures calculated by LAPW. But, the simulated annealing prediction of the ground state demonstrates that CuPt is also the lowest energy configuration out of an astronomical number of possible configurations (without symmetry, there are $\sim 2^N$ possible configurations that the algorithm could explore, where N=4096). For our cluster expansion of $\square$CoO$_2$, the simulated annealing algorithm also finds the layered CuPt as the lowest energy substitutional configuration. As we have already shown above, non-substitutional configurations are even lower in energy for the $\Box CoO_2$ system (e.g., the AAA stacking).

Order-Disorder Transitions for Li/Co Ordering

For $LiCoO_2$, the order-disorder transition between the low-temperature "layered" CuPt phase and the high-temperature disordered (rocksalt) phase is predicted to occur at ~5100 K, well above the melting point of this material. Our calculations indicate that a synthesized [15] disordered rocksalt phase of $LiCoO_2$ *is not thermodynamically stable*, but is rather only stabilized kinetically, consistent with the fact that the disordered phase can only be grown at low temperatures. Upon complete removal of Li, the order-disorder transition of $\Box CoO_2$ drops to ~4400 K, still much too high to be experimentally accessible.

Properties of Disordered and Partially Ordered Cation Arrangements

We show the cluster expansion energetics of several phases with varying degrees of disorder in Table I. The random cation arrangement is shown as is the energetics of a disordered phase with short-range order (SRO). The energetic effect of SRO is to significantly lower the energy of the random phase in both $LiCoO_2$ and $\Box CoO_2$ by 0.27 and 0.40 eV/formula unit, respectively. In Table I, we also show the energetics of CuPt and D4 structures with partial long-range order corresponding to 6% of Li on the Co sites, and vice versa. The $LiCoO_2$ energies of CuPt and D4 are both raised by 0.16 eV/formula unit relative to the $\eta=1$ fully ordered phases, while the corresponding increases for $\Box CoO_2$ is 0.20 and 0.17 eV/formula unit.

The cluster expansion of voltage can also be used to predict the average voltages of these disordered phases. (Table I). The random alloy (3.99 V) is predicted to have a higher average voltage than the ordered CuPt phase (3.78 V). The increase in voltage due to disorder is significantly reduced when one considers the disordered phase with SRO described above (3.86 V). The voltages of partially long-range ordered CuPt and D4 phases are increased relative to CuPt and D4 by 0.05 and 0.01 V, respectively. Note that for either long- or short-range order, the qualitative effect of disordering is the same: Disorder raises the energy of $\Box CoO_2$ more than $LiCoO_2$, and thus raises the average voltage.

FIRST-PRINCIPLES PREDICTION OF LI INTERCALATION VOLTAGES IN RHOMBOHEDRAL AND CUBIC $LiCoO_2$

The material $LiCoO_2$ ($x=1$) has been synthesized in two ordered forms (see Fig. 2): A rhombohedral form, [2, 1] the "layered" or "CuPt-like" structure, and and cubic form which has been produced by solution growth at low temperatures ("D4"). [12, 13] "Layered" CuPt and "cubic" D4 are extremely similar in terms of atomic coordination sequence: Pair and three-body correlations are equivalent, with the first difference between the two occurring at the four-body correlation. Thus, one might expect the two forms of Li_xCoO_2 to exhibit similar energies and electrochemical potentials. However, electrochemical properties of the two compounds are very different: When used as a cathode material in Li_xCoO_2/Li cells, the cubic D4 structure has a nearly flat voltage plateau at 3.6 V, (for $1/2 < x \leq 1$) which contrasts with the voltage profile of the layered CuPt phase which takes place mostly above 4 V and has several voltage drops and plateaus. This distinction is initially difficult to understand given the structural similarity between the two, and the expected similar stability. Two possible explanations could be offered for this unexpected distinction: (a) Despite the similarities in the nominal (i.e., undistorted) CuPt and D4 structures, the differences in symmetry between the two results in different structural distortions: CuPt is a layered LiO/CoO superlattice along the [111] direction, while D4 is not a superlattice. Thus, the CuPt structure has one extra structural degree of freedom (namely a c/a ratio) that the D4 structure does not have. To the extent that $R_{Li} \neq R_{Co}$, the layered nature of

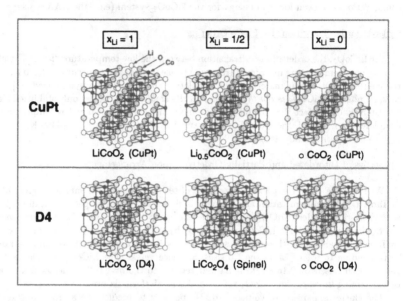

Figure 2: The CuPt and D4 structures of Li_xCoO_2 for $x=1$, $\frac{1}{2}$, and 0. Li, Co, and O atoms are shown as large white, small black, and large grey circles, respectively.

Table II: Calculated average intercalation voltages $\overline{V}(\sigma, x_2, x_1)$ [Eqs. (6) and (5)] for Li_xCoO_2 in the structures $\sigma =$ CuPt and D4.

	CuPt - Structure Preserving	D4 - Structure Preserving	D4 - Non-Structure Preserving
$\overline{V}(\sigma,1,0)$	3.78	3.91	3.91
$\overline{V}(\sigma,1,\frac{1}{2})$	3.37	3.50	3.04
$\overline{V}(\sigma,\frac{1}{2},0)$	4.19	4.32	4.78

the CuPt phase will distort the Li-Co interplanar distances from its ideal value changing both relative total energies, and the x-ray diffraction pattern (e.g., splitting and shifting of peaks) in a measurable way. We have already seen (Table I) that the predicted average voltage between $x=0$ and 1 is larger in the D4 phase than in the CuPt phase, due to the difference in relaxation between the cubic and rhombohedral phases. However, experimental reports [13, 12] find a significantly *lower* voltage (by ~0.5 V) for D4 (for $x > 1/2$), compared with CuPt. (b) Another possibility for the electrochemical distinction between CuPt and D4 is that the structure of these two compounds could change with Li composition x, and correspondingly, relative stability between the two phases might change.

We study the two possibilities (a) and (b) using first-principles total energies to ascertain the structural and electrochemical properties (i.e., Li intercalation energies) of the two compounds. To do so, we explore $x = 1/2$ "derivative Li_xCoO_2 structures" for intermediate compositions.

<u>Results of Total Energy Calculations and Battery Voltages</u>

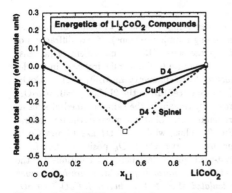

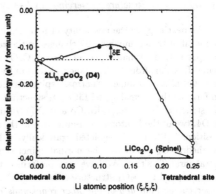

Figure 3: LAPW calculated energetics of Li_xCoO_2 compounds in the CuPt, D4, and spinel structures. The empty (filled) circles show the energetics of "structure preserving" removal of Li from the D4 (CuPt) structures.

Figure 4: Total energy of $LiCo_2O_4$ as the Li atoms are continuously distorted from the T_d sites (Spinel) to the O_h sites ($x=1/2$ D4) along the (ξ,ξ,ξ) path.

$x_{Li}=1/2$: "Structure Preserving" Li Removal

In an attempt to reconcile this discrepancy between calculated and measured voltages of CuPt and D4, we consider several "derivative" structures $Li_{0.5}CoO_2$ formed by partial removal of Li ($x=1/2$) from the parent structures (CuPt or D4). First, we consider partially de-lithiated $x=1/2$ structures formed by "structure preserving" removal of Li from the parent structures (i.e., removal of Li without changes in the positions of the remaining atoms): We have computed the energetics of two $Li_{0.5}CoO_2$ compounds formed from CuPt and D4, respectively, by removing half of the Li atoms. All atomic positions are subsequently relaxed to their minimum energy positions. The CuPt-based $Li_{0.5}CoO_2$ structure (Fig. 2) considered is the ordered vacancy compound observed electrochemically by Reimers et al. [16], which we call "2x1" (see below). For the D4-based $Li_{0.5}CoO_2$ structure, we have simply chosen the only compound which corresponds to removing half the Li from D4 while not increasing the unit cell size (not shown in Fig. 2). The energies of these two structure preserving $x=1/2$ compounds are shown in Fig. 3. The amount by which the energy of $Li_{0.5}CoO_2$ is below the endpoints (indicating the energetic extent of vacancy ordering) is almost identical for the CuPt and D4 cases: Both $x=1/2$ compounds are ~0.2 eV/formula unit below the average of their endpoint compounds. This equality implies that the Li/□ ordering tendency is relatively independent of cation structure.

By using Eq. (5), we may now compute the average voltages in two segments: from $x=1$ to $x=1/2$, and from $x=1/2$ to $x=0$. From Table II, one can see that the average voltage is lower for high Li contents than for low Li contents. However, removal of Li changes the voltages of the CuPt and D4 structures in an almost identical way (due to the similarity of Li/□ ordering tendencies) such that the voltage of the CuPt compound is still lower than D4 for each of the two segments. *Thus, the "structure preserving" removal of Li cannot explain the observed electrochemical differences between CuPt and D4.*

We next explore the possibility of non structure preserving removal of Li from CuPt and D4 (removal of Li accompanied by changes in the remaining atomic positions): X-ray diffraction of the partially de-lithiated ($x_{Li} \sim 1/2$) $Li_x CoO_2$ D4 phase shows [13] the movement of (at least some) Li atoms from their octahedral sites in the D4 structure to the tetrahedral sites in the de-lithiated phase, forming a "normal" spinel structure with stoichiometry $LiCo_2O_4 = 2Li_{1/2}CoO_2$. In forming the normal spinel $LiCo_2O_4$ structure (14 atoms/cell) from the $2LiCoO_2$ ($=Li_2Co_2O_4$) D4 structure (16 atoms/cell), Co and O (12/atoms/cell) remain in the same octahedral sites as in D4, two of the Li atoms are removed, and the remaining two Li atoms move to the tetrahedral positions. [13] This is depicted graphically in Fig. 2. Thus, while $x=1$ D4 has Li and Co in octahedral O_h positions, the normal spinel structure has two Co in O_h positions, but Li in tetrahedral T_d sites. We have also performed first-principles total energy calculations for the normal spinel $LiCo_2O_4$ structure (Fig. 3). *We find that the normal spinel structure has a much lower energy than any of the other compounds calculated at $x=1/2$, including $LiCoO_2 + CoO_2$ in the D4 structure.* This low-energy spinel suggests three conclusions:

(i) Upon extraction of one Li from the $Li_2Co_2O_4$ D4 phase it transforms into the $LiCo_2O_4$ normal spinel. The formation of the spinel phase from D4 can be imagined as follows: one Li atom is removed and the other moves (from O_h to T_d sites); but, the Co and O atoms are not required to move in this transformation. One might ask, however, whether or not CuPt $Li_2Co_2O_4$ (rather than D4) forms the normal spinel structure upon removal of one Li. This is much less likely, since both Li *and* Co atoms would have to change positions to form the normal spinel structure from CuPt: Again, this transformation involves removal of one Li and one Li atom moving from O_h to T_d, however, some of the Co atoms would have to move *substitutionally* from Co O_h sites to Li O_h sites. The only way for Co atoms to make this movement would be to interrupt the close-packed oxygen sublattice. Thus, the formation of the $x=1/2$ normal spinel from the $x=1$ CuPt-based structure, while thermodynamically favored, is presumably kinetically inhibited.

(ii) The low energy of the normal spinel leads [via Eq. (5)] to a significant reduction of the D4 average voltage from $x=1$ to $x=1/2$ (Table II), thus explaining the observed difference in voltage between CuPt and D4. [13, 12] However, our calculations indicate that although the average voltage of the D4 structure is lower than CuPt down to $x=1/2$, subsequent extraction of Li from the spinel structure corresponds to removal of Li from the (energetically favorable) T_d sites, and thus costs a large amount of energy. Thus, the average voltage of the D4 structure is predicted to rise sharply (4.78 V) from $x=1/2$ to $x=0$. Electrochemical measurements [13, 12] of D4 have commonly probed only relatively high Li contents ($x > 0.5$); However, Gummow et al. [13] have extracted Li electrochemically from the spinel $LiCo_2O_4$, observing a sharp increase in voltage (of ~ 1 V) near $x=1/2$, in agreement with our calculations. If the capacity of Li extraction in the spinel phase $LiCo_2O_4$ could be improved, our calculations provide a prediction that this spinel would make a high voltage (4.78 V) battery cathode.

(iii) Because the spinel phase and the D4 are structurally distinct, the D4 $Li_x CoO_2$ system presumably forms a two-phase mixture of spinel+D4 for values of x_{Li} between 1/2 and 1, (as opposed to tolerating a large off-stoichiometry in either of the phases - in other words, when one begins removing Li from D4, one forms small pockets of spinel embedded in the D4 matrix). A two-phase mixture corresponds to straight line (a tie-line) in energy *vs.* composition (the dashed line in Fig. 3). The voltage is proportional to the *slope* of the energy *vs.* composition curve, and hence for a two phase mixture, the voltage is constant. This is consistent with the measured voltage curves for D4 (down to $x_{Li}=1/2$), which show a nearly constant plateau at 3.6 V. Thus, *most of the electrochemical distinctions between the CuPt and D4 phases of $LiCoO_2$ are explained by the low energy $LiCo_2O_4$ spinel.*

Tetrahedral vs. octahedral site preference energies

To examine the *stability* of the Li and Co atoms with respect to distortions from O_h to T_d sites, we have calculated the energetics of the pathway between T_d and O_h sites. We have also mapped the energy of the $LiCo_2O_4$ phase as Li is continuously moved from the octahedral $(0,0,0)$ positions to the tetrahedral $(\frac{1}{4},\frac{1}{4},\frac{1}{4})$ sites along the direct pathway (ξ,ξ,ξ). Along this transition, when Li is at the T_d sites, the structure is the $LiCo_2O_4$ normal spinel structure. When the Li is moved to the O_h sites, the structure becomes the $Li_{0.5}CoO_2$ "structure preserving" version of D4. The energetics along this pathway indicate (Fig. 4) that a small barrier does exist between the O_h and T_d sites, so that the octahedral positions are *metastable*, not unstable. The magnitude of the energy barrier (~ 0.07 eV) corresponds to $\exp(\delta E/kT) \sim 10^{-1} - 10^{-2}$ at room temperature. Thus, Li atoms attempting to cross this barrier (with roughly the Debye frequency) will easily do so on the time scale of seconds and move from octahedral to tetrahedral sites.

INTERCALATION ($\square$/Li) ORDERING: $0\leq x \leq 1$. PREDICTION OF ORDERED VACANCY COMPOUNDS AND THE Li_xCoO_2 PHASE DIAGRAM

We have already seen that vacancy ordered compounds are stable for partially deintercalated Li compositions. Indeed, ordered vacancy compounds of intercalated Li atoms have been observed in several layered lithium-transition metal dichalcogenides, such as Li_xTiS_2, Li_xTaS_2, and $LiCoO_2$. We demonstrate that our first-principles theoretical approach can be used to predict the ground state intercalation ordering, voltage profiles, and ultimately the voltage-temperature phase diagram of Li_xCoO_2. We confirm the stability of the observed ordered vacancy phase at $x=1/2$ for $T < 330K$, and predict several new ordered vacancy compounds, the temperature-dependent voltage profile of this system, and the full Li-intercalation equilibrium phase diagram for the Li_xCoO_2/Li system.

The Li sites in the $R\bar{3}m$ CuPt (layered) structure form close-packed (111) planes which yield a two-dimensional triangular lattice, and these Li planes are stacked in a rhombohedral fashion (ABC). Thus, the problem of intercalation ordering of Li and Li-vacancy (denoted $\square$) is an ordering problem on the stacked triangular lattice. We use the same three-stage procedure described above to both finite temperature and zero temperature properties of intercalation ordering.

We have calculated total energies for a large number of ordered vacancy compounds constructed by removing Li atoms from the $LiCoO_2$ (CuPt) structure. Many of the compounds considered are shown in Fig. 5a. We have also chosen some additional structures to investigate the three-dimensional nature of the ordering (on the ABC triangular stacked planes): For the "2x1" structure, there are two inequivalent ways of stacking 2x1 planes (Fig. 5), and total energies for both these structures were computed. Also, two "staged" structures were considered formed by alternating fully filled and empty 1x1 planes. One structure formed in this way at $x=1/2$ consists of $Li/\square/Li/\square...$ stacking of planes, and an analogous $x=2/3$ structure consists of $Li/Li/\square/Li/Li/\square...$ stacking. In all, total energies for 16 different ordered vacancy compounds were computed.

These 16 energies are mapped onto an Ising-like model with 2, 3, and 4-body interactions, using the cluster expansion (CE) of Eq. (1). In fitting the expansion of Eq. (1), we have used the following 12 figures (and the corresponding interactions): empty, point, first- through fifth-neighbor "in-plane" pairs two "in-plane" triplets (one consisting of three nearest neighbor bonds and the other consisting of two nearest neighbor and one second neighbor bond), one "in plane" quadruplet (five nearest neighbor and one second neighbor bond), and two "out-of-plane" pairs. Figures denoted "in-plane" connect sites in a single triangular plane, while the "out-of-plane" pairs connect two neighboring triangular planes. The 16 energies were fit with these 12 terms with a fitting error of ~ 7 meV/formula unit [typical formation energies in this system (Fig. 6) are 150-200 meV/formula unit]. The resulting expansion is short-ranged as the pair interactions

(a)

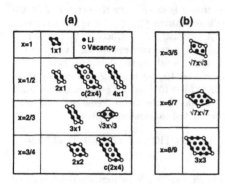

(b)

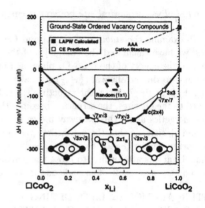

Figure 5: (a) Some of the vacancy ordered compounds in Li_xCoO_2 calculated via LAPW. (b) Other ordered ground state compounds predicted by the cluster expansion.

Figure 6: Predicted ordered vacancy ground state structures of Li_xCoO_2. In the $x=1/2$ phase ($2x1_a$), there are two inequivalent positions where Li atoms may be placed in the neighboring planes (labeled "a" and "b"). We find the "$2x1_a$" structure to be lower in energy than "$2x1_b$".

beyond third-neighbor decay to <1 meV/formula unit. However, the three-, four-body, and "out-of-plane" pair interactions are found to be significant (roughly 20%, 10%, and 20% of the nearest neighbor pair interaction, respectively).

Predicted ordered vacancy ground states in Li_xCoO_2: Fig. 6 shows the first-principles calculated formation energies for the lowest-energy ordered vacancy compounds. Some of the predicted ground states are included in the original set of LDA-calculated energies (filled squares), but some are not (empty squares), and hence are truly "unsuspected" predictions (unit cells shown in Fig. 5b). In addition, we show in Fig. 6 the energy of a completely random arrangement of Li and □ units as a function of composition.

First-principles prediction of Li_xCoO_2/Li voltage profile: The chemical potential difference in Eq. (4) defining the voltage may be computed [9] from (grand canonical) Monte Carlo simulations of the Li_xCoO_2 cluster expansion. This provides a completely parameter-free, first-principles prediction of the Li intercalation voltage of the Li_xCoO_2/Li cell as a function of Li content.

The predicted intercalation voltage profiles are shown in Fig. 7. Shown are results for the equilibrium voltage profile both at zero temperature and $T=300K$. Also shown is the voltage profile of the metastable random solid solution phase. Several aspects of the phase stability of ordered vacancy compounds can be ascertained from these voltage profiles: Two-phase regions, defined in terms of free energies by tie-lines connecting the two phases, correspond to plateaus in the voltage profiles. Likewise, the voltage drops are associated with single phase, ordered regions. As the temperature is increased, drops become more rounded and disappear at order-disorder transitions. The voltage profile of the solid solution phase is completely smooth with no discontinuous voltage drops.

Equilibrium Li-intercalation Li_xCoO_2 Phase Diagram: Via a combination of Monte Carlo simulations (both grand canonical and canonical), thermodynamic integration, and investigation of finite-size effects, one can map the entire chemical potential-temperature, and hence the voltage-temperature, phase diagram for Li intercalation in Li_xCoO_2/Li cells. The predicted phase

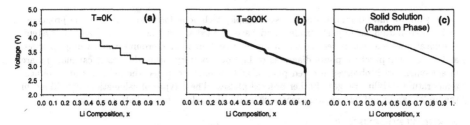

Figure 7: Predicted Li-intercalation voltage of the Li_xCoO_2/Li cell as a function of Li composition, calculated from the chemical potential difference in Eq. (4). (a) Equilibrium profile at T=0K, calculated analytically from the slopes of the ground state hull in Fig. 6. (b) Equilibrium profile at T=300K, calculated from Monte Carlo simulations. (c) Profile of the metastable random solution phase, calculated analytically from the cluster expansion interactions with mean-field (T=300K) entropy.

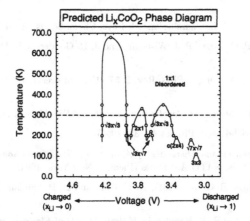

Figure 8: Predicted Li_xCoO_2 voltage-temperature phase diagram. At room temperature (horizontal dashed line), three vacancy compounds remain ordered.

diagram is shown in Fig. 8. Many of the predicted ground state phases undergo order-disorder transitions below room temperature. In equilibrium, three phases are predicted to remain ordered at and above room temperature: the 2x1 (x=1/2) and $\sqrt{3}$x$\sqrt{3}$ (x=1/3 and 2/3) structures. Reimers and Dahn [16] have observed electrochemically and through x-ray diffraction a monoclinic ordered vacancy compound at x=1/2. Their data suggest a "2x1" two-dimensional ordering as we have predicted. Reimers and Dahn [16] have also measured the order-disorder transition temperature for this phase at 60°C, and the width of the 2x1 phase field at room temperature to be ~0.1 V. Our calculations are in excellent agreement with both of these observations. No compounds at x=1/3 or x=2/3 have been experimentally reported in Li_xCoO_2 (although these $\sqrt{3}$x$\sqrt{3}$ phases have been found in other intercalation compounds). It is possible that the formation of these phases is kinetically inhibited in electrochemical experiments. Future electrochemical experiments to investigate the thermodynamic stability of the predicted ordered vacancy phases would therefore be of great interest.

SUMMARY

In sum, we have demonstrated a first-principles technique for predicting ordering properties of Li intercalation battery electrodes, and have applied the method to the Li_xCoO_2 system. Specifically, we have computed the cation and intercalation ordered ground states, voltage profiles, and voltage-temperature phase diagrams of Li_xCoO_2 battery electrodes. Our calculations are in agreement with observed ordered phases, and we have predicted the existence and voltage-temperature stability ranges of other ordered phases. These types of calculations should aid in the understanding of intercalation reactions, and ultimately in future battery design.

ACKNOWLEDGEMENTS

This work was supported by the Office of Energy Research (OER) [Division of Materials Science of the Office of Basic Energy Sciences (BES)], U. S. Department of Energy, under contract No. DE-AC36-83CH10093.

REFERENCES

1. T. A. Hewston and B. L. Chamberland, J. Phys. Chem. Solids **48**, 97 (1987).

2. K. Mizushima, P. C. Jones, P. J. Wiseman, and J. B. Goodenough, Mat. Res. Bull. **15**, 783 (1980).

3. C. Wolverton and A. Zunger, Phys. Rev. B **57** (1998, in press); J. Electrochem. Soc. (1998, submitted)

4. D. J. Singh, *Planewaves, Pseudopotentials, and the LAPW Method*, (Kluwer, Boston, 1994).

5. D. de Fontaine, Solid State Physics **47**, 33 (1994).

6. A. Zunger, in *Statics and Dynamics of Alloy Phase Transformations*, edited by P. E. A. Turchi and A. Gonis, NATO ASI Series (Plenum, New York, 1994).

7. K. Binder, Z. Phys. B **45**, 61 (1981); K. K. Chin and D. P. Landau, Phys. Rev. B **36**, 275 (1987).

8. W. R. McKinnon and R. R. Haering, in *Modern Aspects of Electrochemistry*, edited by R. E. White, J. O'M. Backris, and B. E. Conway (Plenum, New York, 1983), Vol. 15, p. 235.

9. J. N. Reimers and J. R. Dahn, Phys. Rev. B **47**, 2995 (1993).

10. *Solid State Batteries: Materials Design and Optimization*, C. Julien and G. -A. Nazri, eds. Kluwer, Boston (1994), Chap. 1.

11. M. K. Aydinol, A. F. Kohan, G. Ceder, K. Cho, and J. Joannopoulos, Phys. Rev. B **56**, 1354 (1997).

12. E. Rossen, J. N. Reimers, and J. R. Dahn, Solid State Ionics **62**, 53 (1993).

13. R. J. Gummow, D. C. Liles, and M. M. Thackeray, Mat. Res. Bull. **28**, 235 (1993).

14. G. G. Amatucci, J. M. Tarascon, and L. C. Klein, J. Electrochem. Soc. **143**, 1114 (1996).

15. M. Antaya, K. Cearns, J. S. Preston, J. N. Reimers, and J. R. Dahn, J. Appl. Phys. **76**, 2799 (1994).

16. J. N. Reimers, J. R. Dahn, and U. von Sacken, J. Electrochem. Soc. **140**, 2752 (1993); J. Electrochem. Soc. **139**, 2091 (1992).

Reactions of Lithium with Small Graphene Fragments: Semi-Empirical Quantum Chemical Calculations

Marko Radosavljević, Peter Papanek and John E. Fischer

Department of Materials Science and Engineering and Laboratory for Research on the Structure of Matter
University of Pennsylvania, Philadelphia, PA 19104-6272

Abstract

Semi-empirical and *ab initio* calculations [1], as well as inelastic neutron spectroscopy [2], demonstrate that Li can bind to protonated "edge carbons" to create a moiety analogous to the organolithium monomer $C_2H_2Li_2$. This provides a possible additional channel for Li uptake in high capacity Li-ion battery anodes based on low-T pyrolyzed soft carbons. Here we show that similar reactivity is exhibited by polyaromatic hydrocarbons with the protons removed (taken as surrogates for the structural units in hard carbons). In the deprotonated PAH'es the Li serves to saturate dangling bonds, maintaining sp^2 hybridization, whereas Li added to PAH'es creates sp^3 carbons at the edges. In both cases this extra reactivity occurs in parallel with the usual intercalation. These findings have implications for further development in Li-ion rechargeable battery technology.

Introduction

The advances in information and computer technology enable us to perform numerical simulations on large arrays consisting of many atoms. Such calculations parallel physical characterization and synthesis in importance as means of progress in understanding structure of complex new materials. For that reason it is not unusual that computer modeling provides initial push and steers experimental efforts toward more rapid advances. This project is one such example.

Recent breakthroughs in rechargeable Li-ion batteries include the use of lithium-carbon compounds as anodes. Current production of about 15 million batteries per month is dominated by highly graphitic carbons which are well understood. The crystalline norm LiC_6 has been known and extensively studied [3]. In graphite intercalation compounds (GIC) the Li is centered on the highly symmetric interstitial site between adjacent graphene sheets. Lithium occupies only 1/3 of the available sites (second nearest neighbors) forming the ordered $\sqrt{3} \times \sqrt{3}$ in-plane superlattice. The Li-Li spacing (4.2Å) is mainly determined by Coulomb repulsion of the negative ions, while the maximum LiC_2 configuration (first neighbor occupancy) has been achieved in high pressure synthesis [4].

On the other hand, disordered carbons obtained by pyrolyzing organic precursors at low temperatures have some features that compare favorably to those of graphite in terms of battery technology. Because these carbons are synthesized at relatively low temperature, much of the hydrogen remains in the material, with typical H/C ratio ranging between 0.05 and 0.30 [5]. It has been established that hydrogen pacifies the dangling bonds at the perimeter with only one H atom per unsaturated bond [2]. The importance of these materials in terms of rechargeable batteries lies in the fact that they exhibit lithium uptake that rivals that of high-pressure synthesized LiC_2. Several hypotheses have been made concerning the Li location in disordered carbons. The "house of cards" model envisages random stacking of independent graphene fragments which thus can bind "interstitial" Li on both sides and hence twice the capacity of graphite. Other models invoke Li_2 "molecules" which are epitaxially formed over first-neighbor hexagons [6] or involve multiple monolayers of Li over the same graphitic fragment [7]. Alternatively, we have proposed [1] that residual hydrogen enhances the lithium affinity; here we extend that idea to show that fragment edges, whether protonated or not, provide for "excess" Li capacity. Most recently, a group from MIT and Tokyo has reported simulations which closely resemble ours in character but with slightly different results [8].

In this work, we study the Li interaction with the disordered carbohydrogen alloys modeled as aromatic rings with some unsaturated edge valencies. All the simulations were performed at the semiempirical MNDO/AM1 level, the reliability of which was checked in earlier work by comparison with *ab initio* results on small PAH'es [1].

Computational Methods

Over the years, several semiempirical methods have been developed to speed up and simplify calculations for larger molecules. Such approximations most often consider only the valence electrons, while parametrizing the ionic cores. Further improvements such as zero differential overlap (ZDO) and modified neglect of differential overlap (MNDO) [9] are included as well. AM1 is just a newer version of MNDO, in which the core-core repulsion functions are modified and fitting functions for two center integrals are reparametrized [10]. AM1 has proven to be reliable for aromatic carbon systems, however Li has not been parametrized within this system. Thus, we use the older MNDO set of parameters for Li only, a compromise which has produced relatively good qualitative results in related systems [11]. All of these semiempirical calculations were carried out with the molecular orbital software MOPAC [12].

In order to check reliability of our AM1 calculations, we compared some of our results to *ab initio* values for the same geometries obtained from the Gaussian-94 program [13]. As discussed elsewhere [1], our methods suffer from overestimating steric repulsions, due to the small basis set. Also, the potential energy of CLi_2 group is not handled properly, but this should not be a problem since our calculations will not encounter such local insulated geometries. Finally, it is also worth noting that the systems under study exhibit many local minima which makes the search for the global ground state extremely time consuming and difficult. Instead, we start with many initial configurations and draw our conclusions independent thereof.

Results

For completeness, we first summarize results of our simulations on hydrogen-terminated graphite clusters published elsewhere [1]. The smallest interesting test molecule is the 4-ring hydrocarbon $C_{16}H_{10}$ (pyrene) because it contains two distinct types of hexagons terminated by either two or three hydrogens. We shall refer to these as "2H" and "3H" hexagons, though one can apply the terminology of carbon nanotubes and refer to "2H" hexagons as "armchair" edges. Introduction of one lithium atom at either intercalation site reproduces the local structure of crystalline GIC's. However, once pairs of Li atoms are added (one above and one below a hexagon of choice) a distinct difference can be observed between the "3H" and the "2H" case, as shown in Figure 1. If we choose the interstitial positions above and below the "3H" ring the lithium stays put for the most part; it is about 2Å above (and below) the pyrene molecule closely resembling the GIC. On the other hand, starting with interstitials of the "2H" ring, both Li atoms move to the periphery and bind to the carbons. This can only happen if both carbons become sp^3 hybridized which destroys the conjugation of that ring. This effect is known as 1,2-bis-localization. We confirm that both carbons are indeed sp^3 hybridized by observing the increased bond length of 1.52Å between them [1]. This configuration is also closely related to the *ab initio* isomer of $C_2H_2Li_2$ which has the lowest energy and compatibility with the carbon ring structure [14].

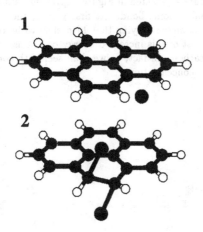

Figure 1: Semiempirically relaxed structures obtained by adding two Li's to the opposite sides of the 4-ring pyrene molecule. In the second configuration we observe evidence of Li bonding via formation of sp^3 hybridized carbon (see text).

We found many other local minima for $Li_2C_{16}H_{10}$, some of which have smaller heats of formation than the isomers shown in Figure 1. However, since the heats of formation of **1** and **2** are approximately the same (99.8 kcal/mol and 100.2 kcal/mol, respectively) we expect that these locally-bonded configurations will be abundant for high Li concentrations. Further explorations of stacked pyrenes, coronene (7-ring) and ovalene (10-ring) verify this observed differentiation between "2H" and "3H" rings while providing much richer spectrum of local minima. In coronene we observe first-nearest neighbor occupancy which is facilitated by the distortion of the originally planar molecule. In ovalene we test the "1H" ring (known as zig-zag edge for carbon nanotubes) which also shows signs of tetrahedral coordination when one Li is introduced. In general, we observe a limit in Li uptake approaching LiC_3, equivalent to "coating" both sides of isolated graphene flakes with second-neighbor occupancy [1].

Hard carbons also consist largely of small graphene fragments but contain insufficient hydrogen to saturate all the dangling bonds. We studied Li uptake in surrogates for such materials simply by leaving some of the edge carbons unsaturated in the starting material. We justify this approach by noting that the average fragment size observed by neutron radial distribution function analysis is in the range 6-20Å for various carbons [2,3,5]. If truly isolated, this implies that the average fraction of edge carbons ($\sim 35\%$) is much larger than typical H/C ratios for hard carbons, leaving a significant number of edge carbons with dangling bonds. We begin with the 4-ring PAH pyrene and start plucking hydrogen atoms away. At first we take away one H atom and start simulations with lithium at each of the 4 possible interstitial positions. As can be seen in Figure 2, some structures(**3**) relax to give Li covalently bound in place of the hydrogen, while others(**4**) remain with interstitial Li. For structure **3**, we observe that Li lies in the plane of the molecule, 2.1Å away from the carbon atom to which it is bound. The Li partial charge is 0.51, typical of a partially covalent bond. The heat of formation is 85.8 kcal/mol, comparable to those of intercalated Li-pyrene compounds [1]. The structure **4** depicted in Figure 2 gives interstitial Li with partial charge 0.65 characteristic of a mainly ionic bond. At first glance, the heat of formation of this molecule (146.3 kcal/mol) is huge implying that Li prefers to be covalently edge-bound to this fragment. However, most of this increased energy cost is associated with the dangling bond of the unsaturated carbon. In fact, as we shall see, both configurations should have similar abundances at high concentration.

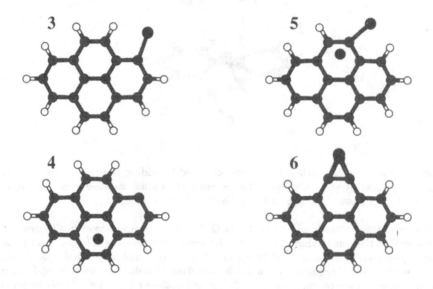

Figure 2: Metastable configurations of one (structures **3** and **4**) and two (molecules **5** and **6**) lithium interactions with partially deprotonated 4-ring pyrene. We observe the difference between interstitial and covalently bound lithium.

Next we introduce a second lithium atom, with either one or two hydrogens missing. For the two relaxed configurations (**5,6**) depicted in Figure 2, all the dangling bonds are saturated and any leftover Li has settled into interstitial positions. It is interesting to observe that in **6** Li's do not bind individually to separate carbon atoms, but instead share the bonds. In traditional chemistry this would require that carbon atoms change hybridization from sp^2 to sp^3, which does not appear to be the case. The carbon-carbon bond length is 1.45Å which is typical for aromatic carbons, and the bond angles resemble much more those of sp^2 hybridized atoms. These facts support our belief that the Li-C bonds are partially covalent.

The partial charges on the covalently bound Li atoms i n the two configurations in Figure 2 are 0.50 on **5** and 0.46 on **6**, implying that the bonds in the second configuration are stronger than that in the first one. The heats of formation, 106.5 and 114.5 kcal/mol respectively, again imply the approximate equality of the Li saturating edge carbons and interstitial Li. Further, these numbers are comparable to those in Li-H-C entities and intercalated lithium pyrene as previously reported [1]. The spectroscopic experiments are underway to search for evidence of the above described bonding patterns.

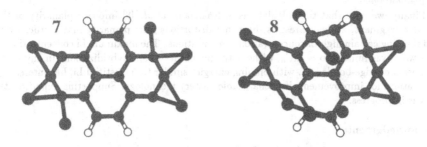

Figure 3: AM1-optimized structure of deprotonated pyrene with excess lithium. Structure **7** shows six lithium covalently bound, while structure **8** has 10 lithiums and is significantly distorted from planarity. Stochiometry of structure **8** fits the observed properties of disordered carbons.

Additional richness of structure is observed as more H is judiciously removed and more Li is added. $Li_6C_{16}H_8$ (not depicted) has two Li's saturating dangling carbons on "3H" rings while the other four create a pair of $Li_2C_2H_2$ structures on "2H" rings. The "isomer" of this molecule with stochiometry $Li_6C_{16}H_4$ (see Figure 3, structure **7**) has very similar heat of formation with some Li's very close to each other. The partial charges of 0.31, 0.35 and 0.45 (reading down on the right side of the molecule or up on the left side) show that indeed some Li's are more tightly bound than others. In the extreme limit, we attempt to merge the last two configurations and end up with a quite disordered non-planar molecule (see Figure 3, structure **8**) $Li_{10}C_{16}H_4$ which nicely matches the experimental observation of Li/C = 0.5 and H/C = 0.25 in a soft carbon. This molecule has no intercalated Li because there are no interior rings; the peripheral bonding mechanism alone is enough to explain the significant increase in capacity.

In order to allow the possibility of intercalation in parallel with edge bonding, some calculations have been performed on the 10-ring system $C_{32}H_{14}$ (ovalene). To date, we have found local minima which stabilize up to 12 Li's, with some fraction occupying interstitial positions but most bonded at the perimeter. We believe it is likely that the ovalene surrogate can eventually accommodate Li in the same proportion as was shown for pyrene.

Conclusions

The results of our semiempirical calculations strongly suggest that hydrogenated or dehydrogenated edges of polyaromatic carbon rings provide the additional mechanism of Li uptake in disordered carbons. Our results provide a natural explanation for high lithium uptake of these materials. The *ab initio* verification of some of these results gives us confidence in their validity.

In hydrogenated flakes, the maximum capacity has stochiometry of LiC_3 due to the flake distortion from planarity, as well as 1,2-bis localization on "2H" rings. The removal of hydrogen atoms provides an additional channel for covalent binding of Li which retains sp^2 hybridization on carbons atoms. In this way LiC_2 is readily obtainable. In addition, the similar energy costs of both interstitial and two sorts of covalent bonds bodes well for rechargeable batteries as it implies that recharge should occur at approximately the same applied voltage.

Finally, we note that the calculations of Nakadaira *et al.* [8] imposed planarity of the lithiated fragment, while our results show that distortions from planarity play a crucial role in stabilizing certain high Li concentration configurations. The maximum Li concentration in their work was limited to $\sim LiC_6$, without the possibility of sp^3 hybridization. In any event the notion of edge-bonded Li, with binding energies similar to interstitial Li, has interesting implications for improvements in rechargeable battery technology. Supporting experimental work is in progress.

Acknowledgments

This work was supported by the MRSEC Program of the National Science Foundation under award Number DMR96-32598, and by the Department of Energy, DE-FC02-86ER45254.

References

[1]. P. Papanek, M. Radosavljević and J. E. Fischer, Chemistry of Materials **8**, 1519 (1996).

[2]. P. Zhou, P. Papanek, R. Lee, J. E. Fischer and W. A. Kamitakahara, J. Electrochem. Soc. **144**, 1744 (1997); P. Zhou, P. Papanek, C. Bindra, R. Lee and J. E. Fischer, J. Power Sources (in press).

[3]. J. E. Fischer in *Chemical Physics of Intercalation*, A. P. Legrand and S. Flandrois, Editors, p. 59, Plenum, New York (1987).

[4]. V. V. Avdeev, V. A. Nalimova, and K. N. Semenenko, High Press. Res. **6**, 11 (1990).

[5]. T. Zheng, Y. Liu, E. W. Fuller, S. Tseng, U. von Sacken, and J. R. Dahn, J. Electrochem. Soc. **142**, 2581 (1995); J. R. Dahn, T. Zheng, Y. Liu, and J. S. Xue, Science **270**, 590 (1995).

[6]. K. Sato, M. Noguchi, A. Demachi, N. Oki, and M. Endo, Science **264**, 556(1994).

[7]. R. Yazami and M. Deschamps, *13th International Seminar on Primary and Secondary Battery Technology and Applications*, Florida Educational Seminar, Boca Raton, FL, March 4-7, 1996.

[8]. M. Nakadaira, R. Saito, T. Kimura, G. Dresselhaus, and M. S. Dresselhaus, J. Mater. Res. **12**, 1367 (1997).

[9]. M. J. S. Dewar and W. Thiel, J. Am. Chem. Soc. **99**, 4899(1977).

[10]. M. J. S. Dewar, E. G. Zoebish, E. F. Healy, J. J. P. Stewart, J. Am. Chem. Soc. **107**, 3902(1985).

[11]. E. Kaufmann, J. Gose, and P. v. R. Schleyer, Organometallics **8**, 2577(1989).

[12]. J. J. P. Stewart, Program MOPAC 6.0; Quantum Chemistry Program Exchange, No. 581.

[13]. M. J. Frisch, et al., Program Gaussian 94, Revision B.1; Gaussian, Inc.: Pittsburgh, PA, 1995.

[14]. U. Röthlisberger and M. L. Klein, J. Am. Chem. Soc. **117**, 42(1995).

COMPUTATIONAL, ELECTROCHEMICAL AND 7Li NMR STUDIES OF LITHIATED DISORDERED CARBONS ELECTRODES IN LITHIUM ION CELLS

G. Sandí, [†]R. E. Gerald II, [††]L. G. Scanlon, K. A. Carrado, and R. E. Winans
Chemistry and [†]Chemical Technology Divisions, Argonne National Laboratory, Argonne, IL 60439; [††]Aero Propulsion and Power Directorate, Wright Laboratory, Wright-Patterson Air Force Base, OH 45433.

ABSTRACT

Disordered carbons that deliver high reversible capacity in electrochemical cells have been synthesized by using inorganic clays as templates to control the pore size and the surface area. The capacities obtained were much higher than those calculated if the resultant carbon had a graphitic-like structure. Computational chemistry was used to investigate the nature of lithium bonding in a carbon lattice unlike graphite. The lithium intercalated fullerene Li_n-C_{60} was used as a model for our (non-graphitic) disordered carbon lattice. A dilithium-C_{60} system with a charge and multiplicity of (0,1) and a trilithium-C_{60} system with a charge and multiplicity of (0,4) were investigated. The spatial distribution of lithium ions in an electrochemical cell containing this novel disordered carbon material was investigated in situ by Li-7 NMR using an electrochemical cell that was incorporated into a toroid cavity nuclear magnetic resonance (NMR) imager. The concentration of solvated Li^+ ions in the carbon anode appears to be larger than in the bulk electrolyte, is substantially lower near the copper/carbon interface, and does not change with cell charging.

INTRODUCTION

High capacity (855 mAh/g), reversible, lithium intercalated carbon anodes were prepared using pillared clays as templates and pyrene, styrene, trioxane/propylene, propylene and ethylene [1,2], that exceed the capacity for stage 1 lithium intercalated carbon anodes (372 mAh/g). Dresselhaus refers to stage 1 Li intercalated graphite compounds as a $\sqrt{3}$ x $\sqrt{3}$ structure of Li ions on a honeycomb lattice [3]. The spacing between the lithium ions in such a lattice is 4.2 Å. In order to account for or rationalize the high capacity lithium carbon anode materials synthesized in our laboratory, we chose to investigate lithium bonding and spacing in endrohedral lithium complexes of C_{60}. Optimized geometries for these systems suggest two types of lithium atoms within the C_{60} lattice. Ionic lithium is obtained for dilithium-C_{60} and covalent lithium for trilithium-C_{60}. In both cases, the calculated lithium-lithium separation of 2.96 Å or less is consistent with that required to achieve a specific capacity greater than that obtained in a stage 1 lithium intercalated graphite.

To study the spatial distribution of lithium ions in an electrochemical cell containing this novel carbon material, in situ Li-7 nuclear magnetic resonance (NMR) was used. In situ NMR methods have been developed to study mobilities, distributions, redox reactions, physicochemical environments, and other properties of the charge carriers and support materials in functioning electrochemical cells [4]. The toroid cavity/electrochemical cell NMR probe used in the present study is capable of recording high resolution spectra for fluid samples and wide line spectra for solid materials, at any position from the central conductor of the toroid cavity (the working electrode of the electrochemical cell). The radiofrequency (rf) magnetic field, B_1, generated by the toroid cavity in the sample volume is inhomogeneous, but precisely defined as $B_1 = A/r$, where A is a constant and r is the radial position from the central conductor [5]. Therefore, a single rf pulse of fixed duration results in a 90° rotation of the nuclear magnetization located only at one radial position. We take advantage of the well-defined B_1 inhomogeneity to form one-dimensional images. These images represent radial concentration profiles of the chemical species in the electrochemical cell that are associated with the NMR resonances in a conventional NMR spectrum. For example, the 5% dispersion in Li-7 nutation frequencies across a 100 μm layer at the surface of the carbon coated working electrode reported here is used to record ^{7}Li NMR spectra at three different radial positions within the 100 μm layer. Since the gradient of the B_1 field ($\nabla B_1 = -A/r^2$) is not constant, the radial spacing between NMR spectra is also not constant. A one-dimensional radial concentration image or profile for any chemical species observed in the NMR spectrum is obtained by plotting the intensity or integral of the resonance versus the radial position at which the spectrum was recorded.

The capacity and power output of an electrochemical cell relies, in part, on the presence of electrolyte at all of the accessible surface area of the carbon particles. In this work, we report the *in situ* ^{7}Li NMR image of solvated lithium ions in an electrolyte wetting a carbon electrode that was derived from pillared clays as templates and pyrene, styrene, trioxane, propylene and ethylene as the organic precursor [1,2]. The carbon was coated onto the central conductor of a toroid cavity detector and served as the cathode electrode in the electrochemical cell.

EXPERIMENTAL

The synthesis of the calcined pillared clays (PILCs) as well as the loading with pyrene, styrene, trioxane/pyrene, ethylene and propylene carbon precursors, has been described in detail elsewhere [1,2]. Electrochemical characterization of the carbon anodes was performed using an Arbin 2400 station cell cycler. Details regarding the preparation of the electrode were described by Sandí et al. [6].

Gaussian 94 was used for all ab initio calculations [7]. Geometry optimizations were performed using the Hartree-Fock method with a 3-21G basis set. However, for the Li_3-C_{60} complex, the restricted open-shell-Hartree-Fock (ROHF) method was used in order to avoid spin contamination.

For the NMR experiments, approximately 1.0 g of electrolyte (1M $LiPF_6$ in ethylene carbonate:diethyl carbonate) was placed in a cylindrical electrochemical cell that formed part of the toroid cavity NMR detector. The cell was assembled from a glass tube with an inside diameter of 10.4 mm and length 22.0 mm. Circular rubber septa were used to seal the tube at both ends. A gold wire counter electrode 1.0 mm in diameter was formed into a 7-turn helix and place against the inner wall of the glass tube. A 0.62 mm diameter copper wire served as a working electrode and passed through the center axis of the cell, piercing both septa. Each end of the working electrode was soldered to a copper rod for mounting the cell in the NMR detector. The working electrode also functioned as the central conductor of the NMR detector. The carbon-based material was made into a slurry as described previously [1,2,6]. The sealed end of a capillary tube was used to coat the working electrode, as it was rapidly rotated in a horizontal orientation, with a thin layer of the cathode material. Between thin-layer coatings the electrode was heated for 20 minutes in air at 170 °C. The final diameter and length of the carbon electrode was 2.0 mm and 10.8 mm, respectively. The total mass and area of the active material was 10 mg and 0.7 cm^2, respectively. The carbon electrode was heated at 80 °C under vacuum prior to incorporation into the electrochemical cell. The cell was charged under potentiostatic control from 0.5 to 4.0 V vs. a gold electrode using an EG&G 273A potentiostat. High-resolution ^{7}Li NMR spectra and images were recorded with a home-built toroid cavity probe using a Varian^{UNITY} I NOVA-300WB spectrometer at the following settings: spectrometer frequency, 116.567 MHz; spectral width, 10 kHz; 4096 data points; 15.0-s recycle delay; 1 transients per spectrum; 128 spectra were recorded with a pulse width increment of 5 μs (spectral width, 100 kHz). The two-dimensional data sets were processed with 200-Hz and 2000-Hz line broadening in the F2 and F1 dimensions, respectively.

RESULTS

Computational Studies

As discussed by Sandí et al.[8], the average reversible capacities obtained from the different carbon precursors are very similar within the limits of the standard deviation, but pyrene exhibited the lowest irreversible capacity and capacity fade. Overall, the performance of these cells in terms of delivered capacity is significantly higher than graphite (855 mAh/g compared to 370 mAh/g at 200-300 mV vs. Li) and some other alternative materials currently under study. While some hysteresis occurs in the discharge-charge voltage profiles, these carbon electrodes deliver very high and stable capacities (>50 cycles).

The electronic energies at 0 K, E_e°, were calculated for the molecular species shown in Table I. These are results from the geometry optimizations which are gas phase calculations. The calculated ΔE_e° shown in Table II represents the difference in energy between the product and reactants at 0 K. A Mulliken population analysis was used to determine the atomic charge distributions in the molecules. The results of the geometry optimization for the C_{60} carbon lattice show that there are two different carbon-carbon bond lengths within the lattice. One carbon-carbon bond length is 1.45 Å and the other is 1.37 Å. The carbon-carbon-carbon bond angles are either 120.0° or 108.0°. There are two triply degenerate lowest unoccupied molecular orbitals [LUMO's] at -0.63eV and +1.06eV. The negative value at -0.63eV implies overlap of the LUMO with the Fermi level and therefore, perhaps, metallic conduction behavior. This may also imply ease of reduction when compared to other carbon lattice systems. For example, the geometry optimization of coronene [$C_{24}H_{12}$] using RHF/3-21G shows two doubly degenerate LUMO's at +1.83 eV.

Structures for the two lithium-endrohedral C_{60} complexes are shown in Figures 1 and 2. The Li_2-C_{60} complex was initially formed by inserting a dilithium cluster with a total charge of 0 and multiplicity of 1 (paired electrons) inside C_{60}. The initial lithium-lithium distance for the dilithium cluster was 2.82 Å.

The distance between the two lithiums in the geometry optimized Li_2-C_{60} complex is 2.96 Å. Typical lithium ion-carbon lattice bond distances vary from 2.09 to 2.55 Å for the Li_2-C_{60} complex. There is very little disruption of the C_{60} lattice as some bonds may increase by 0.01 to 0.05 Å due to lithium intercalation. Carbon-carbon-carbon bond angles of 120° are reduced less than 3°. Table II shows the calculated energy difference at 0 K to be -16.2 kcal/mole. This suggests a thermodynamically stable product.

Table I. Calculated Electronic Energies at 0 K for C_{60} and the Endohedral Lithium Complexes of C_{60}.

Cluster	Method Basis Set	Charge, Multiplicity	E_e° (Hartrees)*
C_{60}	RHF/3-21G	0,1	-2259.05
Li_2 Cluster	UHF/3-21G	0,1	-14.77
Li atom	UHF/3-21G	0,2	-7.38
[Li_2 - C_{60}]	UHF/3-21G	0,1	-2273.85
[Li_3 - C_{60}]	ROHF/3-21G	0,4	-2281.27

* 1 Hartree = 627.51 kcal/mole

Table II. Calculated ΔE_e° for Endohedral Lithium – C_{60} Complexes.

	ΔEe° kcal/mole	Li-Li separation Å	Comments
Li_2 Cluster $+ C_{60} \iff [Li_2 - C_{60}]$	-16.2	2.96	Li⁺ charges: 1.14, 1.17
$3Li^\circ + C_{60} \iff [Li_3 - C_{60}]$	-47.9	2.60, 2.59, 2.74	Li⁺ charges: 0.91, 0.91, 0.98 Li atomic densities: 0.35, 0.36, 0.28

The Li_3-C_{60} complex was initially formed by inserting three lithiums inside C_{60}. After geometry optimization, it can be seen from Figure 2 that the lithiums form a triangle within the C_{60} complex. As shown in Table II, the lithium-lithium distances vary from 2.59 to 2.74 Å. This is much less than that found for the spacing between lithiums in a stage 1 lithium intercalated graphite anode and is consistent with the spacing required for high capacity carbon anodes. There is very little distortion of the carbon C_{60} lattice in the Li_3-C_{60} complex. There is one notable difference, however, between the two lithium complexes. The dilithium complex appears to be ionic whereas for the trilithium complex an electron is shared between the three lithium ions. Each of the lithium ions in the trilithium complex has a value of about 0.3 for the total atomic spin density, which implies covalent bonding.

⁷Li NMR Studies

Figure 3 is a radial concentration profile (a one-dimensional image) of the solvated Li⁺ ions in the electrochemical cell prior to charging the cell. Each data point represents the integrated intensity of the lithium resonance in the high-resolution NMR spectrum recorded at that radial position. The spatial resolution is greatest next to the copper/carbon interface, and decreases in the

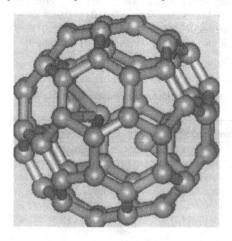

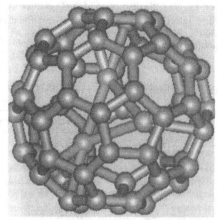

Figure 1. Dilithium C_{60} Endohedral Complex.

Figure 2. Trilithium C_{60} Endohedral Complex.

region of the bulk electrolyte, near the counter-electrode helix. The presence of the metal helix distorts the B_1 field direction and magnitude to varying degrees in the region between 1.5 and 5.5 mm. This region is occupied by the bulk electrolyte and is included to clarify the spatial relationship between the one-dimensional image and the cross section of the electrochemical cell. The profile was normalized to the integral of the resonance recorded at 1.2 mm, which was assigned a lithium ion concentration of 1.0 M, *vide supra*. It appears that the Li^+ ion concentration in the carbon material adjacent to the bulk electrolyte is higher than in the bulk electrolyte itself. This seems highly unusual since the electrolyte in the carbon material and in the bulk should exchange readily. Perhaps the close proximity of the lithium ions to the carbon conductor enhances the NMR sensitivity, as has been observed for water on copper surfaces [9]. Near the copper electrode there is a 0.25 mm zone depleted of lithium ions. The spatial transition is quite abrupt, but the concentration profile directly adjacent to the copper conductor is not clear because the signal-to-noise ratio decreases rapidly in that region. The lithium ion concentration profile was measured again after 7.12 C of charge was passed through the cell, lithiating the carbon cathode to a theoretical C:Li ratio of 11.3:1.0, and revealed very little change. Although the ^{7}Li- NMR spectrum and concentration profile were nearly identical before and after charging, the spin-lattice relaxation time constants differed by a factor of 3.64. The shorter value ($T_1 = 0.65 \pm .01$ s) was measured after the cell was charged and resulted from radical ion byproducts from oxidation reactions that occurred at the counter electrode.

The carbon NMR data provide details about lithium on the carbon surface, but there is not enough evidence to describe the lithium ion intercalation in the carbon lattice itself. It is our goal to incorporate a Li reference electrode in the electrochemical cell to achieve better control of the applied potential and also to provide an additional source of Li^+. The spatial distribution and bonding of Li in the disordered carbon lattice can then be more precisely correlated with the simulated Li_n-C_{60} complexes.

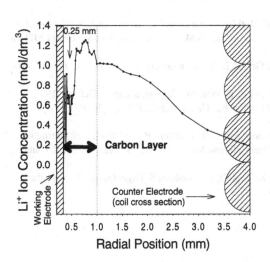

Figure 3: Radial concentration profile of the solvated Li^+ ions in the electrochemical cell.

CONCLUSIONS

It appears as though a curved carbon lattice with low energy LUMO's can be important for lithium cluster formation as reflected by the very short lithium-lithium bond distances. This may account for the very high lithium capacities in carbon anodes. Preliminary lithium ion radial concentration profiles reveal some unusual characteristics of solvated lithium ions adsorbed into the carbon material of the cathode in an electrochemical cell. The corresponding concentration profiles for the electrolyte solvent will be measured in order to clarify the lithium results and provide a direct measure of solvent distribution in the cathode. We plan to increase the surface area of the cathode in order to measure *in situ* wide line ^{7}Li NMR spectra [10] and spatial distributions of intercalated lithium ions in an electrochemical cell at various stages of the charge/discharge cycle.

ACKNOWLEDGMENTS

Dr. Christopher Johnson and Dr. Robert Klingler from the Chemical Technology Division at Argonne National Laboratory are gratefully acknowledged for their helpful comments and assistance. This work has been performed under the auspices of the Basic Energy Sciences, Division of Chemical Sciences, U. S. Department of Energy, under contract number W-31-109-ENG-38.

REFERENCES

1.   G. Sandí, R. E. Winans, and K. A. Carrado, *J. Electrochem. Soc.* **143**, L95 (1996).
2.   G. Sandí, K. A. Carrado, R. E. Winans, J. R. Brenner, and G. W. Zajac, *Mater. Res. Soc. Symp. Proc., Macroporous and Microporous Materials* **431**, 39 (1996).
3.   M. Nakadaira, R. Saito, T. Kimura, G. Dresselhaus, and M. S. Dresselhaus, *J. Mater. Res.* **5**, 1367 (1997).
4.   R. E. Gerald II, R. J. Klingler, J. W. Rathke, G. Sandí and K. Woelk, in *Spatially Resolved Magnetic Resonance*, Proceedings of the 4th ICMRM, Albuquerque, New Mexico (1997). In press.
5.   J. W. Rathke, R. J. Klingler, R.E. Gerald II, K. W. Kramarz, and K. Woelk, *Progr. NMR Spectrosc.*, 30 (1997)
6.   G. Sandí, K. Song, K. A. Carrado and R. E. Winans, *Carbon*, accepted October 1997.
7.   M. J. Frisch, *et al.*, Gaussian 94, Revision E2, Gaussian Inc., Pittsburgh PS (1995).
8.   R. E. Gerald II, unpublished results.
9.   G. Sandí, R. E. Winans, K. A. Carrado, and C. S. Johnson, Proceddings, 2nd International Symposium on New Materials for Fuel Cells and Modern Battery System, Montreal, Canada, 415 (1997) .
10.   K. Tatsumi, T. Akai, T. Imamura, K. Zaghib, N. Iwashita, S. Higuchi, and Y. Sawada, *J. Electrochem. Soc.* **143**, 6 (1996).

This work was performed under the auspices of the Office of Basic Energy Sciences, Division of Chemical Sciences, U.S. Department of Energy, under contract number W-31-109-ENG-38

SIMULATION STUDIES OF POLYMER ELECTROLYTES FOR BATTERY APPLICATIONS

J. W. HALLEY and B. NIELSEN

School of Physics and Astronomy, University of Minnesota, Minneapolis, Minnesota 55455,
woods@jwhp.spa.umn.edu

ABSTRACT

We report modeling studies of polyethylene oxide which are carried out with the goal of elucidating the mechanism of ion conduction in the temperature range of interest to battery applications.
We review our previous work in which the amorphous regions of the polymer between its glass and
melting temperatures is modeled by a molecular dynamics algorithm in which the model system is
polymerized from a model monomeric liquid. We describe new work in which the hydrogen centers
are added to the model in order to permit comparison with recent neutron work. We compare our
simulations of frequency dependent conductivity with experiment and end with a brief discussion
of possibilities for improved conductivity which our current understanding suggests.

INTRODUCTION

In the lithium polymer advanced battery programs of the Department of Energy, the relatively
low conductivity of the polymer electrolytes which are being used in battery prototypes are a
serious problem limiting the technological promise of the devices. The stated target of 10^{-3} (Ohm-
cm)$^{-1}$ at room temperatures is to our knowledge not attained by any of the polymer electrolytes in
current use in prototypes. One approach is to add plasticizers[1] to the polymer matrix. Though
this raises the conductivity, it threatens to compromise the dimensional stability of the electrolyte
which was a primary reason for using polymer electroytes in the first place. In the program on
which progress is reported here, we are trying to elucidate the mechanism of ion transport in one
of the prototype electrolytes, in the expectation that a better understanding will reveal promising
directions for the solution of this engineering problem.

We have chosen polyethylene oxide for the study because of the extensive experimental information[1] (mostly on macroscopic scales) which is available about it. We have used molecular
dynamics techniques to build a model for the polymer[2] and the ions in it and are collaborating
with quantum chemist Larry Curtiss, who provide force fields for our models and Marie-Louise
Saboungi and David Price of Argonne, who are doing neutron scattering studies. Though molecular dynamics studies have been done before [3- 6] our approach has some unique features which
we believe permit some of the central difficulties in this problem to be addresssed.

In the next section we briefly describe our simulation model of the electrolyte, emphasizing
motivations and the methods by which we have recently added hydrogen atoms to the model. The
third section describes some results on ions in the polymer model, including a new comparison of
our simulated conductivity with experimental results. In the last section we discuss some possible
implications for design of improved polymer electrolytes for battery applications.

Mat. Res. Soc. Symp. Proc. Vol. 496 © 1998 Materials Research Society

POLYETHYLENE OXIDE SIMULATION MODEL

Details of our simulation model are described in reference [2] . We chose to develop our model by a crude form of 'computational polymerization' from a model of liquid dimethyl ether[7] because the proposed operating temperatures for polyethylene oxide electrolytes lie between their melting and its glass transition temperatures. This means that the equilibrium state of the polymer is crystalline, but as shown by Armand and coworkers [8] , the ionic conduction is occuring through the amorphous regions which are therefore out of equilibrium. Thus we need a simulation of a nonequilibrium state to study the relevant conductivity problem. There may be many metastable nonequilibrium states and we face the question of how to select the one (or ones) which correspond to the experimental systems being used in the applications. Our approach was to produce our model by a procedure somewhat like what is actually occurring during experimental polyermization of real polyethylene oxide [9]. This does not guarantee that the resulting model will be like the electrolytes used in applications, so we are comparing the resulting model structure with the results of neutron scattering measurements made by our Argonne collaborators as a check. Because our 'polymerization' algorithm contains a time like parameter τ , roughly related to the reaction rate for the polymerization, we can systematically produce a variety of structures for comparison with the experimental structures.

Neutron scattering experiments at room temperatures are difficult to interpret because one must sort out the signal from crystalline and amorphous parts of the sample. Also, we chose in our model to describe methyl and ethyl groups with the united atom model because including explicit account of hydrogen motion in the simulation would significantly increase the computational cost with little corresponding improvement in the accuracy of the structural or low frequency dynamical features of the model. However, neutrons scatter strongly from hydrogen and deuterium, so the absence of hydrogen creates some complications in the interpretation of the experiments. To get around the problem of separating crystalline from amorphous contributions to neutron scattering in the initial stages of the work, our neutron scattering colleagues have chosen to work at 400K, above the melting temperature. To obtain an account of the effects of deuterium (which is substituted for hydrogen in the neutron scattering samples) we add deuterium to the model, not as a part of the full dynamical model, but only for the purposes of calculating the neutron scattering structure factors. To do this, for each configuration in the model we determine 'tetrahedral positions' for deuterium sites (Figure 1).

In chain

    1. Find CCO plane.

    2. Bisect CCO angle in plane

    3. Place 2 H's at tetrahedral angles and 1.08A + random deviation

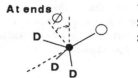

At ends

    1. Extend last CO bond.

    2. Choose random azimuth $\emptyset$

    3. Place three H's

Figure 1. How deuterium was included. Open circles: O; filled circles: CH_2 or CH_3.

Then we choose a position for each deuterium which differs from this by selecting a random displacement with an isotropic Gaussian probability distribution of width 0.1 Å . We are working on implementation of a more sophisticated version of the last step in which the gaussian distribution is not isotropic but is fixed by the assumption that each hydrogen is in its quantum mechanical ground harmonic oscillator states. However the primitive isotropic gaussian distribution works quite well for describing the data which our Argonne collaborators are producing, as illustrated in Figure 2. (The experimental data are preliminary. A complete description of these results is in preparation[10]. Full account of instrumental broadening is not included in the comparison in Figure 2.)

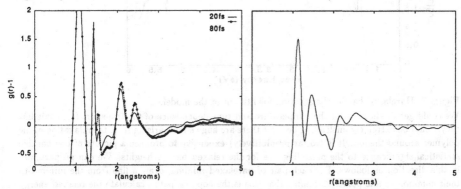

Figure 2. Comparison of the calculated composite radial distribution function(left, for two values of r) with preliminary neutron data (right).

ION CONDUCTION

Our model for the interaction of the elements of this model with a lithium ion is described in detail in [11] . We used force fields derived from first principles calculations by our collaborator L. Curtiss and coworkers at Argonne. We agree with earlier workers in finding that the interactions of the lithium ion with the polymer matrix are so strong that, in a straightforward simulation of a lithium ion in any reasonable model of the the interaction, the lithium ion is essentially observed not to diffuse at all on calculable time scales of molecular dynamics simulation (100 ps). How do we deduce the behavior on long time scales from the short, 100ps , behavior calculable directly, when no diffusion is observed on the shorter, available scales? We briefly discuss three approaches we have taken to this problem

Our first approach is to consider a hopping model and use the molecular dynamics model to determine the free energy barrier heights for hopping between potential minima for the charge carriers. We expect a distribution of barrier heights, rather than a single, rate determining, one because the system is amorphous. Before the polymer relaxes around them, the potential energy surface for lithium ions contains very deep minima(around 100 kcal/mode or 4 eV) relative to vacuum. We can easily obtain the unrelaxed barrier height distribution from simulation as shown in Figure 3.

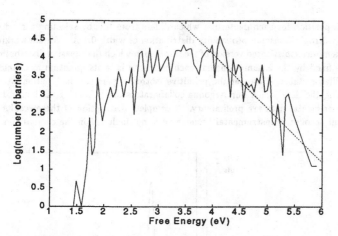

Figure 3. Unrelaxed barrier distribution for lithium in the model.

We could get predictions for the temperature and time dependence of the transport from this but they would be extremely misleading because there are huge effects arising from the relaxation of the polymer around the ion. It is (so far prohibitively) expensive to produce a picture of comparable statistical significance to the preceding one for the relaxed barrier heights. We can go part way in this direction as follows: For each pair of unrelaxed minima, draw a line from the minima to their common saddle point. Assuming that this is the hopping path , calculate the relaxed energy along this path. Do this for all (or as many as you can afford) of the pairs of unrelaxed pairs. Use the resulting barriers to calculate a barrier distribution. This program runs into the following problems: 1) Not many paths can be afforded 2) One finds that a clear barrier is not observed for many of these paths, suggesting that the assumption that the unrelaxed paths approximate the relaxed ones is wrong. Nevertheless, we can use the resulting data to get a rough idea regarding the barrier distribution for a relaxed lithium ion in our model of the electrolyte. It is shown below:

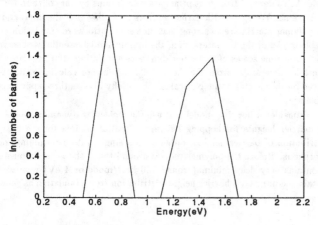

Figure 4. Preliminary relaxed barrier distribution for lithium in the PEO model.

The relaxed barriers are much lower than the unrelaxed ones and are in order of magnitude

104

agreement with first principles results obtained for small clusters by Larry Curtiss. It appears that the ion makes itself more mobile by distorting the environment of the polymer around it. Tentatively, there appear to be two peaks in the relaxed barrier height distribution, unlike the unrelaxed distribution which looked like a very broad gaussian. A high priority in the near future will be to explore whether this double peak survives a more thorough statistical study and, if so, the chemical meaning of the two peaks.

A second approach[11] is to calculate the self correlation of a charge carrier defined for our case of a single lithium in the simulation as

$$S_{s,Li}(\vec{k},\omega) = \frac{1}{2\pi} \int e^{-i\omega t} < e^{i\vec{k}\cdot(\vec{r}_{Li}(t)-\vec{r}_{Li}(0))} > dt$$

and try to deduce the long wave length low frequency diffusive behavior from it by extrapolation. If the motion were diffusive this would have a Lorentzian form as a function of frequency ω centered at zero frequency and with width Dk^2 where D is the diffusion constant whereas if the motion were harmonic it would be a Lorentzian centered at a finite resonant frequency. We fit[11] a linear combination of these two forms to the simulation data. The harmonic motion, corresponding to trapping at oxygen sites, is not dominant. To attempt an extrapolation to long wavelengths (small k values) we looked at the values of D obtained from the fit. They are not independent of k, which means that the motion is not diffusive on these scales . Data on the effective D as a function of k are shown in reference [11]. A unique extrapolation to long wave lengths does not appear to be possible. We learned, however, that the length scale at which a crossover to diffusive behavior occurs is 6 Å or greater, suggesting that the diffusing entity is at least this big.

The third approach is to calculate the frequency dependent electrical conductivity $\sigma(\omega)$ and try to extrapolate to zero frequency. We do this by computing the current response of the system to an imposed, time dependent electric field $E(t) = \sum_{n=1}^{4n_0} E_n \sin 2\pi nt/T$. The current response is $J(t) = \frac{1}{V}\sum_i Q_i v_i$ with Fourier transform $J(\omega_n) = 2/N \sum_{m=1}^{N} J(t_m) \sin\omega_n t_m$. Here $\omega_n = 2\pi n/N\Delta t$ and $t_m = m\Delta t$. We showed earlier[11] that the frequency dependent conductivity is then $\sigma(\omega_n) = J(\omega_n)/E(\omega_n)$. We show some results below, with scales adjusted for comparison with some experimental results on PEO alone and PEO containing a sodium salt.

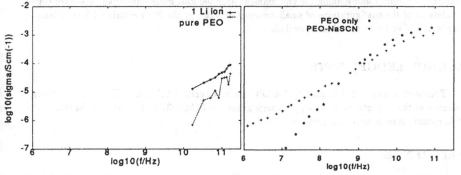

Figure 5. Comparison of simulated conductivity data(left) with experimental results on PEO with and without salt [13] (right).

We see that the calculated $\sigma(\omega)$ is an order of magnitude lower than the experimental one for the case of PEO alone. This is not bad agreement considering that the simulation sample has dimensions of about $(20\text{Å})^3$. The slope of the frequency dependence is also not very different from the experimental one. The smoothness of the experimental data may suggest that there is not a lot of new physics in the decades of frequency between 10^{10} Hz and 10^6 Hz , so extrapolation may be possible. Nevertheless, one also sees that $\sigma(\omega)$ for the neat polymer and the polymer loaded with salt are experimentally almost indistinguishable down to 10^9 Hz which is outside the range of the molecular dynamics simulation. This motivates our effort to find algorithms to extend our calculations to longer times.

CONCLUDING REMARKS

Future work will address the question of the mechanism of ion pairing and the transference number. We hope to develop algorithms which will make lengthier simulations possible to study lower frequencies. We also plan studies of the mechanical properties of our model, since these are important for battery applications.

With what we have learned so far, it appears that the lithium charge carrier should be regarded as an entity which significantly perturbs the polymer matrix within a sphere of radius about 10 Å. (Further evidence for this is cited in reference[11] .) This suggests that entanglements or cross linking on scales much larger than this will not have a serious effect on the conductivity and could be manipulated or introduced in order to prevent crystallization and to increase mechanical stability. This appears to be consistent with reports in the literature that crosslinking often does not have a big effect on conductivity. It is possible to manipulate cross linking or entanglement in several ways: manipulating the structure of the polymer, as in the MEEP family of polymers [14] , by manipulating structure through the polymerization rates or methods, by cross linking with gamma rays[15] or by manipulation of thermal history, for example by rapid quenching from the melt.

In addition, however, it is desirable to increase the conductivity in the amorphous regions. One would like to make the system more liquid like on the short scales (less than or of order 10 Å) and/or to reduce the magnitude of the interaction of the ion with the polymer. One possible approach to the latter problem is to introduce small, controlled amounts of water, which are known to increase the conductivity of PEO at high temperatures [16- 17] . Water is expected to solvate the lithium ions, reducing the electric field which the lithium ion produces on the polymer (through dielectric screening) and making the environment near the ions more liquid-like. We have begun simulations of the configurations of small amounts of water in the PEO matrix with containing a lithium ion in order to test these speculations.

ACKNOWLEDGEMENTS

This work is supported in part by the Division of Chemical Sciences, Office of Basic Energy Sciences, of the Department of Energy through grant DE-FG02-93ER14376 and by the University of Minnesota Supercomputing Institute.

REFERENCES

1. F. Gray, Solid Polymer Electrolytes, VCH, NY, 1991 pp 108-111
2. B. Lin, J.W. Halley and P.T. Boinske, J. Chem. Phys. 105, 1668 (1996)
3. D. Rigby and R-J Roe, J. Chem. Phys. 87, 7285 (1987)
4. J.I. Mckechnie, D. Brown and J.H.R. Clarke, Macromolecules, 25, 1562 (1992)
5. D. Brown, J.H.R. Clarke, M. Okuda and T. Yamazaki, J. Chem. Phys. 100, 6011 (1994)
6. S. Neyertz and D. Brown, J. Chem. Phys. 102, 9725 (1995)

7. B. Lin and J. W. Halley, J. Physical Chemistry 99,16474 (1995

8. C. Berthier, W. Gorecki, M. Minier, M. B. Armand, J. M. Chabgno and P. Rigaud, Solid State Ionics 11, 91 (1983).

9. F.E. Bailey and J.V. Koleske, " Poly(ethylene oxide)" (Academic Press, New York, 1976), pp. 24

10. M-L. Saboungi, D. Price and J. W. Halley (unpublished)

11. P. T. Boinske,L. Curtiss, J. W. Halley, B. Lin and A. Sujianto, Journal of Computer-Aided Materials Design 3, 385 (1996)

12. The problem is well known but not widely discussed in the literature. Some workers changed the model unrealistically until the ion diffused on calculable scales.

13. T. Wong, M. Brodwin, B. L. Papke and D. F. Shriver, Solid State Ionics 5, 689 (1981)

14. Reference [1] p. 97

15. Reference[1], p. 104

16. Reference[1] p. 136

17. R. Huq and G. C. Farrington, J. Electrochem. Soc. 135, 524 (1988)

MOLECULAR DYNAMICS STUDY OF LITHIUM DIFFUSION IN LITHIUM-MANGANESE SPINEL CATHODE MATERIALS

RANDALL T. CYGAN[*], HENRY R. WESTRICH[*], AND DANIEL H. DOUGHTY[**]
[*]Sandia National Laboratories, Geochemistry Department, Albuquerque, NM 87185-0750
[**]Sandia National Laboratories, Lithium Battery Research and Development Department, Albuquerque, NM 87185-0613

ABSTRACT

A series of molecular dynamics computer simulations of the self-diffusion of lithium in pure and several doped lithium-manganese spinel materials has been completed. The theoretical approach is part of an effort to understand the mechanisms and rates of lithium diffusion, and to evaluate the structural control of the cathode materials upon lithium intercalation (charge-discharge) process. The molecular dynamics approach employs a fully ionic forcefield that accounts for electrostatic, repulsive, and dispersion interactions among all ions. A reference unit cell comprised of 56 ions ($Li_8Mn^{3+}_8Mn^{4+}_8O_{32}$) is used to perform the simulations under constant volume and constant pressure constraints. All atomic positions are allowed to vary during the simulation. Simulations were completed for the undoped and doped $LiMn_2O_4$ at various levels of lithium content (based on the number of lithium ions per unit cell and manganese oxidation state). The molecular dynamics results indicate an activation energy of approximately 97 kJ/mole for self-diffusion of lithium in the undoped material. Lithium ion trajectories from the simulations provide diffusion coefficients that decrease by a factor of ten as the cathode accumulates lithium ions during discharge. Molecular dynamics results for the doped spinel suggest a decrease in the diffusion rate with increasing dopant ion.

INTRODUCTION

Molecular modeling and atomic-based energy calculations have recently been used to supplement the synthesis and testing of new oxide materials for lithium ion rechargeable batteries[1,2]. The ability to derive a predictive model is critical to the development of new cathode materials and the improvement of battery performance. Of interest is the investigation of the effects of doping on the crystal chemistry, lattice constants, and electrochemical performance of the lithium manganese oxide spinels[3]. The $LiMn_2O_4$ spinel is one of the best oxide phases for a cathode material, having a voltage plateau of 4 V, a high specific capacity, high thermal stability, low cost, and no or little impact on the environment[4]. Lattice expansion and contraction during, respectively, lithiation and delithiation creates a buildup of stress in the cathode material, and can lead to significant degradation in the battery performance. By providing an atomistic description of lithium ion diffusion through the bulk $LiMn_2O_4$ crystal lattice, a molecular model will be able to evaluate possible diffusion mechanisms, and determine the relative diffusion rates of the lithium for dopant metal and dopant amount, and different levels of lithium intercalation.

Our theoretical approach includes the use of an empirically-derived set of interatomic forcefield parameters to evaluate the stability and crystal structure of pure $LiMn_2O_4$ spinel and several metal-doped derivative compounds. The $LiMn_2O_4$ structure is characterized by a cubic unit cell of 56 atoms ($Li_8Mn^{3+}_8Mn^{4+}_8O_{32}$) and a space group symmetry of Fd3m. Oxidation of

Table I. Lennard-Jones energy parameters for ion interactions.

Interionic Pair	A (kJ Å^{12})	B (kJ Å^6)
Li$^+$ - O^{2-}	141382	0
Mn^{3+} - O^{2-}	195681	13
Mn^{4+} - O^{2-}	525395	8728
Al^{3+} - O^{2-}	200700	156
Co^{2+} - O^{2-}	719640	4863
Ni^{2+} - O^{2-}	385447	0
O^{2-} - O^{2-}	4138809	2832

the cathode material during battery charging creates the lithium-absent spinel λ-MnO$_2$ as lithium ions are transported through the open channels along the [110] direction of the lattice. Conversely, diffusion of lithium back through the channels is important during battery discharge in order to obtain the fully-lithiated and reduced state (LiMn$_2$O$_4$) of the cathode. The reversibility of this process is linked to the fade of the specific capacity and the overall battery performance. A molecular dynamics method is used in this study to evaluate lithium ion diffusion in the pure compound and several doped lithium manganese spinels.

THEORETICAL APPROACH

The interaction energy E of two ions is based on the summation of electrostatic, repulsive, and van der Waals (dispersion) energies as a function of the distance r between the ions[1]:

$$E(r_{ij}) = \frac{z_i z_j e^2}{r_{ij}} + \frac{A_{ij}}{r_{ij}^{12}} - \frac{B_{ij}}{r_{ij}^6} \qquad (1)$$

The electrostatic term includes the electron charge e and the ionic charge z. The interaction parameters A (repulsive) and B (van der Waals) are derived from the observed structures, elastic constants, and dielectric properties of simple binary oxides or from molecular orbital calculations (Table 1)[5]. The total energy of the crystal at 0 K is obtained by the summation of the interactions of the atoms of a reference cell (usually based on the unit cell) with each other and all ions in the other cells. The summations are carried out partially in reciprocal space in order to achieve

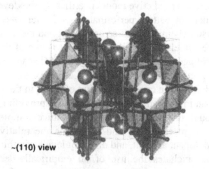

~(110) view

Figure 1. Energy-minimized crystal structure of LiMn$_2$O$_4$ showing slight distortions in the manganese octahedra and the channels for lithium ion diffusion.

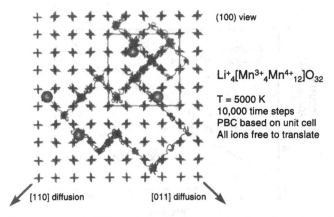

Figure 2. Trajectories of lithium ions from a molecular dynamics simulation of $Li_4Mn^{3+}_4Mn^{4+}_{12}O_{32}$ indicating the preferred diffusion path along the (110) directions. The reference unit cell used in the periodic simulation is denoted by the gray outline in the upper right of the lattice.

proper energy convergence. The ionic model assumes a rigid ion approximation by representing each ion as a point or hard sphere of charge; all ions are assigned their full formal charge. Temperature and transient effects are simulated by a molecular dynamics (MD) method. Atomic velocities are initially assigned based on a Boltzmann distribution of thermal energy. The classical equations of motion are then solved for successive time steps to obtain trajectories of atomic motion as the assembly evolves during the simulation.

Energy minimization and MD calculations were performed using the Discover energy program (Molecular Simulations Inc., San Diego) with periodic boundary conditions (PBC). Initial structures of the $LiMn_2O_4$ and doped derivatives are based on the observed structure parameters and asymmetric unit of $LiMn_2O_4$[6]. A unit cell was generated and then converted to P1 symmetry, thereby, allowing all atoms to freely translate during each simulation. Doped structures were created by substituting Al^{3+}, Co^{2+}, or Ni^{2+} for Mn^{3+} on the octahedral (16d) site and, if necessary, increasing the amount of Mn^{4+} to maintain a neutral unit cell. MD simulations were performed for constant volume (NVT ensemble) and constant pressure (NPT ensemble) conditions, and for temperatures of 2000 K, 3000 K, 4000 K, and 5000 K. Simulations were completed for times up to 30 psec using a one fsec time step. Lithium ion trajectories were monitored after an initial equilibration period of 100 fsec, and were obtained for unit cells having one, four, or seven lithium ions. The elevated simulation temperatures are required to ensure statistically significant transport of lithium during the computationally-limited MD calculation.

RESULTS AND DISCUSSION

The energy-minimized structure of pure $LiMn_2O_4$ obtained from a constant pressure calculation is provided in Figure 1. All of the theoretical cell parameters agree to within 1.5 % of the experimental values. The optimized structure remains cubic and exhibits slight distortions among all of the manganese octahedra. The excellent structural agreement supports the use of the ionic model for simulating the structure and dynamics of the spinel cathode materials and

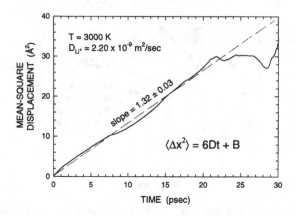

Figure 3. Mean-square displacement as a function of time for all lithium ions from the 3000 K molecular dynamics simulation of $Li_4Mn^{3+}_4Mn^{4+}_{12}O_{32}$. The value for the optimum slope is obtained from a fit to the data from the first 21 psec of the simulation.

underscores the quality of the energy parameters used in describing the ionic interactions.

Figure 2 exhibits the lithium ion trajectories obtained from a 5000 K MD simulation of $Li_4Mn_{16}O_{32}$ superimposed on a large representation of the initial spinel structure. Although all atoms all allowed to translate, only the lithium ions are energetic enough and are able to overcome the energy barriers between the 8a tetrahedral sites. Lithium diffusion is controlled by the crystallography and occurs preferentially by way of zigzag paths along the family of [110] channels. The integrated mean-square displacement for four lithium ions was calculated as a function of simulation time (Figure 3). The slope of this function is directly related to the self-diffusion coefficient of lithium. The excursions of the mean-square displacement for this example after 25 psec reflect the transient motion of one or several of the lithiums back toward their original positions. Diffusion coefficients derived from several constant pressure MD simulations are in agreement with the constant volume calculations. Cell parameters vary by less than 2 % and produce a cubic lattice that has cell lengths that are within 1.5 % of those used in the constant volume simulations.

The results of the MD simulations for the pure spinel compound at the various levels of lithium content can be compared in an Arrhenius plot as provided in Figure 4. Calculations for the lowest lithium content ($LiMn^{3+}Mn^{4+}_{15}O_{32}$) provide the fastest lithium diffusion coefficients, whereas those for the unit cells with the larger number of lithiums ($Li_7Mn^{3+}_7Mn^{4+}_9O_{32}$) have the slowest rates. This result is related to the availability of vacant tetrahedral sites for diffusion to occur. Of course, no lithium diffusion was observed for MD simulations performed for the fully-lithiated (fully-occupied tetrahedral sites) $Li_8Mn^{3+}_8Mn^{4+}_8O_{32}$ spinel. Activation energies derived from the Arrhenius plot are similar for all the three compositions with a mean value of 97 kJ/mole. Assuming that the lithium diffusion mechanism remains the same at lower temperatures, the extrapolation of these data to room temperature provides a very slow diffusion rate of 8×10^{-24} m^2/sec. However, the spinel will be subjected to a voltage and redox effects when performing as a cathode in a real battery, and one would expect significantly faster diffusion rates and ionic mobilities. Nonetheless, the pathway for lithium transport will be equivalent and the MD results will provide a convenient test for relative material performance.

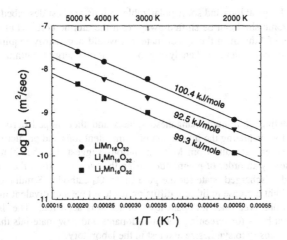

Figure 4. Arrhenius plot of the self-diffusion coefficients for lithium ion in $LiMn_2O_4$ obtained from the molecular dynamics simulations.

MD simulations for the doped $LiMn_2O_4$ compounds suggest a general decrease in lithium diffusion with increasing dopant amount (Figure 5). The doping of the spinel with low amounts of Co^{2+} and Ni^{2+} results in lithium diffusion values that are slightly enhanced relative to the pure material but decrease with increasing dopant amount. The faster lithium diffusion rates observed for the single and double-doped Co^{2+} and Ni^{2+} spinels are related to the reduced electrostatic interactions associated with the lower-charged metals compared to the Mn^{3+}. Further increases in dopant amounts tend to reduce the diffusion rates. Substitution of a similarly-charged dopant such as Al^{3+} does not appear to increase the lithium diffusion rate. No significant lattice distortions are observed for the doped spinel materials.

Due to the limitations of the ionic model in performing the MD simulations, the theoretical diffusion coefficients and activation energies can only be interpreted as relative values. Electronic polarization processes have been ignored and can significantly enhance the rates of

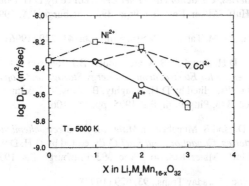

Figure 5. Effect of type and amount of dopant metal on the lithium self-diffusion coefficient as derived from MD simulations at 5000 K.

lithium diffusion. Large anions and some polarizable cations are best described by a shell model in which the electronic shell can be shifted away from the atomic core[1]. Future models of the dynamic properties of lithium in the spinel materials would necessarily require this refinement for deriving diffusion coefficients and analyzing the activation energy for lithium migration.

CONCLUSIONS

The ionic modeling of lithium manganese spinels and their doped derivatives provides a fundamental basis for evaluating the interaction of component ions and predicting their structural and dynamic properties. Results from MD simulations suggest that the lithium self-diffusion coefficient decreases by an order of magnitude during the transition from the oxidized (charged) state to the reduced (discharged) state for the pure $LiMn_2O_4$ cathode. Simulations for the doped spinels indicate a slight increase in lithium diffusion for low levels of divalent metals, however, a general decrease is observed once a dopant level of 13 % is achieved. The theoretical models provide a convenient basis for screening potential dopants and new materials that would be both time-consuming and costly to synthesize and test in the laboratory.

ACKNOWLEDGMENTS

The authors are appreciative of Tim Boyle, David Ingersoll, Bryan Johnson, Mark Rodriguez, Cory Tafoya, and Jim Voigt for their helpful discussions and comments on the material synthesis and battery testing. Behnam Vessal is gratefully acknowledged for his help in deriving the Lennard-Jones potentials. Suggestions by Diana Fisler, Harlan Stockman, and an anonymous reviewer helped to improve the final manuscript. This work was supported by the U.S. Department of Energy, Office of Basic Energy Sciences, Chemical Sciences Program under contract DE-AC04-94AL85000.

REFERENCES

1. R.T. Cygan, H.R. Westrich, D.H. Doughty, in *Materials for Electrochemical Energy Storage and Conversion–Batteries, Capacitors and Fuel Cells*, edited by D.H. Doughty, B. Vyas, T. Takamura, and J.R. Huff, (Mater. Res. Soc. Proc. **393**, Pittsburgh, PA, 1995) pp. 113-118.

2. K. Miura, A. Yamada, and M. Tanaka, Electrochim. Acta. **41**, 249 (1996).

3. J.A. Voigt, T.J. Boyle, D.H. Doughty, B.A. Hernandez, B.J. Johnson, S.C. Levy, C.J. Tafoya, and M. Rosay, in *Materials for Electrochemical Energy Storage and Conversion–Batteries, Capacitors and Fuel Cells*, edited by D.H. Doughty, B. Vyas, T. Takamura, and J.R. Huff, (Mater. Res. Soc. Proc. **393**, Pittsburgh, PA, 1995) pp. 101-106.

4. L. Xie, D. Fouchard, D., and S. Megahed, in *Materials for Electrochemical Energy Storage and Conversion–Batteries, Capacitors and Fuel Cells*, edited by D.H. Doughty, B. Vyas, T. Takamura, and J.R. Huff, (Mater. Res. Soc. Proc. **393**, Pittsburgh, PA, 1995) pp. 285-304.

5. J. D. Gale, J. Chem. Soc., Faraday Trans., **93**, 629 (1997).

6. A. Mosbah, A. Verbaere, and M. Tournoux, Mater. Res. Bull. **18**, 1375 (1983).

STRUCTURE AND ELECTROCHEMICAL POTENTIAL SIMULATION FOR THE CATHODE MATERIAL $Li_{1+x}V_3O_8$

R. BENEDEK *, M. M. THACKERAY *, and L. H. YANG **
*Chemical Technology Division, Argonne National Laboratory
**Condensed-Matter Physics Division, Lawrence Livermore National Laboratory

ABSTRACT

The structure and electrochemical potential of monoclinic $Li_{1+x}V_3O_8$ were calculated within the local-density-functional-theory framework by use of plane-wave-pseudopotential methods. Special attention was given to the compositions $1+x=1.2$ and $1+x=4$, for which x-ray diffraction structure refinements are available. The calculated low-energy configuration for $1+x=4$ is consistent with the three Li sites identified in x-ray diffraction measurements and predicts the position of the unobserved Li. The location of the tetrahedrally coordinated Li in the calculated low-energy configuration $1+x=1.5$ is consistent with the structure measured by x-ray diffraction for $Li_{1.2}V_3O_8$. Calculations were also performed for the two monoclinic phases at intermediate Li compositions, for which no structural information is available. Calculations at these compositions are based on hypothetical Li configurations suggested by the ordering of vacancy energies for $Li_4V_3O_8$ and tetrahedral site energies in $Li_{1.5}V_3O_8$. The internal energy curves for the two phases cross near $1+x=3$. Predicted electrochemical potential curves agree well with experiment.

INTRODUCTION

Transition-metal oxides that intercalate Li are under close scrutiny for their suitability as cathode materials in high-energy-density rechargeable batteries [1]. Among the desired properties of such components are high capacity, which is related to extensive intercalability, and long cycle life, which is related to structural stability upon Li insertion and extraction. A detailed characterization of atomic structure and phase stability as a function of Li insertion would be a valuable complement to electrochemical measurements for assessing the capabilities of a given candidate electrode material, as well as for suggesting strategies for improving the materials. It is frequently difficult, however, to obtain a comprehensive atomic structure characterization by experiment alone. Computational theory is a valuable additional tool for elucidating structural issues. As usual, different levels of computational theory may be considered, ranging from molecular dynamics [2] to semi-empirical tight-binding [3] to *ab initio* methods. Implementation of local density functional theory (LDFT) is of considerable interest, since it is the most rigorous treatment currently feasible. Local density functional theory calculations for lithiated transition metal dioxides have been performed by others recently [4,5].

In the present work, we apply plane-wave pseudopotential methods to the monoclinic trivanadate $Li_{1+x}V_3O_8$, a candidate Li-battery cathode [6], which intercalates between $1+x=1.2$ and $1+x=5$. The selection of this system was motivated, in part, by the availability of x-ray diffraction measurements [7] for the two monoclinic phases (which we refer to as α and β), whose most standard compositions are $1+x=1.2$ and $1+x=4$, which enable tests of the validity of our treatment. Compared to the dioxides [4,5], the trivanadates are lower symmetry, larger unit cell systems, and therefore require greater computational effort. It is encouraging that local density functional theory appears to provide an accurate description of the trivanadate system. The structural information and and physical insights provided by these calculations are not readily obtained by other means.

METHOD

Details of our plane-wave pseudopotential calculations are presented in a recent paper [8]. We employ Troullier-Martins pseudopotentials, a plane-wave basis set with 70 ry cutoff energy, and Gaussian-broadened Brillouin-zone sampling (with 16 special k-points in most cases). The Kohn-Sham equations for occupied and a few unoccupied orbitals are solved iteratively by a band-by-band conjugate gradient algorithm, stabilized by charge density mixing. The internal (reduced) atomic coordinates are relaxed to equilibrium under the guidance of the Hellmann-Feynman forces. The cell parameters (lattice constants) were held constant at the experimental values [7] for $Li_{1.2}V_3O_8$ ($Li_4V_3O_8$) in calculations for the low (high)-Li phase. The error introduced by the constant-lattice-parameter assumption is not expected to alter significantly the basic trends. For one composition, LiV_3O_8, lattice constants as well as internal coordinates were optimized self-consistently at constant energy cutoff. The calculated (experimental for $Li_{1.2}V_3O_8$) lattice constants (in a.u.) are $a=12.66$ (12.46), $b=6.78$ (6.73), $c=22.31$ (22.41). The level of agreement between theory and experiment is comparable to that typically achieved with LDFT. The energy difference between the calculation with experimental lattice constants and optimized lattice constants is approximately 0.1 eV per formula unit. Zero-point energy for Li is neglected, since it is not expected to vary widely for different Li configurations at the same composition.

RESULTS

Reference Systems

The transition metal oxides under consideration are electronically strongly correlated systems that pose a severe test of LDFT. Although the trivanadate materials are known to be insulating, the energy bands calculated within LDFT show no energy gap at the Fermi level (except at the stoichiometric composition $x=0$, which does not occur in nature). Nevertheless, structural properties, which average over all valence electrons, are still reliably calculated within LDFT for many strongly correlated materials. Available atomic structure measurements for the reference compositions of the two monoclinic structures, $1+x=1.2$ and $1+x=4$, provide convenient tests of the LDFT framework for the trivanadates.

We discuss first the reference composition of the high-Li structure, $Li_4V_3O_8$. Anion and cation sublattices of this defect-rocksalt structure each have eight sites per formula unit, with one of the cation sites vacant. In principle, five different Li configurations come into consideration, since there are five non-vanadium cation sites on which the vacancy may reside. We label the configurations 438-1, 438-2, ...,438-5, according to whether the vacant cation site is Li(1), Li(2), Li(3), $S_{oct}(1)$ or $S_{oct}(2)$, in the notation of de Picciotto et al. [7] The calculated energies (in parentheses, eV per formula unit) relative to the most stable are 438-5 (0.0), 438-4 (0.27), 438-3 (0.33), 438-2 (0.37), 438-1 (0.40).

Calculations have also been performed for the low-Li structure, whose standard composition is $Li_{1.2}V_3O_8$. In the unit cell for this system with two formula units, the octahedrally coordinated Li(1) sites are fully occupied, and the tetrahedrally coordinated Li(2) sites have occupancy of 0.2, according to XRD measurements [7]. Four other tetrahedral sites in this structure labeled $S_{tet}(1),..., S_{tet}(4)$ are unoccupied. To verify that LDFT accurately describes the energetics of this structure, calculations were performed for a system with composition $Li_{1.5}V_3O_8$, in which both (octahedral) Li(1) sites are occupied but only one tetrahedrally coordinated Li site per unit cell is occupied. Five configurations were considered, in which the singly occupied site is either Li(2), $S_{tet}(1)$, $S_{tet}(2)$, $S_{tet}(3)$, or $S_{tet}(4)$. We label these configurations 1.538-1, 1.538-2, 1.538-3, 1.538-4 and 1.538-5. 1.538.1 is energetically favored over the others, whose relative energies are 1.538-1 (0.0), 1.538-4 (0.17) 1.538-2 (0.29), 1.538-5 (0.34), 1.538-3 (0.62).

The most stable of the calculated Li configurations for both $Li_{1.5}V_3O_8$ and $Li_4V_3O_8$ are found consistent with experiment. In the case of $Li_4V_3O_8$, only three of the four Li locations have been identified experimentally. These three sites are among the four that are occupied in the two lowest energy configurations 438-5 and 438-4. The configuration 438-5, however, has a substantially lower energy than 438-4, and our calculation therefore predicts that the fourth Li resides at the site $S_{oct}(1)$. For $Li_{1.5}V_3O_8$ we find that the partially occupied site Li(2) is favored over the other possible tetrahedrally coordinated sites.

Figure 1 superposes theoretically predicted (circles) and experimentally determined (crosses) coordinates in a single (010) layer of $Li_4V_3O_8$. The rms deviation between theoretical and experimental positions is about 0.15 Å. A qualitative feature of the structure is the curvature of the rows of oxygen, which reflects the stronger bonding to the vanadium than to lithium.

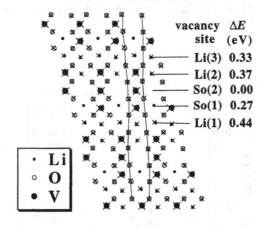

vacancy site	ΔE (eV)
Li(3)	0.33
Li(2)	0.37
So(2)	0.00
So(1)	0.27
Li(1)	0.44

• Li
○ O
• V

Fig. 1. Calculated (open and filled circles) and experimentally determined atomic positions [2] in a (010) layer of $Li_4V_3O_8$ in the 438-5 configuration, with a Li vacancy at $S_{oct}(2)$. The calculated energies for other possible Li vacancy locations are indicated. Vertical rows of oxygen bend toward V trimers and away from Li dimers.

Composition Dependence of Internal Energy

A phase transformation occurs during lithiation between the low-lithium and the high-lithium structures. Substantially different lattice parameters [7] for the two structures ($Li_{1.2}V_3O_8$ and $Li_4V_3O_8$) suggest a first-order transition. To elucidate the relative stability of the two monoclinic structures, calculations were performed at intermediate x. In the absence of experimental atomic structure information except at the reference compositions, additional assumptions are required in order to proceed. Although evidence from both experiment and theory indicates that the reference compositions $Li_{1.2}V_3O_8$ and $Li_4V_3O_8$ are well ordered, configurational disorder is likely at intermediate compositions. We believe, however, that basic trends may be discerned from considerations of internal energy alone and disorder will be neglected.

The calculations for the reference system suggest that the energy cost of creating a Li vacancy in the hypothetical system $Li_5V_3O_8$ is in ascending order for a vacancy on sites $S_{oct}(2)$, $S_{oct}(1)$, $Li(3)$, $Li(2)$, and $Li(1)$. To delithiate the high-Li structure, we therefore depopulate sites in this order. Similarly we lithiate the low-Li structure by filling tetrahedrally coordinated sites in the order suggested by our results for $Li_{1.5}V_3O_8$. Although these prescriptions are somewhat arbitrary, similar qualitative features would most likely emerge from a more precise treatment.

Figure 2 shows the calculated energies per inserted Li relative to the hypothetical reference system LiV_3O_8 (i.e., $[E(1+x)-E(1)]/x$), as a function of x, for the low- and high-Li structures. (Some of the values are preliminary, and may shift slightly as better convergence is achieved) The high-Li structure shows an energy minimum at the standard composition $1+x=4$. We attribute the preference for this composition to the corresponding V^{4+} valence: local lattice relaxation stabilizes the single d-electron on vanadium, analogous to the Jahn-Teller effect.

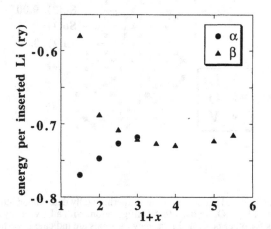

Fig.2. Calculated internal energy per inserted Li, relative to LiV_3O_8, for the low-Li (α) and high-Li (β) structures.

Although free energy curves typically show positive curvature, the x-dependence of internal energy of the α (unlike the β) phase in Fig. 2 appears to have negative curvature. We note that the energies at certain compositions may be overestimated because of the selection of non-optimal Li-configurations. A more precise treatment would perhaps shift the crossing of the two curves, but probably only slightly.

The crossing of the low-Li and high-Li phase internal energy curves occurs in the vicinity of $1+x=3$. Since a line drawn between the circle at $1+x=1.5$ and the triangle at

1+x=5 lies below the curves for α and β, equilibrium thermodynamics would predict two-phase coexistence throughout. Owing to activation barriers, however, the nucleation of β upon lithiation is delayed. The crossing of the internal energy curves (at 1+x=3) occurs close to the composition at which a two-phase (see the voltage plateau in Fig. 3) process is thought to commence during electrochemical lithiation [6,9]. This proximity may be accidental, however, if kinetics is the controlling factor.

Our calculations provide clues regarding transformation paths. As Li is removed from the high-Li structure, features characteristic of the low-Li structure start to develop, while the exclusively octahedral Li coordination characteristic of the high-Li structure remains intact. Specifically, the curvature of the rows of oxygen becomes increasingly pronounced, as does the curvation of the vanadium trimers. The effect of the increased curvatures is to open corridors for Li movement, thereby lowering activation energies for transforming to the preferred low-Li structure at depleted Li compositions.

Electrochemical Potential

We have analyzed the calculated energies for the two trivanadate phases to predict electrochemical potentials, relative to a pure-Li anode. The chemical potential of Li in the oxide is estimated from the difference between total energies for cells in which the lithiation differs by the increment $\Delta x = 0.5$ or 1.0. Thus, electrochemical potentials are predicted from the relation $V(1+x+\Delta x/2) = [E(1+x+\Delta x) - E(1+x)] / \Delta x - \mu(Li)$, where $\mu(Li)$ is the calculated chemical potential for bcc Li. Results are plotted in Fig. 3, along

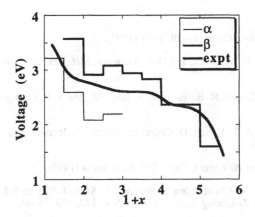

Fig. 3. Calculated electrochemical potential for $Li_{1+x}V_3O_8$ versus lithium filling, for the α (low-Li) and β (high-Li) phases. Voltage drops sharply at 1+x=5, since all vacancies in the defect rocksalt structure are filled.

with an experimental cell voltage curve at low current [9]. It is encouraging that the *ab initio* calculations, with no adjustable parameters, are relatively close to experiment. To make a more detailed interpretation requires a hypothesis regarding the structure in the region

119

between the standard compositions ($1+x = 1.2, 4$). The most straightforward interpretation of the experimental electrochemical potential curve is that pure α persists up to $1+x = 3$, where a two phase regime begins. In this case, the α-phase predicted electrochemical potential at $1+x > 2$ is underestimated. The fixed lattice-parameter approximation, the non-optimal selection of Li configurations, and the neglect of entropic contributions to the free energy may all contribute to this discrepancy. Incidentally, the sharp drop in voltage above $1+x = 5$ is attributed to the lack of additional vacancies for Li to fill in the defect rocksalt structure. Our calculation for $1+x = 5.5$ treats a single Li in an interstitial site, which results in a lower voltage than those that correspond to filling vacancies.

CONCLUSION

Ab initio LDFT calculations for the trivanadate compositions $Li_{1.5}V_3O_8$ and $Li_4V_3O_8$ yield structural coordinates in excellent agreement with experiment, as well as predictions of Li coordinates missing from the XRD refinements. Internal energy curves for the low- and high-Li phases cross in the vicinity of $x=3$. Electrochemical potential simulations (Fig. 3) for the α (β) phase are in close accord with experiment, particularly in the interval $1+x<2$ ($1+x>2$), but the detailed interpretation is obscured by the lack of experimental structural information in the region between the standard compositions ($1+x = 1.2, 4$).

ACKNOWLEDGMENTS

This work was supported at ANL by the U. S. Department of Energy, Office of Basic Energy Sciences. L.Y. is supported at LLNL by the U. S. Department of Energy under contract no. W-7405-ENG-48.

REFERENCES

1. P. G. Bruce, Chem. Commun. **19**, 1817 (1997).

2. R. T. Cygan, H. R. Westrich, and D. H. Doughty, Mat. Res. Soc. Symp. Proc. **393**, 113 (1995).

3. A. Stashans, S. Lunell, R. Bergstroem, A. Hagfeldt, and S.-E. Lindquist, Phys. Rev. **B53**, 159 (1996).

4. M. K. Aydinol, A. F. Kohan, G. Ceder, K. Cho, and J. Joannopoulos, Phys. Rev. **B56**, 1354 (1997).

5. C. Wolverton and A. Zunger, Phys. Rev. B., in press (1997).

6. K. West, B. Zachau-Christiansen, S. Skaarup, Y. Saidi, J. Barker, I. I. Olsen, R. Pynenburg, and R. Koksbang, J. Electrochem. Soc. **143**, 820 (1996).

7. L. A. de Picciotto, K. T. Adendorff, D. C. Liles, and M. M. Thackeray, Solid State Ionics **62**, 297 (1993).

8. R. Benedek, M. M. Thackeray, and L. H. Yang, Phys. Rev. **B56**, 10707 (1997).

9. V. Battaglia, A. Jansen, and A. Kahaian, unpublished.

Ab INITIO CALCULATION OF THE Li$_x$CoO$_2$ PHASE DIAGRAM

ANTON VAN DER VEN, MEHMET K AYDINOL and GERBRAND CEDER
Department of Materials Science and Engineering, Massachusetts Institute of Technology,
77 Massachusetts Ave., Cambridge, MA, 02139

ABSTRACT

The electrochemical properties of the layered intercalation compound LiCoO$_2$ used as a cathode in Li batteries have been investigated extensively in the past 15 years. Despite this research, little is known about the nature and thermodynamic driving forces for the phase transformations that occur as the Li concentration is varied. In this work, the phase diagram of Li$_x$CoO$_2$ is calculated from first principles for x ranging from 0 to 1. Our calculations indicate that there is a tendency for Li ordering at x = 1/2 in agreement with experiment [1]. At low Li concentration, we find that a staged compound is stable in which the Li ions selectively segregate to every other Li plane leaving the remaining Li planes vacant. We find that the two phase region observed at high Li concentration is not due to Li ordering and speculate that it is driven by a metal-insulator transition which occurs at concentrations slightly below x < 1.

INTRODUCTION

The layered transition metal oxide intercalation compound Li$_x$CoO$_2$ has many of the materials properties that define a good cathode material. It exhibits a relatively high voltage with respect to metallic Li and undergoes only small structural changes as it is deintercalated to a Li concentration of approximately 0.3. Furthermore, Li$_x$CoO$_2$ exhibits an exemplary cyclability when the Li concentration is varied between x = 0.5 and x = 1.0. It is, therefore, useful to investigate the fundamental thermodynamic and phase stability properties of Li$_x$CoO$_2$ to obtain a clear understanding of the characteristics that go into making a good cathode. In this respect, first principles methods are a useful tool, as they can complement and often deepen the understanding obtained experimentally about a particular material.

The layered form of Li$_x$CoO$_2$ at high Li concentration belongs to the space group $R\bar{3}m$ and has an ABC oxygen stacking with Co and Li residing in alternating planes between the close packed oxygen layers [2]. As Li is removed from Li$_x$CoO$_2$, vacancies are created within the Li planes. This can lead to Li-vacancy ordering or to changes of a structural nature of the CoO$_2$ host framework. Electrochemical and diffraction studies have indicated that a wide range of phase transformation phenomena occur in Li$_x$CoO$_2$ [1, 3, 4]. For example, at high Li concentrations, a two-phase coexistence region is observed between x = 0.75 and 0.93 where the coexisting phases both have the $R\bar{3}m$ symmetry and differ only in their lattice constants and Li concentration [1]. At around x = 0.5 Reimers and Dahn [1] observed that the Li ions order within the Li planes below approximately 60° C. Ohzuku and Ueda [3] and Amatucci et al [4] found that upon deintercalating below x = 0.21, the layered host transforms to a new phase. To our knowledge, this new phase has not yet been characterized. Fully deintercalated Li$_x$CoO$_2$ was found [4] to maintain its layered structure but with a change in oxygen stacking order to ABAB instead of the ABC stacking order observed at high Li concentration. To further understand and clarify the nature of these phase transformations, we have performed a first principles study of the phase diagram of the layered form of Li$_x$CoO$_2$.

METHOD

To study the thermodynamics of Li$_x$CoO$_2$ we consider the Li sites and assign an occupation variable σ_i which is +1 if Li occupies site i and -1 if site i is vacant. It can be shown

121

[5] that the dependence of the energy on the configuration of Li can then be exactly expanded in terms of polynomials of these discrete occupation variables according to

$$E = V_o + V_{pair} \cdot \sigma_i \cdot \sigma_j + V_{triplet} \cdot \sigma_i \cdot \sigma_j \cdot \sigma_k + \dots \qquad (1)$$

The polynomial basis functions correspond to products of occupation variables belonging to different clusters of lattice sites such as a pair, a triplet etc. For this reason, eqn. 1 is often referred to as a cluster expansion. The sum in equation (1) includes polynomials corresponding to all possible clusters on the lattice, but in practice, the coefficients $V_{cluster}$ of eqn (1) converge to zero as the size and/ or distance between the points of the cluster increase. Therefore, equation (1) can be truncated after some maximal cluster. The coefficients $V_{cluster}$ implicitly account for the interaction between different Li ions, the interaction between Li ions and the CoO_2 host structure and the effect of relaxations of both the host and the Li ions. For more details on cluster expansions and their applications to metals, oxides and semiconductors, we refer the reader to the references [6-9]

Numerical values for the coefficients $V_{cluster}$ for a particular system must be obtained with an accurate first principles total energy method. This can be done by calculating the energy of many different Li-vacancy arrangements within a particular host structure, after which the $V_{cluster}$ can be obtained by either a least squares fit [10] or a more elaborate linear programming fit [11] of eqn 1 to the calculated energies. In this work, we used the pseudopotential method in the local density approximation to calculate the energy of different Li-vacancy arrangements. The application of this method to oxides is well established [12] and the errors incurred by this method are well understood. It suffers from errors as all local density approximations, namely a systematic underprediction of equilibrium lattice parameters, and an over-prediction of cohesive energies.

To investigate the relative stability between different host structures, a separate cluster expansion of the form (1) must be constructed for each host. The linear form of eqn. 1 in terms of the polynomial basis functions makes it ideal to be used in combination with statistical mechanics methods, such as Monte Carlo simulations [13], to find the most stable Li ordering and to obtain finite temperature thermodynamic properties such as free energies. Once the free energies as a function of Li concentration of different host structures are known, the relative stability between these hosts can be determined with the common tangent construction technique.

RESULTS

In this work, we considered three different host structures. These are illustrated in figure 1. The first host structure, conventionally called O3 [14], has an ABC oxygen stacking and is observed to be the stable layered form of Li_xCoO_2 for x larger than 0.30. The second host that we considered is called O1 and has an ABAB oxygen stacking. This host has been identified both experimentally [4] and from first principles [15] as the stable phase when Li_xCoO_2 is completely deintercalated. We also considered a hybrid (H1-3) host structure that has features of both O3 and O1. Every other plane between O-Co-O sheets of H1-3 has a local environment identical to that of O3 whereas the remaining planes have an environment identical to that of O1. Since Li prefers the octahedral sites of O3 to that of O1, we assumed that in H1-3, the Li ions only occupy those alternating planes with an O3 environment. In view of the assumed distribution of Li ions between the O-Co-O sheets, this hybrid host structure can be thought of as a stage II compound similar to that observed in graphite intercalation compounds [16].

We constructed a separate cluster expansion for both the O3 and the H1-3 host structures using the energy values of different Li-vacancy arrangements on the two respective host structures. These energy values were calculated with the VASP pseudopotential program [17, 18] which solves the Kohn Sham equation within the local density approximation using ultra-soft pseudopotentials [19, 20]. All crystallographic degrees of freedom were optimized such that the minimal ground state energy was obtained. For the O3 host we constructed a cluster expansion containing 19 terms (these included 6 nearest neighbor pair clusters within the Li planes, 6 nearest neighbor pair clusters connecting different Li planes and five three point clusters) based on the energy values of 44 different Li-vacancy arrangements within this host. For the H1-3 host, a cluster expansion containing 5 terms (including only in plane first and second nearest neighbor pairs and an in plane triplet cluster) was constructed based on the energies of 5 different Li-vacancy arrangements. Both cluster expansions for the O3 and the H1-3 hosts respectively were implemented in Monte Carlo simulations in the grand canonical ensemble. The details of the pseudopotential calculations, the cluster expansions and the Monte Carlo simulations will be presented in a forthcoming paper. Figure 2 illustrates the calculated phase diagram based on our Monte Carlo simulations. In constructing this phase diagram, we assumed that the O1 host is a line compound at $x = 0$ and set its free energy equal to its energy as calculated with the pseudopotential method. This is a reasonable assumption, since O1 is observed experimentally to be unstable immediately upon lithium insertion [4].

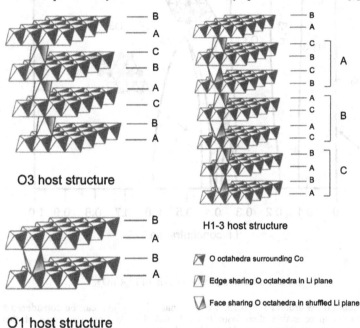

Figure 1. Schematic illustration of the three host structures O3, O1 and H1-3. The edges of the octahedra correspond to oxygen ions. Upper case letters describe the stacking of the close packed oxygen layers.

As can be seen in figure 2, our calculations predict that the O3 host is stable for Li concentrations above $x = 0.3$. For $x > 0.6$, we find that the Li ions remain disordered within the O3 host below room temperature. At Li concentrations around $x = 0.5$, our cluster expansion predicts that the Li ions will order in rows alternated by rows of vacancies as illustrated in the inset of figure 2. The order-disorder transition temperature of this phase was found to be around 160° C. This agrees qualitatively with the experimental observations of Reimers and Dahn [1] who observed Li ordering around $x = 0.5$. Nevertheless the calculated transition temperature is about 100° C higher than the experimentally measured order disorder transition temperature. Such over-prediction of the transition temperature is typical for first principles calculations of phase diagrams in the local density approximation. We find that the Li ions also order at $x = 1/3$

in the O3 host with a configuration in which the Li ions are spread apart as far as possible. The order-disorder transition temperature for this 1/3 Li ordered phase is about 80° C. In view of the of the over-estimation of transition temperatures expected from a first principles phase diagram calculation in the local density approximation, this phase may be expected to occur around 0° C in reality. This is a possible reason why it has not been observed experimentally. At zero Li concentration, we find that the O1 host is stable as opposed to the O3 host, in agreement with the experimental observations [4].

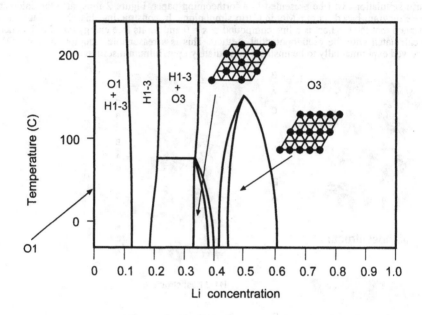

Figure 2. Calculated phase diagram of Li_xCoO_2.

For Li concentrations around x = 0.15, we find that H1-3, which can be considered a stage II compound, is more stable than both the O1 and the O3 host structures at this concentration. Experimentally, Ohzuku and Ueda [3] and Amatucci et al [4] observed that the O3 host undergoes a transformation to a new phase below x = 0.21 which is stable as a single phase at x = 0.148. The predicted stability of the stage II compound around x = 0.15 is consistent with this experimental finding. We found that the calculated diffraction patterns of the H1-3 phase are in good qualitative agreement with those observed experimentally [4] at low concentrations. We believe, therefore, that the experimentally observed phase transformation below x = 0.21 is to a stage II compound. The crystallographic details of this phase along with its X-ray powder diffraction patterns will be published in a forth coming paper.

As can be seen in figure 2, we do not find a two-phase coexistence region between x = 0.75 and x = 0.93 as is observed experimentally. We believe that this two phase region is unlikely to be the result of an ordering reaction, since we do not find any tendency for Li ordering above room temperature for x > 0.6. Instead, we believe that the two phase region is related to the metal-insulator transition that occurs at high Li concentration. Stoichiometric $LiCoO_2$ is a semiconductor [21] whereas Li_xCoO_2 is a metal for x < 0.75. Therefore, there exists a concentration range above 75 % Li, in which the electronic nature of the O3 host changes from insulating behavior to metallic behavior. Our first principles calculations and previous first principles calculations within the local density approximation [15, 22, 23] are in qualitative agreement with this picture. Within the local density approximation, $LiCoO_2$ is predicted to be an insulator while Li_xCoO_2 is predicted to have partially filled valence bands for x less than 1. For every Li removed from $LiCoO_2$, an electron hole is created within the valence band. For x < 0.75, we can expect that there are sufficient holes to allow for a significant degree of screening, and in this regime, the hole states in the valence bands are likely to be delocalized such that Li_xCoO_2 exhibits metallic electronic properties. Under these conditions, the local density approximation can be expected to yield accurate results. For dilute hole concentrations, on the other hand, the hole states are likely to be insufficiently screened causing them to remain localized. In this case, electron correlations are important and first principles investigations in the independent electron approximation such as those performed in the local density approximation break down. A correct study of strongly correlated systems requires the use of many body theories which are currently intractable for systems with large unit cells. The reason that we do not predict a two phase region between x = 0.75 and 0.93 is therefore likely to be the result of an incorrect description within the local density approximation of the electronic states at dilute hole concentrations. The electronic contributions to the free energy will be very different depending on whether the holes are localized or delocalized and the differences will both be of an energetic and entropic nature. These thermodynamic differences are likely to be strong enough in Li_xCoO_2 to cause a two phase region between a metallic phase and a semiconducting phase with localized electrons.

CONCLUSION

In this work we have performed a first principles investigation of the Li_xCoO_2 phase diagram. Our results indicate that first principles investigations can give practical information and insights about the thermodynamic and phase stability properties of transition metal oxide intercalation compounds. We find that at x = 1/2, the Li ions order in agreement with the observations of Reimers and Dahn [1]. We do not find the two phase region observed experimentally at high Li concentration [1, 3, 4]. At 15 % Li concentration, we identify a new stage II compound to be stable.

ACKNOWLEDGMENTS

This work was supported by the Department of Energy, Office of Basic Energy Sciences under contract No. DE-FG02-96ER45571. We thank the San Diego Supercomputing Center for access to their C90 computers and the Center for Theoretical and Computational Materials Science of the National Institute of Science and Technology for generously providing us with computing resources. One of the authors (AVDV) gratefully acknowledges fellowship support from the DOE Computational Science Graduate Fellowship Program.

REFERENCES

[1] J. N. Reimers and J. R. Dahn, J. Electrochem. Soc. **139**, 2091 (1992).

[2] H. J. Orman and P. J. Wiseman, Acta Cryst. **C40**, 12 (1984).

[3] T. Ohzuku and A. Ueda, J. Electrochem Soc. **141**, 2972 (1994).

[4] G. G. Amatucci, J. M. Tarascon and L. C. Klein, J. Electrochem. Soc. **143**, 1114 (1996).

[5] J. M. Sanchez, F. Ducastelle and D. Gratias, Physica **128A**, 334 (1984).

[7] D. de Fontaine, in *Solid State Physics* (eds. Ehrenreich, H. & Turnbull, D.) 33 (Academic Press, 1994).

[8] A. Zunger, in *Statics and Dynamics of Alloy Phase Transformations* 361 (1994).

[9] P. D. Tepesch, A. F. Kohan, G. D. Garbulsky, G. Ceder, C. Coley, H. T. Stokes, L. L. Boyer, M. J. Mehl, B. Burton, K. Cho and J. Joannopoulos, J. Am. Ceram. Soc. **79**, 2033 (1996).

[10] J. W. D. Connolly and A. R. Williams, Phys. Rev. B **27**, 5169 (1983).

[11] G. D. Garbulsky and G. Ceder, Phys. Rev. B **51**, 67 (1995).

[12] A. F. Kohan and G. Ceder, Computational Materials Science **8**, 142 (1997).

[13] K. Binder and D. W. Heermann, *Monte Carlo simulation in statistical physics* (Springer-Verlag, Berlin, 1988).

[14] C. Delmas, C. Fouassier and P. Hagenmuller, Physica B **99**, 81 (1980).

[15] C. Wolverton and A. Zunger, Phys. Rev. B **57**, 2242 (1998).

[16] S. A. Safran, Solid State Physics **40**, 183 (1987).

[17] G. Kresse and J. Furthmuller, Phys. Rev. B **54**, 11169 (1996).

[18] G. Kresse and J. Furthmuller, Computational Materials Science **6**, 15 (1996).

[19] D. Vanderbilt, Phys. Rev. B **41**, 7892 (1990).

[20] G. Kresse and J. Hafner, J. Phys.: Condens. Matter **6**, 8245 (1994).

[21] J. van Elp, J. L. Wieland, H. Eskes, P. Kuiper, G. A. Sawatzky, F. M. F. de Groot and T. S. Turner, Phys. Rev. B **44**, 6090 (1991).

[22] M. T. Czyzyk, R. Potze and G. A. Sawatzky, Phys. Rev. B **46**, 3729 (1992).

[23] M. K. Aydinol, A. F. Kohan, G. Ceder, K. Cho and J. Joannopoulos, Phys. Rev. B **56**, 1354 (1997).

Part III

Fuel Cells

MIXED IONIC-ELECTRONIC CONDUCTION IN NI DOPED LANTHANUM GALLATE PEROVSKITES

N.J. Long[1], and H.L.Tuller[2]
[1]Industrial Research Limited
Gracefield Rd, PO Box 31-310
Lower Hutt, New Zealand

[2]Crystal Physics and Electroceramics Laboratory
Department of Materials Science and Engineering
Massachusetts Institute of Technology
Cambridge, MA 02139, U.S.A.

ABSTRACT

Lanthanum gallate is a promising material for "monolithic" fuel cells or oxygen pumps, i.e. one in which the electrolyte and electrodes are formed from a common phase. We have investigated $La_{1-x}Sr_xGa_{1-y}Ni_yO_3$ ($LSGN_{x-y}$) with x=0.1 and y=0.2 and 0.5 as a potential cathode material for such an electrochemical device. The $\sigma(PO_2,T)$ for $LSGN_{10-20}$ points to a p-type electronic conductivity at high PO_2 and predominantly ionic conductivity at low PO_2. $LSGN_{10-50}$ has an electronic conductivity suitable for SOFC applications of approximately 50 S/cm in air at high temperature. AC impedance spectroscopy on an electron blocking cell of the form M/LSG/LSGN/LSG/M was used to isolate the ionic conductivity in the $LSGN_{10-20}$ material. The ionic conductivity was found to have a similar magnitude and activation energy to that of undoped LSG material with σ_i= 0.12 S/cm at 800°C and E_A= 1.0 ± 0.1 eV. Thermal expansion measurements on the LSGN materials were characterized as a function of temperature and dopant level and were found to match that of the electrolyte under opeating conditions.

1. INTRODUCTION

Developing electrodes for solid oxide fuel cells or oxygen pumps is a difficult task due to the need to simultaneously satisfy electrical, chemical and thermo-mechanical requirements. A promising approach is to dope the electrolyte material with multivalent cations in order to introduce electronic conductivity[1]. If the structure of the electrolyte is not radically altered, then an electrode may be obtained which retains a similar thermal expansion to the electrolyte, is chemically compatible, and is a mixed conductor. A mixed ionic and electronic conducting (MIEC) electrode is expected to have lower overpotential losses than a purely electronic conductor, due to the expansion of the triple phase boundary to the whole surface of the mixed conductor. At temperatures of 800-1000° C, an electronic conductivity (σ_e) of the order of 10 - 100 S/cm and an ionic conductivity (σ_i) at least as large as that in the electrolyte, i.e. of the order of 0.01 -0.1 S/cm is desirable.

This approach for forming MIECs has been investigated in pyrochlores [1-2] and in electrodes for zirconia fuel cells [3]. The recently discovered perovskite oxygen ion conductor La$_{1-x}$Sr$_x$Ga$_{1-y}$Mg$_y$O$_3$ [4,5] is an ideal candidate for this approach given the flexibility to substitute ions on either the A or B sites to enhance the ionic or electronic conductivity. Other perovskite oxides such as La$_{1-x}$Sr$_x$MnO$_3$ and (La,Sr)(Co,Fe)O$_3$ are already known to exhibit either high electronic or mixed conductivities. Solid solutions with the gallate can be expected to exhibit high MIEC as well.

In this paper we study the effects of Ni as a B site dopant in La$_{1-x}$Sr$_x$GaO$_3$. We expect to introduce metallic or polaronic conductivity with increasing Ni content. The strontium dopant, written in conventional notation [6] as, Sr_{La}', is compensated largely by oxygen vacancies [$V_O^{\cdot\cdot}$] giving rise to the high ionic conductivity in the parent material [4, 5]:

$$\sigma_i = 2[V_O^{\cdot\cdot}]q\mu_{V_O^{\cdot\cdot}} \tag{1}$$

The Ni may be in either of the two oxidation states, Ni^{+2} or Ni^{+3}, written, Ni_{Ga}' and Ni_{Ga}^x. The Ni_{Ga}' may be compensated either by further oxygen vacancies or by holes. LaNiO$_3$ with Ni in the +3 oxidation state is known to be a metallic conductor with a narrow conduction band made up of e$_g$ type 3d metal orbitals with an admixture of oxygen 2p [7]. As we add Ni in LSGN$_{x-y}$, we expect to get a hopping conductivity among the Ni sites and eventually metallic conductivity as we form a Ni-O band of sufficient width. The $\sigma(Po_2, T)$ dependence of the Ni doped samples can be correlated with a defect model to better understand the behavior of the material, in particular the relative changes in ionic and electronic conductivity.

Measurement of the ionic conductivity of a material with a high σ_e is known to be problematic. We use an AC impedance technique with electron blocking electrodes for this purpose. We have previously used this method to measure the ionic transference number in Mn doped gadolinium titanate[8]. This method is superior to more common techniques such as concentration cell measurements to measure the transference number, and is particularly useful when $\sigma_e \geq \sigma_i$. However AC blocking experiments can be unreliable due to misinterpretation of the features in the impedance spectrum and oxygen leakage. These results therefore require careful analysis.

For the blocking technique to be valid we must have R$_e^{SE}$ >> R$_i^{MIEC}$ [9]. This ensures that the SE is blocking the electronic current in the mixed conductor, rather than the mixed conductor blocking the ionic current in the SE. There is some uncertainty in the literature as to whether the blocking electrode we have chosen, La$_{0.9}$Sr$_{0.1}$GaO$_3$, is a mixed conductor at high Po_2, with the total conductivity equal to 3.6x10^{-2} S/cm and ionic transference number above 0.9 [10]. Recent diffusion results confirm the predominant ionic character[11]. Our own experiments with Pt blocking electrodes suggest that the electronic conductivity in air must be small [12]. We use the results of our blocking cell to confirm the validity of the above criterion. The use of a lanthanum gallate compound as the blocking SE allows us to co-sinter the MIEC and SE powders. This means leaks or contact resistance at the SE/MIEC boundary is minimized. These are often a problem when YSZ has been used as the blocking SE.

2. EXPERIMENT AND THEORY

High purity (99.99%) oxide powders of La_2O_3, $SrCO_3$, Ga_2O_3, and NiO were used as the starting materials. The powders were mixed, then pelletized and fired 3 times at 1300°C for 16hrs with intermediate regrinding. The pre-reacted powders were isopressed at 350MPa and fired at 1500°C for 48hrs. This preparation was sufficient to avoid the formation of the impurity phases $SrLaGaO_4$ and $SrLaGa_3O_7$ that were previously sometimes visible in XRD and SEM analysis of our samples. The XRD patterns of the Ni samples were obtained using a Philips automated diffractometer and the lattice parameters were determined with the assistance of a computer program.

Specimens used for four probe conductivity measurements were typically of dimensions 2 x 2 x 8 mm. The different oxygen pressures were obtained by mixing oxygen and nitrogen, or CO and CO_2, in appropriate proportions. The oxygen partial pressure was checked with a zirconia sensor. Components of the blocking cell:

$$O_2, M/ La_{0.9}Sr_{0.1}GaO_3/ La_{0.9}Sr_{0.1}Ga_{0.8}Ni_{0.2}O_3 / La_{0.9}Sr_{0.1}GaO_3/M, O_2$$

were prepared by lightly co-pressing the pre-reacted powders, isopressing at 350MPa, and then sintering in air at 1500°C for 48 hrs. The samples were approximately 7 mm in diameter and 10 mm in length. SEM analysis showed the samples to be of high density. Either Pt or Ag paste electrodes (M in the above cell) were used for the AC measurements. A Solartron low frequency 1260 or 1250 impedance analyzer measured the impedance spectrum and fitting was done using the program ZPlot (Scribner Associates, Inc.). Thermal expansion measurements were carried out on a standard dilatometry rig.

The impedance spectrum for an electron blocking cell of the form SE/MIEC/SE has been derived theoretically by Macdonald [13] and Maier [14]. Where SE = solid electrolyte, and MIEC = mixed ionic-electronic conductor. A typical spectrum obtained from our samples is shown in Fig. 1b, along with the equivalent circuit. The high frequency intercept corresponds to the bulk resistance. At intermediate frequencies we see a depressed semicircle resulting from interfacial characteristics, i.e. the Pt/SE and/or SE/MIEC interface. At very low frequencies we see a Warburg diffusion impedance due to O^{2-} diffusion in the MIEC. A grain boundary contribution is not resolvable at the temperatures of interest here. The Warburg impedance is fitted to the equation

$$Z_w = R_w \frac{\tanh\left[\sqrt{iS\omega}\right]}{\sqrt{iS\omega}} \tag{2}$$

This is the solution to a finite length Warburg impedance, where $S = L^2 / \tilde{D}$, L is the diffusion length (i.e. the thickness of the MIEC), and $\tilde{D}$ is the chemical diffusion coefficient of the MIEC. Z_W reflects the diffusion of oxygen ions in the MIEC at low frequencies. Physically the SE blocks the electrons and the oxygen ions are driven by a concentration gradient across the MIEC. From the circuit model given by Macdonald (13) and Maier (14) we also have

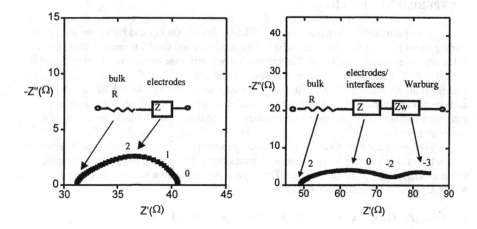

Figure 1. AC impedance spectra for (a, left) Ag/LSG/Ag at 800°C, and (b, right) an electron blocking cell Ag/LSG/LSGN10-20/LSG/Ag at 800°C. The numbers above the data are log (frequency).

$$R_w = t_e R_i \tag{3}$$

where R_i is the ionic resistance of the MIEC and t_e is the electronic transference number. By combining the low frequency fitting with the measured total conductance,

$$1/R_T = 1/R_i + 1/R_e \tag{4}$$

where R_e is the electronic resistance of the MIEC, we are able to separate the ionic conductivity from the total conductivity[8].

3. RESULTS

We present results for two Ni-doped compositions, $La_{0.9}Sr_{0.1}Ga_{0.8}Ni_{0.2}O_3$ and $La_{0.9}Sr_{0.1}Ga_{0.5}Ni_{0.5}O_3$. XRD analysis shows that the $La_{0.9}Sr_{0.1}Ga_{0.8}Ni_{0.2}O_3$ has a hexagonal structure with cell parameters a=5.507 Å and c=6.657 Å, $La_{0.9}Sr_{0.1}Ga_{0.5}Ni_{0.5}O_3$ has the same hexagonal structure with a=5.497 Å and c=6.634 Å. This is the same distorted perovskite structure as $LaNiO_3$. Without the Ni dopant, $La_{0.9}Sr_{0.1}GaO_3$ has an orthorhombic structure[15]. There is some evidence from the XRD patterns for a superlattice peak which would give a unit cell with a doubled c parameter.

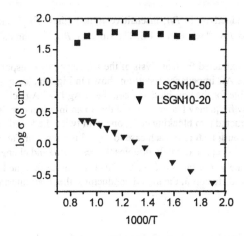

Figure 2. Total conductivity of $LSGN_{10-20}$ and $LSGN_{10-50}$ as a function of temperature.

The conductivity as a function of temperature for these two compounds is shown in Fig. 2. In air $LSGN_{10-20}$ has a semiconductor-like conductivity with an activation energy of $E_A = 0.21$ eV. The conductivity saturates above 800°C with $\sigma_t \approx 2.5$ S cm^{-1}. $LSGN_{10-50}$ has a nearly temperature independent metallic conductivity with $\sigma_t \approx 50$ S cm^{-1}.

The $\sigma(Po_2, T)$ results are shown in Fig. 3. At high Po_2 $LSGN_{10-20}$ has a Po_2 dependence of approximately $\sigma_t \approx Po_2^{1/10}$ at 800°C becoming less Po_2 dependent as T decreases. At low Po_2 the conductivity is largely Po_2 independent. The greater temperature dependence at low Po_2 is reflected in the activation energy which at 10^{-20} atm is $E_A = 0.77$ eV. For $LSGN_{10-50}$,

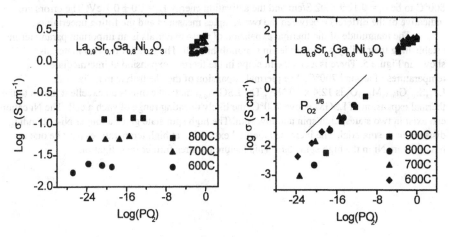

Figure 3. Total conductivity, $\sigma(Po_2, T)$ of $LSGN_{10-20}$ (left) and $LSGN_{10-50}$ (right).

the nearly temperature independent conductivity is Po_2 insensitive at the highest Po_2's but at lower Po_2 we have approximately $\sigma_t \propto Po_2^{1/n}$ where n = 4-6. The slope of $Po_2^{1/6}$ is indicated on the graph.

The ionic conductivity in air is extracted from analysis of the AC impedance response of the blocking cell LSG/LSGN/LSG. The AC impedance spectrum shown in Fig. 1b clearly shows a low frequency feature that is not visible for the non-blocking cell, i.e. a Ag/LSG/Ag cell shown in Fig 1a. To ensure that this low frequency feature was not due to an interfacial resistance at the LSG/LSGN boundary, we constructed two blocking cells, one with twice the length of $LSGN_{10-20}$. The result of fitting the Warburg feature in each case to Eqn. 2 is shown in Table I. As expected, doubling the length of the MIEC approximately doubles the resistance and changes the fitting factor S by a factor of approximately four (as $S = L^2 / \tilde{D}$). This is evidence that the low frequency part of the spectrum is due to diffusion in the mixed conductor rather than interfacial effects.

Table I. Length of $LSGN_{10-20}$ in blocking cell and fitting parameters for Warburg.

Length	$R_w(\Omega)$	S
1.57mm	3.2	74
3.14mm	7.1	284

We therefore fit the Warburg part of the spectrum to Eqn. 2. We then use Eqns. 3 and 4 to solve for the ionic conductivity of the $LSGN_{10-20}$. Figure 4 shows the temperature dependence of the ionic conductivity for the shorter sample of $LSGN_{10-20}$ in air. We find the ionic conductivity at 800°C to be $\sigma_i = 0.12 \pm 0.02$ S/cm and the activation energy $E_A = 1.0 \pm 0.1$ eV. The errors are indicative of the difference between the two samples measured and the fitting uncertainty.

The magnitude of the thermal expansion of these materials is an important parameter in establishing their utility as electrodes in a monolithic cell. The data for the two materials is shown in Figure 5. There is a change in slope in the thermal expansion at intermediate temperatures of around 700°C. The thermal expansion of the electrolyte material $La_{0.8}Sr_{0.2}Ga_{0.85}Mg_{0.2}O_3$ is 12.4 x 10^{-6} K^{-1}. The $LSGN_{10-50}$ mateial exhibits an excellent match to the thermal expansion of LSGM above 700°C, the likely opeating range of such a cell. The Ni cation can exist in two states, high spin and low spin. The high spin state gives a longer Ni-O bonding distance in the material. The increased thermal expansion at high temperatures may be due to having more Ni in the high spin state. This feature requires further investigation.

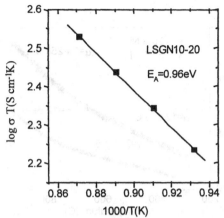

Figure 4. The ionic conductivity of LSGN$_{10\text{-}20}$ in air, extracted from the AC blocking experiment.

5. DISCUSSION

As expected, the addition of Ni to La$_{0.9}$Sr$_{0.1}$GaO$_3$ introduces significant electronic conductivity at high P$_{O_2}$. The conductivity at 800°C of $\sigma_t \approx$ 50 S cm^{-1} is adequate for electrode purposes. At low P$_{O_2}$ as the Ni^{+3} reduces to Ni^{+2} the p-type electronic conductivity decreases. The level to which it decreases depends on how Ni_{Ga}' is compensated. In LSGN$_{10\text{-}20}$ the conductivity at low P$_{O_2}$ is constant with a value of σ_i = 0.13 S/cm, which suggests ionic conductivity. The activation energy of 0.77eV also suggests ionic conduction although this is a slightly lower activation energy than is usual for these compounds[15]. The conductivity is higher than in La$_{0.9}$Sr$_{0.1}$GaO$_3$ and comparable to La$_{0.9}$Sr$_{0.1}$Ga$_{0.8}$Mg$_{0.2}$O$_3$ as can be seen in Table II. This suggests that the Ni_{Ga}' is compensated by additional oxygen vacancies.

There also remains a chance that a phase change occurs at intermediate P$_{O_2}$, that being the source of the change in the magnitude and mechanism of conductivity. That possibility is now being pursued by annealing specimens at low P$_{O_2}$, quenching and examining the x-ray patterns.

From the AC blocking experiment we obtain for LSGN$_{10\text{-}20}$ at 800°C and in air σ_i = 0.12 ± 0.02 S/cm and E$_A$ = 1.0 ± 0.1 eV. The oxygen ion conductivity therefore remains unchanged from the low P$_{O_2}$ region. The oxygen vacancy concentration is expected to decrease as Ni_{Ga}' oxidizes to Ni_{Ga}^x at higher P$_{O_2}$. The mobility at high P$_{O_2}$ would therefore have to increase to compensate the lower $[V_O^{\bullet\bullet}]$. While increased numbers of conduction electrons at high P$_{O_2}$ would increase screening of the compensating cation charge, resulting in lowered potential barriers to anion motion, the larger value of E$_A$ appears to go contrary to this expectation. Blocking cell experiments are being extended to lower P$_{O_2}$ to clarify this trend.

Returning to the question of the suitability of LSG to serve as the electron blocking material, we now find that σ_i(MIEC) = 0.12 S/cm at 800°C. This compares to σ_e(LSG) $\approx$ 0.1σ_{total} $\approx$ 0.004 S/cm[15]. Therefore the criteria of R$_e^{SE}$ >> R$_i^{MIEC}$ is satisfied.

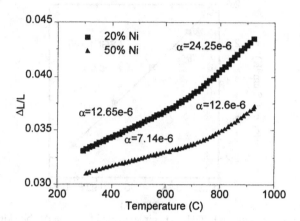

Figure 5. Thermal expansion of LSGN$_{10\text{-}20}$ and LSGN$_{10\text{-}50}$, two expansion coefficients are given for each curve, one for the low temperature region and one for the high. The coefficient is in the units of K^{-1}.

We next discuss the $\sigma(P_{O_2}, T)$ illustrated in Fig. 3. It is well known that in solids such as YSZ where $[V_O^{\bullet\bullet}]$ is fixed by an acceptor (e.g. Sr_{La}') that the oxygen vacancies are fixed and that the holes follow a $P_{O_2}^{1/4}$ dependence. Aside the weaker P_{O_2} dependence of holes observed for LSGN$_{10\text{-}20}$ at high P_{O_2}, it would appear that this material is operating in the vicinity of its electrolytic domain. On the other hand, LSGN$_{10\text{-}50}$ exhibits a much higher p-type conductivity and so one can not see the ionic contribution. Instead we observe a $P_{O_2}^{1/n}$ dependence (n ~ 4-6) which saturates at the highest P_{O_2}'s. This is consistent with a transition from a neutrality equation given by $[Sr_{La}'] = 2[V_O^{\bullet\bullet}]$ at lower P_{O_2} to $[Sr_{La}'] = p$ at higher P_{O_2} as observed in (La$_{2-x}$Sr$_x$)CuO$_4$[16].

Table II. Comparison of ionic conductivities of LaGaO$_3$ compounds.

	$\sigma_{800°C}$	E$_A$ (eV)	
La$_{0.9}$Sr$_{0.1}$GaO$_3$	0.0365	0.81	(ref. 15)
La$_{0.9}$Sr$_{0.1}$Ga$_{0.8}$Mg$_{0.2}$O$_3$	0.121	1.13	(ref. 15)
La$_{0.9}$Sr$_{0.1}$Ga$_{0.8}$Ni$_{0.2}$O$_3$ (high P_{O_2})	0.12	1.0	
La$_{0.9}$Sr$_{0.1}$Ga$_{0.8}$Ni$_{0.2}$O$_3$ (low P_{O_2})	0.13	0.77	

6. SUMMARY AND CONCLUSIONS

The substitution of Ni for Ga in the solid electrolyte $La_{0.9}Sr_{0.1}Ga_{1-y}Ni_yO_3$ results in a strong enhancement of the electronic conductivity at high P_{O_2} reaching values as high as 50 S/cm in the temperature range of 600-900°C for y=0.5. Evaluation of the Warburg impedance in $LSGN_{10-20}$ in air showed the material to retain ionic conductivities comparable to the best Lanthanum gallate solid electrolytes. These results suggest that LSGN could make a good SOFC cathode compatible with the LSG or LSGM electrolyte. Thermal expansion measurements show good matches with LSG solid electrolytes in some temperature ranges. Proper cell design must take into account conditions where TCE values do not match as well.

ACKNOWLEDGMENTS

Prof. Harry L. Tuller thanks Basic Energy Sciences, U.S. Department of Energy for support under contract #DE-FG02-86ER45261.
N.J. Long, thanks the NZ Foundation for Research Science and Technology for financial support. We thank Martin Ryan of IRL, for help with the XRD analysis.

REFERENCES

1. M. Spears and H.L. Tuller, in Solid State Ionics III, G.A. Nazri, J.M. Tarascon and M. Armand, Editors, The Materials Research Society, Pittsburgh, PA, 1993, p. 301.
2. O. Porat, M.A. Spears, C. Heremans, I. Kosacki and H.L. Tuller, Solid State Ionic **86-88**, 285 (1996).
3. S. S. Liou and W.L. Worrel, Appl. Phys. A, **49**, 25 (1989).
4. T. Ishihara, H. Matsuda, and Y. Takita, J. Amer. Chem. Soc., **116**, 3801-3803 (1994).
5. M. Feng and J.B. Goodenough, Eur. J. Solid State Chem. **31**, 663-672 (1994).
6. F.A. Kroger, in The Chemistry of Imperfect Crystals. North Holland, Amsterdam, 1974, p. 14.
7. P. A. Cox, in Transition Metal Oxides: An Introduction to their Electronic Structure and Properties, (Oxford University Press, 1992), p.220
8. N.J. Long, O. Porat and H.L. Tuller, in Proc. Third Int. Symp. on Ionic and Mixed Conducting Ceramics, The Electrochemical Soc., Paris, 1997, in press.
9. I. Riess, in Second Int. Symp. on Ionic and Mixed Conducting Ceramics, T.A. Ramanarayanan, W.L. Worrell and H.L. Tuller, Editors, The Electrochemical Society, Pennington, NJ (1994) p. 286.
10. T. Ishihara, H. Matsuda, and Y. Takita, Solid State Ionics **79**, 147-151 (1995).
11. T. Ishihara, J. A. Kilner, M. Honda, and Y. Takita J. Amer. Ceram. Soc. **119**, 2747-2748 (1997).
12. N.J. Long unpublished results.
13. J.R. Macdonald, J. Chem. Phys. **61**, 3977 (1974).
14. J. Maier, Z. Phys. Chem. (NF) **140**, 191 (1984).
15. P. Huang and A. Petric, J. Electrochem. Soc, **143**, 5, 1644-1648 (1996).
16. E.J. Opila and H.L. Tuller, J. Amer. Ceram. Soc. **77**, 2727-37 (1994).

6. SUMMARY AND CONCLUSIONS

ACKNOWLEDGEMENTS

REFERENCES

Carbon and Fluorinated Carbon Materials for Fuel Cells

D. Wheeler, F. Luczak, R. Fredley, and N. Cipollini
International Fuel Cells Corporation

ABSTRACT

Carbon and fluorinated carbon materials are major constituents of phosphoric acid fuel cells and PEM fuel cells and the stability of these materials is critical for long life operation. Laboratory corrosion studies of separator plate materials were correlated with separator plate changes in commercial PAFC fuel cells. The addition of thin films of Teflon™ to the separator plates extends the life of the separator plates to 60,000+ hours through the formation of a temporary hydrophobic barrier. ESCA studies show the loss of hydrophobicity with time of PAFC electrodes to be a result of delamination of the Teflon from the carbon and not corrosion of the Teflon by phosphoric acid. The projected life of PAFC power plants has been confirmed by commercial operation of power plants for over 40,000 hours.

INTRODUCTION

Carbon, graphite, and fluorinated hydrocarbons, such as Teflon, are the primary materials used in the construction of the PAFC and PEM fuel cells. These two acid fuel cells operate at low to moderate temperatures: the low temperature PEM fuel cell having an operating range of $50°C - 80°C$ and the phosphoric acid fuel cell, PAFC, operating in the temperature range $150°C - 200°C$. Both fuel cells can be operated at pressurized conditions, although optimum power plant efficiency is obtained at atmospheric pressure. The development of advanced carbon–graphite structures with selective Teflon additions is the enabling technology for advancing the commercialization of fuel cells.

Long life is a critical issue for commercial fuel cell power plants. A design of five years, or 40,000 operating hours is desirable for stationary power plants. The corrosion of the carbon graphite composites is the life limiting factor for the fuel cell components where the reactions follow the form:

$$C + 2H_2O \longrightarrow CO_2 + 4H^+ + 4e^- \tag{1}$$

Water is the product of both the PAFC and PEM fuel cells and for the PEM fuel cell water is also a constituent of the perfluorinated sulfonic acid membrane electrolyte. In addition, reaction (1) is favored at the oxidizing potentials typical of the oxygen reduction electrode of the fuel cell, e.g., $0.6V - 1.2V$ vs. SHE.

Two approaches to increase the stability of the carbon components are used: graphitization and application of Teflon to form a hydrophobic barrier on the surface of the carbon composite. The hydrophobic barrier prevents water, and phosphoric acid in the case of the PAFC, from contacting the carbon component. Only low levels of Teflon can be used in the carbon composite because Teflon is an electronic insulator and a fundamental requirement of the fuel cell component is high electronic conductivity.

The unit cell for the PAFC fuel cell is schematically given in Figure 1 and the major carbon containing structures are identified. This unit cell is repeated up to several hundred times in a commercial PAFC cell stack and the cross–sectional area of the stack can be up to one square meter. The anode electrode, cathode electrode, and integral separator plate (ISP) shown in Figure 1 are three components of interest reported here. All three components are carbon–graphite–Teflon composite structures designed to control the distribution of phosphoric acid electrolyte in the PAFC. In particular, the ISP is designed as an impermeable barrier to the transmission of phosphoric acid electrolyte between adjacent cells.[1] The structure of the ISP is schematically given in Figure 2 which shows a graphite sheet with thin films of Teflon bonded to each surface of the graphite sheet. These thin bonded Teflon sheets are discontinuous allowing carbon–carbon contact between the separator graphite sheet and the adjacent respective anode or cathode flow fields as shown in Figure 1. A critical requirement of the ISP is to fill the pores of the up to 25% porous graphite separator to create a hydrophobic barrier to acid penetration while at the same time permitting the carbon – carbon contact to assure optimum electronic conductivity.

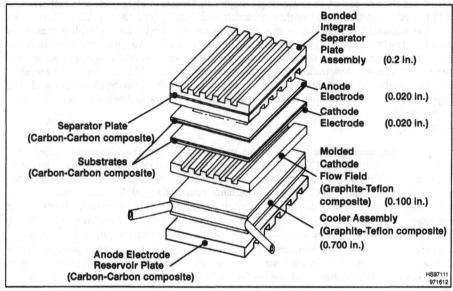

Figure 1. PAFC Cell Stack Assembly Configuration.

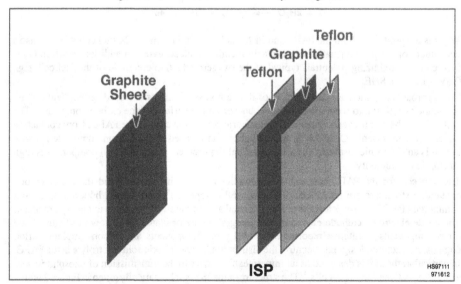

Figure 2. Schematic configuration of an ISP.

140

EXPERIMENT and RESULTS

Anodic polarization studies of graphitized separator plates identified the time dependence of the corrosion of the separator plates over the potential range 0.9V to 1.2V, vs. a hydrogen electrode in the same solution. Data were obtained using the potentiostatic method in 99% H_3PO_4 at 200°C with the data collected continuously over a 1000 minute interval. Typical data are shown in Figure 3. At potentials below 1000 mV the corrosion current, expressed as $\mu A/mg$ carbon, continually decreases from 2 $\mu A/mg$ C to nearly 0.1 $\mu A/mg$ cm – an order of magnitude decrease in corrosion current. On the other hand, the corrosion current of the graphite separators tested at potentials greater than or equal to 1000 mV showed an initial decrease and then an increase in corrosion rate. At the 1000 minute time, all of the corrosion currents of the separator plates tested at 1100 mV or greater were larger than the initial corrosion currents; the samples tested at the higher potentials gave the largest corrosion current as shown in Figure 3. For the three test potentials 1100 mV, 1050 mV, and 1200 mV, the corrosion current appears to be leveling off at times greater than 200 minutes, although a constant value was not reached at 1000 minutes. For the two test conditions 1000 mV and 1050 mV, the corrosion current continued to increase even at the 1000 minute time.

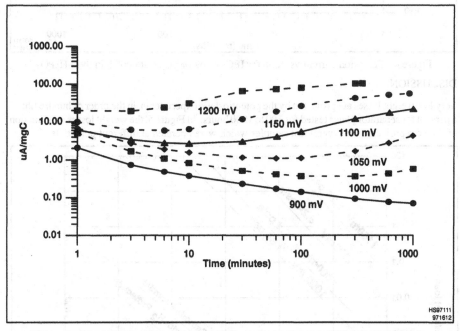

Figure 3. Corrosion current vs. time for IFC separator plates at 400° F in 99% H_3PO_4.

Consistent with the change in corrosion current with time was the change in surface area of the tested samples. The surface area was determined by measuring the double layer capacitance of the sample both at the beginning and the end of the potentiostatic corrosion test. In Figure 4, the corrosion current normalized for the surface area is given; at the higher potentials the normalized corrosion currents are approaching a steady state condition. For the lowest potential test, 900mV, steady state has still not been reached at 1,000 minutes. From reaction (1) the corrosion of the separator plate represents a loss of carbon with a concurrent increase in surface area and porosity of the separator plate.

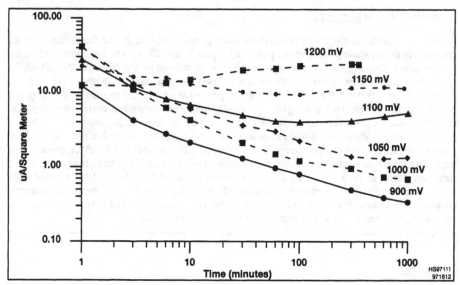

Figure 4. Corrosion current vs. time for IFC separator plates at 400° F in 99% H_3PO_4.

DISCUSSION

Very low weight losses are projected by the potentiostatic experiments, in the order of hundredths of a percent for separator plates tested for 100 to 1000 hours. In Figure 5, the weight losses as a function of time are projected for several test samples which were operated at different potentials.

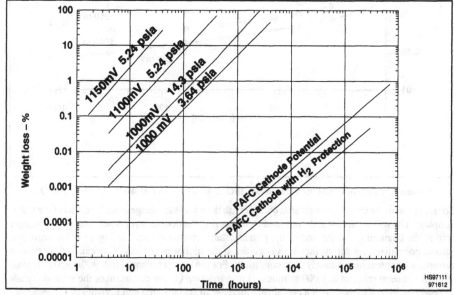

Figure 5. Projected weight loss vs. time for PAFC separator plates at indicated conditions.

142

The effect of the partial pressure of water, as indicated in reaction (1), is also shown. Decreasing water vapor pressure at constant potential results in a decrease in the corrosion rate and lower separator plate weight loss. Also consistent with reaction (1), is the projected decrease in separator weight loss with the protection of the separator plate by hydrogen. For comparison, projected weight losses at the operating potentials of commercial phosphoric acid fuel cells are included and identified as PAFC cathode potential data.

Based on this analysis, the weight loss for separators at actual operating potentials will be on the order of micrograms per gram of separator plate. While very low, the separator plate corrosion results in a change in surface area and also an increase in the porosity of the separator plate. The original porosity is in the range 10–25%. This increase in surface area and porosity results in intrusion of the phosphoric acid into the separator plate and eventually transport of the acid through the separator plate. The transport of the acid through the separator plates, i.e., inter–cell acid flux, is due to the potential gradient across the stack. The inter–cell transport of acid decreases the volume of electrolyte in one area of the stack and increases the volume in another. The maldistribution of electrolyte can lead to stack failure due to electrolyte depletion and reactant crossover or electrolyte flooding of the anode catalyst, i.e., blocking hydrogen from reacting at the anode catalyst sites.

The flux of acid across a separator plate is identified in Figure 6 as a function of "equivalent stack operation hours". The concept of equivalent stack operation hours permits accelerated testing of the separator plates with the primary control potential. As shown in Figure 3, the greater the potential the more rapid the corrosion and the greater the weight loss of the sample. Using the information as expressed in Figure 4, accelerated laboratory tests were established where weight loss of the separator plate samples operated at high potentials was related to actual PAFC cell stack data. Acid flux data of laboratory accelerated test separator plates were also correlated to acid flux data of separator plates operated in commercial 200kW PAFC power plants. The correlation established a direct relationship between accelerated laboratory tests and actual separator plate corrosion at commercial operating conditions.

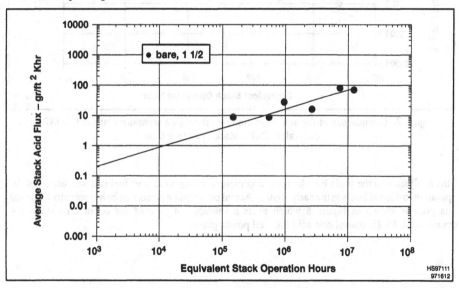

Figure 6. Dependence of the acid migration rate through a separation plate on equivalent PAFC operation time.

143

In Figure 7, the acid flux is shown for a bare separator plate and an ISP. The ISP improvement due to the addition of the Teflon films is an order of magnitude decrease in the acid flux and hence an increase in separator plate life and cell stack life. The two linear regression fits, bare plate and ISP, are essentially parallel clearly indicating that both components follow the same carbon corrosion controlling mode of failure. The impact of the Teflon films used in the ISP appears to delay the onset of corrosion, i.e., acid flux.

The delay in the onset of acid flux is believed to be the result of the hydrophobic nature of the Teflon films and the improvement in cell stack life as seen in Figure 7 suggests that a cell stack could operate for over 40,000 hours. Further, the data base from accelerated testing leads to the conclusion that a phosphoric acid cell stack operating up to 60,000 hours and in an extreme case, 100,000 hours may be obtained.

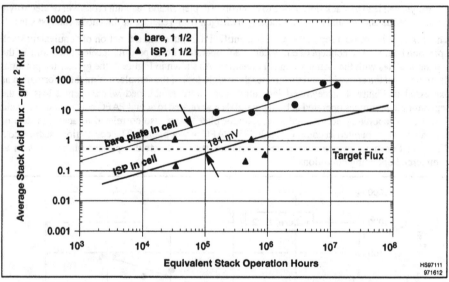

Figure 7. Comparison of the acid migration rate through a separator plate and an ISP vs. equivalent PAFC stack operating time.

This analysis was the basis by which the commercial phosphoric acid fuel cell was designed for operation to 40,000 hours in the early 1990's. Actual power plant performance has recently achieved this goal as shown in Figure 8 which gives a snapshot in time of the operational status for commercial, PC25, phosphoric acid fuel cell power plants.

144

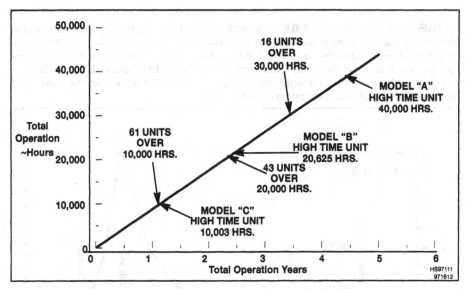

Figure 8. PC25 Fleet durability – unit operation – October 16, 1997.

The loss of hydrophobicity of an ISP with the concurrent onset of acid flux, controlled by the corrosion process was studied in an effort to further improve the cell stack life. Initially it was suspected that corrosion of the Teflon in hot phosphoric acid reduced the hydrophobic nature of the Teflon films. ESCA studies conducted on PAFC electrodes clearly show that corrosion of the Teflon was not the cause of wetting of the samples. Rather delamination of the Teflon from the carbon is observed. The ESCA data shown in Figure 9 of an as prepared PAFC anode shows a peak associated with CF_2 at approximately 292.2eV. Deconvolution of the peak shows two peaks, a strong one at 292 eV and a weak, broad peak at 293.5 eV. We associate the peak at 293.5 eV with the delamination of Teflon from carbon, i.e., charging of the sample where the Teflon is no longer in contact with the conducting carbon. This is consistent with our analysis of pure Teflon–30 where a large CF_2 peak was observed at 305.3 eV; the shift to higher binding energy is the result of charging of the sample.

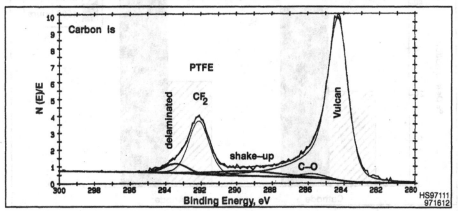

Figure 9. ESCA of a new PAFC anode containing carbon Teflon composite.

145

For PAFC electrodes with over 1500 hours operation in a test fuel cell, the CF_2 peak has shifted to higher energies and an increase in delaminated Teflon is observed as shown in Figure 10. Analysis of the percent delamination shows a two–fold increase in delaminated Teflon as a result of the cell operation and is shown in Figure 11. Based on the ESCA data on PAFC electrodes, we believe the loss of hydrophobicity of the ISP is the result of delamination of the Teflon from the carbon with subsequent wetting of the carbon by the phosphoric acid.

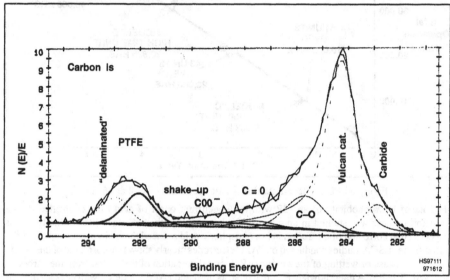

Figure 10. ESCA of used PAFC anode containing carbon Teflon composite.

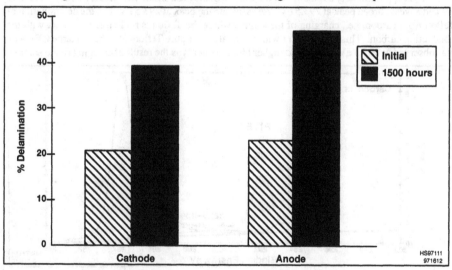

Figure 11. Teflon delimination anode and cathode results.

146

CONCLUSIONS

Corrosion tests of carbon–graphite materials used in phosphoric acid fuel cell separator plates were successfully correlated with the performance of commercial fuel cell separator plates. The corrosion of the separator plates leads to increased acid flux through the separator plates and maldistribution of the phosphoric acid electrolyte. Failure of PAFC cell stacks could result from this maldistribution of electrolyte. The analysis of separator plates bonded with thin films of Teflon, ISPs, showed cell stack life, and hence PAFC power plant life, would exceed 40,000 hours and possibly 60,000 to 100,000 hours of life.

ESCA studies of PAFC electrodes containing carbon, graphite and Teflon showed delamination of Teflon from the carbon–graphite as the cause of decreased hydrophobicity. No corrosion of the Teflon was observed.

The validity of these analyses has been established by commercial 200kW PAFC fuel cells that have achieved over 40,000 hours of operation.

REFERENCES

(1) Robert P. Roach, U.S. Patent No. 5 268 239 (7 December 1993)

HYDROTHERMAL SYNTHESIS AND PROPERTIES OF CERIA SOLID ELECTROLYTES

M. GREENBLATT*, W. HUANG, P. SHUK
Department of Chemistry, Rutgers, the State University of New Jersey, 610 Taylor Rd, Piscataway, NJ 08854-8087, martha@rutchem.rutgers.edu

ABSTRACT

The structure, thermal expansion coefficients and ionic/electronic conductivity of $(Ce_{1-x}Sm_x)_{1-y}(Tb/Pr)_yO_{(2-x/2)+\delta}$ (x= 0-0.30; y= 0-0.10) and $Ce_{1-x}Ca_xO_{2-x}$ (x= 0-0.17) solid electrolytes prepared hydrothermally were investigated. The uniformly small particle size (7-68 nm) of the hydrothermally prepared materials allows sintering of the samples into highly dense ceramic pellets at 1400°C, a significantly lower temperature, compared to that at 1600-1650°C required for samples prepared by solid state techniques. The maximum ionic conductivity in $Ce_{1-x}(Sm/Ca)_xO_{2-\delta}$ was found at x= 0.17 for the Sm ($\sigma_{600°C}$ = 5.7×10^{-3} S/cm). In $(Ce_{0.83}Sm_{0.17})_{1-y}(Tb/Pr)_yO_{1.915+\delta}$ the maximum conductivity was found at y= 0.10 for the Pr and Tb substituted ceria ($\sigma_{600°C}$ = 7.6×10^{-3} S/cm, E_a= 0.55 eV and $\sigma_{600°C}$ = 10^{-2} S/cm, E_a= 0.72 eV respectively) with electronic contribution to total conductivity around 20-30 %. When the Tb or Pr substitution in $Ce_{0.83}Sm_{0.17}O_{1.915}$ is reduced, the conductivity becomes more ionic, and is purely ionic at 2 %. However the conductivity at this lower level doping is not significantly lower.

INTRODUCTION

In the past several years, CeO_2-based materials have been intensely investigated as catalysts, structural and electronic promoters of heterogeneous catalytic reactions and oxide ion conducting solid electrolytes in electrochemical cells [1,2]. The solid electrolyte is a key component of solid-state electrochemical devices, which are increasingly important for applications in energy conversion, chemical processing, sensing and combustion control [3-5]. Presently available oxide ion conducting solid electrolytes are mainly derived from solid solutions based on ZrO_2 (zirconia) [6,7]. Many studies have been made on other, better conductive solid electrolytes as alternatives to zirconia, e.g. Bi_2O_3 [8,9] and CeO_2 [2] based materials. However, at present, the application of Bi_2O_3-based materials is hindered by their limited electrolytic domain [9].

The oxygen vacancy concentration, and concomitant oxide ion conductivity, in cerium oxide can be increased by the substitution of a lower-valent metal ion for cerium. In the past many investigations have been carried out on various aspects of ceria solid electrolytes mostly prepared by conventional ceramic methods [2,10]. Yttrium and samarium doped ceria solid electrolytes have been successfully prepared by hydrothermal method, providing low-temperature preparation and morphological control in ultrafine particles of uniform crystallite dimension [11,12]. Our systematic study of the hydrothermally prepared $Ce_{1-x}Sm_xO_2$ show maximum ionic conductivity for the $Ce_{0.83}Sm_{0.17}O_{1.915}$ composition [13]. Recently Mericle et al. [14] demonstrated that by co-doping small quantities of praseodymium in $Ce_{1-x}Gd_xO_{2-x/2}$ solid solutions, the application region (electrolytic domain) of the electrolyte is shifted by two orders of magnitude to lower oxygen partial pressure.

In this paper we present a systematic study of the structure, ionic and electronic conductivities and thermophysical properties of hydrothermally prepared samarium substituted ceria solid electrolytes doped with small amounts of Pr or Tb as electron traps to extend the oxygen partial pressure application of ceria as solid electrolyte (ion transfer numbers >0.99).

EXPERIMENT

Solid solutions $Ce_{1-x}(Sm/Ca)_xO_{2-\delta}$ (x=0-030) and $(Ce_{0.83}Sm_{0.17})_{1-y}(Tb/Pr)_yO_{1.915+\delta}$ (y=0-0.10) were synthesized by the hydrothermal method as previously reported for the Sm doped ceria solid electrolytes [13]. The appropriate quantities of cerium (III) nitrate hexahydrate $(Ce(NO_3)_36H_2O, 99.9\%$ Aldrich), samarium (III) nitrate hexahydrate $(Sm(NO_3)_36H_2O, 99.9\%$ Aldrich) or calcium nitrate hexahydrate $(Ca(NO_3)_26H_2O, 99.9\%$ Aldrich), praseodymium (III) nitrate hexahydrate $(Pr(NO_3)_36H_2O, 99.9\%$ Alfa) or terbium (III) nitrate pentahydrate $(Tb(NO_3)_35H_2O, 99.9\%$ Alfa) were dissolved separately in water, mixed and coprecipitated with ammonium hydroxide at pH= 10. The precipitated gels were sealed into teflon-lined steel autoclaves and hydrothermally treated at 260°C for several hours. The autoclaves were quenched and the crystallized powders were repeatedly washed with deionized water and dried in air at room temperature or 200°C.

The room/high temperature powder X-ray diffraction patterns (PXD) of the ultrafine powders were obtained with a SCINTAG PAD V diffractometer equipped without/with a high temperature attachment with monochromatized Cu $K\alpha$ radiation at a 2θ scan of 0.5°/min. Cell parameters were calculated by fitting the observed reflections with a least-squares program. The reflection from the (422) plane was used for the determination of average crystallite size [12]. The average crystallite size, D, of the hydrothermally prepared powders was calculated from the Scherrer formula:

$$D= 0.9\ \lambda/(\beta \cos \theta), \tag{1}$$

where λ is the wavelength of the X-rays, θ is the diffraction angle, $\beta= (\beta^2_m - \beta^2_s)$ is the corrected halfwidth of the observed halfwidth, β_m, of the (422) reflection in samples and β_s is the halfwidth of the (422) reflection in a standard sample of CeO_2 (D ~ 100 nm). Differential thermal analysis (DTA) and thermogravimetric analysis (TGA) measurements were carried out in the temperature range 25-750°C with a TA Instruments DSC 2910 and TGA 2050 with a heating and cooling rate of 2°C/min.

The powder samples were pelletized and sintered at 1400-1450°C for 10 h with a programmed heating and cooling rate of 5°C/min. The sintered samples were over 95 % of the theoretical density in all cases. The microstructure of sintered samples was studied with a Field Emission Scanning Electron Microscope (FESEM, Model DSM 962, Gemini, Carl Zeiss, Inc., Germany) equipped with an Energy Dispersive Spectroscopy system (EDS, Princeton, Gamma Tech Inc., NJ) for semiquantitative analysis of Ce, Sm, Ca, Pr or Tb. The electrical conductivity of the materials was measured on sintered ceramic pellets. Silver paste was painted onto two faces of the pellets, using GC Electronics paste. The sample was then dried and fired at 650°C. The ionic conductivity measurements were performed by the complex impedance method at frequencies ranging from 0.1 Hz to 20 KHz (Solartron 1280 Frequency Response Analyser) on isothermal plateaus one hour long, in air on heating and cooling every 25-50°C up to 650°C. In order to determine the oxide ion transfer numbers, the EMF of the following oxygen concentration cell was measured:

$$O_2, p_{O_2}^{\ a}, Ag \mid solid\ electrolyte \mid Ag, O_2, p_{O_2}^{\ c} \tag{2}$$

If the conduction in ceria solid electrolyte is predominantly ionic, the theoretical EMF (E_{th}) is given by the Nernst equation:

$$E_{th}= (RT/4F) \ln [p_{O_2}^{\ c}/p_{O_2}^{\ a}], \tag{3}$$

where R, T, and F are the gas constant, the temperature and the Faraday constant, respectively, and $p_{O_2}^{\ a} = 1.01 \times 10^5$ Pa and $p_{O_2}^{\ c} = 0.21 \times 10^5$ Pa are oxygen partial pressures at the anode (pure oxygen) and cathode (air), respectively. If the specimen has some electronic conduction, the measured EMF (E) will be lower than E_{th} because of the discharge of the electrochemical cell due to the electronic conduction. If the electrodes are sufficiently reversible the oxide ion transfer numbers, t_O, can be calculated as follow:

$$t_O= \sigma_i/(\sigma_i + \sigma_e)= E/E_{th}, \qquad (4)$$

where σ_i and σ_e are ionic and electronic conductivities, respectively.

The electrolytic domain boundary (EDB) was determined by measuring the ceria solid electrolyte impedance as a function of oxygen partial pressure, p_{O_2} and temperature [14]. The electrolytic domain boundary was defined as p_{O_2}, at which the ionic and electronic conductivities are equal. A mixture of O_2, N_2 and H_2 gasses passed through a water separator, set at 25-85°C, was used to fix different oxygen partial pressures determined from the equilibrium of the chemical reactions of the gases. The oxygen partial pressure in the gas mixture was measured by a solid electrolyte oxygen sensor before and after the gasses passed through the closed ceramic cell where the ceria solid electrolyte sample was placed for the conductivity measurements.

RESULTS

The PXD data in Fig. 1 show that $(Ce_{1-x}Sm_x)_{1-y}(Tb/Pr)_yO_{1.915+\delta}$ (x=0-0.30, y=0-0.10), prepared by the hydrothermal synthesis for the first time, forms solid solutions with the fluorite structure in the investigated range y= 0-0.10 for Tb or Pr substitution.

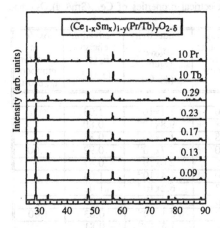
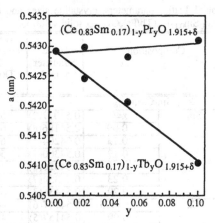

Fig. 1 Powder X-ray diffraction patterns of $(Ce_{1-x}Sm_x)_{1-y}(Tb/Pr)_yO_{2-\delta}$ solid solutions.

Fig. 2 Lattice constants of $(Ce_{0.83}Sm_{0.17})_{1-y}(Pr/Tb)_yO_{1.915+\delta}$.

The fluorite-phase evolution of $Ce_{0.83}Sm_{0.17}O_{1.915}$ as a function of heat treatment time at 260°C in hydrothermal process was examined by XPD: after 90 min., the PXD of the product is that of pure CeO_2. Heat treatment for 120 min. improves the crystallinity of the sample. The unit cell parameter a increases linearly with increasing Sm and Ca content (Table 1) as expected from effective ionic radii ($r_{Ce^{4+}} = 0.1110$ nm; $r_{Sm^{3+}} = 0.1219$ nm; $r_{Ca^{2+}} = 0.1260$ nm) considerations [15]. The unit cell parameter a remains the same with increasing Pr content within the experimental error, and decreases linearly with increasing Tb content (Fig. 2). Based on effective ionic radii considerations ($r_{Ce^{4+}/Sm^{3+}} \sim 0.1128$ nm, $r_{Pr^{4+}/Pr^{3+}} \sim 0.118$ nm, $r_{Tb^{4+}/Tb^{3+}} \sim 0.110$ nm) the trend in the a lattice parameter variation with increasing substitution of Pr/Tb suggests that these ions are in the mixed valent 3+/4+ state. Although the PXD pattern for $Ce_{1-x}Ca_xO_{2-x}$ shows solid state solution formation for 0<x<0.29, the lattice parameter does not change beyond x~0.17. The X-ray map of elements of the sample with nominal composition "$Ce_{0.71}Ca_{0.29}O_{1.71}$" from EDS shows clear nonhomogenity of Ca in the sample. The un-incorporated CaO shows up as an amorphous phase at the grain boundaries or on the grain surface. Thus the solubility limit of Ca is assumed to be x= 0.17.

The average crystallite size of ceria powders, calculated by the Scherrer formula from the XRD data were between 40 to 68 nm after drying at 200°C and between 7 to 14 nm after drying at room temperature (Table 1). The fine samarium, calcium, praseodymium or terbium substituted ceria powders were sintered into pellets with apparent densities over 95 % of the theoretical value even at 1400 °C, whereas $Ce_{1-x}(Ca/Sm)_xO_{2-\delta}$, prepared by conventional ceramic techniques require over 1600°C for sintering. The SEM micrographs of ceria samples sintered at 1400°C in Fig. 3 indicate relatively small particles of uniform size, ~1-5 μm and a very dense microstructure.

Fig. 3 Scanning electron micrograph of the surface of densified (i.e. annealed at 1400°C for 4 h) $(Ce_{0.83}Sm_{0.17})_{0.95}Pr_{0.05}O_{1.915+\delta}$

Table 1 Lattice parameters, crystallite sizes and electrical properties of $Ce_{1-x}(Sm/Ca)_xO_{2-5}$ and $(Ce_{0.83}Sm_{0.17})_{1-y}(Tb/Pr)_yO_{1.915+\delta}$ solid solutions

Composition		lattice parameter a (nm)	average crystallite size D(nm)	conductivity $\sigma_{600°C}$ (S/cm) (±5 %)	activation energy E_a (eV) (±0.05)	oxide ion transfer numbers t_i, 600°C
$Ce_{1-x}(Sm/Ca)_xO_{2+\delta}$						
Sm	x=0	0.54077(2)	50	$1.1x10^{-5}$	1.03	
	0.09	0.54186(8)	68	$9.8x10^{-4}$	0.80	
	0.13	0.54204(5)	68	$1.3x10^{-3}$	0.88	
	0.17	0.54300(7)	50	$5.7x10^{-3}$	0.92	1.00
	0.20	0.54326(6)	40	$2.0x10^{-3}$	0.87	
	0.23	0.54334(7)	50	$1.4x10^{-3}$	1.04	
	0.29	0.54436(3)	68	$6.2x10^{-4}$	1.03	
Ca	x= 0.05	0.54147(6)	40	$1.4x10^{-3}$	0.93	
	0.09	0.54157(5)	40	$2.1x10^{-3}$	0.83	1.00
	0.17	0.54200(4)	50	$1.5x10^{-3}$	0.81	
	0.23	0.54203(6)	50	$4.2x10^{-4}$	0.93	
$(Ce_{0.83}Sm_{0.17})_{1-y}(Tb/Pr)_yO_{1.915+\delta}$						
Tb	y= 0	0.54291(4)	10	$2.9x10^{-3}$	0.74	1.00
	0.02	0.54246(9)	14	$5.6x10^{-3}$	0.80	0.99
	0.05	0.54206(2)	14	$7.1x10^{-3}$	0.76	0.90
	0.10	0.54105(7)	10	10^{-2}	0.72	0.85
Pr	0.02	0.54298(3)	8	$2.1x10^{-3}$	0.80	0.99
	0.05	0.54281(4)	11	$7.0x10^{-3}$	0.62	0.92
	0.10	0.54310(2)	7	$7.6x10^{-3}$	0.55	0.78

As expected these particles are considerably larger than those of the "as-prepared" powder. Temperature dependent X-ray measurements show that the lattice constant increases linearly with temperature; the thermal expansion coefficient determined from this data is 8.6×10^{-6} K^{-1} for the $Ce_{0.83}Sm_{0.17}O_{1.915}$ and slowly increases/decreases with increasing Pr/Tb content. The ionic conductivities of $Ce_{1-x}(Sm/Ca)_xO_{2-\delta}$ solid electrolytes are significantly enchanced compared to CeO_2

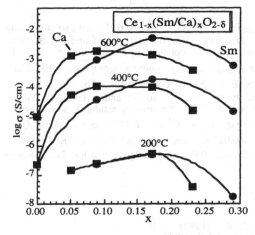
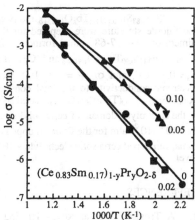

Fig. 4 Concentration dependence of the ionic conductivity of $Ce_{1-x}Sm_xO_{2-x/2}$ and $Ce_{1-x}Ca_xO_{2-x}$ solid solutions.

Fig. 5 Arrhenius plots of the ionic conductivity of $(Ce_{0.83}Sm_{0.17})_{1-x}Pr_xO_{1.915+\delta}$ solid solutions.

by increasing the oxygen vacancies ($V_O^{..}$). As previously reported [13], the ionic conductivity of $Ce_{1-x}Sm_xO_{2-x/2}$ increases systematically with increasing samarium substitution and reaches a maximum for the composition $Ce_{0.83}Sm_{0.17}O_{1.915}$ (Fig. 4). The decrease in the ionic conductivity for higher x is ascribed to defect associations of the type $\{Sm_{Ce}'V_O^{..}\}$ at higher concentrations of $V_O^{..}$. In the $Ce_{1-x}Ca_xO_{2-x}$ system, Ca-substitution introduces twice the number of oxygen vacancies as that of Sm-substitution for the same value of x, and the ionic conductivity reaches a maximum already at x= 0.09. Again, the conductivity decreases for x> 0.09 due to the formation of complex defect associations $\{Ca_{Ce}''V_O^{..}\}$.

In the $(Ce_{0.83}Sm_{0.17}O)_{1-y}(Tb/Pr)_yO_{1.915+\delta}$ system, Pr- or -Tb-substitutions do not introduce additional oxygen vacancies. However, by the possible redox process $Tb^{3+}(Pr^{3+}) \leftrightarrow Tb^{4+}(Pr^{4+})+e^-$ electronic contributions to the total conductivity might be introduced in the system. At low Tb- or Pr-content, because the redox potentials of Tb^{3+}/Tb^{4+} and Pr^{3+}/Pr^{4+} are slightly higher than that of Ce^{4+}/Ce^{3+}, terbium or praseodymium will trap mobile electrons, as previously demonstrated by Maricle et. al. [12] in Pr substituted gadolinium doped ceria solid electrolytes. As can be seen in Fig. 5, the total conductivity (i.e., ionic and electronic) in $(Ce_{0.83}Sm_{0.17})_{1-y}Pr_yO_{1.915+\delta}$ increases with increasing Pr substitution due to the increasing contribution of electronic conductivity (~ 22 % for the Pr-doped sample at 600°C and x= 0.10) to the total conductivity. In both systems (Pr and Tb) the activation enthalpy systematically decreases with increasing Pr or Tb substitution (Table 1) due to the increasing contribution of electrons to the transport process. The highest conductivity was found for the $(Ce_{0.83}Sm_{0.17})_{1-x}Tb_xO_{1.915+\delta}$ composition, $\sigma_{600 °C} \sim 10^{-2}$

S/cm. The ionic conductivity of $(Ce_{0.83}Sm_{0.17})_{0.98}Tb_{0.02}O_{1.915+\delta}$ and $(Ce_{0.83}Sm_{0.17})_{0.98}Pr_{0.02}O_{1.915+\delta}$ solid electrolytes is identical to that of the best conducting ceria solid electrolyte, $Ce_{0.83}Sm_{0.17}O_{1.915}$. Preliminary results of the conductivity as a function of the oxygen partial pressure indicate that the electrolytic domain of ceria solid electrolyte increases from $\sim 10^{-21}$ atm for $Ce_{0.83}Sm_{0.17}O_{1.915}$ to $\sim 10^{-23}$-10^{-24} atm for the x=0.02 Pr or Tb doped samples.

CONCLUSION

$(Ce_{0.83}Sm_{0.17})_{1-x}Tb_xO_{1.915+\delta}$ and $(Ce_{0.83}Sm_{0.17})_{1-x}Pr_xO_{1.915+\delta}$ (x=0-0.10) solid solutions with the fluorite structure were prepared for the first time. Ultrafine particles of uniform crystallite dimension, ~ 7-68 nm formed. The highest conductivity was found for the $(Ce_{0.83}Sm_{0.17})_{0.90}Tb_{0.10}O_{1.915+\delta}$ and $(Ce_{0.83}Sm_{0.17})_{0.90}Pr_{0.10}O_{1.915+\delta}$ compositions ($\sigma_{600°C} = 10^{-2}$ S/cm, E_a= 0.72 eV, and $\sigma_{600°C} = 7.6 \times 10^{-3}$ S/cm, E_a= 0.55 eV respectively). The contribution of electronic conductivity to the total conductivity was ~ 22 % for Pr and ~ 15 % for Tb for $(Ce_{0.83}Sm_{0.17})_{0.90}(Tb/Pr)_{0.10}O_{1.915+\delta}$ solid solutions at 600°C. Preliminary results show an increase of the electrolytic domain of ceria solid electrolyte from $\sim 10^{-21}$ atm at 700°C for $Ce_{0.83}Sm_{0.17}O_{1.915}$ to $\sim 10^{-23}$-10^{-24} atm for the Pr or Tb doped (x= 0.02) samples. The high ionic conductivity of the samarium doped ceria solid electrolyte ~ 10^{-2} S/cm at 600°C is unchanged at the low Pr/Tb doping level.

REFERENCES

1. A. Trovarelli,. Catal. Rev. - Sci. Eng. **38**, 439 (1996).
2. H. Inaba and H. Tagawa, Solid State Ionics **83**, 1 (1996).
3. T. Takahashi and A. Kozawa, eds. Application of Solid Electrolytes (JEC Press, Ohio, 1980).
4. T. Takahashi, ed. High Conductivity Solid Ionic Conductors (World Scientific, Singapore, 1989).
5. W. Göpel, T. A. Jones, M. Kleitz, I. Lundström and T Seiyama, eds. Chemical and Biochemical Sensors , Vol. 2, 3 (VCH, Weinheim, 1991/1992).
6. T. H. Etsell and S. N. Flengas, Chem. Rev. **70**, 339 (1970).
7. S. P. S. Badwal, ed. Science and Technology of Zirconia (Technomic, Lancaster, 1993).
8. T. Takahashi and H. Iwahara, Mat. Res. Bull. **13**, 1447 (1978).
9. P. Shuk, H.-D. Wiemhöfer, U. Guth, W. Göpel and M. Greenblatt, Solid State Ionics **89**, 179 (1996).
10. K. Egushi, T. Setoguchi, T. Inoue, H. Arai, Solid State Ionics **52**, 165 (1992).
11. Y. C. Zhou and M. N. Rahaman, J. Mater. Res. **8**, 1680 (1993).
12. K. Yamashita, K. V. Ramanujachary, M. Greenblatt, Solid State Ionics **81**, 53 (1995).
13. W. Huang, P. Shuk and M. Greenblatt, Chem. Mater. **9**, 2240 (1997).
14. D. L. Maricle, T. E. Swarrm and S. Karavolis, Solid State Ionics **52**, 173 (1992).
15. R. D. Shannon and C. T. Prewitt, Acta Crystallogr. **32A**, 751(1976).

THIN FILM SYNTHESIS OF NOVEL ELECTRODE MATERIALS
FOR SOLID-OXIDE FUEL CELLS

Alan F. Jankowski and Jeffrey D. Morse
University of California - Lawrence Livermore National Laboratory
P.O. Box 808, Livermore, CA 94550

ABSTRACT

Electrode materials for solid-oxide fuel cells are developed using sputter deposition. A thin film anode is formed by co-deposition of nickel and yttria-stabilized zirconia. This approach is suitable for composition grading and the provision of a mixed-conducting interfacial layer to the electrolyte layer. Similarly, synthesis of a thin film cathode proceeds by co-deposition of silver and yttria-stabilized zirconia. The sputter deposition of a thin film solid-oxide fuel cell is next demonstrated. The thin film fuel cell microstructure is examined using scanning electron microscopy whereas the cell perfomance is characterized through current-voltage measurement and corresponding impedance spectroscopy.

INTRODUCTION

A solid-oxide fuel cell (SOFC) device consists of manifolded stacks of cells which combine a fuel and oxidant at elevated temperatures to generate electric current. The basis of each fuel cell is an anode and cathode separated by an electrolyte layer. SOFCs provide an efficient and enviromentally clean method of energy conversion. SOFCs are routinely made using bulk ceramic powder processing.[1] A traditional synthesis approach uses a cermet electrode on which an electrolyte is layered, for example, by tape casting. The unit cell is completed by lamination to the counterpart electrode. SOFCs are traditionally operated at temperatures exceeding 900°C.

The common approach to incorporate thin film technologies in SOFCs is based on coating the cermet electrodes with an electrolyte layer. To decrease the thickness below that obtainable using tape calendaring, thin ceramic films can be deposited by a variety of techniques including variations of sol-gel and colloidal deposition.[2] In addition, thin electrolyte layers can be dc magnetron sputtered from metal alloy targets to optimize ion conductivity and rf magnetron sputtered from oxide targets to form continuous coatings on porous substrates.[3-5]

A thin electrolyte reduces the path for oxygen ion diffusion. A decrease in the electrolyte thickness from ~100 μm to <10 μm can potentially lower the fuel cell operating temperature by several hundred degrees centigrade.[6] Although a thin electrolyte layer can be deposited onto an electrode, a bonding procedure is still required to attach the counterpart electrode and form the unit cell of the anode-electrolyte-cathode trilayer. This process always leads to polarization losses at the electrode-electrolyte interface(s). To address this problem, interfacial layers can be added at both the anode and cathode sides of the electrolyte to reduce interfacial reaction resistances.[7,8] For example, yttria-stabilized Bi_2O_3 is added at the cathode-side and yttria-doped CeO_2 is added at the anode-side of the yttria-stabilized zirconia (YSZ) electrolyte layer.

Thin film processing can be used to produce alternative cermet electrodes.[9,10] Anodes for a YSZ electrolyte can be synthesized as thin wafers. For example, the anode wafers can be sintered compacts of Ni-coated zirconia powder. Metal coating the zirconia powder uniformly distributes the conducting element of the electrode. Sufficient electrical conduction is provided while reducing the metal content to <10 volume percent ensuring a near match in coefficient of thermal expansion to the electrolyte. In comparison, bulk cermet processing involves sintering a powder composite that often contains >30 volume percent metal. The high metal content is needed to yield sufficient conductivity at elevated temperature but often leads to a delamination failure when subjected to routine thermal-cycling.

Chemical and physical vapor deposition methods are shown to be advantageous to produce either the electrolyte layer or the electrodes. A clear advantage exists to explore additional methods to form the anode-electrolyte-cathode trilayer through an integrated synthesis process.

Mat. Res. Soc. Symp. Proc. Vol. 496 © 1998 Materials Research Society

In addition to the use of vacuum deposition to form electrolyte thin films, we investigate the use of sputter deposition to form both electrodes as thin films and to provide a composition graded electrode-electrolyte interface for improved kinetics. We will examine electrode materials for a SOFC based on a YSZ electrolyte that is compatible with low temperature (< 700°C) operation. Specifically, nickel (Ni)-YSZ is selected for the anode and silver (Ag)-YSZ for the cathode.

EXPERIMENTALS

The use of thin film deposition enables the fuel cell unit to be synthesized in a continuous process. A sequential deposition starting with the anode and concluding with the cathode bridges the difficulties of joining electrolyte-coated electrodes (e.g. anodes) to counterpart electrodes (e.g. cathodes). The electrodes are formed as metal and ceramic composites by the co-sputter deposition of targets operated in the dc and rf modes, respectively. The ceramic material chosen for the electrode matrix is the same as that used for the vapor deposition of the electrolyte layer. Synthesis of a dense, thin electrolyte layer is established by the rf-sputter deposition of a $(Y_2O_3)_{5.6}(ZrO_2)_{94.4}$ target.[5] A defect-free electrolyte layer of cubic YSZ is confirmed through transmission electron microscopy (TEM) using plan-view bright-field imaging and selected-area electron diffraction.

To synthesize a thin film fuel cell with an exposed free-standing region suitable for subsequent testing, a brief review of the substrate platform preparation is necessary. A thin layer of silicon nitride is grown using a low pressure chemical vapor deposition process. The 0.22 μm thick film is formed at 800°C using a 112 cm³m⁻¹ flow of dichlorosilane and a 30 cm³m⁻¹ flow of NH_3 at 33 Pa. The backside of the substrate wafer is patterned by standard photolithographic techniques and etched to reveal windowed regions of silicon nitride with areas that range from 0.14 to 16 mm² in size. The substrate is etched at a rate of 1 μmm⁻¹ using 44% KOH at 85°C. A reactive ion etch rate of 0.42 μmm⁻¹ is achieved during nitride removal using 500 W of power with a 2.7 Pa pressure at a 40 cm³m⁻¹ flow of CHF_3 and a 80 cm³m⁻¹ flow of CF_4.

The sequence of deposition process steps to synthesize the fuel cell unit is reviewed as follows. The sputter deposition chamber is evacuated to a base pressure of 5.3 x 10⁻⁶ Pa. The substrate is positioned 10 cm from an array of three planar magnetron sources. The 5 cm diameter YSZ source is positioned between two 3 cm diameter sources that are positioned 60° apart. One 3 cm source has a Ag target and the other a Ni target. To minimize residual stress effects that arise when depositing YSZ using high sputter-gas pressures at elevated substrate temperatures, a low (<50°C) temperature deposition is followed by an air anneal at the cell operating temperature for several hours. A 10 Pa sputter gas pressure at a constant flow of 30 cm³m⁻¹ is used to operate the magnetron sources. Each step of the following process adds an increment of thickness (Δt) to the total thickness (t_f) at each respective stage of the deposition. (Step-1) The deposition begins by sputtering the Ni target at 4.2 Wcm⁻² power yielding a 9.3 nmm⁻¹ rate to give a Δt of 0.3 μm (t_f = 0.3 μm). (Step-2) The deposition continues as the YSZ target is sputtered at 38.2 Wcm⁻² power yielding a 10.6 nmm⁻¹ rate. The Ni and YSZ are co-deposited to a Δt of 0.5 μm (t_f = 0.8 μm). (Step-3) The deposition continues as the YSZ target is sputtered alone to give a Δt of 0.7 μm (t_f = 1.5 μm). (Step-4) The deposition continues as the Ag target is sputtered at 4.2 Wcm⁻² power yielding a 33.3 nmm⁻¹ rate. The YSZ and Ag are co-deposited to a Δt of 1.2 μm (t_f = 2.7 μm). (Step-5) The deposition concludes as the Ag target is sputtered alone to a Δt of 0.8 μm (t_f = 3.5 μm).

RESULTS & DISCUSSION

A thin-film solid-oxide fuel cell (TFSOFC) consisting of a Ni-(YSZ) anode, a YSZ-electrolyte, and a Ag-(YSZ) cathode (Fig. 1) is formed through the continuous deposition process. Scanning electron microscopy indicates that the electrolyte and electrode layers appear continuous both in-plane and through the growth direction. As mentioned, interfacial layer additions or a composition-graded electrolyte-electrode interface (as is featured in this sample) can be added to improve the fuel cell performance by enhancing catalytic activity.

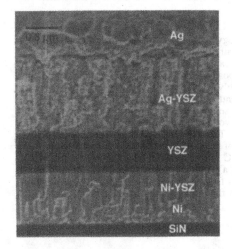

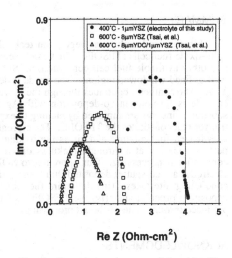

Fig. 1. A thin film solid-oxide fuel cell as deposited by planar magnetron sputtering is viewed in cross-section.

Fig. 2. Impedance spectra of the thin film solid-oxide fuel cell indicate low electrolyte-electrode polarization losses.

The current-voltage (i-V) output of a Ni-(YSZ)/YSZ/Ag-(YSZ) fuel cell is measured as the current across the solid electrolyte is controlled by a galvanostat. After the fuel cell is heated to 400°C in dry N_2, air is supplied as the oxidant to the cathode and a humidified 6% hydrogen (Ar-balance) fuel mixture is supplied to the anode. The measurement of power generation requires that the active area of the thin film cell be free-standing. After the sputter deposition process, the thin support windows of nitride are removed by reactive-ion etching to expose the fuel cell anode over a well-defined area. The i-V output measurements are conducted over square regions that range from 2 to 9 mm² in size. An open circuit voltage (OCV) of ~1.0 V is estimated in preliminary testing along with a short-circuit current of ~240 mAcm[-2] that corresponds to a resistance of ~4 Ωcm². The OCV is less than the 1.1 V expected but is similar to that obtained for other SOFCs of Ni-YSZ-Ag that are formed using a sputtering process for deposition of the electrolyte layer.[8] The cell exhibits somewhat of a non-linear i-V response. The non-ohmic behavior is seen as the resistance decreases with increasing current. The preliminary testing at 400°C generates a maximum power density of 39 mWcm[-2] for the TFSOFC. This cell output is comparable with the maximum power density generated for SOFCs composed of a Ni-YSZ anode and a $La_{.85}Sr_{.15}MnO_3$-YSZ cathode.[11] A SOFC with an electrolyte bilayer of 8μm thick yttria-doped ceria (YDC) and 1μm thick YSZ yields 43 mWcm[-2] at 500°C and 155 mWcm[-2] at 600°C as tested using a 10% hydrogen fuel mixture. The power generated at 600°C decreases to 110 mWcm[-2] for a SOFC with a single 8 μm thick YSZ electrolyte layer.

The electrochemical impedance spectra (Fig. 2) provide a measure of the polarization losses that are found within the cell. A separate Ni-(YSZ)/YSZ/Ag-(YSZ) fuel cell is measured after it is heated to 400°C. Air is supplied to the cathode and the humidified 6% hydrogen (Ar-balance) fuel mixture to the anode. A plot of the real vs imaginary components of the complex impedance spectra forms a semicircular arc for the TFSOFC of Ni-(YSZ)/YSZ/Ag-(YSZ). Results at 600°C are also shown for the SOFCs referenced in the i-V comparison.[11] The real-axis intercepts give the total ohmic losses for the fuel cells. The high frequency intercept indicates that the electrolyte area-specific resistance for the TFSOFC at 400°C is only ~2 Ωcm². This resistance value compares favorably with the output of the SOFC with the 8μmYDC-1μmYSZ electrolyte bilayer operated at a higher temperature of 500°C.[11] An equivalent resistance value of 2 Ωcm² is also reported at 500°C for measurements of a 9μm thick YSZ electrolyte layer.[12] In all, the fuel cell i-V output at low temperature proves fairly consistent with the impedance spectra results.

SUMMARY

Thin film and multilayer deposition technology provides a means to synthesize thin-film solid-oxide fuel cells (TFSOFCs) that can potentially yield greater specific power than found in any other available fuel cell configuration. Substrates were patterned and processed to reveal well-defined windowed regions. The anode is deposited first by sputtering Ni and co-depositing YSZ. The synthesis continues through the electrolyte layer with deposition of only YSZ. The cathode is formed via co-deposition with Ag and the final layer concludes with Ag. The substrate windows are removed by etching to expose a self-supporting anode-electrolyte-cathode trilayer that constitutes the TFSOFC. The current-voltage output from the cell is measured using a hydrogen fuel mixture and air as the oxidant. The TFSOFC yields an acceptable fuel cell performance but at a through thickness orders of magnitude less than that achievable using conventional processing. Our approach to SOFC synthesis is novel in several ways: (i) the electrodes are co-sputter deposited thin films; (ii) a provision exists for the deposition of mixed conducting interfaces; (iii) the entire fuel cell is formed as a thin film through a continuous deposition process; and (iv) the TFSOFC is half the thickness of just the electrolyte layer as produced through state-of-the-art processing.

ACKNOWLEDGMENTS

We thank J. Hayes for his asssistance in sample fabrication, J. Ferreira for the scanning electron microscopy imaging and Q. Pham for his assistance during initial impedance spectroscopy measurements. This work was performed under the auspices of the United States Department of Energy by Lawrence Livermore National Laboratory under contract #W-7405-Eng-48.

REFERENCES

1. N. Minh, J. American Ceramic Society **76**, 563 (1993).

2. S. deSouza, S. Visco and L. DeJonghe, Solid State Ionics **98**, 57 (1997).

3. E. Thiele, L. Wang, T. Mason and S. Barnett, J. Vac. Sci. Technol. A **9**, 3054 (1991).

4. A. Jankowski and J. Hayes, Surface and Coatings Technology **76-77**, 126 (1995).

5. A. Jankowski and J. Hayes, J. Vac. Sci. Technol. A **13**, 658 (1995).

6. S. Barnett, Energy **15**, 1 (1990).

7. L. Wang and S. Barnett, J. Electrochemical Society **139**, 2567 (1992); ibid., **139**, L89 (1992).

8. L. Wang and S. Barnett, Solid State Ionics **61**, 273 (1993).

9. A. Jankowski, Internat. J. Environmentally Conscious Design & Manufacturing **5**, 39 (1996).

10. A. Jankowski, in Ionic and Mixed Conducting Ceramics III, edited by T. Ramanarayanan (The Electrochemical Society Proceeding Series PV-24, Pennington, NJ, 1997) in press.

11. T. Tsai, E. Perry and S. Barnett, J. Electrochemical Society **144**, L130 (1997).

12. A. Atkinson, in Proceedings of the Second European Solid-Oxide Fuel Cell Forum, edited by B. Thorstensen (The Electrochemical Society, Pennington, NJ, 1996) p. 707.

Analysis of Mixtures of Gamma Lithium Aluminate, Lithium Aluminum Carbonate Hydroxide Hydrate, and Lithium Carbonate

M.T. Nemeth, R.B. Ford, T.A. Taylor
Cyprus Foote Mineral Company, 348 Holiday Inn Drive, Kings Mountain, NC 28086
mnemeth@cyprus.com

ABSTRACT

Lithium aluminate, $LiAlO_2$, is a ceramic powder which is used as the porous solid support for the electrolyte in molten carbonate fuel cells (MCFCs). It has previously been reported that gamma $LiAlO_2$ will convert to lithium aluminum carbonate hydroxide hydrate, $Li_2Al_4(CO_3)(OH)_{12} \cdot 3H_2O$ and Li_2CO_3 when exposed to water vapor and carbon dioxide. We compare three techniques, weight gain, carbonate content and x-ray diffraction to measure the amount of conversion. The reaction may involve amorphous intermediates and no one technique by itself is satisfactory to study the conversion.

INTRODUCTION

Lithium aluminate, $LiAlO_2$, is a ceramic powder which is used as the porous solid support for the electrolyte in molten carbonate fuel cells (MCFCs). Changes in the physical properties of gamma-$LiAlO_2$ have been discussed as a potential problem in MCFC life and performance [1]. More recently, the use of beta- and alpha-$LiAlO_2$ for in MCFCs has been studied [2]. We have previously observed that gamma $LiAlO_2$ will convert to lithium aluminum carbonate hydroxide hydrate, $Li_2Al_4(CO_3)(OH)_{12} \cdot 3H_2O$ (LACHH), and lithium carbonate when exposed to moisture and carbon dioxide at 25° C [3]. In light of this, the phase distributions of $LiAlO_2$ and LACHH are important. The degree of conversion of gamma-$LiAlO_2$ to LACHH in the presence of 70% RH is measured by weight gain, carbonate absorption, and x-ray diffraction. One technique alone does not completely characterize the reaction.

EXPERIMENT

The instrumentation, materials, synthesis and procedures have been described previously [3,4]. Gamma $LiAlO_2$ was placed in weighing boats (VWR part 12577-027) to give a bed depth of ~1 inch. This gamma $LiAlO_2$ that was exposed to 70% RH for 43, 86 and 122 days in the previous study [3] was analyzed by weight gain, Knorr carbonate and the multiple x-ray diffraction techniques.

To eliminate any traces of LACHH or alpha $LiAlO_2$, gamma $LiAlO_2$ was heated at 1000° C until no trace of either phase was visible in the XRD pattern. Gamma $LiAlO_2$ was mixed with equimolar concentrations of LACHH and Li_2CO_3 to give standards that were 0, 20, 40, 60, 80, and 100% gamma $LiAlO_2$ by weight. For internal standard analysis, the standards described in the previous sentence were mixed 1:1 by weight with corundum (Aldrich 23,474-5). LACHH and gamma $LiAlO_2$ pack and segregate when mixed with a mortal and pestle or a mixer mill. Therefore, standards were mixed by slow rolling for several hours in jars held by a clamp connected to a Cole-Parmer Mixmaster. The sides of the jar were tapped with a spatula to assure complete mixing.

XRD spectra were recorded from 5° to 70° 2θ with a step size of 0.02° and a count time of 1 s. Primary and secondary Sollar slits, 0.5° aperture, 0.5° scattered radiation and 0.3° detector diaphragms, a monochromator and a copper tube running at 40 kV and 50 mA were used.

For the single peak analysis, the 2θ ranges used were: 10.8 to 12.5° for LACHH; 21.6 to 22.9° for gamma $LiAlO_2$; 31 to 32.3° for Li_2CO_3; and 51.7 to 53.4° for corundum. Conversions were calculated using the integrated areas for the samples, or the integrated area ratio relative to corundum.

The Reference Intensity Ratio (RIR) was calculated for each phase by preparing 1:1 mixtures of corundum with LACHH, Li_2CO_3, and $LiAlO_2$ and recording the XRD spectra. The spectra were processed with the computer program Jade+ ver. 3.1 (MDI). The spectra were smoothed, background corrected, and theta calibrated to the internal standard corundum before calculation of the RIR. The (pattern simulation) quantitative analysis function of Jade+ was used to analyze both the standards described earlier in this section and the samples.

Patterns were also analyzed with Peakfit™ ver. 4.0 (SPSS) and the Pearson VII amplitude function. Data were smoothed, background corrected and converted to ASCII with Siemens Diffrac AT ver 3.3, and normalized to the total corundum area for import into Peakfit™ Patterns for the phases were then simulated with Lotus 1-2-3 rel. 5. Total integrated area for each simulated phase was then calculated. Plotting and least squares were performed with Axum® 4.0 for Windows (Trimetrix).

RESULTS

The conversion of gamma $LiAlO_2$ to LACHH is given in eq. (1). The progress of this reaction can be followed by weight gain, increase in carbonate content, or quantitative XRD analysis.

$$4LiAlO_2 + 9H_2O + 2CO_2 \rightarrow Li_2Al_4(CO_3)(OH)_{12} \cdot 3H_2O + Li_2CO_3 \qquad (1)$$

Figure 1 shows the three spectra that were used to calculate the RIR and the XRD sprectum of the sample that was exposed to 70% RH for 86 days. Table I and Figure 2 summarize the RIR analysis. Accuracy is good. However, for the two samples exposed to 70% RH, recovery of the corundum was higher than the expected value of 50%. The significant broadening for the 100% peak for the LACHH formed *in situ* compared to the synthesized LACHH should be noted.

Calibration curves for the single peak internal standard analysis and the whole pattern simulations are given in Figure 3. Figure 4 shows the simulated patterns of LACHH extracted from the sample that had been exposed to 70% RH for 86 days compared to the pattern extracted from the 26% standard. Note the significant broadening of the 100% peak for LACHH formed *in situ*. Quantitative results obtained by all analytical methods are given in Table II.

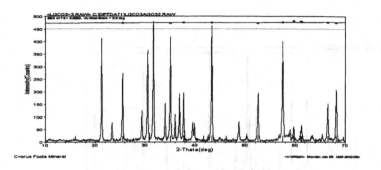

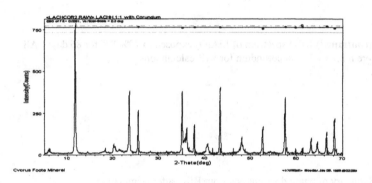

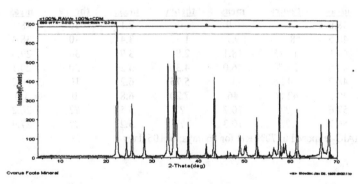

Figure 1. XRD spectra of Li_2CO_3, LACHH, and $LiAlO_2$ (in order from top, continued on next page)

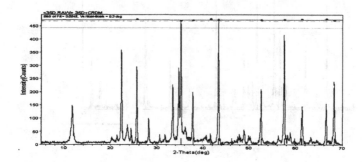

Figure 1. (continued) XRD spectrum of $LiAlO_2$ exposed to 70% RH for 86 days. All samples were mixed 1:1 with corundum for RIR calculations.

Table I. Summary of phase distributions from RIR pattern simulation.

Al_2O_3		LACHH		Li_2CO_3		$LiAlO_2$	
theory	meas	theory	meas	theory	meas	theory	meas
50	48.6	0	0.9	0	2.1	50	48.5
50	48.5	8.55	9.7	1.45	3.4	40	38.4
50	49.1	17.15	18.1	2.85	3.9	30	29.0
50	48.6	25.70	26.0	4.30	5.2	20	20.1
50	48.2	34.25	35.5	5.75	6.2	10	10.2
50	49.5	42.80	46	7.20	6.8	0	1.8
50[a]	57.6	??	10.7	??	5.5	??	26.2
50[b]	56.7	??	10.3	??	5.2	??	27.9

a. gamma $LiAlO_2$ exposed to 70% RH for 86 days; b. 105 days.

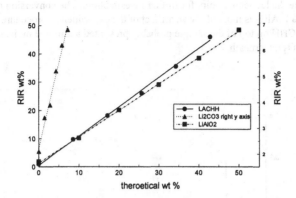

Figure 2. Measured RIR phase concentrations vs. theoretical from data in Table 1.

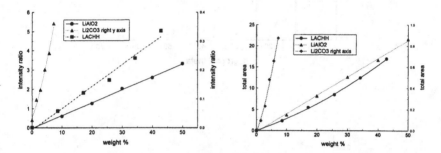

Figure 3. Calibration curves for single peak internal standard XRD and total area simulation.

The results in Table II are calculated from weight gain and CO_3 according to the stoichiometry given in eq. (1). It is assumed that all the weight gain results in the formation of the products and the CO_3 increase results in an equimolar amount of LACHH and Li_2CO_3. The RIR results were normalized to 100% from the distribution given in Table I. The external and internal XRD standard analyses, gravimetric, CO_3, and pattern simulation results give similar results for the concentration of Li_2CO_3 and $LiAlO_2$ in the samples. LACHH results are significantly lower and total recovery is not 100%. RIR results differ markedly from the other techniques. An insufficient amount of CO_3 has been absorbed to give the RIR distribution obtained.

The XRD spectrum in the right of Figure 1 and the pattern simulation in Figure 4 for the LACHH formed *in situ* provide a the launching point for an explanation of the poor agreement. The LACHH formed *in situ* is different from the LACHH prepared for use as a standard. Its particle size and degree of crystallinity differ markedly from the standard material. Another possibility is that amorphous intermediates of the newly formed lithium aluminum species are present that will not give diffraction spectra. The Pearson VII profile fitting parameters from Peakfit[TM] provide the launching point for further investigation. The conversion plateau seen in Table II of ~50% LiAlO$_2$ is most likely an artifact of the experimental procedure. A protective bed depth of LACHH on top of the sample probably prevented additional moisture from reaching the gamma LiAlO$_2$ underneath.

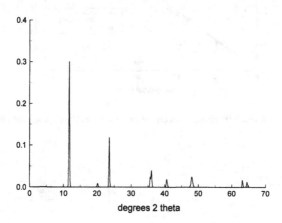

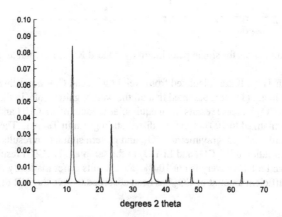

Figure 4. Pearson VII simulated spectra for LACHH extracted from the standard containing 26% LACHH (upper trace) and extracted from LACHH formed upon exposure to 70% RH for 86 days (lower trace).

Table II. Summary of phase analysis of gamma $LiAlO_2$ exposed to 70% RH, by technique.

Time, days[a]	Method	Li_2CO_3 %	LACHH %	$LiAlO_2$ %
43	gravimetric	8.1	48.5	43.4
86	gravimetric	8.0	48	44
	CO_3	6.9	40.8	52.3
	XRD_{ext}	5.8	20.4	48.7
	XRD_{int}	5.3	24	45.8
	RIR	13.0	25.2	61.8
	simulated	5.7	31.0	49.8
122	gravimetric	8.2	48.5	43.3
	CO_3	6.8	40.8	52.4
	XRD_{ext}	5.3	18.3	51.2
	XRD_{int}	5.5	23.2	48.2
	RIR	12.0	23.7	64.3

a. days = number of days @ 70% RH, XRD_{ext} and XRD_{int} are single peak analyses.

CONCLUSIONS

The decomposition of gamma $LiAlO_2$ to $Li_2Al_4(CO_3)(OH)_{12}\cdot3H_2O$ (LACHH), and Li_2CO_3 were studied by weight gain, increase in CO_3 content and XRD. Consistent results are not obtained from the three techniques. The presence of amorphous intermediates or the use of a LACHH standard which did not match the LACHH formed *in situ* are possible explanations for the discrepancy. The peak profiles obtained from the pattern simulation serve as a launching point for further study.

ACKNOWLEDGMENTS

The authors wish to acknowledge Susan Jacob Beckerman for many technical discussions and Mario Fornoff for assistance with x-ray diffraction.

REFERENCES

1. H. Sotouchi, Y. Watanabe, T. Kobayashi and M. Murai J. Electrochem. Soc. **139**, pp. 1127-1130 (1992).

2. K. Nakagawa, H.Ohzu, Y. Akasaka and N. Tomomatsu, N Denki Kagaki oyobi Kogyo Butsuri Kagaku, **64** (6), pp. 478-481 (1996).

3. S.J. Beckerman, R.B. Ford and M.T. Nemeth, M. T. Powder Diffraction, **11** (4), p. 312-317 (1996).

4. E.T. Devyatkina, N.P. Kotsupalo, N.P. Tomilov, and A.S. Berger, Russ. J. Inorg. Chem., **28**, pp. 2774-2778 (1983).

PHASE STABILITY OF THE MIXED-CONDUCTING Sr-Fe-Co-O SYSTEM

B. Ma*, U. Balachandran*, J.P. Hodges†, J.D. Jorgensen†, D.J. Miller†, and J.W. Richardson, Jr.‡
*Energy Technology Division, Argonne National Laboratory, Argonne, Illinois 60439
†Materials Science Division, Argonne National Laboratory, Argonne, Illinois 60439
‡Intense Pulsed Neutron Source, Argonne National Laboratory, Argonne, Illinois 60439

ABSTRACT

Mixed-conducting ceramic oxides have potential uses in high-temperature electrochemical applications such as solid oxide fuel cells, batteries, sensors, and oxygen-permeable membranes. The Sr-Fe-Co-O system combines high electronic/ionic conductivity with appreciable oxygen permeability at elevated temperatures. Dense ceramic membranes made of this material can be used to separate high-purity oxygen from air without the need for external electrical circuitry, or to partially oxidize methane to produce syngas. Samples of $Sr_2Fe_{3-x}Co_xO_y$ (with x = 0, 0.6, 1.0, and 1.4) were prepared by solid-state reaction method in atmospheres with various oxygen partial pressures (pO_2) and were characterized by X-ray diffraction, scanning electron microscopy, and electrical conductivity testing. Phase components of the sample are dependent on cobalt concentration and pO_2. Electrical conductivity increases with increasing temperature and cobalt content in the material.

INTRODUCTION

Recently, oxides with mixed electronic and ionic conductivities have begun to attract attention because of their applications in high-temperature electrochemical processes [1-5]. For example, they are used as electrodes in solid oxide fuel cells, and if their oxygen permeability is high enough, they can be used as oxygen-separation membranes. Sr-Fe-Co-O has not only high combined electronic and oxygen ionic conductivity but also oxygen permeability that is superior to that of other reported ceramic oxides [6-11]. It is a good candidate material as a dense membrane for separation of high-purity oxygen. In an oxygen separation reactor, oxygen can be transported from the high oxygen partial pressure (pO_2) side to the low-pO_2 side under the driving force of the pO_2 difference, $\Delta log(pO_2)$, without the need for external electric circuitry. Balachandran et al. [7] demonstrated that extruded membrane tubes made of Sr-Fe-Co oxide can be used for partial oxidation of methane to produce syngas (CO + H_2) in a methane conversion reactor that operates at ≈850°C. In this type of reactor, oxygen on one side of the membrane was separated from air on the other side of the membrane. Methane conversion efficiencies of >98% were observed and the reactor tubes have operated for >1000 h. Moreover, the oxygen flux obtained from the separation of air in this type of conversion reactor is considered commercially feasible. The use of this technology can significantly reduce the cost of oxygen separation [12].

To obtain sufficient oxygen flux, the gas separation reactor must be operated at high temperatures and substantial pO_2 differences. Therefore, stability of the ceramic membranes is an important issue. Certain compounds of $SrFe_{1-x}Co_xO_{3-\delta}$ were found unsuitable for use in gas separation because they lack stability in highly reducing environments [13]. Thermodynamic calculations can normally give us accurate data on the stability of a system. However, certain

167

examples of discrepancies between the calculated thermodynamic stability and the actual results indicate the importance of obtaining supporting data from real systems [14].

In this paper, we report our studies on the structural and phase stability of $Sr_2Fe_{3-x}Co_xO_y$ materials in air and reducing argon environments. In-situ electrical conductivity and ex-situ room-temperature X-ray diffraction (XRD) and scanning electron microscopy (SEM) experiments were used to characterize the material.

EXPERIMENTAL

The $Sr_2Fe_{3-x}Co_xO_y$ (with x = 0, 0.6, 1.0, and 1.4) powders were made by a solid-state reaction method, with appropriate amounts of $SrCO_3$, $Co(NO_3)_2 \cdot 6H_2O$, and Fe_2O_3; mixing and grinding were conducted in isopropanol with zirconia media for 10 h. After drying, the mixture was calcined in air at 850°C for 16 h, with intermittent grinding. After final calcination, the powder was ground with an agate mortar and pestle to an average particle size of ≈7 μm. The resulting powder was pressed uniaxially with a 1200-MPa load into bars 7.5 mm wide, 44 mm long, and ≈5 mm thick. The bars were then sintered in air at ≈1200°C for 5 h. The true density was measured on powder with an AccuPyc 1330 pycnometer. Bulk density was determined by the Archimedean method, with isopropanol as the liquid medium. Calculated relative density of air-sintered $SrFeCo_{0.5}O_x$ samples was ≈95% of theoretical value. Thin bars ≈ 1 x 5 x 20 mm^3 were cut from the samples and subjected to postannealing heat treatment in various low-pO_2 environments at 900-1100°C.

Room-temperature X-ray powder diffraction analysis was carried out with a Scintag XDS-2000 diffractometer. Data were taken with Cu K_α radiation. A high-purity intrinsic Ge energy-dispersive detector was used to minimize background due to sample fluorescence. A continuous scan with a 2θ scanning rate of 1°/min and chopping step of 0.03° was used to collect data. SEM observations were conducted with a JEOL JSM-5400 scanning microscope at an accelerating voltage of 20 KeV.

Conductivity of the sample was determined by the four-probe method. Platinum wires were attached to the specimen bar to serve as current and voltage leads. Resistance of the specimen was measured with an HP 4192A LF impedance analyzer at 23 Hz. Details of the experimental configuration for measuring conductivity were reported earlier [10]. A K-type thermocouple was attached to an yttria-stabilized zirconia (YSZ) plate on which the sample bars were placed for the heat treatments. The thermocouple was used for both controlling and detecting the temperature of the programmable electric furnace. Temperature tolerance within the uniform hot zone of the furnace is ±1°C. Desirable gaseous environments were obtained by flowing premixed gases through the system during the experiments.

RESULTS AND DISCUSSIONS

Room-temperature XRD patterns of air-sintered $Sr_2Fe_{3-x}Co_xO_y$ are plotted in Fig. 1 for different x values. The powder XRD pattern of the x = 0 composition (Fig. 1a) agrees with the result reported for a single crystal $Sr_4Fe_6O_{13}$ sample [15]. For comparison, the XRD pattern of the perovskite phase $SrFe_{0.8}Co_{0.2}O_x$ is plotted in Fig. 1e. Air-sintered $Sr_2Fe_2CoO_y$ (x = 1, Fig. 1c) is a multiple phase material, with ≈40% perovskite phase, as obtained from Rietveld analysis [16]. Air-sintered $Sr_2Fe_{2.4}Co_{0.6}O_y$ (x = 0.6, Fig. 1b) contains less than 3% perovskite phase, while air-sintered $Sr_2Fe_{1.6}Co_{1.4}O_y$ (x = 1.4, Fig. 1d) has more than 90% perovskite phase. Traces of cobalt

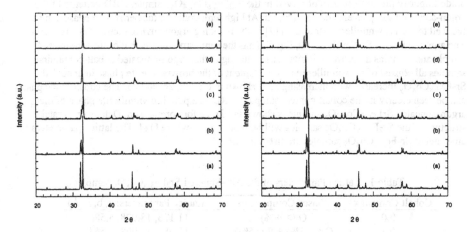

Fig. 1. Room-temperature X-ray diffraction patterns of air-sintered samples: (a) $Sr_2Fe_3O_y$, (b) $Sr_2Fe_{2.4}Co_{0.6}O_y$, (c) $Sr_2Fe_2CoO_y$, (d) $Sr_2Fe_{1.6}Co_{1.4}O_y$, and (e) $SrFe_{0.8}Co_{0.2}O_y$.

Fig. 2. Room-temperature X-ray diffraction patterns of argon-annealed samples: (a) $Sr_2Fe_3O_y$, (b) $Sr_2Fe_{2.4}Co_{0.6}O_y$, (c) $Sr_2Fe_2CoO_y$, (d) $Sr_2Fe_{1.6}Co_{1.4}O_y$, and (e) $SrFe_{0.8}Co_{0.2}O_y$.

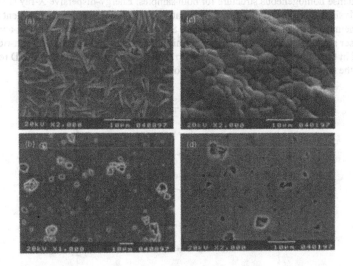

Fig. 3. SEM morphology of $Sr_2Fe_2CoO_x$ sample: (a) original surface of air-sintered sample, (b) polished surface of air-sintered sample, (c) original surface of argon-annealed sample, (d) polished surface of argon-annealed sample.

169

oxide impurity phase were also observed in the $Sr_2Fe_{1.6}Co_{1.4}O_y$ sample. XRD patterns of the argon-annealed samples are shown in Fig. 2. At high temperatures, the perovskite phase can be reduced to the brownmillerite structure [17] in the reducing argon environment ($pO_2 \approx 10^{-6}$ atm). The argon-annealed $Sr_2Fe_3O_y$ (x = 0, Fig. 2a) has the same structure as that of the air-sintered sample and contains no brownmillerite phase, although other argon-annealed cobalt-containing samples all consist of brownmillerite phase. Content of the brownmillerite phase in annealed $Sr_2Fe_{3-x}Co_xO_y$ increases with increasing x, as shown in Fig. 2b, 2c, and 2d. The cobalt oxide phase can be seen clearly in the cobalt-content samples. XRD of a pure brownmillerite phase of the argon-annealed $SrFe_{0.8}Co_{0.2}O_x$ sample is shown in Fig. 2e for comparison. Details on crystal structure of the $Sr_2Fe_{3-x}Co_xO_y$ samples will be reported elsewhere [18]. The lattice parameters of air-sintered $Sr_2Fe_{3-x}Co_xO_y$ samples are listed in Table 1.

Table 1. Lattice Parameters of the Air -Sintered $Sr_2Fe_{3-x}Co_xO_y$ samples

Cobalt Content x	Phases Components	Lattice Parameters: a, b, c (Å)
0.0	O (>99%)	11.126, 18.978, 5.5864
0.3	O (>95%) + P (<5%)	11.081, 19.008, 5.5649
0.6	O (>90%) + P (<10%)	11.065, 19.008, 5.5618
1.0	O (>75%) + P (<25%)	11.017, 19.030, 5.5463

SEM images of the air-sintered and argon-annealed $Sr_2Fe_2CoO_y$ sample are shown in Fig. 3; they reveal a dense homogeneous structure for both samples. Energy-dispersive X-ray (EDX) elemental analysis showed that the overall atomic ratio of the metal elements is consistent with that given in the chemical formula. The polished-surface SEM images showed that the argon-annealed sample is denser and contains fewer pores than the air-sintered sample. Spherical cobalt-rich phase was identified in the argon-annealed sample. By correlating SEM observation with XRD results, we identified the spherical phase as cobalt oxide (CoO, JCPD No. 9-402).

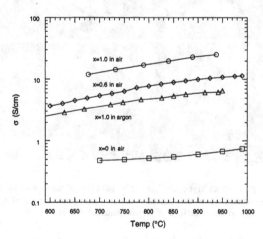

Fig. 4. Temperature dependence of $Sr_2Fe_{3-x}Co_xO_y$ samples.

The total conductivity of $Sr_2Fe_{3-x}Co_xO_y$ sample is shown in Fig. 4 as a function of temperature. Conductivity increases as temperature and cobalt content increase. At 900°C, the conductivity of $Sr_2Fe_2CoO_y$ is ≈20 and ≈6 S·cm⁻¹ in flowing air and argon environments, respectively. Conductivities of $Sr_2Fe_{2.4}Co_{0.6}O_y$ and $Sr_2Fe_3O_y$ are ≈10 and ≈0.6 S·cm⁻¹, respectively, at 900°C in air. Figure 5 shows the transient conductivity of the $Sr_2Fe_2CoO_y$ sample at 950°C after its surrounding atmosphere was changed. The conductivity recovered to its original value after the surrounding atmosphere was switched back to air from the reducing argon environment. This indicates that a phase transition in $Sr_2Fe_2CoO_y$ is reversible after the surrounding atmosphere is changed.

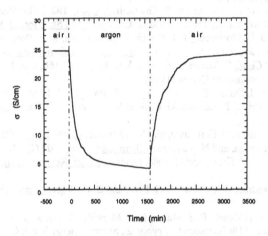

Fig. 5. Conductivity transient behavior of $Sr_2Fe_2CoO_y$ sample at 950°C.

CONCLUSIONS

Phase components of $Sr_2Fe_{3-x}Co_xO_y$ are strongly dependent on the cobalt content in the sample. The $Sr_2Fe_3O_y$ sample is single-phase material, while the high-cobalt-content samples (x > 0.6) contain multiple phases. Perovskite phase in the high-cobalt-content samples increases with increasing cobalt content. In a reducing argon environment, the single-phase $Sr_2Fe_3O_y$ sample has the same crystal structure as the air-sintered sample, while the brownmillerite phase was observed in the cobalt-containing samples. Cobalt oxide has also been found in the argon-annealed cobalt containing samples.

Total conductivity of $Sr_2Fe_{3-x}Co_xO_y$ increases with increasing temperature and cobalt content in the sample. At 900°C, the conductivity of $Sr_2Fe_2CoO_y$ is ≈20 and ≈6 S·cm⁻¹ in flowing air and argon environments, respectively. The conductivities of $Sr_2Fe_{2.4}Co_{0.6}O_y$ and $Sr_2Fe_3O_y$ are ≈10 and ≈0.6 S·cm⁻¹, respectively, at 900°C in air. Transient conductivity behavior of the sample, after surrounding atmosphere is switched back and forth, indicates that the phase transition in the materials is reversible.

171

ACKNOWLEDGMENT

Work at Argonne is supported by the U.S. Department of Energy, Federal Energy Technology Center, under Contract W-31-109-Eng-38.

REFERENCES

1. T. Takahashi and H. Iwahara, Energy Convers., 11, 105 (1971).
2. B. C. H. Steele, Mater. Sci. Eng. B-Solid State M., 13, 79 (1992).
3. N. Q. Minh, J. Am. Ceram. Soc., 76, 563 (1993).
4. R. DiCosimo, J. D. Burrington, and R. K. Grasselli, J. Catal., 102, 377 (1992).
5. K. R. Kendall, C. Navas, J. K. Thomas, and H.-C. Loye, Solid State Ionics, 82, 215 (1995).
6. U. Balachandran, S. L. Morissette, J. T. Dusek, R. L. Mieville, R. B. Poeppel, M. S. Kleefisch, S. Pei, T. P. Kobylinski, and C. A. Udovich, Proc. Coal Liquefaction and Gas Conversion Contractor Review Conf., S. Rogers et al., eds., Vol. 1, pp. 138-160, U.S. Dept. of Energy, Pittsburgh Energy Technology Center (1993).
7. U. Balachandran, T. J. Dusek, S. M. Sweeney, R. B. Poeppel, R. L. Mieville, P. S. Maiya, M. S. Kleefisch, S. Pei, T. P. Kobylinski, C. A. Udovich, and A. C. Bose, Am. Ceram. Soc. Bull., 74, 71 (1995).
8. Y. Teraoka, H. M. Zhang, S. Furukawa, and N. Yamozoe, Chem. Lett., 1985, 1734 (1985).
9. Y. Teraoka, T. Nobunaga, and N. Yamazoe, Chem. Lett., 1988, 503 (1988).
10. B. Ma, J.-H. Park, C. U. Segre, and U. Balachandran, Mater. Res. Soc. Symp. Proc., 393, 49 (1995).
11. B. Ma, U. Balachandran, C.-C. Chao, J.-H. Park, and C. U. Segre, Ceram. Trans. Series, 73, 169 (1997).
12. U. Balachandran, J. T. Dusek, P. S. Maiya, R. L. Mieville, B. Ma, M. S. Kleefisch, and C. A. Udovich, presented at 11th Intersociety Cryogenic Symp., Energy Week Conf. & Exhibition, Houston, Jan. 28-30, 1997.
13. S. Pei, M. S. Kleefisch, T. P. Kobylinski, J. Faber, C. A. Udovich, V. Zhang-McCoy, B. Dabrowski, U. Balachandran, R. L. Mieville, and R. B. Poeppel, Catal. Lett., 30, 201 (1995).
14. H. Iwahara, H. Uchida, K. Morimoto, and S. Hosgoi, J. Appl. Electrochem., 19, 448 (1989).
15. A. Yoshiasa, K. Ueno, F. Kanamaru, and H. Horiuchi, Mater. Res. Bull., 21, 175 (1986).
16. H. M. Rietveld, J. Appl. Cryst., 2, 65 (1969).
17. C. Greaves, A. J. Jacobson, B. C. Tofield, and B. E. F. Fender, Acta Cryst., B31, 641 (1975).
18. J. P. Hodges, J. D. Jorgensen, D. J. Miller, B. Ma, U. Balachandran, and J. W. Richardson, Jr., this proceedings.

CRYSTAL STRUCTURES OF MIXED-CONDUCTING OXIDES PRESENT IN THE Sr-Fe-Co-O SYSTEM

J.P. HODGES*, J. D. JORGENSEN*, D. J. MILLER*, B. MA**, U. BALACHANDRAN**
AND J. W. RICHARDSON, JR.***
*Materials Science Division, Argonne National Laboratory, Argonne, IL 60439.
**Energy Technology Division, Argonne National Laboratory, Argonne, IL 60439.
***Intense Pulsed Neutron Source, Argonne National Laboratory, Argonne, IL 60439.

ABSTRACT

The potential applications of mixed-conducting ceramic oxides include solid-oxide fuel cells, rechargeable batteries, gas sensors and oxygen-permeable membranes. Several perovskite-derived mixed Sr-Fe-Co oxides show not only high electrical-conductivity but also appreciable oxygen-permeability at elevated temperatures. For example, dense ceramic membranes of $SrFeCo_{0.5}O_{3-\delta}$ can be used to separate oxygen from air without the need for external electrical circuitry. The separated oxygen can be directly used for the partial oxidation of methane to produce syngas. Quantitative phase analysis of the $SrFeCo_{0.5}O_{3-\delta}$ material has revealed that it is predominantly composed of two Sr-Fe-Co-O systems, $Sr_4Fe_{6-x}Co_xO_{13}$ and $SrFe_{1-x}Co_xO_{3-\delta}$. Here we report preliminary structural findings on the $SrFe_{1-x}Co_xO_{3-\delta}$ ($0 \leq x \leq 0.3$) system.

INTRODUCTION

Mixed-conducting ceramic oxides are of great interest because of their technological importance in high-temperature electrochemical applications. The mixed Sr-Fe-Co oxide systems possess high electronic (σ_e) and oxygen ionic (σ_i) conductivities, together with structural stability over a wide range of oxygen partial pressures [1-4]. These materials have found applications in high-temperature electrodes, solid-oxide fuel cells, gas sensors and also hold particular promise as oxygen-permeable ceramic membranes. Teraoka et al. [2] have investigated the mixed electronic-ionic conductivity of the $Sr_{1-x}La_xFe_{1-y}Co_yO_{3-\delta}$ perovskite system and have shown that $SrFe_{0.2}Co_{0.8}O_{3-\delta}$, with $\sigma_e \approx 100$ and $\sigma_i \approx 1$ S.cm^{-1} at 800°C in air, has a particularly high oxygen-permeability. Recently, Balachandran et al. [4] have demonstrated that a material of nominal composition $SrFeCo_{0.5}O_{3-\delta}$ possesses a significantly higher oxygen-permeability ($\sigma_e \approx 10$ and $\sigma_i \approx 7$ S.cm^{-1} at 800°C in air). Unlike most of the known mixed conductors, which generally have much higher electronic than ionic conduction, $SrFeCo_{0.5}O_{3-\delta}$ has comparable conductivities. This combination of high electronic and oxide ion conductivities in $SrFeCo_{0.5}O_{3-\delta}$, means that it can act as an selective oxygen-permeable membrane without the need for external electrical circuitry. A promising application of $SrFeCo_{0.5}O_{3-\delta}$ oxygen permeable membranes, is in reactors that directly convert methane to syngas [5].

When used as a ceramic membrane in gas separation, $SrFeCo_{0.5}O_{3-\delta}$ is exposed to a large oxygen partial pressure (pO_2) difference (the resulting chemical potential gradient drives the transport of oxygen from the high to low pO_2 sides of the membrane). Under such a range of pO_2 and at typical operational temperatures of 600-800°C, the oxygen stoichiometry and crystal structure may vary significantly through the membrane.

In order to develop and understand an oxygen permeation mechanism for $SrFeCo_{0.5}O_{3-\delta}$ precise information concerning the crystal structures formed and oxygen nonstoichiometry at various temperatures and oxygen partial pressures is required. We have therefore initiated a wide ranging crystallographic study of the $SrFeCo_{0.5}O_{3-\delta}$ material and related phases.

Quantitative analysis of powder X-ray and neutron diffraction data from $SrFeCo_{0.5}O_{3-\delta}$, performed at Argonne, has revealed its multiphase nature. The major phase, present at a wt. fraction of ~ 70%, is of the form $Sr_4(Fe,Co)_6O_{13}$ and is isotypic in crystal structure to $Sr_4Fe_6O_{13}$ [6,7]. Present also are, an oxygen deficient perovskite of the form $Sr(Fe,Co)O_{3-\delta}$ with a wt. fraction of ~ 25% and CoO as a minor phase at ~ 5% wt. fraction. We have therefore undertaken a study of the two Sr-Fe-Co oxide systems present in the $SrFeCo_{0.5}O_{3-\delta}$ material. In this paper we report preliminary structural findings on the $SrFe_{1-x}Co_xO_{3-\delta}$ ($0 \leq x \leq 0.3$) system.

EXPERIMENTAL

Synthesis

Polycrystalline samples of $SrFe_{1-x}Co_xO_{3-\delta}$ (x = 0, 0.1, 0.2 and 0.3) were prepared by solid-state reaction of $SrCO_3$, Fe_2O_3 and Co_3O_4. The starting materials were mixed under n-amyl alcohol in a ball-mill for 3 hr. After drying, these mixtures were initially fired at 900°C for 12 hr., cooled to room temperature and reground in an agate mortar. Further firings at 1050°C, 1100°C and 1150°C for 12 hr. durations were carried out to complete the reaction. Subsequently, each $SrFe_{1-x}Co_xO_{3-\delta}$ sample was split into three and annealed under the following conditions, flowing Ar at 1050°C, air at 550°C or flowing O_2 at 350°C, all for 8 hr. periods. Overall twelve samples in the $SrFe_{1-x}Co_xO_{3-\delta}$ system were prepared with x = 0, 0.1, 0.2 and 0.3 combined with $\delta \approx 0.12$, 0.25 and $\delta = 0.5$ from the O_2, air, and Ar annealing stages, respectively.

Characterization

Powder X-ray and time-of-flight neutron diffraction data were collected on all samples using a Scintag diffractometer and the Special Environment Powder Diffractometer at Argonne National Laboratory's Intense Pulsed Neutron Source. Rietveld profile analysis of the powder neutron diffraction data was performed using the GSAS suite of programs [8].

RESULTS AND DISCUSSION

For the end member $SrFeO_{3-\delta}$, Takeda et al. [9] have determined that four distinct phases, with differing crystal structures, exist. The phases are defined by the ideal compositions $SrFeO_{3-1/n}$, where n = ∞, 8, 4 and 2. The n = ∞ and 2 members, $SrFeO_3$ and $SrFeO_{2.5}$, possess simple primitive cubic perovskite and brownmillerite crystal structures respectively see Figure 1. In the brownmillerite structure, oxygen vacancies are ordered into lines along the c-axis, which is the $[110]_p$ direction of the parent cubic perovskite. These lines of vacancies are located in every other $(001)_p$ plane, giving rise to alternate layers of FeO_4 tetrahedra and FeO_6 octahedra.

Electron diffraction measurements performed on the n = 8 and 4 members, $SrFeO_{2.88}$ and $SrFeO_{2.75}$, have revealed oxygen vacancy ordering to give perovskite based superstructures [9,10], however the precise crystal structures were unknown. Takano et al. [10] proposed that $SrFeO_{2.88}$ and $SrFeO_{2.75}$ possessed body centred tetragonal ($a = 10.934Å$ and $c = 7.705Å$) and C-face centred orthorhombic ($a = 10.972Å$, $b = 7.700Å$ and $c = 5.471Å$) unit cells, respectively. Furthermore, the volume of these unit cells are related to the parent cubic perovskite, where $a_p \approx 3.85Å$, by $(2\sqrt{2}a_p)^2 \times 2a_p$ and $2\sqrt{2}a_p \times 2a_p \times \sqrt{2}a_p$.

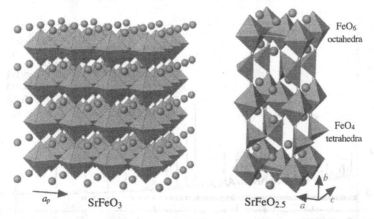

Figure 1. Crystal structures of cubic perovskite $SrFeO_3$ and brownmillerite $SrFeO_{2.5}$.

In this study we have prepared $SrFe_{1-x}Co_xO_{2.88}$, $SrFe_{1-x}Co_xO_{2.75}$ and $SrFe_{1-x}Co_xO_{2.5}$ systems via annealing of $SrFe_{1-x}Co_xO_{3-\delta}$ samples under O_2, air, and Ar, respectively. Powder X-ray diffraction revealed that for all the $SrFe_{1-x}Co_xO_{3-1/n}$ samples, the desired superstructures were formed in single phase. From a consideration of the proposed vacancy ordering schemes and resulting atomic relaxations, we have determined that the crystal structures of $SrFe_{1-x}Co_xO_{2.88}$ and $SrFe_{1-x}Co_xO_{2.75}$ may be described with orthorhombic Fmmm ($a \approx b \approx 4a_p$ and $c \approx \sqrt{2}a_p$) and monoclinic C2/m ($a \approx 2\sqrt{2}a_p$, $b \approx 2a_p$, $c \approx \sqrt{2}a_p$ and $\beta \approx 90°$) cell symmetries, respectively. Schematic representations of these crystal structures are shown in Figure 2. In $SrFe_{1-x}Co_xO_{2.88}$ the oxygen vacancies are located at the vertices and face centres of the unit cell, giving rise to adjacent $(Fe,Co)O_5$ square pyramids. Like brownmillerite, the vacancies are located in every other $(001)_p$ plane such that a layered structure, with $a \neq b$, results.

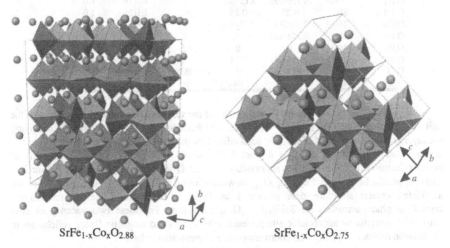

Figure 2. Crystal structures of the $SrFe_{1-x}Co_xO_{2.88}$ and $SrFe_{1-x}Co_xO_{2.75}$ systems.

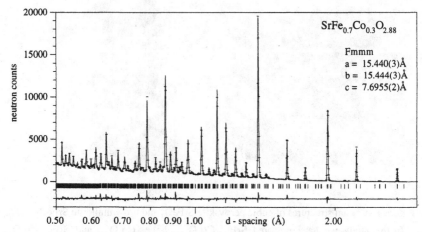

Figure 3. Final observed, calculated and difference profiles for $SrFe_{0.7}Co_{0.3}O_{2.88}$.

Table I. Refined crystallographic parameters for $SrFe_{0.7}Co_{0.3}O_{2.88}$.
Space group Fmmm $a = 15.440(3)$Å, $b = 15.444(3)$Å and $c = 7.6955(2)$Å
$R_{wp} = 0.061$, $R_p = 0.043$, $R_e = 0.019$ and $R_I = 0.067$.

Atom	Site	x	y	z	B_{iso}	Occ.
Sr (1)	16(k)	0.25	0.1278(5)	0.25	0.4(1)	1.0
Sr (2)	16(m)	0	0.1249(5)	0.2341(6)	0.6(1)	1.0
Fe/Co (1)	8(g)	0.1261(5)	0	0	0.4(1)	0.74(2)
Fe/Co (2)	8(g)	0.3755(5)	0	0	0.5(1)	0.68(2)
Fe/Co (2)	16(o)	0.1250(3)	0.2501(3)	0	0.4(1)	0.71(2)
O (1)	16(o)	0.1298(5)	0.1278(8)	0	1.7(2)	1.0
O (2)	8(e)	0.25	0.25	0	1.1(2)	1.0
O (3)	16(o)	0.3742(4)	0.1230(6)	0	0.6(1)	1.0
O (4)	8(g)	0.2533(8)	0	0	2.2(3)	1.0
O (5)	4(b)	0	0	0.5	0.8(2)	1.0
O (6)	16(l)	0.1233(4)	0.25	0.25	0.3(1)	1.0
O (7)	16(n)	0.1132(4)	0	0.2283(7)	1.4(1)	1.0
O (8)	8(h)	0	0.2660(5)	0	0.3(1)	1.0

In $SrFe_{1-x}Co_xO_{2.75}$ the vacancies are located at the vertices and centre of the c faces of the unit cell. Furthermore, this structure can be derived from $SrFe_{1-x}Co_xO_{2.88}$ simply by removing the layers composed only of $(Fe,Co)O_6$ octahedra. For the $SrFe_{1-x}Co_xO_{2.88}$ and $SrFe_{1-x}Co_xO_{2.75}$ systems there exists three and two distinct Fe/Co sites, respectively. Rietveld analysis of powder neutron diffraction data has successfully refined these crystal structures. In Figure 3, a Rietveld profile fit to the $SrFe_{0.7}Co_{0.3}O_{2.88}$ powder neutron diffraction data is shown, with the associated crystal structure data presented in Table I. While in Table II, the refined crystallographic parameters for $SrFe_{0.8}Co_{0.2}O_{2.75}$ are given. For these two vacancy ordered perovskite systems no appreciable site preference by Co and no magnetic Bragg reflections in the neutron diffraction profiles (at room temperature) were detected.

Table II. Refined crystallographic parameters for $SrFe_{0.8}Co_{0.2}O_{2.75}$.
Space group C2/m $a = 10.9762(9)$Å, $b = 7.6729(3)$Å, $c = 5.4736(5)$Å & $\beta = 90.06(2)°$
$R_{wp} = 0.101$, $R_p = 0.068$, $R_e = 0.021$ and $R_I = 0.084$.

Atom	Site	x	y	z	B_{iso}	Occ.
Sr (1)	8(j)	0.1212(6)	0.2369(3)	0.756(2)	0.4(1)	1.0
Fe/Co (1)	4(i)	0.1231(5)	0	0.245(2)	0.4(1)	0.81(1)
Fe/Co (2)	4(i)	0.3735(5)	0	0.744(2)	0.4(1)	0.80(1)
O (1)	8(j)	0.1140(6)	0.2267(4)	0.237(2)	0.9(1)	1.0
O (2)	4(i)	0.2614(8)	0	0.006(3)	0.9(2)	1.0
O (3)	2(b)	0	0.5	0	0.7(3)	1.0
O (4)	2(d)	0	0.5	0.5	2.9(3)	1.0
O (5)	2(c)	0	0	0.5	1.0(2)	1.0
O (6)	4(i)	0.2648(8)	0	0.493(3)	0.4(1)	1.0

The brownmillerite $SrFe_{1-x}Co_xO_{2.5}$ system was refined in the magnetic space group $Icm'm'$ using initial starting parameters reported by Battle et al. [11]. In Figure 4, a Rietveld profile fit to the $SrFe_{0.8}Co_{0.2}O_{2.5}$ neutron diffraction data is shown ($R_{wp} = 0.054$, $R_p = 0.038$, $R_e = 0.019$ and $R_I = 0.040$). In $SrFe_{1-x}Co_xO_{2.5}$ the magnetic moments were found to remain ordered in an antiferromagnetic fashion directed along the c-axis for x = 0 to 0.3. Differing from the $SrFe_{1-x}Co_xO_{2.88}$ and $SrFe_{1-x}Co_xO_{2.75}$ systems, Co was consistently found to show a degree of selectivity between the available Fe/Co sites. To illustrate, in $SrFe_{0.7}Co_{0.3}O_{2.5}$ the layer composed of $(Fe,Co)O_4$ tetrahedra was found to contain 36(1)% Co, this is greater than the 30% expected from a random distribution.

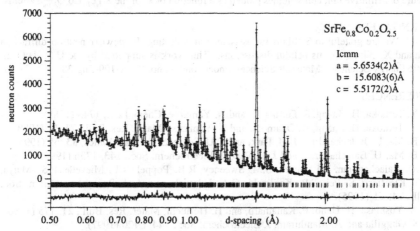

Figure 4. Final observed, calculated and difference profiles for $SrFe_{0.8}Co_{0.2}O_{2.5}$.

For ceramic membranes to operate for long periods under changing temperatures and large pO_2 gradients, they must be structurally stable. If the crystal structure of the membrane material changes significantly between different operating conditions, then this may lead to structural instability, for example, cracking. Large changes in unit cell volume as one varies the pO_2 are particularly undesirable in ceramic membrane materials. In Figure 6, we have plotted the primitive perovskite unit cell volumes for the $SrFe_{1-x}Co_xO_{3-1/n}$ systems as a function of x. It can be seen that Co doping has only a small effect on the unit cell volumes, however the oxygen content has a much greater influence. In moving from the oxygen rich $SrFe_{1-x}Co_xO_{2.88}$ to the $SrFe_{1-x}Co_xO_{2.75}$ system, the average unit cell volume increases by ~ 1%. The situation is clearly different on moving to the $SrFe_{1-x}Co_xO_{2.5}$ system where the unit cell volume increases by ~ 6%. This relatively huge volume increase is a consequence of significant changes in crystal structure and bonding. For the n = 8, and 4 members of $SrFe_{1-x}Co_xO_{3-1/n}$, there exists significant bonding in all 3-directions. In the n = 2 brownmillerite system, however, the alternate layers composed of $(Fe,Co)O_4$ tetrahedra can be considered to be without bonding in the a axis direction, and this results in a large relative expansion of the unit cell. These unit cell volume considerations, in part suggest that oxygen-permeable ceramic membranes composed only of $SrFe_{1-x}Co_xO_{3-\delta}$ may not possess the structural stability required in high-temperature reactors.

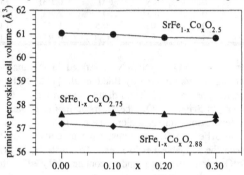

Figure 6. Primitive unit cell volumes plotted as a function of x for the $SrFe_{1-x}Co_xO_{3-1/n}$ systems.

ACKNOWLEDGMENTS
We are grateful to S. Short for assistance in collecting the powder neutron diffraction data and X. Xiong for his helpful discussions. This work is supported by the U.S. D.O.E., Basic Energy Sciences - Materials Sciences, under contract no. W-31-109-Eng-38.

REFERENCES

1. Y. Teraoka, H. Zhang, S. Furukawa, and N. Yamazoe, Chem. Lett., 1743 (1985).
2. Y. Teraoka, H. Zhang, K. Okamoto, and N. Yamazoe, Mater. Res. Bull., 23, 51 (1988).
3. B. Ma, U. Balachandran, J.-H. Park, and C.U. Segre, Solid State Ionics, 83, 65 (1996).
4. B. Ma, U. Balachandran, and J.-H. Park, J. Electrochem. Soc., 143, 1736 (1996).
5. U. Balachandran, T.J. Dusek, S.M. Sweeney, R.B. Poppel, R.L. Mievelle, P.S. Maiya, M.S. Kleefisch, S. Pei, T.P. Koblinski, C.A. Udovich, and A.C. Bose, Am. Ceram. Soc. Bull., 74, 71 (1995).
6. A. Yoshiasa, H. Ueno, F. Kanamaru, and H. Horiuchi, Mater. Res. Bull., 21, 175 (1986).
7. S. Guggilla and A. Manthiram, J. Electrochem. Soc., 144, L120 (1997).
8. A.C. Larson and R.B. Von Dreele, General Structure Analysis System, Los Alamos National Laboratory, Los Alamos, NM (1985, 1994).
9. Y. Takeda, K.Kanno, T. Takada, O. Yamamoto, M. Takano, N. Nakayama, and Y. Bando, J. Solid State Chem., 63, 237 (1986).
10. M. Takano, T. Okita, N. Nakayama, Y. Bando, Y. Takeda, O. Yamamoto, and J.B. Goodenough, J. Solid State Chem., 73, 140 (1988).
11. P.D. Battle. T.C. Gibb, and P. Lightfoot, J. Solid State Chem., 76, 334 (1988).

APPLICATION OF EUTECTIC CERAMIC MIXTURES FOR THE FUNCTIONAL COMPONENTS OF HIGH TEMPERATURE SOFC's

Ch. Gerk, M. Willert-Porada, Dept. Chem. Eng., Div. of Mater. Sci., University of Dortmund, D-44221 Dortmund, F.R. Germany, wipo@chemietechnik.uni-dortmund.de

ABSTRACT

A novel design for a high temperature SOFC, based on lamellar electrode-electrolyte segments obtained by solidification of an oxidic eutectic melt on an electrolyte substrate is presented. Such „composite" electrodes contain NiO or MnO - 8Y-ZrO_2 lamellae, which after reduction /oxidation yield electrode-electrolyte lamellae with 1-2 µm width and a vertical dimension of > 100 µm, depending upon the amount of eutectic melt solidified on a polycrystalline substrate. The nucleation of the eutectic on a polycrystalline substrate followed by a semi-directional crystallization of the two phases yields a gradient of 3-phase boundaries over the height of such an electrode, with the number of 3-phase boundaries increasing towards the substrate.

INTRODUCTION

Different geometrical concepts of high temperature Solid Oxide Fuel Cells (SOFC), based on ZrO_2-electrolytes, were tested over the last decade. In planar as well as tubular arrangements the improvement of cell performance and of the mechanical stability of the electrode-electrolyte units was achieved by implementation of thicker supporting elements or by increasing the dimensions of the electrodes, made up from porous composites of the electrocatalytic and the ionic conductor material [1-4]. However, the materials concept underlying the modern SOFC is still based on separate electrode-electrolyte materials, with relatively small areas of „overlapping" functionality based on the overall cell volume, even in most recent composite electrodes [5]. A further increase of the number of 3-phase contacts in such composite electrodes by, e.g., application of finer powders, is limited due to grain coalescence and growth as well as post sintering, which reduces the porosity under service conditions. Therefore alternative concepts for an electrode-electrolyte contact area should not only satisfy the long term high temperature stability of the microstructure but also the mechanical stability of the device.

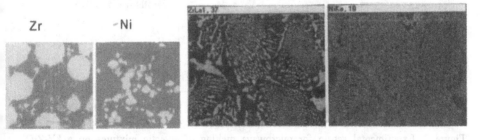

Figure 1: Comparison of the Ni- (anode)/Zr- (electrolyte)-distribution between a modern composite electrode (after [5]) and a Ni-8Y-ZrO_2-electrode from an eutectic mixture (EDX/SEM-results)

179

In order to insure short transport paths for the electrochemical reaction and a sufficient mechanical stability by a reasonable thickness of the cell, eutectic mixtures in the system ZrO_2-NiO-(R_2O_3) and ZrO_2-MnO -(R_2O_3), R=Gd, are of particular interest. From directionally solidified melts of NiO-ZrO_2 [6] and MnO-ZrO_2 [7] as well as the Gd-system [8] it is known, that a fine lamellar microstructure is developed, with ZrO_2 and NiO-lamellae of 1-2 μm thickness. Such a structure can be visualized as a composite electrode, in which a further refinement of mixing between Ni and the elctrolyte is achieved, as indicated in Figure 1.

In the following, first results on the real microstructure of samples obtained from eutectic melts solidified on commercial zirconia substrates will be presented. The microstructure of these materials is analyzed in terms of fuel cell requirements. DC-conductivity of eutectic and composite Ni-ZrO_2 will be discussed.

EXPERIMENT

Commercial powders were employed for the eutectic melt: 8Y-ZrO_2 (0.3 μm d_{50}, TOSOH, Japan); MnO, NiO, La_2O_3, Gd_2O_3 (99.9% each, 0.5-1 μm, ChemPur, Germany). The substrates were obtained either from commercial 50x50x 0.5 mm^3 green foils of 8Y-ZrO_2 (Kerafoil, Germany), sintered at 1450°C to full density prior to use or from 8Y-ZrO_2 bulk 20x20x5 mm^3 samples sintered at 1400°C. The eutectic composition was adjusted to +/- 5% above and below the eutectic [6,9], in order to adjust for compositional changes during melting on a substrate, e.g., 20 +/-5Mol% NiO in ZrO_2 and 38 +/-%wt MnO in ZrO_2. Gd_2O_3-additions varied from 10 - 20 Mol%, replacing ZrO_2. All powders were milled in ethanol and calcined several times prior to the melting step. Finally, discs with ϕ 13 mm were pressed from the eutectic powder mixture and placed on top of the substrate. Melting and crystallization was performed by heating the composite in a microwave furnace (microwave frequency 2.45 GHz, power 2.5 kW, cylindrical applicator, Puls Plasma Technik GmbH, Dortmund). In order to increase the thermal load, a CSZ-cylinder was used as suszeptor. A scheme of the experimental set-up is shown in Figure 2.

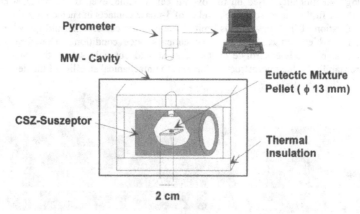

Figure 2: Experimental set-up for microwave melting of eutectic mixtures on a 8Y-ZrO_2 polycrystalline substrate.

The anode-electrolyte composite was fired at 1750-1800°C in air, the cathode-electrolyte composite at 1550-1600°C in Ar. NiO was reduced to Ni at 800 - 1000°C in H_2. For a cathode two processing routes were applied: starting from the MnO-8Y-ZrO_2-eutectic an oxidation of the

Mn^{2+} in air was performed, followed by infiltration with La_2O_3 or from the reduced NiO-8Y-ZrO_2-eutectic Ni was removed by acid-etching and the porosity was filled with a MnO-La_2O_3 - slurry. After firing in air, $LaMnO_3$ is abtained.

Conductivity measurements were performed at 20 - 800°C using a HP 4284 A impedance analyzer, operating at 20 Hz - 1 MHz. The DC-conductivity was extracted from the low frequency semicircle. Measurements on NiO-containing samples were performed in air.

RESULTS

Melting by Microwave Heating

Limitations due to the thermal stability of conventional heating elements exist when firing temperatures > 1750°C in air are necessary. Upon microwave heating the material to be processed acts as „heating element", therefore temperatures as high as the melting point of the material can be achieved, providing the electrical conductivity of the material does not limit the penetration depth for the radiation [10.11]. Furthermore, when local composition differences exist, heat is deposited selectively within the area of higher dielectric loss. Applied to a solid 8Y-ZrO_2-substrate covered by a pellet composed of an, e.g., eutectic NiO-8Y-ZrO_2-mixture, at temperatures close to the melting temperature higher dielectric losses due to „lattice softening" can be assumed [11] at the interface between the pellet and the substrate as compared to the bulk of the substrate. A temperature gradient can be established by this heating method, which facilitates the melting process and also governs the solidification process of the eutectic lamellae on to the substrate. Results of a calculation for the temperature gradient achievable at conditions close to the electric brake down are shown in Figure 3, together with a scheme of the heat flow.

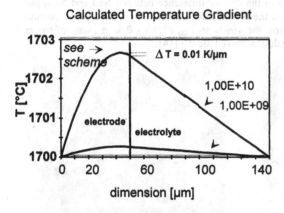
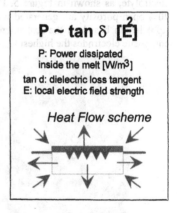

Figure 3: Temperature gradient at the melt-substrate interface, calculated for microwave power density values typical in microwave melting experiments

A temperature gradient of 0.01 K/μm is typically the highest value of a temperature gradient used in directional solidification of single crystals grown from a melt. Due to the polycrystalline nature of the „seeding" substrate, a polycrystalline eutectic is grown on the electrolyte substrate, however, with only one direction of the eutectic lamellae, as shown in Figure 4 for the NiO-8Y-ZrO_2-eutectic solidified on a 8Y-ZrO_2-substrate. The number of lamellae per surface is highest at the substrate-eutectic interface, with a sufficient percolation between the „eutectic"-8Y-ZrO₂-lamellae and the electrolyte substrate, as shown in the insert of Figure 4.

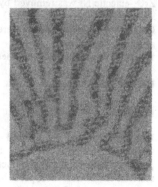

Figure 4: Microstructure of a „directionally" solidified NiO-8Y-ZrO₂ -eutectic, substrate 8Y-ZrO₂ (SEM)

<u>Anode and Cathode Preparation</u>

The reduction of NiO to Ni is the preferred method to convert a NiO-8Y-ZrO₂-composite into Ni-8Y-ZrO₂. Within the eutectic microstructure the reduction reaction proceeds very smoothly, with nearly no NiO-lamellae left. Upon reduction, the Ni crystals remain attached to the electrolyte, as shown in Figure 5. Due to the density difference between NiO and Ni approx. 20% open porosity are generated within the Ni-lamellae. Calculation of the 3-phase boundary (tpb) length based on the real microstructure yields an L_{tpb} of 0.3 - 0.6. In „state of the art" composite electrodes the highest L_{tpb}-values achieved are 0.3 [14].

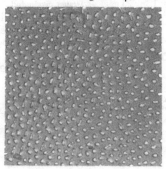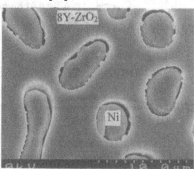

Figure 5: Microstructure and electrode-electrolyte contacts within a lamellae-electrode generated from the eutectic oxidic melt.

For the cathode side application of the Ni-8Y-ZrO₂-eutectic microstructure has proven to yield the best results. In this case Ni is removed by etching and the empty lamellae are filled with a MnO/La₂O₃ slip. After firing, the desired LaMnO₃. The microstructure of such a cathode is shown in Figure 6.

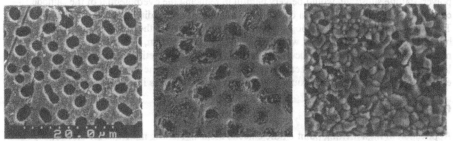

Figure 6: Microstructure of the lamellar cathode „preform" and of the LM-filled and fired cathode.

Electrical Conductivity at the Anode

The electrical conductivity of a lamellae-composite electrode as compared to a „dispersion"-type composite electrode is shown in Figure 7. The real NiO-8Y-ZrO$_2$-eutectic does not achieve the calculated value of a perfectly directional lamellae electrode, due to some misalignment of the polycrystals. Further optimization of the solidification process is necessary.

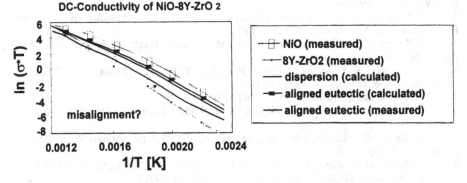

Figure 7: dc-conductivity of the anode-electrolyte components as a function of microstructure.

DISCUSSION

In view of recent model calculations by Virkar and co-workers, regarding the electrocatalytic performance of SOFC with a „lamellar" microstructure[13], the suggested eutectic electrode-electrolyte systems would, due to the thickness of the NiO-lamellae of 1-5 μm, insure a complete independence of the effective resistance of the electrode on the thickness, even when the thickness of the lamellar electrode would be > 100μm. Therefore the eutectic electrode could be used as a supporting element for a SOFC-unit. In this case, the volume ratio of active, 3-phase boundary and 2-phase boundary contacts on the overall volume of a SOFC device would be significantly higher than in planar devices. Furthermore, the gradient of 3-phase boundaries, with the number of 3-phase contacts increasing towards the electrode due to solidification on a polycrystalline substrate, would be very helpful in reducing the overpotential of the cathode, caused by the decrease of oxygen concentration over the vertical dimension of the cathode [14]. Therefore, in spite of the technological problems connected with the preparation of the desired high melting oxide eutectics and with the directional solidification, necessary to insure maximum percolation and conduction of the desired phase, theoretical predictions support the lamellae-composite electrode-electrolyte concept. Furthermore, the application of a melt also enables casting as a processing route to achieve large area devices in a very short time.

CONCLUSION

Application of high melting point oxidic eutectic melts as starting materials for electrode-electrolyte segments of SOFC-units has been shown to be practical, using selective microwave melting and directional crystallization on a polycrystalline electrolyte substrate. Transformation of the eutectic oxide lamellae to the desired electrode materials is achieved by reduction of NiO to Ni for the anode side or by removal of Ni and replacing with $LaMnO_3$ for the cathode side. In view of recent theoretical models the achieved microstructure is very promising for improvement of the SOFC performance.

REFERENCES

1.   K. Ogasawara, I. Yasuda, Y. Matsuzaki, T. Ogiwara and M. Hishinuma, Electrochem. Proc. Vol. 97-40 (1997) 143-152
2.   H.P. Buchkremer, U. Diekmann, L.G.J. de Haart, H. Kabs, U. Stimming, and D. Stöver, Electrochem. Proc. Vol. 97-40 (1997) 160 - 170
3.   S. Kakigami, T. Kurihara, N. Hisatome, and K. Nagata, Electrochem. Proc. Vol. 97-40 (1997) 180-186
4.   S. Kawasaki, K. Okumura, Y. Esaki, M. Hattori, Y. Sakaki, J. Fujita, and S. Takeuchi, Electrochem. Proc. Vol. 97-40 (1997) 171-179
5.   H. Itoh, J. Electrochem. Soc., Vol. 144 [2], 1997
6.   A. Revcolevschi, G. Dhalenne and F. d'Yvoire, J. Phys. Colloq., C4 (46), S. 441-447 (1985)
7.   W. Kurz, P.R. Sahm, Gerichtet erstarrte eutektische Werkstoffe, Springer Verlag Berlin-Heidelberg, 1975, p. 120 ff
8.   B. Dubois, G. Dhallene, A. Revcolvschi, J. Am. Ceram. Soc., 69 [1], C6 - C8 (1986)
9.   Schultz and A. Muan, J. Am. Ceram. Soc., 54 [10] 504 (1971)
10.   M. Willert-Porada, in „Ceramic Processing Science and Technology", ed. H. Hausner, G. Messing, Shin-ichi Hirano, Amer. Ceram. Soc., Westerville, OH, Ceram. Trans. Vol. 51, S. 501-506 (1995)
11.   M. Willert-Porada, T.Gerdes, H. Kolaska, K. Rödiger, Metall, 50 [11], 744-752 (1996), Metall, 51 [1/2], 57-65 (1997)
12.   D.J. Duval, Mat. Res. Soc. Symp. Proc. Vol. 430, 125 - 129 (1996)
13.   C.W. Tanner, K.-Z. Fung, and A. Virkar, „The Effect of Porous Composite Electrode on Solid Oxide Fuel Cell Performance", J. Electrochem. Soc., Vol. 144 [1], (1997) 21-30
14.   H. Fukunaga, M. Ihara, K. Sakaki, and K. Yamada, Solid State Ionics, Vol. 86-88, 1179-1185 (1996)

Financial support of DFG, SPP Gradientenwerkstoffe, contract Wi 856/8-2, is gratefully acknowledged.

A COMPOSITE ELECTROLYTE FOR AN SOFC CONSISTING OF A CERIA SHEET AND A ZIRCONIA FILM DEPOSITED BY THE SOL-GEL METHOD

Reiichi Chiba*, Fumikatsu Yoshimura* and Junichi Yamaki**
*NTT Integrated Information & Energy Systems Laboratories, 162, Tokai-Mura, Ibaraki-Ken, 319-11, JAPAN, chiba@iba.iecl.ntt.co.jp
**Institute of Advanced Material Study, Kyushu University, Kasuga Koen 6-1, Kasuga 816, Japan

ABSTRACT

We investigated a composite electrolyte for solid oxide fuel cells prepared by coating a ceria sheet ($Ce_{0.8}Gd_{0.2}O_{2-d}$ or GDC) with a scandia alumina doped zirconia ($0.850ZrO_2$-$0.110Sc_2O_3$-$0.04Al_2O_3$) film by the sol-gel method. The sol-gel film annealed at 1200°C was examined by X-ray diffraction analysis and found to be in a cubic phase at room temperature. The ionic conductivity of this film is comparable to that of bulk sintered at 1620°C. Scanning electron microscope observations revealed that the film forms a good interface with the electrolyte of the ceria sheet, even though the annealing temperature is as low as 1200°C.

We fabricated a single cell consisting of a composite electrolyte, a $La_{0.8}Sr_{0.2}MnO_3$ cathode and a Ni-YSZ anode. The composite electrolyte consisted of zirconia film about one micron thick deposited by the sol-gel method and a 0.2 mm thick ceria sheet. A cell operated with moist H_2 and O_2 gas exhibited an open circuit voltage of 1.00 V at 800°C. This value is much closer to the value of 1.13 V expected from the Nernst equation than the value of 0.76 V for a cell with a ceria sheet but without the sol-gel film.

INTRODUCTION

Solid oxide fuel cells (SOFC) operating at about 1000°C have been frequently studied [1,2]. However, such high temperature operation causes material problems including electrode sintering and interfacial diffusion between electrolyte and electrode [1]. One possible way to overcome this problem is to reduce the SOFC operating temperature around 800°C [1,3]. To achieve this, the ohmic loss in the electrolyte must be reduced. This is because it accounts for most of the power generation loss of an SOFC at such low temperature, and the ionic conductivity decreases when the operating temperature is reduced. Gd or Sm doped ceria (GDC or SDC) are candidates as the electrolyte material for low temperature operating SOFCs, because of their high ionic conductivity [4].

However, ceria-based materials have several problems originating from their instability in a reduced atmosphere. The ionic transport number for ceria is less than 1.0 [4]. This means that the electrolyte short circuits electrically and this reduces the electromotive force and results in a drop in the open circuit voltage from the value expected by the following Nernst equation.

$$\eta = \eta_0 + (RT/nF)\cdot \ln\left[\left(P_{O_2} \text{ in anode}\right)\middle/\left(P_{O_2} \text{ in cathode}\right)\right]$$

Moreover, the thermal expansion coefficients of ceria in a reduced atmosphere are larger than those in an oxidized atmosphere. This produces stress in the electrolyte and in the interfaces with the cathode or anode [5].

These problems can be overcome, if the ceria electrolyte surface in a reduced atmosphere (anode side) is covered with a film made of zirconia-based material [6]. In such a case, the film should be as thin as possible in order to retain the low ohmic loss in the electrolyte.

Therefore, we prepared our electrolyte thin films by the sol-gel method, because it can provide films of submicron order thickness. In addition, the coating apparatus of the sol-gel method is much simpler than other commonly used methods such as electrochemical vapor deposition or

185

Mat. Res. Soc. Symp. Proc. Vol. 496 © 1998 Materials Research Society

vacuum processes including rf sputtering. We selected zirconia doped with Sc^{3+} Al^{3+} as the film material to cover the ceria, because it exhibits the highest ionic conductivity among zirconia doped with rare-earth ions [7].

In this study, we fabricated a single cell with a composite electrolyte composed of a ceria sheet and zirconia sol-gel film to investigate the potentialities of this kind of electrolyte.

EXPERIMENT

The sol-gel film samples were prepared by the conventional sol-gel method [8]. We used an alkoxide of Zr, Al and a water solution of $Sc(NO_2)_3$ mixed in alcohol. The 0.7 μm-thick films, which consisted of ten individually spin-coated layers, were deposited on a sintered alumina substrate for characterization and on GDC sheets. Each layer was dried at 150°C and calcined at 400°C before the next layer was deposited. Once this process was complete, the films were annealed in air for two hours at 800°C-1600°C for crystallization.

An X-ray diffractometer was used to investigate the crystal structure at room temperature. We investigated the morphology of the films and the interfaces between the electrolytes and the electrode by examining cross-sections with a scanning electron microscope (SEM). We evaluated the ionic conductivity of the samples with the DC four terminal method. Here, the film samples were about 0.8 μm x 0.3 cm x 1.8 cm. Platinum wires were attached to the samples with platinum paste by firing at 800°C. A constant current of 0.1-10 μA was supplied along the substrate plane and the potential drop was measured with a high input impedance digital multi-meter. We analyzed the grain boundary resistance factor for the films by the AC impedance method (two terminal). The measurements were performed in the 0.1-10 MHz frequency range at 600°C in air with an impedance analyzer.

The geometry of the cell we used is shown in **Fig.1**. We used Ni-YSZ, $La_{0.8}Sr_{0.2}MnO_3$, and porous Pt for the anode, cathode and reference electrode, respectively. First, sol-gel film was spin coated on the 0.2 mm-thick GDC sheet and annealed at 1200°C for two hours. Then NiO-YSZ powder with PVA was painted on the film coated side of the GDC sheet and fired at 1200°C for 1 hour, then $La_{0.8}Sr_{0.2}MnO_3$ powder with ethylene glycol and Pt paste were painted on the plate and fired at 1100°C in air for 1 hour. The $La_{0.8}Sr_{0.2}MnO_3$ powder had a mean diameter of about 3.0 μm. Pt mesh was used as the current collector. The typical area of the cathode and anode was about $0.283cm^2$ (0.6 mm in diameter). The firing conditions and materials for the cell are listed in **Table 1**.

Table 1. Cell components and their materials.

component	materials
electrolyte	
ceria sheet	$Ce_{0.8}Gd_{0.2}O_{2-\delta}$ + MgO (0.2 mm t)
sol-gel film	$0.85ZrO_2$-$0.11Sc_2O_3$-$0.04Al_2O_3$ (1.0 μm t)
cathode	$La_{0.8}Sr_{0.2}MnO_3$ (fired at 1100°C)
anode	Ni-YSZ (fired at 1200°C)
reference	Pt paste (fired at 1100°C)
effective area	0.283 cm^2 mmφ

Figure 2 shows the schematic structure of the fuel cell. Oxygen gas (99.9% purity) was supplied to the cathode compartment and reference electrode at flow rates of 150 cc/min and 300 cc/min, respectively. The hydrogen gas was supplied to the anode compartment at a flow rate of 300 cc/min. The cell current was varied between 0-200 mA with a galvanostat and the cell voltage (between cathode and anode) was measured with a digital multi-meter.

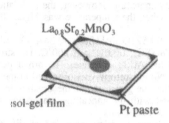

$La_{0.8}Sr_{0.2}MnO_3$

sol-gel film   Pt paste

Fig.1 Cell geometry.

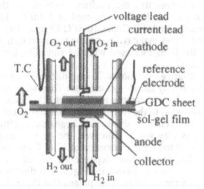

voltage lead
current lead
O_2 out O_2 in   cathode
T.C
reference
electrode
O_2
GDC sheet
sol-gel film
anode
collector
H_2 out
H_2 in

Fig.2 Fuel cell structure.

RESULTS

Figure 3 shows the annealing temperature dependence of the morphology of sintered sol-gel films. The film annealed at 800°C (**Fig.3** (a)) has a very fine texture. The layer structure resulting from the spin-coating process still remains. The film annealed at 1000°C (**Fig.3** (b)) no longer has a layer structure, and the grain size has increased. The grain size in the film annealed at 1200°C (**Fig.3** (c)) is comparable to the film thickness. Although relatively large cavities developed in this film, it has a continuous, isotropic and well-sintered structure.

The large cavities stem from the small cavities observed between the grains in film annealed at lower temperatures. The thickness of the film annealed at 800°C was almost the same as that of the film annealed at 1200°C, while adjacent grains fitted together with almost plane boundaries.

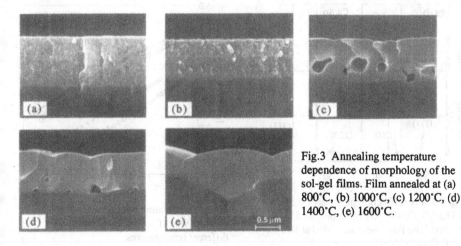

Fig.3 Annealing temperature dependence of morphology of the sol-gel films. Film annealed at (a) 800°C, (b) 1000°C, (c) 1200°C, (d) 1400°C, (e) 1600°C.

187

This means that the cavities observed in the film annealed at 800°C converged and still remained in the film annealed at 1200°C. The number and size of the cavities in the films annealed at 1400°C and 1600°C (**Fig.3** (d),(e)) decreased with the annealing temperature. This shows that cavities can be removed from the film by annealing. However, the undulations on the films increased with the annealing temperature, when the temperature was higher than 1200°C.

Figure 4 shows X-ray diffraction patterns of $0.85ZrO_2$-$0.11Sc_2O_3$-$0.04Al_2O_3$ films prepared by the sol-gel method. They were annealed at 800°C, 1000°C and 1200°C. The strong peaks for the sample annealed at 800°C are those of the Al_2O_3 substrate. The broad peaks correspond to the films and are indexed as cubic phase. No anisotropy was observed in the X-ray diffraction analysis. The large peak width shows that the grain size of the crystals in the film is very small. However, the peak widths decreased with the annealing temperature and those of the sample annealed at 1200°C are comparable to that of bulk.

The temperature dependence of the ionic conductivity of films annealed at different temperatures is shown in **Fig.5**. No discontinuous change in conductivity was observed for any of the samples. This means that these films are completely stabilized in the cubic phase. The conductivity of the film annealed at 800°C is less than 1/10 of bulk. In contrast, the conductivity of the film annealed at 1200°C is 86% of the bulk value. This shows that the sol-gel method can considerably reduce the process temperature.

We analyzed the resistance of the films by the AC impedance method. Cole-Cole plots of the films annealed at 800°C, 1000°C and 1200°C are shown in **Fig.6**. The measurements were made at 600°C. The Cole-Cole plot for film annealed at 1200°C consists of a large symmetrical semi-circle in the high frequency region and a small elliptic semi-circle in the low frequency region. The elliptic semi-circle originates from grain boundary resistance and the large semi-circle originates from grain resistance [9]. This means that grain boundary resistance is less than 5% of total resistance in the film annealed at 1200°C. The Cole-Cole plot for the film annealed at 1000°C includes a degree of asymmetry. This curve should be the superposition of a symmetrical semi-circle and an elliptic semi-circle of similar sizes. The plot for the film annealed at 800°C has a greater degree of asymmetry. This curve should be the superposition of a small symmetrical semicircle and a large elliptic semi-circle. These results suggest that grain boundary resistance accounts for most of the resistance in films annealed at 800°C and 1000°C. Therefore, the improvement in the conductivity seems to be caused by a decrease in the grain boundary effect.

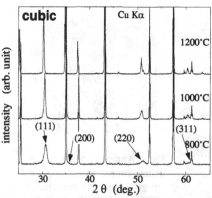

Fig.4 X-ray diffraction patterns of $0.85ZrO_2$-$0.11Sc_2O_3$-$0.04Al_2O_3$ films prepared by sol-gel method. The films were annealed at 800°C, 1000°C and 1200°C.

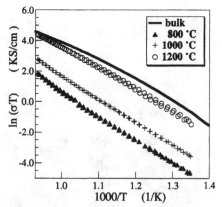

Fig.5 Temperature dependence of ionic conductivity of sol-gel films annealed at different temperatures.

Figure 7 (a) and (b) are SEM images of the cross-sections of a porous GDC sheet and fine GDC sheet spin-coated with sol-gel film. Both sheet were made by the doctor blade technique. The sol-gel solution soaked into the porous sheet and did not cover its surface, because the viscosity of the solution was very low. By contrast, the fine sheet (**Fig.7(b)**), which contained MgO as a sintering agent, was covered with sol-gel film in spite of the surface undulations. The sol-gel film annealed at 1200°C formed a good interface with the fine GDC sheet. We therefore used the fine GDC sheet for the cell.

Figure 8 shows SEM image of the cross-section of a cell with a composite electrolyte consisting of a GDC sheet and sol-gel film. The sol-gel film was spin-coated 10 times. In this figure, only a portion of the anode side is shown. While the sol-gel film remains between the GDC sheet and the Ni-YSZ anode, they form good interfaces.

We placed the cell, whose cross-section is shown in **Fig.8**, in the fuel cell shown in **Fig.2** and measured the I-V characteristics.

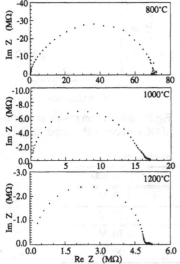

Fig.6 Cole-Cole plots of sol-gel films annealed at 800°C, 1000°C and 1200°C, measured at 600°C

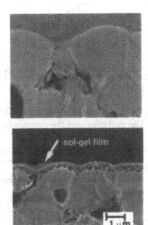

Fig.7 SEM images of cross-sections of porous and fine GDC sheets spin-coated with sol-gel film, and annealed at 1200°C.

Ni-YSZ

sol-gel film

$Ce_{0.8}Gd_{0.2}O_2$ +MgO

Fig.8 SEM image of cross-section of cell with composite electrolyte of GDC sheet and sol-gel film (10 times spin-coating).

The cell voltage at 800°C is plotted against cell current in **Fig.9**. The cell voltage at zero current or OCV was 1.00 V. This value is about 0.24 V higher than that of a cell with only a GDC sheet (0.76 V). The OCV values are listed in **Table 2**. This means that the sol-gel film functions as an electrolyte. This value is lower than 1.13 V which is the OCV for the cell with zirconia electrolyte at 800°C. However, when we take the sol-gel film thickness into consideration, the ionic transport number is close to 1.0.

Figure 10 shows the film thickness dependence of the OCV of the cell with a composite electrolyte. When the film thickness was small (0.2 μm), we observed no improvement in the OCV. At thicknesses above 0.5 μm the OCV improved abruptly and approached the YSZ value. This suggests that the formation of pinholes in the sol-gel film causes the OCV to drop from 1.13 V. If pinholes are the main cause of the OCV drop, the ionic transport number of the film should be close to 1.0.

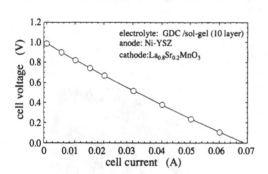

Fig.9 Cell voltage dependence on cell current at 800°C. The cell contains the composite electrolyte.

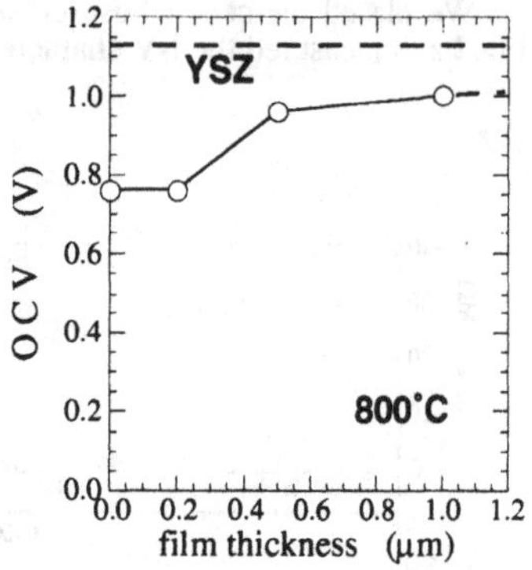

Fig.10 Sol-gel film thickness dependence of OCV of cell with composite electrolyte.

Table 2. OCV for cells with different electrolytes.

electrolyte	OCV at 800°C
YSZ	1.13 V
Sc_2O_3, Al_2O_3 doped zirconia	1.13 V
GDC only	0.76 V
GDC and *sol-gel film composite	1.00 V

* $0.85ZrO_2$ -$0.11Sc_2O_3$,-0.04 Al_2O_3 , annealed at 1200°C.

CONCLUSIONS

Cubic stabilized $0.85ZrO_2$-$0.11Sc_2O_3$-$0.04Al_2O_3$ films can be obtained by using the sol-gel method. The film annealed at 1200°C had a continuous, isotropic structure and the ionic conductivity of the film was 7.6 x 10^{-2} S/cm at 800°C. This is comparable to that of bulk samples prepared by a solid state reaction at 1620°C. The cavities in the films grow with annealing temperature. These cavities can be removed by high temperature annealing. The grain boundary resistance of the film annealed at 1200°C was estimated to be 1/20 of the total resistance at 600°C. The improvement in the ionic conductivity of the film with annealing temperature originated from a reduction in grain boundary resistance. The cell with the composite electrolyte exhibited an open circuit voltage of 1.00 V at 800°C. This value is much closer to the value of 1.13 V expected from the Nernst equation than the value of 0.76 V for a cell with a ceria sheet but without the sol-gel film. The OCV improves sharply when the sol-gel film thickness exceeds 0.5 μm. This suggests that the existence of pinholes in the film causes the OCV to fall below the value expected for zirconia electrolyte.

REFERENCES

[1] Nguyen Q. Minh, J. Am. Ceram. Soc. 76 [3], p. 563 (1993).

[2] S. C. Singhal, in Proceedings of the Third International Symposium on Solid Oxide Fuel Cells, Hawaii, p. 665, (1993).

[3] K. Krist, J. D. Wright, in Proceedings of the Third International Symposium on Solid Oxide Fuel Cells, Hawaii, p. 782, (1993).

[4] T. Kudo, H. Obayashi, J. Electrochem. Soc. Vo.123, No.3, p. 415 (1976)

[5] I. Yasuda and M. Hishinuma, in Proceedings of the 63rd Meeting of The Electrochemical Society of Japan, p. 176 (1996).

[6] F.P.F van Berkelin, G.M. Christie, F.H. van Heuveln, J.P.P. Huijsmans, Proceedings of the Fourth International Symposium on Solid Oxide Fuel Cells, Yokohama, p. 1062, (1995).

[7] T. Ishii, T. Iwata, Y. Tajima, in Proceedings of the Third International Symposium on Solid Oxide Fuel Cells, Hawaii, p. 59, (1993).

[8] T. W. Keuper, S. J. Visco, L. C. De Jonghe, Solid State Ionics, Vol.52 (1-3), p. 251 (1992).

[9] J.E. Bauerle and J.Hrizo, J. Phys. Chem. Solids, Vol.30, p565 (1969).

ACKNOWLEDGMENTS

We express our gratitude to Dr. Yonezawa, Dr. Ogata and Ms. Endou at the Central Research Center of Mistubishi Material Corporation for their help in coating the sol-gel films. We also express our gratitude to Mr. Ohki in NTT Advanced Technology for his help with the scanning electron microscope observation of the film samples.

MOLECULAR DYNAMIC SIMULATION AND ELECTRICAL PROPERTIES OF $Ba_2In_2O_5$

MASAMI KANZAKI,AKIHIKO YAMAJI AND KAZUYA KAWAKAMI
DEPARTMENT OF MECHANICAL ENGINERRING, TOKYO INSTITUTE
OF TECHNOLOGY,OOKAYAMA,MEGUROKU,TOKYO 152,JAPAN

ABSTRACT

Brownmillerite($Ca_2Al_2O_5$-$Ca_2Fe_2O_5$ solid solution) structure can be regarded as an oxygen-ion deficient perovskite structure. Because of high proportion of the oxygen vacancies in the structure, this material could be a candidate of fast oxide-ion conductor. Goodenough et al. indeed observed a first-order transition to a fast oxide-ion conductor at 930° C for $Ba_2In_2O_5$ which adapts brownmillerite structure at ambient temperature. Molecular dynamics simulation was employed to study oxygen ion diffusion and phase transition of $Ba_2In_2O_5$. The structure was well simulated at 300 K. When the system was heated, the original orthogonal cell transformed to a tetragonal cell at 2300 K. Inspection of the structure revealed that oxygen ions started to migrate from their original sites to nearest vacant oxygen sites at this temperature. The diffusion was restricted for the oxygen sites around In-tetrahedron, resulting highly anisotropic diffusion on the ac plane. At 4600 K it further transformed to an oxygen vacancies-disordered cubic perovskite structure. Although predicted transition temperature were apparently overestimated, the transition way to the phases with high oxygen ion diffusivity is consistent with the experimental results from electrical conductivity measurements. The high temperature cubic phase shows large ion conductivity. It is of interest to examine whether or not the cubic phase stabilizes in the low temperature region by making solid solution of another elements. We found that the cubic phase is stabilized below 500° C without any decrease of conductivity in $BaIn_{1.9}Ce_{0.1}O_y$ and $Ba_2In_{1.8}Nb_{0.2}O_5$.

INTRODUCTION

Brownmillerite (Ca_2AlFeO_5) structure can be regarded as an oxygen-deficient perovskite structure [1]. In the structure, oxygen vacancies are ordered in lines parallel to <101>, resulting tetrahedral coordination for a half of (Al,Fe) ions. Because of a sixth of oxygen sites in the corresponding perovskite structure are vacant, this material could be a candidate for fast oxygen ion conductor. Goodenough et al. [2] have indeed observed a first-order transition to a fast oxide-ion conductor at 930 C for $Ba_2In_2O_5$ which adapts brownmillerite structure at ambient temperature. More recently Sr_2ScAlO_5 [3] and $Ba_2GdIn_{1-x}Ga_xO_5$ [4] have been reported as fast oxide-ion conductors. Shin et al. [5] have reported that $Sr_2Fe_2O_5$ with brownmillerite structure transforms to an oxygen disordered cubic perovskite structure above 700° C. However the details of the diffusion mechanism and its relation with structural transition of the compound have not been reported to date.

The purpose of present study is to obtain the insight of the mechanism of the oxygen ion diffusion in $Ba_2In_2O_5$. We employed molecular dynamics (MD) simulation technique for this purpose. The MD simulation has been used to study the diffusion behavior of the superionic conductors [6]. However no MD study has been reported for the compounds with brownmillerite structure. In order to examine the effect of composition on the transition behavior, we extended the simulation to $A_2In_2O_5$ (A=Ba, Sr, Ca; B=Al, Fe, In) systems as well. These MD simulation results were compared with the experimental results.

EXPERIMENTAL

1. Molecular dynamic simulation

A MD simulation program (MXDORTO and MXDTRICL) developed by Prof. K. Kawamura of Tokyo Institute of Technology was employed [7]. The MD simulation procedure was similar to our previous study of $ACuO_2$ [8]. We used full ionic two-body potential with Born-Mayer-type short-range repulsion term shown below.

$$u_{ij}=q_iq_j/r_{ij}+f_0(b_i+b_j)\exp[(a_i+a_j+r_{ij})/(b_i+b_j)]\qquad(1)$$

First term is for electrostatic potential between ions i and j separated by r_{ij}, having formal charges of q_i and q_j, respectively. Second term is for repulsive potential and fo is a constant. Repulsion parameters (a_i, b_i) for ions used in this simulation were taken from Kunz and Armbruster [9].

Total number of atoms in a simulated cell was 324 (containing 9 unit cells; 3a, 1b, 3c). Initial internal coordinate for each ion was taken from that of Ca_2AlFeO_5 [1], and was relaxed at 300 K. In order to study the structure at high temperature, the system was heated up to 5000 K with a heating rate of 0.5 K/step (total 104 steps). At temperatures near the transitions, constant temperature MD runs were also performed in order to obtain details of oxygen ion diffusion behavior.

2. Sample preparation and measurements
The sample were prepared by the following procedure. The starting materials were high purity $BaCO_3, In_2O_3$ and others. The mixed powders were pressed into cylindrical pellets at 2 ton/cm^2 and sintered at 1300° C to 1350° C for 10 to 24 hours in air. The specimens obtained were examined by means of X-ray diffractometry. The texture of the specimen was observed with a scanning electron microscope. The ion conductivity of the specimens was evaluated from their complex impedance in the frequency range of 10-10^6 Hz with an impedance analyzer. A platinum electrode was baked onto the surface of the specimens at 1000° C.

RESULTS AND DISCUSSION

1 Molecular dynamic simulation
We performed a 5000-steps simulation at 300 K to relax the initial structure [1] with fixed cell parameters as observed for $Ba_2In_2O_5$ (a=0.6111, b=1.6816 and c=0.5992 nm) [2]. The structure was stable during the simulation and each ion slightly moved from the initial position. We simulated a powder x-ray diffraction pattern using the obtained structure. The pattern matched quite well with that of observed, supporting the MD-derived structure.
In order to study the structure at high temperature, the system was heated up to 5000 K. Fig. 1 shows the reduced cell parameters of corresponding cubic perovskite (i.e., a'=a/$\sqrt{2}$, b'=b/4, and c'=c/$\sqrt{2}$) during heating. At temperatures below 2300 K, the cell parameters linearly increased with increasing temperature.
At around 2300 K, the cell parameters changed drastically, and above the temperature coalescence of a'-and c'-axes was observed (i.e., a tetragonal cell). Above 4600 K b'-axis further coalesced with other axes (i.e., a cubic cell).
In order to observe the oxygen ion diffusion behavior, the mean-square-displacement (msd) of oxygen ions was monitored during the heating. Figure 2 shows the simulated results of the mean-square-displacement of oxygen and the Arrhenius plot of oxygen diffusion coefficient calculated from the msd data. We noted a sudden increase in the msd at the orthorhombic/tetragonal transition, suggesting start of the oxygen ion diffusion at the transition. Goodenough et al. [2] observed a first-order transition to a fast oxide-ion conductor for this compound at 930° C. Although the simulated transition temperature is about 1000 K higher than that of observed, the transition to the phase with oxygen ion diffusion is consistent with the observation. Figure 3 shows DTA and dilatometric analytical results of $Ba_2In_2O_5$ ceramics. We can observe the peaks corresponding to phase change at about 920° C. Also in fig.4 which shows the temperature dependence of ionic conductivity for $Ba_2In_2O_5$ ceramics at temperatures 500° C to 1000° C, a sharp discontinuity in conductivity was observed at around 920° C and this compound undergoes a phase transition from insulator to a good ionic conductor. Although we carried out high temperature X-ray diffraction analysis for $Ba_2In_2O_5$ ceramics, we can not confirm the crystal structure changes from orthorhombic to cubic at this

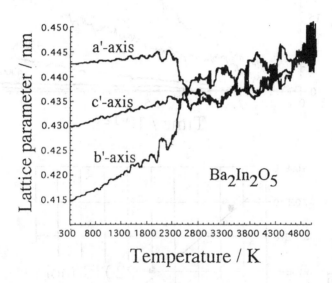

Temperature / K

Fig.1. Reduced cell parameters vs temperature.
Initial orthrhombic cell transform to a tetragona cell above 2300 K, and to a cubic cell above 4600 K.

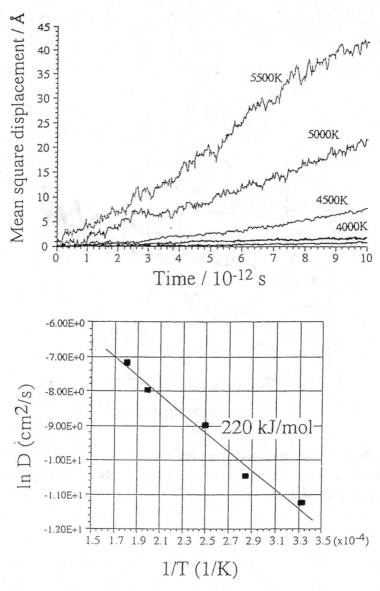

Fig.2.MSD of oxygen vs time at 3000-5500 K(top)
Arrhenius plot of diffusion coefficient for oxygen(bottom)

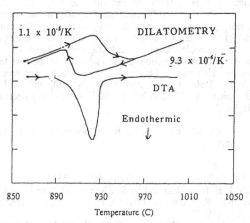

Fig.3.DTA and dilatometric traces for a sintered $Ba_2In_2O_5$

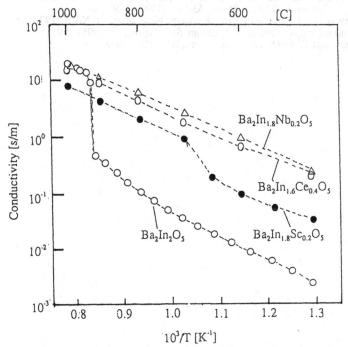

Fig.4.Ionic conductivity of $Ba_2In_{2-x}M_xO_5$ as a function of reciprocal temperature. M:Nb,Sc,Ce.

transition temperature because of the chemical alteration of sample surface. Activation energy obtained from conductivity data is 0.75 eV and this value is very different from the MD simulated value of 2.3 eV. This difference may result in the fact that experimentally determined value is only migration energy of oxygen; but on the contrary the simulated activation energy includes the vacancy creation energy and migration energy.

Since the high temperature cubic phase of $Ba_2In_2O_5$ shows large ion conductivity, it is of interest to examine whether or not the cubic phase stabilizes in the low temperature region by solid solution of another elements. Figure 4 shows the temperature dependence of ion conductivity for $Ba_2In_2O_5$, $BaIn_{1.9}Ce_{0.1}O_y$ and $Ba_2In_{1.8}Nb_{0.2}O_5$ and $Ba_2In_{1.8}Sc_{0.2}O_5$ ceramics. We found that the cubic phase is stabilized below 500 C without any decrease of conductivity in $BaIn_{1.9}Ce_{0.1}O_y$ and $Ba_2In_{1.8}Nb_{0.2}O_5$.

Acknowledgment The authors thank Prof. K. Kawamura for supplying his molecular dynamics simulation programs.

References
[1] A. A. Colville and S. Geller, Acta Cryst., B26 (1971) 2311.
[2] J. B. Goodenough, J. E. Ruiz-Diaz and Y. S. Zhen, Solid State Ionics,44 (1990) 21.
[3] Y. Takeda, N. Imanishi, R. Kanno, T. Mizuno, H. Higuchi, O. Yamamoto and M. Takano, J.Solid State Chem., 63 (1992) 53.
[4] M. Schwartz, B. F. Link and A. F. Sammuells, J. Electrochem. Soc., 140   (1993) L62.
[5] S. Shin, M. Yonemura and H. Ikawa, Mat. Res. Bull., 13   (1978) 1017.
[6] C. R. A. Catlow, Solid State Ionics, 53-56   (1992) 955.
[7] K. Kawamura, Japan Chem. Prog. Exchange Newsletter, 6 No. 4 (1995) 91.
[8] M. Kanzaki and A. Yamaji, A, J. Ceram. Soc. Japan,103 (1995) 529.
[9] M. Kunz and T. Armbruster, Acta Cryst., B48 (1992) 609.

DYNAMICS OF THE PEROVSKITE-BASED HIGH-TEMPERATURE PROTON-CONDUCTING OXIDES: THEORY AND EXPERIMENT

C. KARMONIK[1], T. YILDIRIM[1,2], T.J. UDOVIC[1], J.J. RUSH[1] and R. HEMPELMANN[3]

[1] NIST Center for Neutron Research, National Institute of Standards and Technology, Gaithersburg, Maryland 20899-0001, USA
[2] University of Maryland, College Park, Maryland, 20742, USA
[3] Physikalische Chemie, Universität Saarbrücken, Germany

ABSTRACT

The dynamics of the doped perovskite-based high-temperature protonic conductors (HTPC) were studied by means of neutron vibrational spectroscopy (NVS) and first-principles pseudopotential supercell calculations. Vibrational spectra from hydrogen-charged samples with different rare-earth dopants revealed three well-defined vibrational bands in the energy ranges 20-60, 60-90, and 100-140 meV. The two lowest-energy bands were insensitive to the dopants. First-principles phonon calculations indicate that they are mainly associated with oxygen modes. In contrast, the high-energy band was very sensitive to the dopant, and in this case, calculations indicate that it is associated with OH bending modes.

INTRODUCTION

In recent years, scientific interest in perovskite-based high-temperature protonic conductors (HTPC) has increased strongly because of the promising industrial applications of these materials. [1] For large-scale stationary applications, intense materials research is underway on solid oxide fuel cells (SOFC) operating at temperatures considerably lower than those oxygen-conducting, yttrium-stabilized zirconia as the solid electrolyte. With decreasing temperature, which is one of the main aims of current developments, proton-conducting oxides become competitive. To have a better understanding of the proton dynamics and thus the proton conductivity mechanism, we have performed first-principles supercell phonon calculations in combination with an inelastic neutron scattering study of the vibrational dynamics of various HTPC's. As a model system, we chose the extensively studied , doped strontium cerates $SrCe_{0.95}M_{0.05}H_xO_{3-\delta}$, where M=Nd, Ho, and Sc. Together with the experimental results, preliminary calculations for the similar system $SrZrO_3$ and its Sc-doped derivative are presented, since Zr^{4+} ions are much easier to model theoretically than Ce^{4+} ions. Extension of these calculations to the doped strontium cerate analogs are in progress and will be reported elsewhere.

EXPERIMENTAL DETAILS

The preparation of the ceramic powder and the charging with hydrogen were carried out as described elsewhere [2]. The powder samples (approx. 35 g each, $\sim$ 90 % neutron transmission) were sealed under a helium-containing atmosphere in aluminum cans with a diameter of approximate 25 mm. The hydrogen concentrations were determined using prompt-gamma activation analysis [3] at the cold neutron PGAA spectrometer situated on the Neutron Beam Split-core Reactor (NBSR) at the NIST Center for Neutron Research (NCNR).

Sample	Sc	Ho	Nd	Undoped
mol % H	1.77 ± 0.13	1.70 ± 0.09	1.06 ± 0.08	0.2 ± 0.1
Ionic radii	$r(Sc^{3+}) = 0.732$ Å	$r(Ho^{3+}) = 0.894$ Å	$r(Nd^{3+}) = 0.995$ Å	$r(Ce^{4+}) = 0.920$ Å

Table 1: Hydrogen contents of the different HTPC samples and ionic radii of the dopants.

The PGAA results of the hydrogen content in $SrCe_{0.95}M_{0.05}H_xO_{3-\delta}$ shown in Table 1 revealed dopant-dependent differences. While the Sc- and Ho-doped material contained 1.77 ± 0.13 and 1.70 ± 0.09 mol % hydrogen, respectively (which is typical of these kinds of doped HTPC's), the Nd-doped material incorporated considerably less hydrogen, i.e., 1.07 ± 0.08 mol % under the same charging conditions. This result can be correlated with the different ionic radii of the dopants, i.e., Sc^{3+} (0.732 Å) and Ho^{3+} (0.894 Å) are smaller than Ce^{4+} (0.92 Å) while Nd^{3+} (0.995 Å) is considerably larger [4]. The existence of some H (~ 0.2 mol %) in the undoped sample suggests the presence of protons in these kinds of compounds as intrinsic defects; however, as expected, the amount is considerably smaller than the concentrations achieved by charging the doped materials.

The NVS experiments were carried out at the BT4 spectrometer at the NCNR. All vibrational spectra were measured using the Cu (220) monochromator with either the low-resolution beryllium filter analyzer (with an assumed final energy of 3 meV) or the high-resolution beryllium-graphite-beryllium filter analyzer (with an assumed final energy of 1.2 meV). Collimations of 40'-40' were used before and after the monochromator, respectively, for the low-resolution experiments, and 60'-40' for the high-resolution experiments. During the measurements, temperature regulation was accomplished with a closed-cycle He refrigerator.

FIRST-PRINCIPLES PHONON CALCULATIONS

To achieve a qualitative understanding of the NVS-results for the $SrCeO_3$ system, we apply, for the first time, the first-principles force-constant approach to calculate zone-center (i.e. $q=0$) phonons in $SrZrO_3$ and its Sc-doped derivative. $SrZrO_3$ and a number of other perovskite oxides are isostructural with $GeFeO_3$ representing a structural type with space group *Pbnm*. The crystal-structure information [5] for $SrZrO_3$ is summarized in Table 2.

For the zone-center phonon calculations, we consider a superlattice obtained from the conventional cell given in Table 2 by the following transformation:

$$\mathbf{a} \to (\mathbf{a} + \mathbf{b}) \qquad \mathbf{b} \to (\mathbf{a} - \mathbf{b}) \qquad \mathbf{c} \to \mathbf{c} \ . \tag{1}$$

Hence, the superlattice is $\sqrt{2} \times \sqrt{2} \times 1$ times larger than the primitive cell and contains eight ($SrZrO_3$) units. For the doped system, we replaced one of the Zr atoms in the supercell by (Sc + H), which yields a cell formula $SrZr_{1-x}Sc_xH_xO_3$ with $x = 1/8$. Before calculating the dynamical matrix at the zone center, both the structure and atomic coordinates are fully optimized by the conjugate-gradient method such that the total RMS force on each of the atoms is less then 0.005 eV/Å. In table 2, we give the theoretical values obtained from this structure-optimization calculation for the undoped system which are in good agreement with the experimental values.

Properties	Experiment	Theory
Cell Formula	$Sr_4Zr_4O_{12}$	$Sr_4Zr_4O_{12}$
a (Å)	5.786	5.738
b (Å)	5.815	5.864
c (Å)	8.196	8.179
Sr (1)	(0.003,0.526,0.25)	(0.012, 0.545, 0.25)
Zr (1)	(0,0,0)	(0,0,0)
O (1)	(-0.073,-0.018,0.25)	(-0.105, -0.040, 0.25)
O (2)	(0.217,0.284,0.035)	(0.207, 0.292, 0.056)

Table 2. Various experimental crystal structure information for $SrZrO_3$ (Space Group # 62 P b n m) and the corresponding optimized values obtained from the first-principles pseudopotential total-energy calculation.

The elements of the dynamical matrix, $D(kk')$ at the zone-center were found from the Hellmann-Feynman forces generated when the (k')th ion is displaced a small amount $(u_\beta = 0.05$ Å$)$ from equilibrium along the $\pm\beta$ directions:

$$D_{\alpha\beta}(kk') = -\frac{1}{\sqrt{m(k)m(k')}} \frac{F_\alpha^+(k) - F_\alpha^-(k')}{u_\beta} \qquad (2)$$

where $F_\alpha^\pm(k)$ are the components of the resulting Hellmann-Feynman forces due to a positive/negative displacement. By calculating $D(kk')$ in this way, we eliminate all force constants that depend on odd powers of the atomic displacements; in particular, the linear and the lowest anharmonic cubic terms have been eliminated. When complete, the dynamical matrix $D(kk')$ is then diagonalized to find the zone-center phonons.

Our calculations of the total energy and the forces have been performed using the first-principles code CASTEP [6], and the results have been obtained within the local-density functional approximation using a pseudopotential approach. The electronic wave functions were represented as plane waves with a cut-off energy of 800 meV. To describe the ion-electron interaction, the optimized norm-conserving pseudopotentials have been used. Brillouin-zone integrations have been carried out using only the Γ point, which is sufficient due to the large size of our superlattice.

RESULTS AND DISCUSSION

In the NVS experiments, we measured high-resolution vibrational spectra at low temperature (17 K) for all $SrCeO_3$-samples. The results are summarized in the left panel of Fig. 1. All spectra can be roughly divided into three regions: a low-energy-transfer region I (20 - 60 meV) dominated by complex, high-intensity modes; a medium-energy-transfer region II, (60 - 90 meV) which exhibits similar scattering features for all four samples; and a high-energy-transfer region III (90 - 140 meV) where considerable differences in the spectra exist. As the samples in the present study differ only by the kind of dopant cation, it is clear that any observed differences are attributable to the influence of that cation. Furthermore, as the neutron scattering cross sections for all the atoms in these compounds are low compared to hydrogen, the observed scattering features above 100 meV are mainly reflective of the vibrational modes involving large-amplitude hydrogen motions.

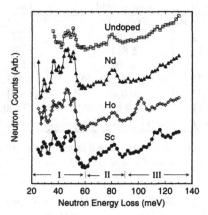

Figure 1: High-resolution vibrational spectra for the different doped samples at 17 K.

The H features near 112 meV and 102 meV for the Sc-doped and Ho-doped HTPC's, respectively, suggest a dopant-related site for the hydrogen in these materials. In contrast, the absence of this feature in region III for the Nd-doped material suggests the lack of any additional Nd-related hydrogen. These results are consistent with the relatively lower H concentration for the Nd-doped sample as obtained from PGAA (see Table 1).

This behavior is consistent with the results of IR investigations of $SrZrO_3$ compounds doped with different cations [7]. These authors also found a dependence of the hydrogen concentration on the dopant radius by investigating the OH stretching modes near 400 meV. The intensity of the dopant-related modes dropped approximately fivefold for dopant radii exceeding the radius of Zr^{4+}.

A comparison of the spectra for the Sc- and Ho-doped materials suggests that a smaller dopant cation yields a higher-energy vibrational feature in region III. We are currently extending our first-principles total-energy calculations to help us understand the relationship between dopant size and the proton-dopant interaction [8].

The equilibrium geometries and neutron vibrational spectra obtained from the zone center phonon calculations for undoped $SrZrO_3$ and Sc-doped $SrZr_{1-x}Sc_xH_xO_3$ (x=1/8) are shown in Fig. 2. We first note that even though resulting structures are different for doped and undoped ones, the vibrational spectra are quite similar. Comparing Fig. 2 with the experimental spectra in Fig. 1, we identify the modes in regions I and II as mainly due to oxygen motions. This identification is also consistent with the fact that these modes are very similar for all four samples. For example the modes in the energy range 20–30 meV are the modes in which whole ZrO_6-octahedra (or $HScO_6$-octahedra) move together with respect to the surrounding ions in the solid. As shown in the inset to the right panel of Fig. 2, introducing H into the O_6-octahedron distorts the geometry significantly. In particular, due to Coulomb interactions, the Sc atom moves away from the proton while the O ions move towards the proton. In the equilibrium configuration, we find that the H forms an OH bond with one of the oxygen ions, yielding a bond length of about 1.06 Å. The O–H–O is 159.7° and the distance between H and the other O ion is 1.403 Å. As expected, the OH stretching mode is the highest-energy mode (greater than 300 meV), whereas the OH bending modes are around 150-160 meV.

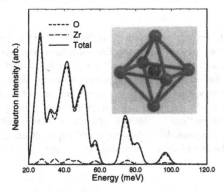 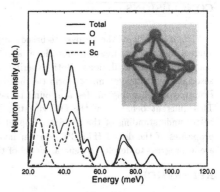

Figure 2: Left: Vibrational spectrum of the Zr-O_6 cluster shown in the inset. **Right:** Vibrational spectrum of the Sr-H-O_6 cluster shown in the inset.

In Fig. 3 we compare the experimental and calculated spectra. We assume that the total vibrational spectrum for a sample with $x < 1/8$ (as in the real systems) is a superposition of the vibrational spectrum of the undoped $SrZrO_3$ (left panel in Fig. 2) and the doped supercell structure of $SrZr_{1-x}Sc_xH_xO_3$ (x=1/8, right panel in Fig. 2) with appropriate weighting factors. However, as the experimental system, $SrCe_{1-x}Sc_xH_xO_3$, is different than the system used in the calculations we do not expect full quantitative agreement. Nevertheless, the similarities between the two spectra for the oxygen modes are quite noticeable. For the hydrogen modes, our calculations predict that the OH bending modes are much stiffer in $SrZr_{1-x}Sc_xH_xO_3$ than in $SrCe_{1-x}Sc_xH_xO_3$. Currently we are preparing $SrZr_{1-x}Sc_xH_xO_3$ samples to verify this.

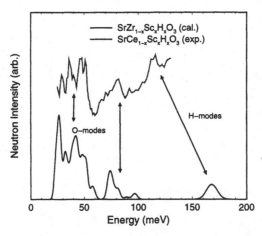

Figure 3: Comparison of the neutron vibrational spectrum of $SrZr_{1-x}Sc_xH_xO_3$ (calculated) and $SrCe_{1-x}Sc_xH_xO_3$ (measured). Even though the experimental system is different than the one used in the calculations, the vibrational modes associated with the oxygen atoms are remarkable similar. However, the proton dynamics are quite sensitive to the structure and dopant.

CONCLUSIONS

The dynamics of the perovskite-based, high-temperature protonic conductors (HTPC) were studied by means of neutron vibrational spectroscopy combined with first-principles pseudopotential supercell calculations. Both experiment and theory indicate that the vibrational spectra associated with oxygen modes form two bands below 90 meV and are insensitive to the dopant. However, OH bending modes are found to be very sensitive to both dopant and temperature. This different behavior of the OH vibrational modes emphasizes the importance of gaining a better understanding of the hydrogen-dopant interactions in order to optimize the conductivity properties of the doped HTPC's. Further investigations, both theoretical and experimental [8], are in progress to develop a general model of the proton dynamics in these compounds.

REFERENCES

1. P. Colomban (Ed.), Proton Conductors, Cambridge University Press (1992).
2. T. Matzke, U. Stimming, C. Karmonik, M. Soetratmo, R. Hempelmann, and F. Guethoff, Solid State Ionics **86-88**, p. 621 (1996).
3. R. L. Paul, Analyst **122**, 35R (1997).
4. R. C. Weast, M. J. Astle (Ed.), Handbook of Chemistry and Physics, 62nd Edition, CRC Press, Inc., Boca Raton, Florida (1981-82).
5. S. Geller and E. A. Wood, Acta Crystallogr. **9**, p. 563 (1996).
6. M. C. Payne, M. P. Teter, D. C. Arias, and J. D. Joannopoulos, Rev. Mod. Phys. **64**, p. 1045 (1992).
7. H. Yugami, Y. Shibayama, S. Matsu, M. Ishigame, S. Shin, Solid State Ionics **85**, p.319 (1996).
8. T. Y. Yildirim et al., to be published.

NEW INTERCONNECTIONS FOR PLANAR ALLOY-SEPARATOR SOFC STACKS

Y.-T. YAMAZAKI, T. NAMIKAWA, T. IDE, N. OISHI, O. SUZUKI
Department of Innovative and Engineered Materials, Tokyo Institute of Technology,
Nagatsuta, Midori-ku, Yokohama, 226 Japan, yamazaki@iem.titech.ac.jp

ABSTRACT

A new design of interconnections between air electrodes and alloy separators in planar stacks is presented. The edges of the air electrodes supporting the electrolyte films are connected, through thin $LaCrO_3$ layers, with the edges of adjacent alloy separators. The electric current in the cell flows radially in the porous thick air electrodes of LSM. The $LaCrO_3$ layers are connected with the alloy separators through nickel felt in the fuel gas atmosphere. By using these interconnections, we can overcome the corrosion problems caused by contacting air electrodes with alloy separators in conventional planar stacks.

INTRODUCTION

High temperature fuel cells utilizing solid oxide electrolytes are collecting increased attention because of (1) their ability to utilize hydrocarbon fuels, (2) potential for long term operations due to the absence of liquids, and (3) higher current densities thereby higher power densities. On the other hand, the insufficient durability for thermal cycles limits the expansion of their applications. The durability of planar SOFC stacks against thermal cycles increases by utilizing alloy separators with sliding seals and flexible interconnections [1-4]. The use of alloy separators, however, brought about the corrosion problems which lowered the quality of the electric connections between the air electrodes and separators. Farthermore corrosion proceeds at the interfaces between LSM cathodes and the surfaces of alloy separators. In this study, we present a new design of interconnections between air electrodes and alloy separators in planar SOFC stacks, which solves these problems based on the direct contact of the LSM cathodes with the alloy separators in oxidative atmosphere.

INTERCONNECTIONS THROUGH EDGES OF PLANAR CELLS

Current flow in a thick air electrode

The concept of the air electrode supported (AES) cell, which was developed for tubular cells in Westinghouse, can be extended to planar cells. When we stack AES planar cells, currents in the air electrodes flow radially toward the edges of the cells. Therefore each air electrode can be connected to the adjacent alloy separator in the outside of the air compartment, that is, the air electrodes and alloy separators can be connected in the reducing atmosphere of fuel gas. We call this planar stack "CE (cell edge) stack" in this paper. Figure 1 shows the current path in a cross flow CE stack. The two sides of a CE stack exposed to air and fuel gas are illustrated in Fig.2. The figure indicates that the LSM supports (air electrodes) are connected with separators in the reducing atmosphere of fuel gas. Figures 3 and 4 show cross sectional views of a CE stack. The fringes of the LSM supports are coated with LSC $(La(Sr)CrO_3)$. Figure 3 illustrates currents passing through the nickel felts which connect the LSC layers with the alloy separators. The LSC coating protects the LSM support from reducing atomspher of fuel gas. Ceramic sheets must be inserted to isolate the air electrodes from adjacent separators in the fuel sides of the cells as shown in Fig.3. Sliding seals and flexible interconnections are necessary in a planar stack, so as to release thermal stresses . The interfaces of the sliding seals between the alloy and ceramics, and the flexible interconnections using nickel felts are indicated in Fig.3 and 4, respectively.

Mat. Res. Soc. Symp. Proc. Vol. 496 © 1998 Materials Research Society

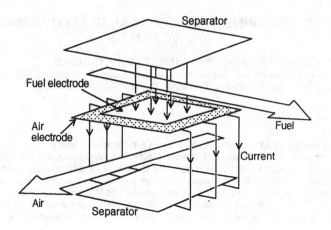

Fig. 1 Conceptual drawing of a CE stack. The air electrode is connected with the separator edge, in the outside of the air compartment of the cell.

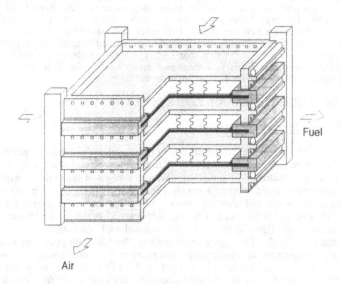

Fig. 2 Schematic drawing of square planar AES cells which are stacked with the proposed interconnections.

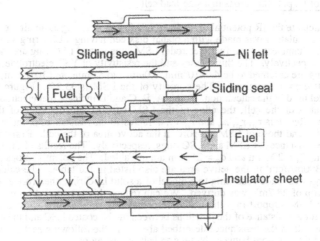

Fig. 3   Schematic diagram of a CE stack showing the current in air electrodes, Ni - felt interconnections and alloy separators.

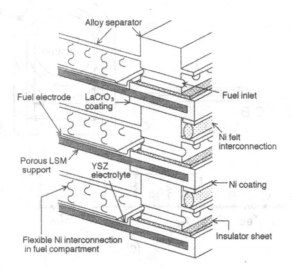

Fig. 4   Schematic diagram showing sliding seals and Ni - felt flexible interconnections in a CE stack.

Estimation of IR potential drops in a side-lead cell

We calculate the IR potential drops in an LSM support acting as an air electrode in a CE stack. The calculation was made with a planar disk cell having a LSC ring shown in Figs.5a and 5b. The diameters of the fuel electrode, LSM support and LSC ring are indicated as d_A, d_M and d_C, respectively. The thicknesses and the resistivity of YSZ electrolyte, LSM support and LSC ring are assumed to be 4.0×10^4mm, 3.0mm and 5.0mm, and 10.0Ωcm, $5.0 \times 10^{-3}\Omega$cm and $2.5 \times 10^2 \Omega$cm, respectively. The porosity of the LSM is 33.3%. Figure 5c shows the circuit model used in the calculation. Points A, B, C and D indicate fuel electrode, the edge of the active area of the cell, the edge of the LSM support and the edge of the LSC ring, respectively. Resistors r_{p1}, r_{p2}, ..., and r_{s1}, r_{s2}, ... represent the resistances of the small segments of electrolyte, and those of LSM support, in the active area of the cell. Resistors r_{BC} and r_{CD} indicate the resistances of LSM and LSC rings, respectively. The results of the calculation for $d_A = 192$mm, $d_M = 200$mm and $d_C = 210$mm are in Table I. The potential drops for the current density of 300mA/cm^2 in the active area are also listed in the table. The calculation of the resistance r_{CD} was made according to a model of a solid LSC ring, therefore a relatively high potential drop of 28.7mV was derived. We can reduce this potential drop by extending the edge of the LSM support into the LSC interconnecting ring, as it is shown in Fig.3. It is assumed that the resistance of the junctions between the Ni coated LSC and the alloy separator is much lower than the resistances described above. In the following section, we reduce the resistance of the junctions between Ni felts and alloy separators.

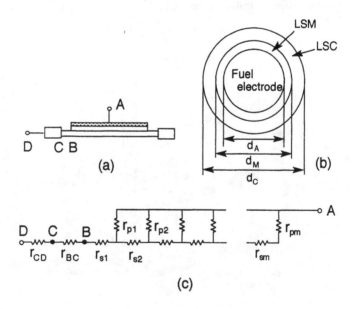

Fig. 5   A disk cell with a LSC connecting ring, used in the calculation of IR losses in the proposed cell stack; (a) side view, (b) top view, (c) equivalent circuit.

Table I Resistances calculated for the segments in the model shown in Fig. 5, and the potential drops for the current density of 300mA/cm^2 in the active area of the cell.

	Resistance (Ω)	IR drop for 300mA/cm^2 (mV)
r_{AB}	1.38 x 10^{-4}	12.0
r_{BC}	1.62 x 10^{-4}	14.1
r_{CD}	3.3 x 10^{-4}	28.7

REDUCTION IN THE RESISTANCE OF THE JUNCTIONS BETWEEN HEAT RESISTANT ALLOYS AND NICKEL FELT

Reduction in the resistance by Ni coating

Heat resistant alloys form passive oxide films when they are heated to high temperatures. Typically, these films are Cr$_2$O$_3$ and Al$_2$O$_3$. The Cr and Al atoms are contained in the alloys, and they precipitate on the surfaces of alloys at high temperatures and then form impermeable and insulating oxide films. When the atmosphere contains water vapor, the passive films are formed even in a fuel gas. We have developed a method to control the development of passive oxide films in fuel gases by coating the surfaces of alloys with thin Ni layers before the alloys are heated.

Figure 6 schematically shows the test piece used for the resistance measurement at 900°C. A square sample of 20 x 20 mm was cut out from an HA214 alloy plate with a thickness of 2 mm (Mitsubishi materials Co.). The both sides of the plate were polished with a diamond paste, and Ni was evaporated using an electron beam evaporation in a vacuum of 6.0x10^{3}Pa. The substrate temperature was 600°C. The deposition rate and the thickness of the Ni layer were 0.5 μm/min and 4 μm, respectively. The sample was sandwiched with Ni felt pieces and pressed with alumina plates. Pt wires are connected to the Ni felts and then the sample was heated in hydrogen gas containing 3% H$_2$O. The heat treatment was continued for 200 h at 900°C. The resistivity change was measured during the heat treatment. The resistivity change was also measured with samples which were not coated with Ni.

Results and discussion

The resistance changes measured with the Ni coated and uncoated samples are shown in Fig.7. The low resistance of the Ni coated sample demonstrates that the formation of the passive oxide films was suppressed by the Ni coating. This result suggests that the Ni layer on the surface of the alloy retards the accumuration of the passive atoms such as Cr or Al on the surface of the alloy. When the uncoated alloy surface came into contact with Ni felt and was heated, the formation of the oxide film was suppressed only at the contact points between the alloy surface and the fine fibers of the Ni felt, and consequently, the progress of Ni felt sintering with the alloy was blocked by the oxide layer on the surface of the alloy. The oxide layer which was developed in the same heat treatment condition, showed a high resistance of 3kΩ at 900°C.

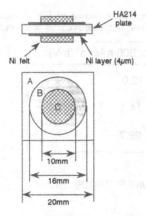

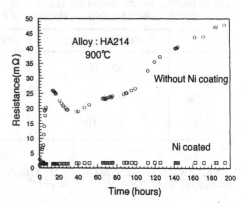

Fig. 6 Cross section and top view of the sample used for the measurement of the resistance in the junctions between Ni coated HA214 alloy and Ni felts.

Fig. 7 The resistances of the junctions between the HA214 alloy and Ni Ni felt at 900°C in H_2 (3% H_2O).

CONCLUSIONS

A new design of interconnections between air electrodes and alloy separators in planar SOFC stacks is developed. It solves the corrosion problems caused by the direct contacts between LSM air electrodes and alloy separators at high temperatures in oxidizing atmosphere. A concrete example of the new interconnection in a planar stack was presented. The calculation of the ohmic losses on the proposed stack design proved that the losses in the stack were not substantial. An effective method to make good contacts between alloy separators and Ni felt, which was essential to materialize the new stack, was presented.

ACKNOWLEDGMENT

A part of this work carried out as a research project of The Japan Petroleum Institute commissioned by the Petroleum Energy Center with the support of the Ministry of International Trade and Industry.

REFERENCES

1. S. C. Singhal, "Advances in Tubular Solid Oxide Fuel Cell Technology", in *Proc. 4th Int. Symp. on SOFC*, (M. Dokiya, O. Yamamoto, H. Tagawa, S. C. Singhal, eds.), The Electrochemical Society, Inc., Pennington, NJ, Vol. 95-1, pp.195-207(1996).
2. Y. Yamazaki, T. Namikawa, N. Oishi, and T. Yamazaki, "Relaxation of Thermal Stresses in Planar SOFC Stacks", *ibid.*, pp.236-244.
3. T. Yamazaki, N. Oishi, T. Namikawa and Y. Yamazaki, "Frictional Forces in an SOFC Stack with Sliding Seals", *Denki Kagaku*, Vol. 64, pp.634-637(1996).
4. N. Oishi, T. Namikawa and Y. Yamazaki, "Thermal Cycle Tests of a Planar SOFC Stack with Flexible Interconnections and Sliding Seals", *Denki Kagaku*, Vol. 64, pp.620-623(1996).

Development of Electrolyte plate for Molten Carbonate Fuel Cell

C. Shoji*, T. Matsuo*, A. Suzuki*, Y. Yamamasu*
*Research Institute, Ishikawajima-Harima Heavy Industries Co., Ltd., 3-1-15
Toyosu, Kotou-Ku, Tokyo, Japan

ABSTRACT

It is important for the commercialization of molten carbonate fuel cell (MCFC) to improve the endurance and the reliability of the electrolyte plate. The electrolyte-loss in the electrolyte plate increases the cell resistance and deteriorates the cell voltage. The formation of cracks in the electrolyte plate causes a gas cross leakage between the fuel gas and the oxidizer gas. The pore structure of electrolyte plate must be stable and fine to support liquid electrolyte under MCFC operation. It is necessary to prevent the formation of cracks in electrolyte plate during thermal cycling. We have improved the stability of electrolyte plate using advanced $LiAlO_2$ powder and improved the durability of electrolyte plate for thermal cycling by the addition of the ceramic fiber .

The initial cell voltage using electrolyte plate with advanced $LiAlO_2$ powder was 820mV at current density $150mA/cm^2$ and the decay rate of cell voltage was under 0.5%/1000h for 8,800h. According to the post analyses, the pore structure of the electrolyte plate did not change. The stability of advanced $LiAlO_2$ powder was confirmed. It was proved that the electrolyte plate reinforced with ceramic fiber is effective for thermal cycling.

INTRODUCTION

A fuel cell is expected to become a new electric power-generation device in near future. Because it has high efficiency to convert the chemical energy of the fuel gas directly into the electrical energy and has low emission to minimize environmental pollution. Especially MCFC is capable of large capacity and has a merit of utilizing a variety of fuels such as abundant coal. In order to commercialize MCFC, the goal of the lifetime of MCFC is more than 40,000h at point of profit, the initial cell voltage is higher than 0.8 V and the decay rate is less than 0.25%/1000h (as for the initial cell voltage). The most important item is to accomplish the decay rate in our development program .

MCFC is composed of an anode, a cathode, an electrolyte plate sandwiched between them and a separator. An anode, a cathode and an electrolyte plate are porous media and their pores are filled with molten carbonate depending on a balance in capillary pressures. Especially, the electrolyte plate plays an important role to transport ions and prevent the gas cross-leakage between anode side and cathode side. Therefore, it is very important to stabilize the structure of the electrolyte plate. Electrolyte impregnated in the pore of the electrolyte plate can carry ions and prevent the gas leakage.

Recently the particle growth of the commercial $LiAlO_2$ powder, which is a material for matrix of the electrolyte plate, was reported[1,2]. The stability of $LiAlO_2$ in MCFC seems to be unreliable for 40,000h. Many researchers say that particle growth of $LiAlO_2$ is due to a dissolution-reprecipitation mechanism in molten carbonate[1-3]. Namely $LiAlO_2$ forming the smaller particles tend to dissolve and precipitate onto larger particles. The coarsening of particle deteriorates electrolyte-retention capability. P.A. Finn reported that γ-$LiAlO_2$ was the most stable phase in molten carbonate[3]. However phase transformation from γ-phase to α-phase in molten carbonate was recently reported. The specific gravity of γ-phase is different from α-phase. Thus the pore structure of electrolyte plate may change. K. Nakagawa et.al. have insisted that the use of α-phase for electrolyte plate is more reasonable than that of γ-phase under MCFC operation[4]. Therefore we attempted to develop more stable $LiAlO_2$ powder in MCFC with Nippon Chemical Industrial Co., Ltd. Our objective is to fabricate the advanced γ-$LiAlO_2$ powder which is uniform in particle size and in shape, and which is low content of α-$LiAlO_2$. In this report we

211

discuss the evaluation result of the advanced powder, the commercial γ-LiAlO$_2$ powder, and α-LiAlO$_2$ under MCFC operation.

On the other hand, the formation of cracks in electrolyte plate can cause gas cross leakage between fuel gas and oxidizer gas. The drop of open circuit voltage is caused by the gas cross leakage and deteriorates cell performance. In this paper the effects on the electrolyte plate reinforcement with ceramic fiber was evaluated.

EXPERIMENTAL

Electrolyte plates of various types of LiAlO$_2$ powder were analyzed after cell test. HSA10 and LSA15 (Cyprus Foote Mineral Co., Ltd.) are commercial γ-LiAlO$_2$ powder. TYPE2 (Nippon Chemical Industrial Co., Ltd.) is an advanced γ-LiAlO$_2$ powder. As for the reinforce material, the high purity α-Al$_2$O$_3$ fiber (Mitsui Mining Material Co., Ltd.) was used and the rod-shaped γ-LiAlO$_2$ (made in our laboratory) was used[5].

Fig.1 shows the process of the fabrication of electrolyte plate. Either LiAlO$_2$ powder or the mixture of LiAlO$_2$ powder and ceramic fiber is dispersed in the organic solvent which contains dispersant, binder, plasticizer to form

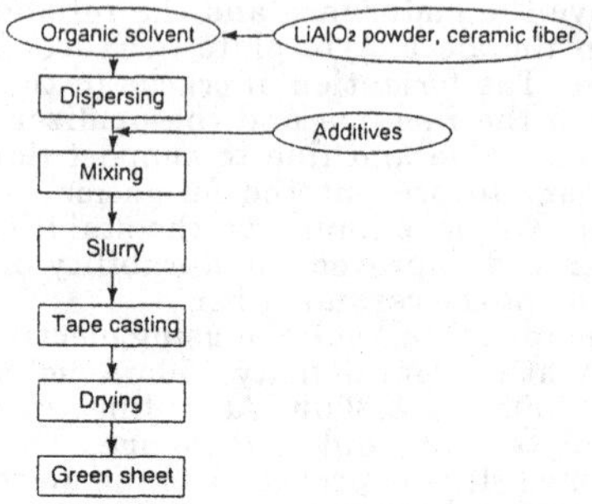

Fig.1 Process of fabrication of electrolyte plate

slurry. The viscosity of the slurry is adjusted within a suitable range. Then it is cast into tape using the doctor-blade method.

Many tests on a single cell with the electrode area of 100 cm^2 were carried out. Each cell was operated at 650 °C, 1 atm and the current density was 150 mA/cm^2. The fuel gas utilization was 75% and the oxidizer gas utilization was 50%. The composition of gasses was H$_2$/CO$_2$/H$_2$O=64/16/20 (Fuel) and Air/CO$_2$=70/30 (Oxidizer).

The electrolyte plate after cell test was immersed in the 1:1 solution of acetic acid and anhydrous acetic acid to remove carbonate. Then the sample was cleaned with ethanol and dried at about 100 °C.

The surface area, the pore size distribution, the particle shape, and the crystalline form of LiAlO$_2$ after removing carbonate were analyzed by BET type surface area measuring unit, Mercury porosimetry, scanning electron microscope (SEM), and powder X-ray diffraction, respectively.

Many thermal cycling tests on a single cell with electrode area of 100 cm^2 were carried out. The temperature was 650°C during cell operation. Thermal cycling test was conducted between 650°C and below 350°C. The gas cross-leakage was determined by measuring nitrogen gas concentration in the anode outlet at a pressure difference of 100 mm H$_2$O.

RESULTS AND DISCUSSION

Long term operation

Figure2 shows the cell performance using the advanced LiAlO$_2$ powder (TYPE2). Its initial cell voltage was about 820 mV at current density 150 mA/cm^2. The steadiness of the open circuit voltage (OCV) suggests that the cracks were not significant in the electrolyte plate during cell operation. The cell voltage at the current density of 150 mA/cm^2 and the internal resistance (IR)-loss were very steady, the decay rate was under 0.5%/1000h. This cell performance suggests that the electrolyte-retention capability of carbonate of the electrolyte plate was kept. Fig. 3 shows the voltage of the cells using the electrolyte plate made of TYPE2, HSA10 and α-LiAlO$_2$ respectively at the current density of 150 mA/cm^2. The voltage of the cell using the electrolyte plate made of α-LiAlO$_2$ remarkably deteriorated, the decay rate was over 2%. The decay was mainly caused by the increase of internal resistance-loss. The voltage of a cell using the electrolyte plate made of HSA10 slightly decreased at the rate of 1.2%/1000h.

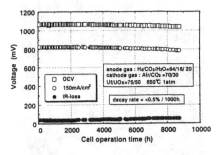

Fig.2 Cell performance using electrolyte plate of TYPE2

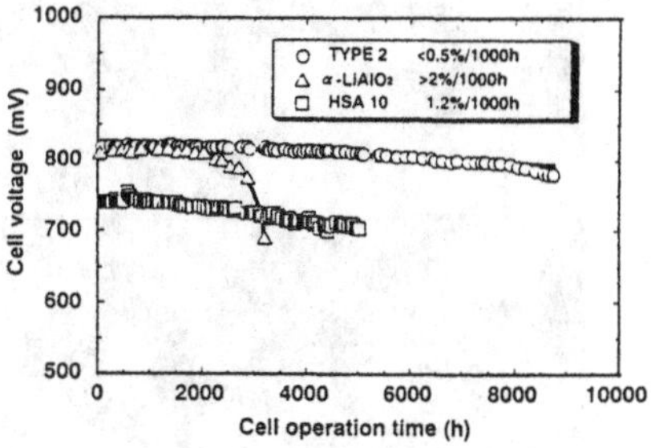

Fig.3 Change of cell voltage at current density of 150 mA/cm^2

Post-test analyses

Table 1 shows the change of surface area of LiAlO$_2$ powder after cell test. The decrease in surface area indicates the deterioration of electrolyte-retention capability. The surface area of the advanced LiAlO$_2$ (TYPE2) did not change much within 8,800h. The surface area of HSA10 decreased by 87% in comparing with pre-test. On the other hand, the surface area of α-LiAlO$_2$ decreased by 36% in comparing with pre-test.

SEM photographs of before and after cell test are shown in Fig. 4. According to SEM photograph before cell test, the particle of the advanced LiAlO$_2$ powder (TYPE2) is uniform in size and its surface is smooth. In the case of HSA10, there are some large particles among fine particles before cell test. The particle surface of α-LiAlO$_2$ is rough and remarkably irregular. In the case of HSA10 and α-LiAlO$_2$, there were clear differences between before and after cell test. On the other hand, the shape and size of the advanced LiAlO$_2$ powder (TYPE2) did not change much. These results suggest that the particle growth does not simply depend on the crystal structure. The physical properties such as particle size distribution, shape might be more important than crystal structure in the particle growth.

The ratio of MPS (Mean Pore Size) of these electrolyte plates before and after cell test is shown in Fig. 5. The coarsening of mean pore size (MPS) is considered to indicate the deterioration of electrolyte-retention capability. The pore structure of the electrolyte plate made of α-LiAlO$_2$ coarsened with the particle growth. In the case of HSA10, the pore structure slightly coarsened. On the other hand, the pore structure of the electrolyte plate of the advanced LiAlO$_2$ powder (TYPE2) was stable using both Li/K carbonate and Li/Na carbonate.

Many researchers reported that the phase transformation from γ-LiAlO$_2$ to α-LiAlO$_2$ occurred under MCFC operation. It was reported that initial content of α-LiAlO$_2$ affects the rate of phase transformation[2]. Fig. 6 shows the change of content of α-LiAlO$_2$ after cell test. In cases of both HSA10 and the advanced LiAlO$_2$ powder (TYPE2), content of α-LiAlO$_2$ increased in spite of the initial low content. All

Table.1 Change of surface area of LiAlO$_2$ after cell operating

	Operating time (h)	Change ratio of surface area (%,Post-test/Pre-test)
TYPE 2	8800	104
HSA 10	5000	87
α-LiAlO$_2$	3200	36

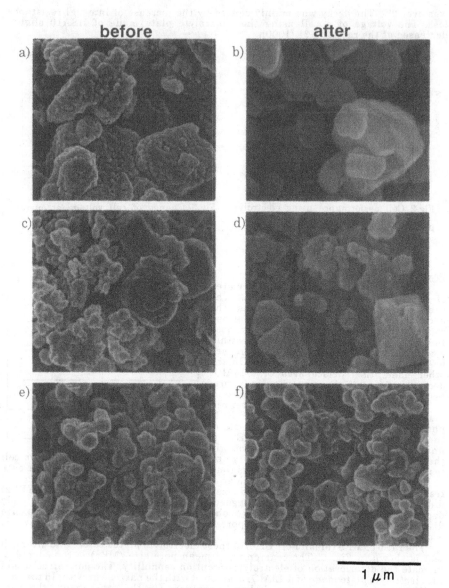

before   after

a)   b)

c)   d)

e)   f)

$1 \mu m$

Fig.4 SEM photographs of before and after cell test
a) α-LiAlO$_2$ 0h b) α-LiAlO$_2$ 3200h
c) HSA10 0h d) HSA10 5000h
e) TYPE2 0h f) TYPE2 8800h

electrolyte plates used in those cell tests were fabricated by ball milling using α-Al₂O₃ balls. The small amount of α-Al₂O₃ worn from the balls mixed into the electrolyte plates. There is a possibility that the α-Al₂O₃ would promote the transformation from γ-LiAlO₂ to α-LiAlO₂. The difference of the content of α-LiAlO₂ between HSA10 and the advanced LiAlO₂ (TYPE2) after cell tests might be due to the initial content of α-LiAlO₂.

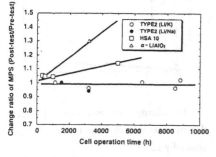

Fig.5 Change of mean pore size of electrolyte plate

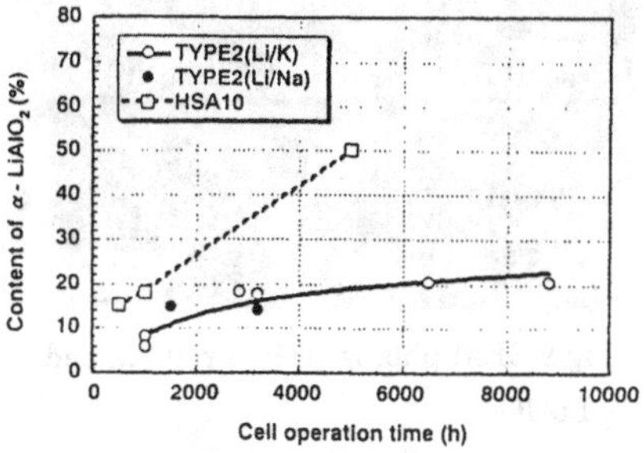

Fig.6 Change of content of α-LiAlO₂

Fig. 7 shows SEM photograph of α-Al₂O₃ fiber. This diameter is about 10μm. Fig. 8 shows the change of the gas cross leakage during thermal cycling test using the electrolyte plate reinforced with and without α-Al₂O₃ fibers. For the electrolyte without Al₂O₃ fibers, the gas cross leakage increased remarkably after three times. In the case of the electrolyte plate reinforced with Al₂O₃ fiber, the gas cross leakage did not increase over 5%. However Al₂O₃ fibers tend to react with Li₂CO₃ and give rise to change electrolyte composition. Therefore, we are developing the rod-shaped γ-LiAlO₂. Fig. 9 shows SEM photograph of the rod-shaped γ-LiAlO₂[5]. Fig. 10 shows the change of the gas cross leakage during thermal cycling test using the electrolyte plate reinforced with rod-shaped γ-LiAlO₂. The addition of rod-shaped γ-LiAlO₂ was effective to prevent the formation of crack in the electrolyte plate.

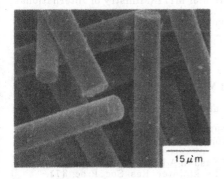

Fig.7 SEM photograph of Al₂O₃ fiber

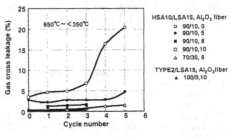

Fig.8 Change of gas cross leakage during thermal cycling test with Al₂O₃ fiber

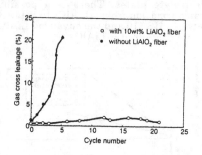

Fig.9 SEM photograph of rod-shaped γ-LiAlO₂

Fig.10 Change of gas cross leakage during thermal cycling test with rod-shaped γ-LiAlO₂.

SUMMARY

1. The decay rate of cell voltage using electrolyte plate made of the advanced LiAlO₂ powder (TYPE2) was under 0.5%/1000h for 8,800h.
2. The advanced powder (TYPE2) was more stable than the commercial LiAlO₂ powder under MCFC operation.
3. The stability of the electrolyte plate depends on the uniformity of LiAlO₂ in particle size and in shape.
4. The electrolyte plate reinforced with ceramic fiber had durability for thermal cycling.

ACKNOWLEDGMENTS

This work has been conducted under a contract from NEDO (New Energy and Industrial Technology Development Organization) and MCFC Research Association (Technology Research Association for Molten Carbonate Fuel Cell Power Generation System) as a part of the New sunshine Program of MITI (Ministry of International Trade and Industry). We appreciate their advice and support.

We would like to appreciate Nippon Chemical Industrial Co., Ltd. because they have managed to fabricate the advanced LiAlO₂ powder corresponding to our requests.

REFERENCES

1. H.Sotouchi, Y.Watanabe, T.Kobayashi and M.Murai, J.Electrochem.Soc., **139**,1127(1992)
2. K.Hatoh, J.Niikura, N.Taniguchi, T.Gamo and T.Iwaki, Denki Kagaku, **57**, 728(1989)
3. P.A..Finn, J.Electrochem.Soc.,**127**, 236(1980)
4. K.Nakagawa, H.Ohzu, Y.Akasaka and N.Tomimatsu, Denki Kagaku, **65**, 231(1997)
5. K.Suzuki, T.Kakihara, Y.Yamamasu and T.Sasa in Advances in Porous Materials, edited by S.Komarneri, D.M.Smith and J.S.Beck(Mater. Res. Soc. Proc. **371**, Pittsburph, 1994) pp.297-302

NOVEL PROTON EXCHANGE MEMBRANE FOR HIGH TEMPERATURE FUEL CELLS

M. BHAMIDIPATI, E. LAZARO, F. LYONS, R.S. MORRIS, Cape Cod Research, Inc., 19 Research Road, E. Falmouth, MA 02536

ABSTRACT

This research effort sought to demonstrate that combining select phosphonic acid additives with Nafion could improve Nafion's high temperature electrochemical performance. A 1:1 mixture of the additive with Nafion, resulted in a film that demonstrated 30% higher conductivity than a phosphoric acid equilibrated Nafion control at 175^0C. This improvement to the high temperature conductivity of the proton exchange membrane Nafion is without precedent. In addition, thermal analysis data of the test films suggested that the additives did not compromise the thermal stability of Nafion. The results suggest that the improved Nafion proton exchange membranes could offer superior electrochemical performance, but would retain the same degree of thermal stability as Nafion. This research could eventually lead to portable fuel cells that could oxidize unrefined hydrocarbon fuels, resulting in wider proliferation of fuel cells for portable power.

INTRODUCTION

Adams[1] has identified six criteria that define the "ideal" hydrocarbon/air fuel cell electrolyte. The "perfect" electrolyte would:

1) be a good medium for the electrochemical reaction of hydrocarbons with air.

2) facilitate good ionic charge transport.

3) be a suitable medium for good diffusion of reactants to catalytic sites and removal of reaction products from those sites.

4) be physically, chemically and electrochemically stable over the operating temperature and potential range.

5) not interfere with catalytic reactions.

6) not react with fuel cell structural materials.

Conventional PEM meet all but two of these requirements. Criteria 1 and 4 are the primary areas where improvement in PEMs needs to be attained to realize the full developmental potential of PEM fuel cells. Modern PEM fuel cells operate at high rate on hydrogen fuel, but only in a very narrow temperature range set by the delicate balance between water production and water removal. Any imbalance between electroformation of water and evaporation can result in either electrode flooding or membrane dehydration. The problem is made worse by the need to diffuse reactants to reaction sites in the presence of water vapor. As the temperature rises, the need to prevent dehydration of the membrane tends to reduce the partial pressures of the reactants. This factor leads to designs requiring cells which operate at several atmospheres

217

total pressure at low load levels. While humidification and pressure operation solve the problem for hydrogen operation, they do nothing for cells operating on hydrocarbon fuel. Attempts to substitute even simple hydrocarbons for hydrogen have, to date, failed largely because efficient hydrocarbon electrooxidation will probably always involve operation at temperatures above 150^0C. All state of the art PEM dry out rapidly when heated above 150^0C. The result is local cell failure. This reduces the rate of water formation which only makes the situation worse. PEMs are very difficult to rehydrate once dehydrated.

This research sought to explore the feasibility of raising the useful operating temperatures of PEMs above 100^0C through changes in the chemistry of the polymer. The significance of the research lies in the possibility that not only would the system constraints (pressure levels and humidification requirements) be relaxed on hydrogen/oxygen systems, but it is also likely that improvements in electrooxidation of simple hydrocarbons would result.

Recently, Yeager et al[2] equilibrated Nafion 117 with ortho-H_3PO_4 in a successful attempt to demonstrate improvement in the high temperature operating characteristics of the PEM Nafion. This study's findings echoed those of others in that H_3PO_4 acted like a Brönsted base in conjunction with the sulfonic acid groups and thereby improved the dissociation of the sulfonic acid as demonstrated below:

$$-CF_3SO_3H + H_3PO_4 \longrightarrow H_4PO_4^+ + -CF_3SO_3^-$$

This effect together with what these authors found to be better water retention by the phosphoric acid-equilibrated Nafion probably explains the improved ionic conductivity of these modified membranes at elevated temperatures. The proposed research sought to build on this initial study and offer further improvement to modified perfluoroalkane sulfonic acid membranes through addition of fluorinated phosphonic acid monomers. Perfluorinated phosphonic acid should be a far more powerful Brönsted base which will accentuate the dissociation of sulfonic acids while maintaining high temperature water retensive properties. The result should be a modified Nafion membrane with excellent high temperature conductivity that can serve as a model for a future polymeric counterpart with similar properties. Such an approach will be necessary since a solid state, polymeric membrane is one sure way to guarantee sustainable improvement to the properties of Nafion.

EXPERIMENTAL

Perfluorinated phosphonic acids were synthesized by following a literature[3] procedure. The structures of the perfluoroalkylphosphonates and their designations are shown in Table I. The current research focused on compounds 1, 2 and 5. Compounds 1 and 2 were chosen because they are difunctional and offered the possibility of greatest success. Compound 5 was chosen as an extreme case, since it was the most hydrophobic of this group.

Two basic approaches were taken to combine the individual test compounds with Nafion. In one instance, separate perfluorophosphonates were combined with the 5% Nafion solution (Nafion 117, Aldrich Chemical Co.) in specific weight ratios. The second approach involved ex situ hydrolysis of the separate esters using 37% HCl which converted them to the corresponding phosphonic acids followed by combination of the acid with Nafion solution in a specific weight ratio. Of the two techniques, the phosphonate/Nafion combinations were more successful in forming homogeneous mixtures. The acid/Nafion approach was successful with all but the 2- fluorocarbon di-phosphonic acid (2D) compound. Once hydrolyzed, the 2D phosphonic acid could not be extracted from water.

Table I. Structure and nomenclature of perfluoroalkyl phosphonates

Structure	Nomenclature
1. (Et_2O)-PO-$(CF_2)_2$-$PO(OEt_2)$	2D
2. (Et_2O)-PO-$(CF_2)_4$-$PO(OEt_2)$	4D
3. CF_3-$(CF_2)_3$-PO-(OEt_2)	4M
4. CF_3-$(CF_2)_5$-PO-(OEt_2)	6M
5. CF_3-$(CF_2)_7$-PO-(OEt_2)	8M

Spin casting of the separate Nafion/test compound mixtures onto four probe test chips was done using a Headway Research (Garland, TX) Model 1-EC101DT-R485 Photo Resist Spinner. The four probe test chips were composed of alumina onto which a gold pattern is applied. The pattern has four individual gold lines for contact with the test electrolyte which are each, in turn, connected to their own gold pad for effecting electrical contact to external sensing and regulating devices. The gold leads were masked with Teflon tape throughout the spin casting procedure. The film thickness for all the films is about 5 μm.

Two approaches were used to hydrolyze the perfluoroalkyl phosphonates in the phosphonate/Nafion films thus converting them to their corresponding acids. The first technique involved immersing the test films in 27^0C concentrated H_3PO_4 for one hour followed by wiping off the excess acid. A Nafion control film was prepared in the same manner for later comparison with the test films. The second approach involved *in situ* hydrolysis accomplished by incubating the phosphonate/Nafion films on their respective test chips in a vapor of 37% HCl for one hour. This was achieved by placing concentrated HCl in a 50 ml beaker, placing the test chip on top of the beaker, covering both with a second 500 ml beaker and placing the entire assembly on a hot plate to reflux the HCl. Again, a control film of Nafion was prepared in the same manner for comparison. Conductivity tests were performed on all three types of films; phosphonic acid/Nafion, H_3PO_4-hydrolyzed phosphonate/Nafion and HCl-hydrolyzed phosphonate/Nafion.

Conductivity measurements using the four probe chips were taken in a test chamber where the available water vapor was tightly controlled. A separate moisture generator provided moisture to the test chamber. This moisture generator consisted of a round bottom flask held at a prescribed temperature into which a stream of nitrogen gas was bubbled. The moist nitrogen was then carried over to the test chamber using insulated glass plumbing. The water vapor activity in the membrane (held within the test chamber) was determined from the partial pressure of water in the carrier gas and the available water in the membrane at saturation $P_{H_2O}/P_{sat.}$ The saturated water content of the film was considered to be that of Nafion and phosphoric acid. Conductivity measurements consisted of applying specific currents (1,5,10,50, and 100μA) across two of the four lines in the test chip at each temperature, while measuring the voltage drop at each current applied across the film using the other set of lines. Triplicate runs were performed for each compound tested, and a newly prepared chip was used for each run. Measurements were taken after allowing the apparatus to equilibrate for up to thirty minutes at each temperature, corresponding to a specific water vapor activity. To execute these measurements, a Keithley Model 224 Progammable Current Source together with a Simpson Model 474 multimeter were used. Conductivity measurements were then taken at various vapor contents at several test chamber temperatures ranging from 25°C to 175°C. The voltage and current measurements were each tabulated and the resistance of the test film was determined using Ohm's law. The conductivity of the film was then determined using the following relationship:

$$\sigma = l / R A$$

$$\text{where}: \sigma = \text{film conductivity in S/cm}$$
$$l = \text{distance between the reference electrodes in cm}$$
$$R = \text{film resistance in ohms}$$
$$A = \text{film cross sectional area in cm}^2$$

RESULTS

Generally speaking, combining Nafion with the fluorophosphonates resulted in films with conductivities higher than those of either of the acid-equilibrated Nafion controls. Our results concurred with those of Yeager et al[2] in that the conductivity of the test and control films was temperature dependent and strongly dependent upon the water vapor activity (WVa) in the test cell. Contrary to Yeager's results (0.0068 S/cm @ 0.36 WVa 130^{0}C), we found the conductivity of the H_3PO_4-Nafion control to be considerably less (0.0016S/cm @ 0.38Wva 130^{0}C). As explained previously, we used Sone[4] et al's procedure for calculating the film conductivity which calls for considering the film cross sectional area in the calculation. Prior to applying the correction for this cross sectional area to our data, we noted that our conductivity readings for the Nafion controls were very close to those reported by Yeager. The improvement in conductivity demonstrated by one of the better perfluoroalkyl phosphonic acid/Nafion films in comparison to Nafion controls is illustrated in Figure I.

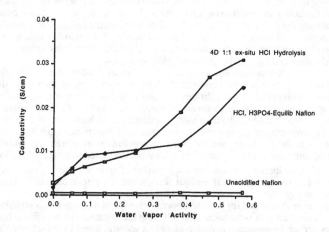

Figure I. Comparison of HCl and H_3PO_4 Equilibrated Nafion with Ex-Situ HCl Hydrolyzed Films @ 175^{0}C

These test results confirm the feasibility of the proposed approach. These results are for a film composed of a 1:1 weight mixture of the perfluorobutyl diphosphonic acid (4D) compound plus Nafion versus pure Nafion films that were equilibrated in HCl and H_3PO_4. The 4D compound in this test was hydrolyzed in HCl and then combined in a 1:1 weight ratio with 5% Nafion solution, and then tested the conductivity at 175^{0}C. As illustrated in Figure I, as the water vapor activity increases, the 4D phosphonic acid/Nafion film retains water better than either the H3PO4 equilibrated Nafion or the HCl-equilibrated Nafion. Above 0.3 WVa, the 4D/Nafion film showed conductivities of up to 30% greater than the Nafion controls, highly significant in comparison to the standard deviation in the readings of this test; 1×10^{-3}S/cm. It would appear that the 4D material is better able to retain water at this temperature which helps the sulfonic acid groups and the phosphonic acid groups to remain protonated. The greater acidity of the fluorophosphonic acid over that of H3PO4 is probably playing a role in the higher conductivity of the test film.

Two other perfluoroalkyl phosphonates, 2D and 8M were hydrolyzed and tested in this manner are shown in Figure II. The general trend for all of these materials is higher conductivities than Nafion at water vapor activities above 0.2WVa. The 8M acid/Nafion mixture appears to show higher conductivities at lower water contents indicating that perhaps this material forms a more homogeneous mixture with the Nafion by virtue of its longer fluorocarbon chain. This same trend appears to occur when the 2D and 4D film mixtures are compared. The 4D appears to out perform the 2D across the range of water content. This may mean that the available water in the film would be more evenly partitioned between the sulfonic acid and phosphonic acid moieties in the 4D and 8M combination films resulting in higher conductivities for these combinations over the entire water vapor range. This result argues for compounds with even longer perfluoroalkyl chains to possibly improve the conductivity further.

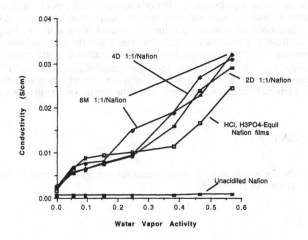

Figure II. Conductivities of Various HCl Hydrolyzed Films @ 175°C

In addition, to conductivity measurements, thermal analysis was performed on mixtures of Nafion and the perfluoroalkyl phosphonates (Figure III). Thermogravimetric analysis (TGA) was performed by Polymer Solutions, Inc., Blacksburg ,Va, using a Perkin-Elmer model TGA 7 thermogravimetric analyzer. The samples were heated from room temperature to 300⁰C under nitrogen atmosphere. Results indicated that the thermal stability of the mixtures was not significantly different from that of either acid treated or untreated Nafion.

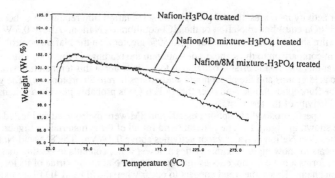

Figure III. TGA of Nafion and Nafion/Perfluoroalkyl Phosphonic Acid Mixtures

CONCLUSIONS

This research effort sought to demonstrate that combining select perfluoroalkyl phosphonic acids with Nafion (a perfluoroalkyl sulfonic acid membrane) could improve Nafion's high temperature electrochemical performance. The original premise for this effort centered on utilizing the better high temperature water retentive properties (and high pK_as) of the perfluoroalkyl phosphonic acids to augment the performance of the sulfonic acid groups in the Nafion. The results of this work have successfully vindicated the proposed premise. A 1:1 mixture of a perfluoralkyl phosphonic acid with Nafion, resulted in a film which demonstrated 30% higher conductivity than a phosphoric acid equilibrated-Nafion control at 175^{0}C. This result was clearly dependant upon the availability of water, strongly suggesting that the phosphonic acid retained water at high temperatures, thereby improving the conductivity of the mixed film.

ACKNOWLEDGMENTS

The authors would like to acknowledge the financial support from National Science Foundation under the SBIR Award DMI-9561693.

REFERENCES

1. Adams, A.A., and Barger, H.J., *J. Electrochem. Soc.*, 121, 987-990 (1974).

2. Savinell, R., Yeager, E., Tryk, D., Landau, U., Wainwright, J., Weng, D., Lux, K., Litt,M. and Rogers, C., *J. Electrochem. Soc.*, 141, (4), L46-48 (1994).

3. Nair, H.K., Guneratne, R.D., Modak, A.S., Burton, D.J., *J. Org. Chem.*, 59, 2393 (1994).

4. Sone, Y. Ekdunge, P. and Simonsson, D., *J Electrochem. Soc.*, 143, (4), 1254, (1996).

PULSED FIELD GRADIENT NMR INVESTIGATION OF MOLECULAR MOBILITY OF TRIMETHOXYMETHANE IN *NAFION* MEMBRANES

Y. WU *[†], T.A. ZAWODZINSKI [†], M. C. SMART [‡],**, S.G. GREENBAUM *,[‡], G.K.S. PRAKASH**, and G.A. OLAH**
*Physics Department, Hunter College of CUNY, New York, NY 10021, steve.greenbaum@hunter.cuny.edu
[†] Electronic Materials Division, Los Alamos National Laboratory, Los Alamos, NM 87545
[‡] Electrochemical Technologies Group, NASA Jet Propulsion Laboratory, California Institute of Technology, Pasadena, CA 91109
** Loker Hydrocarbon Research Institute, University of Southern California, Los Angeles CA 90089

ABSTRACT

Molecular mobility of water and trimethoxymethane (TMM) in *NAFION* membranes of two different equivalent weights (EW), 1100 and 1500, were investigated. Self-diffusion coefficients were determined by the NMR pulsed field gradient method, using the water and methyl protons NMR signals, in saturated *NAFION* samples containing concentrations of TMM in water varying between 0.5 and 14 M, and at temperatures varying from 30°C to 80°C. Diffusion of molecular species containing methyl protons is more than a factor of two slower in the 1500 EW membrane than in the 1100 EW membrane at 30°C and 1 M concentration. The difference rises to about a factor of four at 80°C and 14 M concentration. These differences are attributed to the lower extent of plasticization of the higher EW material as well as the greater effective distance between acid functional groups. *NAFION* samples containing methanol/water mixtures were also investigated. Comparison with the methanol results and the permeation behavior, as characterized by gas chromatographic methods, show that more than half of the TMM is hydrolyzed to methanol as it passes through the acidic membrane. The implications of these findings for alternative fuels in direct oxidation fuel cells are discussed.

INTRODUCTION

Direct methanol oxidation fuel cells are hampered by high methanol crossover rates. This has led to consideration of alternative fuels, one of which is trimethoxymethane (TMM - $[CH_3O]_3CH$). TMM is similar to methanol in terms of overall electrochemical activity but is a volumetrically efficient fuel. Other advantages that TMM has as a fuel include a substantially higher boiling point and lower vapor pressure than methanol. TMM is a considerably larger molecule than methanol, and previous studies suggested reduced molecular crossover compared to methanol.[1] However, TMM is a rich fuel, yielding 20 electrons per complete oxidation reaction compared to six electrons for methanol. Therefore, the crossover "penalty" for TMM is higher than for methanol on a per mole basis.

Another strategy for reducing cross-over which has been described is the use of higher equivalent weight membranes. In that case, the rationale has been the lower uptake of methanol (and water), which leads to lower permeability. The lower uptake results in a decrease in the overall molecular mobility of fuel in the polymer. Since the conductivity of the membrane also

223

Mat. Res. Soc. Symp. Proc. Vol. 496 © 1998 Materials Research Society

drops as EW is increased, there is a trade-off associated with increased EW membranes--increased resistive loss in the cell versus decreased fuel permeation rates.

In this investigation, we studied the mobility of water, methanol and TMM in *NAFION* membranes. We also assessed the effect of changing the equivalent weight of the proton exchange membrane on molecular crossover for both methanol and TMM. We carried out permeation and self diffusion measurements, by gas chromatography and pulsed field gradient (PFG) NMR techniques respectively, in *NAFION* membranes equilibrated in various solutions of water/TMM and water/methanol. The PFG NMR method is a powerful technique which yields directly the self-diffusion coefficient, D*, of a particular nucleus (e.g. [1]H) associated with a diffusing molecule.[2] In certain favorable cases, as in the present investigation, it is possible to distinguish between several species (e.g. water and methanol) containing the same NMR nucleus, and thus determine their D* values separately. Inspection of the NMR results in light of known electrochemical crossover measurements led us to the chromatographic studies.

EXPERIMENTAL

NAFION membranes were obtained from E.I. DuPont de Nemours, Inc. in two ew forms, EW = 1100, which is commercially available as *NAFION-117*, and EW = 1500. The membranes were pre-treated in aqueous solutions of H_2O_2, and then H_2SO_4, and washed with deionized water, as described previously.[2,3] The PFG NMR measurements were performed on a Bruker AMX-400 NMR spectrometer equipped with a microimaging probe with gradient coils which can sustain a maximum gradient strength of 4.0 mT/cm. Gradients were calibrated with a water standard and temperatures were controlled to ± 1K with a Bruker VT1000 temperature controller. For the NMR measurements, slivers of membrane of dimensions 8 mm X 15 mm were immersed for several days in aqueous solutions of TMM (obtained from Aldrich), with the TMM concentration varying from 0.5 to 14 M. The membranes were then removed from solution and quickly blotted dry to remove surface liquid, loaded into 10 mm OD glass NMR tubes and sealed with parafilm. The self diffusion coefficients, D* were determined by the pulsed gradient, stimulated spin echo sequence (PGSSE), utilizing the proton NMR signals originating from both water and methyl protons.

The determination of methanol permeability in polymer electrolyte membranes was accomplished by measuring the change in concentration as a function of time, by gas chromatographic methods, of two vessels separated by membrane samples. The procedure involved preparing a solution of the permeate and placing it in one vessel, typically 3.0M solutions, while the other vessel (B) contained de-ionized water. Samples were taken from the vessel B at specified times and analyzed with a Varian 3400 gas chromatograph equipped with a Carbowax column. An internal standard, such as a higher molecular weight alcohol, was added to the samples to determine the concentration.

The direct electrochemical oxidation of methanol and trimethoxymethane was investigated in liquid feed fuel/oxygen cells (25 cm^2 electrode area) which contained membrane electrode assemblies (MEAs) utilizing Nafion 1100 e.w. (7 mil.) and Nafion 1500 e.w. (5 mil.) membranes as the solid polymer electrolytes (MEAs were supplied by Giner, Inc., Watham, Mass.). In this design, an aqueous solution of the organic fuel is fed to an unsupported Pt/Ru anode , whereas oxygen is supplied to an unsupported Pt cathode containing ~ 4.0 mg cm^{-2} Pt electrocatalyst. The cell was operated at temperatures ranging from 20°C to 90°C, oxygen pressures of 20-30 psig, oxygen flow rates ranging between 1.0L/min. to 5.0L/mim., and fuel concentrations of 0.5 to 2.0M. The cells were operated at current densities in the range of 1-400 mA cm^{-2}. The methanol crossover rates present in operating fuel cells was measured by analyzing the CO_2 content of the cathode exit stream. This was accomplished by utilizing an on-line analyzer, purchased from Horiba Co., which measures the CO_2 volume percent in the cathode stream by passing the sample through an infra-red detector. Before each measurement, the instrument was calibrated with gases of known CO_2 content.

RESULTS

To assess the overall permeation rate from individual measurements, we must know both the TMM uptake and the diffusion rate. The former could also yield some information concerning any selective partioning of TMM into or out of the membrane. The ^1H NMR spectrum of solutions of TMM in water and in the membrane consisted primarily of two lines, from methyl protons and water protons. To determine the partitioning of TMM into the membrane, the water to methyl group peak intensity ratio was determined from the NMR spectrum. Intensity ratios were essentially identical inside the membrane to that observed in the immersion solution. Thus, no selective partitioning occurs.

Self-diffusion coefficients extracted from the PGSSE signal intensities for methanol and water in both *NAFION* 117 (N117) and the 1500 ew membrane (N1500) are plotted in Fig. 1a and 1b, respectively, under conditions of varying temperature and methanol concentration. The methanol diffusion results for N117 are essentially the same as previously reported. At similar methanol concentrations and temperatures, both water and methanol D* values drop by a factor of two to three in the higher EW material relative to those in N117. The lower water and methanol diffusion rates in N1500 are attributed to the lower overall 'plasticization' of the N1500 membrane (i.e. lower solvent uptake) as well as the greater average spacing between SO_3^- functional groups, as compared to the N117. Although this result has important implications concerning methanol crossover, the proton conductivity is also lower in the higher EW material, most likely due to the same factor that limits the methanol and, more importantly, the water diffusivity. As noted previously for N117, there is a weak dependence of the diffusion coefficient on concentration in both membranes.

NMR diffusion coefficients of water and TMM in both membranes, equilibrated in water/TMM solutions, are plotted as a function of concentration for several temperatures in Figure 2a and 2b, respectively. The close similarity in TMM and methanol behavior is immediately apparent, and a more detailed comparison between the methanol and TMM data shows that the D* values are also quite similar. It should be emphasized that the NMR spectroscopic method cannot easily distinguish between CH_3 protons in CH_3OH and TMM since they have very similar electronic environments (both are in O CH_3 moieties). The diffusion results are apparently at odds with electrochemical permeation results. With similar diffusion coefficients and partitioning for methanol and TMM,we would expect a factor of >3 (the ratio of number of electrons harvested from TMM to that from methanol) increase in TMM electrochemical cross-over current. This led us to hypothesize that TMM can hydrolyze into methanol and other products within the highly acidic Nafion medium. Thus, gas chromatographic permeation measurements, presented below, were performed in order to test this hypothesis.

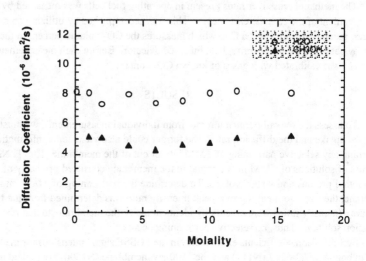

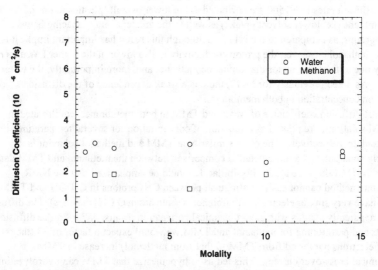

Figure 1a (top): Diffusion coefficients of water and methanol in a N117 membrane at 30°C; 1b (bottom): Diffusion coefficients of water and methanol in a N1500 membrane at 30°C

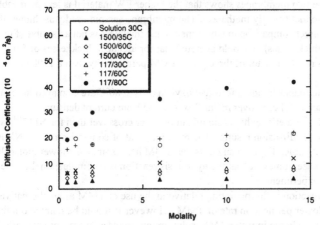

Water Diffusion Coefficient in N1500, N117 and TMM Solutions for Various TMM Concentrations at Elevated Temperatures (Immersed Membrane)

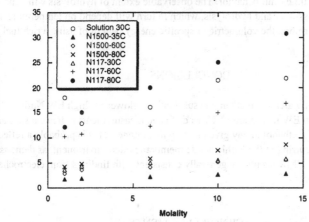

TMM Diffusion Coefficient in N1500, N117 and TMM Solution for Various TMM Concentrations at Elevated Temperatures (Immersed Membrane)

Figure 2a (top) Diffusion coefficients of water in N117 and N1500 membranes at various temperatures; 2b (bottom): Diffusion coefficients of TMM in N117 and N1500 membranes at various temperatures.

227

Methanol permeation data obtained by GC for both membranes are shown in Figure 3. Even though TMM is being studied as the permeant, the clear detection of methanol demonstrates that, indeed, the TMM hydrolyzes by the time the molecules reach the other side of the membrane. There is also evidence for the presence of about 5% of an unidentified secondary product, most likely methyl formate. Thus it appears that the NMR diffusion measurements reflect methanol rather than TMM diffusion in the equilibrated membranes. Comparison between the two membranes shows that the higher EW material is less permeable to the TMM hydrolysis product (mostly methanol). The membrane sample used was thinner than the N117 sample. For direct comparison in which membrane thickness is not a variable (Figure 3 indicates two different thicknesses), it should be noted that for identical thicknesses of 5 mil, a 1500 ew membrane has nearly a factor of three lower TMM permeability than a 1100 ew sample.

Crossover measurements in liquid feed fuel/oxygen cells at 60°C are displayed in Figure 4. Figure 4 shows a comparison of crossover molar flux, derived from current density measurements. The N1500 material exhibits significantly lower crossover than for N117[1,4], which is consistent with the NMR diffusion results. The relevant point of comparison to the NMR results is the open circuit value. Figure 4 suggests that TMM has a somewhat lower molar flux than methanol. In this case, it may be that the hydrolysis reaction is incomplete on the time scale of the permeation experiment.

The hydrolysis reaction is disappointing relative to the use of TMM as an alternative fuel based on the expected lower permeation rate of TMM. However it should be pointed out that the equilibration of the membrane in water/TMM solution represents a "worst case scenario". That is, it is expected that some of the fuel will be electro-oxidized at the anode before hydrolysis can occur within the bulk of the membrane in an operating fuel cell. The experiments reported here represent the long time limit behavior. The observable extent of hydrolysis will depend on the relative rates of cross-over and hydrolysis, which in turn will depend on temperature and membrane EW. Furthermore, the volumetric or specific energy benefit of using a rich fuel such as TMM still accrues.

CONCLUSIONS

Self-diffusion of water and methanol is substantially slower for high EW Nafion membranes than for low EW membranes. TMM diffusion measurements indicate a similar diffusion rate to that of methanol in any given Nafion membrane. This is probably a reflection of hydrolysis of TMM occurring in the highly acidic membrane microenviroment, as demonstrated via GC measurements. NMR results are generally consistent with findings from electrochemical permeation measurements.

ACKNOWLEDGMENTS

We gratefully acknowledge DARPA, ONR and the DOE Office of Transportation Technology for support of this work. One of the authors (S.G.) acknowledges the NASA/JOVE program and the National Research Council for sabbatical leave support.

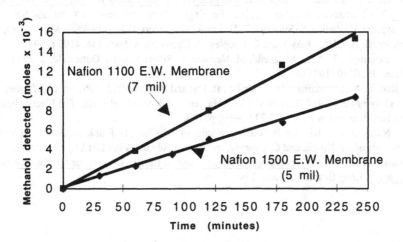

Figure 3: GC detection of methanol permeation from a TMM-fed Nafion membranes. T=25°C.

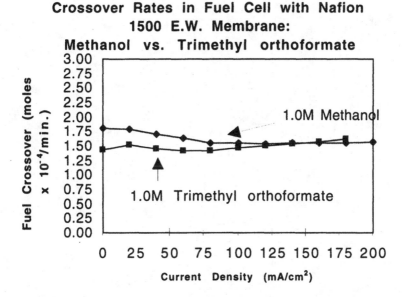

Figure 4: Electrochemically determined molar fluxes of methanol and TMM through a N1500 membrane, 60°C.

REFERENCES

1. M.C. Smart, G.K.S. Prakash, G.A. Olah, S.R. Narayanan, H. Frank, S. Surampudi, G. Halpert, J. Kosek and C. Cropley, in The Electrochemical Society Meeting Abstracts, vol. 96-2, The Electrochemical Society, Inc., Pennington, NJ, 1996, abstract # 789; and S.R. Narayanan, E. Vamos, S. Surampudi, H. Frank, G. Halpert, G.K.S. Prakash, M.C. Smart, R. Knieler, G. A. Olah, . Kosek and C. Cropley, *J. Electrochem. Soc.*, **144**, 4195 (1997).

2. see, for example, T. A. Zawodzinski, M. Neeman, L. Sillerud, and S. Gottesfeld, *J. Phys. Chem.*, **95**, 6040 (1991).

3. X. Ren, T. A. Zawodzinski Jr., F. Uribe, H. Dai and S. Gottesfeld in *Proton Conducting Membrane Fuel Cells I*, S. Gottesfeld, G. Halpert and A. Landgrebe eds., The Electrochem. Soc. Inc, Pennington NJ, PV95-23 (1995), p. 284.

4. S.R. Narayanan, A. Kindler, B. Jeffries-Nakamura, W. Chun, H. Frank, M. Smart, T.I Valdez, S. Surampudi, J. Kosek, and C. Cropley, in "Recent Advances in PEM Liquid Feed Direct Methanol Fuel Cells", Proceedings of the 11th Annual Battery Conference on Applications and Advances, Long Beach, CA, Jan. 1996.

ACCELERATED CORROSION OF STAINLESS STEELS IN THE PRESENCE OF MOLTEN CARBONATE BELOW 923 K

Ken-ichiro OTA, Katsuya TODA, Naobumi MOTOHIRA, Nobuyuki KAMIYA
Department of Energy Engineering, Yokohama National University
79-5 Tokiwadai, Hodogaya-ku, Yokohama 240, Japan

ABSTRACT

The high temperature corrosion of stainless steels (SUS316L and SUS310S) in the presence of molten carbonate [$(Li_{0.62}K_{0.38})_2CO_3$ and $(Li_{0.52}Na_{0.48})_2CO_3$] has been studied in a CO_2-O_2 atmosphere by measuring the weight gain of the specimens.

The corrosion of SUS316L significantly depended on the reaction conditions. With the carbonate coating (both Li/Na and Li/K carbonates), severe corrosion occurred during the initial period of the corrosion test below 923 K, especially around 823 K. The initial severe corrosion was a local corrosion which produced through holes in the metal specimens and occurred more clearly at low Pco_2 with the Li/Na coating than with the Li/K coating. The corrosion became more severe at higher CO_2 pressures and lower O_2 pressures. In a pure CO_2 atmosphere (without O_2), the corrosion rate significantly increased at 823 K. The steel was corroded uniformly at that time.

INTRODUCTION

Generally the energy conversion processes utilizing a molten salt have many advantages, such as a high reaction rate and high conversion efficiency. However, the processes always have material problems for commercialization. The molten carbonate fuel cell (MCFC) is expected to be one of the most promising power generation systems for the coming century due to its high efficiency, excellent environmental characteristics and the ability to utilize a wide variety of fuels. Besides the MCFC, molten carbonate could be used as the electrolyte for other electrochemical systems, such as the high temperature carbon dioxide separator or the high temperature electrochemical heat pump using the water gas shift reaction[1]. In these systems, material durability is quite important for long term operation[2]. Although the long term operation of 40,000 h was proved by using a 100 cm^2 cell[3], several materials should be improved before commercialization. In present MCFCs, the stainless steels such as SUS316L or SUS310S, which have been used for the separator alloy or the current corrector, have some corrosion in an oxygen-containing atmosphere (cathode side) the presence of molten carbonate. The corrosion separator or the current collector causes an increase in the ohmic resistance due to the formation of oxide scale and the loss of the electrolyte due to the corrosion reaction. The decrease in the metallic corrosion in the presence of molten carbonate is thought to be very important in order to decrease the decay rate of the cell. Besides this, SUS316L has been reported to have severe corrosion with the presence of the Li/Na carbonate melt below 923 K[4] that causes trouble during the start of MCFCs.

In this paper, the corrosion of stainless steels (SUS310S and SUS316L) has been studied in the presence of molten carbonates [$(Li_{0.62}K_{0.38})_2CO_3$ and

Mat. Res. Soc. Symp. Proc. Vol. 496 © 1998 Materials Research Society

$(Li_{0.52}Na_{0.48})_2CO_3]$. The major interest is in the corrosion below 923 K and the reaction conditions for the low temperature corrosion.

EXPERIMENTAL

Commercially available stainless steels (SUS316L and SUS310S) were used for the corrosion tests. The metal sheets (thickness=0.5 mm) were cut to a rectangular size (6x12mm) and used for the corrosion test. The corrosion tests were carried out with the presence of a carbonate melt coating (up to 60 mg/cm²). For the coating, the carbonates were put on the surface of the specimen by dipping in the carbonate-ethanol mixture and then dried. The amount of carbonate on the metal was determined by the weight change before and after the coating process. The corrosion tests were carried out mainly at 923 K and at lower temperatures in the CO_2-O_2 atmosphere with $(Li_{0.62}K_{0.38})_2CO_3$ [Li/K carbonate] or $(Li_{0.52}Na_{0.48})_2CO_3$ [Li/Na carbonate]. The weight gain of the specimen was continuously monitored by TGA (Shimadzu DT-40). The heating rate of the TGA from room temperature to the experimental temperature was normally 40 K/min. In order to determine the effect of the heating rate, the heating rate was changed from 0.7 to 40 K/min. After the corrosion test, the oxide scales of some specimens were removed by etching and the metal consumption due to the corrosion reactions was determined.

RESULTS AND DISCUSSION

The high temperature corrosion of stainless steels in air follows the parabolic rate law in most cases and the rate determining step is the diffusion of reacting species through the corrosion scale[5]. The corrosion scale of stainless steels in an oxygen-containing atmosphere consists of at least 2 layers; an Fe rich layer and a Cr rich layer. The characteristics of the stable Cr rich layer is important for the corrosion protection. Generally, the corrosion rate of SUS310S is smaller than that of SUS316L because of the higher Cr content in the steel[6].

In the presence of molten carbonate, the corrosion characteristics of SUS316L significantly depended on the reaction conditions (temperature, atmosphere, melt composition, etc.) compared to those of SUS310S and the corrosion rate of SUS316L was larger in most cases than that of SUS310S. However, SUS316L is mainly used for the current collector or the separator in MCFCs, since

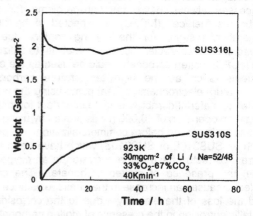

Fig.1 Weight gain curves for the corrosion of stainless steels with the Li/Na carbonate melt coating.

the electrical conductivity of the corrosion scale is higher than that of SUS310S. This instability of SUS316L should be solved before its commercialization.

Figure 1 shows the weight gain curves of the stainless steels with a Li/Na carbonate eutectic melt coating at 923 K. The corrosion of SUS310S follows the parabolic rate law and the corrosion rate is very small under this condition. On the other hand, a sharp weight increase is observed during the very initial period of the SUS316L corrosion which is followed by a weight decrease. With the Li/K carbonate melt coating at the same temperature and in the same atmosphere, the corrosion of SUS316L followed the parabolic law and the corrosion rate was very low like SUS310S. The big difference appears with the Li/Na carbonate melt coating. The difference would come from the lower Cr content of SUS316L. If the Cr content was low, the formation of a stable Cr oxide layer would be seriously affected by the reaction conditions.

Since the initial weight increase of SUS316L was very similar to that of pure Fe, the initial weight increase is mainly due to the formation of Fe oxides. The weight decrease of a specimen after the sharp increase might be caused by the reaction of Fe oxide and Li carbonate followed by the CO_2 liberation. The large weight increase means that the severe corrosion takes place during the initial period of the corrosion of SUS316L with the carbonate melt coating.

In order to confirm the severe corrosion, the metal loss due to the corrosion was measured after the test. Figure 2 show the metal loss of SUS316L with Li/K and Li/Na carbonate melt coatings. Compared to the Li/K carbonate coating, a large metal loss was obtained with the Li/Na carbonate coating especially during the initial period of the corrosion test. These trends were almost same as those of

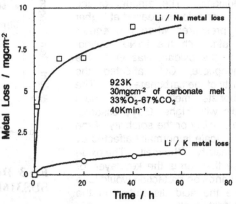

Fig.2 Metal loss plots for the corrosion of SUS316L with a carbonate melt coating.

the TGA. This means that the TGA results approximate the corrosion characteristics, although some reaction takes place between the metal and carbonate coating. The initial severe corrosion was also confirmed by measuring the metal loss of a specimen after the corrosion test.

After removing the corrosion scale, several through holes were observed in the metal specimens when a severe initial corrosion took place. The corner or the side edge of a specimen was sometimes severely damaged. This severe initial corrosion might be local corrosion. The corrosion took place within 1 h, which meant that the through hole was made in a very short time. Considering the role of the separator, this corrosion should be avoided or SUS316L could not be used for the separator of an MCFC.

Severe initial corrosion was observed with the Li/K carbonate melt coating as well as with the Li/Na carbonate. Figure 3 shows a comparison of the weight increase of SUS316L at 50 h between the Li/K and Li/Na carbonate melt coatings. The weight increase was mainly due to the initial weight change. With the Li/K carbonate coating, the amount of corrosion significantly depended on CO_2 pressure. When the CO_2

pressure is less than 0.5 atm, the corrosion with the Li/K carbonate coating is lower than that with the Li/Na coating. On the other hand, the amount of corrosion with the Li/K carbonate coating is higher when the CO_2 pressure is greater than 0.75 atm.

The initial corrosion also depended on the amount of the carbonate coating. The severe initial corrosion took place when the coating amount was between 10 and 30 mg/cm^2 for the Li/K carbonate melt. When the carbonate melt was small, there was not enough meet to dissolve the oxide scale. The dissolution of oxide scale generally accelerates the corrosion reaction. On the other hand, the supply of oxidant from the gas phase through the melt might be limited when a large amount of the melt was present on a metal specimen. The corrosion rate reached a maximum with the intermediate amount of melt coating.

The severe initial corrosion also depended on the atmospheric conditions. The amount of the initial corrosion increased at higher CO_2 pressures when Po_2 equals 0.1 atm with the Li/Na coating. Since the oxidant was O_2 in the atmosphere, CO_2 affected the quality of the carbonate melt. The carbonate melt becomes more acidic with higher CO_2 pressures. The stability or the solubility of the oxide scale is normally affected by the basicity of the melt. The oxide scale that forms the key layer for the corrosion protection would take place the acid dissolution in the carbonate melt.

The characteristics of the severe corrosion also depended on Po_2. The severe initial corrosion took place at a Po_2 lower than 0.3 atm in a Pco_2 of 0.1 atm at 923 K with the Li/Na coating. The severe initial corrosion did not occur at Po_2 higher than 0.5 atm. If the oxidant potential was high enough to form the stable oxide scale, severe corrosion would not take place.

In a pure CO_2 (without O_2) atmosphere, the rapid corrosion rate of SUS316L was observed with the Li/Na carbonate coating around 823 K. The corrosion follows the linear rate law rather than the parabolic rate law. This

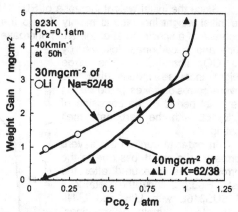

Fig.3 Dependence of weight gain of SUS316L at 50h on Pco_2.

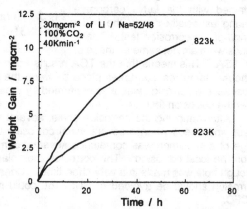

Fig.4 Weight gain curves for the corrosion of SUS316L with the Li/Na carbonate melt coating.

is another type of severe corrosion related to the Li/Na carbonate melt. This severe corrosion did not depend on the coating amount. Figure 4 shows the corrosion of SUS316L with the Li/Na coating in a pure CO_2 atmosphere. The corrosion was a general corrosion and a metal specimen was uniformly corroded. Again, the corrosion rate at 823 K is higher than that at 923 K. Generally, the corrosion rate should decrease at lower temperatures. This is not a case. The stable Cr oxide layer is very important for corrosion protection in an oxygen containing atmosphere. The stable oxide scale could not be formed at this temperature in low oxygen potential such as in the CO_2 atmosphere. The small addition of oxygen to CO_2 could suppress this severe corrosion.

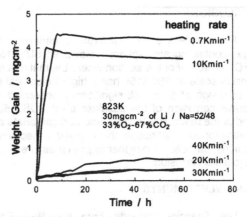

Fig.5 Weight gain curves for the corrosion of SUS316L with the Li/Na carbonate melt coating.

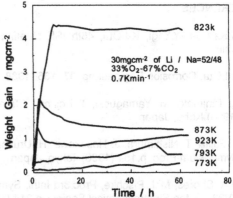

Fig.6 Weight gain curves for the corrosion of SUS316L with the Li/Na carbonate melt coating.

The severe initial corrosion was affected by the heating rate of the TGA from room temperature to the desired temperature. Figure 5 shows the effect of the heating rate on the weight gain curves for the corrosion of SUS316L with a Li/Na carbonate melt coating at the desired temperature of 823 K where the most severe initial corrosion took place. This severe initial corrosion is observed at a heating rate slower than 10 K/min. At the heating rates of 0.7 and 10 K/min when the severe initial corrosion occurred, several through holes in the specimens were found after removing the corrosion scale from the specimen. The severe initial corrosion was a local corrosion that makes a hole. Considering the starting conditions of a MCFC, the low temperature corrosion should be eliminated.

Figure 6 shows the corrosion curves of the different temperatures at the heating rate of 0.7 K/min. This heating rate is close to the heating-up rate of on actual MCFC. At this heating rate, the largest initial corrosion is observed at 823 K. Below this temperature, the severe initial corrosion is not observed. Above this temperature, an initial weight increase is observed, but the amount is smaller than that at 823 K. Considering the start-up condition of a MCFC, this low temperature corrosion should be avoided.

CONCLUSIONS

The stainless steel SUS310S is generally more stable than SUS316L in an oxygen-containing atmosphere. However, SUS316L is utilized for the separator of a MCFC, especially at the section where the material meets the carbonate melt, since the corrosion scale of SUS316L has a higher electrical conductivity than that of SUS310S. The corrosion of SUS316L significantly depended on the reaction conditions and the abnormal corrosion of the stainless steel SUS316L was observed during the initial period of the reaction with Li/Na and Li/K carbonate coatings and also without oxygen in the atmosphere. Although the corrosion mechanism is not clear, the atmospheric conditions, namely, the higher oxygen pressure and lower CO_2 pressure, are important for avoid the corrosion.

ACKNOWLEDGMENTS

The financial supports from the New Energy and Industrial Technology Development Organization (NEDO) of Japan and the Molten Carbonate Fuel Cell Association of Japan are greatly appreciated.

REFERENCES

1. N.Kamiya, Y.Yagi, K-I.Ota, 46th ISE Meeting Abstract No.I5-41 (1995) Xiamen, China.

2. K-I.Ota, Corrosion Engineering, **37,** 135 (1988).

3. K. Tanimoto, M. Yamaguchi, T. Kojima, Y.Tamiya, Y. Miyazaki, Proc.2nd IFCC p3. (1996) Kobe, Japan.

4. Y. Fujita, T. Nishimura, J. Hosokawa, H. Urushibata, A.Sasaki, 3rd FCDIC Fuel Cell Symposium Proc. p.154 (1996) Tokyo, Japan.

5. D.A. Shores, M.J. Pischke, Proc.3rd Intnl. Symp. Carbonate Fuel Cell Technology, PV93-16, the Electrochemical Society, p.214 (1993).

6. S. Nakagawa, J. Isozaki, S. Kihara, Boshoku Gijutu, **36,** 438 (1987).

PREPARATION OF Pt/WO₃ POWDERS AND THIN COATINGS ON CARBON BLACK AND METAL SUPPORTS BY THE COMPLEX SOL-GEL PROCESS (CSGP)

A. DEPTUŁA*, W. ŁADA*, T. OLCZAK*, B. SARTOWSKA*, A. CIANCIA** L. GIORGI** AND A. Di BARTOLOMEO**
*Institute of Nuclear Chemistry and Technology, ul. Dorodna 16, 03-195 Warsaw, Poland, adept@orange.ichtj.waw.pl
**ENEA-CRE-Casaccia, AD 00100 Roma, Italy, giorgil@casaccia.enea.it

ABSTRACT

In this work the CSGP was applied for preparation of Pt-WO$_x$ and its coatings on carbon black, silver and titan substrates. Saturated tungsten sol (0.15M) was prepared by dissolving $(NH_4)_4W_5O_{17} \cdot 2.5 \, H_2O$ in ascorbic acid solution. To this solution H_2PtCl_6 was added to obtain the molar ratio Pt:W=1. After evaporation and heating at 700°C for 2h homogeneous yellow-green powders were obtained. Another part of this solution was diluted with ethanol and after ultrasonic mixing used for preparation of coatings on Ag and Ti substrates by the immersion technique. Gel layers were dried at 200°C for 20h and then calcined at 700°C for 2h. The resulting white colored layers (thickness 10-65μm) were very adherent. Ascorbate aqueous tungsten-platinum sol diluted to 0.03 or 0.06 M ΣMe was used for preparation of Pt-W (10 and 20 %) catalyst by impregnation of carbon black Vulcan XC-72, followed by thermal treatment in air, argon and hydrogen atmospheres.

INTRODUCTION

Direct Methanol Fuel Cells (DMFC) have been recognized as one of the most promising systems of energy sources for automotive area [1-4]. Oxidation of methanol in these cells is catalyzed by Pt (often mixed with other metals or/and oxides e.g. Ru, Au, Os, Ir, Pd, Sn, Pb, Bi, Cr, Cu, Fe and W) on carbon black support [1-12]. The catalysts can be applied either to electrodes (generally Teflon bonding) or to membranes. The methods of catalyst preparation for DMFC applications described in the above cited papers can be systematized in three groups:
1. Adsorption of active components by carbon black from their aqueous solutions (often complex or colloidal) [1,6,8-10],
2. Precipitation on the support from similar solutions [6,7,11],
3. Precipitation by electrodeposition from solutions [5].
Freshly prepared catalysts are subjected to thermal treatment in nitrogen (argon) or /and hydrogen atmosphere.

For preparation of similar type catalysts the method of impregnation has been widely used [13]. However there is no information on its application to the preparation of catalyst for DMFC. There are two main advantages of the impregnation process: the relative simplicity and possibility of precisely obtaining materials with the desired composition.

The goal of the present work was the application of this method for the preparation of Pt-WO$_x$ catalysts on carbon black. Sols prepared with a very strong complexing agent (ascorbic acid-ASC) were used as impregnating solutions. The application of ASC to sol preparation by the Complex Sol Gel Process (CSGP) has been patented by INCT (Poland) and ANL (USA) teams [14] and successfully applied to the synthesis of Li$_x$Mn$_2$O$_{4\pm\delta}$ [15]. Because this process was also successfully used for preparation of ceramic coatings e.g. hydroxyapatite on Ag or Ti supports [16] we decided to apply it for preparing thin layers of Pt/WO₃ on metallic supports. Similar coatings were obtained by electrodeposition [17].

EXPERIMENTAL

Starting saturated tungsten solutions were prepared by dissolving $(NH_4)_4W_5O_{17} \cdot 2.5 \, H_2O$ (POCh, Gliwice, Poland) in concentrated aqueous solution of ASC containing 200 g/dm³, (solution No I) or in the solution diluted 10 times (solution No II). In both cases the concentration of W (0.15M) was similar. To both solutions an aqueous solution of H_2PtCl_6

237

(11gPt/dm^3) was added to the molar ratio of Pt:W=1:1. The resulting products are respectively denoted I+Pt and II+Pt. The solutions were immediately subjected to the next steps because after several minutes the precipitation was observed, despite the complexing action of ASC. Solution I+Pt, diluted with ethanol, was used for preparing coatings on silver or titanium substrates by the immersion technique with ultrasonic mixing described in detail in [16]. Homogeneous yellow-green powders were obtained from this solution by evaporation at boiling point and further heating at 700°C for 2h.

Solution No II+Pt was used for impregnation (method IM-2) of carbon black Vulcan XC-72 (Carbochrom, Milan, Italy). The amounts of starting sols were calculated as to obtain 10 and 20% of Pt+W in the final product. Dilution of the starting solutions with water to a total volume of ~170 cm^3 per 10g of carbon black was necessary for complete wetting of the carbon black. The impregnation was carried out in an ultrasonic bath supported by mechanical stirring until black, creamy, homogeneous, very viscous slip was obtained. In some experiments (denoted method IM-1) W-ASC solution (II) diluted with water was added first to the carbon black followed by the addition of respective Pt solutions. The obtained slips were initially dried under an infrared lamp and then at 110°C for 24h. The thermal treatment at higher temperatures was carried out in air, argon and hydrogen atmosphere.

Samples of materials were characterized by:
-thermal analysis (TG, DTA) using a Hungarian MOM Derivatograph (sample weight -100 mg, heating rate -5°C/min., atmospheric air, reference material- Al$_2$O$_3$, sensitivity: TG-200, DTG-1/5, DTA-1/5),
-X-ray diffraction (XRD), Co K$_\alpha$ (Positional Sensitive Detector, Ital Structure)
-SEM using a scanning electron microscope (Tesla BS340). The samples were coated a thin layer (~30nm) of Au.

Evaluation of chloride and oxide bonding on carbon black was carried out by leaching (0.5g catalyst/ 25gH$_2$O) at room temperature.

RESULTS AND DISCUSSION

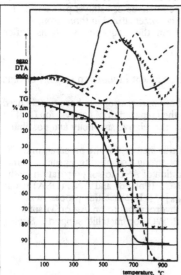

Fig.1.Thermal analysis of carbon black (--) and catalyst 10% (—) and 20% Pt+W ($^{\infty\infty}$) dried at 110°C

It has been observed that in contrast to water the ascorbate sols perfectly wet the carbon black. No crystallization of the added metallic compounds was observed. Consequently the slips were perfectly homogeneous.

The results of thermal analysis of the samples dried at 110°C as well as of the carbon black are shown in Fig.1. Only samples prepared by variant IM-2 are shown because those prepared by IM-1 technique are almost identical. A significant weight loss of carbon black starts at 500°C due to oxidation. Weight losses of impregnated samples start earlier (150°C) due to losses of volatiles (H$_2$O, ammonia and Cl$^-$). Decomposition of ASC begins at 400°C and is accompanied by a large exothermic effect in the region 400-700°C. The shoulder on the DTA curve near 700°C corresponds to the final oxidation of pure carbon black.

The results of the calcination experiments at various temperatures and atmospheres and the leaching experiments of the dry residues are summarized in Table I. At a temperature of 110°C a part of the compounds added during the impregnation process is loosely bonded and is easily leached with water. Unexpectedly after leaching of the 20% Pt-W catalyst the extracted amounts of chlorides were 10 times lower than for the 10% Pt-W catalysts.

238

TABLE I. Weight changes (Δm) and leaching results

Sample No, (%) Pt+W and impregnation method used	Δm, % (+ in the case of weight increase), in ( ) % of leached Cl⁻ and inorganic species (calcined at 400°C)				
	Air			Ar	H₂
	110°C, 24h*	400°C, 0.5h, Δm after 1h	900°C, 10 h	900°C 1h**	700°C 1h
K1-10 (only W)-1	(0;0.5)				
K2-10-IM-1	(2.5;1.4)	9(0.3;0.16);13	90	10 (0;0)	+15
K3-10-IM-2	(3.4;0.7)	11(0.4;0.13);17	87	12 (0;0)	+9
K4-20-IM-1	(0.3;5.1)	10(0.1;0.14);12	82	14 (0;0)	+13
K5-20-IM-2	(0.3;5.6)	10(0.1;0.16);12	82	15 (0;0)	+7

*- starting samples for treatment at higher temperature in air and argon
**- after these experiments parts of the samples were heated in hydrogen

In contrast to chlorides solid elements pass to water in amounts several times higher (approximately 25% of the added amounts). After heating at 400°C for 0.5h the leached amounts of solid species are significantly lower (<0.2%) and similar for all types of catalysts. Since further heating in air (Fig. 1) is connected with high carbon losses the final thermal treatment of the catalysts was carried out in argon atmosphere at 900°C, as in [6]. In those conditions only a limited weight loss is observed and no substances are leached. This indicates that the bonding of catalytic elements on carbon is effective. Unexpected weight increases were always observed after heating the catalyst in hydrogen atmosphere at 700°C. Since the only possible reaction is that with carbon black it means that approximately 0.5 and 1 mole of H₂ per mole of C, respectively for IM-2 and IM-1 technique, reacts with carbon. This suggests that in the latter technique the activity of the catalyst is higher presumably due to the deposition of Pt on the surface of the formed tungsten species. However in order to avoid hydrogenation it seems to be necessary to decrease the temperature of the thermal treatment e.g. to 500°C (as suggested in [1, 10]).

SEM micrographs of the carbon black and catalysts are shown in Fig. 2. In contrast to flocky surface of the carbon black the catalyst surfaces are relatively smooth. No significant differences were observed between samples prepared under various conditions.

Thin films on Ag and Ti substrate were prepared from solution (I+Pt). Fig. 3 shows micrographs of Pt/WO₃ on Ag. It can be observed that after the thermal treatment at 200°C then at 700°C the obtained coatings are very adherent. After the treatment at 200°C grains of different size can be seen while after heating at 700°C those layers exhibit characteristic modular appearance. The cross sections show that the layers have uniform thickness. As can be seen from the data presented in Table II the diameter of the films depends on the preparation conditions.

TABLE II. Pt/WO₃ coating deposited on silver (withdrawal rate 5 cm/sec).

sol- ethanol volume ratio	immersion numbers	firing temp.°C and time, h	thickness, μm
1 : 1	1	200, 20	22
1 : 1	3	700, 2	63
1 : 2	1	200, 20	15
1 : 2	1	700, 2	11

black carbon
VULCAN XC-72

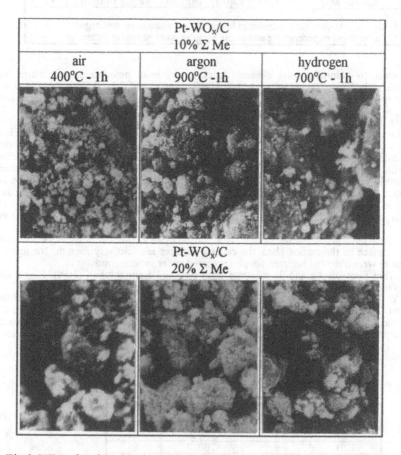

Pt-WO$_x$/C 10% Σ Me		
air 400°C - 1h	argon 900°C -1h	hydrogen 700°C - 1h

Pt-WO$_x$/C 20% Σ Me		

Fig.2 SEM of carbon black and Pt-WO$_x$/C catalysts prepared under various conditions. (⊢20μm⊣)

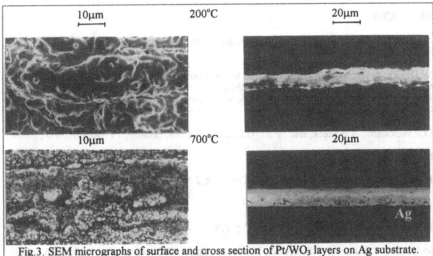

Fig.3. SEM micrographs of surface and cross section of Pt/WO₃ layers on Ag substrate.
Sol : ethanol volume rate 1:2.

In order to obtain Pt-WO₃ powders the solution I+Pt was quickly evaporated and dried at 200°C. After further heating at 700°C for 2h homogeneous yellow-green powders were obtained. Their XRD patterns, shown in Fig.4, indicate that Pt is homogeneously distributed in WO₃.

The catalytic properties of the obtained catalysts and their application for DMFC will be subject to further studies

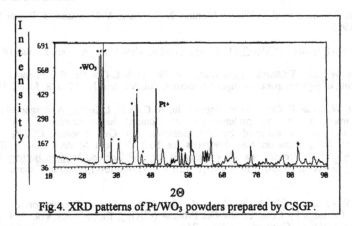

Fig.4. XRD patterns of Pt/WO₃ powders prepared by CSGP.

CONCLUSIONS

1. Pt-WO$_x$ catalysts well bonded on carbon black were obtained by the impregnation technique using CSGP aqueous ascorbate solutions of ammonium pyrowolframate and chloroplatinic acid.
2. Adherent coatings of the Pt-WO₃ catalysts on Ag and Ti substrates of various thickness (10-65μm) were obtained.
3. Homogeneous powders of Pt-WO₃ were synthesized by CSGP.

REFERENCES

1. L.Giorgi, A.Pozio, Celle a combustibile ad elettrolita polimerico solido, RT/ERG/95/05, ENEA, Rome, Italy, 1995.

2. W.Dönitz, G.Gutmann, P.Urban, Second International Symposium on New Materials for Fuel Cell and Modern Battery Systems, June29-July3,1997.Proseedings. New Materials for Fuel Cell and Modern Battery Systems II, Editors, O.Savadogo and P.R. Roberge, ISBN 2-553-00624-1, Montreal, Canada, 1997, p.14.

3. D.P.Wilkinson, A.E.Steck, ibid., p.27. G.Lalande, R.Côte, D.Guay, J.P.Dodelet, ibid., p.768

4. C.Lamy, J-M.Leger, ibid., p. 477.

5. F.Gloaguen, T.Napporn, S.Donon, M-J.Croissant, J-M.Leger, C.Lamy, ibid., p.510.

6. A.Fischer, M.Götz, H.Wendt, ibid., p.489.

7. R.Borkowska, R.Lipka, J.Przyłuski, ibid., p.778.

8. M.Uchida, Y.Fukuoka, Y.Sugawara, N.Eda, A.Ohta, J. Electrochem. Soc. 143, 2245 (1996).

9. M.Uchida, Y.Aoyama, M.Tanabe, N.Yanagihara, N.Eda, A.Ohta, J. Electrochem. Soc. 142, 2572 (1995).

10.G.Tamizhmani, G.A.Capuano, J.Electrochem. Soc. 141, 968 (1994).

11.A.Hamnett, B.J.Kennedy, Electrochimica Acta 33, 1613 (1988).

12.A.S.Arico, Z.Poltarzewski, H.Kim, A.Morana, N.Giordano, V.Antonucci, J. Power Sources 55, 159 (1995).

13.J.R.Anderson Structure of Metallic Catalysts, London, New York, Academic Press 1975

14.A.Deptuła, W.Łada, T.Olczak M.Lanagan, S.E. Dorris, K.C.Goretta, R.B.Poeppel, Method for preparing of high temperature superconductors, Polish Patent No 172618, June 2 1997.

15.A.Deptula, W. Łada, F, Croce, G.B. Appetecchi, A. Ciancia, L.Giorgi, A. Brignocchi and A. Di Bartolomeo. Synthesis and preliminary electrochemical characterization of $Li_xMn_2O_4\pm$ d) (x=0.55-1.1) powders obtained by the Complex Sol Gel Process (CSGP), Second International Symposium on New Materials for Fuel Cell and Modern Battery Systems, June29-July3, 1997. Proceedings. New Materials for Fuel Cell and Modern Battery Systems II, Editors, O.Savadogo and P.R. Roberge, ISBN 2-553-00624-1,Montreal,Canada, 1997,p. 732.

16.Deptuła, W.Łada, T. Olczak, R. Z. LeGeros, J. P. LeGeros, Preparation of Calcium Phosphate Coatings by Complex Sol-Gel Process (CSGP), Bioceramics Vol. 9, Pergamon (ISBN 0 08 0426840), Cambridge, 1996, p.313.

17.P.K.Shen, A.C.C. Tseung, J.Electrochem. Soc. 141, 3082 (1994).

CARBON COMPOSITE FOR A PEM FUEL CELL BIPOLAR PLATE

T. M. BESMANN, J. W. KLETT, AND T. D. BURCHELL
Metals and Ceramics Division, Oak Ridge National Laboratory, P. O. Box 2008
Oak Ridge, TN 37831-6063, tmb@ornl.gov

ABSTRACT

The current major cost component for proton exchange membrane fuel cells is the bipolar plate. An option being explored for replacing the current, nominal machined graphite component is a molded carbon fiber material. One face and the volume of the component will be left porous, while the opposite surface and sides are hermetically sealed via chemical vapor infiltration of carbon. This paper will address initial work on the concept.

I NTRODUCTION

The first use of a proton exchange membrane (PEM) fuel cell was in General Electric fuel cells supplied to the Gemini Space Program in the 1960s. These fuel cells use a perfluorinated ionomer polymer membrane electrolyte that conducts protons. It was not until the mid-1980s, however, that the PEM fuel cell began to see significant research and development that is continuing to this day [1].

The advantage of PEM fuel cells over other types is its low-temperature operation (<100°C). This allows for rapid start when hydrogen is the fuel and the use of a wide variety of materials such a polymers and elastomers. The disadvantage of PEM fuel cells are their intolerance to unreformed fuels. Yet, the almost instantaneous availability of power from PEM fuel cells have made them the primary candidate for the fuel cell effort under the Partnership for Next Generation Vehicle program. The objective of the program is to develop a cost-effective power system for a passenger vehicle that will obtain high mileage (34 km/l) with near-zero emissions.

The challenges for PEM fuel cell technology for vehicular use lie in reducing the cost of the fuel stack, with a goal of $35/kW for the system. While there has been impressive progress in reducing catalyst loading and in the use of very thin membranes, the cost of the materials of construction remain a barrier. The current key issue is the cost of the bipolar plate, which is the electrode plate that separates individual cells in a stack [2]. The current reference design requires the bipolar plate be of high density graphite with machined flow channels. Both the material and machining costs are thus prohibitive. The bipolar plate requirements include low cost materials and processing, lightweight, thin (<3mm), sufficient mechanical integrity, high areal electronic conductivity, low permeability (boundary between fuel and air), and corrosion resistance (in the moist atmosphere of the cell) [3].

The above requirements eliminate conventional materials from consideration. Polymers have insufficient conductivity and therefore require substantial conductive fillers. Inexpensive metals suffer corrosion and would need protection. These constraints have thus led to the concept described in this paper.

The bipolar plate approach developed at ORNL is to use a low-cost slurry-molding process to produce a carbon fiber preform. The molded carbon fiber component would have an inherent volume for diffusing fuel or air to the electrolyte surface and could be molded with entrance and exit channels. Thus it would serve as a diffuser or flow field, as well as a bipolar plate. The surface of the bipolar plate separating cells is made hermetic through chemical vapor infiltration (CVI) with carbon. The infiltrated carbon would also serve as a conductive electrode.

EXPERIMENTAL

The porous carbon fiber preform was manufactured via the slurry molding technique illustrated in Fig. 1, originally developed to produce high temperature thermal insulation. Milled Amoco DKD-x mesophase pitch carbon fibers (~400 m in length) were slurried in a water solution with phenolic DUREZ® resin (Occidental Chemical Corp.). The slurry was then vacuum molded into a flat plate ~2 mm in thickness and dried at 50°C in a convection oven. After curing for 14 hours at 130°C, the as-molded part was carbonized under flowing nitrogen at 1050°C and graphitized at 2800°C. Sample disks 25 mm in diameter were machined from the molded plate. The resultant material had a 15 vol.% fiber loading.

In order to seal one surface of the disks as well as their sides, the disks were placed horizontally in a 4.6-cm diameter vertical graphite furnace with one face resting on graphite base. The face in contact with the base will not be infiltrated since the gases cannot easily transport to that surface. The disks were then exposed for one hour to the CH_4 infiltrating gas under conditions described in part by Delhaes [4] and Bammidipati et al. [5]: Temperature: 1400°C; Pressure: 5 kPa; Flow Rate: 600 cm^3/sec at STP.

The surface of the samples were examined using scanning electron microscopy (SEM) (Hitachi S-800) and polished cross-sections were viewed using optical microscopy.

RESULTS

The random network of carbon fibers in the preform can be seen in Fig. 2, which is an image of the uninfiltrated preform after slurry-molding and heat treatment. The network structure, while quite open, retains its shape and is easily handled.

After CVI with carbon, the fibers on the exposed surface are highly coated and the surface appears to be sealed. This is confirmed by a hydrogen permeability test which indicated a leakage rate of <1 cm^3 hydrogen under a 2 bar differential. In the low-magnification optical image of Fig. 3, it is apparent that one side of the disk is well-coated, while the majority of the volume as well as the other side remain porous. Figure 4, a plan-view SEM image of the infiltrated surface and Fig. 5, a high magnification optical image of the cross-section, reveal that the surface is coherent. Polarized light views of the deposited carbon indicate that it is highly graphitic, implying that it will also be highly electrically conductive. This is confirmed by several 4-point probe measurements on the dense side yielding values ranging from 374 to 388 S/cm. The range on the porous side was from 219 to 250 S/cm.

DISCUSSION

The cross-section of the carbon fiber bipolar plate material shows an apparently sealed surface and edge, with a porous volume and surface containing bonded carbon fibers. This structure is well-suited for the PEM fuel cell application in that it may be sufficiently hermetic with regard to leakage of fuel or oxidant between half-cells, yet provides a flow field for fuel or oxidant to transport to the catalyst/electrolyte surface. Producing a single component that provides both the bipolar plate and diffuser components, and into which can be molded entrance

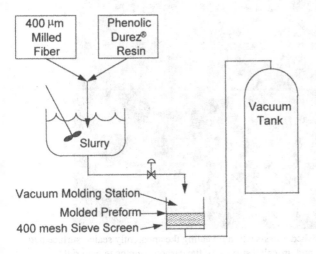

Fig 1. Schematic of slurry-molding apparatus.

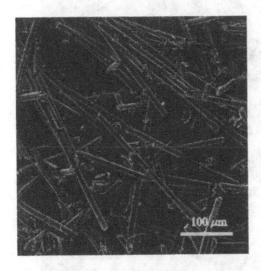

Fig. 2. SEM image of a slurry-molded preform.

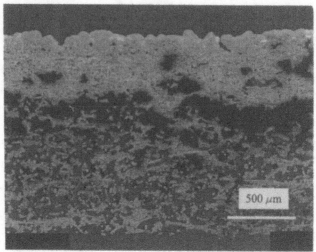

Fig. 3. Optical image of a polished cross-section revealing the apparently sealed surface and porous interior volume and unsealed surface of the carbon composite material.

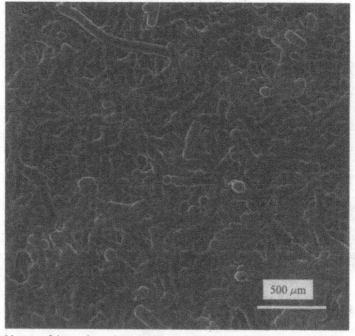

Fig. 4. SEM image of the surface of the carbon fiber preform infiltrated with chemical vapor deposited carbon.

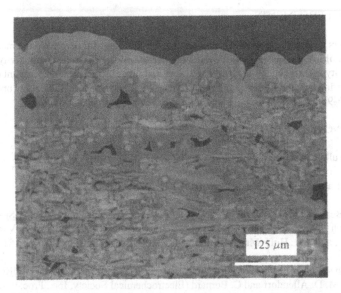

Fig. 5. Optical image of a polished cross-section revealing the apparently sealed surface of the carbon composite material.

flow channels solves several problems. First, the component is exceptionally thin and light, allowing for lighter and more compact fuel cell stacks. Second, as the component is a fiber-reinforced material it is relatively tough compared to monolithic graphite, which must be sufficiently thick to withstand the stresses within the fuel stack. Third, the electronic efficiency of the material is high, offering significant efficiencies since the component will be superior to that of any combination of bipolar plate and diffuser which has an interface that acts a source of ohmic loss.

Issues that remain in the current approach to the bipolar plate/diffuser described in this work are those of scaling to full-size components, molding –in channels and holes, simplification of the preform processing, and more rapid densification of the surfaces by CVI. The latter two issues speak to the need for low-cost production. Thus a more streamlined, preferably continuous process, is required for preform fabrication, and possibly a continuous and rapid CVI step.

CONCLUSIONS

A unique and possibly very low cost bipolar plate/diffuser has been demonstrated in the current work. The concept of a molded preform with inherent entrance/exit channels is conceptually easy to foresee. The material development to date has shown that appropriate preforms are relatively easily prepared, densification provides for a highly conductive material, and that the necessary surfaces may be made hermetic leaving highly porous the volume and the face of the component in contact with the electrolyte.

ACKNOWLEDGEMENTS

The authors would like to thank D. F. Wilson and T. N. Tiegs for reviewing the manuscript. M. S. Wilson of Los Alamos National Laboratory performed the electrical conductivity and permeability measurements. This research was supported by the U. S. Department of Energy, Energy Efficiency and Renewable Energy, Office of Transportation Technology under contract DE-AC05-96OR22464 with Lockheed Martin Energy Research Corporation.

REFERENCES

1. T. F. Fuller, Interface, Fall 1997, p. 26 (1997).

2. T. R. Ralph, Platinum Metals Rev., **41** (3), p. 102 (1997).

3. R. L. Borup and N. E. Vanderborgh in Materials for Electrochemical Energy Storage and Conversion – Batteries, Capacitors and Fuel Cells, edited by D. H. Doughty, B. Vyas, T. Takamura, and J. R. Huff (Mater. Res. Soc. Proc. 393, Pittsburgh, PA 1995), p.151-155.

4. P. Delhaes, Chemical Vapor Deposition. Proc. Fourteenth Intl. Conf. and EUROCVD 11, edited by M. D. Allendorf and C. Bernard (Electrochemical Society, Inc., Proc. 97-25, Pennington, NJ 1997), p. 486-495.

5. S. Bammidipati, G. D. Stewart, J. R. Elliott, Jr., S. A. Gokoglu, and M. J. Purdy, AIChE Journal **42** (11), p. 3,123 (1996).

DETERMINATION OF OXYGEN PERMEATION KINETICS IN A CERAMIC MEMBRANE WITH THE COMPOSITION SrFeCo$_{0.5}$O$_{3.25-\delta}$

S. KIM, Y.L. YANG, R. CHRISTOFFERSEN AND A. J. JACOBSON
University of Houston, Department of Chemistry, Houston, TX 77204-5641.

ABSTRACT

The oxygen permeation through an oxide membrane with bulk composition SrFeCo$_{0.5}$O$_{3.25-\delta}$ has been measured as a function of both oxygen partial pressure and temperature. The results of the pressure dependence of the permeation indicate that the oxygen transport in this membrane is dependent primarily on the bulk diffusion rate. Although the permeation experiments were carried out at temperatures within, or very close to, the range where SrFeCo$_{0.5}$O$_{3.25-\delta}$ is stable as a pure single phase, the membrane was found to consist of SrFe$_{1.5-x}$Co$_x$O$_{3.25-\delta}$ (x = ~0.42) together with fractions of Sr(Co,Fe)O$_{3-\delta}$ perovskite and Co-Fe oxide that formed as stable phases during densification of the membrane at high temperature (1090°C). These additional phases persisted in the membrane during the permeation measurements.

INTRODUCTION

Mixed-conducting oxides with high oxygen ion conductivities can form the basis for ceramic membranes that separate oxygen from air. Such membranes are also of interest for their potential use in membrane reactors that can produce synthesis gas (CO+H$_2$) by direct conversion of hydrocarbons such as methane. For this application, materials are required to have structural stability in reducing atmospheres as well as high oxygen permeability.

Recently a non-perovskite oxide material SrFeCo$_{0.5}$O$_{3.25-\delta}$ has been of interest as a promising candidate for membrane reactors [1-4] because it has greater structural stability in reducing atmospheres than many other cobalt containing perovskite oxides currently being studied for membrane applications. The crystal structure of SrFeCo$_{0.5}$O$_{3.25-\delta}$ consists of a perovskite-type octahedral layer that alternates with an edge-sharing non-perovskite layer containing the additional B-site cations [5].

The properties of SrFeCo$_{0.5}$O$_{3.25-\delta}$ membranes, including their stability, have been well documented. The oxygen permeation kinetics of SrFeCo$_{0.5}$O$_{3.25-\delta}$ membranes, however, have not been completely defined, particularly with respect to the importance of the surface exchange kinetics relative to the bulk diffusion rate. We have shown previously that these two effects can be separately evaluated by measuring the pressure dependence of the oxygen flux through the membrane [6-8]. In this paper, we report the oxygen transport kinetics for a tubular membrane with the bulk composition SrFeCo$_{0.5}$O$_{3.25-\delta}$.

A recent phase stability study of compositions within the solid solution SrFe$_{1.5-x}$Co$_x$O$_{3.25-\delta}$ indicated that cobalt solubility in this phase is dependent on temperature [9]. At high temperature (1150 -1200°C), an iron rich SrFe$_{1.5-x}$Co$_x$O$_{3.25-\delta}$ phase is apparently in equilibrium with a Sr(Co,Fe)O$_{3-\delta}$ perovskite phase and CoO. These observations are relevant to our permeation studies because although membrane operating temperatures are lower, we have used comparably high temperatures to fabricate dense tubular membranes. Consequently, in addition to the permeation experiments, we have also investigated the microstructure of membrane after the permeation measurements using scanning electron microscopy (SEM) and electron-probe microanalysis (EPMA) techniques. The results from these studies have been combined with data from experiments that investigate the phase stability relations of the membrane composition under the conditions of our permeation experiments.

EXPERIMENTAL

The ceramic powder for the membrane was produced by freeze-drying nitrate solutions. To synthesize $SrFeCo_{0.5}O_{3.25-\delta}$, pre-dried $SrCO_3$ (Aldrich, 99.995%) was dissolved in nitric acid together with Fe metal powder (Aldrich, 99.99+%) and Co metal powder (Aldrich, 99.9+%) in the required stoichiometric ratio. The nitric acid solution was then sprayed using an atomizer (Sonotek) into liquid nitrogen at a rate of 2.5 ml/min. The resulting nitrate "snow" was freeze-dried using an FTS Dura-Dry II MP freeze-dryer. The resulting nitrate powder was gradually heated from 90°C to 300°C to remove residual acid and then heated at 800°C for 2h to decompose the nitrates. Finally the powder was fired in a platinum crucible in air at 1090°C for 3h and cooled to room temperature by shutting off the furnace. Characterization of this and subsequent samples by X-ray powder diffraction was carried out with a Scintag XDS 2000 powder diffractometer using CuKα radiation.

To fabricate a dense tubular membrane the final powder was ball milled for 48 h in ethanol and then mixed with 0.20 wt % stearic acid lubricant (Aldrich, 99+%). This mixture was formed into a tube by cold isostatic pressing at 40,000 psi using a custom-designed rubber mold. The green tube was heated at 0.1 °C/min to 390°C in oxygen to first remove the stearic acid, followed by heating at 0.5°C/min to 1090°C. After 5 h at 1090°C the tube was cooled to room temperature at 0.5°C/min. The tube had a density of >97% of theoretical. The ends of the tube were then cut square and ground on 600 grit SiC paper for final installation into the permeation apparatus.

The details of the measurement apparatus for the tubular membranes have been described in a previous paper [7]. Leakage was found to be less than the detection limit of the GC (<5 ppm) through the entire series of permeation measurements.

The phase relations for the composition $SrFeCo_{0.5}O_{3.25-\delta}$ in air were investigated using pieces of pelletized starting material prepared by two different methods as follows. In the first method, portions of the final powder prepared by the freeze-dried route were isostatically pressed into pellets under the same conditions used to prepare the ceramic tube but without lubricant. In the second method, the predried components $SrCO_3$ (Aldrich, 99.995%), Fe_2O_3 (Aldrich, 99.998%) and Co_3O_4 (Aldrich, 99.995%) were mixed and fired in powder form at 1150°C in air for 10 h, then reground and fired an additional 9 h in powder form at 1150°C. This powder was pelletized and heated at 1150°C in air for 16 h then pulled directly from the furnace and allowed to cool in air. Once prepared, both types of pellets were cut into pieces that were isothermally annealed at 900°C in air for various times followed by drop-quenching into liquid N_2. The samples were characterized by X-ray powder diffraction, SEM and EPMA.

EPMA techniques (JEOL JXA-8600 EPMA) were used to characterize the tubular membrane ceramic as well as the pellet samples used for the phase stability experiments. Samples were embedding in epoxy and polished in the usual way. The microstructure of the membrane ceramic prior to the permeation experiments was studied using the pieces that had been trimmed from the tube ends. After the permeation experiment EPMA study was performed on cross-sections cut across the cylindrical axis of the tube.

RESULTS

Oxygen Transport kinetics

In reporting the permeation data we use previously derived relations that provide information about the relative roles of surface exchange versus bulk diffusion in the oxygen transport [7]. Under the assumptions discussed by Kim *et al.* [7], if the oxygen transport is limited by the rate of oxygen exchange between the gas and the membrane surface the following relation holds:

$$F = \frac{\pi r_1 r_2 w c_i k_{io}}{r_1 + r_2} \left(\sqrt{p_1/p_0} - \sqrt{p_2/p_0} \right) \qquad (1)$$

where r_1 and r_2 are the radial coordinates of the outer and inner tube walls respectively, w is the tube length, c_i is the density of oxygen ions, k_{io} is the surface exchange coefficient and p_o is 1 atm oxygen pressure. Eq. (1) predicts that under purely surface limited transport conditions a plot of the oxygen flow rate versus the parameter $(p_1/p_o)^{0.5}-(p_2/p_o)^{0.5}$ will be linear.

For the alternative case in which transport is limited by the bulk diffusion rate through the membrane the flow rate F is given by:

$$F = \frac{\pi w c_i D_a}{2\ln(r_1/r_2)}\ln(p_1/p_2) \tag{2}$$

where D_a is the ambipolar oxygen ion-electron hole diffusion coefficient and the flow rate is now linearly proportional to $\ln(p_1/p_2)$. In Figs. 1a and 1b our measured oxygen flow rates are plotted against the pressure terms in Eqs. (1) and (2) respectively.

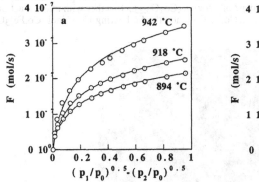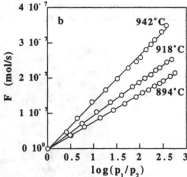

Fig.1: Oxygen flow plotted vs. (a) $(p_1/p_0)^{0.5}-(p_2/p_0)^{0.5}$ and (b) $\log(p_1/p_2)$ in a tubular membrane with the composition $SrFeCo_{0.5}O_{3.25-\delta}$ measured at the indicated temperatures. The inlet pressure p_1 (outside tube) varied from 0.001 to 1 atm and the outlet pressure p_2 (inside tube) varied from 0.0003 to 0.0027 atm. The dimension of tube is $r_1 = 0.66$ cm, $r_2 = 0.50$ cm, and $w = 2.87$ cm.

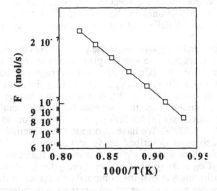

Fig.2. The temperature dependence of oxygen permeation in a tubular membrane with the composition $SrFeCo_{0.5}O_{3.25-\delta}$.

The non-linear dependence in Fig. 1a when compared with the linear behavior in Fig. 1b demonstrates that transport is bulk diffusion limited with little or no contribution from the surface exchange kinetics. Based on the slopes of the plots in Fig. 1b, the values of an effective diffusion coefficient (see below) were determined from eqn. 2 to be 4.48×10^{-8}, 3.18×10^{-8} and 2.58×10^{-8} cm^2/sec at 942, 918, and 894°C, respectively.

The temperature dependence of the oxygen flow through the $SrFeCo_{0.5}O_{3.25-\delta}$ membrane in the temperature range 796°C to 942°C is shown in Fig.2. The apparent activation energy based on these data is 73 kJ/mol. An oxygen flow of 8.49×10^{-8} mol/s was obtained at 796°C with oxygen partial pressures of 0.21 and 7.0×10^{-4} atm at the high and low pressure sides, respectively.

Membrane Microstructure and Phase Stability

Backscattered electron imaging and X-ray microanalysis of the pre- and post-experimental membrane showed that both samples consist of $SrFe_{1.5-x}Co_xO_{3.25-\delta}$ together with major fractions of two additional phases, the perovskite $Sr(Co,Fe)O_{3-\delta}$ and the Fe-bearing Co-spinel $(Co,Fe)_3O_4$ (Fig. 3).

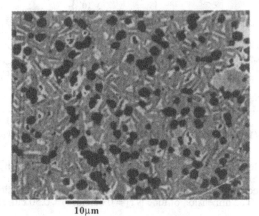

10μm

Fig. 3. Backscattered electron image for $SrFeCo_{0.5}O_{3.25-\delta}$ membrane after oxygen permeation measurements. Contrast is proportional to the average Z of the phases present which are $Sr(Co,Fe)O_{3-\delta}$ (light gray to white), $SrFe_{1.5-x}Co_xO_{3.25-\delta}$ (medium gray) and $(Co,Fe)_3O_4$ (black).

Phase volume proportions and compositions in the pre- and post-experimental ceramics are not significantly different. The average phase volume proportions are 68% for $SrFe_{1.5-x}Co_xO_{3.25-\delta}$, 17% for $Sr(Co,Fe)O_{3-\delta}$ and 15% for $(Co,Fe)_3O_4$. The average compositions of the different phases are $SrFe_{1.08}Co_{0.42}O_{3.25-\delta}$, $SrCo_{0.20}Fe_{0.80}O_{3-\delta}$ and $Co_{2.5}Fe_{0.5}O_4$ for the Co-oxide. These data are compared to the bulk composition of the membrane on a ternary composition diagram in Fig.4. The same three-phase assemblage that we observe in the membrane ceramic from the persent study was reported previously [9] for $SrFe_{1.5-x}Co_xO_{3.25-\delta}$ compositions with x = 0.425 to 0.525 synthesized at 1150° to 1200°C. We have also found this assemblage in a sample prepared by the mixed oxide synthesis route described above and annealed at 1150°C in air. The phase compositions in this sample show a much tighter distribution than that for the membrane, with average values of $SrFe_{1.13}Co_{0.37}O_{3.25-\delta}$ and $SrCo_{0.18}Fe_{0.84}O_{3-\delta}$ for the two perovskite related phases (Fig. 4). These results suggest that the composite character of the membrane developed initially when the membrane ceramic underwent final sintering at high-temperature (1090°C) during its preparation.

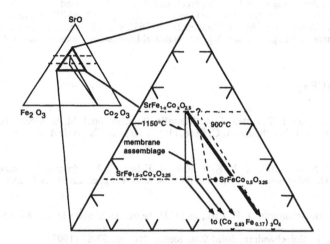

Fig. 4. Ternary composition diagram showing approximate phase relations based on phase compositions in a mixed-oxide sample prepared at 1150°C, this sample annealed at 900°C, and a freeze dried sample heated to 900°C.

In order to place constraints on the stability relations expected for the membrane's bulk composition at temperatures similar to the permeation experiment, we annealed a pelletized sample of the membrane precursor powder at 900°C in air for 10 days followed by drop quenching into liquid N_2. Electron microprobe data showed that the pellet consisted of $SrFe_{1.02}Co_{0.48}O_{3.25-\delta}$ with only very minor amounts of Co-oxide. This suggests that the temperature for stability of the composition $SrFeCo_{0.5}O_{3.25-\delta}$ as a single phase in air is very close to the temperature range of our permeation experiments. Based on the phase relations we would expect the membrane to convert back to a single phase of $SrFeCo_{0.5}O_{3.25-\delta}$, but this did not occur. Therefore it is concluded that during the permeation experiment the membrane slowly re-equilibrates to a single phase but never completed the transition due to slow kinetics. The permeation properties measured and reported here are those for a composite which, although it consists mostly of $SrFeCo_{0.5}O_{3.25-\delta}$, is not a homogeneous single-phase composition and the diffusion coefficients discussed above are only effective diffusion coefficients for the three phase composite membrane.

CONCLUSIONS

The oxygen transport in an oxide membrane with the bulk composition $SrFeCo_{0.5}O_{3.25-\delta}$ was determined to be bulk diffusion limited with an apparent activation energy of 73kJ/mol. Under the sintering conditions (1090°C) that were necessary to fabricate the dense tubular membrane used for permeation experiments, the $SrFeCo_{0.5}O_{3.25-\delta}$ phase was found to be unstable with respect to a more iron rich composition. Microstructural evaluation of membranes as prepared and after permeation studies gave average compositions and volume fractions for the phases present: $SrFe_{1.08}Co_{0.42}O_{3.25-\delta}$ (68 vol%), $SrCo_{0.2}Fe_{0.80}O_{3-\delta}$ (17 vol%) and $Co_{2.5}Fe_{0.5}O_4$ (15 vol%). The kinetics of re-equilibration of the membrane at lower temperatures (800-900°C), after the high temperature densification, to single phase $SrFeCo_{0.5}O_{3.25-\delta}$ are slow.

ACKNOWLEDGMENTS

We thank the Texas Center for Superconductivity and the Robert A. Welch Foundation for financial support of this work. The work was supported by the MRSEC program of the National Science Foundation under Award Number DMR-9632667. SK acknowledges support from the US Department of Energy through Argonne National Laboratories under Contract W-31-109-Eng-38.

REFERENCES

1. U. Balachandran, J. T. Dusek, R. L. Mieville, R. B. Poeppel, M. S. Kleefisch, S. Pei, T. P. Kobylinski, C. A. Udovich, and A. C. Bose, (1995) *Applied Catalysis A: General*, v. 133, p. 19-29, (1995).

2. S. Pei, M. S. Kleefisch, T. P. Kobylinski, J. Faber, C. A. Udovich, V. Zhang-McCoy, B. Dabrowski, U. Balachandran, R. L. Mieville, and R. B. Poeppel, *Catalysis Letters*, v. 30, p. 201-212 (1995)

3. B. Ma, U. Balachandran, J.-H. Park, and C.U. Segre, Solid State Ionics, **83**, 65 (1996).

4. B. Ma and U. Balachandran, Solid State Ionics, **100**, pp. 53-62 (1997)

5. A. Yoshiasa, K. Ueno, F. Kanamaru and H. Horiuchi, Mat. Res. Bull., **21**, pp. 175-181 (1986).

6. T.H. Lee, Y.L. Yang, A.J. Jacobson, B. Abeles and M. Zhou, Solid State Ionics, **100**, 77 (1997)

7. S. Kim Y.L. Yang, A.J. Jacobson and B. Abeles, Solid State Ionics, in press.

8. S. Kim, Y.L. Yang, A.J. Jacobson and B. Abeles, submitted to Solid State Ionics Proc., Hawaii, USA, 1997.

9. S. Guggilla and A. Manthiram, J. Electrochem. Soc., **144**, pp. L120-L122 (1997).

Part IV

Lithium Ion Rechargeable
Batteries – Cathode Materials

NOVEL SYNTHESIS PROCESS AND CHARACTERIZATION OF Li-Mn-O SPINELS FOR RECHARGEABLE LITHIUM BATTERIES

Toshimi Takada, Hirotoshi Enoki, Etsuo Akiba, Takenori Ishizu*, and Tatsuo Horiba*
National Institute of Materials and Chemical Research, Ibaraki, JAPAN; takada@nimc.go.jp
* Shin-Kobe Electric Machinery Ltd, Saitama, JAPAN.

ABSTRACT

A new process has been developed for the synthesis of well-crystallized Li-Mn-O spinels with a homogeneous composition $Li[Li_xMn_{2-x}]O_4$ ($0.0 \leq x \leq 0.333$) using the stoichiometry mixtures of lithium acetate and manganese nitrate as starting materials. The crystal structure of these compounds was studied with Rietveld refinements of the X-ray diffraction profiles. The lattice parameter of the spinels shows a strong dependence on the composition and manganese oxidation state. SEM micrographs indicate that the crystallites appear as single crystals. The size of the crystallites are in the range of $0.1 - 2\mu m$, depending on the synthesis conditions. Samples with $x \leq 0.125$ show good electrode performance for the cell $Li/Li_{1+x}Mn_{2-x}O_4$ in the 4V region, whereas $Li_4Mn_5O_{12}$ ($x=0.333$) spinels show good cyclability with a rechargeable capacity of over 100mAh/g in the 3V region.

1. INTRODUCTION

The high energy density (120Wh/g) and good cycle life (500-1000 cycles) of commercial lithium ion batteries, which use $LiCoO_2$ as cathode and graphite or carbon as anode, make them very attractive for use in energy storage and powering emission free vehicles in the near future. In order to overcome the resource limitation and thus the high cost of Co, Li-Mn-O spinels are being studied extensively as a replacement for $LiCoO_2$. Recent research has clarified that the stoichiometric spinels, $Li[Li_xMn_{2-x}]O_4$ ($0.0 \leq x \leq 0.333$) with end members $LiMn_2O_4$ and $Li_4Mn_5O_{12}$, are of particular significance among the Li-Mn-O compounds because they can be used as electrodes in rechargeable 3V and 4V lithium batteries(1-3).

In this study, a new synthesis process has been developed to obtain well-crystallized lithium manganese oxide spinels with a homogenous composition $Li[Li_xMn_{2-x}]O_4$ ($0 \leq x \leq 0.333$). Redox titration, powder X-ray diffraction and Rietveld refinement were conducted to clarify the relations between the spinel lattice parameter, the Mn occupancy at $16d$ sites, and the average oxidation state of manganese. The electrode performance of these samples was examined using the cell $Li/Li[Li_xMn_{2-x}]O_4$ in both the 3V and 4V regions.

2. EXPERIMENTAL

Synthesis of Well Crystallized Li-Mn-O Spinels. 99.9% pure lithium acetate $LiOAc \cdot 2H_2O$, and manganese nitrate $Mn(NO_3)_2 \cdot 6H_2O$ (from WAKO Pure Chemical Industries, Ltd.) were used as starting materials. The raw materials with a molar ratio of Li/Mn $1/2 - 4/5$ were first heated at 100°C to obtain a uniform solution, and then slowly oxidized around 250°C under flowing O_2 to convert the solution to a solid Li-Mn-O precursor. The precursor was then ground and pelletized (10mm in the diameter and height) before heating at a temperature ranging from 400 to 900°C for $1 - 3$ days. Samples with a composition $Li[Li_xMn_{2-x}]O_4$ ($0 \leq x \leq 0.125$) were obtained by heat treatments, first at 650°C, then at 800°C for 1 day each. Details of the preparation process were described in our previous report (4).

Characterization. X-ray powder diffraction data were collected between $2\theta = 15° - 120°$ with a step interval of 0.02° at room temperature on a Rigaku RAX-I X-ray diffractometer with monochromatized Cu-Kα radiation ($\lambda=1.5406$ Å) at 40kV, 30mA. Structural refinements were carried out with a Rietveld refinement program, *RIETAN-97* β-version (5). Scanning electron micrographs were taken at room temperature on a Hitachi S-800 microscope equipped with a field emission gun, at 10kV. The average oxidation state of Mn (Mn valency hereafter) in the final spinel products was determined from the active oxygen content which was measured by the standard volumetric method (6).

Electrode Performance. The cyclic voltammogram of Li/Li[Li$_x$Mn$_{2-x}$]O$_4$ (0≤x≤0.333) was collected at room temperature with a conventional electrochemical cell using Li metal as negative and reference electrodes, and 1.0 M LiPF$_6$ dissolved in ethylene carbonate (EC) and dimethyl carbonate (DMC) (1:1 in volume) as the electrolyte. The slurry was pressed onto an aluminum mesh current collector (10 × 10mm square). The charge/discharge capacity and cyclability were evaluated using the 2430 coin cell Li/Li[Li$_x$Mn$_{2-x}$]O$_4$.

3. RESULTS AND DISCUSSION
3.1 Synthesis and Structural Changes of Li-Mn-O Spniels

Figure 1 shows a typical SEM image of the spinel Li$_{1+x}$Mn$_{2-x}$O$_4$ crystallites in the samples with a Li/Mn=3/5 (x=0.125). At temperatures below 800℃, the size of the crystallites ranged from 0.1 to 0.3μm. At higher temperatures, well-developed polyhedra (mainly octahedra bounded by eight (111) planes, were clearly observed, and the size of the crystallites increased to 1-2μm in the samples prepared at 900℃.

Fig. 1. Typical SEM micrographs for the spinel Li$_{1+x}$Mn$_{2-x}$O$_4$ samples (x=0.125).

Rietveld refinements were carried out for all samples synthesized at temperatures between 400 and 900℃ using the space group $Fd\bar{3}m$. For 800-900℃ samples with a Li/Mn=5/7 (x=0.25) or 4/5 (x=0.333), the refinements included two crystalline phases: spinel Li[Li$_x$Mn$_{2-x}$]O$_4$ and Li$_2$MnO$_3$ (C/2m). Details of the procedure are described in our previous reports (7,8). The refined lattice parameter **a**, the occupancy of manganese at the 16d site, **g(Mn)**, and the Mn valency are plotted against the synthesis temperature and the Li/Mn ratio in the precursors in Figs. 2(a), 2(b), and 2(c), respectively. Clearly, synthesis temperature strongly affects the spinel structure in all samples with Li/Mn ranging from 1/2 (x=0.0) to 4/5 (x=0.333). As a result, a higher synthesis temperature gives a larger lattice parameter. The g(Mn) at 16d sites also increases with the increase of the synthesis temperature, whereas the Mn valency decreases accordingly. The increase of the lattice parameter with synthesis temperature, can therefore be ascribed to the reduction of a part of Mn^{4+} to Mn^{3+} at elevated temperatures because Mn^{3+} ions are more stable than Mn^{4+} at temperatures above 580℃, and the effective ionic radii of octahedrally coordinated Mn^{3+} ions (0.645, at high spin state) is larger than that of Mn^{4+} (0.56 Å).

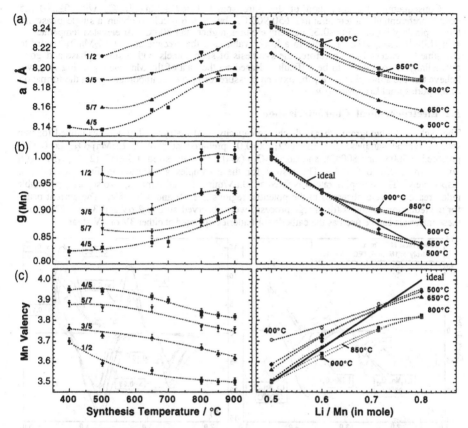

Fig.2. Plot of (a) the lattice parameter **a**, (b) the refined Mn occupancy at 16d sites, **g(Mn)**, and (c) Mn valency for spinel $Li_{1+x}Mn_{2-x}O_4$ against the synthesis temperature and the Li/Mn ratio in the precursors.

The spinel lattice parameter decreases in proportion to the Li/Mn ratio. Accordingly, the refined g(Mn) decreases with increasing Li/Mn ratio. The line designated as **ideal** for g(Mn) was calculated from the ratio of Li/Mn corresponding to the stoichiometric $(Li)_{8d}[Li_xMn_{2-x}]_{16d}O_4$ ($0 \leq x \leq 0.333$) spinels. Clearly, for Li/Mn$\leq$0.6, g(Mn) is very close to that of single phase $(Li)_{8d}[Li_xMn_{2-x}]_{16d}O_4$ with $0 \leq x \leq 0.125$, except for those samples synthesized at a temperature below 650°C. This can be ascribed to the formation of the minority phase Mn_2O_3 in the samples synthesized at/under 650°C as detected by X-ray diffraction. On the contrary, when Li/Mn>0.6, g(Mn) is very close to that of the pure spinel $(Li)_{8d}[Li_xMn_{2-x}]_{16d}O_4$ with $0.125 < x \leq 0.333$ at temperatures below 650°C, but deviates from the line-**ideal** as synthesis temperature is raised above 800°C. Strict correspondence of Mn valency with g(Mn) was obtained (Fig. 2(c)), which verifies the validity of the refined g(Mn) from XRD data by the Rietveld method, and the formation of pure spinel phase with a stoichiometric composition $Li[Li_xMn_{2-x}]_{16d}O_4$ ($0 \leq x < 0.33$) under certain conditions. Note that the Mn valency was found to be somehow lower than 4.0 in those samples with Li/Mn=4/5, which indicates the difficulty of synthesizing the stoichiometric $Li_4Mn_5O_{12}$ with all Mn in the 4+ state.

Consequently, the formation of the pure spinel $(Li)_{8a}[Li_xMn_{2-x}]_{16d}O_4$ with $0 \leq x \leq 0.125$ occurs preferentially at temperatures 650-900°C, but it is difficult to obtain a single phase for those spinels with $0.125 < x \leq 0.333$ at temperatures higher than 800°C. At elevated temperatures, x in $(Li)_{8a}[Li_xMn_{2-x}]_{16d}O_4$ shifts to a lower value and the precipitation of Li_2MnO_3 occurs to consume the excess Li. Therefore, the synthesis of such spinels with a single phase and good crystallinity is considered to be difficult. Based on the data of the Mn valency in Fig.2(c), we believe that the atmosphere. i.e., the oxygen pressure also plays an important role in the formation of the pure spinel $Li_{1+x}Mn_{2-x}O_4$.

3.2 Electrochemical Characterization

The cyclic voltammogram, charge-discharge capacity, and the cyclability were examined for two sets of typical samples: (1). $LiMn_2O_4$ (x=0.0, prepared at 800°C), $Li_9Mn_{15}O_{32}$ (x=0.125, prepared at 500°C and 800°C), and $Li_4Mn_5O_{12}$ (x=0.33) synthesized at 400°C. (2) $Li_{1+x}Mn_{2-x}O_4$ with x from 0.0 to 0.125. The results for these samples are shown in Fig 3(a) and (b), respectively. The sweeping rate was fixed at 20mV/min. For all samples, two separate reversible redox reactions were observed in the potential regions 2-3.5V and 3.5-4.5V. The current peaks for the reduction reactions (discharge process) were observed around 2.6V and 3.8V. Therefore, these materials can be used as the cathode for batteries operated at either 3V or 4V stage.

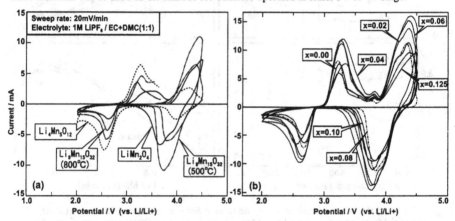

Fig.3. Cyclic voltammogram of the cells $Li/Li_{1+x}Mn_{2-x}O_4$ ($0.0 \leq x \leq 0.333$) for the first cycle.

The discharge profiles for these samples in both the 4V and 3V region are shown in Figure 4(a) and (b), respectively. In the 4V region, voltage plateaus were observed between 4.1V and 3.8V, and the discharge capacity was in the order $LiMn_2O_4 > Li_9Mn_{15}O_{32}$ (800°C) $> Li_9Mn_{15}O_{32}$ (500°C) $> Li_4Mn_5O_{12}$. Similar results were obtained for the samples $Li_{1+x}Mn_{2-x}O_4$; the discharge capacity decreased as x increased from 0.0 to 0.125. Also, for samples with $x \geq 0.06$, the voltage changed linearly from 4.1 to 4.0V, indicating the difference in the structure changes during the discharge process between these samples. In the 3V region, voltage plateaus were observed at 2.8V, the discharge capacity was in the order $Li_4Mn_5O_{12} > Li_9Mn_{15}O_{32}$ (800°C) $> LiMn_2O_4 > Li_9Mn_{15}O_{32}$ (500°C). For $Li_{1+x}Mn_{2-x}O_4$ samples, the discharge capacity varies with the cut-off voltage. In the region 3.6-1.5V, samples with a higher x give larger discharge capacity. However, in the voltage range of 3.6-2.5V, the discharge capacity varies independently with x. Apparently, the discharge process in the 3V region is complex. For $x \geq 0.06$ samples, it seems that two reactions occur in the ranges 3.6-2.3V and 2.3-1.5V. Therefore, it would be interesting to further investigate how the structure of these samples changes during the discharge process in both the 3V and 4V regions.

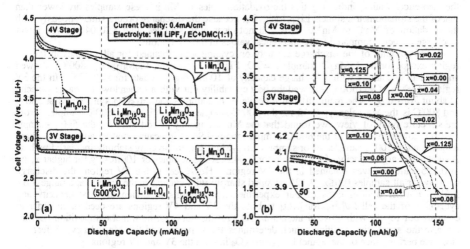

Fig.4. Discharge profiles of the cells Li/Li$_{1+x}$Mn$_{2-x}$O$_4$ (0.0≤x≤0.333) in the 4V region: 4.5-3.6V, and in the 3V region: 3.6-2.5V or 3.6-1.5V, for the first cycle.

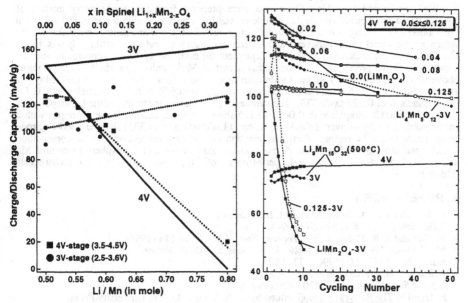

Fig.5. Discharge capacities of the cell Li/Li$_{1+x}$Mn$_{2-x}$O$_4$ (0.0≤x≤0.333).

Fig.6. Cyclability of the cell Li/Li$_{1+x}$Mn$_{2-x}$O$_4$ (0.0≤x≤0.333).

The discharge capacities of these samples in the 3V and 4V region are plotted in the Fig.5, along with the theoretical values for the stoichiometric spinels Li$_{1+x}$Mn$_{2-x}$O$_4$ (0.0≤x≤0.333). Clearly, in the 3V region, the experiment values are rather low, which suggests that it is not possible to reach 100% occupancy of the 16c sites (1.0 Li per formula Li$_{1+x}$Mn$_{2-x}$O$_4$). On the contrary, in the 4V region, the experiment values for the samples with a x≥0.06 were higher than the theoretical values, indicating that the oxidation states of Mn in these samples are lower than that of the stoichiometric spinels $Li_{1+x}Mn_{2-x}O_4$, because the discharge occurs simultaneously with the oxidation of Mn^{3+} to Mn^{4+}. It is still not clear why samples with $x \le 0.04$ showed lower discharge capacity compared to the theoretical values in the 4V region .

Good stability of the discharge capacity on cycling was obtained for all samples in the 4V region except for the $LiMn_2O_4$ sample (x=0.0, Fig.6). Especially, $Li_9Mn_{15}O_{32}$ (800°C) showed an excellent cyclability with a discharge capacity of 100mAh/g (see data for x=0.125). In the 3V region, $Li_9Mn_{15}O_{32}$ (500°C) showed the best cyclability but with a rather low discharge capacity of about 70mAh/g. The main difference between the samples $Li_9Mn_{15}O_{32}$ synthesized at 500°C and 800°C is the Mn valency, 3.72 for the 500°C-sample, and 3.65 for the 800°C-sample. From the structure point of view, we found that the Mn occupancy at 16d sites for the 500°C-sample (0.89) is lower than that of the 800°C-sample (about 0.94), indicating that an amorphous phase remains in the 500°C sample. For samples $Li_{1+x}Mn_{2-x}O_4$, the cycling stability increased with x. A stable discharge capacity of 120mAh/g for $0.04 \le x \le 0.08$ samples or 100mAh/g for higher x, up to 0.125 samples, was obtained in the 4V region. However, in the 3V region, the discharge capacity for all samples $0.0 \le x \le 0.125$ dropped sharply with cycling in the cell voltage range 3.6 - 2.5V. Further optimizations of our experimental conditions to obtain the best capacity and cyclability of the cell $Li/Li_{1+x}Mn_{2-x}O_4$ in both the 3V and 4V regions are needed, in order to compare our results with those in literature. As a result, not only the spinel composition (Li/Mn) but also the synthesis condition which determines the structure and the Mn valency, affect the electrode performance of the spinel $Li_{1+x}Mn_{2-x}O_4$ in both the 3V and 4V regions.

4. CONCLUSION

Well crystallized $Li_{1+x}Mn_{2-x}O_4$ spinels were prepared from the stoichiometry mixture of $LiOAc \cdot 2H_2O$ and $Mn(NO_3)_2 \cdot 6H_2O$. Single-crystal-like crystallites with a size of 0.1 - 2.0 μm were observed. Using powder X-ray diffraction and Rietveld refinements, we were able to determine the spinel lattice parameter and the Mn occupancy at 16d sites, g(Mn). It was found that the lattice parameter and g(Mn) are dependent on both the synthesis temperature and the Li/Mn ratio in the precursor. We found that a part of Mn^{4+} reduces to Mn^{3+} at temperatures above 650°C, independently of the Li/Mn ratio in the precursors. Formation of $Li_{1+x}Mn_{2-x}O_4$ spinels with $0 \le x \le 0.125$ occurs at temperatures ranging from 650 to 900°C, but below 800°C for those spinels with $0.125 < x < 0.333$. The synthesis temperature for stoichiometric $Li_4Mn_5O_{12}$ has to be under 650°C. Samples with $0.04 \le x \le 0.125$ showed good cyclability in the 4V region with a rechargeable capacity of over 100mAh/g, whereas $Li_4Mn_5O_{12}$ (x=0.333) showed good electrode performance in the 3V region. Particular attention should be given not only to the composition (Li/Mn) but also to synthesis conditions, such as temperature and atmosphere to control Mn valency and therefore, the electrode performance of $Li_{1+x}Mn_{2-x}O_4$ spinels for rechargeable lithium batteries.

5. REFERENCES

1. R. J. Gummow, A. de Kock, and M. M. Thackeray,
   Solid State Ionics, 69, 59 - 67 (1994).
2. Y. Gao and J. R. Dahn, J. Electrochem. Soc., 143, 100 - 114 (1996).
3. M. M.Thackeray, M. F. Mansuetto, D.W. Dees, and D. R. Vissers,
   Mater. Res. Bull., 31 (1996) 133- 140.
4. T. Takada, H. Hayakawa, and E. Akiba,
   J. Solid State Chem., 115, 420 - 426 (1995).
5. F. Izumi in The Rietveld Method, edited by R. A. Young, Oxford University Press,
   New York (1993), pp. 236-253.
6. Japan Industrial Standard (JIS) M8233-1982, "Determination of Active Oxygen
   Content in Manganese Ores".
7. T. Takada, H. Hayakawa, E. Akiba, F. Izumi, and B. Chakoumakos,
   J. Solid State Chem., 130, 74-80 (1997).
8. T. Takada, H. Enoki, H. Hayakawa, and E. Akiba, J.Solid State Chem., submitted.

NOVEL CATHODE MATERIALS BASED ON ORGANIC COUPLES FOR LITHIUM BATTERIES

N. RAVET[a], C. MICHOT[b], M. ARMAND[a]
a Département de Chimie, Université de Montréal , Montréal QC H3C 3J7, Canada;
b LEPMI (UMR 5631 INPG-CNRS), Institut National Polytechnique de Grenoble, B.P. 75, 38402 Saint Martin d'Hères, France.

ABSTRACT

The electrochemical reduction of the oxocarbons: squarate, croconate and especially rhodizonate lithium salts have been studied in all solid state lithium batteries. Lithium rhodizonate cells were tested on cycling in the 1.5 - 3.5 V potential range The reduction of lithium rhodizonate occurs in two waves of two electrons. The number of electrons transferred in reduction on the first cycle was around 3.5 based on a capacity of 515 mA.h.g⁻¹ and a discharge depth of 87 %. This process is quite reversible but we observed a fast decline of the capacity on cycling. This loss of capacity may be attributed to residual water in the salt. The reduction of the lithium croconate occurs at a potential of 1.8 V in a quasi-irreversible process. We could not observe the reduction of lithium squarate which occurs in the potential range where the lithium is inserted in carbon black. We also report an investigation on rhodizonate salts of transition metals. The best results, in term of capacity, on the 1.5 - 3.5 V potential range, were obtained with copper rhodizonate which exhibits a capacity of 579 mA.h.g⁻¹ on the first discharge.

INTRODUCTION

A wealth of studies dealing with advanced battery systems [1] have appeared since the early 80's. This interest is driven by the perspective of practical applications and by innovative concepts in this field. A milestone was the introduction of the solid-state insertion cathode in 1973. The polymer electrolytes has then offered the possibility of practical solid rechargeable lithium battery [2]. Later, the insertion concept applied to the negative electrode ("rocking chair") has led rapidly to the commercialisation of the so called lithium-ion technology [3]. Cathode materials fall into the main divisions of chemistry, organic or inorganic: redox polymers and transition metal intercalation compounds.

Conjugated polymer used as cathode materials, such as polyacetylene, polypyrrole or polyaniline are p-dopable materials. The charge-discharge process is a simple redox reaction of the polymer compensated by anion intercalation. However, low specific energy and stability make them less attractive. More promising materials were proposed by Visco [4-5] based on solid redox polymerization electrodes. In this kind of cathode, the active materials are polymeric disulfides that undergo reductive depolymerization upon discharge and oxidative polymerization upon charge. Very high energy densities and decent cycling behavior are obtained with the polydisulfide derived from dimercapto thiadiazole [6], active around 2.8 - 2.9 V and the maximum discharge depth is 80% if the capacity is less than 1 C.cm⁻² (which corresponds to a overall capacity of about 100 mAh.g⁻¹). However there is evidence for solubility of the reduction intermediates and of the dithiolate obtained

Transition metal based intercalation compounds constitute the bulk of studies in this area [1, 7-8]. Today, the most popular are $LiCoO_2$, $LiNiO_2$ and $LiMn_2O_4$. $LiCoO_2$ and NiO_2 [8-9] have a layered structure and provide good capacities in cycling (130 - 150 mAh.g^{-1} respectively), with a discharge plateau being near 4V. The high cost of $LiCoO_2$ and the difficulty to synthesize $LiNiO_2$ in its electrochemically active structure make $LiMn_2O_4$ more attractive.

$LiMn_2O_4$ is generally described as an inexpensive and rather "green" material. Thus the $LiMn_2O_4$ spinel phase has been extensively studied and considerable amount of work has been focused on the optimization of this material. Today, the reversible capacity on the 4V range is almost the same as for $LiCoO_2$ [10]. A synthesis of $LiMnO_2$ having the same layered structure than $LiCoO_2$ has been recently reported [9-11]. The initial capacity announced is 270 mAh.g^{-1} but the cycling performance of this material is unsatisfactory. Mn^{2+} can dissolve in the electrolyte and all 4V materials exceed the stability windows of known electrolytes, raising safety issues.

An interesting recent development is the use of compounds with olivine structure such as $LiFePO_4$ [12] or with a NASICON framework such as $Li_xFe_2(SO_4)_3$ [13]. In both cases, lithium insertion occurs close to 3.6 V vs. Li^+/Li° with a reversible capacity ≈ 100 mAh.g^{-1} at low discharge current densities. Increasing the current densities leads to a loss of capacity.

Quinones are well known to give reversible redox couples in aqueous solutions and probably the most ubiquitous mediators in biological systems. In nonprotic media, radical anionic species are involved but the potentials tend to be rather low (≈ 2.2 V vs. Li^+/Li°) and in the neutral form, they tend to be soluble in the electrolyte media [14].

In the present paper, we report on our studies on the electrochemical behavior in all solid state lithium cells of a novel family of cathodic materials based on organic compounds : the oxocarbons.

The term oxocarbon designates a series of compounds in which all, or by extension nearly all, the carbon atoms are bonded to carbonyl or enolic oxygens as well as their hydrated or deprotonated equivalents. The most important oxocarbons are the cyclic compounds $C_nO_n^{x-}$ such as the four, five and six carbons rings : squarate, croconate and rhodizonate.

squarate   croconate   rhodizonate

The stability of these compounds decreases when the number of carbons increase. This is the reason why the seven or eight member rings have not been reported. The chemistry and the main features of oxocarbons are summarized in reviews [15 - 18]. Oxocarbons are known since the early years of the 19th century and probably predate the transition from inorganic to organic chemistry by Wohler [19]. Though, the aromatic nature of the monocyclic oxocarbons anions was recognized less than 30 years ago [20]. More recent studies [21] tend to question this point and claimed that for monocyclic anions the aromaticity decreases rapidly with increasing ring size. Nevertheless, these anions are stabilized by delocalization of π electrons at the ring periphery. Crystallographic and spectroscopic studies show a symmetrical planar structure for squarate, croconate and rhodizonate anions [15], with the absence of the typical C=O stretching band at 1800 cm^{-1}.

Only a few electrochemical studies are reported in the literature on oxocarbons solutions and most of them in aqueous electrolyte [22-24]. More recently, some studies in aprotic electrolytes referring to the oxidation process of croconates [25] or to the oxidation reduction process of squaric acid [26] have appeared.

To our knowledge, the only studies on the oxydo-reduction process of rhodizonic acid and its salts have been carried out in aqueous media, though the existence of transient odd-electron radical anions produced electrochemically have been observed at low temperature in CH$_2$Cl$_2$ [27].

Potentiometric titrations have shown that rhodizonic acid can be reduced reversibly in a two steps sequence where each step involves the transfer of two electrons and two protons.

$$\text{rhodizonic acid} \quad \xrightleftharpoons[-2H^++2e^-]{+2H^++2e^-} \quad \text{tetra hydroxy quinone} \quad \xrightleftharpoons[-2H^++2e^-]{+2H^++2e^-} \quad \text{hexahydroxybenzene}$$

The standard reduction potentials reported are 0.41 V$_{SHE}$ for the first wave and 0.35 V$_{SHE}$ for the second one. The complete delocalization of the electronic charges around the ring stabilizes both anionic and radical structures. The electrochemical behavior of rhodizonic acid in aqueous media depends strongly on the pH of the solutions [22].

These reduction potentials and the high theoretical capacity (589 mAh.g^{-1} for lithium rhodizonate) make this material interesting to be tested in all solid state lithium batteries. It was also expected that these materials would show negligible solubility in non-protic electrolytes. We report on our investigations on the lithium insertion in squarate, croconate and rhodizonate salts used as cathode materials. We have also tested some transition metal rhodizonates derivatives of Cu, Pb and Ba.

EXPERIMENTAL

All the salts were dried under dynamic vacuum at room temperature for a few hours and then at 55 °C overnight.

Lithium rhodizonate and squarate were obtained by neutralization of the commercial acid with a stoichiometric amount of LiOH in methanol. Methanol was removed at 35 °C on a rotary evaporator leaving a greenish black powder with metallic luster (C6) or a white powder (C4).

265

Lithium croconate is synthesized from barium croconate obtained as described in reference [28] using lithium carbonate instead of sodium carbonate (yellow powder).

Copper, barium, and lead rhodizonates are obtained by precipitation of the corresponding salts in aqueous media from solutions of potassium rhodizonate and stoichiometric quantities of $CuSO_4$, $BaCl_2$ and $Pb(CO_2CH_3)_2$ respectively.

Electrochemical characterization of the materials were made in coin type cells. The electroactive material was ground and blended with Ketjen black® and PEO (400,000 g/mol) in the weight ratio 34 / 10 / 56. Acetonitrile was added to dissolve the PEO and the mixture was spread on stainless steel current collectors (0.8 cm^2). The composite cathode was dried under dynamic vacuum at 55 °C overnight. Electrolyte was a $PEO_{20}LiTFSI$ film (σ ≈ 10^{-3} S.cm^{-1} at 80 °C, TFSI = [(CF$_3$SO$_2$)$_2$N]) and the negative electrode was punched from a thin lithium layer pressed on a nickel foil. Coin cells were assembled in a helium atmosphere dry box. Cells performances were investigated at 80 °C using slow scan voltammetry (MacPile®). The cells were allowed to stabilize one hour at 80 °C before experiments.

DSC and TGA measurements were carried out on Perkin-Elmer apparatus Pyris 1 and TGA 7 respectively. For the DSC experiments, samples were tested in open alumina dishes and platinum crucibles were used for the TGA analyses. All analyses were carried out under nitrogen atmosphere with a heating rate of 10°C min-1.

RESULTS AND DISCUSSION

<u>Redox behavior of lithium squarate, croconate and rhodizonate</u>

The electrochemical behavior of these cells is reported on figure 1.

Investigations performed on squaric acid in dimethyl formamide have shown that this compound is reduced in two 1-electron step [26]. The two reduction process occur with 1 volt difference. In our case, the squarate did not show any electroactivity on the potential range investigated. The current wall observed at 1.3 V, also recorded in the case of croconate, is attributed to the lithium insertion in acetylene black.

For the croconic acid a complex reduction mechanism has been reported in aqueous media [24]. Depending on the pH, reductions of 1, 2, 3 or 4 electrons were observed with, in some cases, a dimerisation step . The described mechanism include reactions of protonation.

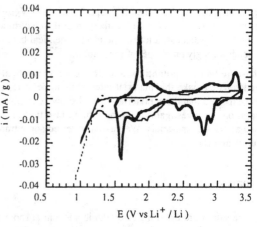

Figure 1 : First voltammetric scan on lithium salt of
· · · · · squarate
——— croconate
——— rhodizonate
(T = 80 °C, v = 20 mV / h)

In solid state the reduction of croconate occurs in two peaks of one electron. On cycling the croconate on the first redox couple it appears that the reduction is not reversible.

On decreasing the ring size the injection of electrons is inhibited due to the aromatic stabilization of electrons in the ring: compounds become less prone to electron injection.

The most interesting material in term of electroactivity is lithium rhodizonate which has deserved further attention.

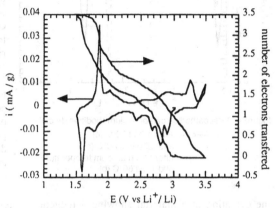

Figure 2 : First voltammetric scan on a lithium rodizonate cell
(T = 80 °C, v = 20 mV / h)

Lithium rhodizonate cells

Lithium rhodizonate cells were cycled in the 1.5 - 3.5 V potential range (figure 2). Within this range, the number of electrons transferred in reduction, on the first cycle, was near to 3.5. This yield with an actual capacity of 515 mA.h.g^{-1} and a discharge depth of 87 %. Reduction occurs in two distinct potential ranges, each corresponding to the transfer of two electrons. The first one between 3 - 2.2 V corresponds to the reduction of lithium rhodizonate to tetrahydroxy benzoquinone tetra-lithium salt and the second one between 1.9 - 1.5 V corresponds to the reduction of the tetrahydroxy benzoquinone tetra-lithium salt to hexahydroxy benzene hexa-lithium salt. The reduction of the lithium rhodizonate seems to be fully reversible. The same capacity is exchanged during the reduction and the oxidation scans.

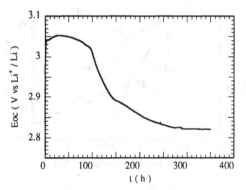

Figure 3 : Evolution of the open-circuit potential with the time of a lithium rhodizonate cell at a temperature of 80°C

The shape of the discharge curves depends strongly on the initial state of the cells. We have followed the open circuit potential of a lithium rhodizonate cell at 80 °C during 350 hours.

It first stabilizes at a value near 3.05 V but after 100 hours drops to 2.82 V before a second stabilization (figure 3).

On figure 4 one can see the volumetric trace of this cell heated at 80 °C for 350 hours (a) and the curve obtained for a cell maintained at room temperature for the same time (b). The third curve was obtained by cycling a freshly prepared cell which was allowed to stabilize at

267

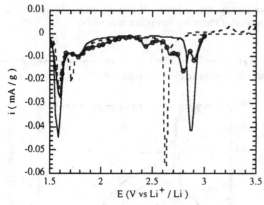

Figure 4 : First cathodic scan on lithium rhodizonate cells
——●—— After 1 h at 80°C (c)
- - - - - after 350 h at 80°C (a)
———— after 350 h at room temperature
and 1 h a 80°C (b)

80°C for one hour (c). In order to check that the decrease of the open circuit potential could not be attributed to a self discharge, the cell (a) was first subjected to an anodic scan. The oxidation phenomenon observed in this case is similar to the one obtained on a freshly made cell. The reduction curve of the cell c shows four peaks on the upper potential range. On this range, only one of these peaks is observed for each of the cells a and b. The cell stabilized at room temperature shows only the 2.85 V peak and the one maintained at 80 °C only the 2.65 V peak. We have to point out that in both cases, the discharge capacity obtained was around 3.5 electrons. The observed evolution cannot be attributed to a degradation of the active material. The time evolution of a material showing 4 reduction peaks to another one showing only one reduction peak, whose position depends on the temperature, is suggestive of a crystal reorganization occuring with time and/or temperature.

In order to point out a possible evolution of the morphology with the temperature, DSC and TGA studies were made on a carefully dried lithium rhodizonate sample (figure 5).

DSC scans do not show any phenomenon which could be associated with a recrystallization or a reorganization of the crystal structure. In the investigated temperature range, an endotherm is observed at 143 °C. This phenomenon is associated with the onset of a weight loss which starts above 50 °C as seen from the TGA scan. Degradation occurs at 240 °C on the DSC scan and the weight loss then recorded corresponds to the extrusion of one CO_2 molecule. We have observed that a sample heated at 150°C looses only about 6 % of its weight but returns to its original weight on cooling, only a few minutes after being exposed to ambient atmosphere.

We suggest that the reversible weight loss reported until 140-150 °C could be attributed to an hydratation- deshydratation process occuring without any morphological change. The regular weight loss indicates that water molecules are not strongly bonded in the crystal. The endothermal peak recorded on the DSC scan is probably due to the evaporation of crystallization water.

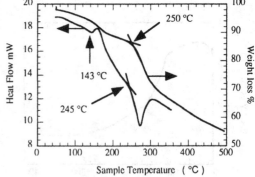

Figure 5 : TGA and DSC scans of lithium rhodizonate performed at 10 °C / min under nitrogen atmosphere

268

Indeed, we have observed by IR spectroscopy that water was always present whatever the drying process used. On a sample of potassium rhodizonate (Lancaster), the presence of water was not detected by IR spectroscopy.

After the drying process and before to be stored in the dry box, the cathodes were exposed to the room atmosphere. We could not exclude that they reabsorb small quantities of water.

In order to verify if the shift of the first reduction peak observed between the a and b cells was linked to the stabilization temperature, we have assembled a cell using a liquid electrolyte PC-DME-0.5M LiTFSI. The open circuit of this cell is somewhat lower than for other cells but no systematic evolution with time was observed. The first reduction peak of a lithium rhodizonate cell tested with a PC-DME liquid electrolyte at room temperature appears at the same value than for the 80°C cells.

The LiTFSI salt used in all the electrolytes is very hygroscopic and could act as a getter for the water left in the active material, eventually reacting with Li°.

The rate of the water diffusion in the lithium salt of rhodizonate depends on the crystal structure. There is only a few studies available on the literature on this subject. Oxocarbons salts of monovalent cations are described as layered compounds. Cations are located between the planes of the oxocarbons rings [29-31]. In some cases, a columnar structure has been clearly mentioned (similar to the stacking of TCNQ charge transfer complexes).

This low dimensionality would provide easy access for water and ion diffusion in lithium rhodizonate, taking into account the small size of the lithium ion leaving empty galleries. Bulkier cations (K^+) would hinder the diffusion process or occupy the larger sites.

The release of moisture not strongly bonded to the structure could explain the drop of the open circuit potential of figure 3 . This argument matches well with the quite abrupt decrease of the open circuit potential which would be initially a mixed potential (cathode + water redox couples). The differences observed on the first reduction peak potentials between the heated cell and the one kept at room temperature could be attributed to the lithium passivation and to the deshydratation of the active material. We suggest that the potential shift observed is probably due to an evolution of the water content in the structure of the rhodizonate salts.

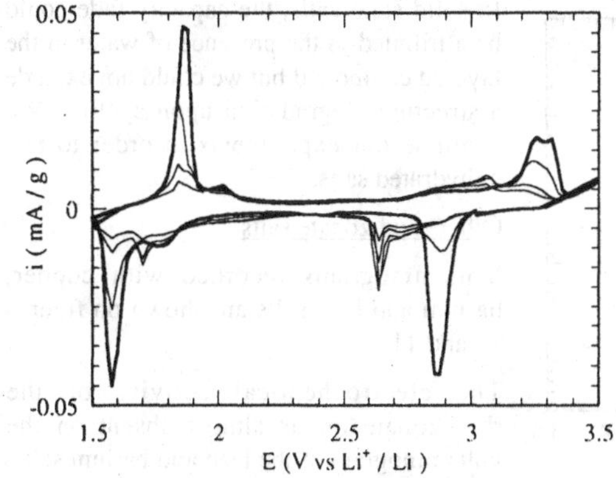

Cycling results on a lithium rhodizonate cell on the extended range 1.5 - 3.5 V and on 2.5 - 3.4 V are reported on figures 6 and 7 respectively. These results were obtained after a stabilization of 350 hours at room temperature. The cycling behavior of a cell made of potassium rhodizonate tested on the first redox couple is presented in figure 8.

Figure 6 :   Cycling voltammograms of a lithium rhodizonate cell on the 1.5-3.5 range (T= 80°C v=20 mV.h^{-1})

An noticeable decrease of the capacity is observed in all cases (figure 9). Though, the number of electrons transferred in the reduction and oxidation processes is almost the same (C/D = 1) for all the lithium rhodizonate cells investigated.

For the potassium salt, the charge passed during the oxidation process always exceeds that for reduction. Oxidation of the rhodizonate potassium salt may occurs in this case to explain the loss of capacity of this material.

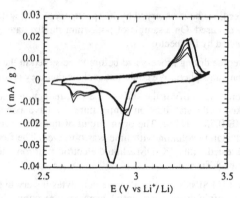

Figure 7 : Cycling traces of a lithium rhodizonate cell within the 2.5-3.4 V range (T= 80°C v=20 mV.h^{-1})

An anodic shift of the reduction peak is observed on the second scan as the probable result of the exchange of potassium ions for lithium in the structure.

For lithium rhodizonate, the first redox couple seems to be less reversible than the second one. After the first cycle the 2.85 V peak is almost disappeared. On the second cycle, we noted the apparition of the 2.65 V reduction peak and a 3 V oxidation peak for the cell tested on the extended range. The same behavior was observed with the unstabilized cell. No shift in peak positions was observed for the cell cycled after 350 hours at 80 °C.

For the cell tested on the first redox couple, a separation of the reduction peak was recorded. No significant evolution is observed for the oxidation process.

Peak shift from 2.85 to 2.65 V could again be the consequence of the release of water from the electroactive salt.

In aqueous electrolyte, an irreversibility of the rhodizonic acid have been reported only in acidic media where the acid is not ionized and two C=O group form a gem-diol .

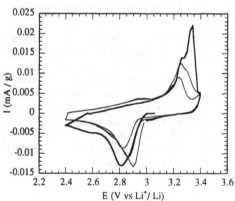

Figure 8 : Cycling traces of a potassium rhodizonate cell on the 2.5-3.4 range (T= 80°C v=20 mV.h^{-1})

In solid state cells, the capacity fade could be attributed to the presence of water in the layered compound but we could not exclude a structural degradation upon cycling. We continue our experiments in order to test unhydrated salts.

Other rhodizonate salts

Voltammograms recorded with copper, barium and lead salts are shown on figures 10 and 11.

The electrochemical activity of the rhodizonate ion is almost absent on the voltammogram of the lead and barium salts. In the case of the lead salt the peaks recorded are probably due to the reduction of Pb^{++}. With barium ions, which are not

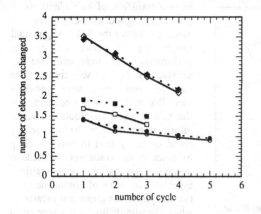

Figure 9 Evolution of the number of transferred electron during the charge process (dashed lines) and the discharge process (solid lines) for potassium rhodizonate (squares), lithium rhodizonate on the 1.5 - 3.5 V range (losanges) and on the 2.5 - 3.4 V range (circles)

electroactive on this range of potential, almost no redox activity is apparent.

Crystallographical studies carried out on hydrated complexes of croconate [32] and squarate [33] ions with divalent metals have shown that these compounds have an infinite polymeric structure. Metals are in the ring plane. Lead and barium could fill up the structure and inhibit the lithium diffusion. This inhibition is not observed with copper which is smaller.

The copper salt voltammogram is quite similar to that obtained for the lithium derivative. In the discharge process, there is only one peak added near 3V. This peak is likely due to signal the reduction of copper ions. During the first discharge, the capacity involved is 580 mA.h^{-1} which corresponds to 5 exchanged electrons. After oxidation, the same capacity is recovered. The 3V peak

disappears on the second scan (figure 10). This behavior suggest that a part of the charge evolved during the oxidation process does not lead to the re-oxidation of the copper.

As for several studied cells, the capacity involved during the oxidation process is larger than for reduction. It was interesting to determine whether an irreversible anodic degradation of the active material would occur. In the solid state and aprotic media, we cannot rule out the formation the neutral species cyclohexanehexone C_6O_6. This compound could diffuse on the electrolyte. The oxidation product of squarate are C_4O_4·$^-$ and transient C_4O_4 [26]. Electrochemical oxidation of rhodizonic acid in aqueous media leads to the transient triquinone C_6O_6. This compound rapidly reacts with water to a non-reductible octahydrate containing only $C(OH)_2$ entities $C_6H_{16}O_{14}$.[34].

However, scanning anodically to 3.5 V a lithium salt did not show any evidence of an important anodic phenomenon.

CONCLUSIONS

Lithium rhodizonate is a promising cathode material for lithium secondary or primary batteries. The reduction process is quasi reversible and occurs

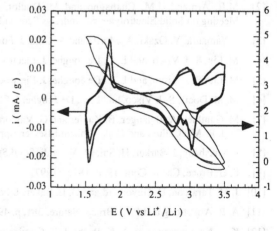

Figure 10 : Cycling traces of a copper rhodizonate cell (T= 80°C v=20 mV.h^{-1})

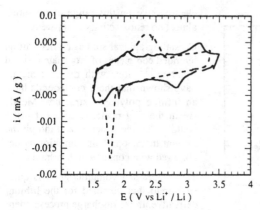

Figure 11 : Cycling traces of a barium (solid line) and lead (dashed lines) rhodizonate salts cells. (T= 80°C v=20 mV.h⁻¹)

in two transfers of two electrons. It provides a good capacity with 515 mA.h.g⁻¹ within the 3 - 1.5 V potential range. Up to now, the cycling performances of these cells are not satisfactory, but the sensitivity of the salt to water has been recently identified and this was unexpected as the salt is quasi insoluble in water; improvement of the drying process could probably lead to better cycling behavior. Also, no attempt was made to optimize the cathode composition, especially in terms of homogeneity of the composite, an important parameter when electron diffusion is expected to be rate limiting in this type of molecular compound. Copper rhodizonate is a candidate at least for in primary batteries because of its outstanding capacity of 580 mA.h.g⁻¹. It should be emphasized that rhodizonate salts could be made very inexpensively directly from the natural compound inositol [(CHOH)$_6$] and is expected to be environmentally benign.

ACKNOWLEDGEMENTS

We wish to thank ACEP Inc (Hydro-Québec and Yuasa C°) for financial support for this work. One of us (CM) is currently a recipient of a CNRS-Hydro-Quebec joint post-doctoral fellowship.

REFERENCES

[1] Handbook of batteries second edition, Edited by D. Linden McGraw-Hill, (1994).

[2] M.B. Armand, J.M. Chabagno and M. Duclot, extended abstracts, 2nd International Meeting on Solid Electrolytes, St Andrews Scotland, (1978)

[3] J. Yamaura, Y. Ozaki, A. Morita and A. Ohta, J. Power Sources 43-44, p. 233, (1993)

[4] M. Liu, S. J. Visco, and L. C. De Jonghe, J. Electrochem. Soc. 138, p. 1891, (1991)

[5] M. Liu, S. J. Visco, and L. C. De Jonghe, J. Electrochem. Soc. 138, p. 1896, (1991)

[6] M. M. Doeff, S. J. Visco, and L. C. De Jonghe, J. Electrochem. Soc. 139, p. 1808, (1992)

[7] M. Gauthier, A. Belanger, B. Kapfer and G. Vassort, Polymer Electrolyte reviews 2, Edited by J. R. MacCallum and C. A. Vincent Elsevier Applied Science, London (1989)

[8] R.Koksbang, J. Barker, H. Shi, M. Y. Saïdi, Solid State Ionics. 84, p. 1, (1996)

[9] P. G. Bruce, Chem. Com. 19, p. 1817, (1997)

[10] J. M. Tarascon, D. Guyomard and G. L. Baker, J. Power Sources 43-44, p. 689, (1993)

[11] A. R. Armstrong and P. G. Bruce, Nature. 381, p. 499, (1996)

[12] K. S. Nanjundaswamy, A. K. Padhi, J. B. Goodenough, S. Okada, H. Ohtsuka, H. Arai and J. Yamaki, Solid State Ionics. 92, p. 1, (1996)

[13] A. K. Padhi, K. S. Nanjundaswamy, and J. B. Goodenough, J. Electrochem. Soc. **144**, p. 1188, (1997)

[14] L. Miller, T.H. Josefiak Synth. Metals **23** (**3-4**), p. B431, (1988).

[15] R. West and J. Niu, The Electrochemistry of the carbonyl group **2**, Edited by J. Zabicky Interscience publishers, p. 241 (1970)

[16] Oxocarbons Edited by R. West, Academic Press (1980)

[17] A. H. Schmidt, Janssen Chim. Acta **4**, p. 3, (1986)

[18] G. Seitz and P. Imming, Chem. Rev. **92**, p. 1227, (1992)

[19] L. Gmelin, Ann. Phys. **4**, p. 1, (1825)

[20] R. West, H.-Y. Niu, D. L. Powel and M. V. Evans, J. Am. Chem. Soc. **82**, p. 6204, (1960)

[21] J. Aihara, J. Am. Chem. Soc. **103**, p. 1633 (1960) ; A. Moyano and F. Serratosa, J. Molec. Structure **131**, p. 90, (1982) ; K. Jug, J. Org. Chem. **48**, p. 1344, (1983)

[22] M. B. Fleury and G. Molle, Electrochim. Acta **20**, p. 951, (1975); J. Moiroux, D. Escourrou and M. B. Fleury, Electrochim. Acta **25**, p. 785, (1980); *ibidem* Bioelectrochem. Bioenerg. **7**, p. 333, (1980).

[23] B. Carré, J. Paris, L. Fabre, S. Jourdannaud, P. Castan, D. Deguenon and S. Wimmer, Bull. Soc. Chim. Fr. **127**, p. 367, (1990).

[24] M. B. Fleury , P. Souchay and M. Gouzerh, Bull. Soc. Chim. Fr. **6**, p. 2562, (1968) ; D. Sazou and G. Kokkinidis, Can. J. Chem. **65**, p. 397, (1987).

[25] L. M. Doane and A. J. Fatiadi, J. Electroanal. Chem. **135**, p. 193, (1982)

[26] G. Farnia and G. Sandonà, J. Electroanal. Chem. **348**, p. 339, (1993)

[27] E. V. Patton and R. West, J. Phys. Chem. **77**, p. 2652, (1973).

[28] A. J. Fatiadi, H. S. Isbell and W. F. Sager, J. Res. Natl. Bur. Stand., Sec A **67**, p. 153, (1963)

[29] W. M. Macintyre and M. S. Werkema, J. Chem. Phys. **40**, p. 3563, (1964)

[30] N. C. Baenziger and J. J. Hegenbarth, J. Am. Chem. Soc. **86**, p. 3250, (1964)

[31] E. Krogh Andersen and I. G. Krogh Andersen., Acta Cryst. B **31**, p. 379, (1975)

[32] R. West and H.-Y. Niu, J. Am. Chem. Soc. **85**, p. 2586, (1963)

[33] R. West and H.-Y. Niu, J. Am. Chem. Soc. **85**, p. 2589, (1963)

[34] G. Kokkinidis, D. Sazou and I. Moumtzis, J. Electroanal. Chem. **213**, p. 135, (1986)

IN-SITU X-RAY CHARACTERIZATION OF LiMn$_2$O$_4$: A COMPARISON OF STRUCTURAL AND ELECTROCHEMICAL BEHAVIOR

MARK A. RODRIGUEZ, DAVID INGERSOLL, & DANIEL H. DOUGHTY
PO Box 5800, Sandia National Laboratories, Albuquerque, NM 87185-1405

ABSTRACT

Li$_x$Mn$_2$O$_4$ materials are of considerable interest in battery research and development. The crystal structure of this material can significantly affect the electrochemical performance. The ability to monitor the changes of the crystal structure during use, that is during electrochemical cycling, would prove useful to verify these types of structural changes. We report in-situ XRD measurements of LiMn$_2$O$_4$ cathodes with the use of an electrochemical cell designed for in-situ X-ray analysis. Cells prepared using this cell design allow investigation of the changes in the LiMn$_2$O$_4$ structure during charge and discharge. We describe the variation in lattice parameters along the voltage plateaus and consider the structural changes in terms of the electrochemical results on each cell. Kinetic effects of LiMn$_2$O$_4$ phase changes are also addressed. Applications of the in-situ cell to other compounds such as LiCoO$_2$ cathodes and carbon anodes are presented as well.

INTRODUCTION

A large number of highly applied electrochemical systems, *e.g.* batteries, employ crystalline materials. Examples of some of these include electrolytic MnO$_2$ used in alkaline cells, LiCoO$_2$ found in lithium ion rechargeable cells, and Zn/ZnO found in a number of cell chemistries.[1] The performance of many of these materials is highly dependent on their crystal structure. For example, only the spinel polymorph of LiMn$_2$O$_4$ exhibits a 4V and a 3V plateau, each of which have theoretical capacities of 148 mAh/g. However, the spinel structure does not appear stable along the 3V plateau. The ability to monitor the changes of the crystal structure during electrochemical cycling, should prove useful to verify these types of structural changes as well as aid in the development of structure-activity relationships. Furthermore, the ability to perform the measurements in a cell configuration that allows for use of an electrode structure similar to or identical to that employed in commercial devices has the potential to make significant impact on future cell development. Finally, the ability to monitor these changes real time, could provide useful insight in the dynamics of the system.

The choice of material used to contain the cell or to serve as a window into the cell exerts a significant effect on the overall performance of the final device. For example, Richard, *et. al.*[2] recently described an *in-situ* XRD cell containing a system based on the Ultralife gel technology. In this experimental setup a beryllium window was used in close proximity to the cathode electrode material. Consequently, the window was required to be physically separated from the electrode and electrolyte, in this case by aluminum spacers, to ensure that it did not participate in

275

the electrochemical reactions. In fact, as described, only the gel system could be used in this cell since liquid electrolyte would lead to wetting of the beryllium window and allow its subsequent participation in the redox reactions. Over 20 diffraction peaks are seen from the cell components themselves over the 2θ range studied. This number of artifact diffraction peaks makes observation of the diffraction pattern from the sample of interest problematic at best. Additionally, very long data acquisition times are required (on the order of 12 hours) for a single pattern. This severely limits real-time data collection on samples to only the slowest charge or discharge rates.

For *in-situ* XRD electrochemical studies, the material used for housing the cell must meet several criteria. First, it must be relatively transparent to the X-ray radiation so as to minimize the absorption of both the incident and diffracted rays. Second, it should have few if any diffraction lines at, or near angles where the sample diffracts. Third, it should be chemically stable to the solvent system utilized, as well as when in contact with the electrode materials. Fourth, it must be electrochemically stable and preferably inert, so as not to interfere with the electrochemical characteristics of the active electrode materials. Fifth, it should be impervious to the solvent system utilized so it will not dry out during the course of the experiment. Sixth, the mechanical characteristics of the material should be such that compression can be applied to the cell to ensure good electrical contact and minimum interfacial resistance. Finally, a method of sealing the cell material to itself and other materials, such as the electrical feedthroughs, must be available. The two materials that were selected for the in-situ cell were polypropylene and polyethylene (1mil thickness).

In order to determine unit cell parameters, high quality X-ray data must be collected. This seemingly simple requirement is not so easily met in an electrochemical cell primarily due to the fact that the electrode can and usually does swell after cell activation and upon charging and discharging. As a result of this swelling the electrode surface can move out of the focal plane of the X-ray source. This movement will likely result in displacement of the diffracted energy and incorrect determination of cell parameters. In the system described here, an internal standard is present that ensures precise determination of the diffraction angles and hence the unit cell parameters.

EXPERIMENTAL

Cell Design

Porous electrodes were prepared and consisted of 83% active material, 8% Teflon used as a binder, and the balance as carbon. The electrodes were prepared using a standard mixing, knead, and roll technique. The final electrode thickness was 11 mil and was approximately 50% porous. The active material used was $LiMn_2O_4$ obtained from FMC and Chemetals and $LiCoO_2$ obtained from FMC. The porous electrodes were pressed into an aluminum current collector grid in such a way so that the aluminum grid and porous electrode are at the same height. In this way the aluminum grid served as the internal standard for all XRD measurements. Lithium metal was used as the negative electrode for these studies and was pressed onto a nickel grid which served as the current collector. A 1 mil Celgard separator was used in these cells. The supporting electrolyte solution used was 1M LiPF$_6$ in ethylene carbonate/diethylene carbonate (70:30). The 12 cm^2 electrodes (6cm X 2cm) were assembled into a stack and placed into a small plastic bag constructed using 1 mil polypropylene (or polyethylene). The bag was assembled by impulse heating of the bag materials. The current collectors were also sealed to the bag assembly using impulse heating. The cell was then vacuum filled with the electrolyte through a small fill hole left in the cell, and allowed to soak for approximately 5 minutes. At the end of this time period the cell was sealed while a vacuum was applied, thus ensuring compression on the cell components. A picture of the cell dis-assembled into its various parts and fully assembled is shown in Figure 1.

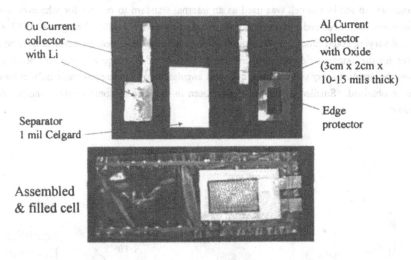

Cu Current collector with Li

Separator 1 mil Celgard

Al Current collector with Oxide (3cm x 2cm x 10-15 mils thick)

Edge protector

Assembled & filled cell

Figure 1. In-situ cell dis-assembled into its respective parts and fully assembled within polyethylene bag.

<u>XRD Data Collection</u>

In-situ measurements can be classified into two categories, *static* or *dynamic* measurements.[3] The varying parameter of interest in this investigation is voltage. In the case of a static in-situ measurement, data is collected on a sample cell at constant voltage. In this type of measurement, the cell is brought to a specific voltage value and held for a long time to reach a relatively equilibrated state. During this time the current is allowed to decay to its equilibrium value. The diffraction pattern is then collected on this sample. For the static measurement the scan speed is not a critical issue since the structure should not be changing appreciably during the analysis and so the scan rates tended to be longer in length to improve the data quality. In a dynamic measurement the voltage is being varied constantly while the cell is being processed. In this scenario the scan speed for data collection is much more important. In order to obtain useful information, the rate of data collection for a pattern should be much faster than changes occurring in structure of the lattice. This important issue for the dynamic experiments required modification to smaller 2θ ranges and shorter counting times.

All XRD patterns were collected using a Siemens automated θ–θ powder diffractometer equipped with Cu Kα radiation, a diffracted-beam graphite monochromator, and scintillation counter. The cell was loaded into the diffractometer in such a way as to keep the cell level and attention was paid to avoid sample displacement error as much as possible. Figure 2 shows a schematic view of the experimental setup used. The in-situ cell was connected by lead wires to a computer-controlled potentiostat. Voltage profiles were programmed into the PC for variable voltage experiments or held at a constant voltage during static measurements. The current collector within the in-situ cell was used as an internal standard to correct for whatever sample displacement error occurred. Parameters for typical scans were a 34-80° 2θ range, 0.05° 2θ step-size, and varying count times ranging from 0.75 for some dynamic measurements to as long as 5 sec/step for the longer static-type measurements. The 34-80° 2θ range was chosen since the cell showed the greatest x-ray transparency over this angular range where reasonable diffraction data could be obtained. Smaller 2θ ranges were chosen in the case of some dynamic measurement scenarios.

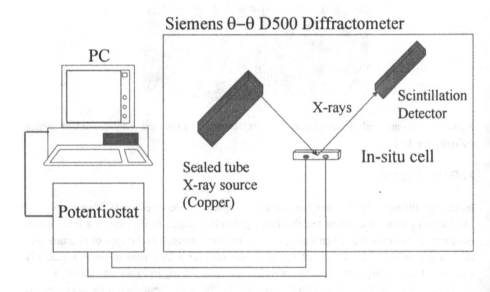

Figure 2: Schematic diagram of experimental setup for XRD employing in-situ cell.

RESULTS & DISCUSSION

Cell X-ray Transparency

The X-ray transparency of the in-situ cell is illustrated in Figure 3. This figure compares three diffraction patterns. The bottom pattern is that of a commercially available (FMC) $LiMn_2O_4$ spinel powder. This pattern is shown for comparison purposes. The middle pattern illustrates what the diffraction data looks like for a cathode prepared of this powder which has been placed within the in-situ cell bag but without any electrolyte added. It is clear that the pattern has not changed very much at all over the 34-80° 2θ range. This results indicates the ease of obtaining diffraction data from the material within the bag of the cell. Additional peaks from the aluminum current collector appear in the diffraction data since it has been placed on top of the cathode and is also detected during the scan. Since the structure and pattern for Al metal is well known, these peaks are very useful as an internal standard for the experiment. The top diffraction pattern was collected on an in-situ cell which was filled with electrolyte. In this case there does appear to be some attenuation of the diffraction data with the addition of the electrolyte. However, the diffraction peaks for both the aluminum and the spinel phase are easily detected above background with sufficient intensity to be used in both qualitative and quantitative type diffraction measurements.

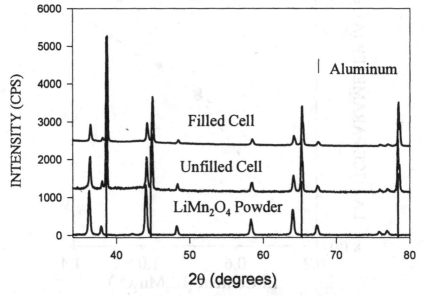

Figure 3: Illustration of X-ray transparency of in-situ cell. Bottom diffraction pattern shows typical XRD data for $LiMn_2O_4$ powder. Middle pattern shows diffraction data of cathode in the in-situ cell (no electrolyte); top pattern is diffraction data taken on in-situ cell with electrolyte.

Static Measurements on $LiMn_2O_4$

A series of diffraction scans on samples equilibrated at various voltage values are shown plotted with their cubic spinel lattice parameter vs. Li content (see Figure 4). The Li content was determined by first determining the quantity of Li in the initial starting powder by chemical analysis and then making the assumption that all the Li present is found as the spinel. The Li content was determined to be $Li_{1.2}Mn_2O_4$ by inductively coupled plasma (ICP). Using the initial value of the Li content, the reduction of Li in each cell was based on the capacity of each cell and the charge passed to reach equilibrium. Cells prepared using this powder were charged to various voltage values and then held constant at that voltage while diffraction data was collected. The peaks were fit to obtain d-spacings for the respective hkl's and a least square refinement was performed to obtain a lattice parameter for the cubic spinel structure. Figure 4 shows the results of these static measurements. It is evident that the unit cell shrinks nearly linearly with reduction of Li content. There does, however, appear to be a slight break in the linearity at about 0.8 Li; below 0.8 the slope is higher at about 0.24 A/Li while above 0.8 Li the slope decreases to 0.14 A/Li. This suggests that the behavior of Li intercalation is different in the two regions. Study of the behavior in $Li_xMn_2O_4$ is ongoing. Diffraction data for the lower Li contents (below 0.4 Li content) were difficult to obtain due to the breakdown of the solvent in the in-situ cell.

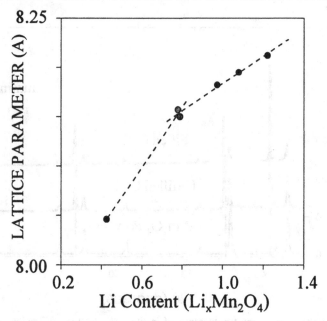

Figure 4. Dependence of lattice parameter on Li content in $LiMn_2O_4$ spinel.

Dynamic Measurements on LiMn$_2$O$_4$

As noted earlier, dynamic experiments can be used to detect changes of a material under non-equilibrium conditions. In one such experiment involving analysis of LiMn$_2$O$_4$, the diffraction data demonstrated some rather interesting results when the cathode was charged and analyzed too quickly for equilibration at the designated voltage. Figure 5 shows the diffraction data for the 35-40° 2θ range on a cell charged to a series of voltage values but not allowed sufficient time to come to the equilibrium Li content for that voltage. What is observed in this case is the presence of varying compositions of Li within the spinel structure which gives rise to a broadening of the Li$_x$Mn$_2$O$_4$ peaks and ultimately an apparent phase separation. The phase separation is illustrated by formation of two peaks for the (311) reflection at about 4.06V (and possibly even three peaks by 4.24V). The phase separation concept was determined to be the best explanation for the observed data since other cubic spinel peaks recorded (but not shown) display the same peak separation behavior independent of their indexed (hkℓ) assignment. This is in contrast to the possibility that a phase transition occurred (e.g. change in lattice symmetry) in which case the peak splitting would have a marked dependency on a the (hkℓ) of each x-ray reflection.

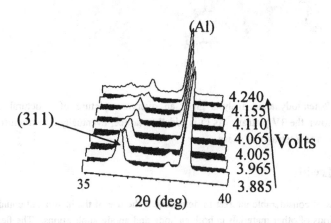

Figure 5. 3-D view of diffraction data for LiMn$_2$O$_4$ cathode monitored as a function of voltage. The (311) reflection for the spinel shows peak splitting consistent with a kinetically-controlled phase separation of the spinel structure into varying regions of Li content within the lattice.

Another example of a dynamic measurement is shown in Figure 6. In this experiment the cell was discharged from about 3.8V down the 3 volt plateau to 2V potentiodynamically while a series of diffraction scans were collected. This analysis proved quite revealing; a phase transition is clearly seen occurring as the cell is forced to intercalate more and more Li. The observed reflections for the new structure match well with the tetragonal structure $Li_2Mn_2O_4$ reported by Goodenough, *et.al.*[5]   This information concerning the structural transition of the spinel structure is very important in understanding the potential usefulness and likely problems posed by employing the 3 volt plateau in battery development.

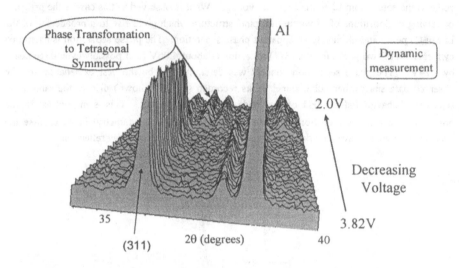

Figure 6.   Potentiodynamic analysis gives an overview picture of structural distortions in $Li_{1+x}Mn_2O_4$ down the 3 V plateau. As Li is added to lattice, eventually the structure undergoes a phase transition to a tetragonal phase.

Additional Materials

It was of considerable interest to determine if the use of the in-situ cell could be extended to investigations of other materials in both cathode and anode applications. The flexibility of the cell design makes analysis of either anode or cathode as easy as flipping the cell over and remounting it in the diffractometer. Just as the aluminum current collector could be used as an internal standard for the cathode, the anode current collector can also be used as the internal standard on the anode side. Figure 7 illustrates diffraction data collected on an anode of Lonza graphite. This graphite material displays a high degree of preferred orientation along the c-axis and hence has very strong $(00l)$ reflections dominating the diffraction data. The bottom diffraction scan shows data for an initial cell prior to Li intercalation. The data shows very strong (002) and (004) graphite peaks and additional peaks for nickel which was used as the anode current collector in this cell. Employing the nickel as an internal standard, the c-axis lattice parameter was calculated for the graphite material. Next, the cell was biased and Li was intercalated into the graphite. The middle diffraction pattern shows the diffraction data for the Li-intercalated graphite. This pattern shows a very pronounced shift in the (00ℓ) graphite peaks to lower 2θ or higher d-spacings consistent with Li intercalating between the graphite sheets. The c-axis lattice parameter was expanded by more than 8% by the Li intercalation. Next, the cell was reversed biased and Li was de-intercalated from the graphite. The top diffraction pattern in Figure 7 shows the diffraction data after Li was removed. Recovery of the Li from the graphite lattice looked promising in that the c-axis lattice parameter calculated for final state was nearly identical the starting value.

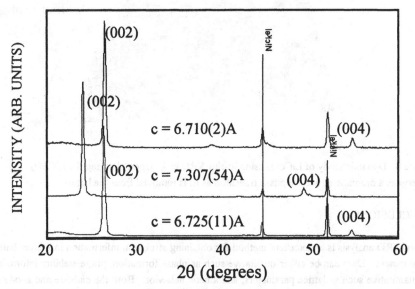

Figure 7. Diffraction data for graphite anode analyzed via the in-situ cell. Bottom pattern is initial scan of initial anode (Ni peaks from current collector). Middle scan is data on Li-intercalated graphite. Top scan is from anode after Li was de-intercalated back out of graphite.

Other cathode materials of interest in battery materials can be investigated using this in-situ cell. Successful in-situ XRD measurements have been performed on cells prepared with $LiCoO_2$ as the cathode material. Analysis of this cathode material is slightly more difficult since the cobalt has a tendency to display fluorescence during analysis which increases the background of the diffraction pattern. However, very useful results have been obtained like the dynamic study shown in Figure 8. This 3-D graph shows the behavior of $LiCoO_2$ during charging. The initial peaks shown in the 48-62° 2θ range are consistent with the typical $LiCoO_2$ layered-type lattice.[5] Upon charging up the 4V plateau, Li is being removed from the lattice and eventually the structure undergoes a phase transition indicated by the very dramatic peak shifts. The nature of this new lattice has not been fully determined at this time but the results of a phase transition at low Li contents in $LiCoO_2$ have been predicted in computer modeling studies.[6]

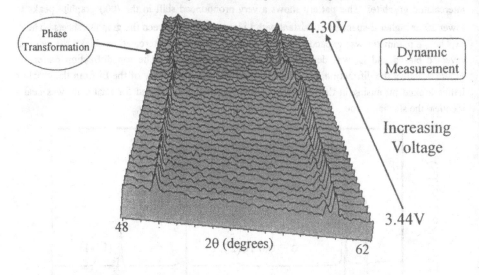

Figure 8. Dynamic study of $LiCoO_2$ using in-situ XRD cell. Upon charging the $LiCoO_2$ undergoes a dramatic structural phase transition as Li is removed from the lattice.

CONCLUSIONS

In-situ XRD analysis is an excellent method for obtaining structural information in lithium battery development. Data can be either qualitative such as phase formation, phase stability information or quantitative such as lattice parameters, and kinetic behavior. Both the cathode and anode can be analyzed easily and the in-situ cell transparency to both light and x-ray wavelengths makes analysis straight-forward. In-situ studies can be expanded to investigate new cathode/anode materials quickly and accurately.

ACKNOWLEDGMENTS

The authors would like to thank Jill Langendorf for her help with in-situ cell preparation. Sandia is a multiprogram laboratory operated by Sandia Corporation, a Lockheed Martin Company, for the United States Department of Energy under Contract DE-AC04-94AL85000.

REFERENCES

1. D. Linden, editor, Handbook of Batteries, 2nd ed., McGraw Hill, New York, 1995.
2. M. N. Richard, I. Koetschau, and J. R. Dahn, J. Electrochem. Soc. 144 554-557 (1997).

3. E. M. Levin, C. R. Robbins, H. F. McMurdie, Phase Diagrams for Ceramists, The American Ceramic Society, Columbus, OH, 1964, pp. 21.

4. J. Goodenough, M. Thackeray, W. David, and P. Bruce, *Rev. Chem. Miner.* **21** 435 (1984).

5. R. J. Gummow, M. M. Thackeray, W. I. F. David, and S. Hull, *Mat. Res. Bull.* **27** 327-337 (1992).

6. C. Wolverton, private communication.

PHASE TRANSITION OF LiMn$_2$O$_4$ SPINEL AND ITS APPLICATION FOR LITHIUM ION SECONDARY BATTERY

Junji Tabuchi *, Tatsuji Numata *, Yuichi Shimakawa **, Masato Shirakata ***
*Material Development Center, NEC Corporation, Kawasaki, Japan
**Fundamental Research Laboratories, NEC Corporation, Tsukuba, Japan
***Nippon Moli Energy Corporation, Toyama, Japan

ABSTRACT

LiMn$_2$O$_4$ has a phase transition at room temperature, which is caused by Jahn-Teller distortion. DC resistivity of LiMn$_2$O$_4$ shows an anomaly at the transition temperature, while no such anomaly is observed in samples with excess lithium. X-ray diffraction patterns of LiMn$_2$O$_4$ reveal that the crystal structure changes from cubic at higher temperature to orthorombic, as a first approximation, at lower temperature. However, no differences in initial charge-discharge curve are observed, which means that the Jahn-Teller distortion has no effect on electrochemical characteristics. The authors have succeeded in mass-producing lithium ion secondary batteries with a manganese spinel cathode.

INTRODUCTION

LiMn$_2$O$_4$ is a promising cathode material for lithium ion secondary batteries, because of its advantages such as low cost and low-toxicity. Recently, it has been reported that LiMn$_2$O$_4$ shows a phase transition at room temperature, which is caused by Jahn-Teller distortion. On contrary, manganese spinel with excess lithium has no such phase transition [1] [2]. Lithium ion secondary batteries using a manganese spinel cathode with excess lithium is reported to show good cycle life performance. [3] In order to make clear the effect of the phase transition on cell performance, LiMn$_2$O$_4$ and manganese spinel with excess lithium are synthesized and evaluated by X-ray diffraction, resistivity measurement and electrochemical test.

EXPERIMENT

LiMn$_2$O$_4$ and excess lithium samples with α=1.05, 1.10 (Li/Mn=α/2.00) were synthesized using a solid state reaction. Stoichiometric LiMn$_2$O$_4$ was α=1.00. The detailed synthesis procedure for these three samples was reported elsewhere. [2] [4] The samples were confirmed to be single phase by X-ray diffraction. Lattice constants from 310K to 20K were determined by a full-profile fit of X-ray diffraction data using the Rietveld analysis program RIETAN. [5] Resistivity of the samples was measured by a four-probe method. Rectangular bars of the polycrystalline samples were used. Discharge curves and capacity were measured in coin cells (2320) using Li metal anodes. Cathodes were prepared by mixing manganese spinel powders with conductive carbon (7wt%) and PVDF binder (3.5wt%). Electrolyte used was 1M LiPF$_6$ in

Mat. Res. Soc. Symp. Proc. Vol. 496 © 1998 Materials Research Society

an EC(Ethylene Carbonate)/DEC(Diethyl Carbonate)=30/70 mixed solvent. The electrochemical tests were performed at a constant current, 40 hour rate, with a voltage range from 3.0 to 4.3 V. Details of the cell assembly method were reported elsewhere. [4]

RESULTS AND DISCUSSION

Jahn-Teller Distortion

The X-ray diffraction patterns of the $\alpha=1.00$ and 1.10 samples at 300K and 250K are shown in figure 1, respectively. Both of the $\alpha=1.00$ and 1.10 samples at 300K can be indexed with cubic symmetry. The X-ray diffraction peaks of the $\alpha=1.00$ sample at 250K were obviously split. Yamada reported that the phase transition of $LiMn_2O_4$ was from cubic, $Fd3m$, at higher temperature to mixed phase of cubic and tetragonal, $I4_1/amd$ at lower temperature. [6] In order to confirm the phase transition, Rietveld refinement for the 250K phase was carried out. The best fit for the 250K phase is obtained assuming orthorombic symmetry, $Fddd$, as a fist step.

The calculated lattice constants from 310K to 20K for $\alpha=1.00$ and 1.10 are shown in figure2 (a) and (b), respectively. The calculated lattice parameters of the $\alpha=1.00$ sample at lower

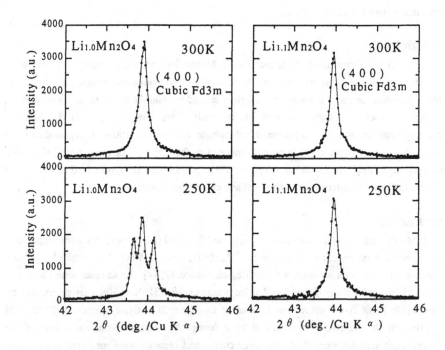

FIG. 1.   X-ray diffraction patterns of $Li_{1.0}Mn_2O_4$ (left) and $Li_{1.1}Mn_2O_4$ (right) at 300K (top) and 250K (bottom)

temperatures were based on the orthorombic symmetry model. In figure 2(a), we see that the lattice parameter, b, increases with decreasing temperature, in contrast a and c decrease with decreasing temperature. This result also suggests that the 250K phase is not a mixed phase, but a single phase with the orthorombic symmetry. Strictly speaking, the 250K phase may have lower symmetry than orthorombic symmetry because the fitting is not enough.

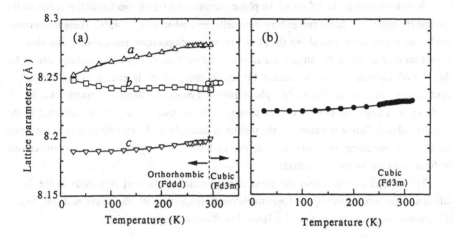

FIG. 2. Temperature dependence of lattice parameters of (a) $Li_{1.0}Mn_2O_4$ and (b) $Li_{1.1}Mn_2O_4$.

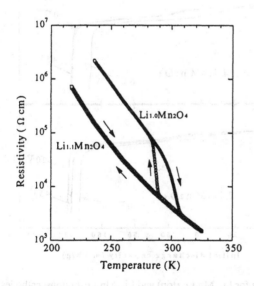

FIG. 3. Temperature dependence of the resistivity for $Li_{1.0}Mn_2O_4$ and $Li_{1.1}Mn_2O_4$

Figure 3 shows temperature dependence of the resistivity for the α=1.00 and 1.10 samples. The temperature dependence of the resistivity for the α=1.10 and α=1.00 sample show semiconductor-like behavior. With decreasing temperature, the resistivity for the α=1.00 sample increases by nearly an order magnitude at 290K. In contrast, for the α=1.10 sample, no such anomaly is observed.

In order to clarify the effects of the phase transition on the electrochemical property of the manganese spinel, discharge curves from coin cells were taken at 0, 20, 45°C. These temperature are below, intermediate and above the phase transition temperature, respectively. The obtained curves for α=1.00 and 1.10 samples are shown in figure 4 (a) and (b). All discharge curves for the α=1.00 cathode are similar except for an IR drop, which is caused by the electrolyte conductivity. No obvious effects of the phase transition on electrochemical properties was found. In figure 4, a larger IR drop at 0°C is observed for the α=1.00 cathode as compared to the α=1.10 cathode. This was caused by the resistivity anomaly in the α=1.00 sample at the phase transition. Considering cell performance, the α=1.10 cathode gives better low temperature performance than the α=1.00 cathode.

The α=1.05 sample shows the same characteristics as the α=1.10 sample in the X-ray diffraction pattern, resistivity and electrochemical properties, which means that 5% excess lithium is enough to suppress the Jahn-Teller distortion in manganese spinel.

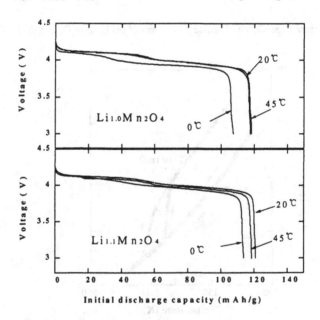

FIG. 4. Discharge curves for $Li_{1.0}Mn_2O_4$ (top) and $Li_{1.0}Mn_2O_4$ (bottom) cathodes at 0, 20, 45°C

Prismatic Cell Performance

Nippon Moli Energy Corporation has been successfully producing prismatic lithium ion secondary batteries with manganese spinel cathode since July 1996. Various prismatic cell designs are now in the market. Cell performance for IMP260848, one of the products, whose nominal capacity is 600mAh, is shown in figures 5-7. The cell is charged at 20°C at a constant current 1C up to 4.2V and then hold at constant voltage for 2.5 hours in total. Discharge curves at various C rates at 20°C are shown in figure 5. Since the anode material is graphite, flat discharge curves are observed. Temperature characteristics are shown in figure 6. The discharged capacity at various temperatures are normalized to the discharged capacity at 0.2C rate at 20°C which is set equal to 100%. The cell can be discharged even at -20°C at a 1C rate. Cycling performance at 20°C is shown in figure 7. The initial capacity is more than 600 mAh. The capacity at cycle 1000 is more than 70% of the initial capacity, which is comparable to performance of lithium ion batteries using $LiCoO_2$ cathodes.

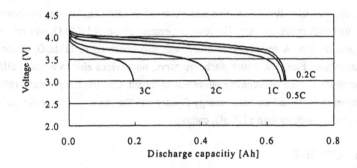

FIG. 5. Discharge curves of IMP260848 cell at 20°C.

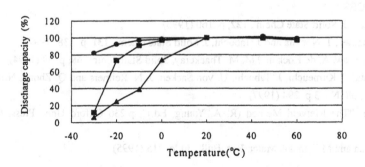

FIG. 6. Temperature characteristics for IMP260848.

291

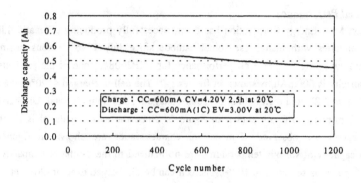

FIG. 7. Cycle performance for IMP260848

CONCLUSION

The effect of the Jahn-Teller distortion in manganese spinel on electrochemical properties was investigated. The Jahn-Teller distortion was observed in $LiMn_2O_4$, but was not observed in manganese spinel with excess lithium. The low temperature phase of $LiMn_2O_4$ was orthorombic as a first approximation. A resistivity anomaly was also observed in $LiMn_2O_4$ at the phase transition temperature. From the initial discharge curve, no obvious electrochemical difference between cubic phase and the orthorombic phase was observed. Commercial prismatic cell with a manganese spinel cathode had the same energy density and the same cycling performance as a lithium ion secondary battery with a $LiCoO_2$ cathode.

ACKNOWLEDGEMENTS

The prismatic cell has been jointly developed with Moli Energy (1990) Ltd. The authors thank Mr. Tomioka and Mr. Kumeuchi for preparing the samples.

REFERENCES

1. A. Yamada, J. Solid State Chem., **122**, p. 160 (1996)

2. Y. Shimakawa, T. Numata and J. Tabuchi, J. Solid State Chem., **131**, p. 138 (1997)

3. R. J. Gummow, A. de Kock and M. M. Thackeray, Solid State Ionics, **69**, p. 59 (1994)

4. T. Numata, T. Kumeuchi, J. Tabuchi, U. von Sacken, J. N. Reimers and Q. Zhong, NEC Res. & Develop., **38**, No. 3 p. 294 (1997)

5. F. Izumi, "The Rietveld Method (R. A. Young, Ed.)", p.236 Oxford Univ. Press, Oxford (1993)

6. A. Yamada and M. Tanaka, Mater. Res. Bull., **30**, p. 715 (1995)

LiCoO$_2$ Thin Films Grown by Pulsed Laser Deposition on Low Cost Substrates

D.S. Ginley, J.D. Perkins, J.M. McGraw, and P.A. Parilla
National Renewable Energy Laboratory , Golden CO
M.L. Fu and C.T. Rogers, *University of Colorado, Boulder CO*

Abstract

We report on the use of pulsed laser depositon (PLD) to grow thin films of LiCoO$_2$ on a number of low cost substrates including SnO$_2$ coated Upilex, stainless steel and SnO$_2$ coated glass. Highly textured (00l) films grown on CVD deposited SnO$_2$ films on 7059 glass, were obtained at 200 to 500 mTorr O$_2$ and a temperature of 500 C. Similar texture was not obtained on the stainless or Upilex however dense films from crystalline to amorphous were obtained. The films were characterized by x-ray diffraction and Raman spectroscopy.

Introduction

LiCoO$_2$ is one of the key materials being looked at as a cathode for rechargable Li secondary batteries. These materials are key for 4 V and 3 V batteries. LiCoO$_2$ is also of interest for electrochromic devices [1,2]. It is part of the general familiy of materials including LiCoO$_2$, LiNiO$_2$ and LiMnO$_2$. These materials all have layered rock salt structure and need to be synthesized in the lithiated state unlike the vanadium oxide materials which are classic bronze formers and can be lithiated after synthesis. It is key to understanding these materials to have good control of the crystallinity and morphology (porosity) on the substrates of interest. Thin films of LiCoO$_2$ have been synthesized by a variety of techniques including sputtering [3-6], spray deposition [7-12], reaction of metals [13] sol gel synthesis [14,15] and pulsed laser deposition [16-19]. Typically the films were deposited on stainless steel or Ta substrates. These techniques produce a wide spread in crystal quality from amorphous to highly crystalline and of porosity from very porous aerogel and sol gel structures to highly dense films.

In this work we report on the pulsed laser deposition of thin films of LiCoO$_2$ on a variety of potentially low cost substrates including Upilex (a polyimide), stainless steel and tin oxide coated 7059 glass. In all cases dense films were grown. In the case of the SnO$_2$ coated glass substrates the highly textured films were obtained. On the other substrates the films were predominately amorphous but some polycrystalline films were grown. Films were characterized by a variety of techniques inc1udeing x-ray diffraction, Raman spectroscopy and inductively coupled plasma analysis.

Experimental

Synthesis: The films are grown by laser ablation of a stoichiometric 2.54 cm ceramic LiCoO$_2$ target obtained from Target Materials or synthesized in-house (most of the results) in a controlled atmosphere vacuum chamber. A schematic of the apparatus is shown in Figure 1. The target and heater are on-axis with each other at a distance of approximately 8.5 cm. The target is rotated about its axis at between 1-10 rotations per minute. Typically three 1 cm x 4 cm substrates are mounted on the heater for each run, thus creating three sister samples at each set of deposition conditions. The typical substrates included SnO$_2$-coated glass (11 Ω-cm) obtained by CVD of tetramethyl tin on 7059 glass substrates, stainless steel (430) foil, and

InSnO$_2$ (ITO) (12-16 Ω-cm) coated Upilex provide from ITN corporation. After the chamber is pumped down to a base pressure of at least 10^{-6} Torr, the substrates are heated to the chosen deposition temperature (65C-600C) and O$_2$ is bled into the chamber at a flow rate of 24.8 sccm to maintain a set background oxygen pressure. With a frequency of 10 Hz and a typical fluence of 360 mJ/pulse, the 248 nm excimer laser strikes the target at 45 degrees, ablating a plume of material normal to the surface. The ablated material settles on the substrates and a thin film is created. Deposition runs of 20,000 pulses result in film thicknesses ranging from 1000 Å to 4000 Å depending on ambient conditions.

Characterization: Average thickness measurements were made with a Dektak III profilometer. Corners of the SnO$_2$/7059 samples were covered with TiO$_2$ as a mask which could be easily removed after growth. Inductively coupled plasma analysis (ICP) was determined on a Varian Liberty system. Sample were initially dissolved in 1 ml 1:1 HCl:H2O and then diluted to 10 volume % HCl. Using a Scintag X-ray diffraction machine that emits Cu K$_\alpha$ radiation, we determined the phase and crystallininty of the films. Raman scattering measurements were performed in a 180° back-scattering configuration with an Ar-ion laser operating at 514.5 nm and a Spex 1877 Triplemate spectrometer with a CCD detector. A polarized incident beam, ~4 mW, was focused through a microscope objective to a spot size of ~2μm. Analysis was performed in air. The FTIR was determined on a Nicolet spectrophotometer. This information was consistent with spectroscopy of Raman-active phonons in cobalt oxides.

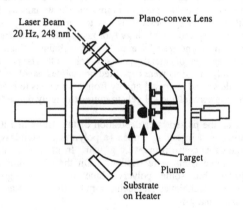

Figure 1. Pulsed laser deposition vacuum system.

Results and Discussion:

Table 1 tabulates the growth conditions for the films in this study. All samples with the same major sample number (Co-L-MFxx) were grown in the same run. The same analytical studies may have been performed on one or all of the films. All films were grown by pulsed laser deposition in an oxygen ambient with a laser energy of approximately 350 mJ per pulse and a total of approximately 20,000 pulses per growth run.

Sample thicknesses were typically .1 and .5 μm. The key factor controlling the thickness was the oxygen partial pressure with the thicker films deposited at the lower oxygen partial pressures. The film thickness was nearly temperature independent. The thicknesses reported are from the $SnO_2/7059$ control that was run in each run. The thickness was difficult to determine directly on the stainless steel because of slight non-planarity on the non polished surface and on the Upilex because it is inherently to soft for the DekTac.

Table 1: Growth Conditions for PLD Films

Film ID	Thickness (avg)	Substrate	Temp. (C)	Pressure/b kgd. gas	#pulses/ rep. rate	avg. laser power
Co-L-MF7-3		ss/citranox				
Co-L-MF8-1	1911.5	111 $SnO_2/7059$	500	200 mTorr O_2	20k/10 Hz	225 mJ/pulse
Co-L-MF8-2	2576.5	SnO_2 Cherry Prod.				
Co-L-MF9-1	4255	112 $SnO_2/7059$	550	500 mTorr O_2	20k/10 Hz	325mJ/p to 275mJ/p =300mJ/p
Co-L-MF9-2		ss/acid etch				
Co-L-MF9-3	3039	111 $SnO_2/7059$				
Co-L-MF11-1	5408.5	112 $SnO_2/7059$	65	200 mTorr O_2	20k/10 Hz	350mJ/p to 300mJ/p =325mJ/p
Co-L-MF11-2	4973	112 $SnO_2/7059$				
Co-L-MF13-1		ITO 12/Upilex	300	no bkgd. gas.	20k/10Hz	275mJ/p to 225mJ/p =250mJ/p
Co-L-MF15-1		ss/acid etch	300	500 mTorr O_2	20k/10Hz	
Co-L-MF15-2	2614	112 $SnO_2/7059$				
Co-L-MF16-1	2017	ITO 12/Upilex	300	100 mTorrO_2	21k/10Hz	250mJ/p to 225mJ/p =238mJ/p
Co-L-MF16-3	1824	ITO 16/Upilex				

Note that there is more than one type (112, 113, etc.) of SnO_2 on the 7059 glass which are from different batches. These samples had slightly different CVD growth parameters resulting in slightly different film texture as is reflected in the x-ray data. The origin of these textural differences is currently under investigation.

Table 2 is a summary of the analytical data for the same films as in Table 1. Note that not all samples were characterized the same way. For some samples ICP was run as a control for the amount of Li in the film. In nearly all cases the ratio of Li:Co is nearly 50:50 indicative of nearly stoiciometric incorporation of Li into the films. Also, FTIR was performed on the Upilex films since they luminesce strongly at the Raman wavelength.

Table 2: Summary of analytical data for the films in Table 1

Film ID	XRD	Raman (Number of active modes)	ICP Li:Co Ratio	FTIR (phonons at cm^{-1})
Co-L-MF7-3		4 phonons		
Co-L-MF8-1	(003), (104)	2 phonons	54.050:45.950	
Co-L-MF8-2		2 phonons		
Co-L-MF9-1	(003), (006), (009), oriented	2phonons	53.276:46.724	538, 687
Co-L-MF9-2	polycrystalline	2 phonons		
Co-L-MF9-3	(003), (101),(006), (012), (104)	2 phonons		
Co-L-MF11-1				
Co-L-MF11-2	amorph	4 phonons	48.627:51.373	
Co-L-MF13-1		2 phonons		
Co-L-MF15-1	(003), (101), (104), (110), less than Co-L-MF9-2	3 phonons		
Co-L-MF15-2	(003) weak orientation		45.72:54.28	
Co-L-MF16-1	questionable (101), (104) peaks	Upilex/ITO photoluminesence		548, 677
Co-L-MF16-3	small (101), (104) peaks	Upilex/ITO photoluminesence		

As illustrated in the x-ray in Figure 2 highly textured (003) films are obtained [18,20,21]. Note that the peaks are intentionally shifted for more ready viewing. In all cases, the films on the 7059 glass were preferentially textured with this orientation. The degree of orientation depended on the SnO_2 growth conditions. The film Co-L-MF17-2 was grown on an SnO_2 substrate that had less initial texture than the other two in the figure. The small peaks at 29 and 65 degrees are indicative of the presence of other orientation. Figure 3 illustrates the morphology of films grown on the three substrates. In all three cases the average grain size is approximately 100 nm. This is very uniform in the case of the glass substrate and there is evidence of some larger grain growth on the stainless and Upilex substrates.

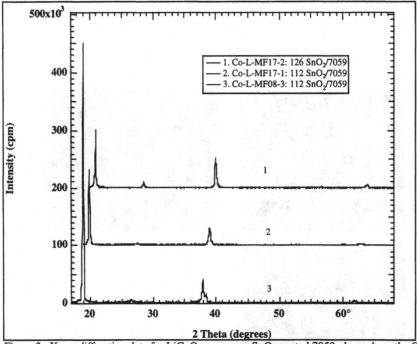

Figure 2: X-ray diffraction data for $LiCoO_2$ grown on SnO_2 coated 7059 glass where the CVD growth parameters for the SnO_2 were different for the 3 substrates.

In the case of the film on glass which is highly textured by x-ray as shown in Figure 4 the small grain size is somewhat surprising. It is interesting that the growth habit is very similar in all of the cases even though as is shown below the crystallinity is quite different. This morphology is particullarly well suited to battery electrodes due to the high surface area and small grains.

Figure 4 illustrates the x-ray data for a set of films grown under different growth conditions. The key indicator is the peak at 19° which is a probe of the degree of (003) texturing. Note first that the curves 7 and 8 which were grown under vacuum and low temperature respectively show no peak indicative of very little crystallinity. Curve 6 grown at 300 C shows substantial texture and crystallinity. The maximum intensity for the x-ray signal is obtained for a temperature of 400 C and an O_2 pressure of 200 mTorr. A further increase in temperature toe 500 C causes very little difference in the x-ray intensity. Growth at increasing temperatures causes a loss in x-ray signal. Note that increasing O_2 pressure seems to be required to maintain the texture (compare curves 1 and 2). This might be expected because of the volatility of Li and

LiO$_x$. The oxygen partial pressure may drive the surface to the LiCoO$_2$ and maintain stoichiometry and crystallinity. Figure 5 is the Raman spectra for films similar to those in curves 5 and 6 in Figure 4. The bands are a sharp set of two peaks which match the data in the literature. These peaks are the 486 cm^{-1} (E$_g$) and the 596 cm^{-1} (A$_{1g}$) transitions as previously identified by Inaba et.al. [22,23]. Both the x-ray and Raman are supportive of a highly crystalline film.

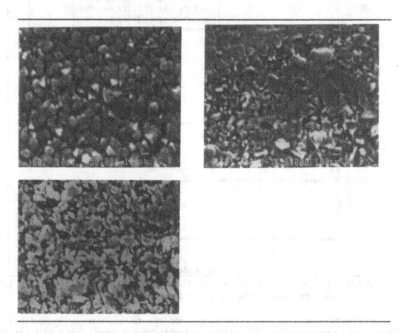

Figure 3: SEM micrographs of a LiCoO$_2$ film on SnO$_2$ glass at 400 C/200 mTorr, a film on stainless at 300 C/500 mTorr and a film on ITO coated Upilex at 300 C/100 mTorr.

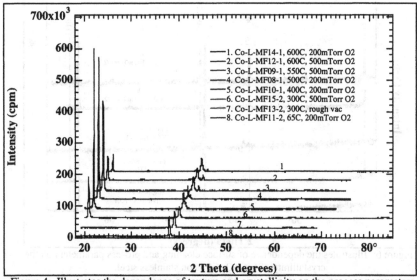

Figure 4: Illustrates the dependence of texture and crystallinity on the process parameters for films grown on SnO_2/7059 substrates.

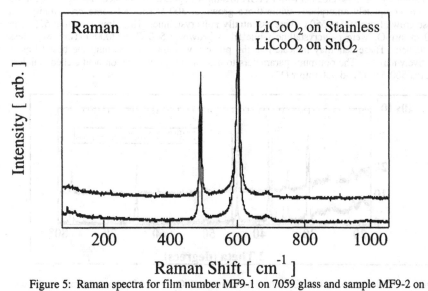

Figure 5: Raman spectra for film number MF9-1 on 7059 glass and sample MF9-2 on stainless steel.

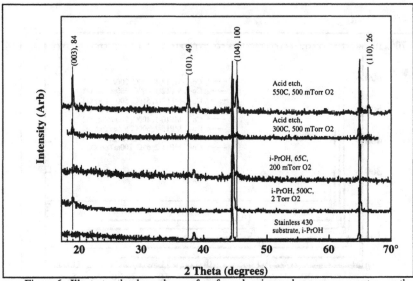

Figure 6: Illustrates the dependence of surface cleaning and process parameters on the crystallinity of LiCoO₂ films on stainless steel.

For films grown on stainless steel substrates Figure 6 presents data for the dependence of film crystallinity on the process parameters. Light acid etches (HCl or HNO_3) resulted improved crystallinity over those cleaned with i-PrOH to simply degrease the surface. Films grown at 65 C were completely amorphous while those grown at 300 C showed some crystallinity and those grown from 500 to 550 C were essentially polycrystalline. The films grown at 550 C in 500 mTorr O_2 were polycrystalline while films grown at 500 C in 2000 mTorr were less crystalline. These data may indicate that the process window for obtaining the best films is relatively narrow. The optimum parameters to date are to grow films on acid etched stainless steel at 500-550 C and 500 mtorr O2.

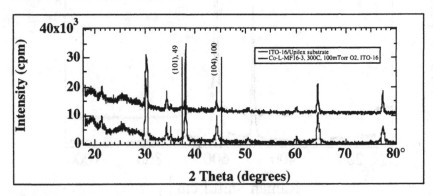

Figure 7: XRD data for a LiCoO₂ film grown on a InSnO₂ coated Upilex substrate vs the data for the virgin substrate (bottem).

Figure 7 illustrates the x-ray diffraction data for an approximately .2 μm LiCoO₂ film grown on the ITO coated Upilex. Note that there is essentially no difference between the bottom spectrum which is the substrate alone and the top one containing the film. This indicates that the film is predominately amorphous. Figure 8 is the FTIR data for the same film compared to the bare substrate and a LiCoO₂ film on SnO₂ coated 7059 glass. The data suggests that there may be some ordered regions in the film on the Upilex. Indicating at least incipient crystallinity.

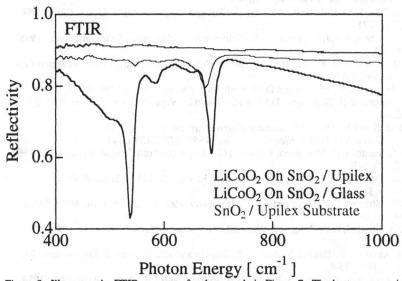

Figure 8: Illustrates the FTIR spectrum for the sample in Figure 7. The bottom curve is the FTIR spectrum for the LiCoO₂ film on SnO₂ coated 7059 glass in Figure 2.

Conclusions:

Pulsed laser deposition was used to grow LiCoO₂ films on a variety of substrates including SnO₂ coated 7059 glass, 430 stainless steel and Upilex. In all cases the Li stoichiometry was maintained during growth. The films on SnO₂ coated glass were highly crystalline and the degree of texture correlated to the initial texture of the SnO₂. Films on stainless steel varied from amorphous to crystalline depending on process parameters including surface cleaning, pressure and process temperature. All of the films on ITO coated Upilex were amorphous by x-ray but showed some evidence of crystallinity by FTIR. Interestingly, films on SnO₂ coated glass at 300 C show substantial crystallinity. This may indicate as in the case of the stainless steel that the LiCoO₂ growth is very sensitive to the surface conditions. Future work is focused on improving the materials growth and evaluating the electrochemical properties.

Acknowledgment:
This work was supported by the U.S. Department of Energy, Office of Basic Energy Sciences under contract #DE-AC36-83Ch10093

References:

[1]   G. Wei, T.E. Haas and R.B. Goldner *Proc. - Electrochem. Soc.* **1990**, *90-2*, 80-8.
[2]   G. Wei Thesis, 1991.
[3]   F.K. Shokoohi and J.M. Tarascon In U.S., 1992; pp.
[4]   E.J. Plichta and W.K. Behl *Proc. - Electrochem. Soc.* **1993**, *93-24*, 59-67.
[5]   E.J. Plichta and W.K. Behl *J. Electrochem. Soc.* **1993**, *140*, 46-9.
[6]   B. Wang, J.B. Bates, C.F. Luck, B.C. Sales, R.A. Zuhr and J.D. Robertston *Proc. - Electrochem. Soc.* **1996**, *95-22*, 183-92.
[7]   C. Chen, E.M. Kelder, P.J.J.M. van der Put and J. Schoonman *J. Mater. Chem.* **1996**, *6*, 765-771.
[8]   C.H. Chen, A.A.J. Buysman, E.M. Kelder and J. Schoonman *Solid State Ionics* **1995**, *80*, 1-4.
[9]   C.H. Chen, E.M. Kelder, P.J.J.M. van der Put and J. Schoonman *Proc. - Electrochem. Soc.* **1996**, *96-14*, 43-57.
[10]   P. Fragnaud, T. Brousse and D.M. Schleich *J. Power Sources* **1996**, *63*, 187-191.
[11]   P. Fragnaud, R. Nagarajan, D.M. Schleich and D. Vujic *J. Power Sources* **1995**, *54*, 362-6.
[12]   D.M. Scheich "Thin film inorganic electrochemical systems," 1995.
[13]   I. Uchida and H. Sato *J. Electrochem. Soc.* **1995**, *142*, L139-L141.
[14]   D.G. Fauteux, A. Massucco, J. Shi and C. Lampe-Onnerud *J. Appl. Electrochem.* **1997**, *27*, 543-549.
[15]   L.A. Dominey, T.J. Blakley, V.R. Koch, D.R. Vujic and D.M. Schleich *Proc. Int. Power Sources Symp.* **1992**, *35th*, 290-3.
[16]   K.A. Striebel, C.Z. Deng and E.J. Cairns *Mater. Res. Soc. Symp. Proc.* **1995**, *393*, 85-90.
[17]   K.A. Striebel, C.Z. Deng, S.J. Wen and E.J. Cairns *J. Electrochem. Soc.* **1996**, *143*, 1821-1827.
[18]   M. Antaya, J.R. Dahn, J.S. Preston, E. Rossen and J.N. Reimers *J. Electrochem. Soc.* **1993**, *140*, 575-8.
[19]   M. Antaya, K. Cearns, J.S. Preston, J.N. Reimers and J.R. Dahn *J. Appl. Phys.* **1994**, *76*, 2799-806.
[20]   G.G. Amatucci, J.M. Tarascon and L.C. Klein *J. Electrochem. Soc.* **1996**, *143*, 1114-23.
[21]   J.N. Reimers and J.R. Dahn *J. Electrochem. Soc.* **1992**, *139*, 2091-7.
[22]   M. Inaba, Y. Iriyama, Z. Ogumi, Y. Todzuka and A. Tasaka *J. Raman Spectrosc.* **1997**, *28*, 613-617.
[23]   C.K. Huang, S. Surampudi, A.I. Attia and G. Halpert In U. S. Pat. Appl., 1993; pp.

SYNTHESIS AND CHARACTERIZATION OF LiNiO₂ AS A CATHODE MATERIAL FOR PULSE POWER BATTERIES

H.S. Choe and K.M. Abraham*
EIC Laboratories, Inc., 111 Downey Street, Norwood, MA 02062, USA

ABSTRACT

The optimum conditions for preparing $LiNiO_2$ with good electrochemical performance are presented. Three different methods were investigated:

- LiOH and NiO at 700 °C in air (Batch A),
- LiOH and NiO between 150 and 750 °C in oxygen (Batch B), and
- $LiNO_3$ and $NiCO_3$ between 600 and 750 °C in oxygen (Batch C).

Batch C exhibited the highest first discharge capacity of about 150 mAh/g between 2.0 and 4.0V with a solid state Li/LiNiO₂ cell containing a PAN-based polymer electrolyte. Solid state carbon/LiNiO₂ cells and two-cell bipolar batteries demonstrated high rate capabilities at room temperature with an ability to sustain pulsed discharge currents as high as 50 mA/cm². About 22 500, 13 000, and 1000 pulses were obtained at current densities of 10, 20, and 50 mA/cm², respectively, with a 10 ms pulse width followed by a 50 ms interpulse rest period.

INTRODUCTION

Rechargeable Li ion batteries have received widespread interest due to an increasing demand for safe, lightweight, high energy, high power batteries for both consumer and military applications. To optimize specific power and energy, several high voltage cathodes have been extensively studied, which include $LiCoO_2$, $LiNiO_2$, $LiNi_{1-x}Co_xO_2$, and $LiMn_2O_4$. Almost all Li ion batteries commercially available today use $LiCoO_2$ due to its good electrochemical performance, ease of manufacture, and safety. However, its main drawback is cost. Hence, low cost is the key emphasis in developing cathode materials for practical applications.

Lithiated nickel oxide ($LiNiO_2$) is an attractive cathode material for Li ion rechargeable batteries due to low cost, high working voltage, and a high theoretical capacity of 274 mAh/g. Extensive studies on $LiNiO_2$ [1-3] have, however, shown that its electrochemical performance is strongly dependent on the stoichiometry of the material. Stoichiometric $LiNiO_2$ consists of a cubic close-packed oxygen array with lithium and nickel ions occupying the octahedral sites. It has been found that depending on the method of preparation, it is very easy to obtain a non-stoichiometric material, $Li_{1-x}Ni_{1+x}O_2$, with extra nickel ions within the lithium sites. Even a small amount of structural disorder due to the displacement of nickel and lithium ions from their respective sites can affect the working voltage and reversible capacity of the cathode material.

In this paper, we present our investigation of three different methods for the preparation of $LiNiO_2$, and the electrochemical performance of the resulting products. Solid state mono- and bipolar carbon/LiNiO₂ cells containing highly conductive poly(acrylonitrile) (PAN)-based polymer electrolytes as separators were fabricated and tested. The rate capabilities of the cells were evaluated by carrying out pulse discharges at current densities ranging between 10 and 50 mA/cm².

---

*Present address: Covalent Associates, Inc., 10 State Street, Woburn, MA 01801.

EXPERIMENTAL

Three different methods were investigated to prepare stoichiometric LiNiO$_2$:

(a) Batch A	LiOH (Aldrich, WI) and NiO (CERAC, WI) heated at 700 °C for 24 hours in air.
(b) Batch B	LiOH and NiO heated at 150 °C for 12 hours in air, followed by 550 °C for 16 hours in O$_2$, and then 750 °C for 21 hours in O$_2$.
(c) Batch C	LiNO$_3$ (Aldrich, WI) and NiCO$_3$ (Alfa, MA) heated at 600 °C for 16 hours in O$_2$, followed by 750 °C for 12 hours in O$_2$.

Stoichiometric amounts of nickel and lithium salts (Li:Ni = 1:1 by molar ratio) were mixed thoroughly before heating in alumina boats. The intermediate products were ground with a mortar and pestle before reheating. The final products were stored in the dry-box. The structure and purity of the different batches of LiNiO$_2$ were characterized by X-ray diffraction analysis with a Rigaku diffractometer using monochromated Cu-K$_\alpha$ radiation (λ=1.5405 Å).

Polyacrylonitrile-based polymer electrolyte having a composition 10.4 w/o PAN (Polyscience, PA), 46.9 w/o ethylene carbonate (EC, Baxter, IL), 33.3 w/o propylene carbonate (PC, Scientific Products, IL), and 9.4 w/o LiN(CF$_3$SO$_2$)$_2$ (3M, Minneapolis) was used as the battery separator. The polymer electrolytes were prepared by first dissolving the salt in a mixture of EC and PC. PAN was then added and the entire mixture was heated in a closed vial at ~ 135 °C until the polymer dissolved and formed a homogenous mixture. The viscous solution was then placed between two metal sheets covered with Teflon release films, and rolled between two rollers to obtain a thin film.

Both graphite (Aldrich, WI) and Conoco petroleum coke (Conoco, OK) were evaluated as anode materials. The carbon composite electrode comprised 52.5 w/o carbon and 47.5 w/o polymer electrolyte binder. The composite electrode was prepared by first making the polymer electrolyte binder as described earlier, and then adding the carbon. The mixture was then heated at ~135 °C and the hot melt was pasted onto a nickel foil by extruding between two rollers. The composite cathode was prepared in a similar way. It comprised 51 w/o LiNiO$_2$, 7 w/o carbon, and 42 w/o polymer electrolyte binder. Stainless steel was used as the current collector.

The carbon/LiNiO$_2$ cells were fabricated by sandwiching a polymer electrolyte between a carbon anode and LiNiO$_2$ cathode. The cell was sealed under vacuum in a metallized plastic bag. The overall dimension of a cell was 10 cm^2. A bi-polar carbon/LiNiO$_2$ electrode was fabricated by spot-welding a SS current collector coated with LiNiO$_2$ composite electrode and a Ni current collector coated with carbon composite electrode, back-to-back. A bi-polar battery was constructed by placing a cathode or an anode on each side of the bi-polar electrode, separated by the polymer electrolyte. The battery was vacuum sealed in a metallized plastic bag.

Galvanostatic cycling of carbon/LiNiO$_2$ cells was carried out using an Arbin battery test system (Bryan, TX). The mono-polar cells were cycled between 2 and 4V at 0.1 and 0.2 mA/cm^2, and the bi-polar batteries were cycled between 6 and 8V at 0.25 mA/cm^2. The pulse power performance of the cells were tested by first charging them to 4V (mono-polar) and 8V (bi-polar) at 0.1 mA/cm^2, followed by pulse discharging to 2V (mono-polar) and 4V (bi-polar) at current densities ranging between 10 and 50 mA/cm^2. Pulse discharging was carried out using a software developed in-house on an IBM-compatible 486-computer. Each pulse typically lasted for 10 ms, followed by an interpulse rest period of 50 ms.

RESULTS

Synthesis of LiNiO₂: It has been demonstrated that [3] the Bragg intensity ratio $R_{(003)} = I_{(003)}/I_{(104)}$ can serve as a reliable quantitative criterion for the stoichiometry of $Li_{1-x}Ni_{1+x}O_2$. Figure 1 represents the XRD patterns of Batches A, B, and C synthesized in this laboratory. The peak positions appear to correspond to the diffraction data of stoichiometric $LiNiO_2$ given by Powder Diffraction File 9-63 [4]. However, the Bragg intensity ratio $R_{(003)}$ was calculated to be 0.76, 1.15, and 1.3 for Batches A, B, and C, respectively. Another difference between the different batches is the splitting of the peak at around $2\theta = 64.5°$. For stoichiometric $LiNiO_2$, there should be two peaks of equal intensity at around $2\theta = 64.5°$, clearly observed for Batches B and C.

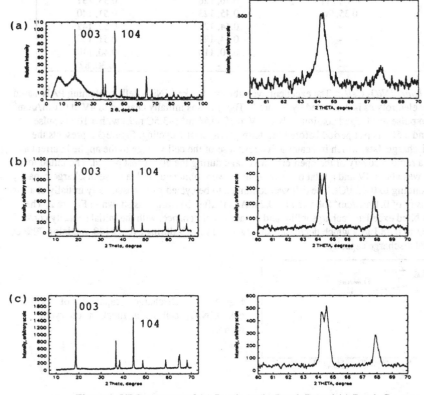

Figure 1. XRD patterns of (a) Batch A, (b) Batch B, and (c) Batch C.

The electrochemical performance of the different $LiNiO_2$ batches were evaluated in $Li/SPE/LiNiO_2$ cells. The cells were cycled between 2 and 4V at 0.2 mA/cm², and some of the discharge capacities for each cell are listed in Table I. Cell A containing Batch A shorted during the fourth charge. It exhibited a 46% first cycle irreversibility and a discharge capacity of 0.35 Li/NiO_2 (96 mAh/g) during the third discharge. Cell B containing Batch B performed about 60 cycles before shorting, with a capacity corresponding to 0.4 Li/NiO_2 (110 mAh/g). A first cycle irreversibility of 28% was observed. Cell C containing Batch C exhibited a first cycle irreversibility of about 30%, with the first discharge capacity corresponding to 0.55 Li/NiO$_2$ (151 mAh/g). The capacity was observed to slowly decrease to 0.51 Li/NiO$_2$ during 20th discharge and 0.30 Li/NiO$_2$ (82.2 mAh/g) during 120th discharge. Hence, it is observed that LiNiO$_2$ prepared in an oxygen environment (Batches B and C) have superior electrochemical properties in terms of capacity and first cycle reversibility. It appears that the Bragg intensity ratio R$_{(003)}$ should be greater than 1 and the peak at 2θ = 64.5° should be a doublet for LiNiO$_2$ with good electrochemical properties.

Table I. Discharge Capacities of Cells Containing Different Batches of LiNiO$_2$.

Cycle No.	Discharge Capacity(Li/NiO$_2$, mAh/g)		
	Cell A (Batch A)	Cell B (Batch B)	Cell C (Batch C)
1	0.41, 112	0.46, 126	0.55, 151
3	0.35, 96	0.45, 123	0.51, 140
10	-	0.44, 121	0.52, 142
20	-	0.44, 121	0.51, 140
60	-	0.40, 110	0.40, 110
120	-	-	0.30, 82

Carbon/LiNiO$_2$ Cells: The excellent cycleability of coke/LiNiO$_2$ cells containing PAN-based polymer electrolytes is illustrated by Figure 2. The cell was initially charged to 4V at 0.1 mA/cm^2 and then pulse discharged continuously to 2V at 10 mA/cm^2 (~3.5C rate) with a 10 ms pulse width and a 50 ms rest period before long-term galvanostatic cycling. Figure 3 represents the pulse discharge data, which indicates a fast response of the cell voltage to the applied current as well as a rapid recovery of the open-circuit voltage during the 50 ms rest period. The initial load voltage was about 3V and a total of 33 000 pulses were obtained, with a pulse discharge capacity corresponding to 0.3 Li/C$_6$. The cell was continued to be cycled galvanostatically initially at a current density of 0.1 mA/cm^2 followed by 0.2 mA/cm^2 after 20 cycles, as shown in Figure 2. The cell exhibited excellent rechargeability and capacity maintenance, with the initial capacity corresponding to 0.48 Li/C$_6$ which decreased slowly to 0.34 Li/C$_6$ during cycle #140, retaining 70% of the initial discharge capacity.

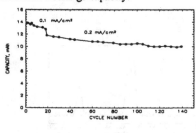

Figure 2. Discharge capacity of a coke/LiNiO$_2$ cell as a function of cycle number.

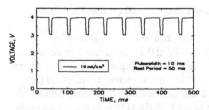
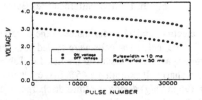

Figure 3. Pulse discharge data of a coke/LiNiO$_2$ cell at a current density of 10 mA/cm^2.

Graphite was investigated as an alternate anode in our search for carbon anodes with higher intercalation capacity and rate capability. It is well established that graphite composite anodes intercalate reversibly about 1 Li/C_6 in liquid electrolytes. However, PC-based electrolytes are observed to reduce on graphite electrodes, interfering with Li intercalation. The incorporation of a small amount of 12-crown-4-ether has been observed to suppress the electrolyte reduction.

A graphite/$LiNiO_2$ cell was charged to 4V at 0.1 mA/cm^2 before pulse discharging to 2V at current densities of 10, 20, 30, and 40 mA/cm^2, which correspond to rates ranging between 7C and 26C. Figure 4 represents the pulse discharge data. At a pulse rate of 10 mA/cm^2 (7C), the initial load voltage was 3.3V, and the cell delivered about 27 000 pulses corresponding to a pulse discharge capacity of 0.5 Li/C_6. When the current density was increased to 20 mA/cm^2 (13C), the load voltage decreased to 3.1V and the cell delivered 13 000 pulses with a pulse discharge capacity of 0.47 Li/C_6. At 30 mA/cm^2 (19C), the load voltage decreased to 2.6V and 3000 pulses were obtained. Only 3 pulses were obtained at the 26C rate (40 mA/cm^2). Hence, it shows that even at a high pulse rate of 13C, an excellent capacity utilization of 0.5 Li/C_6 was exhibited by graphite.

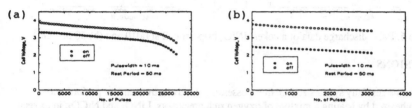

Figure 4. Pulse discharge data of a graphite/$LiNiO_2$ cell at (a) 10 and (b) 30 mA/cm^2.

Carbon/LiNiO₂ Bi-polar Batteries: The power capability of a battery can be increased in several ways, such as using thin electrodes, electrolytes with high conductivity, and bi-polar electrodes. Power is related to the operating cell potential (V) and total cell resistance (R) by

$$P = Vi = V^2/R \qquad (1)$$

Hence, high cell voltage and low cell resistance are important to achieve high power capabilities. The internal resistance of a battery can be lowered by using bi-polar electrodes, which provides the shortest current path between cells, eliminating the need for intercell connectors.

Polymer electrolytes are ideal for the construction of bi-polar batteries. Laminates of these electrolytes and bi-polar electrode plates can be easily stacked without the problem of electrolyte leakage. In this work, two-cell carbon/$LiNiO_2$ bi-polar batteries were fabricated with PAN-based polymer electrolytes, as described earlier. It is very important to match the capacities of the two cells in a bi-polar configuration, since the cell with a lower capacity will be the limiting cell that controls the overall performance of the bi-polar battery.

Figure 5 represents the galvanostatic cycling data for a coke/$LiNiO_2$ bi-polar battery, cycled between 6 and 8V at 0.25 mA/cm^2. Very little fading was observed, with a capacity corresponding to about 0.4 Li/C_6 at cycle #26, after which the cell shorted due to cycler malfunctioning. A typical pulse data for a coke/$LiNiO_2$ bi-polar battery is given by Figure 6. At a current density of 10 mA/cm^2 (10C), the load voltage was about 7V and a total of 22 500 pulses with a capacity corresponding to 0.44 Li/C_6 were obtained. At a much higher current density of 50 mA/cm^2 (50C), the load voltage was about 5V and 1000 pulses were obtained, see Figure 6. Hence, a bi-polar coke/$LiNiO_2$ battery also exhibited high pulse rate capabilities.

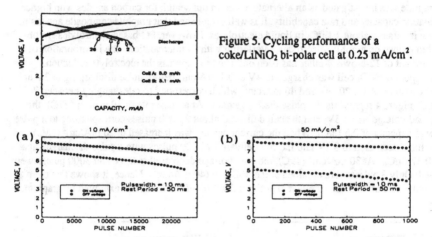

Figure 5. Cycling performance of a coke/LiNiO₂ bi-polar cell at 0.25 mA/cm².

Figure 6. Pulse discharge data of a coke/LiNiO₂ bi-polar cell at (a)10 and (b) 50 mA/cm²

CONCLUSIONS

We have successfully synthesized $LiNiO_2$ possessing good electrochemical performance. $LiNiO_2$ synthesized by heating a mixture of oxygen rich precursors, $LiNO_3$ and $NiCO_3$, in oxygen exhibited a first discharge capacity of about 150 mAh/g corresponding to about 0.55 Li/NiO_2.

The carbon/$LiNiO_2$ cells demonstrated good rechargeability between 2 and 4V and high rate capabilities at room temperature. Graphite/$LiNiO_2$ cells exhibited high pulse rate capabilities, where at a 13C rate, 0.5 Li/C_6 was utilized.

Two cell coke/$LiNiO_2$ bi-polar batteries also demonstrated good rechargeability between 6V and 8V and high pulse rate capabilities. At a 10C rate (10 mA/cm²), the cell could deliver 22 500 pulses, which corresponds to a capacity of 0.44 Li/C_6.

ACKNOWLEDGMENTS

This work was carried out with financial support from the Air Force, Contract F29601-92-C-0099.

REFERENCES

1. J. Morales, C. Peres-Vicente, and J. Tirado, Mater. Res. Bull., 25, 623 (1990).
2. R.V. Moshtev, P. Zlatilova, V. Manev, A. Sato, J. Power Sources, 54, 329 (1995); S. Yamada, M. Fujiwara, and M. Kanda, ibid., 54, 209 (1995).
3. T. Ohzuku, A. Ueda, and M. Nagayama, J. Electrochem. Soc., 140, 1862 (1993).
4. Powder Diffraction Files, Joint Committee on Powder Diffraction Standards, Swarthmore, PA (1975).

A New Polymer Cathode-Conducting Polymer Interconnected With Sulfur-Sulfur Bonds: Poly(2, 2'-Dithiodianiline)

Katsuhiko NAOI,* Ken-ichi KAWASE, Mitsuhiro MORI and Michiko KOMIYAMA
Cooperative Research Center, Tokyo University of Agriculture & Technology
Koganei, Tokyo 184, Japan

Abstract

Poly(2, 2'-Dithiodianiline)(Poly(DTDA)), a conducting polymer having disulfide bond in it, has been proposed as a new class of high energy storage material. DTDA has one S-S bond interconnected between two moieties of anilines. The structure of the poly(DTDA) was similar to that of polyaniline in addition that the S-S bond is preserved after electropolymerization of DTDA. The poly(DTDA) has some advantages because of its high theoretical energy density, faster kinetics and higher electrical conductivity than other organosulfur cathodes

Introduction

A series of compounds having -SH or -SS- groups within the molecules are being considered as energy storage materials(1-11), whereby energy exchange occurs based on the reversible polymerization-depolymerization process(2SH <--> S-S).

According to the previous studies(12-18), disulfide materials show a redox reaction which accompanies structural changes of the monomers, dimers, and polymers upon oxidation and reduction processes.   However, the cleavage of S-S bonds and the recombination efficiency, or charge retention, is generally low(16,17). The recombination efficiency can be improved if the material has a confined polymer structure or conformation which is interconnected with S-S bonds.

Here the authors present dithiodianiline(DTDA) as a new conducting polymer compound having S-S bonds confined in-between the chains of polyaniline(see Figure 1). The poly(DTDA) is expected to show higher electrochemical activity and higher electrical conductivity by virtue of its intramolecular electrocatalytic effect(13-15) as well as the conducting nature of the material. Theoretical charge density of poly(DTDA) is 330 Ah/kg assuming that this polymer shows self-doping process.

Experimental

Poly(DTDA) was electropolymerized by cycling potential between -0.2 and +1.2 V (vs. Ag/AgCl) at glassy carbon electrode in a solution containing 20 mM of DTDA monomer + 0.2M $HClO_4$ + 0.4M $LiClO_4$ / CH_3CN-H_2O(1:1).

A standard three-electrode, two-compartment electrochemical cell was used for all the electrochemical experiments. Cyclic voltammograms were obtained on a normal potentiostat PS-07 (Toho Giken Ltd.). All experiments were run under a nitrogen atmosphere at room temperature.

Results and Discussion

As shown in Fig. 2, the polymeric film continued to grow by means of the cycled-potential electropolymerization of DTDA monomer.   The potential was scanned between -0.2 and +1.2 V (vs. Ag/AgCl) in an acidic solution of CH_3CN-H_2O. The cyclic voltammogram showed a sharp redox in which the oxidation-reduction of the polyaniline

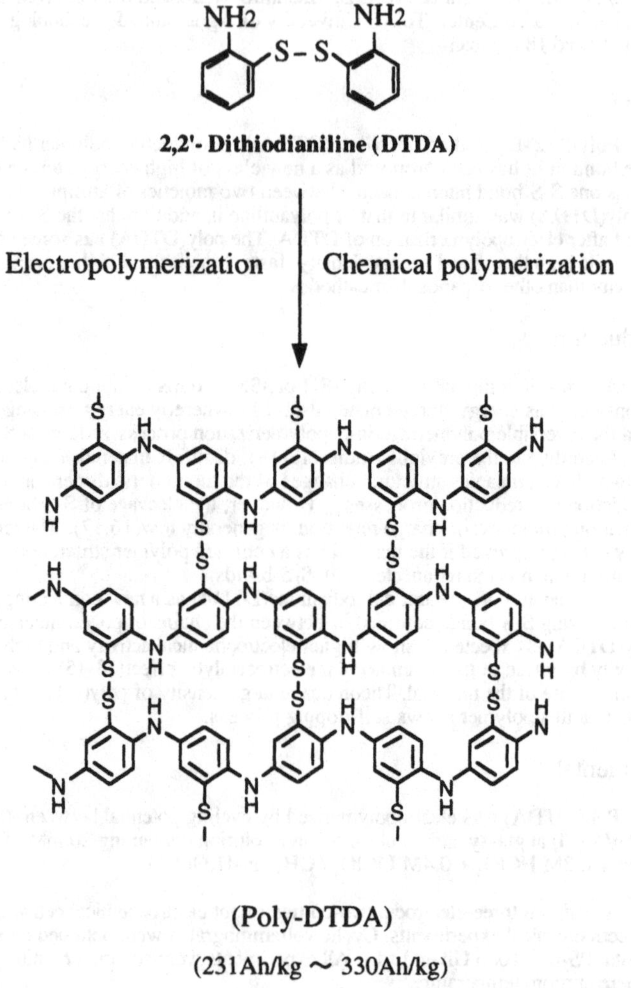

NH2   NH2

2,2'- Dithiodianiline (DTDA)

Electropolymerization | Chemical polymerization

(Poly-DTDA)

(231Ah/kg ~ 330Ah/kg)

Fig. 1   poly(DTDA) and its theoretical energy density.

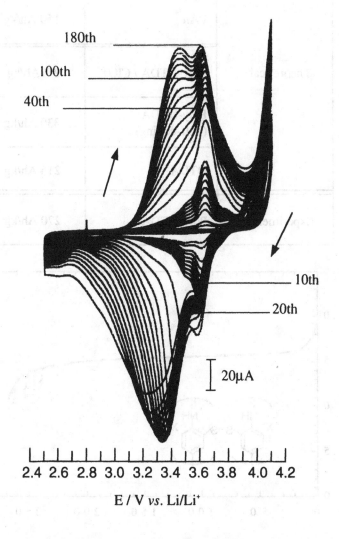

180th

100th

40th

10th

20th

20μA

2.4 2.6 2.8 3.0 3.2 3.4 3.6 3.8 4.0 4.2

E / V *vs.* Li/Li⁺

Fig.2   Cyclic voltammograms of 50 mM DTDA in 0.5 M
CF₃COOH, 0.5 M NaClO₄ / MeCN at ITO. Scan
rate , 50 mV/s.

	PAn	150 Ah/kg
Theoretical	Poly-DTDA / ClO4⁻	231Ah/kg
	Poly-DTDA (Self-doping)	330 Ah/kg
	DTDA	216 Ah/kg
Experimental	Poly-DTDA	270 Ah/kg

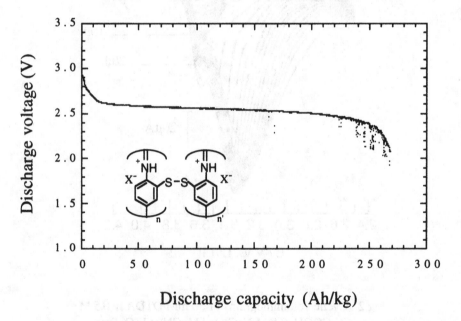

Fig.3　Discharge curve of Li/poly-DTDA cell (at 20℃). Discharge current : 0.1 mA/cm². Termination voltage : 2.0V.

at N sites occurs and that the charge is compensated by a self doping/undoping process of thiolate(S^-) anions.

The FT-IR spectrum for poly(DTDA) film indicated that the structure is almost the same as that of the quinonoid form (22-24) of polyaniline(emeraldine base) except for the strong absorption peaks at 750 cm^{-1} (C-S) and no absorption peak responsible for S=O (at 1100 cm^{-1}), indicating that S-S bonds were not further oxidized to form any kind of sulfone or sulfonate groups(S=O).

In-situ Surface Enhanced Raman Spectroscopy(SERS) analysis(25) indicated that the electrochemical cleavage and recombination process of S-S is catalyzed or enhanced by polyaniline side chains in poly(DTDA). The behavior is similar to the electrocatalytic effect that has been observed for polyaniline/2,5-Dimercapt-1,3,4 thiadiazole(DMcT)(13-15).

The discharge test of the lithium cells with the poly(DTDA) was performed at a current density of 0.1 mA cm^{-2}. Figure 3 shows a discharge performance of the cell consisted of lithium anode, $LiClO_4$/EC:PC(1:1)/polyacrlonitrile-based gel electrolyte, and composite cathode containing poly(DTDA)(20wt%), $LiClO_4$/PEO(20wt%), carbon(60%), and binder(5%). Termination voltage was 2.0 V. From this flat and single-step discharge curve, the energy density was calculated to be 675 Wh/kg-cathode. The specific capacity calculated 270 Ah/kg corresponds to the cathode utilization to be around 81%. Further studies on cyclability and rate capability of poly(DTDA) cathode are currently underway.

Conclusions

The electropolymerized poly(DTDA) showed an enhanced redox process which is based on the intramolecular electrocatalytic effect of aniline moiety and thiolate anions. The discharge test of Li/poly(DTDA) cell showed flat curve with a charge capacity of 270 Ah/kg-cathode and an energy density of 675 Wh/kg-cathode.

REFERENCES

1.   S. J. Visco, C. C. Mailhe, L. C. De Jonghe and M. B. Armand, J. Electrochem. Soc., 136, 661 (1989).
2.   S. J. Visco and L. C. De Jonghe, Mat. Res. Symp. Proc., 135, 553 (1989).
3.   M. Liu, S. J. Visco and L. C. De Jonghe, J. Electrochem. Soc., 136, 2570 (1989).
4.   S. J. Visco, M. Liu and L. C. De Jonghe, J. Electrochem. Soc., 137, 1191(1990).
5.   M. Liu, S. J. Visco and L. C. De Jonghe, J. Electrochem. Soc., 137, 750 (1990).
6.   M. Liu, S. J. Visco and L. C. De Jonghe, J. Electrochem. Soc., 138, 1891 (1991).
7.   M. Liu, S. J. Visco and L. C. De Jonghe, J. Electrochem. Soc., 138, 1896 (1991).
8.   M. M. Doeff, M. M. Lerner, S. J. Visco and L. C. De Jonghe, J. Electrochem. Soc., 139, 2077 (1992).
9.   M. Ue, S. J. Visco and L. C. De Jonghe, Denki Kagaku., 61, 1409 (1993).
10.   S. J. Visco, M. Liu, M. M. Doeff, Y. P. Ma, C. Lampert and L. C. De Jonghe, Solid State Ionics, 60, 175 (1993).
11.   E. M. Genies and S. Picart, Synth. Met., 69, 165 (1995).
12.   K, Naoi, M. Menda, H. Ooike and N. Oyama, Proc. 31st Bat. Symp. Jpn., p.31 (1990).
13.   K. Naoi, M. Menda, H. Ooike and N. Oyama, J. Electroanal. Chem., 318, 395

(1991).
14. T. Sotomura, H. Uemachi, K. Takeyama, K. Naoi and N. Oyama, Electrochimica Acta., 37, 1851(1992).
15. K. Naoi and N. Oyama, Proc. Symp. on Lithium Battery, Honolulu, Hawaii, May 16-21, (1993).
16. K. Naoi, Denki Kagaku, 61, 135 (1993).
17. K. Naoi, Y. Oura and A. Kakinuma, Extd Abst., ISE Meeting(Berlin) (1993).
18. (a) K. Naoi, K. Kawase, Y. Inoue, Abst. IBA Meeting, Chicago(1995); (b) K. Naoi, K. Kawase, Y. Inoue Proc. Symp. Lithum Batteries of ECS meeting ,San Antonio, in press ; (c) K. Naoi, K. Kawase, Y. Inoue, J. Electrochem. Soc., submitted.
19. K. Naoi, K. Kawase, Y. Inoue, M. Komiyama, Abst. Battery Symp in Japan. (1996).
20. K. Naoi, K. Kawase, Y. Inoue, M. Komiyama, Abst. Electrochem. Soc. Meeting in Japan (1996).
21. K. Naoi, K. Kawase, Y. Inoue, Extd. Abst. Electrochemical Society, San Antonio, TX (1996).
22. D. K. Moon, K. Osakada, T. Maruyama and T. Yamamoto, Makromol. Chem., 193, 1723 (1992).
23. J. Tang, X. Jing, B. Wang and F. Wang, Synth. Met., 24, 231 (1988).
24. Y. Wei, R. Hariharan and S. A. Patel, Macromol., 23, 758 (1990).
25. K. Naoi, K. Kawase, M. Mori, and M. Komiyama, J. Electrochem. Soc., 144, L173 (1997).

LiMn$_{2-x}$Cu$_x$O$_4$ Spinels - 5 V Cathode Materials

Yair Ein- Eli, W. F. Howard, Jr. and Sharon H. Lu
Covalent Associates, Inc.,
10 State Street, Woburn, MA 01801, USA
Sanjeev Mukerjee and James McBreen
Brookhaven National Laboratory
Upton, NY 11973-5000, USA
John T. Vaughey and Michael M. Thackeray
Argonne National Laboratory
9700 South Cass Avenue, Argonne, IL 60439, USA

Abstract

A series of electroactive spinel compounds, LiMn$_{2-x}$Cu$_x$O$_4$ (0.1 $\leq$ x $\leq$ 0.5) has been studied by crystallographic, spectroscopic and electrochemical methods and by electron-microscopy. These LiMn$_{2-x}$Cu$_x$O$_4$ spinels are nearly identical in structure to cubic LiMn$_2$O$_4$ and successfully undergo reversible Li intercalation. The electrochemical data show slight shifts to higher voltage for the delithiation reaction that normally occurs at 4.1 V in standard Li$_{1-x}$Mn$_2$O$_4$ electrodes (1 $\geq$ x $\geq$ 0) corresponding to the oxidation of Mn^{3+} to Mn^{4+}. The data also show a remarkable reversible electrochemical process at 4.9 V which is attributed to the oxidation of Cu^{2+} to Cu^{3+}. The inclusion of Cu in the spinel structure enhances the electrochemical stability of these materials upon cycling. The initial capacity of LiMn$_{2-x}$Cu$_x$O$_4$ spinels decreases with increasing x from 130 mAh/g in LiMn$_2$O$_4$ (x=0) to 70 mAh/g in "LiMn$_{1.5}$Cu$_{0.5}$O$_4$" (x=0.5). Although the powder X-ray diffraction pattern of "LiMn$_{1.5}$Cu$_{0.5}$O$_4$" shows a single-phase spinel product, neutron diffraction data show a small, but significant quantity of an impurity phase, the composition and structure of which could not be identified. X-ray absorption spectroscopy was used to gather information about the oxidation states of the manganese and copper ions. The composition of the spinel component in the LiMn$_{1.5}$Cu$_{0.5}$O$_4$ was determined from X-ray diffraction and XANES data to be Li$_{1.01}$Mn$_{1.67}$Cu$_{0.32}$O$_4$ suggesting, to a best approximation, that the impurity in the sample was a lithium-copper-oxide phase.

Introduction

Materials that reversibly intercalate lithium form the cornerstones of the emerging lithium-ion battery industry. The spinel LiMn$_2$O$_4$ is an inexpensive, environmentally benign intercalation cathode that is the subject of intense development (1), although it is not without faults. The achievable electrode capacity (120 mAh/g) is 15-30% lower than that which can be obtained from Li(Co,Ni)O$_2$ cathodes. Moreover, an unmodified LiMn$_2$O$_4$ electrode exhibits an unacceptably high capacity fade. Several researchers have stabilized the LiMn$_2$O$_4$ electrode structure to lithium insertion/extraction reactions at ~4 V by substituting a small fraction (~2.5%) of the manganese ions with other metal cations (2-4).

Recently, we reported a preliminary account of the preparation and electrochemical behavior of a copper substituted spinel, LiMn$_{1.5}$Cu$_{0.5}$O$_4$ (5). In this paper, we present electrochemical, structural and spectroscopic data obtained from an examination of various compounds in the LiMn$_{2-x}$Cu$_x$O$_4$ system (0 $\leq$ x $\leq$ 0.5) which provide a much greater understanding of the behavior of these materials than initially reported (5). Structural properties and variations in the cation charge distribution are used to explain the electrochemical behavior of LiMn$_{2-x}$Cu$_x$O$_4$ electrodes. X-ray absorption studies have been used to determine the oxidation states of the manganese and copper ions. The structure of the spinel component, as determined by a Rietveld refinement of the powder X-ray diffraction pattern, is presented.

Experimental

$LiMn_{2-x}Cu_xO_4$ ($0 \leq x \leq 0.5$) cathode materials were prepared by conventional solid state and sol-gel methods. In the solid state syntheses, $LiOH \cdot H_2O$ was intimately mixed with the required amounts of CuO and MnO_2 for a given stoichiometry, and then heated for 18 hours in air at 750° C. The product was free-flowing and did not require milling.

Nearly phase-pure $LiMn_{1.5}Cu_{0.5}O_4$ was prepared by a sol-gel process by dissolving stoichiometric amounts of CH_3COOLi, $Cu(OOCCH_3)_2 \cdot H_2O$, and $Mn(OOCCH_3)_2$ in deionized water, and adding a 4 -times molar amount of NH_4OH. The mixture was stirred with gentle heating for 2 hours, then concentrated to dryness on a rotary evaporator.

Cyclic voltammograms were obtained with an EG&G/PAR potentiostat, model 263A; they were recorded at a slow sweep rate of 15 $\mu V/s$. Cycling data were collected on either a Maccor series 4000 or Starbuck multi-channel cyclers. Cathode materials were studied using a lithium foil anode, separated with Whatman BS-65 glass microfibers in a 1 cm^2 parallel-plate configuration. Cathode films were prepared from a slurry of $LiCu_xMn_{2-x}O_4$ with 10% PVDF and 10% acetylene black (w/w) dissolved in N-methyl-2-pyrrolidinone. The mixture was doctor-bladed onto aluminum foil, dried at 140° C under vacuum for several hours and then roll-compressed. Cathode discs with an average weight of 6 mg of active material were obtained. The electrolyte composition was 1.2 M $LiPF_6$ in a mixture of ethylene carbonate (EC) and dimethyl carbonate (DMC) in a volume ratio of 2EC:3DMC. Cells were charged and discharged galvanostatically at a current density of 0.25 mA/cm^2 between 3.3 and 5.1 V.

X-ray absorption data were collected at Beam Line X11A of the National Synchrotron Light Source at Brookhaven National Laboratory with the storage ring operating at 2.584 GeV and an electron current between 110 and 350 mA. The monochromator was operated in the two crystal mode using Si (111) crystals. Since the main interest was the determination of the oxidation state of the Mn and Cu ions, only the X-ray absorption near edge fine structure (XANES) was analyzed in detail.

The X-ray patterns of $LiCu_{0.5}Mn_{1.5}O_4$ samples showed a single-phase spinel product. By contrast, neutron diffraction data of these two samples collected by the Intense Pulsed Neutron Diffractometer (IPNS) at Argonne National Laboratory and by the High Resolution Powder Diffractometer (HRPD) at the Rutherford Laboratory (U.K.) showed evidence of a small, but significant impurity phase that could not be identified. Because of problems encountered with a two-phase refinement of the neutron data, a best approximation of the structure of the spinel component was determined from the X-ray data by the Rietveld profile refinement technique.

Results and Discussion

Sample Preparation: Attempts to prepare $LiMn_{2-x}Cu_xO_4$ cathode materials via conventional solid-state reaction techniques resulted in persistent MnO_x/Li_2MnO_3 impurities in the products, whereas sol-gel methods produced purer materials. The X-ray diffraction patterns of typical products (obtained with Fe K_α radiation) made by the two routes are shown in Figure 1. Three subsequent firings of the products made by the solid state reaction route reduced only marginally the Mn_2O_3/Li_2MnO_3 impurity levels. For the sol-gel route a firing sequence produced what appeared to be essentially phase-pure materials (Figure 1). The X-ray data show a small, but significant [220] peak at 39° 2θ. The intensity of this peak indicates that a small amount of copper resides on the 8a tetrahedral sites of the spinel structure.

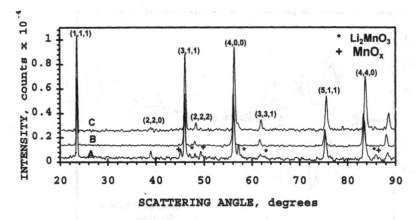

Figure 1:XRD patterns obtained from $LiCu_{0.5}Mn_{1.5}O_4$ spinel prepared by solid state (A) and sol-gel (C) preparation methods. The XRD pattern of unmodified $LiMn_2O_4$ spinel (B) is shown for comparison.

The neutron diffraction pattern of a $LiMn_{1.5}Cu_{0.5}O_4$ sample shows clear evidence of a spinel phase and an impurity phase. The composition of the spinel component in the sample as determined by X-ray diffraction analysis (described in a following section) indicated that the impurity phase was a lithium-copper-oxide compound. The appearance of the impurity only in the neutron diffraction pattern was attributed to the fact that copper is a very strong scatterer of neutrons ($b=0.78 \times 10^{12}$ Å). The observation of the impurity phase only in the neutron diffraction pattern highlights the possible dangers of misinterpreting X-ray diffraction patterns.

Electrochemical studies: Figure 2 shows the cyclic voltammograms of the spinel series. Note that the spinel composition with the lowest copper content ($LiMn_{1.9}Cu_{0.1}O_4$) provides a voltage profile and cyclic voltammogram which closely resembles those obtained from standard $LiMn_2O_4$ (4). As the copper content of the spinel increases, the peaks which are located at 4.05 and 4.16 V in the cyclic voltammogram of $LiMn_2O_4$, (attributed to a two-step extraction of lithium from the tetrahedral 8a sites), shift to higher voltages. The higher voltage peak splits into a doublet. These features are particularly noticeable in the $LiMn_{1.5}Cu_{0.5}O_4$ sample in which the higher voltage peak (at 4.16 V in $LiMn_2O_4$) is shifted and split into two peaks at approximately 4.22 and 4.30 V. Although the reasons for this behavior are not yet fully understood, it is believed that the peak shifts are associated with the presence of some copper on the tetrahedral sites of the spinel structure. It is possible that one of the higher voltage peaks of the doublet may be due to oxidation of the lithium-copper-oxide impurity phase. The high voltage reaction at 4.9 V that increases with increasing copper content is attributed to the oxidation of Cu^{2+} to Cu^{3+} on the octahedral (16d) sites of the spinel structure. The cyclic voltammograms indicate that all these reactions appear to be reversible, at least on the first cycle. Table 1 summarizes the relative amounts of capacity obtained in the high voltage region (5.1-4.5 V) and low voltage region (4.5-3.3 V) during the third cycle for the various $Li/LiMn_{2-x}Cu_xO_4$ cells ($0 \leq x \leq 0.5$). The total capacity of $LiMn_{2-x}Cu_xO_4$ electrodes drops with increasing copper content from 119 mAh/g at x = 0.1 to 71 mAh/g at x = 0.5.

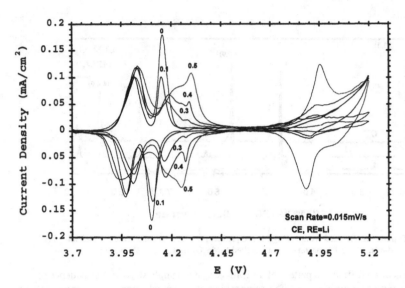

Figure 2:Cyclic voltammogram obtained from LiCu$_x$Mn$_{2-x}$O$_4$ (x = 0, 0.1, 0.3, 0.4 and 0.5) cycled in the potential limits of 3.75 - 5.2 V at scan rate of 15 μV/s. Li metal served both as counter and reference electrode in EC(2):DMC(3)/1.2M LiPF$_6$.

It is clear that copper substitution has two major effects on the electrochemistry of the spinel electrode that can be interpreted in terms of the coordination and oxidation state of the copper ions. If it mimicked Amine's nickel-substituted analog, Li[Mn$_{1.5}$Ni$_{0.5}$]O$_4$ (6), the copper ions would all be divalent and the manganese ions would be fully-oxidized in a tetravalent state. In this circumstance, no charge capacity would be expected in the 3.9 - 4.3 V region, which originates from Mn^{3+} → Mn^{4+} oxidation. The electrochemical data are clearly inconsistent with a spinel electrode with the simple cation arrangement in Li[Mn$_{1.5}$Cu$_{0.5}$]O$_4$; the data imply a different cation arrangement and charge distribution in the structure of the spinel electrode.

Table I: Relative Capacities of Empirical "LiCu$_x$Mn$_{2-x}$O$_4$" Electrodes

x in LiCu$_x$Mn$_{2-x}$O$_4$	Capacity, mAh/g at 5.1-4.5V	Capacity, mAh/g at 4.5-3.3V	Cu:Mn Capacity Ratio
0.1	7	112	1:16
0.2	10	96	1:10
0.3	13	79	1:6
0.4	19	63	1:3
0.5	23	48	1:2

Figure 3 shows the cycling performance in terms of discharge/charge capacity obtained between 5.1 and 3.3 V, expressed in mAh/g, *vs.* cycle number for the series of $LiMn_{2-x}Cu_xO_4$ electrodes $(0 \leq x \leq 0.5)$. The available capacity decreases as the amount of copper in the spinel increases. However, electrodes with higher copper content showed significantly improved capacity retention on cycling.

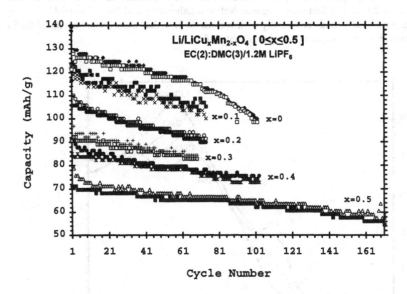

Figure 3: The cycle life behavior (charge/discharge capacity, expressed in mAh/g *vs.* cycle number) of $LiCu_xMn_{2-x}O_4$ $(0 \leq x \leq 0.5$, in steps of x = 0.1). Li metal served as a counter electrode in EC(2):DMC(3)/1.2 M $LiPF_6$.

XANES measurements: Figure 4 shows the XANES spectra at the Cu K-edge for Cu foil, Cu_2O, CuO and $LiCu_{0.5}Mn_{1.5}O_4$. Theoretical calculations predict that the white line should shift by 10 eV in going from Cu^{2+} to Cu^{3+} (7). Measurements on $KCuO_2$, in which the Cu^{3+} ions have square planar coordination, show a peak shift of only 4 eV (8). Since no shift is seen in the peak, and there is no shoulder in the edge, the XANES data would appear to indicate only Cu^{2+} ions with a symmetric coordination.

Structure Refinements: In a spinel structure with the cation distribution $Li[Mn_{1.5}Cu_{0.5}]O_4$, charge neutrality is obtained when the manganese ions are all tetravalent and the copper ions are all divalent. Replacement of copper by lithium on the octahedral sites results in $Li_4Mn_5O_{12}$ (or in spinel notation $Li[Mn_{1.67}Li_{0.33}]O_4$) in which charge neutrality is achieved by increasing the Mn^{4+} content to compensate for the monovalent lithium ions. These two compounds are, in principle the end-members of a possible solid solution system $Li[Mn_{1.5+x}Cu_{0.5-3x}Li_{2x}]O_4$ $(0 \leq x \leq 0.167)$. In this solid solution system, it would also be possible for lithium and copper ions to exchange on the tetrahedral sites. Structural refinements of such complex systems are difficult, particularly when three different cation-types are disordered over one crystallographically independent site. Despite the difficulties that were encountered in obtaining a meaningful fit to the neutron diffraction data of a two-phase reaction product, these analyses provided valuable information about the structure that could not be determined from the X-ray diffraction patterns. This information was then used for the final structure refinement with the "single-phase" X-ray diffraction data. The structure was refined using the prototypic cubic spinel space group Fd3m.

Structure analyses of the spinel component in two independent $LiMn_{1.5}Cu_{0.5}O_4$ samples using neutron diffraction data from Argonne National Laboratory and the Rutherford Laboratory showed unequivocally, and consistently with various models, that some copper occupied the 8a tetrahedra with a site occupancy of approximately 0.1.

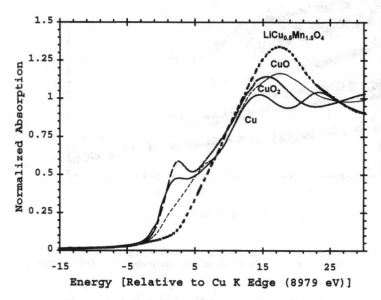

Figure 4:Normalized Cu K edge XANES for Cu foil (_____), Cu_2O (- - -), CuO (-----), and $LiCu_{0.5}Mn_{1.5}O_4$ (·····).

Refinements were carried out on the system $(Li_{0.9}Cu_{0.1})_{8a}[Mn_{1.67}Cu_{0.33-\delta}Li_{\delta}]_{16d}O_{4-\alpha}$ using the input from the neutron data. The parameter, δ, was used to control the stoichiometry of the lithium and copper on the 16d octahedral sites. A second parameter, α, was refined to determine if there was any non-stoichiometry in the oxygen content.

The results of the refinement which yielded the best fit to the data (R_p=8.7%) are summarized in Table 2. Lattice parameter was refined to 8.1923(2) Å. Refinements of two separate samples did not favor an oxygen-deficient structure; both refinements showed that the oxygen ion positions were fully occupied (α=0). The structure analysis showed a cation distribution $(Li_{0.9}Cu_{0.1})_{8a}[Mn_{1.67}Cu_{0.22}Li_{0.11}]_{16d}O_4$; the overall composition is $Li_{1.01}Mn_{1.67}Cu_{0.32}O_4$. The results strongly suggest that the impurity detected in the neutron diffraction profile is a lithium copper oxide phase.

Table II: Crystallographic parameters of $(Li_{0.9}Cu_{0.1})_{8a}[Mn_{1.67}Cu_{0.33-\delta}Li_\delta]_{16d}O_4$
(Space group Fd3m, $a = 8.1923(2)$ Å; Vol.= 549.82(2) Å^3)
[Rp=8.7%]

Atom	Wyckoff Notation	x	y	z	U (x100)	Site Occupancy, n
Li (1)	8a	0.125	0.125	0.125	2.42(6)	0.9
Cu (1)	8a	0.125	0.125	0.125	2.42(6)	0.1
Mn (1)	16d	0.5	0.5	0.5	2.42(6)	0.835
Cu (2)	16d	0.5	0.5	0.5	2.42(6)	0.112(6)
Li (2)	16d	0.5	0.5	0.5	2.42(6)	0.053(6)
O (2)	32e	0.2652(2)	0.2652(2)	0.2652(2)	3.63(7)	1

Although several lithium copper oxide compounds are known to exist, particularly $Li_2O\cdot nCuO$ compounds with divalent copper, such as $Li_2Cu_2O_3$ (n=2) (9) and Li_2CuO_2 (n=1) (10), none of these compounds could be indexed satisfactorily to all the peaks of the impurity phase in the neutron diffraction pattern.

Conclusions

Novel electroactive materials, $LiMn_{2-x}Cu_xO_4$, (0.1 ≤ x ≤ 0.5) have been prepared and evaluated in lithium cells. The data show that lithium can be extracted from the spinel structures in two main potential regions: 3.9 to 4.3 V and 4.8 to 5.0 V, attributed to the oxidation of Mn^{3+} to Mn^{4+} and Cu^{2+} to Cu^{3+}, respectively. The reactions are reversible. Stable electrochemical cycling has been observed for electrode compositions with high values of x, but with low capacity (60-70 mAh/g), for example, in an electrode with overall composition "$LiMn_{1.5}Cu_{0.5}O_4$" (x=0.5). Analysis of X-ray and neutron diffraction data, and XANES spectra have revealed that it is difficult to synthesize single-phase lithium-copper-manganese-oxide spinel compounds with a predetermined composition. The stability and relatively low reactivity of CuO (compared to Li_2O) appears to restrict the complete incorporation of copper into the spinel structure. The cation distribution in these spinel structures is highly complex, which makes it difficult to perform detailed structure analyses with X-ray and neutron diffraction data with a high degree of accuracy. Although the XANES data show that the oxidation state of copper in these spinel compounds is divalent, the possibility of a small amount of Cu^{1+} (in tetrahedral sites) or Cu^{3+} (in octahedral sites) in the starting spinel structures should not be entirely discounted.

Acknowledgments

This work was performed under an SBIR Phase I DoD program sponsored by the U.S. Army CECOM, administered by the Army Research Laboratory, Fort Monmouth, NJ, under Contract No. DAABO7-97-C-D304. Support for Argonne National Laboratory from the U.S. Department of Energy's Advanced Battery Program, Chemical Sciences Division, Office of Basic Energy Sciences, under Contract No. W-31-109-Eng-38 is gratefully acknowledged. The XAS measurements were done at Beam Line X11A at NSLS. This work was supported by the Assistant Secretary for Energy Efficiency and Renewable Energy, Office of Transportation

Technologies, Electric and Hybrid Propulsion Division, USDOE under Contract Number DE-AC02-76CH00016. Dr. W. I. F. David and Dr. R. M. Ibberson are thanked for collecting neutron diffraction data at the Rutherford Appleton Laboratory and for undertaking some preliminary structural refinements. The authors would like to thank Professor J. B. Goodenough for stimulating discussions.

References

1. M. M. Thackeray, Progress in Batteries and Battery Materials, Vol. 14, R. J. Brodd, ed., ITE Press, Inc., Brunswick, OH, p.1 [1995], and references therein.
2. R. J. Gummow, A. de Kock, D. C. Liles, and M. M. Thackeray, *Solid State Ionics*, **69**, 59 [1994].
3. G. Pistoia, C. Bellito, and A. Antonini, 190th Electrochem. Soc. Meeting, San Antonio, TX, 6-11 October 1996.
4. A. D. Robertson, S. H. Lu, and W. F. Howard, Jr., *J. Electrochem. Soc.*, **144**, 3505 [1997].
5. Y. Ein-Eli and W.F. Howard, Jr., *J. Electrochem. Soc.*, **144**, L205 [1997].
6. K. Amine, H. Tukamoto, H. Yasuda, and Y. Fujita, 188th Electrochem. Soc. Meeting, Chicago, IL 8-13 October 1995; K. Amine, H. Tukamoto, H. Yasuda, and Y. Fujita, *J. Electrochem. Soc.*, **143**, 1607 [1996].
7. E. E. Alp, G. K. Shenoy, D. G. Hinks, D. W. Capone II, L. Soderholm, H. B. Schuttler, J. Guo,D. E. Ellis, P. A. Montano, and M. Ramanathan, *Phys. Rev.* **B 35**, 7199 [1987].
8. J. M. Tranquada, S. M. Heald, and A. R. Moodenbaugh, *Phys.Rev.*, **B 36**, 5263 [1987].
9. JCPDS Data File 36-661.
10. J. Barker and M. Yazid Saidi, U.S. Patent 5,670,277 [1997].

Application of potentially biodegradable polyamide and polyester containing disulfide bonds to positive active materials for lithium secondary batteries

H. Tsutsumi*, S. Okada, K. Toda, K. Onimura, and T. Oishi
Department of Applied Chemistry & Chemical Engineering, Faculty of Engineering,
Yamaguchi University, 2557 Tokiwadai, Ube 755, Japan

ABSTRACT

Potentially biodegradable polyamide (poly(imino-3, 3'-dithiodipropionylimino-2, 2'-dithiodiethyl), -($NHCOCH_2CH_2$-S-S-CH_2CH_2CO-$NHCH_2CH_2$-S-S-CH_2CH_2)$_n$-, PIDI) and polyester (crosslinked poly(oxy-3, 3'-dithiodipropionyloxymethylene-1,4-phenylenemethylene, -($OCOCH_2CH_2$-S-S-CH_2CH_2COO-CH_2-C_6H_5-CH_2)$_n$- c-PODOM)) were prepared by interfacial polymerization between 3,3'-dithiodipropionic acid and cystamine, or 1, 4-benzenedimethanol. PIDI and c-PODOM showed the electrochemical responses based on reduction of the disulfide bonds in the polymer chains and re-oxidation of the produced thiolate anions. Degradation of PIDI and PODOM was investigated in various pH buffer solution. The apparent rate constant of c-PODOM at 25 °C was 1.3×10^{-3} % day^{-1} at pH 7 and that of PIDI at 37 °C was 3.5×10^{-3} % day^{-1} at pH 7. Estimated period for 90 % degradation of the polymers at pH 7 is about 1800 days (c-PODOM, 25 °C) and 630 days (PIDI, 37 °C).

Introduction

Light and high power secondary batteries are important devices for lightening potable electrical equipment, such as cellular phone, portable computer. Lithium or lithium-ion batteries are powerful candidates. Lithium cobalt oxide ($LiCoO_2$) is used for a positive active material of lithium-ion practical secondary batteries[1]. However, estimated amount of cobalt (25 ppm) in the earth's crust is lower than that of iron (5.63 %), manganese (950 ppm), sulfur (260 ppm), and carbon (200 ppm) [2]. New positive active material not based on cobalt for lithium or lithium-ion batteries is indispensable to construct a practical high power secondary battery in future.

Use of organosulfur compounds, such as 2, 5-dimercapt-1, 3, 4-thiadiazole (DMcT), for positive active material of high power secondary lithium batteries have been also investigated and reported by Visco's [3-5] and Oyama's groups [6-8]. The theoretical capacity of DMcT is 362 Ah kg^{-1}. Furthermore, the redox reaction of organosulfur compounds (R-S-S-R + 2e$^-$ = 2(RS$^-$)) is usually reversible. We have investigated other organosulfur polymers which can be used as a positive active material for lithium batteries [9-12]. We think that the advantage of the organosulfur compounds is not only their high energy capacity but also their biodegradability. About 6 billion cells were produced in Japan on 1995 [13]. Wastes of batteries have also increased. Usual rechargeable batteries, such as lead-acid, nickel-cadmium, lithium-ion battery, contain heavy metals which are expected to pollute our environment. Recycling of the metals has been investigated. However, those materials is not recycled perfectly. Secondary batteries containing biodegradable materials are preferable for the natural environment.

In this paper we propose that polyamide and polyester containing disulfide bonds are not only prefer to the positive active materials of lithium secondary batteries with high capacity and but also to the biodegradative electrode materials. Figure 1 shows their structures and their theoretical capacities. The disulfide bonds in the polyamide or polyester act as redox centers for battery reaction and the amide or ester bonds in the chain are hydrolyzed in the outdoor environment. The polyamide and polyester has intrinsic electric conductivity. Thus polymer-graphite mixtures were prepared and used for all the electrochemical measurements.

323

Mat. Res. Soc. Symp. Proc. Vol. 496 ©1998 Materials Research Society

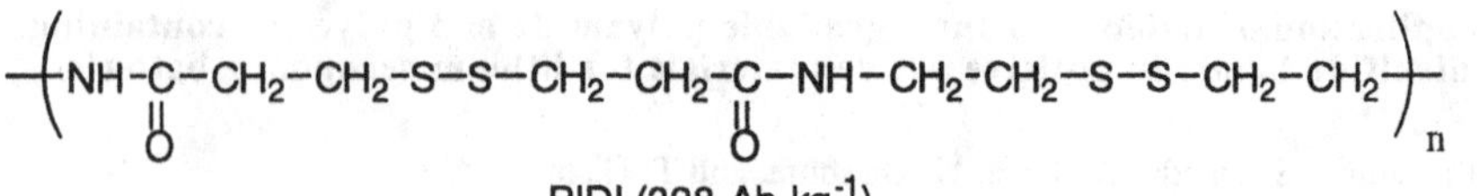

$$-\!\!\left(\!NH\!-\!\underset{\underset{O}{\|}}{C}\!-\!CH_2\!-\!CH_2\!-\!S\!-\!S\!-\!CH_2\!-\!CH_2\!-\!\underset{\underset{O}{\|}}{C}\!-\!NH\!-\!CH_2\!-\!CH_2\!-\!S\!-\!S\!-\!CH_2\!-\!CH_2\!\right)_{\!n}$$

PIDI (328 Ah kg^{-1})

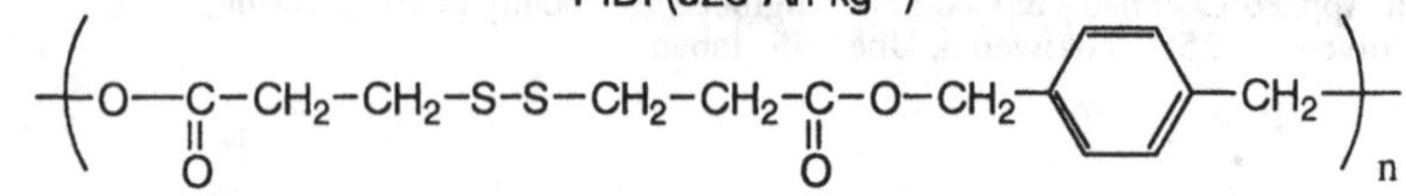

$$-\!\!\left(\!O\!-\!\underset{\underset{O}{\|}}{C}\!-\!CH_2\!-\!CH_2\!-\!S\!-\!S\!-\!CH_2\!-\!CH_2\!-\!\underset{\underset{O}{\|}}{C}\!-\!O\!-\!CH_2\!-\!\!\left\langle\!\!\bigcirc\!\!\right\rangle\!\!-\!CH_2\!\right)_{\!n}$$

PODOM (172 Ah kg^{-1})

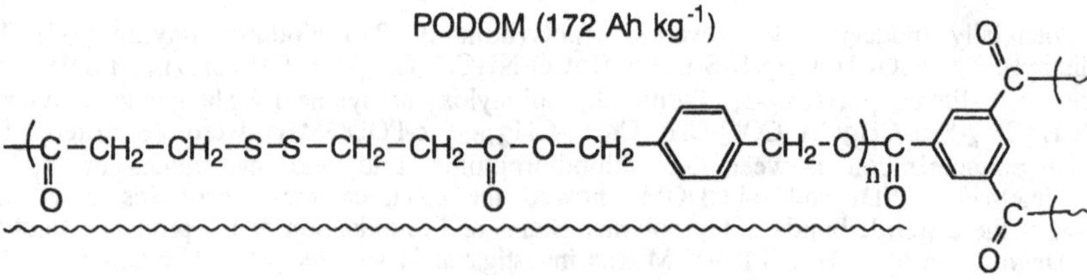

c-PODOM (135 Ah kg^{-1})

Figure 1. Structures of PIDI, PODOM, and c-PODOM, and their theoretical capacities.

EXPERIMENT

Materials

3, 3'-Dithiodipropionic acid (Aldrich), cystamine (Ishizu), 1, 4-benzenedimethanol (Ishizu), and acetonitrile (HPLC grade, Ishizu) were used as received. Propylene carbonate (PC) and 1, 2-dimethoxyethane (DME) were used as received (Tomiyama Pure Chemical Industries, Battery grade). Lithium perchlorate trihydrate (Ishizu) was heated at 150 °C for 24 h under vacuum to remove hydrated water. Graphite powder (Wako) and other reagents were also used as received.

Preparation of polyamide and polyester

Polyamide (poly(imino-3, 3'-dithiodipropionylimino-2, 2'-dithiodiethyl), PIDI) was prepared by condensation between 3, 3'-dithiodipropionic acid and cystamine [9, 12]. Detail procedure of preparation of 3, 3'-dithiodipropionyl dichloride was reported in previous paper [9]. Polyester (crosslinked poly(oxy-3, 3'-dithiodipropionyloxymethylene-1,4-phenylenemethylene, c-PODOM)) was prepared by condensation between 3, 3'-dithiodipropionic acid, 1, 4-benzenedimethanol, and 1, 3, 5-benzenetricarbonyl chloride (cross-linker). Polyester, c-PODOM, was synthesized under similar condition of PIDI preparation. A typical preparation procedure of c-PODOM was as follows: a mixture of 4.205 g of 3, 3-dithiodipropionic acid (20.0 mmol) and 8.60 ml of excess thionyl chloride was heated under reflux for 150 min. After cooling the mixture to room temperature, excess thionyl chloride was removed using a rotary evaporator. The oily acid chloride and 0.531 g of 1, 3, 5-benzenetricarbonyl chloride (2 mmol) was dissolved with 60 ml of CHCl3. 1, 4-Benzenedimethanol (3.18 g, 23 mmol) was dissolved into a sodium hydroxide aqueous solution (1.84 g, 46 mmol). The aqueous solution was stirred (1000 r. p. m., at 25 °C) with a high-speed stirring motor (Three-one motor™, BL1200S, Shintokagaku Co.) to stir the reaction solution. The CHCl3 solution containing the acid dichloride and the cross-linker was gradually added to the stirred aqueous solution. The resulting polymer was collected, washed with acetone, water, and acetone, and dried under vacuum over 24 h at room temperature. PODOM(no crosslinked) powder was also prepared as similar manner except for the cross-linker addition step.

Detailed configuration of the carbon paste electrode for this study was described in our previous paper [9]. A mixture of polymer, graphite powder, and a small amount of the electrolyte solution was blended with a mortar and pestle, and the paste was loaded on a glassy carbon current collector. Typical loading levels of polymer on the electrode was 1 (polymer) : 9 (graphite) by weight. Unless otherwise stated, all electrolyte solutions consisted of 0.1 mol dm^{-3} LiClO$_4$ in acetonitrile. A conventional three-electrode cell with a Pt counter electrode (2 cm × 2 cm × 0.01 cm), (a lithium plate was used in case of a model lithium cell.) was used. Cyclic voltammograms were recorded with a potentiogalvanostat (HA-301, Hokuto), a function generator (HB-104, Hokuto), and an X-Y recorder (WX-1100, Graphtec). The measurements were performed at room temperature (20 - 25 °C). All the electrode potential in the cyclic voltammetry measurements cited in this paper are referred to *sce*.

Degradation of PIDI in aqueous solution

Degradability of PIDI and PODOM was tested *in vitro*. Polymer powder was immersed into the citrate-phosphate buffer solution with various pH, 5, 7, and 9 at 37 °C or 25 °C. Polymer powder (300 mg) was immersed into 20 ml of buffer solution in a bottle with screw cap and the bottles were sealed and placed in a thermostatic bath at 37 °C or 25 °C. Weight loss of the polymer powder was recorded with time. Degradation products were analyzed and identified by NMR and IR measurements.

RESULTS

Electrochemical behavior of PIDI and c-PODOM electrode in acetonitrile electrolyte solution
Figure 2 shows the cyclic voltammograms at fifth cycle for PIDI and c-PODOM electrode in acetonitrile (AN) solution containing 0.1 mol dm^{-3} LiClO$_4$. The loaded amount of both polymer in the electrode was 1 (polymer) : 9 (graphite). As shown in Figure 2(a), cathodic peak at -2.3 V and anodic peak at 0.5 V were observed for PIDI electrode and both peaks are attributed to the reduction and re-oxidation steps of disulfide bonds in PIDI. Electrochemical responses of PIDI are similar to those of polyamides containing 3, 3'-dithiodipropionyl group [9]. c-PODOM electrode showed different electrochemical behavior (Figure 2(b)). The cathodic peaks at 0, -0.6, and -1.1 V and the anodic peaks at -0.5, 0.1, 0.7, and 1.2 V were observed. The peak potential shift and some peaks which were not observed in PIDI electrode may be caused by difference of chemical structure of polymer, some impurity, and the mixing and loading condition of polymer. Under potential cycling electrochemical behavior of both electrodes was different each other.

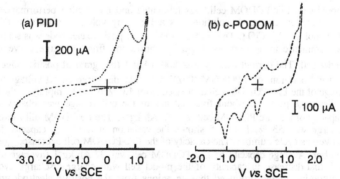

(a) PIDI   (b) c-PODOM

$\boxed{}$ 200 μA $\boxed{}$ 100 μA

-3.0 -2.0 -1.0 0 1.0   -2.0 -1.0 0 1.0 2.0
V *vs.* SCE   V *vs.* SCE

Figure 2. Cyclic voltammograms for PIDI (a) and c-PODOM (b) electrode in acetonitrile contanining 0.1 mol dm^{-3} LiClO$_4$. Weight ratio of polymer to graphite is 1 : 9 in both electrodes. Scan rate 0.5 mV s^{-1}.

Figure 3 shows the variation of the capacity for PIDI and c-PODOM electrode with potential cycling. The capacity was estimated from cathodic peak area of CV for the electrodes. The theoretical capacity of PIDI is 328 Ah kg^{-1} and that of c-PODOM is 135 Ah kg^{-1}. The capacity of PIDI electrode increased with potential cycling and the maximum capacity was 53.4 Ah kg^{-1} at 5 cycle. Average capacity of PIDI electrode was 44.1 Ah kg^{-1} and average utilization was 13.4 %. The capacity of c-PODOM electrode was also increased with potential cycling and the maximum value was 32.4 Ah kg^{-1} at third cycle. Average capacity of c-PODOM electrode was 16.3 Ah kg^{-1} and average utilization was 12.0 %.

Increase in capacity of the PIDI and c-PODOM electrode in the range first to fifth or third cycle was observed. When a PIDI electrode was dipped into the electrolyte solution for 26 h before potential cycling, the observed capacity at first cycle was 37.3 Ah kg^{-1}. Thus penetration of the electrolyte into the electrode is a problem for our electrode. The disadvantage will be improved using a highly porous current corrector or other designs for the electrode.

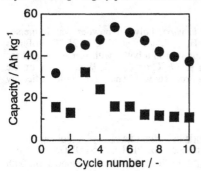

Figure 3. Variation of capacity for PIDI and c-PODOM electrode estimated from peak area of CV curves in 0.1 mol dm^{-3} LiClO$_4$/ acetonitrile solution with cycle number. PIDI (●), c-PODOM (■).

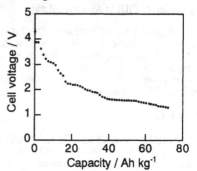

Figure 4. First discharge curve of Li/0.1 mol dm^{-3} LiClO$_4$ PC-DME (1 : 1, by vol.)/c-PODOM model cell. Current density 500 μA g^{-1} cathode.

Discharge-charge behavior of Li/c-PODOM model cell

A model cell, Li/c-PODOM cell, was fabricated and the cycling performance of the cell was demonstrated. The electrolyte solution was a 1 : 1 (by volume) mixture of PC and DME containing 0.1 mol dm^{-3} LiClO$_4$. The configuration of the positive electrode was same as that for the cyclic voltammetric measurements. Figure 4 shows the first discharge curve of the Li/c-PODOM model cell. The current density was 500 μA g^{-1} (per gram of positive electrode). This value is 5 mA g^{-1} per gram of c-PODOM (0.037 C). The discharge cutoff voltage was 1.25 V. The cell voltage of the Li/c-PODOM cell decreased from 4.0 V to 3.2 V, rapidly at the initial stage of the first discharge process. About five plateaus of the cell voltage was observed. The first discharge capacity of the Li/c-PODOM cell was 72 Ah kg^{-1}. The c-PODOM utilization in the cell at first discharge was 53 %. Figure 5 shows the variation of discharge capacity for the Li/c-PODOM cell versus cycle number. The capacity of the Li/c-PODOM cell decreased gradually with the cycle number. Average capacity of Li/c-PODOM cell was 37 Ah kg^{-1} (average utilization 27 %). We think that decrease in capacity of the model cell with charge-discharge cycles may be caused by dissolving of the produced thiolate anions from the positive electrode into the bulk electrolyte. PODOM is soluble into organic solvents, such as AN, PC, and the solubility of the produced thiolate anions from PODOM may be also high.

<u>Degradation of PIDI and PODOM in buffered aqueous solution</u>

When the container of the battery containing PIDI or c-PODOM as a positive active material is broken, degradability of the polymer and effect of the polymer on the environment are very important problem. Degradability of PIDI and PODOM was conducted *in vitro*. The polyamide and polyester powder was immersed into the citrate-phosphate buffer solution with various pH, 5, 7, and 9. Degradation behavior of PIDI under the same condition was reported in our previous paper [12]. Figure 6 shows the weight loss curves of PODOM (at 25 °C) in various pH aqueous buffer solution. The weight of the polyester decreased slowly. The weight loss ratio was 10 % at pH 5 for 56 days. Rate constant of the degradation reaction was estimated on the assumption that this reaction is pseudo-first-order reaction. However, the reaction condition was a heterogeneous system. Therefore, the rate constants are apparent values. The rate constants of c-PODOM at 25 °C are 1.5×10^{-3} % day^{-1} at pH 5, 1.3×10^{-3} % day^{-1} at pH 7, and 1.4×10^{-3} % day^{-1} at pH 9. Those of PIDI at 37 °C is 4.6×10^{-3} % day^{-1} at pH 5, 3.5×10^{-3} % day^{-1} at pH 7, and 4.2×10^{-3} % day^{-1} at pH 9. Degradation of both polymers was slow under neutral condition. Estimated period for 90 % degradation of the polymers at pH 7 is about 1800 days (c-PODOM) and 630 days (PIDI). The slow degradation rate will be improved in the natural environment. Because the biotic degradation by microorganism and the abiotic degradation by heat, light, and oxygen in the natural environment will accelerate the degradation of polyester or polyamide [14]. For practical use of the polyamide or polyester in a battery, more detailed degradation behavior in the real ecosystem should be investigated.

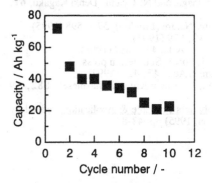

Figure 5. Variation of discharge capacity of Li/c-PODOM model cell (see Fig. 4) with cycle number.

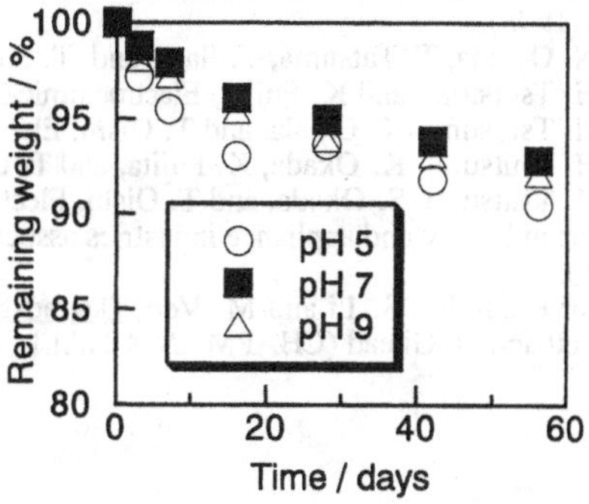

Figure 6. Mass loss data of PODOM in citrate-phosphate buffer solution, pH 5 pH 7, and pH 9 at 25 °C.

CONCLUSIONS

The polyamide (PIDI) and polyester (c-PODOM) containing disulfide bonds were prepared by condensation between 3,3'-dithiodipropionic acid and cystamine, or 1, 4-benzenedimethanol. Both polymers showed the electrochemical responses based on the reduction of the disulfide bonds in the polymer and re-oxidation of the produced thiolate anions. The capacity was estimated from peak area of the cyclic voltammograms for PIDI and c-PODOM. The maximum capacity of PIDI was 53.4 Ah kg^{-1}. Average capacity of PIDI electrode was 44.1 Ah kg^{-1} and average utilization was 13.4 %. The maximum value of c-PODOM was 32.4 Ah kg^{-1}. Average capacity of c-PODOM electrode was 16.3 Ah kg^{-1} and average utilization was 12.0 %.

A Li/c-PODOM model cell was constructed and its discharge-charge performance was tested. Average capacity of Li/c-PODOM cell was 37 Ah kg^{-1} (average utilization 27 %).

Degradability of the polymers in buffered aqueous solution was conducted. PIDI and PODOM showed their degradable properties and the apparent rate constants for degradation were estimated.

ACKNOWLEDGMENTS

The authors thank for the financial support of this investigation from The Iwatani Naoji Foundation's Research Grant and The Electric Technology Research Foundation of Chugoku.

REFERENCES

1. Y. Nishi, K. Sumida, J. Seto, K. Ozawa, S. Oishi, and M. Yokokawa, Denki Kagaku, 62, 578 (1994).
2. S. R. Taylor, Geochim. Cosmochim. Acta, 28, 1273 (1964).
3. S. J. Visco, M. Liu, M. B. Armand, L. C. De Jonghe, Mol. Cryst. Liq. Cryst., 190, 185 (1990).
4. M. M. Doeff, S. J. Visco, and L. C. De Jonghe, J. Electrochem. Soc., 139, 1808 (1992).
5. M. M. Doeff, S. J. Visco, and L. C. De Jonghe, J. Appl. Electrochem., 22, 307 (1992).
6. T. Sotomura, H. Uemachi, K. Takeyama, K. Naoi, and N. Oyama, Electrochimica Acta, 37, 1851 (1992).
7. T. Sotomura, H. Uemachi, Y. Miyamoto, A. Kaminaga, and N. Oyama, Denki Kagaku, 61, 1366 (1993).
8. N. Oyama, T. Tatsuma, T. Sato, and T. Sotomura, Nature (London), 373, 598 (1995).
9. H. Tsutsumi. and K. Fujita, Electrochimica Acta, 40, 879 (1995).
10. H. Tsutsumi, K. Okada, and T. Oishi, Electrochimica Acta, 41, 2657 (1996).
11. H. Tsutsumi, K. Okada, K. Fujita, and T. Oishi, J. Power Sources, in press.
12. H. Tsutsumi, S. Okada, and T. Oishi, Electrochimica Acta, 43, 427 (1997).
13. Japan battery and appliance industries association Ed., Denchi Kigu, (in Japanese) 882, 9 (1995).
14. for example, S. Li and M. Vert, Degradable Polymers Principle & Applications, edited by G. Scott and D. Gilead (CHAPMAN & HALL, London, 1995) pp. 43-87.

RAMAN SCATTERING IN LiCoO$_2$ SINGLE CRYSTALS AND THIN FILMS

J.D. Perkins[*], M.L. Fu[***], D.M. Trickett[**], J.M. McGraw[**], T.F. Ciszek[*], P.A. Parilla[*], C.T. Rogers[***] and D.S. Ginley[*].
[*]National Renewable Energy Laboratory, Golden, CO
[**]Colorado School of Mines, Golden, CO
[***]University of Colorado, Boulder, CO

ABSTRACT

We report Raman scattering measurements for uniaxially textured and randomly oriented polycrystalline LiCoO$_2$ thin films as well as for LiCoO$_2$ and LiCo$_{0.4}$Al$_{0.6}$O$_2$ single crystals. For both the crystalline LiCoO$_2$ thin film samples and the single crystal LiCoO$_2$ samples, well defined phonon modes are observed at Raman shifts of 486 cm^{-1} and 596 cm^{-1} corresponding to the expected E$_g$ and A$_{1g}$ modes of the layered LiCoO$_2$ crystal structure with $R\bar{3}m$ symmetry. Upon Al substitution for Co in LiCoO$_2$, the two phonon modes appear to shift to higher energy, but further work is needed to clarify this point.

INTRODUCTION

While LiCoO$_2$ is one of the most promising cathode materials for rechargeable Li-ion batteries, many questions persist regarding the interrelationship of the electronic, structural and battery properties[1-4]. To better understand the vibrational excitations of LiCoO$_2$ and hence, the associated structural properties, we have begun Raman scattering studies of oriented thin film and single crystal samples of LiCoO$_2$. In addition to providing insight into the properties of stoichiometric LiCoO$_2$, the studies should also provide a baseline for future *in-situ* Raman scattering studies of structural transitions during electrochemical cycling.

LiCoO$_2$, along with LiNiO$_2$, LiVO$_2$ and LiCrO$_2$, has a layered metal-oxygen rock-salt structure with alternating Li and transition metal monolayers along the cubic <111> direction of the metal ion sublattice. The resultant structure, shown in Figure 1, has $R\bar{3}m$ space group symmetry and two Raman active modes are expected, one A$_{1g}$ and one E$_g$[5,6]. Inaba *et al.* [5,7], Julien *et al.*[8] and Huang and Frech [9] have reported Raman spectra for LiCoO$_2$ and LiCo$_{1-x}$Ni$_x$O$_2$ powders and thin films. Suzuki *et al.* [6] have studied Raman scattering from phonon and magnon excitations in LiCrO$_2$ single crystals.

In this work, we report polarized and unpolarized room temperature Raman scattering spectra for uniaxially textured and randomly oriented LiCoO$_2$ thin films as well as for LiCoO$_2$ and LiCo$_{1-x}$Al$_x$O$_2$ single crystals. In all samples, the expected A$_{1g}$ and E$_g$ phonon modes are seen. The polarization dependence of the measured spectra is consistent with the expected phonon symmetries. A Co$_3$O$_4$ impurity phase is identified in films grown at low substrate temperatures.

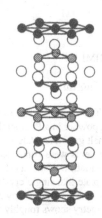

Figure 1: Crystal Structure of LiCoO$_2$. Li(Gray), Co(Black), Oxygen (White).

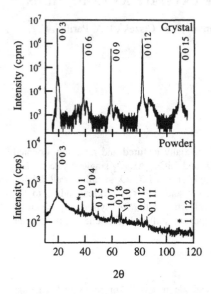

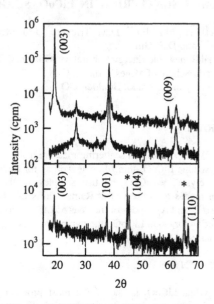

Figure 2: θ/2θ X-ray diffraction spectra of LiCoO$_2$. Top: (001) oriented single crystal plate. Bottom: Powder pattern of crushed single crystal. Prominent peaks are labeled (* denotes an unidentified peak).

Figure 3: θ/2θ X-ray diffraction spectra. Top Panel: Textured LiCoO$_2$ thin film (top curve) and SnO$_2$/glass (bottom curve) substrate. The substrate spectrum has been offset for clarity. Bottom Panel: Randomly oriented LiCoO$_2$ thin film grown on stainless steel. Substrate peaks marked with *.

EXPERIMENT

Sample Growth

The single crystal samples were grown by either direct melting and recrystallization of LiCoO$_2$ powder in a Pt crucible or via flux growth using a Li$_2$O flux in an alumina crucible. The direct melt grown crystals are small (< 1 mm) crystallites embedded in a ceramic matrix. Raman measurements were conducted on an as-grown crystalline facet without removing the crystallite from the matrix. Using a 4:1 ratio of Li$_2$O flux to LiCoO$_2$ solute mixture, the flux growth method yielded plate-like single crystals with lateral dimensions up to 1.5 cm with the thickness ranging from 5 - 400 μm. However, inductively coupled plasma (ICP) analysis of the metals stoichiometry shows roughly 50% Al substitution for Co. As the Al in these flux grown crystals comes from partial dissolution of the alumina crucible by the Li$_2$O flux, there is substantial batch to batch variation in the Al content, $0.45 < x < 0.7$ in LiCo$_{1-x}$Al$_x$O$_2$. Figure 2 shows x-ray diffraction spectra for an Al substituted crystal with nominal stoichiometry LiCo$_{0.4}$Al$_{0.6}$O$_2$. The diffraction spectra for both the single crystal plate (top panel) and the crushed crystal powder pattern (bottom panel) are consistent with $R\bar{3}m$ symmetry. Further work is underway to refine the crystal growth which will be reported separately.[10]

The thin film samples were grown on SnO_2 coated glass and stainless steel substrates via pulsed laser deposition (PLD) from a stoichiometric ceramic $LiCoO_2$ target. Films grown on the SnO_2 at substrate temperatures (T_s) between 400 and 600 °C and at an oxygen pressure (pO_2) of 200 mTorr are highly crystalline and uniaxially textured with a grain size of roughly 100 nm. Films grown at similar deposition conditions on stainless steel substrates are randomly oriented with weaker x-ray diffraction peaks. This is evident from the $\theta/2\theta$ x-ray diffraction spectra shown in Figure 3. The top panel shows the diffraction spectrum for a $LiCoO_2$ thin film grown on a SnO_2 coated glass substrate at T_s = 400 °C and

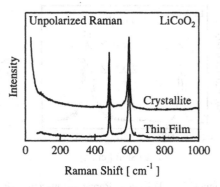

Figure 4: Room temperature unpolarized Raman scattering spectrum of melt grown $LiCoO_2$ crystal and PLD grown crystalline $LiCoO_2$ thin film.

pO_2 = 200 mTorr. The $LiCoO_2$ (003) and (009) peaks are labeled. A spectrum of the substrate is included for reference. Pole figures (not shown) show the textured grains to have randomly oriented a- and b-axis in the plane. These spectra show the film to be highly textured with the (001) planes parallel to the substrate surface. In contrast, as can be seen in the bottom panel of figure 3, for a $LiCoO_2$ film grown on a stainless steel substrate at T_s = 550 °C and pO_2 = 500 mTorr, the x-ray diffraction pattern closely resembles a powder pattern, indicating a randomly oriented polycrystalline film. Films grown on SnO_2 at low temperature (T_s = 65 °C, pO_2 = 200 mTorr) show no discernible x-ray diffraction peaks. For all these PLD grown thin film samples, ICP analysis shows a Li:Co ratio of 1±0.05. The details of the film growth are reported in the related proceedings paper by Ginley *et al.*[11]

<u>Raman Spectroscopy</u>

Raman scattering measurements were performed in an 180° back-scattering configuration with an Ar-ion laser operating at 514.5 nm and a Spex 1877 Triplemate spectrometer with a LN_2 cooled CCD detector. A polarized incident beam, ~4 mW, was focused through a microscope objective to a spot size of ~2 μm. A wedged window used as a beam splitter allows for co-linear incident and collected light through the microscope objective. For polarized Raman spectra, an additional polarizer is placed in the collected light path between the beam splitter and the spectrometer. All samples were measured in air and at room temperature.

RESULTS AND DISCUSSION

Figure 4 shows unpolarized Raman scattering spectra for a melt grown $LiCoO_2$ crystallite and for a uniaxially textured $LiCoO_2$ thin film grown on a SnO_2 substrate. For both samples, there are two well defined phonon peaks with Raman shifts of ~ 486 cm^{-1} and 596 cm^{-1}. For layered $LiCoO_2$, two Raman active modes are expected and the measured energies are consistent with those reported by Inaba *et al.* on powdered and thin film $LiCoO_2$ samples.[5] Note the similarity in both the line width and the intensity of the spectra measured for the thin film and crystallite sample.

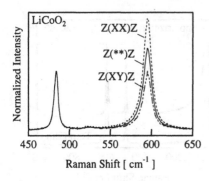

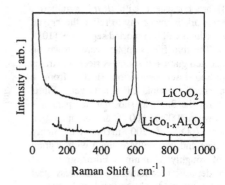

Figure 5: Polarized Raman scattering spectra of c-axis oriented $LiCoO_2$ thin film. Spectra have been normalized to match at the 484 cm^{-1} peak. Notation for labels: k_{inc} (E_{inc}, E_{scat}) k_{scat}.

Figure 6: Unpolarized Raman scattering spectra for melt grown $LiCoO_2$ crystal and flux grown $LiCo_{0.4}Al_{0.6}O_2$ crystals.

Figure 5 shows polarized Raman scattering spectra measured at room temperature on a (001) textured $LiCoO_2$ film. The spectra have been normalized to match at the 486 cm^{-1} peak. The laser is normally incident upon the sample and the scattered light is collected in an 180 degree backscattering geometry. Thus, both the incident and scattered light wave vectors are nominally parallel to the $LiCoO_2$ (001) axis. The 596 cm^{-1} peak is stronger for parallel incident and scattered polarizations (XX) than for crossed (XY) polarizations. For an unpolarized spectrum (**), the peak strength lies midway between the parallel and crossed peak intensities. In the XX polarization, both the E_g and A_{1g} modes are allowed, whereas in the XY polarization only the E_g mode is expected. Hence, based on the smaller relative scattering strength for the 596 cm^{-1} mode in XY polarization, it is assigned to the A_{1g} mode and the 486 cm^{-1} mode to the E_g mode. These assignments are consistent with those of Inaba et al.[5] and in agreement with the expectation that the A_{1g} bond stretching mode should be at higher energy than the E_g bond bending mode. The strength of the A_{1g} mode in XY polarization probably arises from a break down in strict selection rules due to the non-normal incident and collected light inherent to using a high numerical aperture (N.A. = 0.6) microscope objective.

Figure 6 shows unpolarized Raman scattering spectra measured at room temperature for a melt grown $LiCoO_2$ crystal and an Al substituted flux grown crystal with nominal stoichiometry $LiCo_{0.4}Al_{0.6}O_2$. Recent calculations predict and experiments show an increased battery voltage for Al substituted $LiCoO_2$.[12] The effect of Al substitution on cyclability has yet to be determined. We find that for the Al substituted crystal, two reasonably defined peaks are apparent at Raman shifts of 502 cm^{-1} and 624 cm^{-1} along with a broader band at 434 cm^{-1} and possibly one at 550 cm^{-1}. As x-ray diffraction measurements of the Al substituted sample indicate a $LiCoO_2$ crystal structure, the two sharper lines in the Raman spectrum are likely the E_g and A_{1g} phonon modes shifted to higher energy upon random substitution of the lighter element Al for Co.

Figure 7 shows Raman scattering spectra for $LiCoO_2$ thin films grown via PLD on stainless steel and SnO_2 coated glass substrates. The spectrum for the $LiCoO_2$ film grown on stainless steel shows the expected two phonon modes for $LiCoO_2$ with comparable intensity and line width to that observed for the $LiCoO_2$ film grown on SnO_2 coated glass. However, the x-ray diffraction spectra

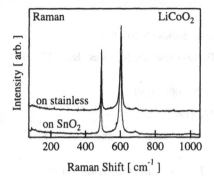

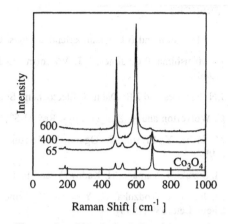

Figure 7: Unpolarized Raman scattering spectra of LiCoO$_2$ films grown by PLD on stainless steel (top curve) and SnO$_2$/glass (bottom curve) substrates.

Figure 8: Unpolarized Raman scattering spectra of LiCoO$_2$ films grown by PLD on SnO$_2$/glass substrates. Substrate temperatures during growth were 600 °C, 400 °C and 65 °C. An additional spectrum for a Co$_3$O$_4$ thin film is shown for comparison.[13]

for the film grown on stainless steel (figure 3) shows a non-oriented powder pattern, indicating a crystalline but randomly oriented film.

Figure 8 shows the Raman scattering spectra for three LiCoO$_2$ films grown at substrate temperatures of 600 C, 400 C and 65 C. All three films are grown on SnO$_2$ coated glass substrates and at an oxygen partial pressure of 200 mTorr. The Raman scattering spectra show two effects of lowering the substrate temperature. First, the two expected LiCoO$_2$ phonon peaks become markedly weaker and second, weak but distinguishable peaks appear at 186, 519, 616 and 690 cm^{-1}. As can be seen in Figure 8, these extra peaks correspond to a Co$_3$O$_4$ impurity phase. This impurity phase, which is not stable at higher growth temperatures, is believed to arise from the target, but further work is needed on this point.

CONCLUSIONS

Raman scattering spectra for unixially textured and randomly oriented LiCoO$_2$ thin films as well as for LiCoO$_2$ and LiCo$_{0.4}$Al$_{0.6}$O$_2$ single crystals have been presented. For both the crystalline LiCoO$_2$ thin film samples and the single crystal LiCoO$_2$ samples, well defined phonon modes are observed at Raman shifts of 486 cm^{-1} and 596 cm^{-1} corresponding to the expected E$_g$ and A$_{1g}$ modes respectively of the layered LiCoO$_2$ crystal structure with $R\bar{3}m$ symmetry. Upon Al substitution for Co in LiCoO$_2$ the two phonon modes appear to shift to higher energy.

ACKNOWLEDGMENTS

This work was supported by the U.S. Department of Energy, Office of Basic Energy Sciences under contract #DE-AC36-83CH10093.

REFERENCES

1. T.A. Hewston and B.L. Chamberland, J. Phys. Chem. Solids **48**, 97 (1987).

2. K. Mizushima, P.C. Jones, P.J. Wiseman and J.B. Goodenough, Mat. Res. Bull. **15**, 783 (1980).

3. J.N. Reimers and J.R. Dahn, J. Electochem. Soc. **139**, 2091 (1992).

4. C. Wolverton and A. Zunger, Phys. Rev. B **57**, 2242 (1998).

5. M. Inaba, Y. Iriyama, Z. Ogumi, Y. Todzuka and A. Tasaka, J. Raman Spect. **28**, 613 (1997).

6. M. Suzuki, I. Yamada, H. Kadowaki and F. Takei, J. Phys.-Cond. Mat. **5**, 4225 (1993).

7. M. Inaba, Y. Todzuka, H. Yoshida, Y. Grincourt, A. Tasaka, Y. Tomida and Z. Ogumi, Chem. Lett. 889 (1995).

8. C. Julien, C. Perez-Vicente, E. Haro-Poniatowski, G.A. Nazri and A. Rougier, Materials for Electrochemical Energy Storage and Conversion II, edited by D. S. Ginley and D. Doughty (Mater. Res. Soc. Proc. Pittsburgh, PA, 1997) pp. XX.

9. W. Huang and R. Frech, Solid State Ionics **86-88**, 395 (1996).

10. D.M. Trickett, to be published (1998).

11. D.S. Ginley, J.D. Perkins, J.M. McGraw, P.A. Parilla, M.L. Fu and C.T. Rogers, Materials for Electrochemical Energy Storage and Conversion II, edited by D. S. Ginley and D. Doughty (Mater. Res. Soc. Proc. **496**, Pittsburgh, PA, 1997) pp. XX.

12. M.K. Aydinol, A. Van Der Ven and G. Cedar, Materials for Electrochemical Energy Storage and Conversion II, edited by D. S. Ginley and D. Doughty (Mater. Res. Soc. Proc. **496**, Pittsburgh, PA, 1997) pp. XX.

13. M.L. Fu, C.T. Rogers, J.D. Perkins, J.M. McGraw, P.A. Parilla, J.G. Zhang, J.A. Turner and D.S. Ginley, submitted to J. Mat. Res. (1997).

NR MEASUREMENTS TO STUDY THE REVERSIBLE TRANSFER OF LITHIUM ION IN LITHIUM TITANIUM OXIDE

T. Esaka *, S. Takai * and M. Kamata **
*Faculty of Engineering, Tottori University, Minami 4-101, Koyamacho, Tottori 680, JAPAN.
**Department of Science Education, Tokyo Gakugei University, Koganei 184, JAPAN.

ABSTRACT

Using Neutron Radiography (NR), the transfer of lithium ions has been investigated in the high temperature-type lithium ion conductor, $Li_{1.33}Ti_{1.67}O_4$. After supplying direct current to the oxides with different isotope ratios ($^6Li/^7Li$), NR images were obtained, which confirmed lithium ion movement in the oxides as changes of bright parts on negative films. Analysis of the NR images clarified that lithium ions in the spinel-type $Li_{1.33}Ti_{1.67}O_4$ were transported almost reversibly according to the polarity of electric field applied. The relations between the lithium contents in the samples and the film gray level showed that the cathode region in the charged sample contains about twice more lithium than in the original sample. Furthermore, NR method was applied to check lithium ion insertion into the TiO_2 (rutile) sample. As a result, an informative Neutron Radiogram showing the lithium ion insertion was obtained.

INTRODUCTION

Neutron Radiography (NR) has not been utilized in the field of electrochemistry. Nevertheless, using NR, we can visualize the movement of ionic species, particularly in lithium ion or proton conductors, because the interaction of these atoms with neutrons is widely different among their isotopes. For example, neutron cross section of ^6Li is approximately four orders of magnitude larger than that of ^7Li. Therefore, if ^6Li is transferred into a lithium ion conductor containing no ^6Li by charging, the movement of ^6Li can be clearly visualized as a brightness movement on a negative film. Considering the practical neutron cross section, of course, the movement of natural lithium (^nLi), which contains 7.4 % ^6Li, could be visualized not only in ^7Li-containing compounds but also in some compounds composed of elements having much smaller neutron cross sections than that of ^6Li.

The authors have already reported that the movement of lithium ions in $Li_{1.33}Ti_{1.67}O_4$ [1-4], $Ca_{0.95}Li_{0.10}WO_4$ [5,6] and in some lithium batteries [7,8] can be directly observed using NR. In the first case, the transport number of lithium ions in $Li_{1.33}Ti_{1.67}O_4$ can be determined by analyzing a film gray level profile of the NR image for electrolyzed samples. According to the previous works, moreover, $Li_{1.33}Ti_{1.67}O_4$ should change into TiO_2 on the anolyte surface after electrolysis [2]. And TiO_2 has been known as a lithium ion insertion material. Therefore, in this work, NR is also employed in order to investigate how lithium ions are transferred in $Li_{1.33}Ti_{1.67}O_4$ depending on the polarity of the electric field applied and thereby how the content of lithium ion is varied in the sample. In addition, same radiographs were used in order to study how lithium ion insertion was carried out into TiO_2.

EXPERIMENT

Preparative and electrolyzing procedure of samples

Two types of cylindrical samples of $Li_{1.33}Ti_{1.67}O_4$ were prepared for electrolysis by the ordinary solid state reaction method using Li_2CO_3 and TiO_2 as starting materials. Both samples have the same chemical compositions. However, one was prepared using Li_2CO_3 synthesized from $^7Li(OH)H_2O$ of Toyama's High Purity Chemicals (99.93 ^7Li%) (shown by "smp^7Li" below) and another was prepared using the starting materials with the natural $^6Li/^7Li$ ratio (Wako Pure Chemicals) (shown by "smp^nLi" below). Although ^7Li exists in both samples, the lithium based on $^7Li(OH)H_2O$ will be denoted as ^7Li in the following text.

After sintering and machining (9 mm diameter by 7 to 9 mm length), gold paste was applied by baking on one plane of each sample. Then, these two samples were piled and placed in an electric furnace as illustrated in Fig.1. Dc current (j = 35 mA cm^{-2}) was passed through the samples for electrolysis at 900°C, where the baked gold on smp^{N}Li was used as the anode and the gold on smp^{7}Li as the cathode. Thereafter, the polarity of electrolyzing field was changed to intent lithium ions, transferred once toward the cathode, to go back to their initial positions. This reverse electrolysis was carried out at 20 to 25 mA cm^{-2}. In order to check the change of sample surfaces after electrolysis, X-ray diffraction with nickel-filtered CuKα radiation was employed.

In order to obtain exact profiles of lithium contents in ^{N}Li$_{1.33}$Ti$_{1.67}$O$_4$ as numeric values, rectangular samples (9×9 mm, 16 mm length) were separately prepared. Dc current (j = 24 mA cm^{-2}) was passed through the gold paste baked on both opposite planes of the samples for electrolysis at 900°C. In this case, the electrolysis with reverse polarity was also carried out. The standard samples were prepared, which have defined contents (0, 20, 40, 60, 80 and 100%) of ^{N}Li in the same sized depth of samples. The "0 %" and "100 %" samples correspond to TiO$_2$ and Li$_{1.33}$Ti$_{1.67}$O$_4$, respectively. In order to convert the film gray levels into numeric data, an image scanner (HP Scanjet 3c) connected to a computer was employed

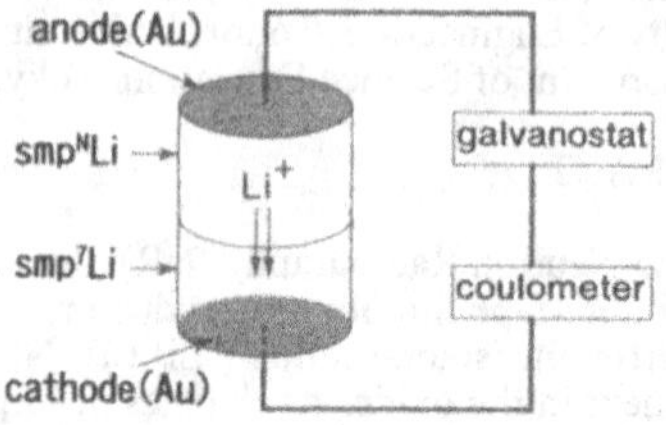

In the case to see the lithium ion insertion into TiO$_2$, a cylindrical sample of TiO$_2$ with 9 mm diameter by 3 mm length was prepared. This was attached to smp^{N}Li, placed in the furnace and electrolyzed at 37 mA cm^{-2} at 900°C, where TiO$_2$ was a catholyte and the smp^{N}Li an anolyte.

Fig.1 Schematic illustration of electrolysis.

Radiographic procedure

After the electrolysis, the samples were taken out from the furnace and fixed on the surface of a vacuum film cassette (12.5μm Gd + Kodak SR for industrial use). Then they were irradiated for defined minutes in the thermal neutron radiography (TNR) facility of Research Reactor Institute, Kyoto University. Irradiated periods were changed occasionally depending on the condition of the Reactor, especially heavy water tank for attenuation of neutrons; 16 minutes for cylindrical samples and 45 minutes for rectangular sample. The NR images of the samples (i.e. smp^{N}Li) containing ^{6}Li should be observed as much more whitish parts on negative films than those of the samples including no ^{6}Li.

RESULTS AND DISCUSSION

Lithium ion transfer in Li$_{1.33}$Ti$_{1.67}$O$_4$

Figure 2 shows a NR image of Li$_{1.33}$Ti$_{1.67}$O$_4$ samples. The smp^{7}Li contains no ^{6}Li and the smp^{N}Li contains a few percent of ^{6}Li. The magnitude of interaction between neutron and each lithium isotope is so different that the presence of lithium in each sample can be distinguished as difference of the brightness; smp^{7}Li looks blackish and smp^{N}Li whitish on the radiogram (negative film), which means that the former is transparent and the latter is opaque to neutron. Here, we can also see the difference of brightness between the sample center and side, which is based on the sample thickness because of usage of cylindrical samples.

The NR images of electrolyzed samples are presented in Fig. 3. The whitish region moved from smp^{N}Li toward smp^{7}Li, which means that the lithium in smp^{N}Li moved

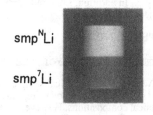

smp^{N}Li

smp^{7}Li

Fig.2 NR image of the sample stack of Li$_{1.33}$Ti$_{1.67}$O$_4$ (smp^{N}Li and smp^{7}Li) before electrolysis.

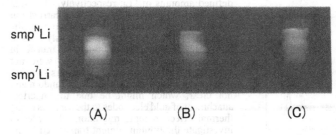

smp^{N}Li

smp^{7}Li

(A)   (B)   (C)

Fig.3 NR images of the sample stacks of Li$_{1.33}$Ti O (smp^{N}Li and smp^{7}Li) after electrolysis. (A):after electrolysis by 650 C, (B):(A) plus the reverse charging by 325 C and (C):(A) plus the reverse charging by 650 C.

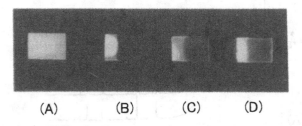

(A)   (B)   (C)   (D)

Fig.4 NR images of the rectangular samples of Li$_{1.33}$Ti$_{1.67}$O$_4$ (smp^{N}Li). (A):before electrolysis, (B):after electrolysis by 1400 C, (C):(B) plus the reverse charging by 700 C, (D):(B) plus the reverse charging by 1400 C.

into smp^{7}Li and the lithium near the anode was fully consumed. When the dc polarity was changed, the white region was going to go back toward the initial condition.   These illustrates that the lithium ions are possible to be reversibly transferred in the oxides according to the direction of dc current.

From the above results, the reversible transfer of lithium ions in Li$_{1.33}$Ti$_{1.67}$O$_4$ is easily recognized even in smp^{N}Li. As the cylindrical samples were used in those cases, however, the brightness difference is appeared between the sample center and side. In order to improve this problem as much as possible and, furthermore, to obtain numeric data on the radiogram, another experiment was carried out using smp^{N}Li with rectangular shape in the followings.

Figures 4 and 5 shows NR images of the rectangular samples of Li$_{1.33}$Ti$_{1.67}$O$_4$ (smp^{N}Li) before and after electrolysis and the rectangular standard samples with the

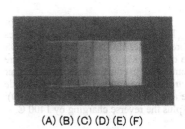

(A) (B) (C) (D) (E) (F)

Fig 5   NR image of the rectangular standard samples with the defined amount of ^{N}Li. (A):0, (B):20, (C):40, (D):60, (E):80 and (F):100 % ^{N}Li sample.

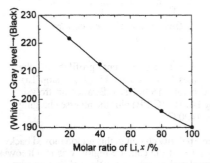

Molar ratio of Li, x /%

Fig.6   Calibration curve showing the relation between film gray level and ^{N}Li content.

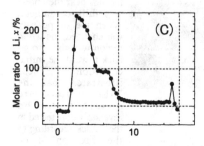

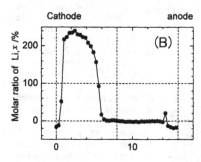

Cathode   anode

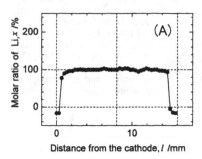

Distance from the cathode, l /mm

Fig.7 Lithium content profiles in the samples of $Li_{1.33}Ti_{1.67}O_4$ (smp[N]Li). (A):before electrolysis, (B):after electrolysis by 1400 C, (C):(B) plus the reverse charging by 1400 C.

defined amounts of [N]Li, respectively. The first charging by 1400 C in Fig. 4 was carried out regarding the left-hand side of sample as the cathode. The white region obviously moved leftward, which denotes the lithium ion transfer in this sample. However, the images were not completely sharp; the boundary regions showing lithium ion migration are somewhat random and not clear, which might be due to imperfect attachment of gold electrodes to the samples or the thermal effect to ionic migration. In order to investigate the lithium content transferred in the samples, the film gray levels along the horizontally centered line in each sample were conveniently converted to numeric data, where was used a calibration curve (Fig.6) obtained from Fig.5. These results are shown in Fig.7. The region having about twice larger lithium content than the initial was found to be present near the cathode after electrolysis by 1400 C. By the

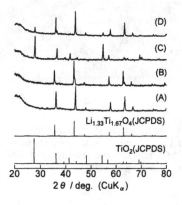

Fig.8 X-ray diffraction patterns of $Li_{1.33}Ti_{1.67}O_4$ (smp[N]Li) near the electrodes before and after electrolysis and the JCPDS data of $Li_{1.33}Ti_{1.67}O_4$ and TiO_2. (A):before electrolysis, (B):on the cathode side after electrolysis by 1400 C, (C):on the anode side after electrolysis by 1400 C and (D): (C) plus the reverse charging by 1400 C.

inverse polarity charging, this region moved backward keeping the high lithium content. In this case, gray level at the cathode region was much lower than that indicated in Fig.6. Therefore, a line extrapolated simply from the calibration curve in Fig.6 was used to get the lithium content. Moreover, the lithium content was confirmed by ICP spectroscopy. This result is indicated in Table 1, where the lithium content is expressed against titanium not responsible to ionic transfer and the value after charging is about twice as much as that before charging.

Table 1 The ratio of lithium to titanium in anode and cathode regions of $Li_{1.33}Ti_{1.67}O_4$ (smp^NLi) measured by ICP spectroscopy.

Li / Ti	Before charging	After charging by 1400 C
Cathode region	0.80	1.53
Anode region	0.80	0.0135

smp^NLi, TiO_2 smp^NLi, TiO_2

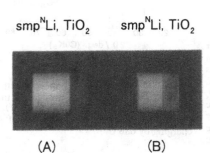

(A)   (B)

Fig.9 NR images of the sample stacks of $Li_{1.33}Ti_{1.67}O_4$ (smp^NLi) and TiO_2. (A):before electrolysis at 25°C and (B):after electrolysis by 680 C at 900°C.

Figure 8 shows the X-ray diffraction patterns of $Li_{1.33}Ti_{1.67}O_4$ (smp^NLi) near each gold electrode before and after electrolysis. The pattern on the cathode side are almost the same as that before electrolysis. On the contrary, there is no peak corresponding to $Li_{1.33}Ti_{1.67}O_4$ on the anode side after electrolysis by 1400 C. The resultant pattern perfectly coincided with that of TiO_2. After further application of inverse electric field, the peaks of $Li_{1.33}Ti_{1.67}O_4$ appeared again. From these results together with the radiograms shown above, it was considered that the following reaction would reversibly proceed in the anode region during electrolysis,

$$3 Li_{1.33}Ti_{1.67}O_4 = 4Li^+ + 5TiO_2 + O_2 + 4e^- \qquad (1).$$

In the cathode region, the following reaction should be considered,

$$Li_{1.33}Ti_{1.67}O_4 + \alpha Li^+ + \alpha e^- = Li_{1.33+\alpha}Ti_{1.67}O_4 \qquad (2).$$

As long as we see the lithium ion distribution in Fig.7(C) and (D), the reaction (2) itself is not likely to be reversible. However, the reaction concerning lithium ion transfer is considered to be reversible. One reason why the reversible reaction is possible would be that as TiO_2 is not a strictly defined insulator but a poor electronic conductor, it works as an electrode material. As long as we checked the X-ray diffraction patterns of TiO_2 formed by electrolysis, the lattice deformation, i.e. expansion or contraction, was not observed. This means that the lithium ion insertion would be possible even into the separately prepared TiO_2. In order to check this phenomenon visually, NR technique was also used.

Figure 9 shows the NR image of the sample stacks of $Li_{1.33}Ti_{1.67}O_4$ (smp^NLi) and TiO_2. Smp^NLi looks whitish on the radiogram before electrolysis. On the contrary, TiO_2 looks blackish, because neutrons can easily penetrate through TiO_2. This result is almost same as that of electrolyzed sample in Fig.2. When the cell stack was electrolyzed, the black region of TiO_2 becomes gray. This means that lithium ion moved to and reacted with TiO_2 as the lithium insertion material. As far as these images are concerned, the right-hand side of TiO_2 seems to be more whitish than that in the TiO_2 bulk, which means the fact that a small part of lithium ion can pass through TiO_2 and accumulate at the electrode. Considering the mass of TiO_2 used, the electric quantity for this electrolysis was not enough to form $Li_{1.33}Ti_{1.67}O_4$ totally from TiO_2. Therefore, the partly formation of $Li_{1.33}Ti_{1.67}O_4$ should be taken in consideration, which would construct the conduction path of lithium in TiO_2 and subsequently make the lithium (or lithium compound) deposit at the electrode.

Figure 10 shows the result of the X-ray diffraction patterns of TiO_2 before and after electrolysis near the electrode. After electrolysis, obviously, the formation of $Li_{1.33}Ti_{1.67}O_4$ is observed, which confirmed that the lithium ion insertion reaction is possible.

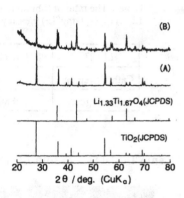

Fig.10 X-ray diffraction patterns of the charged TiO_2 near the electrode and the JCPDS data of $Li_{1.33}Ti_{1.67}O_4$ and TiO_2. (A):before electrolysis and (B):after charging by 680 C.

CONCLUSION

Neutron radiography (NR) was useful in order to study the reversible transfer of lithium ions in the high temperature-type ionic conductor such as $Li_{1.33}Ti_{1.67}O_4$. Analysis of the NR images clarified that lithium ions in the spinel-type $Li_{1.33}Ti_{1.67}O_4$ transferred almost reversibly according to the polarity of the electric field applied. After electrolysis, the cathode region in the sample contains about twice more lithium than that in the initial composition.

Lithium ion insertion to TiO_2 (rutile-type) was also able to be confirmed visually by this method; a beautiful Neutron Radiograms showing the lithium ion insertion was obtained.

ACKNOWLEDGMENTS

This work is partly supported by the Grand-in-Aid for Developmental Scientific Research (No.07555197) and for Scientific Research on Priority Area (No.260) from the Ministry of Education, Science and Culture of Japanese Government.

REFERENCES

1. M.Kamata, T.Esaka, S.Fujine, K.Yoneda and K.Kanda, Nucl. Inst. Methods in Phys. Res. A, **377**, 161 (1996).

2. M.Kamata, T.Esaka, N.Kodama, S.Fujine, K.Yoneda and K.Kanda, J. Electrochem. Soc., **143**, 1866 (1996).

3. M.Kamata, T.Esaka, K.Takami, S.Takai, S.Fujine, K.Yoneda and K.Kanda, Denki Kagaku, **64**, 984 (1996).

4. T.Esaka, M.Kamata, K.Takami, S.Takai, S.Fujine, K.Yoneda and K.Kanda, Key Eng. Mater., **132**, 1393 (1997).

5. T.Esaka, M.Kamata and H.Saito, Solid State Ionics, **86**, 73 (1996).

6. M.Kamata, T.Esaka, K.Takami, S.Takai, S.Fujine, K.Yoneda and K.Kanda, Solid State Ionics, **91**, 303 (1996).

7. M.Kamata, T.Esaka, S.Fujine, K.Yoneda and K.Kanda, Denki Kagaku, **63**, 1063 (1995).

8. M.Kamata, T.Esaka, S.Fujine, K.Yoneda and K.Kanda, J. Power Sources, **68**, 495 (1997).

EFFECTS OF MICROSTRUCTURE ON OXYGEN PERMEATION IN SOME PEROVSKITE OXIDES

K. Zhang, Y. L. Yang, A. J. Jacobson, and K. Salama
Materials Research Science and Engineering Center
University of Houston
Houston, TX 77204

ABSTRACT

The effects of microstructure on the oxygen permeation in $SrCo_{0.8}Fe_{0.2}O_{3-\delta}$ (SCFO) and $La_{0.2}Sr_{0.8}Fe_{0.8}Cr_{0.2}O_3$ (LSFCO) was investigated using disc samples fabricated under different processing conditions. The microstructure of LSFCO remained unchanged when the sintering temperature was increased from 1300 to 1450 °C, but the average grain size of SCFO increased considerably when the sintering temperature was increased from 930 to 1200 °C. The change in grain size was found to have a strong effect on the oxygen permeation flux in SCFO, which increased considerably as the grain size was decreased. This indicates that the contribution of the grain boundary diffusion to the steady state oxygen flux in SCFO is substantial and grain boundaries provide faster diffusion paths in oxygen permeation through the samples.

INTRODUCTION

Mixed type conducting materials have attracted much attention in recent years because of their potential applications in oxygen separation membranes, solid oxide fuel cells and electrocatalytic reactors. One group of these materials, the perovskite oxides have been extensively investigated due to their electronic and ionic conductivities at elevated temperatures as well as their structural stability in both oxidizing and reducing atmospheres [1-9], but limited study has been conducted to understand the effect of microstructure on the electrochemical properties of those materials. Badwal et al [10] investigated the effect of microstructure on the conductivity of yttria-zirconia, and found that the grain size had no measurable effect on the lattice resistivities, while the grain boundary resistivity decreased dramatically with the increase of the grain size. An inflection was observed when the grain boundary resistivity was plotted as a function of grain size. Kawada et al [11] determined the oxygen permeation flux through bulk and grain boundary of $La_{0.7}Ca_{0.3}CrO_3$ by measuring the steady state flux and the tracer diffusion profile respectively. Although they confirmed the existence of fast diffusion paths along grain boundaries, the contribution of the grain boundary diffusion to the steady state oxygen flux was much smaller than expected from the tracer diffusion experiment. No direct experiments, however, has been conducted to study the effect of microstructure on the oxygen permeation properties of pervoskite materials. It is thus interesting to investigate whether or not the grain boundaries in perovskite oxides membranes have a significant effect on the oxygen permeation properties.

In this investigation, the microstructure and its effect on the oxygen permeation in SCFO and LSFCO membranes were studied following our previous work [12].

EXPERIMENTAL

Disc samples with approximate dimensions of 14 mm in diameter and 2 mm in thickness were fabricated using SCFO (Praxair Specialty Ceramics) and LSFCO (freeze dry synthesis) powders. The pellets were pressed using 220 Mpa pressure and sintered in air for 6 hours at temperatures ranging from 930 to 1200 °C for SCFO and 1300 to 1450 °C for LSFCO. The densities of the sintered pellets were determined using Archimedes' method, and were greater than 94% of the theoretical values. To study the sintering characteristics of SCFO, samples were sintered in air at 1100 °C and held from 15 min. to 168 hours. The stability of the microstructure of SCFO was investigated using samples which were sintered at 1100 and 1200 °C for 6 hours, and then exposed at 1050 °C for different times after samples were ground and polished to 1 μm.

The microstructure of sample used in this investigation is determined using a Nikon AFX-II A optical microscope. The grain size of the samples was determined by measuring the mean linear intercept lengths on micrographs of at least three images. The oxygen permeation fluxes were measured using the apparatus described in Ref. 6. In these experiments, disc samples ground using 600 grit paper were sealed between two quartz tubes with gold rings. Air gas was fed at one end of the sample, while helium gas with a flow rate of about 38 cm^3/min was swept through the other end. The flow rate was controlled by a MKS 247C mass flow controller. A computer controlled Antek 3000 gas chromatograph was connected to the exit on the sweep side, where both oxygen and nitrogen concentrations were measured. The gas chromatograph was calibrated using a standard gas of O$_2$ in helium. The test temperature was measured by a thermocouple inserted in the quartz tube at the air side with the accuracy of ± 1 °C. Experiments were conducted in the temperature range of 800 to 900 °C for SCFO (to 930 °C for LSCFO). Samples were first heated up to 900 °C (or 930 °C) and kept at this temperature until a steady state was reached when the oxygen concentration varied less than 1%. The temperature was then dropped to the next measurement point. The oxygen permeation flux at each temperature was monitored periodically and recorded when the steady-state was reached. The reported oxygen permeation fluxes are the steady-state values.

RESULTS AND DISCUSSION

Microstructure Characterization

While the microstructure of LSFCO samples shows little change in the sintering temperature range with the average grain size about 20 μm, the microstructure of SCFO is considerably changed with the sintering temperature. Fig. 1 shows the microstructure of SCFO samples sintered at several temperatures, where a bimodal grain size distribution is observed. The grain size as a function of sintering temperature for SCFO is plotted in Fig. 2. It is seen that the grain size increases from 2.0 to 14.8 μm when the temperature is increased from 930 to 1200 °C. The grain size, however, only changed from 3.4 μm to 12.8 μm when sintering time is increased from 15 min. to 168 hours at 1100 °C as can be seen in Fig. 3. After the initial sharp increase of grain size, the grain size increases only slightly with further increase of the sintering time, which indicates that the microstructure of this material becomes relatively stable after the densification process is completed.

Furthermore, the microstructure of SCFO exhibited good stability as shown in Fig. 4. It can be seen that the microstructure of the sample remains fairly stable at elevated temperatures as long as it is below the sintering temperature. The stability of the microstructure of this material has been confirmed by observing the microstructure of the samples after they were exposed to oxygen chemical potential gradient at high temperatures. The samples retrieved after the oxygen permeation at temperatures ranging from 800 to 900 °C for about 7 days changed little.

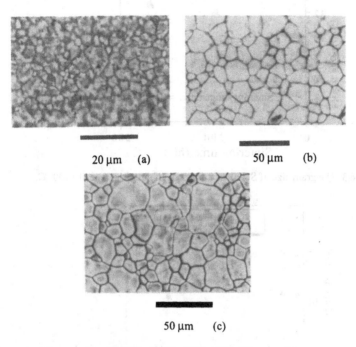

20 μm   (a)   50 μm   (b)

50 μm   (c)

Figure 1. Micrographs of SCFO samples sintered at different temperatures: (a) 950 °C, (b) 1100 °C and (c) 1200 °C.

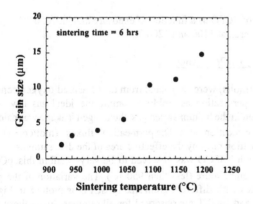

Figure 2. The effect of sintering temperature on the grain size of SCFO.

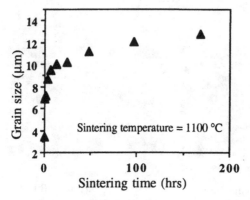

Figure 3. The grain size of SCFO as a function of sintering time at 1100 °C.

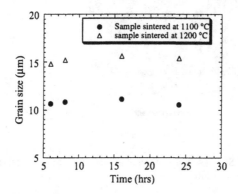

Figure 4. The grain size of SCFO as a function of exposure time at 1050 °C, for sample sintered at 1100 and 1200 °C.

Effect of Microstructure On Oxygen Permeation

The oxygen permeation rates (mol/s) were calculated from the measured oxygen concentrations at the helium exit end of the permeation assembly assuming the ideal gas law. From the concentration of nitrogen detected in the helium sweep gas, the oxygen leakage was calculated and used to correct for the oxygen permeation rates. The permeation fluxes (mol/cm² s) are then calculated by dividing the permeation rates by the effective area of the disc samples.

The oxygen permeation through LSFCO was observed to be very small in this pO$_2$ gradient and almost negligible at the test temperature (less than 950 °C). The variation of the permeation flux with temperature for samples with different grain size of SCFO is plotted in Fig. 5. Linear relationships between log (JO$_2$) and 1000/T are observed for all samples. From these plots, it is clear that the oxygen permeation flux depends on the microstructure of the sample. The oxygen permeation flux is generally increased with the decrease of the grain size of the sample. The microstructure effect on oxygen permeation of SCFO can also be examined when the log (JO$_2$) is plotted as a function of the average grain size of the samples as is shown in Fig. 6. The oxygen permeation flux is increased rapidly as the grain size becomes smaller.

These results reveal that the electrochemical process can be affected by the sample preparation conditions. Since the densities of the samples used for oxygen permeation are similar, the observed difference in oxygen permeation flux is believed to be caused by the microstructure difference. As shown in Fig. 5, the oxygen permeation flux through SCFO is increased with the decrease of the average grain size, indicating that grain boundaries provide a faster path for oxygen penetration. Although the exact mechanism of this effect is not clear at this point, it is possible that the increase of grain boundaries may affect both the surface exchange and the diffusion processes. Previous studies suggested that both of these processes play rate-limiting roles in oxygen permeation through SCFO membranes [8]. Thus, the increase of the oxygen permeation flux through SCFO membranes can be due to the faster diffusion along grain boundary which has higher defect concentrations and / or higher defect mobility than in the bulk. It may also be due to the enhancement of the surface exchange rate by introducing more interfaces at the surface. Work is currently in progress to determine the mechanisms.

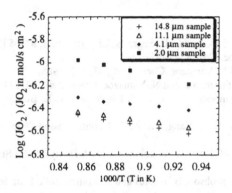

Figure 5. The oxygen permeation flux in SCFO as a function of temperature for different grain size samples.

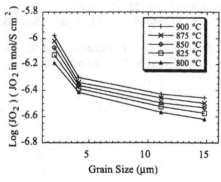

Figure 6. The oxygen permeation flux in SCFO vs grain size at several temperatures.

CONCLUSIONS

The microstructure of LSFCO remains unchanged when the sintering temperature is between 1300 and 1450 °C and the oxygen permeation through this material is negligible up to 930 °C. However, the microstructure of SCFO is affected significantly by sintering temperature such that the average grain size of the sample is considerably increased with the increase of the sintering temperature from 930 to 1200 °C. The effect of microstructure on the oxygen permeation flux through SCFO has been observed. The flux is increased significantly with the decrease of the average grain size, indicating that the grain boundaries provide a faster path for the oxygen diffusion and / or an enhancement of the surface exchange rate.

ACKNOWLEDGMENTS

The authors gratefully acknowledge the support of the National Science Foundation through the MRSEC program under Award Number DMR-9632667.

REFERENCES

1. B. Ma, U. Balachandran, J.-H. Park, and C. U. Segre, Solid State Ionics, **83**, 65 (1996).
2. N. Q. Minh, J. Am. Ceram. Soc., **76**, 563 (1993).
3. Y. Teraoka, T. Nobunaga, and N. Yamazoe, Chem. Lett., **83**, 503 (1988).
4. Y. Teraoka, H. -M. Zhang. S. Furukawa, and N. Yamazoe, Chem. Lett., **80**, 1743 (1985).
5. H. Kruidhof, H. J. M. Bouwmeester, R. H. E. V. Doorn and A. L. Burggraaf, Solid State Ionics, **63-65**, 816 (1993).
6. L. Qiu, T. H. Lee, L. M. Liu, Y. L. Yang, and A. J. Jacobson, Solid State Ionics, **76**, 321 (1995).
7. B. A. Van Hassel, T. Kawada, N. Sakai, H. Yokokawa, and M. Dokiya, Solid State Ionics, **66**, 41 (1993).
8. T. H. Lee, Y. L. Yang, A. J. Jacobson, B. Abeles and S. Milner, Solid State Ionics, **100**, 77 (1997).
9. A. P. Sutton and R. W. Balluffi, in Interfaces in Crystalline Materials (Oxford Science Publications, London, 1995) p. 467.
10. S. P. S. Badwal, and J. Drennan, J. Mater. Sci., **22**, 3231 (1987).
11. T. Kawada, T. Horita, N. Sakai, H. Yokokawa, and M. Dokiya, Solid State Ionics, **79**, 201 (1995).
12. K. Zhang, Y. L. Yang, D. Ponnusamy, A. J. Jacobson, and K. Salama, Effect of Microstructure on Oxygen Permeation in $SrCo_{0.8}Fe_{0.2}O_{3-\delta}$, submitted to J. Mater. Sci..

SYNTHESIS AND PRELIMINARY ELECTROCHEMICAL CHARACTERIZATION OF LiNi$_{0.5}$Co$_{0.5}$O$_2$ POWDERS OBTAINED BY THE COMPLEX SOL-GEL PROCESS (CSGP)

A.DEPTUŁA*, W. ŁADA*, T. OLCZAK*, F. CROCE**, F. RONCI**, A. CIANCIA***, L. GIORGI***, A. BRIGNOCCHI*** AND A. DI BARTOLOMEO***
*Institute of Nuclear Chemistry and Technology, ul. Dorodna 16, 03-195 Warsaw, Poland, adept@orange.ichtj.waw.pl
**Department of Chemistry, University „La Sapienza", P.le A.Moro, 5 - 00137 Roma, Italy, croce@uniroma1.it
***ENEA-CRE-Casaccia, AD 00100 Roma, Italy, giorgil@casaccia.enea.it

ABSTRACT

Complex Sol-Gel Process (CSGP) was applied to the preparation of LiNi$_{0.5}$Co$_{0.5}$O$_2$. Starting sol-solutions were prepared in two different ways: I, in which aqueous ammonia was added to a starting solution of Li$^+$-Ni^{2+}-Co^{2+} acetate-ascorbate, and II in which LiOH was added to a solution of Ni^{2+}- Co^{2+} and NH$_4^+$ acetate-ascorbate. It was found that in the absence of ascorbic acid, or at its lower content ($\leq$0.2 M on 1M Σ Li$^+$- Ni^{2+}- Co^{2+}) precipitation of Ni hydroxides occurred. Regular sols were concentrated ~3 times, gelled and dried at 140°C. Intensive foaming was observed for samples during further heating. Consequently for scaling up to 200g in a run a preliminary long drying procedure followed by self-ignition step (~400°C) was introduced. Thermal transformation of the gel to solid was studied by TG, DTA, XRD and IR. The main feature of this step is carbonate formation. The final structure LiNi$_{0.5}$Co$_{0.5}$O$_2$ is observed after heating for 1h at 800°C. For larger scale production the extension of firing time was necessary. Electrochemical properties of the LiNi$_{0.5}$Co$_{0.5}$O$_2$ compound, prepared by the CSGP were evaluated and considered satisfactory.

INTRODUCTION

Lithiated transition metal oxides having a layered structure and the general formula LiMO$_2$, have been extensively studied as positive electrode active materials for lithium or lithium-ion batteries. Recently LiNi$_x$Co$_{1-x}$O$_2$,especially at x=0.5, is being considered for such applications because of its higher capacity and lower material cost as compared to LiCoO$_2$ [1,2]. LiNi$_{0.5}$Co$_{0.5}$O$_2$ is generally prepared by a solid state processes in which long times, high temperatures and oxygen atmosphere are applied [1-4]. According to the author's best knowledge it has never been synthesized by the sol-gel process.

The goal of the present study was the application of the Complex Sol-Gel Process (CSGP) patented by INCT (Poland) and ANL (USA) teams [6] to the fabrication of high temperature superconductors The process has also been successfully applied to the synthesis of Li$_x$Mn$_2$O$_{4\pm\delta}$ [5].

EXPERIMENTAL

The experimental procedures were similar as in former experiments on the synthesis of Li$_{0.55-1.11}$Mn$_2$O$_4$ by CSGP described in [5]. Starting sol-solutions were prepared in two different ways: I, in which aqueous ammonia (Am) was added to the starting solution of Li$^+$-Ni^{2+}-Co^{2+} acetate (AC)-ascorbate (ASC), and II in which LiOH was added to the solution of Ni^{2+}- Co^{2+} and NH$_4^+$ acetate-ascorbate. For evaluation of the CSGP results some experiments, without addition of ascorbic acid were carried out. All reagents used were of A.R. grade. Concentrated solutions were prepared by dissolving the reagents in deionized water (equipment produced by „Deionizatori" Genoa, Italy). Molar concentrations of the starting solutions were: LiAC-3.429M;

3.429M; NiAC$_2$-0.641M; CoAC$_2$- 1.161M, NH$_4$OH- 12,1; LiOH- 4.25M. Solid ascorbic acid was added to the acetate solutions.

Li concentrations in the starting solutions were determined gravimetrically (accuracy $\pm$ 0.1%); Ni and Co- by complexometric titration (accuracy $\pm$ 0.2%). Potentiometric titrations were carried out with a Labor-pH-Meter, Knick- type 646/647/647-1 using Ingold combined electrodes. Sols were concentrated under vacuum using a Buchi RE 121 Rotavapor. The viscosity of sols was measured using Ubbelohde capillaries.

Gels and the products of their thermal treatment (in a programmed Carbolite furnace type CSF 1200) were characterized by the following methods:

-thermal analysis (TG, DTA) using a Hungarian MOM Derivatograph (sample weight - 100 mg, heating rate -10°C/min., atmospheric air, reference material- Al$_2$O$_3$, sensitivity: TG-200, DTG-1/5, DTA-1/5),

-infrared (IR) spectroscopy (Perkin Elmer Model 983 Spectrometer), using the potassium bromide pellet technique (1/300 mg). Carbonate concentrations were determined by internal standardization [7], using sodium nitride and ν_2 carbonate bands at 875 cm^{-1}

-X-ray diffraction (XRD), Co K$_\alpha$ (Positional Sensitive Detector, Ital Structure).

Apparent density of the powders, was measured picnometrically using CCl$_4$.

The electrochemical properties of the mixed oxide were evaluated by galvanostatic, voltammetric and impedance measurements. All tests were performed at room temperature inside a MBRAUN Labmaster 130 glove box having an argon atmosphere with H$_2$O and O$_2$ content below 1ppm. The electrode membranes were obtained using Poly(vinylchloride), PVC as binder (5% w/o), Acetylene Black carbon (AB) as electronic conductor (5% w/o) and LiNi$_{0.5}$Co$_{0.5}$O$_2$ (90% w/o). A 1M solution of LiClO$_4$ in a Ethylencarbonate (EC) - Dimethylcarbonate (DMC) 1:1 mixture was used as electrolyte, while lithium metal was used as anode and/or as reference electrode.

RESULTS AND DISCUSSION

Potentiometric titrations of the Li-Ni-Co acetate starting solution (pH=6.5) with aqueous ammonia (to final pH=9, Variant I) as well as of the Ni-Co-NH$_4$ (pH=4.7) acetate solution with LiOH (to pH=4.8) and Am (to final pH=8, Variant II) indicate relatively quick hydrolysis (no inflection point on the potentiometric curve). In variant I the addition of NH$_4$OH results in the formation of a green precipitate (Ni^{2+} species) at pH=8. Similar precipitate is formed when a small amount of ASC (less than corresponding to the molar ratio (MR) ASC:ΣLi+Ni+Co=0.15) was added to the above solution. In the presence of larger amounts of ASC (MR $\geq$ 0.2) the molar ratios of Am:ΣMe at the inflection points on the potentiometric curves were proportional to the ASC concentration. It was also observed that a higher concentration results in slower hydrolysis kinetics and moreover inhibits the precipitation due to the strong complexing of the cations by ascorbic acid. Consequently in further experiments ascorbic acid was added in an amount corresponding to MR=0.5. After several days gelation to violet monolithic gels were observed for both variants. It was due to the solvent evaporation; in a closed recipient the sols were liquid for several months.

The gels prepared above (on the scale ~10g of oxides) were dried for 48h at 140°C. Intensive foaming, especially for gels type I was observed. The results of thermal analysis of these dried gels are shown in Fig.1. Thermal analysis of the acetate precursor shows one exothermic effect at ~400°C which can be attributed to the decomposition of acetates. The thermal decomposition of both types of the AC+ASC precursors is more complex. It seems that the first exotherm at ~ 350°C is connected with the decomposition of the metal acetates. The second one, composed of many superimposed effects, observed in the temperature range 460-700°C, is mainly caused by the oxidation of complex compounds formed by ASC decomposition. On the basis of IR spectra (Fig. 2) it can be supposed that also in this step carbonates are formed and partially decomposed. For the temperature region 540-800°C weight losses <15% were observed. During prolonged calcination in stationary conditions (Fig. 2) weight losses were 20-40%. It means that the amounts of carbonates formed in dynamic conditions during the thermal analysis were significantly lower than during the prolonged calcination. However even after firing for 1h at 800°C traces of carbonates are present. As follows from IR spectra distinct traces of

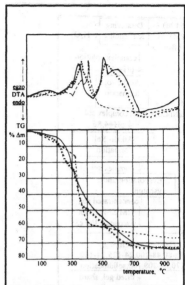

Fig. 1. Thermal analysis of Li-Ni-Co gels dried at 140°C for 48h: AC (--) and AC + ASC, variants: I —; II ˣˣˣ.

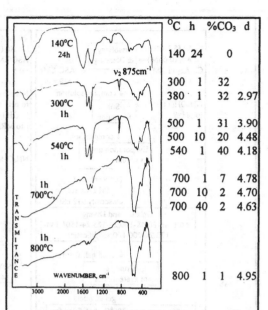

°C	h	%CO₃	d
140	24	0	
300	1	32	
380	1	32	2.97
500	1	31	3.90
500	10	20	4.48
540	1	40	4.18
700	1	7	4.78
700	10	2	4.70
700	40	2	4.63
800	1	1	4.95

Fig.2.Infrared spectra,%CO₃ and density (d,g/cm³) of Li-Ni-Co ASC gels (I) calcined in various conditions.

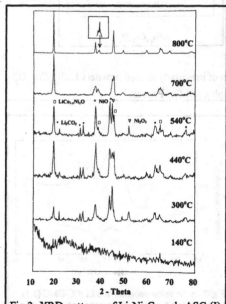

Fig.3. XRD patterns of Li-Ni-Co gels ASC (I) calcined for 1h at various temperatures.

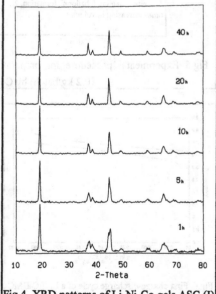

Fig.4. XRD patterns of Li-Ni-Co gels ASC (I) calcined at 700°C for various times (h).

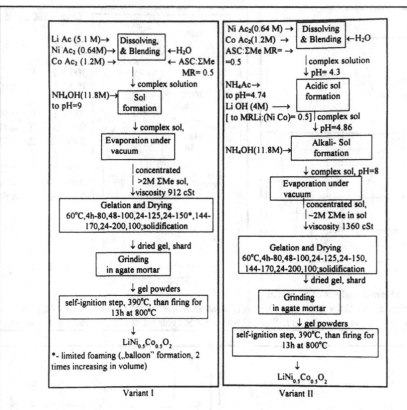

Variant I

Li Ac (5.1 M)→
Ni Ac₂ (0.64M)→ | Dissolving, |
Co Ac₂ (1.2M)→ | & Blending | ←H₂O
⎯⎯ ← ASC:ΣMe
| MR= 0.5
↓ complex solution
NH₄OH(11.8M)→ | Sol |
to pH=9 | formation |

↓ complex sol,
| Evaporation under |
| vacuum |

| concentrated
| >2M ΣMe sol,
↓viscosity 912 cSt

| Gelation and Drying |
| 60°C,4h-80,48-100,24-125,24-150*,144- |
| 170,24-200,100;solidification |

↓ dried gel, shard
| Grinding |
| in agate mortar |

↓ gel powders
| self-ignition step, 390°C, than firing for |
| 13h at 800°C |

↓
LiNi₀.₅Co₀.₅O₂
*- limited foaming („balloon" formation, 2
times increasing in volume)

Variant II

Ni Ac₂(0.64 M) → | Dissolving |
Co Ac₂(1.2M) → | & Blending | ←H₂O
ASC:ΣMe MR= →
=0.5 | complex solution
| ↓ pH= 4.3
NH₄Ac→ | Acidic sol |
to pH=4.74 | formation |
Li OH (4M) ⎯→
[to MRLi:(Ni Co)= 0.5] | complex sol
↓ pH=4.86
NH₄OH(11.8M)→ | Alkali- Sol |
| formation |

↓ complex sol, pH=8
| Evaporation under |
| vacuum |

| concentrated sol,
| ~2M ΣMe in sol
↓viscosity 1360 cSt

| Gelation and Drying |
| 60°C,4h-80,48-100,24-125,24-150, |
| 144-170,24-200,100;solidification |
↓ dried gel, shard

| Grinding |
| in agate mortar |

↓ gel powders
| self-ignition step, 390°C, than firing for |
| 13h at 800°C |

↓
LiNi₀.₅Co₀.₅O₂

Fig.5. Experimental procedure for preparation of irregularly shaped powders LiNi$_{0.5}$Co$_{0.5}$O$_2$ (0.2 kg/batch) by Complex Sol-Gel Process

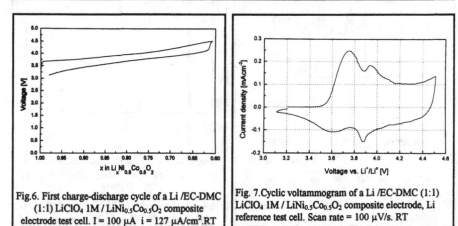

Fig.6. First charge-discharge cycle of a Li /EC-DMC (1:1) LiClO₄ 1M / LiNi$_{0.5}$Co$_{0.5}$O₂ composite electrode test cell. I = 100 μA i = 127 μA/cm². RT

Fig. 7.Cyclic voltammogram of a Li /EC-DMC (1:1) LiClO₄ 1M / LiNi$_{0.5}$Co$_{0.5}$O₂ composite electrode, Li reference test cell. Scan rate = 100 μV/s. RT water exist after calcination at 600°C. According to Pereira-Ramos [8] the presence of some interlayer water in the structure of cathodic material for Li batteries prepared by the sol-gel processes improves the cycle life.

Figure 3 shows XRD patterns of the ASC gels calcined at various temperatures in air for 1h. It can be noted that the final structure $LiNi_{0.5}Co_{0.5}O_2$ occurs for samples fired for 1h at 800°C. The formation of a unique phase with the expected layered (SG=R3m) structure in these conditions indicates the advantage of CSGP over the solid state processes in which longer times, higher temperatures and oxygen atmosphere are commonly used [1-4]. The XRD patterns shown in Fig. 4 indicate that a similar effect can be obtained for samples heated for a longer time at 700°C. However, the shape of the peak at 2Θ~39 suggests the presence of small amounts of unreacted NiO. However even a prolonged heating (up to 40h) does not lead to positive results at 500°C.

On the basis of the results we scaled up both variants to 200g of the final product in a run. However during the preliminary experiments it was observed that the thermal treatment of the resulting homogeneous gel to the desired temperature of several hundred °C results in very intensive foaming (the volume increases several times). Therefore, similarly as in our former work [5] concerned with the fabrication of $LiMn_2O_4$ a long duration preheating procedure followed by self-ignition step (at ~400°C) were applied. The detailed experimental procedures are shown in Fig. 5. It was observed that for this scale a longer heating time was necessary for obtaining the desired final product. SEM analysis of the final powders has shown that the mean particle radius ranged from 1 to 10 μm. It seems that further scaling up to the industrial scale should not be connected with serious difficulties.

The foaming was not observed only for very thick layers (several tens of μm) prepared on metallic substrates (Al, Ag and Ti) from concentrated sols (diluted with ethanol) by the technique described in [9].

Apparent densities of many samples were measured. They are shown in Fig. 3. The materials produced on a large scale, heated for 1h at 800°C had lower densities (I-4.78 g/cm³ and II,-4.83 g/cm³) than those obtained on a laboratory scale. Further heating increases the density to a similar value (4.94 g/cm³).

Electrochemical properties, of the $LiNi_{0.5}Co_{0.5}O_2$ compound prepared on large scale by variant I were evaluated. The first cycle (charge-discharge processes) of a Li / EC-DMC (1:1) $LiClO_4$ 1M / $LiNi_{0.5}Co_{0.5}O_2$ (a 200g batch, variant I) composite electrode test cell, performed with a constant current (I = 100 μA, i = 127 μA/cm²) is shown in Fig. 6. The main feature of this plot is the good reversibility and the high coulombic efficiency of the Li⁺ deintercalation-intercalation process. The specific capacity for Δx=0.4 in $Li_xNi_{0.5}Co_{0.5}O_2$ (about 110 mAh/g) can be considered good at there current densities.

In Fig. 7 is reported the cyclic voltammogram obtained with a three electrode cell using a 100 μV/s scan rate. Once again, the presence of well defined peaks both in the anodic part (positive current, Li⁺ deintercalation) and in the catodic (negative current, Li⁺ intercalation) confirms the good reversibility of these processes.

Finally, in Fig 8 is reported the impedance spectrum obtained with a three electrode cell in the frequency range 65 KHz - 1 mHz. This is the typical spectrum of a thin film intercalation electrode. The semicircle, associated with the charge transfer process, shows a low charge transfer resistance. The diffusion related response is divided in two

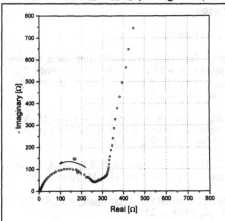

Fig. 8.Impedance spectrum of a Li /EC-DMC (1:1) $LiClO_4$ 1M / $LiNi_{0.5}Co_{0.5}O_2$ composite electrode, Li reference test cell. Frequency range = 65 KHz - 1 mHz. R.T.

parts: a linear part with 45° slope (related to the semi-infinite length diffusion) and a linear part with higher slope (related to the finite length diffusion). This behavior demonstrates high diffusion rates: from the analysis of this spectrum we have obtained a high value of the lithium chemical diffusion coefficient ($D \cong 10^{-7} cm^2 s^{-1}$).

$LiNi_{0.5}Co_{0.5}O_2$ prepared by the Complex Sol-Gel Method has shows a good cyclability if the anodic potential is limited to about 4.2 V. It is suggested that this evidence could be a consequence of the formation of a passivating layer which seems to be responsible for the very low values of the chemical diffusion coefficients calculated on the basis of impedance measurements. The poor kinetics could be the cause of the low specific capacity obtained with high current rates, together with the possible presence of small amounts of unconverted NiO. Considering the fact that still many variables in the sol-gel synthesis can be tuned in order to improve the performance of the compound under investigation and, besides, the use of a different electrolyte could prevent the passivating film formation, or, at least, improve its kinetics.

CONCLUSIONS

1. Addition of ascorbic acid prevents the precipitation during alkalization in aqueous solutions of Li^+- Ni^{2+} Co^{2+} acetates. Consequently the so called Complex Sol Gel Process (CSGP) can be successfully applied for the preparation of homogeneous sols of those elements by hydrolysis with aqueous ammonia.
2. The resulting complex sols can be readily concentrated and gelled to a monolithic homogeneous product.
3. Intensive foaming during thermal treatment of fresh gels was observed. Consequently for scaling up to 200g in a run a preliminary long drying procedure followed by self-ignition step was introduced. For large scale fabrication it is necessary to increase the calcination time.
4. Smooth and adherent thin coatings on Al and Ti substrate were prepared.
5. The overall result of this preliminary electrochemical characterization can be considered good,

REFERENCES

1. R. Koksbang, J. Barker, H. Shi and M.Y. Saidi, Solid State Ionics 84,1-21(1996).

2. W.Li, J.C.Currie, J.Electrochem.Soc. 144, 2773(1997).

3. A.Ovenston, D.Qin, J.R.Walls, J.Mat. Sci. 30, 2496 (1995).

4. Y. Grincourt, G.Brisard, R.Yazami,A. Ovenstone, Second International Symposium on New Materials for Fuel Cell and Modern Battery Systems, June29-July3, 1997. Proceedings, New Materials for Fuel Cell and Modern Battery Systems II, Editors, O.Savadogo and P.R. Roberge, ISBN 2-553-00624-1, Montreal, Canada, 1997,p. 732.

5. A.Deptula, W. Łada, F, Croce, G.B. Appetecchi, A. Ciancia, L.Giorgi, A. Brignocchi, A. Di Bartolomeo. Synthesis and preliminary electrochemical characterization of $Li_xMn_2O_4$ (x=0.55-1.1) powders obtained by the Complex Sol Gel Process (CSGP), ibid.,p. 732.

6. A.Deptuła, W.Łada, T.Olczak M.Lanagan, S.E. Dorris, K.C.Goretta, R.B.Poeppel, Method for preparing of high temperature superconductors, Polish Patent No 172618, June 2 1997.

7. A. L. Smith, Applied Infrared Spectrometry, Fundamentals, Techniques and Analytical Problem- Solving, John Wiley and Sons, New York, 1979, p. 251.

8. J-P.Pereira-Ramos, J. Power Sources 54, 120 (1995).

9. W.Łada, A.Deptuła, T.Olczak, W.Torbicz, D. Pijanowska, A. Di Bartolomeo, Preparation of Thin SnO_2 Layers by Inorganic Sol-gel Process, J. of Sol-Gel Science and Technology 2, 551 (1994).

SILVER-DOPED VANADIUM OXIDES AS HOST MATERIALS FOR LITHIUM INTERCALATION

F. COUSTIER, S. PASSERINI*, J. HILL and W. H. SMYRL
Corrosion Research Center, Department of Chemical Engineering and Material Science, University of Minnesota, Minneapolis, MN 55455; *passerin@cems.umn.edu

ABSTRACT

An improved cathodic material has been obtained by doping vanadium oxide hydrogel with silver. Silver-doped vanadium pentoxides with a silver molar fraction ranging from 0.01 to 1 were synthesized. With the successful doping, the electronic conductivity of V_2O_5 was increased by 2 to 3 orders of magnitude. The electrochemical performance of the silver doped materials is very high, up to 4 moles of lithium per mole of silver-doped V_2O_5 were found to be reversibly intercalated. In addition, the lithium diffusion coefficient is found to be high in the silver-doped material and with a smaller dependence on the lithium intercalation level. These enhancements resulted in high rates of insertion and delivered capacities.

INTRODUCTION

Vanadium pentoxide gels have been studied over the past several decades and especially during the last two decades with the advent of sol-gel process techniques. It has been shown that the gels serve as host materials for a wide variety of metal cations. The large lithium insertion capacity of vanadium pentoxide gel-based cathodes, equal or above two equivalent of lithium per vanadium [1-3], makes them attractive for use as cathodes in high capacity lithium batteries. Thin films or pellets of the xerogel material have been successfully employed in lithium batteries [4] as well as in electrochromic devices either as electrochromic electrodes or as passive lithium reservoirs [5]. Either as thin film or powder, gel-based materials have been shown to intercalate larger amounts of lithium than crystalline V_2O_5 [1-3,6]. Unfortunately, the V_2O_5 gel-based materials are characterized by modest electronic conductivity that affect the rate of insertion of Li^+ ions. They share the low conductivities with other amorphous transition metal host materials, and success here could have an impact for the other systems. The synthesis of silver doped vanadium oxide, prepared starting from the hydrogel precursor, is one the ways we are investigating to enhance the conductivity of gel-based cathode materials.

Amorphous forms of vanadium bronzes have been synthesized from the parent vanadium pentoxide in the gel state. More than twenty different cation bronzes have been prepared in the latter studies [7]. In previous work in this laboratory [8], it was found that the reaction of silver metal with vanadium pentoxide hydrogel was rapid at room temperature. When the product was dried as a spin coated xerogel film, it supported highly reversible Li^+ intercalation as well as highly reversible Ag^0 formation and reoxidation on repeated redox half cycles respectively. The silver vanadium pentoxide formed in this way is amorphous and had higher conductivity than the parent V_2O_5 xerogel. The study of the enhancement in conductivity and its influence on the ability to support reversible high rate of Li^+ is the subject of the present report.

EXPERIMENTAL

V_2O_5 gels were prepared via the sol-gel route, by protonation of sodium metavanadate solution [7]. A homogeneous gel was obtained from the ion exchanged solution after aging for one day.

Silver doped V_2O_5 gels were prepared by mixing the selected stoichiometric amount of silver powder (ALFA 99.9% purity) with V_2O_5 hydrogel. The mixture was vigorously stirred until the silver was completely oxidized and no powder was left unreacted. The amount of silver incorporated (y) corresponds to the equivalent reduction of the formal valency of vanadium from +5 to (+5-y/2). Several silver doped hydrogels were made with the silver molar fraction ranging from 0.01 to 1. Except for the highest silver molar fraction (y=1), the doped hydrogels had a consistency similar to the original, undoped V_2O_5 hydrogel. AgV_2O_5 showed a flocculated consistency instead of a gel.

Thin film samples were prepared by spreading the silver-doped hydrogels onto stainless steel or glass plates. The surface coverage, i.e. the mass of material per unit area of the thin film samples, was varied from 0.02 to 30 mg/cm^2 by changing the amount of coated hydrogel. Due to the flocculant consistency, AgV_2O_5 xerogel samples with a surface coverage below 0.3 mg/cm^2 could not be prepared. All samples were first slowly dried in air overnight to prevent cracking and then dried under vacuum. Active material mass was determined by weighing the samples on a high precision balance ($\pm 1\mu g$).

Film thickness measurements were made by using an Atomic Force Microscope from Topometrix as well as a mechanical profilometer (Dektak). The average density of the dip coated samples was 2.3 g/cm^3, determined by averaging thickness and weigh measurements over several samples.

Powders of composite, silver-doped vanadium pentoxide materials (Hex.$Ag_yV_2O_5$), were prepared through a multistep procedure [9]. The first step consisted in the preparation of an organogel. It was prepared by exchanging the water from the hydrogel with acetone (Aldrich 99.5% used as received). A second solvent exchange with a low surface tension organic solvent (hexane) was then performed to produce the hexane organogel (hexanogel). Hexane needed in the latter step was from Aldrich(98.5% used as received). The double solvent exchange was necessary since water and hexane are not miscible and a single step preparation of the hexanogel was not possible at ambient conditions. The hexanogel still maintain the characteristic bicontinuous network typical of the parent hydrogel. If dried, the Hex.$Ag_yV_2O_5$ powder has a surface area as high as 190 m^2/g.

The following step consisted of the preparation of the composite hexanogel. Selected amounts of Ketjen black (KJB, AKZO NOBEL) and polytetrafluoroethylene (PTFE, Du Pont) were added to the hexanogel in order to obtain the following ratio $Ag_yV_2O_5$:KJB:PTFE = 87%, 9% and 4% w/w. The mixture was stirred until a homogeneous suspension was obtained (10-15 hours). As a final step, the composite hexanogel was dried by evaporation of the solvent in dry air (R.H. <1%) and then under vacuum (10^{-3} Torr) at 70°C. For simplicity, later on the dried hexanogel is indicated as Hex. $Ag_yV_2O_5$. The dried powder was compressed to form pellets. A typical pellet that weighed about 10mg was 0.15mm thick with an average density of 1.9 g/cm^3. For electrochemical tests, the pellets were pressed onto a folded stainless steel X-met that was used as a current collector.

The electrochemical tests were performed in a conventional three-electrode electrochemical cell. A 1M solution of lithium perchlorate in anhydrous propylene carbonate was used as the electrolyte. Lithium metal strips or discs (Foote Mineral Co.) were used as reference and counter electrodes. All chemicals used were reagent grade. LiClO$_4$ (Fluka reagent grade) was vacuum dried at 140°C for 12 hours. Propylene carbonate (Grant Chemicals, H$_2$O<0.002% w/w) was used as received. The cells were assembled and tested in a He-filled glove box(H$_2$O<1ppm).

X-ray diffraction patterns were obtained by using a Siemens D-500 diffractometer. AES (Auger Electron Spectroscopy) measurements were performed with a Physical Electronics instrument(PHI 595).

Conductivity measurements were made by the Van der Pauw method [10] with an in-line four-point configuration experimental set-up. The four probes were placed in contact with the gold dots evaporated on the surface of the dip-coated xerogel samples (subsequently designated as $Ag_yV_2O_5$(XRG)). Current was measured by a potentiostat/galvanostat (model 173 Princeton Applied Research) and voltage was read with a high impedance voltmeter. The measurements were performed in a temperature range from 20°C to 110°C. Each measurement was made after a 24 hours equilibration time at the selected temperature.

Cyclic voltammetry experiments were performed by using a Solartron Electrochemical Interface (ECI 1287). GITT experiments and galvanostatic charge/discharge tests were run by using a customized, fully computer controlled galvanostat.

RESULTS

X-ray diffraction measurements (not shown here) indicate that, with the exception of AgV_2O_5 compound, all materials maintained the layered structure typical of layered vanadium pentoxide gel-based materials. The highest silver content material did show several peaks which positions in the XRD pattern coincided with $Ag_2V_4O_{11}$ peaks reported in the literature [11].

The electronic conductivity of the silver doped samples was measured as a function of temperature. For comparison purposes, the conductivity of the parent material, undoped V_2O_5, is also reported. The presence of silver does affect the electronic conductivity of the materials and such an effect is shown, at several concentrations and temperatures, in the Arrhenius plots reported in Figure 1. In the limit of one mole of silver per mole of V_2O_5, the electronic conductivity was increased by almost 3 orders of magnitude ($\sigma_{RT}=5.6S/cm$) but, as previously pointed out, it is associated with the formation of a crystalline product. $Ag_{0.1}V_2O_5$ showed a substantial increase in conductivity ($\sigma_{RT}=0.8S/cm$) as well, and it was achieved without any disruption of the original layered structure.

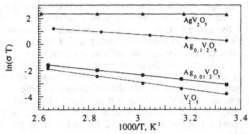

Figure 1. Arrhenius plot for V_2O_5 (circles), $Ag_{0.01}V_2O_5$ (squares), $Ag_{0.1}V_2O_5$ (lozenges), AgV_2O_5 (triangles) xerogel dip-coated samples.

Galvanostatic intermittent titration (GITT) [12] was performed on dip coated $Ag_{0.1}V_2O_5$ and V_2O_5 thin films. The experiment was run by applying a constant cathodic current for an amount of time corresponding to insertion steps of 0.1 equivalents of lithium per mole of $Ag_{0.1}V_2O_5$ up to $x=1$, 0.2 equivalents up to $x=2$, and 0.25 equivalents at higher intercalation levels (x is the molar fraction of lithium). After each cathodic pulse, the electrodes were allowed to relax for approximately 80,000 seconds before measuring the open circuit electrode potential.

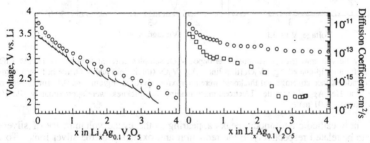

Figure 2. GITT experiment on a dip-coated $Ag_{0.1}V_2O_5$ xerogel sample. Left plot: Open Circuit Potential (vs. Li) vs. x, i.e. equivalents of lithium intercalated, and electrode voltage behavior during constant current pulses vs. x. Right plot: Calculated diffusion coefficient vs. x. For comparison purposes, the diffusion coefficient of lithium in the parent material (V_2O_5) is also shown (square dots). Surface coverage and thickness were approximately 0.2 mg/cm^2 and 1 μm. Pulse current density: 12μA/cm^2. T: 25°C; Electrolyte: 1M LiClO$_4$-PC.

Four equivalents of lithium per mole of silver-doped electrode material were electrochemically intercalated. The quasi-equilibrium electrode potential is shown in Figure 2 (left plot) as a function of the lithium content. The electrode voltage behavior during the galvanostatic pulses is also shown in Figure 2. Only a slight increase of the voltage drop is seen upon the insertion of 4 equivalents of lithium suggesting a good ability of the doped material to intercalate lithium throughout the whole composition range. Such a behavior certainly reflects the high electronic conductivity of the silver-doped material but it also indicates a high diffusivity of the Li^+ ions in the solid phase that is almost independent on the intercalation level. In fact, the lithium diffusion coefficient, calculated according to the procedures introduced for compact non porous materials [12] (Figure 2; right plot), decreases of less than two orders of magnitude upon the insertion of 4 equivalents of lithium per mole of $Ag_{0.1}V_2O_5$. For comparison purposes in the figure is also reported the Li^+ diffusion coefficient in the parent material (V_2O_5). In the same composition range, the lithium diffusion coefficient of the latter material showed a five orders of magnitude decrease.

In Figure 3 (left part) are shown the cyclic voltammograms of $Ag_yV_2O_5$(XRG) (y =0.01 and 0.1) performed at a sweep rate of 0.1mV/s. The samples were thin films with a surface coverage around 0.02 mg/cm^2 corresponding to a thickness near 0.1 μm. The material showed the typical current-voltage behavior of V_2O_5 (XRG) [6] with three broad peaks during the cathodic scan related to different lithium intercalation steps, and three corresponding anodic peaks during the reverse voltage sweep, i.e. the anodic scan. Nevertheless, a deeper investigation shows some differences (evidenciated by the arrows in the figure). In fact, the voltammetric curve of $Ag_{0.01}V_2O_5$ shows a small peak appearing in the anodic sweep at 2.9 V (vs.Li) which counterpart is probably hidden below the first cathodic peak. The voltammetric curve of $Ag_{0.1}V_2O_5$ showed a new cathodic peak at 2.7V (vs.Li). The anodic counterpart of such a peak could be either the shoulder on the left side of the anodic peak above 3.0V (vs.Li) or hidden within the background between the other peaks. Figure 3 (right plot) shows the voltammetric curve of a Hex.$Ag_{0.1}V_2O_5$ composite pellet at a sweep rate of 0.01mV/s. Although the shape of the curve slightly differed from the ones of the thin films, the small cathodic peak at the onset of the largest cathodic peak (2.5V vs.Li) is shown. The position of the small peak (2.7V vs.Li) corresponds very well with the one showed by the thin film.

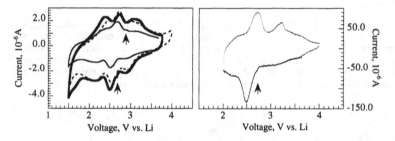

Figure 3. Cyclic voltammetries of various silver-doped V_2O_5 samples at room temperature in 1M LiClO$_4$-PC electrolyte. Left plot: dip-coated $Ag_{0.01}V_2O_5$ (thin line), $Ag_{0.1}V_2O_5$ (thick line) and V_2O_5 (dashed line) xerogel thin samples. Electrode surface coverage and thickness were approximately 0.02 mg/cm^2 and 0.1 μm. Sweep rate: 0.1 mV/s. Right plot: Hex.$Ag_{0.1}V_2O_5$ pellet. Electrode mass loading and thickness were approximately 10mg/cm^2 and 150 μm. Sweep rate: 0.01 mV/s.

The new cathodic and anodic peaks appearing in the cyclic voltammetry of silver-doped materials can be related respectively to the reduction and oxidation of the silver ions. To further investigate this point, Auger Electron Spectroscopy measurements (AES) were performed on $Ag_{0.1}V_2O_5$ thin film to verify the oxidation state of the doping silver upon lithium insertion and deinsertion. The results indicate that, in agreement with the cyclic voltammetry experiments, the silver ions were reduced to a more metallic state upon lithium insertion (cathodic sweep) and, at least, partially reoxidized during the lithium deinsertion (anodic sweep). The reversibility of the process was verified by analyzing the composition of a Hex.$Ag_{0.1}V_2O_5$ pellet after prolonged cycle test (200 cycles). The chemical analysis results showed that no silver loss occurred upon repeated insertion/deinsertion cycles.

The left plot in Figure 4 illustrates the voltage behavior of Hex.Ag$_{0.1}$V$_2$O$_5$ pellets vs. the delivered capacity, as a function of the discharge rate. In the same figure is also illustrated the voltage behavior of Hex.Ag$_{0.3}$V$_2$O$_5$ pellets at 1C rate. The discharge rate is calculated assuming the full capacity of the silver-doped V$_2$O$_5$ to be 4 equivalents of lithium per mole of material (560 mAh/g). A discharge at C rate is then obtained by passing a quite high cathodic current of 560 mA per gram of active material through the cell. The samples were always charged at C/40 rate. in such conditions, the materials were able to deliver specific capacity as high as 380 mAh/g at C/40 discharge rate.

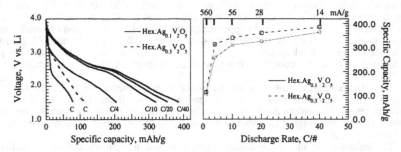

Figure 4. Voltage behavior vs. delivered capacity at different discharge rates (left plot) and delivered capacity vs. discharge rate (right plot) for Hex.Ag$_y$V$_2$O$_5$ pellets. Data shown in the latter plot are averaged over several cells. Electrode mass loading and thickness were approximately 10 mg/cm^2 and 150 µm. Electrolyte: 1M LiClO$_4$-PC.

A more comprehensive picture of the battery performance is shown in the right plot of Figure 4 where is reported the delivered capacity of Hex.Ag$_{0.1}$V$_2$O$_5$ and Hex.Ag$_{0.3}$V$_2$O$_5$ as a function of the discharge rate (data are averaged over several cells). Both materials showed very high electrochemical performances. The specific capacity given at C/40 were about 360 mAh/g and 380 mAh/g, respectively by Hex.Ag$_{0.1}$V$_2$O$_5$ and Hex.Ag$_{0.3}$V$_2$O$_5$. Even at very high rate (1C), corresponding to an actual current in the cell of 5.6 mA or a current density of 16.5 mA/cm^2, capacities of about 80 mAh/g and 115 mAh/g were delivered. Particularly interesting is the behavior of the heavily doped material that in C/4 rate discharges was able to deliver over 310 mAh/g capacity. Such an improvement is certainly related with the higher electronic conductivity within the cathodic material particles due to the larger amount of doping silver.

Figure 5 shows the delivered capacity of Hex.Ag$_y$V$_2$O$_5$ composite pellets (y= 0.1 and 0.3) obtained upon repetitive galvanostatic charge and discharge cycles at C/4 rate. Both materials showed good performances with a modest capacity fading upon cycling.

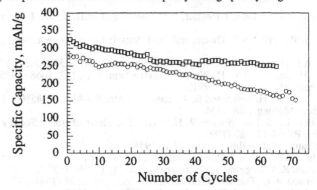

Figure 5. Discharge capacity vs. cycle number of Hex.Ag$_{0.1}$V$_2$O$_5$ (circles) and Hex.Ag$_{0.3}$V$_2$O$_5$ (squares) pellets. Electrode mass loading and thickness were approximately 10 mg/cm^2 and 150 µm. Electrolyte: 1M LiClO$_4$-PC.

Particularly interesting is the behavior of the Hex.$Ag_{0.1}V_2O_5$ composite that showed a decrease of less than 20% of the initial capacity upon the initial 25 cycles. After the 25th cycle the delivered specific capacity remained constant and larger than 250mAh/g.

CONCLUSIONS

The doping of V_2O_5 hydrogel with silver results in a strong improvement of the electrochemical performance of the final materials. Such an effect is certainly caused by the formation of metallic silver during the cathodic cycle as a results of the reduction of the silver ions doping the material. In previous work it was shown that in other materials such as silver vanadium oxide [13] and in silver and copper tungstates [14,15], the formation of metallic silver or copper enhanced the electronic conductivity of the material. The silver-doped V_2O_5 gel-based materials showed in this work appeared to possess this characteristic too, together with a complete reversibility of the silver reduction/oxidation process and without any phase transformation or loss of metallic silver.

The electrochemical properties of the silver doped material were also found to be enhanced. The lithium diffusion coefficient had a very small dependence on the lithium intercalation level. The enhancement is especially important at higher Li composition where undoped V_2O_5 was seen (Figure 2) to show a dramatic drop of the lithium diffusion coefficient of about 5 order of magnitude.

The enhancement of the electronic conductivity and the lithium diffusion coefficient, combined with the reduction of the host solid phase thickness, resulted in the very good host capability showed by the silver-doped Hex.V_2O_5 composite material at high rates. It is in fact under these conditions that high electronic conductivities and ionic diffusivities play an important role in enhancing the rate of intercalation/deintercalation processes in conventional composite configurations.

Further, the material showed a relatively low capacity fading upon cycling.

ACKNOWLEDGMENTS

We are grateful for the support of this work by the Army Research Office through the Illinois Institute of Technology MIFC/B/M Research Center. Contract : DAAH04-93-R-BAA10.

REFERENCES

1.   S. Passerini, D.B. Le, W.H. Smyrl, M. Berrettoni, R. Tossici, R. Marassi, M. Giorgetti, Solid State Ionics, accepted for publication (1997).
2.   D.B. Le, S. Passerini, A.L. Tipton, B.B. Owens and W.H. Smyrl; J. Electrochem. Soc., 142, L102 (1995).
3.   D.B. Le, S. Passerini, J. Guo, J. Ressler, B.B. Owens and W.H. Smyrl; J. Electrochem. Soc., 143, 2099 (1996).
4.   A.L. Tipton, S. Passerini, B.B. Owens and W.H. Smyrl; J. Electrochem. Soc., 143, 3473 (1996).
5.   S. Passerini, A.L.Tipton and W.H. Smyrl; Solar Energy Materials 39 (1995) 167.
6.   H.K. Park W.H. Smyrl and M. D. Ward, J. Electrochem. Soc., 142, 1068 (1995).
7.   J. Livage, Chem. Mater., 3, 578 (1991).
8.   D.B. Le, P. Foong, W.H. Smyrl and R. Atanasoski; Extended Abstract #580, Electrochemical Society Meeting, May 1994.
9.   F. Coustier, D.B. Le, S. Passerini and W.H. Smyrl; The Electrochemical Society Proceeding Volumes, PV 97-13, 180 (1997).
10.   L.J. van der Pauw, Philips Res. Report, 13,1 (1958).
11.   E.S. Takeuchi and P. Piliero; J. Power Sources, 21, 133 (1987).
12.   W. Weppner and R.A. Huggins; J. Electrochem. Soc., 124, 1569 (1977).
13.   E.S. Takeuchi and W.C. Thiebolt; J. Electrochem. Soc., 135, 2691 (1988).
14.   B. Di Pietro and B. Scrosati; J. Electrochem. Soc., 124, 161 (1977).
15.   S. Passerini, S. Loutzky and B. Scrosati; J. Electrochem. Soc., 141, L80 (1994).

PREPARATION AND BATTERY APPLICATIONS OF MICRON SIZED Li₄Ti₅O₁₂

D. PERAMUNAGE* AND K. M. ABRAHAM**
*EIC Laboratories, Inc., 111 Downey Street, Norwood, MA 02062

ABSTRACT

The objective of this study was to highlight the usefulness of micron-sized Li₄Ti₅O₁₂ in three distinctive areas: a) cathode of a low-voltage Li battery, b) insertion type auxiliary electrode to investigate the electrochemistry of oxide cathode materials, and c) anode of a Li-ion cell in conjunction with LiMn₂O₄ cubic spinel cathode. Li cells with Li₄Ti₅O₁₂ exhibited an open circuit voltage of ~1.6V, >90% utilization (in terms of the theoretical capacity) at ~C/10 rate, ~40% utilization 5C rate, and extended full-depth charge/discharge cycling at ≥ 1C rates with virtually no capacity fade. LiMn₂O₄ cathodes, evaluated in Li₍₄₊ₓ₎Ti₅O₁₂ (x = ~1.2)/LiMn₂O₄ cells, exhibited extended full-depth cycling capability with a small capacity fade rate of <0.1% which appeared to slow down with cycling. At a 1C discharge rate, over 190 cycles were demonstrated corresponding to an end utilization of ~90 mAh/g or ~0.6 mole Li per LiMn₂O₄. Balanced Li₄Ti₅O₁₂//solid polymer electrolyte//LiMn₂O₄ full cells of slightly cathode-limited configuration had an open-circuit voltage of ~3.0V and a mid-discharge voltage of ~2.5V showing full-depth extended cycling capability at a utilization of ~90 mAh/g or ~0.6 mole Li per LiMn₂O₄ at the 1C and ~0.45 mole Li per LiMn₂O₄ at the 7.5C discharge rate.

INTRODUCTION

Passivation layers at the electrode/electrolyte interface and diffusion limitations of Li⁺ in the solid state are two factors that limit the performance of an anode in a Li-ion battery at high discharge rates. Mid discharge voltage for Li insertion into Li₄Ti₅O₁₂ is close to 1.56V *versus* Li⁺/Li which is well above the potential regime where either the electrolyte or the solvent could be reduced and consequently, the electrode is free of passivation layer formation on its surface. Insertion or extraction of Li ions to and from Li₄Ti₅O₁₂ is shown in equation [1-3].

$$\frac{1}{3}Li_4Ti_5O_{12} + Li^+ + e^- \xrightarrow[\text{Charge}]{\text{Discharge}} \frac{1}{3}Li_7Ti_5O_{12} \qquad E = \sim 1.5V \qquad [1]$$

Diffusion limitations inside the solid phase could be circumvented by minimizing the time constant for diffusion in the solid state, L^2/D (L = diffusion width, and D = diffusion coefficient) (4) using smaller size particles (meaning smaller L). In a recent study we presented a procedure to prepare micron-sized LiMn₂O₄ which resulted in a substantially improved rechargeability and capability in Li//polymer electrolyte//LiMn₂O₄ cells (5). We showed that particle size control of the product could be achieved primarily by controlling the particle size of the precursor mixture. We adopted a similar concept to synthesize micron-sized Li₄Ti₅O₁₂ which, as will be shown later, exhibited substantial improvement in the rate capability in Li and Li ion cells

Insertion/extraction processes of Li₄Ti₅O₁₂ are known to proceed with no noticeable change in lattice dimensions (1) and, therefore, it is considered as a zero strain insertion process. Absence of any mechanical deformations in Li₄Ti₅O₁₂ during charge/discharge processes may have also resulted

---

**Present address: Covalent Associates, 10 State Street, Woburn, MA 01801.

in the low capacity fade observed in Li/Li$_4$Ti$_5$O$_{12}$ cells (1-3). Utility of Li$_4$Ti$_5$O$_{12}$ will be discussed in terms of three areas of potential applications including; a) cathode of a low-voltage Li battery, b) insertion type auxiliary electrode to investigate the electrochemistry of oxide cathode materials, and c) anode of a Li-ion cell in conjunction with LiMn$_2$O$_4$ cubic spinel cathode.

EXPERIMENTAL

Synthesis of Lithiated Li$_4$Ti$_5$O$_{12}$

Two procedures were used. They both involved solid-state synthesis; a) heating of a well-ground stoichiometric mixture of TiO$_2$ (anatase, APS 32 nm, AESAR) and LiOH (AESAR) at 800°C and subsequently grinding and sieving the product to desired particle size and, b) dispersing submicron size TiO$_2$ and Li$_2$CO$_3$ (Aldrich) in hexane followed, first by removal of the solvent, and then calcination of the residue at 800°C. Both of these high temperature reactions were done in a slow O$_2$ stream. Precursors were mixed to yield a final stoichiometry of Li$_4$Ti$_5$O$_{12}$.

Synthesis of Polymer Electrolyte

Polymer electrolyte was prepared by heating a weighed and thoroughly mixed slurry of PAN, Li salt, and EC/PC mixture at 140C followed by casting on FEP films as described elsewhere (6). Typical electrolyte composition was 21 m/o PAN:36.5 m/o EC:36.5 m/o PC:6.0 m/o LiPF$_6$.

Preparation of Thin Composite Electrodes

Both Li$_4$Ti$_5$O$_{12}$ and LiMn$_2$O$_4$ were made into thin composite electrodes having thickness between 0.002 and 0.005 cm. A slurry prepared by blending a mixture of the active material, a polymeric binder, and high surface Chevron carbon black with enough N-methypyrrolidone was coated on an Al substrate using the doctor-blade method. The solvent was allowed to evaporate at ~100°C under vacuum followed by pressing at 5500 psi pressure.

Electrochemical Characterization of Cells Made of Li$_4$Ti$_5$O$_{12}$ and LiMn$_2$O$_4$

Cells were fabricated by sandwiching the polymer electrolyte between the anode and the cathode composites and hermetically sealing the partially evacuated cell by heat-sealing the metallized plastic envelope at the edges. Electrode preparation, cell construction and sealing were carried out in an Argon-filled glove box.

RESULTS AND DISCUSSION

Synthesis of Li$_4$Ti$_5$O$_{12}$

Spinel material with a nominal composition of Li$_4$Ti$_5$O$_{12}$ can be conveniently synthesized from the two processes outlined in the experimental section. The corresponding reactions are

$$6\,\text{LiOH} + \frac{15}{2}\,\text{TiO}_2 \quad \rightarrow \quad \frac{3}{2}\,\text{Li}_4\text{Ti}_5\text{O}_{12} + 3\,\text{H}_2\text{O} \qquad [2]$$

$$3\,\text{Li}_2\text{CO}_3 + \frac{15}{2}\,\text{TiO}_2 \quad \rightarrow \quad \frac{3}{2}\,\text{Li}_4\text{Ti}_5\text{O}_{12} + 3\,\text{CO}_2 \qquad [3]$$

Li$_4$Ti$_5$O$_{12}$ was obtained as an off-white powder consistent with the previously reported work (1-3)

Lithium battery applications of Li₄Ti₅O₁₂

$Li_4Ti_5O_{12}$ was characterized in solid polymer electrolyte (SPE) cells of the type $Li//SPE//Li_4Ti_5O_{12}$ with a Li reference electrode included in it. The SPE typically used had 21 m/o PAN:36.5 m/o EC:36.5 m/o PC:6.0 m/o $LiPF_6$. Prior to fabrication of the cell, the $Li_4Ti_5O_{12}$ electrode was allowed to soak in a ~1M solution of $LiPF_6$ in 1:1 EC and PC. Capacity utilization at various discharge rates were of primary interest. Since the polymeric binder, e.g. PAN and/or PVDF, used in the composite is soluble in N-methypyrrolidone, the active particles in the dried composite can become coated with a layer of these polymers as the solvent is removed. During cell operation, Li^+ will have to diffuse through these layers to reach the active particles, and unless they are fully swelled with the liquid electrolyte they may create a diffusion barrier for ions. Li^+ may not get to or out of active particles sufficiently fast enough to sustain high rates, resulting in concentration polarization and poor cathode utilization. We could circumvent this problem by soaking the composite electrode with EC-PC/LiPF₆ electrolyte at approximately 140°C, which in effect, converts the polymer coating on the active particles to a solid polymer electrolyte coating with high ionic conductivity. Presented in Table 1 is a comparison of the performance of; a) $Li_4Ti_5O_{12}$ obtained from LiOH and TiO_2 soaked at room temperature, b) the same material soaked at 140°C as described above, and c) the material obtained from Li_2CO_3 and TiO_2 and soaked at 140°C

Table 1. Electrochemical Characteristics of $Li_4Ti_5O_{12}$ Composites

Composition	Capacity, mAh/g during discharge at a current density, mA/cm².										
	0.02	0.05	0.1	0.2	0.5	1.0	2.0	5.0	0.02[i]	0.2[j]	[a]mA/cm²
[b]1 (S)	147	142	139	126	106	78	56	36	147	129	0.40
[c]2 (S)	137	133	129	123	98	68	35	3	139	124	0.51
[d]1 (H)	147	143	141	134	117	93	66	24	148	135	0.46
[e]2 (H)	152	150	147	144	128	107	82	46	152	145	0.60
[f]1' (C,H)	157	154	149	141	127	112	94	69	156	142	0.52
[g]2' (C,H)	155	152	149	142	127	112	81	64	154	144	0.52
[h]3 (C,H)	163	160	156	150	134	121	105	71	164	151	0.61

a) Current density corresponding to the C-Rate, b) 87.5 w/o $Li_4Ti_5O_{12}$:10.0 w/o Chevron C:2.5 w/o PVDF, soaked at room temperature (S), c) 87.5 w/o $Li_4Ti_5O_{12}$:10.0 w/o Chevron C:2.5 w/o PAN, soaked at room temperature (S), d) composition 1 soaked at 140°C (H), e) composition 2 soaked at 140°C (H), f) composition 1 with $Li_4Ti_5O_{12}$ (C) obtained from Li_2CO_3 and TiO_2 soaked at 140°C g) composition 2 with $Li_4Ti_5O_{12}$ (C), soaked at (H), h) composition, 86.5 w/o $Li_4Ti_5O_{12}$ (C):10.0 w/o Chevron C:2.5 w/o PVdF:1.0 w/o PAN soaked at 140°C i) Stability check at low rate following 33 cycles j) Extended cycling (up to 100 cycles) at a moderately high current density.

As evidenced by the data in Table 1, significant improvement in the performance of $Li_4Ti_5O_{12}$ could be realized by soaking the electrode at 140°C instead of at room temperature. Further improvement in performance is possible by substituting the $Li_4Ti_5O_{12}$ material obtained from LiOH and TiO_2 with that obtained from Li_2CO_3 and TiO_2. *The latter is hereafter referred to as $Li_4Ti_5O_{12}$ (C) and it was routinely used for the work described below together with soaking at 140°C.* The actual voltage versus capacity profiles for a typical for $Li/PAN-EC/PC-LiPF_6/Li_4Ti_5O_{12}$ cell based on $Li_4Ti_5O_{12}$ (C) are depicted in Figure 1 at various discharge rates.

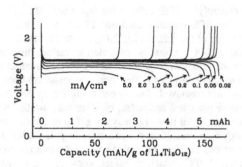

Figure 1. Cycling curves at ambient temperature for Li/PAN-EC/PC-LiPF$_6$/Li$_4$Ti$_5$O$_{12}$ cell based on the composition 3 (C,H) in Table 1. Theoretical capacity was 6.06 mAh. Discharge currents are noted on the figure and charging was done at a constant 20 μA/cm^2 rate throughout.

The initial open-circuit voltage was close to 2.9V. The cells of three electrode configurations were typically cycled between 2.3V to 1.2V. Discharge rate was varied from 20 μA to 5 mA/cm^2 in steps with few cycles for each discharge rate while charging was done at a constant 20 μA/cm^2 rate throughout. During discharge the voltage dropped quickly until the voltage reached a plateau around 1.55V *versus* Li$^+$/Li. Discharge/charge capacities for the cell depicted in Figure 1 is shown as a function of cycle number in Figure 2. .

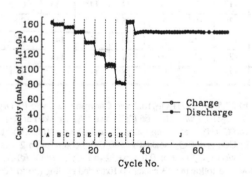

Figure 2. Discharge/charge capacities as a function of cycle number for the cell shown in Figure 1. Charge(discharge) current density, mA/cm^2, corresponding to each marked region is [A] = 0.02(-0.02), [B] = 0.02(-0.05), [C] = 0.02(-0.1), [D] = 0.02(-0.2), [E] = 0.02(-0.5), [F] = 0.02(-1.0), [G] = 0.02(-2.0), [H] = 0.02(-5.0) [I] = 0.02(-0.02), [J] = 0.02(-0.2). (0.5 mA/cm^2 equals to ~0.8 rate). •

Application of Li$_4$Ti$_5$O$_{12}$ as an Auxiliary Anode for Cathode Evaluation

Unlike the Li cells used to evaluate Li$_4$Ti$_5$O$_{12}$, the first cycling step of a Li/LiMn$_2$O$_4$ cell is a charge which involves plating of Li metal onto the Li counter electrode. Because of a thick passivation layer already present on the Li counter electrode, the plating process is usually uneven and very often favors the growth of dendritic Li and cell shorting. This difficulty is circumvented by replacing the Li anode with a partially lithiated Li$_{(4+x)}$Ti$_5$O$_{12}$ (x = ~1.2) anode which is electrochemically prepared in a Li/Li$_4$Ti$_5$O$_{12}$ cell. The capacity of this anode is chosen to be twice or more of the capacity of $LiMn_2O_4$ cathode. Partial lithiation satisfies the slight irreversible capacity of the $Li_4Ti_5O_{12}$ electrode during its first cycle, ~2.5%, and as a result, Li deintercalated from the $LiMn_2O_4$ cathode during its first charge is not wasted for that purpose. Electrochemical potential of the $Li_{(4+x)}Ti_5O_{12}$ (x ~1.2) electrode is fixed on its horizontal charge/discharge plateau (~1.56V *versus* Li^+/Li) and since it has a Li capacity far higher than $LiMn_2O_4$, the electrode remains free of dendrite formation. No solvent or electrolyte reduction is possible at this potential plateau either. In essence, the high capacity $Li_{(4+x)}Ti_5O_{12}$ (x ~1.2) behaves as an ideal passivation free counter electrode with a constant potential and ready to act as the perfect Li "sink" for the $LiMn_2O_4$ electrode. Depicted in Figures 3 and 4 are the voltage *versus* capacity profiles and long-term cycling performance of a $Li_{(4+x)}Ti_5O_{12}/LiMn_2O_4$ cell based on a partially lithiated anode. $LiMn_2O_4$ electrode was processed similar to the $Li_{(4+x)}Ti_5O_{12}$ (x ~1.2) electrode.

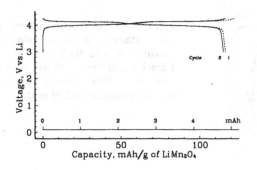

Figure 3. Capacity *versus* voltage profiles at ambient temperature for $Li_{(4+x)}Ti_5O_{12}(x~1.2)//SPE//LiMn_2O_4$ cell. The cathode had a 6.2 mAh theoretical capacity.

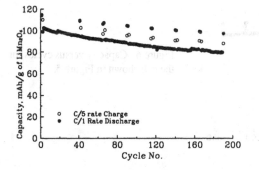

Figure 4. Discharge/charge capacities as a function of cycle number for $Li_{(4+x)}Ti_5O_{12}$ (x~1.2)//SPE//LiMn$_2$O$_4$ Cell shown in Figure 3. Charge(discharge) current density was 0.12 (-0.6) mA/cm^2. Capacity stability was checked intermittently at 0.04(-0.04) mA/cm^2. (0.6 mA/cm^2 equals to ~1.0 C rate).

Utilization observed in initial cycles appeared to be closer to the optimum of 0.8 Li mole per $LiMn_2O_4$. The capacity fade appears to slow down with cycling and at around 75 cycles it has reached 0.14% per cycle.

Application of $Li_4Ti_5O_{12}$ in $Li_4Ti_5O_{12}//SPE//LiMn_2O_4$ Li-ion Cells

Li-ion cells discussed in this section are similar to those used above to evaluate $LiMn_2O_4$ except that they have balanced anode and cathode capacities. Selected capacity *versus* voltage profiles for a typical $Li_4Ti_5O_{12}/LiMn_2O_4$ cell are presented in Figure 5 and its capacity as a function of cycle life is presented in Figure 6. Evolution of the capacity with cycling is similar to that for the cathode (Figure 3). Capacity fade appears to level off with extended cycling and it has reached nearly 0.12%/cycle at around 100 cycles. Slightly diminishing capacity observed during stability check suggests that the capacity loss is real and, most likely, of materials origin and not related to any polarization phenomenon occurring at high rate.

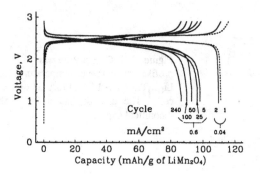

Figure 5. Selected cycling curves at room temperature for a balanced $Li_4Ti_5O_{12}//SPE//LiMn_2O_4$ cell.
Capacity is expressed relative to the cathode. The cathode had a theoretical capacity of 5.91 mAh.

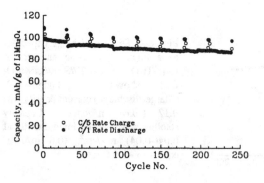

Figure 6. Capacity versus cycles for the cell shown in Figure 5.

Rate capability of a typical $Li_4Ti_5O_{12}//SPE//LiMn_2O_4$ cell is depicted in Figure 7. The cell demonstrated high utilization close to 0.45 Li per $LiMn_2O_4$ at ~7.5 C rate.

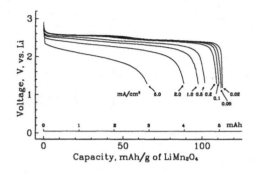

Figure 7. The voltage *versus* capacity profiles at ambient temperature for a balanced $Li_4Ti_5O_{12}//SPE//LiMn_2O_4$ cell as a function of different discharge rates. Capacity is expressed relative to the cathode which had a capacity of 6.67 mAh theoretical capacity

CONCLUSION

$Li_4Ti_5O_{12}$ electrodes prepared with PVDF or PAN binder and soaked at 140°C in the plasticizer solution achieved close to 90% of the theoretical capacity, the highest observed and over 40% of this utilization was still possible at rates higher than 5C. These electrodes underwent extended full-depth charge/discharge cycling at ≥ 1C rates with virtually no capacity fade. Utility of a partially lithiated $Li_4Ti_5O_{12}$ as an auxiliary electrode for $LiMn_2O_4$ evaluation was also demonstrated. Balanced $Li_4Ti_5O_{12}//SPE//LiMn_2O_4$ cell showed full-depth extended cycling capability at the C-rate for 250 cycles corresponding to an end utilization of >80 mAh/g or ~0.55 mole Li per $LiMn_2O_4$. Capacity fade rate was low at <1% and the electrodes appeared to remain free of passivation during extended cycling. The cell exhibited high rate capability as demonstrated by a utilization close to 0.45 Li per $LiMn_2O_4$ at about the 7.5 C rate.

ACKNOWLEDGMENTS

This work was carried out on Contract DE-FG02-96ER82158 from the department of Energy. The authors would like to thank Brian G. Carroll for the preparation of materials.

REFERENCES

1.   T. Ohzuku, A. Ueda, and N. Yamamoto, J. Electrochem. Soc., 142, 1431 (1995).

2.   T. Ohzuku, A. Ueda, and N. Yamamoto, and Y. Iwakoshi, J. Power Sources, 54, 99 (1995).

3.   K. M. Kolbow, J. R. Dahn, and R. R. Haering, J. Power Sources, 26, 397 (1989).

4.   S. Altung, K. West, and T. Jacobsen, J. Electrochem. Soc., 126, 1311 (1979).

5.   Z. Jiang, and K. M. Abraham, J. Electrochem. Soc., 143, 1591 (1996).

6.   H. S. Choe, B. G. Carroll, D. M. Pasquariello, and K. M. Abraham, Chem. Mater., 9, 369 (1997).

Figure 2. The voltage versus capacitance...

CONCLUSION

ACKNOWLEDGMENTS

This work was carried out under Contract No. DE-AC02-76-... from the Department of Energy. The authors would like to thank Arthur D. Carslaw of the Department of Materials

REFERENCES

1. T. Ohachi, A. Ueda and N. Yasunaga, Electrochem. Soc. 118, ... (1971)
2. T. Ohachi, A. Ueda and H. Yasunaga, ... Proc. ... (1984)
3. J. M. Snow, J. H. Dalsgaard, P. J. Bladon ... (1989)
4. F. S. Stone, R. West and J. Electrochem. Soc. 121, 117 (1979)
5. F. Fang, and K. McCarthy, J. Electrochem. Soc. 144, 1511 (1967)
6. H. C. Gatos, S. G. Carroll, D. M. Hargreaves and K. W. Abraham, Thin Solid Films ... (1991)

HYDROTHERMAL SYNTHESIS AND CHARACTERIZATION OF A SERIES OF NOVEL ZINC VANADIUM OXIDES AS CATHODE MATERIALS

Fan Zhang, Peter Y. Zavalij, and M. Stanley Whittingham*,
Chemistry Department and Materials Research Center
State University of New York at Binghamton, Binghamton, New York 13902

ABSTRACT

We report here the hydrothermal synthesis of a series of novel zinc vanadium oxides, using tetramethyl ammonium ion in order to stabilize the layered structure of vanadium oxide during electrochemical redox reactions with lithium. The compounds were synthesized by the reaction of zinc chloride, vanadium (V) oxide, and tetramethyl ammonium hydroxide at 165°C for 60 hours. Four new zinc vanadium oxide compounds were discovered, only one of which contained the organic cation; another was found to have a layered structure with bridging V-O-V groups analogous to that of beta alumina. These compounds were characterized by X-ray powder diffraction, and IR. Three of the compounds reacted readily with lithium, and their electrochemical behavior in lithium cells were determined.

INTRODUCTION

Lithium is a very attractive material for high energy density batteries because of its low equivalent weight and high electrode potential. Our research on materials for advanced lithium batteries has focussed on the hydrothermal synthesis, in tetramethyl ammonium containing solutions, of vanadium oxides. These solutions normally contain lithium and may contain an additional transition metal, such as iron or zinc, in order to stabilize the layered structure of vanadium oxide for electrochemical redox reactions with lithium. Earlier we reported the use of the tetramethyl ammonium ion as a structure-directing cation [1] and showed that a number of new phases of tungsten [2], molybdenum [3], and vanadium could be formed. In the case of vanadium, two new structure types were reported in 1995 [4], the layered $N(CH_3)_4V_4O_{10}$ [5] and a hydrated vanadium dioxide [6,7], also a new structure type formed by iron chloride and vanadium oxide in 1997, the layered $[N(CH_3)_4]_zFe_yV_2O_5 \cdot nH_2O$, where z is 1/6, y ≈ 0.1 and n is 1/6 [8]. Subsequently, several other layer structure vanadium oxides were formed by various groups [9-12]. To date only the tetramethyl ammonium ion has shown the ability to form a range of different structures with differing organic to vanadium ratios [13]; the present count is six structures including two with a string-like morphology formed at pH values of 3 or less such as $N(CH_3)_4V_8O_{20}$ [13], and layer structures like $N(CH_3)_4V_3O_7$ [14] and $[N(CH_3)_4]_5V_{18}O_{46}$ [15].

Vanadium oxides have been extensively investigated as possible cathodes for lithium batteries. In particular, V_2O_5 and V_6O_{13} have been the most extensively studied. Although showing good redox behavior, their cyclability is not as good as the lithium cobalt oxide cathodes and the discharge curve slopes much more, making electronic control more difficult.

Mat. Res. Soc. Symp. Proc. Vol. 496 © 1998 Materials Research Society

This paper describes the formation of new vanadium oxide cathodes for rechargeable lithium batteries by the hydrothermal reaction of zinc chloride and vanadium pentoxide in the presence of the tetramethyl ammonium cation, $N(CH_3)_4^+$. Four new vanadium oxides, $Zn_{0.4}V_2O_5 \cdot 0.3H_2O$, $[N(CH_3)_4]_{0.5}Zn_{0.4}V_2O_5$, $Zn_3(OH)_2(V_2O_7) \cdot H_2O$, and $Zn_2(OH)_3(VO_3)$, with layered structures were found. The tendency of this cation to direct the synthesis toward layered structure has been previously noted [1,3]

EXPERIMENTAL

The zinc vanadium oxides were synthesized by mixing $ZnCl_2$ and V_2O_5 powder from Johnson and Matthey with 25% tetramethyl ammonium hydroxide solution from Alfa in different molar ratios, as the following 1:1:1, 1:1:2, 1:1:3, 1:1:4. The products, synthesized from the above molar ratios, are $Zn_{0.4}V_2O_5 \cdot 0.3H_2O$, $[N(CH_3)_4]_{0.5}Zn_{0.4}V_2O_5$, $Zn_3(OH)_2(V_2O_7) \cdot H_2O$, and $Zn_2(OH)_3(VO_3)$, respectively. When $[N(CH_3)_4]_{0.5}Zn_{0.4}V_2O_5$ was synthesized, the pH of the solution was adjusted to 3.67-4.00 by acetic acid. No acid was added in the other reactions. The resulting solution was transferred to a 125-ml Teflon-lined autoclave (Parr Bomb), sealed, and reacted hydrothermally for 2.5 days at 165°C. The resulting crystals of each compound were filtered and dried in air. The colors of these four compounds are greenish black, black, white and white, respectively.

X-ray powder diffraction was performed using Cu Kα radiation on a Scintag θ–θ diffractometer. The data was collected from 4°2θ to 90°2θ with 0.03°2θ steps and 15 sec per step. The FTIR data was obtained on a Perkin-Elmer 1500 series. Chemical analysis was performed using TGA, a JEOL8900 Electron Microprobe, and an ARL Spectrospan-7 DCP Atomic Emission Spectrometer.

The degree of reduction of the vanadium oxide by lithium was determined by reaction with n-butyl lithium from Aldrich Chemicals, following standard procedures [16]. Initial electrochemical studies were conducted in lithium cells using 1 molar $LiAsF_6$ in a 1:1 propylene/dimethyl carbonate mixture as the electrolyte; the vanadium oxide was mixed with 10% carbon black and 10% Teflon powder, and hot pressed for 20 minutes at 440°F. A MacPile potentiostat was used to cycle the cells.

RESULTS AND DISCUSSION

X-Ray Diffraction Determination

The X-ray diffraction data of the four compounds, $Zn_{0.4}V_2O_5 \cdot 0.3H_2O$, $[N(CH_3)_4]_{0.5}Zn_{0.4}V_2O_5$, $Zn_3(OH)_2(V_2O_7) \cdot H_2O$, and $Zn_2(OH)_3(VO_3)$, are shown in Fig. 1. The pattern of $Zn_{0.4}V_2O_5 \cdot 0.3H_2O$ shows a repeat distance of 10.43Å. Preliminary analysis of the structure indicates a monoclinic unit cell: $a=11.743(2)$Å, $b=3.6149(2)$Å, $c=10.503(1)$Å, and $\beta = 96.69(2)^0$. This is consistent with a double layer sheet of V_2O_5 with zinc ions and water molecules between the layers. It is almost certainly related to the ν and δ phases described by Galy [17]. The product of heating to 150°C showed a contraction in the repeat distance from 10.43 to 9.23Å; on exposure to atmospheric air, the lattice expanded again to 10.48Å and 10.59Å, indicating a variable water content. The powder X-ray diffraction pattern could be readily indexed with a monoclinic lattice system and space group C 2/m. We are presently solving the structure of the compound $[N(CH_3)_4]_{0.5}Zn_{0.4}V_2O_5$.

$Zn_3(OH)_2(V_2O_7) \cdot H_2O$, has lattice parameters of a=6.061(1)Å. c=7.207(1)Å, with space group P-3m1. The lattice system is hexagonal. The structure is sandwich-like, with the layers comprising zinc oxide/hydroxide octahedra, with V-O-V pillars and water molecules between the layers [18]. It is reminiscent of beta alumina with less than half the volume between the layers being occupied in the dehydrated compound.

The powder diffraction pattern of $Zn_2(OH)_3(VO_3)$, was indexed with a hexagonal lattice system and the space group R-3m. The cell parameters of this compound are: a=3.0721(1)Å, c=14.313(3)Å. This compound has repeating edge-shared zinc oxide/hydroxide octahedra and edge-shared VO_4 tetrahedra bilayered structure.

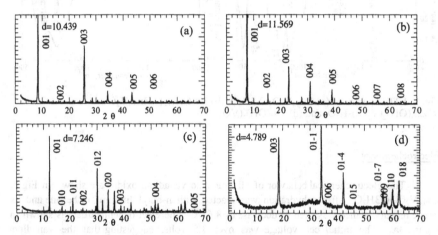

Fig. 1 XRD patterns of (a) $Zn_{0.4}V_2O_5 \cdot 0.3H_2O$, (b) $[N(CH_3)_4]_{0.5}Zn_{0.4}V_2O_5$, (c)$Zn_3(OH)_2(V_2O_7) \cdot H_2O$, and (d) $Zn_2(OH)_3(VO_3)$.

FTIR Spectra

The FTIR spectra of $Zn_{0.4}V_2O_5 \cdot 0.3H_2O$, $[N(CH_3)_4]_{0.5}Zn_{0.4}V_2O_5$, $Zn_3(OH)_2(V_2O_7) \cdot H_2O$, and $Zn_2(OH)_3(VO_3)$, are shown in figure 2. The 1008cm^{-1} peaks of FTIR spectra of compound 1 and 2 indicate the presence of V=O bond in both compounds, those at 1487, 1385 and 946 cm^{-1} are due to the organic species, leaving only those at (900-700 cm^{-1}) and (800-400 cm^{-1}) as being associated with the vanadium oxides network. Those paucity of V-O bands speak against the presence of a vanadium cluster, which have rich IR spectra.

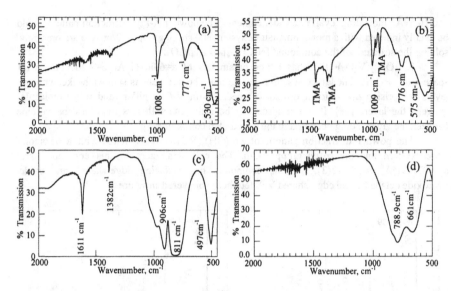

Fig. 2. FTIR spectra of (a) $Zn_{0.4}V_2O_5 \cdot 0.3H_2O$, (b) $[N(CH_3)_4]_{0.5}Zn_{0.4}V_2O_5$, (c)$Zn_3(OH)_2(V_2O_7) \cdot H_2O$, and (d) $Zn_2(OH)_3(VO_3)$.

Electrochemistry

The electrochemical behavior of all four zinc vanadium oxides are shown in Fig 3. $Zn_{0.4}V_2O_5 \cdot 0.3H_2O$ after dehydration was reacted with n-butyl lithium in hexane and its lattice repeat distance was seen to decrease to 8.62Å. The electrochemical data is shown in Figure 3a. The initial cell voltage was over 3.5 volts, suggesting that the vanadium approaches the +5 oxidation state. The lithium insertion then proceeded in a number of stages, reminiscent of that of V_2O_5. These steps were maintained during the subsequent charge and discharge cycles, showing that the structure of this material did not change during the redox reactions.

The compound $[N(CH_3)_4]_{0.5}Zn_{0.4}V_2O_5$, before lithium intercalation, has an initial cell voltage of 3.4. A maximum of 1.2 lithium could be incorporated into the structure on discharge consistent with the vanadium being reduced to the +4 oxidation state.

The curve of the electrochemical intercalation of lithium into $Zn_3(OH)_2(V_2O_7) \cdot H_2O$ compound, after removal of the interlayer water, suggests that the oxidation state of the vanadium is between 5 and 4, as the cell open circuit voltage is 3.38 V. A maximum of 1.05 lithium above a cut-off of 2 volts could be inserted into the structure on discharge suggesting that only a partial reduction of the vanadium took place. The existence of the hydroxide group in this compound only appears to impede the lithium cycling with only 0.55 lithium being discharged and charged after the first cycle.

The essentially three-dimensional bonding in $Zn_2(OH)_3(VO_3)$, has an interlayer spacing of 4.789Å. Thus the possibility for the insertion of lithium is relatively small compared to the first three compounds. This was confirmed by electrochemical cycling, which shows that only 0.013 lithium could be incorporated into the structure of $Zn_2(OH)_3(VO_3)$ above 2 volts.

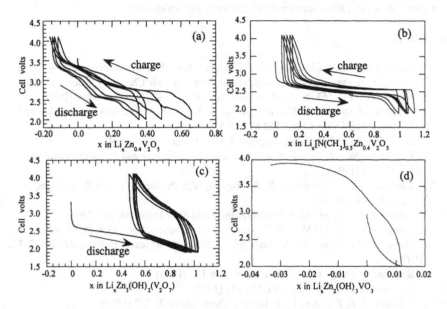

Fig. 3. Electrochemical cycling of (a) $Zn_{0.4}V_2O_5 \cdot 0.3H_2O$, (b) $[N(CH_3)_4]_{0.5}Zn_{0.4}V_2O_5$, (c)$Zn_3(OH)_2(V_2O_7) \cdot H_2O$, and (d) $Zn_2(OH)_3(VO_3)$.

CONCLUSIONS

Mild hydrothermal synthesis has been shown to be a powerful technique in the synthesis of novel compounds. We have successfully synthesized a series of novel zinc vanadium oxides, $Zn_{0.4}V_2O_5 \cdot 0.3H_2O$, $[N(CH_3)_4]_{0.5}Zn_{0.4}V_2O_5$, $Zn_3(OH)_2(V_2O_7) \cdot H_2O$, and $Zn_2(OH)_3(VO_3)$, by using this technique. Several have open crystalline structures. $Zn_{0.4}V_2O_5 \cdot 0.3H_2O$, $[N(CH_3)_4]_{0.5}Zn_{0.4}V_2O_5$ are electrochemically active, readily and reversibly undergoing redox reactions with lithium in non-aqueous cells. $Zn_3(OH)_2(V_2O_7) \cdot H_2O$, is partially electrochemically active due to the reaction between hydroxide group and lithium. Moreover, this very open structure compound may well be the model for forming active cathode materials, when the zinc is replaced by other redox active transition metal ions such as manganese. In contrast $Zn_2(OH)_3(VO_3)$ did not show electrochemical activity.

ACKNOWLEDGMENT

We thank the Department of Energy through Lawrence Berkeley Laboratory and the National Science Foundation through grant DMA-9422667 for partial support of this work. We also thank Professor Richard Naslund for the use of the DCP atomic emission spectrometer, and Bill Blackburn for the electron microprobe studies.

REFERENCES

1. M. S. Whittingham, J. Li, J. Guo and P. Zavalij, "Hydrothermal Synthesis of New Oxide Materials using the Tetramethyl Ammonium Ion," in Soft Chemistry Routes to New Materials, Vol. 152-153, p. 99 Trans Tech Publications Ltd., Nantes, France (1994)

2. P. Y. Zavalij, J. Guo, M. S. Whittingham, R. A. Jacobson, V. Pecharsky, C. Bucher and S.-J. Hwu, J. Solid State Chem. 123 83 (1996)

3. J.Guo, P. Zavalij and M. S. Whittingham, Chem. Mater. 6, 357 (1994)

4. M.S. Whittingham, J. Guo, R. Chen, T. Chirayil, G. Janauer and P.Zavalij, Solid State Ionics 75, 257 (1995)

5. P. Zavalij, M.S. Whittingham, E. A. Boylan, V.K.Pecharsky and R.A.Jacobson, Z. Kryst. 211, 464 (1996)

6. T. Chiryil, P. Zavalij and M. S. Whittingham, Solid State Ionics 84, 163 (1996)

7. T. Chirail, P. Zavalij and M. S. Whittingham, J. Electrochem. Soc 143, L143 (1996)

8. F. Zhang, P. Zavalij and M. S. Whittingham, Materials Research Bulletin, 32, 701-707 (1997)

9. D. Riou and G. Ferey, J. Solid State Chem. 120 173 (1995)

10. D. Riou and G. Ferey, Inorg.Chem. 124 151 (1996)

11. L. F. Nazar, B. E. Koene and J. F. Britten, Chem. Mater. 8, 327 (1996)

12. Y. Zhang, J. R. D. Debord, C. J. O'Connor, R. C. Haushalter, A. Clearfield and J. Zubieta, Angew.Chem. Int. Ed. Engl 35, 989 (1996)

13. T. Chirayil, P. Y. Zavalij, and M. S. Whittingham, J. Mater. Chem. 7, 2193 (1997)

14. T. Chirayil, P. Y. Zavalij and M. S. Whittingham, Chem. Commun., 33, (1997).

15. L. Nazar, University of Waterloo, Personal Communication (1196)

16. M. B. Dines, Mater. Res. Bull. 10, 287 (1975)

17. J. Galy, J. Solid State Chem. 100, 229 (1992)

18. P. Y. Zavalij, F. Zhang and M. S. Whittingham, Acta Cryst. C53, 1738 (1997)

STRUCTURAL PHASE TRANSFORMATIONS IN V_2O_5 THIN FILM CATHODE MATERIAL FOR Li RECHARGEABLE BATTERIES

J.M. MCGRAW*, J.D. PERKINS**, J.-G. ZHANG**, P.A. PARILLA**, T.F. CISZEK**, M.L. FU***, D.M. TRICKETT*, J.A. TURNER**, and D.S. GINLEY**
*Colorado School of Mines, Golden, CO, USA
**National Renewable Energy Laboratory, Golden, CO, USA
***University of Colorado, Boulder, CO, USA

ABSTRACT

We report on investigations of V_2O_5 thin film cathodes prepared by pulsed laser deposition and the phase transformations which occur during electrochemical cycling. Our experimental results on PLD-grown, textured V_2O_5 crystalline films concur with reports in the literature that there is a voltage threshold above which, cycling appears to be completely reversible and below which, cycling appears to be irreversible. Crystalline films discharged beyond the threshold to 2.0 V vs. Li exhibited an immediate and continuous fade in capacity as well as a ~90% decrease in XRD peak intensity and a similar decrease in Raman signal intensity in as few as ~10 cycles. We have made ω-phase material by both electrochemically discharging virgin, crystalline V_2O_5 and by further discharging previously cycled films which showed irreversible structural changes.

INTRODUCTION

The high cell potentials, high power densities, and rechargeability of the bronze-forming transition-metal oxides make them very attractive candidates for cathode materials in Li rechargeable batteries. The ability to intercalate Li depends on the openness of the material structure. For crystalline material, it must be layered, channeled, or both. Electrical conductivity is also required. Assuming these fundamental requirements are met, one of the most critical parameters in evaluating materials is stability to repeated electrochemical cycling. The degree of reversibility of a discharge reaction depends upon the extent of the irreversible structural damage incurred by the host material.[1] The V_2O_5 structure and the phase transformations upon lithiation have been studied extensively.[2-9]

The open circuit voltage of V_2O_5 is ~3.5 V vs. Li. The α- δ- and γ-$Li_xV_2O_5$ phases are shown in Figure 1. The α-$Li_xV_2O_5$ structure can be described as being made up of alternating corner- and edge-sharing VO_5 square pyramids in a layered structure with two pyramids pointing upwards followed by two pyramids pointing downwards. The α-$Li_xV_2O_5$ phase occurs for Li intercalation of x<0.1. In the δ-$Li_xV_2O_5$ phase (0.9<x<1.0), some puckering of the layers occurs, however no major bonds are broken. This puckering recedes upon de-intercalation. Cycling between the α- and δ-phases is completely reversible.[7,8] Intercalating more than one Li results in the γ-$Li_xV_2O_5$ phase (1<x<2). Major, irreversible structural changes occur and charge capacity decreases, indicating poor reversibility. The pristine α-phase can be regained by heating the γ-phase to 340 °C.[5] The poor cycling reversibility of the γ-phase material significantly hinders its performance. In bulk powders, the lower voltage limit of reversible cycling has been determined to be ~2.3 V.[10] Above this threshold voltage, cycling appears to be completely reversible. When discharged below this voltage, charge capacity decreases, indicating irreversibility. When discharged to intercalate 3 Li per V_2O_5, the ω-phase is attained.

Mat. Res. Soc. Symp. Proc. Vol. 496 © 1998 Materials Research Society

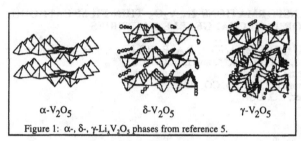

α-V₂O₅   δ-V₂O₅   γ-V₂O₅

Figure 1: α-, δ-, γ-Li$_x$V₂O₅ phases from reference 5.

The ω-Li$_x$V₂O₅ phase (x=3) has a Rocksalt structure, distinct from the aforementioned phases. In bulk powder experiments, the material formed after the first cell discharge is a superstructure characterized by a tetragonal unit cell. Subsequent cycles lead to a cubic system (a_{cub} = 4.11 Å) in which the oxygen are arranged in an fcc array with a statistical distribution of V and Li in the octahedral sites.[11] This structure yields a dense packing fraction of Li in V₂O₅ with 1.5 Li per V. The ω-phase exhibits good cyclability.

We have grown (h00) textured V₂O₅ thin films by pulsed laser deposition (PLD). We present here the results of experiments investigating the cycling properties and structural changes observed in voltage-limited experiments.

We found that, like bulk powders, crystalline V₂O₅ thin films have a voltage threshold above which, cycling is completely reversible and crystal structure is preserved. When discharged below this threshold, the long-range order quickly breaks down, but local order is retained. We have made ω-phase material by both electrochemically discharging virgin, crystalline V₂O₅ and by further discharging previously cycled films which showed irreversible structural changes. In both instances, the cubic form of the ω-phase formed without first forming the tetragonal superstructure.

EXPERIMENTAL PROCEDURE

Thin Film Deposition

Films were deposited by PLD in a 12-inch spherical vacuum chamber. A Lambda Physik KrF (λ=248 nm) EMG 103 laser delivered ~12 J/cm² at 20 Hz to the target. The target was sintered V₆O₁₃ produced in-house and the phase was confirmed by XRD. The substrates were conductive, polycrystalline SnO₂ on float glass (Cherry Display Products Corp.). The deposition temperature was 200 °C. The deposition atmosphere was 20 mTorr O₂.

Film Characterization

Electrochemical Cycling
Electrochemical cycling experiments were performed in an Ar filled glove box. The reference and counter electrodes were Li foil. The electrolyte solution was 1 M LiClO₄ in propylene carbonate (PC). All experiments were performed under constant current conditions with various currents and between various voltage ranges. For further characterization, some samples were cycled, then rinsed with neat PC followed by isopropyl alcohol and dried, after which XRD and Raman spectroscopy measurements were taken. These samples were then returned to the electrochemical cell for further cycling.

X-ray Diffraction

X-ray diffraction was performed on a 4-circle Scintag X-1 Diffractometer with a Cu-Kα anode source operating at 45 kV/40 mA. A 2 mm pin-hole collimator was used to ensure a constant irradiation area of the electrochemically cycled area. Analysis was performed in air.

Raman Spectroscopy

Raman scattering measurements were performed in a 180° back-scattering configuration with an Ar-ion laser operating at 514.5 nm and a Spex 1877 Triplemate spectrometer with a CCD detector. A polarized incident beam, ~4 mW, was focused through a microscope objective to a spot size of ~2 μm. Analysis was performed in air.

EXPERIMENTAL RESULTS AND DISCUSSION

Stepped Voltage Experiment

In order to probe the limit of reversible intercalation in crystalline thin films, a crystalline V_2O_5 film was cycled in three different voltage ranges. The predominantly (h00) textured, orthorhombic V_2O_5 film was first cycled in the range 4.1 - 3.3V, then 4.1 - 3.0V, and finally 4.1 - 2.0V vs. Li. After each 50 cycles, the film was analyzed by XRD and Raman spectroscopy. The Li ion capacity vs. cycle number is plotted in Figure 2. The capacity of the film when cycled to 3.3 V and to 3.0 V does not degrade with cycling. This indicates that cycling in this voltage range is completely reversible. When cycled to 2.0 V, an immediate and continuous decrease in capacity is observed. While cycling to 2.0 V shows a degree of irreversibility, it is not catastrophic. Even after 200 cycles to 2.0 V, the film retains roughly half of the initial charge capacity.

The crystal structure of this film was probed by XRD and Raman after each 50 cycles and plots are shown in Figure 3. In the XRD, unmarked peaks correspond to the SnO_2 substrate and marked peaks correspond to orthorhombic V_2O_5. The predominantly (h00) texturing is clear. After cycling 50 times to 3.3 V plus an additional 50 times to 3.0 V, the XRD pattern of the film is

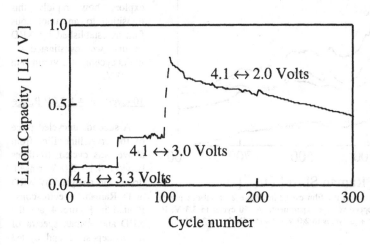

4.1 ↔ 2.0 Volts

4.1 ↔ 3.0 Volts

4.1 ↔ 3.3 Volts

Figure 2: Li Ion Capacity vs. Cycle Number for the stepped voltage experiment. The film was cycled in the indicated voltage ranges at constant current of 100 μA/cm². Capacity is constant when cycled to 3.3 V and to 3.0 V, but decreases immediately when cycled to 2.0 V.

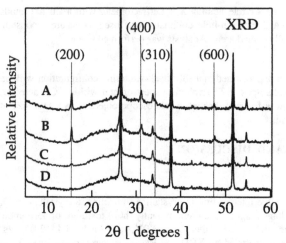

Figure 3a: XRD at various points during cycling in the stepped voltage experiment of Figure 2. Unmarked peaks are associated with the SnO_2 substrate. Marked peaks correspond to V_2O_5, showing the predominantly (h00) texturing. A: As-deposited; B: 50 cycles to 3.3 V + 50 cycles to 3.0 V; C: +50 cycles to 2.0 V; D: +151 cycles to 2.0 V.

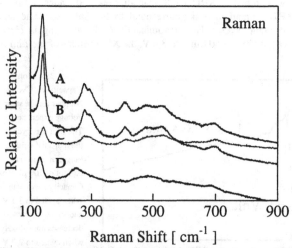

Figure 3b: Raman spectra of the film cycled in Figure 2 at various points during cycling in the stepped voltage experiment. A: 50 cycles to 3.3 V; B: +50 cycles to 3.0 V; C: +50 cycles to 2.0 V; D: +151 Cycles to 2.0 V.

very similar to the as-deposited film, indicating that no significant structural changes occurred in this limited cycling range. The Raman spectra in this same cycling range exhibits no significant difference between the cycling to 3.3 V and cycling to 3.0 V.

Significant differences occurred when the film was cycled to 2.0 V. After the first 50 cycles to 2.0V, nearly 90% of the V_2O_5 XRD peak intensity is lost. After 200 cycles to 2.0 V, there are no distinguishable peaks. The Raman spectra also show large decreases in intensity, however, unlike the XRD peaks, there continues to be a discernible signal even after 200 cycles to 2.0 V. This indicates that some local order remains in spite of the destruction of the gross crystal structure. To explore how rapidly this transition to an amorphous film as established by XRD occurs, we investigated the initial cycling of a virgin film to 2.0 V.

10 Cycles in 4.1-2.0V Range

A second, uncycled piece of the crystalline film from above was cycled 10 times between 4.1 - 2.0 V, rinsed, dried, and analyzed by XRD and Raman spectroscopy. Plotted in Figure 4 are the XRD and Raman spectra of the as-deposited and cycled film. After only 10 cycles, there is ~90% decrease in the XRD peak intensity. The Raman spectrum shows significant decrease in signal intensity to levels similar to those observed in Figure 3. These results suggest that changes in crystal structure occur in fewer than 10 cycles when cycled to 2.0 V. This indicates that films typically cycled to deep discharge develop an amorphous character.

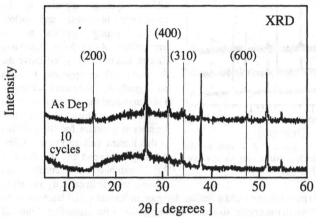

Figure 4a: Raman spectra of the crystalline V_2O_5 film comparing the film as-deposited and after being cycled 10 times between 4.1-2.0 V. Marked peaks correspond to V_2O_5.

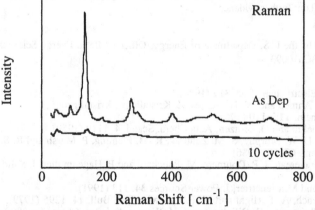

Figure 4b: Raman spectra of the crystalline V_2O_5 film comparing the film as-deposited and after being cycled 10 times between 4.1 - 2.0 V.

ω-Li$_x$V$_2$O$_5$

One piece of crystalline film and the film which had been cycled 10 times between 4.1-2.0 V were each discharged to 1.2 V and the voltage held until the current was <5 μA to ensure equilibrium distribution of Li in the films. Each film was rinsed, dried, and analyzed by XRD. The XRD of the films together with a blank reference SnO$_2$ substrate is shown in Figure 5. Both films exhibit a broad peak around 44° and no other non-substrate peaks are distinguishable. The strongest ω-Li$_3$V$_2$O$_5$ peak is the (200) peak and occurs at a 2θ value of 44.14°[11] with Cu $K\alpha$ radiation. The original films were both (200) orthorhombic textured. It is possible that the resulting material retains its texture transforming into (200) cubic textured ω-Li$_3$V$_2$O$_5$ in which case, only the (200) peak would be expected in this 2θ range. It is consistent that the peak observed in both films is due to ω-phase material; however, this cannot be definitively concluded without further evidence.

377

CONCLUSIONS

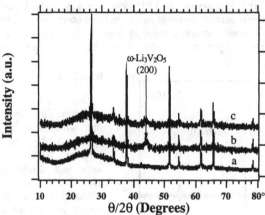

Figure 5: XRD of the films discharged to 1.2 V showing the development of ω-Li$_3$V$_2$O$_5$. (a) SnO$_2$ substrate; (b) As-deposited, crystalline film; (c) Film cycled 10 times, from Figure 4.

Our experimental results on crystalline films grown by PLD concur with reports in the literature that there is a voltage threshold above which, cycling appears to be completely reversible and below which, cycling appears to be irreversible. As evidence, XRD and Raman data of films cycled above the threshold voltage appeared to have no significant structural changes. This supported the capacity vs. cycle number data which showed that capacity is constant for films cycled in this limited voltage range. Films discharged beyond the threshold to 2.0 V exhibited an immediate and continuous fade in capacity as well as a nearly 90% decrease in XRD peak intensity and a similar decrease in Raman signal intensity in as few as 10 cycles. The breakdown in the crystal structure did not result in the immediate failure of the cathode material, only a continuous decline in charge capacity.

Films which were cycled to an essentially amorphous state as well as (200)-textured, crystalline, as-deposited films developed a new peak near 44° after having been discharged to 1.2 V. It is consistent that the peak observed in both films is due to ω- Li$_3$V$_2$O$_5$, however, this cannot be definitively concluded without further evidence.

ACKNOWLEDGMENTS

This work was supported by the U.S. Department of Energy, Office of Basic Energy Sciences under contract #DE-AC36-83Ch10093

REFERENCES

1. M. S. Whittingham, J. Electrochem. Soc. 315 (1976).
2. H. G. Bachmann, F. R. Ahmed and W. H. Barnes, Z. Kristallogr., Kristallgeom., Kristallphys., Kristallchem. **115**, 110 (1961).
3. A. Byström, K. A. Wilhelmi and O. Brotzen, Acta Chem. Scand. **4**, 1119 (1950).
4. R. J. Cava, A. Santoro, D. W. Murphy, S. M. Zahurak, R. M. Fleming, P. Marsh and R. S. Roth, J. Solid State Chem. **65**, 63 (1986).
5. J. M. Cocciantelli, P. Gravereau, J. P. Doumerc, M. Pouchard and P. Hagenmuller, J. Solid State Chem. **93**, 497 (1991).
6. C. Delmas, S. Bréthes and M. Ménétrier, J. Power Sources **34**, 113 (1991).
7. P. G. Dickens, S. J. French, A. T. Hight and M. F. Pye, Mat. Res. Bull. **14**, 1295 (1979).
8. D. W. Murphy, P. A. Christian, F. J. DiSalvo and J. V. Waszczak, Inorg. Chem. **18**, 2800 (1979).
9. B. Pecquenard, D. Gourier and N. Baffier, Solid State Ionics **78**, 287 (1995).
10. A. Shimizu, T. Tsumura and M. Inagaki, Solid State Ionics **63-65**, 479 (1993).
11. C. Delmas, H. Cognac-Auradou, J. M. Cocciantelli, M. Ménétrier and J. P. Doumerc, Solid State Ionics **69**, 257 (1994).

HIGH PERFORMANCE VANADIUM OXIDE THIN FILM ELECTRODES FOR RECHARGEABLE LITHIUM BATTERIES

Ji-Guang Zhang, Ping Liu, C. Edwin Tracy, David K. Benson and John A. Turner,
National Renewable Energy Laboratory, Golden, CO 80401

ABSTRACT

Plasma Enhanced Chemical Vapor Deposition (PECVD) was used to prepare vanadium oxide thin films as cathodes for rechargeable lithium batteries. The reactants consisted of a high vapor pressure liquid source of vanadium ($VOCl_3$) and hydrogen and oxygen gas. Deposition parameters such as the flow rates of H_2, O_2 and $VOCl_3$, the substrate temperature and the Rf power were optimized, and high deposition rate of 11 Å/s was obtained. Vanadium oxide films with high discharge capacities of up to 408 mAh/g were prepared. The films also showed a superior cycling stability between 4 and 1.5 V at a C/0.2 rate for more than 4400 cycles. The films were amorphous up to a deposition temperature of 300°C, however, deposition on to substrates with textured surfaces facilitated the formation of crystalline films. We demonstrate that both the vanadium oxide material and the PECVD deposition method are very attractive for constructing thin-film rechargeable lithium batteries with high capacity and long-term cyclic stability.

INTRODUCTION

The miniaturization of electronic devices has resulted in very low current and power requirements for batteries. Consequently, thin film rechargeable lithium batteries are being developed as the power source for these devices. An advantage of this type of batteries is that they can be incorporated into the same integrated circuit with other electronic elements. Electrode materials with high resistance can also be employed due to the thin layer structure. This fact is evident in the case of vanadium oxide, which will be discussed in detail in this report.

After more than twenty years of extensive research, the choice of cathodes for lithium or lithium ion batteries is now mainly limited to several transition metal oxides within the first series of the periodic table, such as: vanadium oxides (V_2O_5 and V_6O_{13}), manganese oxide and lithiated manganese oxide (MnO_2 and $LiMn_2O_4$), lithium cobalt and nickel oxides ($LiCoO_2$ and $LiNiO_2$) [1-3]. A recent trend is to dope these materials with other elements such as Cu, Al, Cr, Zn, etc., either to improve the cyclability or to generate a higher working voltage [4].

Research efforts on using vanadium oxide as a cathode for rechargeable lithium batteries were started in the late seventies [5]. The lack of wide acceptance of using this material compared to other cathode materials had been due to its low electronic conductivity and poor cyclability. Extensive phase transitions due to lithium intercalation were observed in both V_2O_5 and V_6O_{13} which might have partially contributed to their poor cyclability. However, the high capacity and energy density of vanadium oxide has led to a continuous effort to further improve its electrochemical performance. In terms of thin film batteries, several deposition methods including magnetron sputtering, electron beam evaporation, sol-gel/spin coating and thermal evaporation have been used to prepare vanadium oxide cathodes [6-9] with varying success.

In an attempt to improve the charge capacity and cyclability of thin film vanadium oxide cathodes, we chose to use the plasma enhanced chemical vapor deposition (PECVD) method. Our results show that through the optimization of the deposition parameters, improved

Mat. Res. Soc. Symp. Proc. Vol. 496 © 1998 Materials Research Society

microstructure can be obtained which leads to a high capacity and highly reversible thin film cathode.

EXPERIMENTAL

Vanadium oxide films were prepared using a 13.56 MHz capacitively coupled PECVD process with 24 cm diameter parallel-plate electrodes. Vanadium oxytrichloride ($VOCl_3$) was used as the primary reactant precursor, with argon as the carrier gas. The precursor temperature was controlled at 10°C, and the vapor pressure of the precursor was ~9 torr. The pressure of the argon carrier gas was 15 psig (1406 torr in Golden, Colorado), so that the partial pressure of the vanadium oxytrichloride was ~0.64 % of the Ar/$VOCl_3$ mixture. This mixture was then combined with hydrogen and oxygen before entering the plasma region of the reaction chamber. The flow rates of the reactants were controlled using mass flow controllers, and partial pressures of hydrogen, oxygen, and $VOCl_3$ vapor in the chamber were varied to optimize film properties. The weights of the substrates (SnO_2) were measured before and after film deposition using a high-precision microbalance to determine the net weight of the vanadium oxide films. Substrate temperature (during film deposition) was varied between 30°C and 300°C. Film thicknesses were measured with a DekTak profilometer after deposition. The films were stored and tested inside a controlled-environment glove box under argon atmosphere. The water and oxygen concentrations in the dry-box were measured to be less than 1 ppm and 1.5 ppm, respectively. The samples were then tested in an electrochemical cell with a three-electrode configuration. Lithium metal was used as both the counter and reference electrodes, and a solution of 1M of $LiClO_4$ in propylene carbonate (Merck & Co., Inc.) was used as the electrolyte. Samples with active areas of ~1 cm^2 were cycled between 4.0 volts and 1.8 (or 1.5) volts at 25 °C and tested at discharge/charge rates between C/0.1 to C/1.0 (where the C/1 rate is defined as the insertion of one mole lithium in one mole V_2O_5 per hour). The electrochemical experiments were performed using a computer-controlled battery-testing system (BT2043, Arbin Instruments Corp.). Film crystallinity was characterized by X-ray diffraction (XRD) on a 4-circle Scintag X-1 diffractometer with a Cu-Kα anode source. Film stoichiometry was measured by X-ray photoelectron spectroscopy (XPS).

RESULTS AND DISCUSSION

We started our systematic approach with the optimization of the H_2 flow rate, using the deposition rate as the criterion for guidance. The flow rates of vanadium oxytrichloride ($VOCl_3$) and oxygen (O_2) were held constant at 3.2 and 3.8 sccm, respectively, while the hydrogen flow rate was varied between 14 and 42 sccm. All of the films were deposited at a substrate temperature of 30°C for 10 min, and the RF power was controlled at 50 W. The chamber pressure during deposition was maintained at 0.6 torr.

The film deposition rate exhibits a maximum of 11 Å/s at a H_2 flow rate of 28 sccm and is more than 5 times higher than those reported for vanadium oxide films prepared by thermal evaporation (~2 Å/s) [9]. Another interesting property of these films is their high density. For the films deposited at a H_2 flow rate of 28 sccm, the density of the film varied from 3.1 to 3.3 g/cm^3. For comparison, bulk V_2O_5 and V_6O_{13} have a density of 3.36 and 3.90 g/cm^3 respectively. Considering that the density of thin-film material is usually 10% to 20% less than that of bulk material, our vanadium oxide films have a density close to that of bulk V_6O_{13}. This high density has an important implication since it is believed that energy density (Wh/l) is more important than specific energy (Wh/kg) in many applications, such as batteries for portable computers.

We then fixed the hydrogen flow at 28 sccm and optimized the oxygen flow rate, using charge-discharge capacity as the criterion for guidance. The flow rate of vanadium oxytrichloride (VOCl₃) was held constant at 3.2 sccm, while the oxygen flow rate was varied from 3.8 to 20 sccm. The substrate temperature was again kept at 30°C. The charge-discharge data between 4.0 and 1.5 V indicate that the charge capacity increases with increasing oxygen flow rate up to 10.5 sccm and remains unchanged afterwards when the other deposition conditions are held constant. The discharge capacity of our material exceeds 408 mAh/g (or 1.265 mAh/cm³) under these discharge conditions as shown in Fig. 1. The energy density for this film (~5000 Å thick) is 960.3 Wh/kg and is close to the best value reported for a cathode material in the literature [10]. The featureless shape of the discharge curve indicates the film is amorphous which is confirmed by XRD measurements.

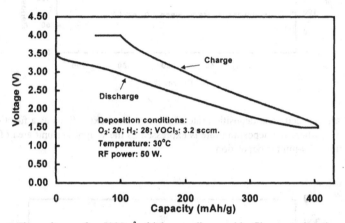

Figure 1   The voltage of a 5000 Å thick vanadium oxide film as a function of discharge capacity.

The effect of RF power on both deposition rate and charge-discharge capacity was also investigated. The flow rates of VOCl₃, hydrogen and oxygen were fixed at 3.2, 28, and 7.5 sccm, respectively, while the RF power applied to the electrode was varied from 10 to 100 W. Figure 2 shows the lithium charge capacity and the deposition rate as functions of the RF power, where the deposition rate is expressed as the weight per unit area of samples deposited in 10 minutes. Although the maximum deposition rate is achieved at a power of 20 W, the charge capacity of the film increases up to a power of 50 W.

Optimized films were prepared using the following gas flow rates: VOCl₃ = 3.2 sccm, O₂ = 15 sccm, H₂ = 28 sccm. The actual partial pressures of the reactant gases in the chamber were measured as follows: P_{VOCl3} = 4 mtorr, P_{O2} = 18 mtorr, P_{H2} = 31 mtorr. The other deposition parameters were: RF power = 50 W, chamber pressure = 600 mtorr, and substrate temperature = 30°C. Figure 3 shows the long-term cyclic stability of an optimized 5000 Å thick VOₓ film prepared under the above deposition conditions. The film was tested at a relatively high charge/discharge rate (C/0.2) and shows negligible degradation after an initial 15% decrease during the first two cycles which is common to vanadium oxide materials [10]. The optimized VOx thin film has been cycled for more than 4400 cycles with no signs of degradation and is still under test.

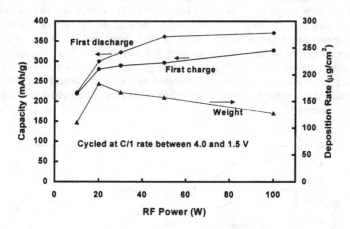

Figure 2   Charge capacity and deposition rate of the vanadium oxide films as a function of RF power, where the deposition rate is expressed as the weight per unit area of samples during a 10-minute deposition.

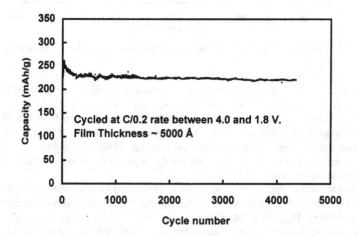

Figure 3   The long-term cyclic stability of an optimized 5000 Å thick vanadium oxide film cycled between 4.0 and 1.8 V in a high charge/discharge rate (C/0.2).

We also investigated the effect of the substrate temperature during deposition. In the pulsed laser deposition process, increasing the temperature above 200°C usually led to the formation of crystalline films [9]. However, in our PECVD process amorphous VOx films were always formed on smooth SnO₂ substrates up to temperatures of 300°C. This fact is evident in both the nature of discharge curve and the results from XRD data. To induce the growth of crystalline films, substrates with textured surfaces of SnO₂ were used. When the substrate temperature was raised above 200°C, the discharge curve began to show several plateaus which are characteristic of crystalline vanadium oxide. Fig. 4 shows the charge-discharge curve for the first two cycles. After the sample is discharged to 1.8V, the following charge and discharge curves are featureless which is in agreement with previous observations on crystalline vanadium oxide [10].

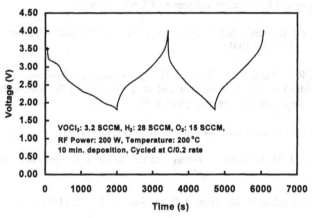

Figure 4   Charge-discharge curves of a crystalline vanadium oxide thin film electrode.

CONCLUSIONS

To our knowledge, we are the first to report the successful use of the PECVD technique to prepare thin films of vanadium oxide. Optimization of the deposition conditions has allowed us to obtain battery cathode materials with very high charge-discharge capacity. These films also show a high stability during repeated cycling. While amorphous films are formed under most conditions, the use of high temperature (> 200°C) and textured-surface substrates induces the crystallization of vanadium oxide. Our results show that both the VOx material and the PECVD method are very attractive choices for use in the construction of rechargeable lithium batteries.

ACKNOWLEDGEMENTS

Dr. Alan Kibbler and David Niles are gratefully acknowledged for valuable discussions and help on the set up of PECVD system and XPS measurements, respectively. This work was supported by the Advanced Battery Program, Office of Basic Energy Sciences, Department of Energy.

REFERENCES

1. A.N. Dey, Thin Solid Films, **43**, 131 (1977).

2. G. Pistoia, J. Power Sources, **9**, 307 (1983).

3. K.M. Abraham, J. Power Sources, **7**, 1 (1982).

4. Y. Ein-Eli and W.F. Howard, Jr., J. Electrochem. Soc., **144**, L205 (1997) and references therein.

5. D. W. Murphy and P.A. Christian, Science, **205**, 651 (1979).

6. J.B. Bates, G.R. Gruzalski, N.J. Dudney, C.F. Luck, X.H. Yu, and S.D. Jones, Solid State Technology, p.59, (July 1993).

7. S.F. Cogan, R. D. Rauh, T. D. Plante, N. M. Nguyen, and J.D. Westwood, in Electrochromic Materials, edited by M. K. Carpenter and D. A. Corrigan (The Electrochemical society Proceedings Series, Pennington, NJ, 1990), p. 99.

8. H.K. Park and W.H. Smyrl, J. Electrochem. Soc., **141**, L25 (1994).

9. J. -G. Zhang, J. M. McGraw, J. Turner and D. Ginley, J. Electrochem. Soc., **144**, 1630 (1997).

10. C. Delmas, S. Brethes, M. Ménétrier, J. Power Sources, **34**, 113 (1991).

ELECTROCHEMICAL LI INSERTION IN LAMELLAR (*BIRNESSITE*) AND TUNNEL MANGANESE OXIDES (*TODOROKITE*)

M J. DUNCAN, F. LEROUX, and L.F. NAZAR[*]
University of Waterloo, Department of Chemistry, Waterloo, Ontario Canada N2L 3G1;
lfnazar@uwaterloo.ca

ABSTRACT

A comparison of Li insertion in manganese oxide phases with a tunnel (todorokite) framework, its two-dimensional layered precursor (birnessite/buserite), and Li-exchanged materials are presented. The results outline the effect of the MnO_6 octahedral arrangement and framework composition on the electrochemical response. The interlayer cations in the lamellar materials are exchangeable for Li, giving rise to a lithiated birnessite that displays a sustainable capacity of 125 mAh/g. For todorokite, molten salt exchange using $LiNO_3$ results in displacement of water from the tunnels, and incorporation of additional Li into the structure. Some of this Li is extractable during charge, resulting in a reversible capacity of 172 mAh/g in the voltage window 4.2-2.0V.

INTRODUCTION

Host structures based on assemblies of Mn oxide polyhedra can be classified into spinels (3D), lamellar phases (2D), and tunnel structures (1D). Within these classes, subtypes can be defined based on the size of path. In the case of the hydrated lamellar phases, two general phases can be defined. Those with one layer of water in the interstitial space such as Na-birnessite give rise to a 7Å interlayer spacing. Exchange of the Na^+ for Mg^{2+} results in accomodation of a second water layer and hence an increase in the basal spacing to ~10Å. Its electrochemical properties have not been examined, in contrast to the 7Å birnessites, where the Li capacities exhibit wide discrepancies, possibly as a result of compositional and structural variations as a result of synthesis method.[1]

1D tunnel structures have also been extensively studied over the past years, although new developments in this area offer intriguing possibilities for revisiting the chemistry. Reversible insertion has been reported in $Na_{0.44}MnO_2$,[2] and α-MnO_2,[3] the latter of which adopts the hollandite structure comprised of edge-sharing MnO_6 octahedra that span "2x2" channels. An initial insertion capacity of ~0.5 Li/Mn was reported for this material, although cycling is not highly reversible. A more spacious tunnel framework is obtained by hydrothermal transformation of lamellar Mg-buserite, whereupon the MnO_6 octahedra reorganize to form a "3x3" channel framework structure analogous to the mineral todorokite.[4] Cations and water molecules are accomodated within these tunnels. There are two synthetic methods for the preparation of todorokite, reported by Golden et al.,[4] and Shen et al.[5] The difference lies in the synthesis of the layered precursor Na-birnessite; namely in the method used to oxidize the Mn^{II} starting material to the birnessite; and whether or not it results in the incorporation of Mg into the birnessite framework. Here, we report the first investigation of the electrochemical properties of Li intercalation exhibited by this large tunnel (3x3) Mn phase, todorokite. The similarities and differences of the electrochemical response between the 2D and 3D materials are presented, in addition to the enhanced perfomance obtained by ion-exchange of the interstitial cations for Li.

EXPERIMENTAL

Synthesis and characterization. Phyllomanganates were prepared by two literature methods, using either O_2-oxidation (O_2-BIR), or conproportionation using $Mg(MnO_4)_2$ as the oxidizing agent (MnO_4-BIR). Ion-exchange of the Na^+ for Mg^{2+} in both Na-birnessites was achieved by dispersing the solid in 1L of 1M $MgCl_2$ solution, to give the resultant Mg-buserites (O_2-BUS and MnO_4-BUS). Li exchange of the interstitial cations was carried out by dispersing the as-synthesized material with a 50-fold molar excess of Li (as a 0.3M LiCl solution). This was repeated thrice, isolating the material each time. Todorokites were prepared by hydrothermal treatment of layered Mg-buserite. Approximately 0.5g of wet O_2-BUS and MnO_4-BUS in their Mg-exchanged forms were resuspended in 16ml of DW in a Teflon liner of a 23mL capacity Parr bomb. These suspensions were autoclaved for 5 days at $170^{\circ}C$, and the black precipitates recovered and washed to yield O_2-TDK and MnO_4-TDK.

Instrumentation. X-ray powder diffraction (XRD) patterns were obtained on random powder mounts in a Siemens D500 diffractometer equipped with a diffracted beam monochromator, using Cu-Kα radiation. The transformation of the layered to tunnel framework was confirmed by high resolution TEM studies. A Varian Liberty 100-ICP-AES spectrometer was used to determine the atomic ratio of alkali (Na, Mg,) to Mn. Oxidation state measurements were performed following the method by Vetter and Jaeger.[6] Thermogravimetric analysis was carried out on a PL Thermal Sciences STA 1500 thermal analysis system at a heating rate of $10^{\circ}C$/min from room temperature up to $700^{\circ}C$ to obtain the full stoichiometry. Surface areas of ~45m^2/g were measured for both the phyllomanganates and tektomanganates by the BET method. Samples were degassed at $80^{\circ}C$ under vacuum for 24 hours prior to analysis.

Electrochemical Li Insertion. Samples were ground through a 50μm mesh to ensure a narrow distribution of particle sizes. The active material was then mixed with 20% acetylene black and 5% KYNAR FLEX® 2820-00 binder, and cyclopentanone was added to form a slurry. This mixture was coated onto an Al disc (1.44 cm^2) and dried by allowing the solvent to evaporate. Typically, each cathode constituted 1-2 mg of active material. Li insertion was performed on active material pretreated at $180^{\circ}C$ prior to battery assembly. To ensure there was no subsequent uptake of water during the preparation, the coated Al discs were then reheated at $180^{\circ}C$ before being introduced into the glove box. Propylene carbonate (PC, Aldrich) was distilled under vacuum before being used to make up the electrolyte, 1M $LiClO_4$.

The cell components were transferred into an Ar atmosphere-filled glove box and assembled in a Swagelock housing-type cell. The electrochemical testing was studied under galvanostatic control using a Mac-Pile™ system. The voltage limits set for galvanostatic cycling were 4.2-2.0V, with a typical cycling rate of C/20 (15-20 μA/cm^2).

RESULTS AND DISCUSSION

Table 1. Stoichiometries (reported per Mn) for MnO_4-and O_2-$180^{\circ}C$ treated materials.

Sample	Composition	Mn^{x+}
MnO_4 -BIR	$Na_{0.18}Mg_{0.15}MnO_{2.06}\cdot0\cdot35H_2O$	3.64
MnO_4 -BUS	$Mg_{0.25}MnO_{2.07}\cdot0\cdot59H_2O$	3.63
MnO_4 -TDK	$Mg_{0.24}MnO_{2.06}\cdot0\cdot52H_2O$	3.64
O_2 -BIR	$Na_{0.25}MnO_{1.97}\cdot0\cdot17H_2O$	3.69
O_2 -BUS	$Mg_{0.18}MnO_{1.94}\cdot0\cdot27H_2O$	3.52
O_2 -TDK	$Mg_{0.15}MnO_{2.02}\cdot0\cdot44H_2O$	3.74

The relationship between the 2D and 1D phases, and their structural transformation is illustrated in **Fig. 1**. The difference between birnessites synthesized using O_2 vs $Mg(MnO_4)_2$ as oxidizing agents is primarily due to incorporation of Mg into the layered framework structure in the latter case. The corresponding compositions for all of the materials, based on TGA, oxidation state and chemical analysis are summarized in **Table 1**. The discharge-charge curves for MnO_4-BIR, -BUS and -TDK which all have a Mn oxidation state of ~3.64 are shown in **Fig. 2**. Birnessite (**Fig. 2a**, upper curve) displays a sloping discharge curve indicative of single phase topotactic insertion of Li, and a reversible capacity of 0.48 Li/Mn. This is the only material from which cations can be electrochemically extracted on charge. In contrast, MnO_4-BUS (**Fig. 2b**) elicits a somewhat different electrochemical response: the first discharge cycle has a well defined plateau region at 2.7V with a sharp drop in voltage at 2.4V. On oxidation, 0.13 moles of Li remain trapped in the

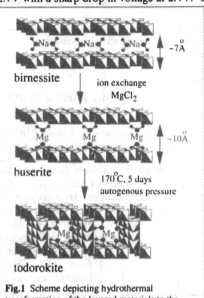

birnessite | ion exchange $MgCl_2$

buserite | $170°C$, 5 days autogenous pressure

todorokite

Fig.1 Scheme depicting hydrothermal transformation of the layered materials to the tunnel todorokite structure

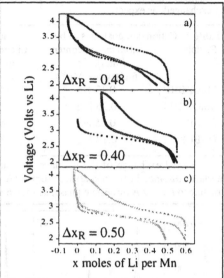

a) $\Delta x_R = 0.48$

b) $\Delta x_R = 0.40$

c) $\Delta x_R = 0.50$

Voltage (Volts vs Li)

x moles of Li per Mn

Fig. 2 Galvanostatic discharge-charge curves of a) MnO_4-BIR; b) MnO_4-BUS; c) MnO_4-TDK, cycled at a rate of C/20 within the voltage limits 4.2-2.0V

structure. Despite the lack of reversibility shown in the first cycle, subsequent cycles show a fully reversible capacity of 0.4 Li/Mn (100mAh/g). The electrochemical response of todorokite, obtained by hydrothermal transformation of this material has a similar profile (**Fig. 2c**), although the initial discharge capacity is higher (0.6 Li/Mn). A sharp voltage drop in the curve near 2.0V is associated with full occupation of a specific site, and approximately 98% theoretical capacity is achieved (assuming reduction of 0.64 Mn^{4+} to Mn^{3+}). This capacity is completely reversible on charge, albeit not fully on subsequent discharge. Nevertheless, todorokite displays the highest reversible capacity of these materials, 130 mAh/g. The effect of oxidation method on the electrochemical behavior of the two todorokite materials is to alter the profile of the discharge-charge curves, without significantly changing the total reversible capacity. Specifically, the initial discharge capacity of O_2-TDK is the same, but on charge, 0.15 moles Li/Mn are irreversibly trapped within the structure. This value coincides with the onset of increased polarization in the MnO_4-derived material. The second discharge cycle is fully reversible, in this case, however, yielding the same reversible capacity as MnO_4-TDK.

It appears that the difference in total vs reversible capacity of about 0.15 Li/Mn is a fundamental characteristic of the tunnel framework, as previous studies of MnO_2 having the hollandite (2x2 tunnel structure) also showed that a similar amount of Li remained after electrochemical oxidation.[3] These ions may become trapped in the 1x1 tunnel cavities which are located between the larger 2x2 tunnels in hollandite or 3x3 tunnels in todorokite (**Fig. 1**). In todorokite, the amount of Li trapped correlates well with what would be predicted on the basis of the structure (0.16 [1x1] tunnel sites/Mn in the unit cell).

Li - exchanged layered materials. We sought to increase the total reversible capacity of these materials by exchanging the interlayer or tunnel cations with Li, using ion-exchange with an excess of LiCl, in aqueous solution. This proved to be successful in the case of the layered materials, but not for todorokite. The cation compositions of the layered materials before and after ion-exchange, (and after heat treatment at 180°C) are summarized in **Table 2.** Only partial exchange of the interlayer cations was achieved in all cases, and framework cations (Mg^{2+}) were not exchanged, as expected.

Table 2. Cation compositions of birnessites and buserite, before and after Li exchange

Sample	Initial cation content (interlayer + framework Mg^{2+})	Li content after exchange (per Mn)	Composition
MnO_4-BIR	0.17 Na	0.09	$Li_{0.09}Na_{0.12}Mg_{0.08}MnO_{2.00} \cdot 0.33H_2O$
MnO_4-BUS	0.25 Mg	0.08	$Li_{0.08}Mg_{0.20}MnO_{1.99} \cdot 0.68H_2O$
O_2-BIR	0.25 Na	0.12	$Li_{0.12}Na_{0.14}MnO_{1.94} \cdot 0.20H_2O$
O_2-BUS	0.18 Mg	0.14	$Li_{0.14}Mg_{0.09}MnO_{1.88} \cdot 0.30H_2O$

Partial exchange of the interlayer Na cations for Li resulted in a slight decrease of the interlayer distance in the birnessites from 7.1Å to 6.9Å, and conversion of the ~10 Å buserite structure to the 6.9 Å birnessite phase, as shown in **Fig. 3**. The decrease in interlayer distance for the buserites on ion-exchange of Mg^{2+} for Li^+ corresponds to ejection of water from the inter-lamellar gap as a result of the lower hydration energy of Li compared to Mg. This was apparent in the XRD patterns of the materials immediately after ion-exchange, with little change in the interlayer distance occuring after partial dehydration at 180°C.

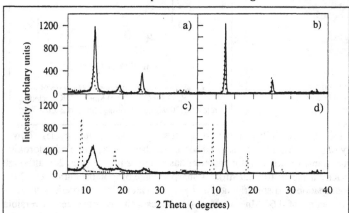

Fig. 3. XRD patterns before and after Li-ion exchange of a) MnO_4-BIR; b) O_2-BIR; c) MnO_4-BUS; d) O_2-BUS. Pristine material (--); Li-exchanged material (-)

As the chronopotentiometric curves for the exchanged buserites were similar to those of the birnessites, only those for the latter are illustrated in **Fig. 4**. *All* of the chemically exchanged Li is extracted during the initial charge cycle, as shown for Li-MnO_4-BIR, **Fig. 4a,** as the increase in charge capacity of 0.09 correlates exactly with the increase in Li content (0.09). The Na cations which remained within the structure during ion-exchange therefore are also apparently trapped during this process. Compared to pristine MnO_4-BIR (**Fig. 2a**), the profile on the following discharge-charge cycle shows less hysteresis, a more well-defined and reversible insertion process, and increased capacity, Q_R of 0.56 Li/Mn. The dx/dV vs V (or electronic density of sites *vs* voltage) plot, **Fig. 4c,** displays a correspondingly defined site for Li insertion at 2.85V (discharge). The improvement on ion-exchange is even more apparent in the profiles for the Li-O_2-prepared birnessite (**Fig. 4b,d**). Here, the distinct plateau region is indicative of a

388

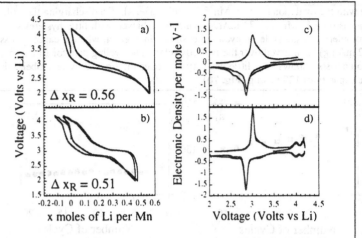

Fig. 4 Galvanostatic discharge-charge curves cycled at a rate of C/20 of a) MnO_4-BIR; b) O_2-BIR; c) differential capacity curve for a; d) differential capacity curve for b.

singular phase transition, with occupancy of sites at 2.85V and little hysteresis on charge (3.0V). The difference in behavior between the two materials is a consequence of the lack of structural Mg in the framework of Li-O_2-BIR, which affects the distribution of Li potential sites.

Li - exchanged todorokite. As the aqueous LiCl route resulted in *no* exchange of the tunnel cations for Li, an alternative route using molten salt exchange ($LiNO_3$) was employed. This resulted in insertion of Li rather than simple exchange, and an increase in the Mn oxidation state, leading to a material with the composition $Li_{0.41}Na_{0.08}Mg_{0.16}MnO_{2.32} \cdot 18H_2O$ for O_2-TDK (Li-O_2-TDK). The diffraction pattern (**Fig. 5a**) shows that the architecture of the fundamental framework is unaltered by this process. The change in the intensity ratio of the 001/002 reflections, which is sensitive to tunnel occupation, suggests that the population of the tunnels is different. This is confirmed by TG analysis (**Fig. 5b**) which clearly indicates a substantial loss of the structural water within the tunnels, which is centered around 300°C in the pristine material ($Mg_{0.15}MnO_{2.02} \cdot 0.44H_2O$).

The chronopotentiometric curve for Li-O_2-TDK on initial charge is shown in **Fig. 6**. It is evident that 0.12 Li can be oxidatively extracted, in

Fig. 5 a) XRD pattern of Li-O_2-TDK before (--), and after (-) Li exchange; b) TGA profile of Li-O_2-TDK before (thin line), and after Li exchange (thick line)

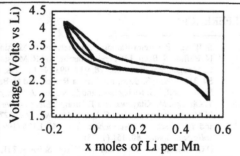

Fig. 6 Galvanostatic charge-discharge curve of Li-O_2-TDK at a cycling rate of C/20 (20 μA/cm²).

good accordance with the oxidation state of $Mn^{3.82+}$ in this material. On discharging the cell, the reversible capacity is 0.66 Li/Mn, or 172 mAh/g, which is achieved with little hysteresis. This capacity is not completely sustainable on cycling, as shown in **Fig. 7a**, where a capacity loss of 5% (down to 152 mAh/g) is observed over the first 10 cycles. The cycling data for Li-exchanged MnO_4-BIR by comparison, shows very little capacity loss over the first 20 cycles, albeit with a lower sustainable capacity of 125 mAh/g (**Fig. 7b**).

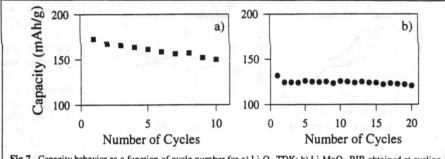

Fig 7. Capacity behavior as a function of cycle number for a) $Li-O_2$-TDK; b) $Li-MnO_4$-BIR obtained at cycling rates of about C/20.

CONCLUSIONS

Lithiated todorokite is a promising material for use as a cathode in lithium rechargeable batteries. Its appealing characteristics include a very low structural water content, a high reversible capacity (up to 170 mAh/g), and lack of conversion to spinel over the voltage window 4.2-2.0V. Capacity fading of approximately 5% occurs over the first 10 cycles, although this may be suppressed by optimizing the material by judicious choice of synthetic method and framework dopants. By comparison, the layered lithiated birnessite exhibits a somewhat lower reversible capacity of 125 mAh/g; however, it is more stable to cycling over the same voltage range.

ACKNOWLEDGMENTS

LFN gratefully acknowledges the generous financial support of NSERC through the strategic and operating programs. The authors thank Dr. V. Karanassios for assistance with the ICP measurements.

REFERENCES

1.   S. Bach, J.P. Pereira-Ramos, N. Baffier, and R. Messina, *Electrochimica Acta*, **36**, 1595 (1991); P. Le Goff, N. Baffier, S. Bach, J.P. Pereira-Ramos, and R. Messina, *Solid State Ionics*, **61**, 309 (1993); P. Strobel and C. Mouget, *Mat. Res. Bull.*, **28**, 93 (1993); F. Le Cras, S. Rohs, M. Anne, and P. Strobel, *J. Power Sources*, **54**, 319 (1995); S. Bach, J.P. Pereira-Ramos, and N. Baffier, *J. Electrochem. Soc.*, **143**, 3429 (1996).
2.   M.M. Doeff, T.J. Richardson, and L. Kepley, *J. Electrochem. Soc.*, **143**, 2507 (1996).
3.   T. Ohzuku, M. Kitagawa, and T. Hirai, *J. Electrochem. Soc.*, **138**, 360 (1991); M.H. Rossouw, D.C. Liles, M.M. Thackeray, and W.I.F. David, *Mat. Res. Bull.*, **27**, 221 (1997); Q. Feng, H. Kanoh, K. Ooi, M. Tani, and Y. Nakacho, *J. Electrochem. Soc.*, **141**, L135 (1994); S. Bach, J.P. Pereira-Ramos, and N. Baffier, *Solid State Ionics*, **80**, 151 (1995).
4.   D.C. Golden, C.C. Chen, and J.B. Dixon, *Science*, **231**, 717 (1986).
5.   Y. F. Shen, R.P. Zerger, R.N. DeGuzman, S.L. Suib, L. McCurdy, D.I. Potter, and C. L. O'Young *Science*, **260**, 511 (1993).
6.   K.J. Vetter and N. Jaeger, *Electrochimica. Acta*, **11**, 401 (1996).

LITHIUM INSERTION BEHAVIOUR OF $Li_{1+x}V_3O_8$ PREPARED BY SOL-GEL METHOD IN METHANOL

J. KAWAKITA, Y. KATAYAMA, T. MIURA, T. KISHI
Department of Applied Chemistry, Faculty of Science and Technology, Keio University, Hiyoshi 3-14-1, Kouhoku-ku, Yokohama 223, JAPAN, kawakita@applc.keio.ac.jp

ABSTRACT

$Li_{1+x}V_3O_8$ (LT-M form) was obtained by sol-gel method in CH_3OH. This form, prepared at 350°C, possessed smaller grain size and better electrochemical performance than the HT form prepared by conventional high temperature synthesis. High discharge capacity (372 $mAh \cdot g^{-1}$: $x = 4.0$) and reversible discharge and charge cycle were attained. When heated at 200°C, CH_3OH molecules remained in the compound and crystallinity became lower by lithium insertion over $x = 2.0$. The lithium de-intercalation was irreversible.

INTRODUCTION

The layered lithium trivanadate, $Li_{1+x}V_3O_8$, has been investigated as a cathode material for lithium secondary batteries [1,2,3]. The main advantage as an active material is its ability to accommodate up to three moles of Li^+ ions per formula unit. Vanadium exists in pentavalent state though Li(I) is contained in the host structure before discharge. A good cycling performance is derived from the fact that the layers composed of $V_3O_8^-$ are linked strongly by an ionic bond between Li(I) ions and octahedrally coordinated oxygen atoms faced to the layers and the host structure changes little on lithium insertion and extraction[4]. Recently, it was reported that amorphous and low crystalline $Li_{1+x}V_3O_8$ (LT form) prepared by sol-gel method in an aqueous solution was superior to one (HT form) obtained by conventional high temperature synthesis in terms of electrochemical property such as the discharge capacity [5,6]. So far, LT form has been synthesized only in an aqueous solution. However, water is one of the hindrances that have to be excluded from the non-aqueous lithium cell. In this work $Li_{1+x}V_3O_8$ (LT-M form) is prepared by sol-gel method using CH_3OH as a solvent. Moreover, the structural characteristics and the electrochemical properties have been investigated and compared with those of HT form.

EXPERIMENT

$Li_{1+x}V_3O_8$ (LT-M form) was prepared by sol-gel method, referring the report by Pistoia et al. [5]. $LiOCH_3$ and CH_3OH were substituted for LiOH and H_2O, respectively. V_2O_5 was slowly added to 0.6 M $LiOCH_3$ / CH_3OH solution and the concentration of it was 0.75 M. This operation was carried out in a glove box filled with argon gas. Then the mixture was stirred at 50°C for two days under argon atmosphere. Greenish brown precipitate was formed, and was heated and evacuated at 90°C for 12 hours to evaporate solvent CH_3OH. A fine powder was obtained by mild grinding of the sample. While this powder was highly hygroscopic, it became non-hygroscopic after the heat treatment at 200 and 350°C in dry air.

Thermogravimetry (TG) was carried out on the samples heated at 200 and 350°C. As the reference sample, $Li_{1.2}V_3O_8$ (HT form) was prepared by melting the mixture of Li_2CO_3 and V_2O_5 in an appropriate molar ratio and cooling slowly down to room temperature.

The crystal structure of both forms was characterized by X-ray powder diffraction (XRD) and infrared (IR) spectroscopic measurements. The content of pre-existing vanadium in tetravalent, V(IV), was determined by combination of atomic absorption analysis and redox titration using $KMnO_4$ / H_2SO_4 solution.

Galvanostatic discharge and discharge / charge cycle tests were performed using a cylindrical glass cell. This cell had three electrodes. The working electrode was prepared as follows. The pellet was obtained by pressing a mixture of active material, acetylene black and PTFE in a weight ratio of 70: 25: 5 on a porous nickel net. The obtained pellet was connected to a lead wire by silver paste on a nickel side of the pellet and insulated electrically with silicone resin leaving counter side of the pellet as electrode surface. Lithium metal rods were used as the reference and the counter electrodes. The chemical diffusion coefficient of lithium ($\tilde{D}_{Li}$) was obtained by the galvanostatic intermittent titration technique (GITT) [7]. The steady rest potential within $0.001 \ V·h^{-1}$ was regarded as the open circuit potential (OCP) after galvanostatic discharge up to each x value.

RESULTS

Synthesis of LT-M $Li_{1+x}V_3O_8$

Figure 1 shows IR spectra of LT-M forms heated at several temperatures, and that of the HT form. CH_3OH molecule existed in the sample heated at 90°C (Fig. 1(a)), as the absorption bands were observed ascribed to the stretching (ν) and deformation (δ) vibrations among oxygen, hydrogen and carbon atoms. When heated at 200°C as in Fig. 1(b), the absorption bands as cited above did not disappear, so it was considered that CH_3OH molecule was still remained in the compound. The result of both chemical analysis and TG showed that the composition of the compound heated at this temperature was represented as $Li_{1.13}V_3O_{7.68}·0.09 \ CH_3OH$. After heat treatment at 350°C (Fig. 1(c)), CH_3OH molecule was removed from LT-M form owing to disappearance of all the absorption bands corresponding to vibrations in CH_3OH. The absorption bands around 1000 and 750 cm^{-1} were attributed to the stretching vibrations between vanadium and oxygen atoms, i.e. ν(V=O) and ν(V-O-V), respectively [8]. The wavenumbers of the vibrations, ν(V=O) and ν(V-O-V), in LT-M form heated at 350°C agreed mostly with those in HT form, although they shifted to higher wavenumbers slightly.

Figure 2 shows XRD patterns of LT-M forms heated at several temperatures, together with that of the HT form. As seen in Fig. 2(a), no peaks were observed except two attributed to Mylar film, which was put on the sample in the XRD measurement to avoid contamination by humid air. Thus, LT-M form heated at 90°C was obtained in an amorphous state. This form, however, was too hygroscopic to prepare as a the pellet for the working electrode in our experiment. When heated at 200°C (Fig. 2(b)), many peaks with weak intensity indicated crystalline to some extent but were not indexed to any kinds of vanadium oxides and lithium vanadate. This phenomenon was also observed in the case of the LT form [6]. All the peaks

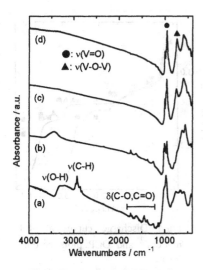

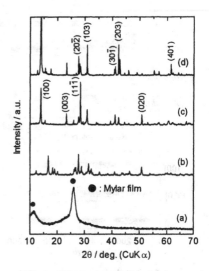

Fig.1 IR spectra of LT-M forms heated at (a) 90°C, (b) 200°C, (c) 350°C, and (d) HT form.

Fig.2 XRD patterns of LT-M forms heated at (a) 90°C, (b) 200°C, (c) 350°C, and (d) HT form.

of LT form heated at 350°C were attributed to $Li_{1+x}V_3O_{7.9}$ (indexed in JCPDS card No. 18-754), as shown in Fig. 2(c). Their relative intensities, however, seemed to be different from those of HT form (Fig. 2(d)). In particular, the ratio of (100) / (020) on LT-M form was smaller than that on HT form. Figure 3(a) showed the relative intensity of (020) and (003) planes, normalizing (100) as 100 %, in the XRD pattern of HT and LT-M forms, and in JCPDS card. Apparently, in the case of LT-M form the relative intensities on (020) and (003) planes to (100) was stronger than that of HT form and were similar to that of JCPDS card. The absolute intensity of (100) plane of LT-M form, however was much smaller than that of HT form. On the other hand, each of grain size in [100], [020] and [003] directions of LT-M form was smaller than that of HT form. Especially, in [100] direction the grain size of the former was as half as that of the latter. As the layers in the host lattice was piled up to [100] direction, it revealed that LT-M form had less stacking of the layer than HT form. This fact might explain shift of the absorption bands of ν(V=O) and ν(V-O-V) in IR spectra to high wavenumber direction (see Fig. 1).

Electrochemical Properties of LT-M $Li_{1+x}V_3O_8$

Figure 4 shows discharge curves and OCP plots for HT and LT-M forms heated at 200 and 350°C. OCPs of LT-M form heated at 350°C agreed with that of HT form up to $x = 3.2$. The plateau region appeared near 2.4 V (vs. Li) for $3.2 < x < 4.0$ in the case of the former. Thus, additional lithium insertion proceeded in this region as single-phase reaction and $Li_5V_3O_8$ formed finally, which Picciotto et al. expected to exist [9]. This phenomenon was confirmed by XRD patterns, as shown in Fig. 5(a). On the other hand, OCPs of LT-M form

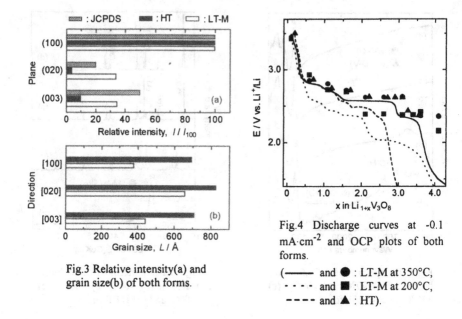

Fig.3 Relative intensity(a) and grain size(b) of both forms.

Fig.4 Discharge curves at -0.1 mA·cm^{-2} and OCP plots of both forms.

(——— and ● : LT-M at 350°C,
- - - - and ■ : LT-M at 200°C,
– – – – and ▲ : HT).

heated at 200°C dropped rapidly at x = 2.1 and kept constant at 2.4 V for x > 2.1. XRD patterns (see Fig. 5(b)) in this range showed lower crystallinity than the starting material at x = 0.1, so it was revealed that LT-M form heated at 200°C had the process of lithium insertion different from either LT-M form heated at 350°C or HT form. In the case of LT-M form heated at 350°C, the two plateaus appeared obviously near 2.55 and 2.4 V, corresponding to constant OCPs observed for 2.0 < x < 3.2 and x > 3.2, respectively. West et al. reported this phenomenon concerning LT form prepared in an aqueous solution [6]. The latter plateau contributed mainly to increase in discharge capacity compared with HT form. This might be explained by improvement in ion kinetics. Presumably, the smaller grain size resulted in a shorter diffusion path of Li$^+$ ions and the less stacking of the layers facilitated transformation of the phase. In fact, chemical diffusion coefficient of lithium, $\widetilde{D}_{Li}$, in LT-M form became evidently larger that HT form in the primary single phase region of x < 2.0, as seen in Fig. 6. While the discharge capacity of LT-M form heated at 200°C was as large as that at 350°C, the potential of the former was lower than the latter, except the early stages of discharge. As low crystallinity and defects in the lattice, i.e. oxygen deficiency in this case, usually gave positive effect to kinetics, the large gap between discharge potential and OCPs indicated that CH$_3$OH molecule in the interlayer prevented Li$^+$ ions from smooth jump from a site to neighbor one.

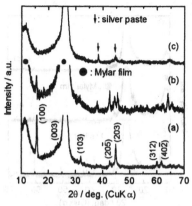

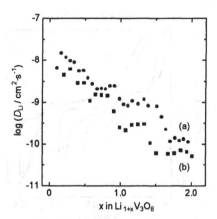

Fig.5 XRD patterns of discharged sample up to (a) $x = 3.2$ on HT, (b) $x = 4.0$ on LT-M heated at 350°C, and (c) $x = 2.5$ on LT-M heated at 200°C.

Fig.6 Dependence of diffusion coefficient of lithium ($\tilde{D}_{Li}$) on x value in (a) LT-M heated at 350°C and (b) HT form.

Figure 7 shows discharge and charge cycle curves of HT and LT-M forms heated at 200 and 350°C. $Li_4V_3O_8$ phase formed in the last stage of discharge ($x = 3.0$) and returned to LiV_3O_8 phase in the case of HT form [10]. On charging of LT-M form heated at 350°C had two plateaus near 2.45 and 2.6 V probably corresponding to each plateau on discharge. Moreover, only LiV_3O_8 phase was observed in the XRD patterns after discharge and charge cycle, as shown in Fig. 8(a). A slight shifting of some diffraction lines in the peak position was due to distortion of the structure caused by the remaining Li^+ ions in it. Accordingly, it was revealed that the crystal structure of LT-M form heated at 350°C was also transformed reversibly, involving extraction of Li^+ ions from $Li_5V_3O_8$. This cycle reversibility was maintained as the structure returned to the original one even after at 10th cycle, as shown in Fig. 8(b). No obvious plateaus and knees were observed on charging curve of LT-M form heated at 200°C. This aspect was a common characteristic of amorphous materials. Beyond $x > 2.0$, crystallinity of this material lowered immediately (see Fig. 5(b)). Thus, lithium vanadate with low crystallinity close to amorphous state was formed in the course of discharge ($x = 2.0$) and additional two Li^+ ions were inserted. On the other hand, after discharge and charge cycle, some weak and broad peaks were observed in the XRD patterns, as shown in Fig. 8(c). It was considered that lithium was extracted from lithium vanadate described above during charging. Consequently, it was found that insertion and extraction of Li^+ ions proceeded irreversibly in the case of LT-M form heated at 200°C.

CONCLUSIONS

The phase $Li_{1+x}V_3O_8$ (LT-M form) was obtained by sol-gel methods in methanol as well as in water. This form, prepared at 350°C, resulted in attainment of both high discharge capacity (372 mAh·g^{-1}: $x = 4.0$) and reversible cycle between discharge and charge. It was considered that these were explained by fast diffusion of Li^+ ions and feasible transformation

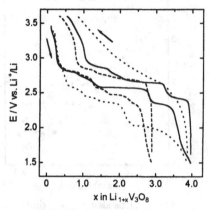

Fig.7 Discharge and charge cycle curves at ±0.1 mA·cm^{-2},

(—— : LT-M at 350°C,
···· : LT-M at 200°C,
---- : HT).

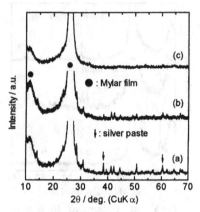

Fig. 8 XRD patterns of LT-M at 350°C after (a) 1st cycle, (b) 10th cycle, and (c) of LT-M at 200°C after 1st cycle.

from the original phase to next one during cycling owing to smaller grain size and less stacking of the layers consisting of vanadium and oxygen atoms. After heat treatment at 200°C, LT-M form contained CH_3OH molecule and oxygen defects in the lattice. It was hypothesized that large polarization was caused by obstruction of movement of Li^+ ions by methanol. Crystallinity became much lower by further lithium insertion beyond $x = 2.0$ and extraction of Li^+ ions from this lower crystalline material like amorphous indicated that charging process was irreversible.

REFERENCES

1. S. Panero, M. Pasquali and G. Pistoia, J. Electrochem. Soc. **130**, p. 1225 (1983).
2. N. Kumagai and A. Yu, J. Electrochem. Soc. **144**, p. 830 (1997).
3. J. Kawakita, Y. Katayama, T. Miura and T. Kishi, Solid State Ionics, in press.
4. G. Pistoia, M. Pasquali, M. Tocci, V. Manev and R.V. Moshtev, J. Power Sources **15**, p. 13 (1985).
5. G. Pistoia, M. Pasquali, G. Wang and L. Li, J. Electrochem. Soc. **137**, p. 2365 (1990).
6. K. West, B. Zachau-Christiansen, S. Skaarup, Y. Saidi, J. Barker, I.I. Olsen, R. Pynenburg and R. Koksbang, J. Electrochem. Soc. **143**, p. 820 (1996).
7. W. Weppner and R.A. Huggins, J. Electrochem. Soc. **124**, p. 1569 (1977).
8. Y. Kera, J. Solid State Chem. **51**, p. 205 (1984).
9. L.A. de Picciotto, K.T. Adendorff, D.C. Liles and M.M. Thackeray, Solid State Ionics **62**, p. 297 (1993).
10. G. Pistoia, S. Panero, M. Tocci, R.V. Moshtev and V. Manev, Solid State Ionics **13**, p. 311 (1984).

DEMONSTRATION OF MIXED METAL OXIDE CATHODE MATERIALS IN PRISMATIC LI-ION CELLS

G. M. Ehrlich and R. L. Gitzendanner
Yardney Technical Products, Inc., Pawcatuck, CT. 06379, USA

Abstract

Recent advances in mixed metal oxide ($LiNi_{1-x}Co_xO_2$) cathode materials for lithium ion (Li-ion) batteries have resulted in a new generation of high capacity cathode materials. High capacity materials are particularly useful for applications where either volume or weight is limited, such as in space applications. These applications also benefit from cell chemistries with high energy efficiency as this permits reduced ancillary energy storage and dissipation apparatus. Mixed metal oxide materials have been evaluated and demonstrated in prismatic Li-ion cells. The capacity of the mixed metal oxide materials is exceptional, reversible capacity of 160mAh/g is demonstrated although the cell voltage is lower than when $LiCoO_2$ materials are used, 3.58V vs. 3.74V, at low ($0.5mA/cm^2$) discharge rates. The high capacity results in a significant improvement in specific energy, further, the sloping discharge curve characteristic of the mixed metal oxide materials facilitates determination of the state of charge based on cell voltage. The design of composite cathode materials using the mixed metal oxide materials has a significant influence on the cell impedance and the rate capability of the material. Results describing the rate capability of these materials is presented.

Introduction

An increasingly diverse array of portable electronic devices require cost effective, energy dense batteries. Further, applications such as computers or radios frequently require high rate pulse discharge capability. Lithium-ion (Li-ion) batteries which use a $LiCoO_2$ [1] cathode material have demonstrated exceptional performance in cost insensitive applications but their applicability has been limited by their high cost, especially when NiCd alternatives may be considered. Recently developed cathode materials, such as $LiNi_{1-x}Co_xO_2$[2], address the need for a low cost Li-ion technology by offering comparable performance at potentially reduced cost.

Commercially available $LiNi_{1-x}Co_xO_2$ materials have been demonstrated both in flat-plate prismatic Li-ion cells with carbon based anodes and laboratory coin cells with lithium metal anodes. The prismatic cell design is attractive as it takes full advantage of the available space in rectilinear battery boxes and provides good rate capability. The technology demonstrated in this paper utilized a rigid plastic cell case which enhanced the specific energy of the cell while providing the reliability and safety of a rigid cell case. Lightweight cells of this type are attractive for satellite applications as the high energy efficiency reduces solar cell and radiator requirements thus lowering satellite weight.

Experiment

<u>3Ah Prismatic Cells</u> Prototype Flat plate prismatic cells were fabricated using a commercial $LiNi_{1-x}CoO_2$ cathode material coated onto Aluminum foil and a graphitic anode material coated onto copper foil. Acetylene black and graphite were added to the cathode material to enhance its conductivity. The cell contained 21 electrodes separated by a microporous polypropylene film and was filled with an electrolyte of 1M $LiPF_6$ in EC:DMC 1:1. A sealed polypropylene cell case was utilized and exposure to moisture minimized. The cell was designed to be cathode limited to prevent over-reduction of the anode material.

<u>Coin Cells</u>   Coin cells with Li metal anodes were used to evaluate the $LiNi_{1-x}CoO_2$ material without effects from the carbon anode material used in the prismatic cell. The electrolyte was a 1M solution of $LiPF_6$ in a 1:1 mixture of ethylene carbonate and dimethylcarbonate.

Results and Discussion

<u>Rate Capability</u>   Cells were charged at 0.55 mA/cm^2 (C/5) at 25°C and discharged at various rates up to 3mA/cm^2 (1.1 C). A plot of the discharge capacity versus current density is presented in Figure 1, a plot of the mean cell voltage versus current density is presented in Figure 2. At current densities between 0.5 and 3 mA/cm^2 the cell demonstrated little loss in discharge capacity with increasing rate. At 3mA/cm^2, the discharge capacity was 95% of that observed at 0.5 mA/cm^2. Similarly, as shown in Figure 2, the cell did not become polarized over the regime investigated. At 0.5 mA/cm^2, the mean cell voltage during discharge was 3.64V. At the highest rate investigated the mean cell voltage during discharge was 3.55V demonstrating the rate capability of the system.

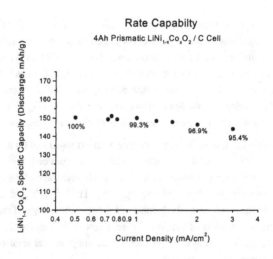

Figure 1. A plot of the positive electrode specific capacity on discharge at various rates for a 4Ah prismatic Li-ion cell.

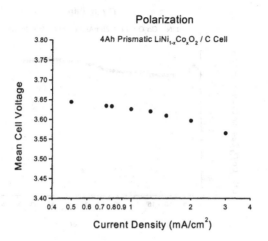

Polarization
4Ah Prismatic LiNi$_{1-x}$Co$_x$O$_2$ / C Cell

Figure 2. A plot of the mean cell voltage for constant current discharge at various rates.

<u>Cycle-ability</u>

To examine the characteristics of the LiNi$_{1-x}$Co$_x$O$_2$ material without effects from a carbon anode material, the LiNi$_{1-x}$Co$_x$O$_2$ material was cycled versus a Li anode. Shown in Figure 3 is a plot of the discharge capacity versus cycle number for a LiNi$_{1-x}$Co$_x$O$_2$ / Li cell cycled at 1.25mA/cm^2 charge/discharge. Although the initial capacity was high, 200mAh/g for the first de-intercalation of the cathode (charge), the reversible capacity is lower, 160 mAh/g.

In a prismatic cells the LiNi$_{1-x}$Co$_x$O$_2$ cathode material demonstrated a high level of stability when cycled at moderate rates, 2mA/cm^2 (0.73C) charge/discharge. As shown in figure 4, a capacity fade rate of 0.020%/cycle was observed over 200 cycles at room temperature. As expected based on the half-cell data, irreversible loss of 27% was observed on the initial cycles, a portion of which (8%) is attributable to the anode material.

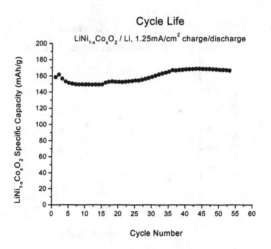

Figure 3. A plot of the specific capacity versus cycle number for a LiNi$_{1-x}$Co$_x$O$_2$ electrode cycled versus Li metal in a coin cell.

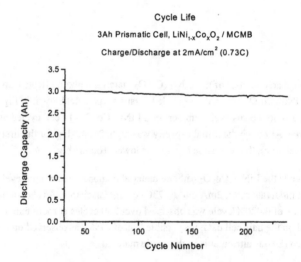

Figure 4. Cycle life of a 3Ah prismatic Li-ion cell which utilized LiNi$_{1-x}$Co$_x$O$_2$ cathode and MCMB carbon anode materials. The cell was cycled at 2mA/cm^2 charge/discharge. The capacity fade rate was 0.020%/cycle.

Energy Efficiency   Shown in Figure 5 is a plot of the coulombic and energy efficiency. These cells offer the high (96%) energy efficiency characteristic of Li-ion technologies. For applications where excess heat is undesirable, such as computers or satellites, this high level of efficiency is desirable.

Specific Energy   Including the cell case and terminals, a 3Ah prismatic cell weighed 114 grams and offered 12.8 Wh on discharge thus a specific energy of 112 Wh/Kg, 14% more than demonstrated with $LiMn_2O_4$ cathode materials in the same prototype cell design.

Conclusions

Mixed metal oxide cathode materials have been demonstrated in both 3Ah prismatic Li-ion cells and coin cells to offer high capacity, good rate capability and good cycle-ability. This technology has demonstrated high energy efficiency and high specific energy. To maximize the energy density at the cell level, minimization of the irreversible capacity observed on the initial cycles is desirable.Future work will evaluate the high and low temperature performance of the system.

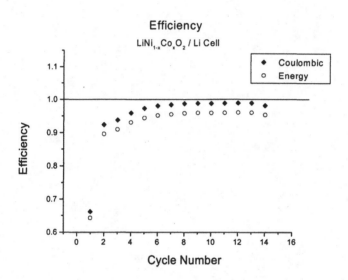

Figure 5. A plot of the coulombic and energy efficiency of cell which utilized a $LiNi_{1-x}Co_xO_2$ positive electrode material.

References

[1] T. Nagaura and K. Tozawa, Prog. Batt. Solar Cells, **9**, p. 209 (1991).
[2] W. Li and J. C. Currie, J. Electrochem. Soc., **144**, p. 2773 (1997).

EFFECTS OF ALUMINUM DOPING ON THE PHASE STABILITY AND ELECTROCHEMICAL PROPERTIES OF LiCoO₂ and LiMnO₂

Y.-I. Jang, B. Huang, H. Wang, Y.-M. Chiang, and D. R. Sadoway
Department of Materials Science and Engineering
Massachusetts Institute of Technology, Cambridge, MA 02139

ABSTRACT

Aluminum is of interest as a constituent for Li battery electrodes due to its low cost and low mass, and because *ab initio* calculations indicate that solid solution of $LiAlO_2$ with $LiMO_2$ (M = transition metal) in the α-$NaFeO_2$ structure can increase intercalation voltage [1]. In this study, we investigated the effect of Al doping on $LiCoO_2$ and $LiMnO_2$. Single phase $LiAl_yCo_{1-y}O_2$ has been synthesized up to y = 0.5 by firing homogenous hydroxide precursors. A systematic increase in the open circuit voltage is observed with Al content. In $LiAl_yMn_{1-y}O_2$, the addition of $LiAlO_2$ stabilizes $LiMnO_2$ in the α-$NaFeO_2$ structure under conditions where neither endmember is stable in the structure. High reversible capacity was obtained over both a 4 V and 3 V plateau, indicating that the compound transforms to a spinel-related structure during cycling, but that the cooperative Jahn-Teller distortion is suppressed.

INTRODUCTION

Compounds of the type $LiMO_2$ (M = Co [2], Ni [3]) with the α-$NaFeO_2$ structure (space group R$\overline{3}$m) have been extensively studied as cathodes for Li-ion batteries. The effects of doping with other transition metal elements have also been studied [4-6]. Recently, a related polymorph of $LiMnO_2$ (space group C2/m) has been obtained by the ion-exchange reaction of lithium salts with α-$NaMnO_2$ [7,8]. Compared with the transition metals, Al has not received wide attention as a dopant, with only a few reports on $LiAl_yM_{1-y}O_2$ (M=Co, Ni) systems to date [9-11]. However, doping with Al is of significant interest for several reasons. *Ab initio* calculations by Aydinol *et al.* [1] have shown that $LiAlO_2$ has a theoretical intercalation voltage of ~5 V *vs.* Li/Li⁺, which is higher than that of any other lithium transition-metal oxide. While pure $LiAlO_2$ is electrochemically inactive, the solid solution of $LiAlO_2$ with lithiated transition-metal oxides can potentially increase the intercalation voltage and energy density. Secondly, the fact that $LiAlO_2$ is stable in the α-$NaFeO_2$ structure at temperatures below ~600°C [12] suggests that it could have a stabilizing effect on other compounds. Finally, its low cost and low density make $LiAlO_2$ attractive as an intercalation compound constituent. In this paper, we report effects of Al doping on the phase stability and electrochemical properties of $LiCoO_2$ and $LiMnO_2$.

EXPERIMENT

SYNTHESIS

$LiAl_yCo_{1-y}O_2$ (y = 0, 0.25, 0.5) and $LiAl_yMn_{1-y}O_2$ (y = 0, 0.25) samples were synthesized by firing homogeneous precursor powders obtained by a co-precipitation and freeze-drying technique [13]. Mixed aluminum-cobalt or aluminum-manganese hydroxides were co-precipitated from mixed aqueous solutions of $Al(NO_3)_3 \cdot 9H_2O$ and $Co(NO_3)_2 \cdot 6H_2O$ or $Mn(NO_3)_3 \cdot 6H_2O$, respectively. A more detailed description of precursor preparation can be found elsewhere [13]. Precursors thus obtained for LiAl$_y$Co$_{1-y}$O$_2$ were fired for 2 h at 800°C in air. In the case of LiAl$_y$Mn$_{1-y}$O$_2$, the valence of Mn ions, firing atmosphere was controlled by firing in a pre-mixed Ar/O$_2$ mixtures for 2 h at 945°C. All samples were furnace-cooled to room temperature after firing. Calcined powders were characterized by TEM, STEM/EDX and X-ray diffraction (XRD) using Cu-K$_\alpha$ radiation.

ELECTROCHEMICAL TESTING

For electrochemical evaluation, oxide powders were mixed with carbon black, graphite and poly(vinylidene fluoride) (PVDF) in the weight ratio of 78:12:0:10 for LiAl$_y$Co$_{1-y}$O$_2$ and 78:6:6:10 for LiAl$_y$Mn$_{1-y}$O$_2$. The electrochemical test cell consisted of two stainless steel electrodes with a Teflon holder. Lithium ribbon of 0.75 mm in thickness was used as the anode. The separator was a film of Celgard 2400^{TM}, and the electrolyte consisted of a 1 M solution of LiPF$_6$ in ethylene carbonate (EC) and diethylene carbonate (DEC). The ratio of EC to DEC was 1:1 by volume. Charge-discharge studies were performed at constant current density in the range of 0.1 to 0.4 mA/cm^2. Additional details of electrochemical testing are discussed elsewhere [14].

RESULTS AND DISCUSSION

LiAl$_y$Co$_{1-y}$O$_2$

XRD patterns of LiAl$_y$Co$_{1-y}$O$_2$ for y = 0, 0.25 and 0.5 are shown in Figure 1(a-c), respectively. All samples are single phase and have the α-NaFeO$_2$ structure, space group R$\bar{3}$m. Miller indices (hkl) are indexed for y = 0 in the hexagonal setting. Peaks for the γ-LiAlO$_2$ phase (tetragonal) are barely distinguishable at y = 0.5 (Figure 1(c)), indicating that the solid solubility limit is exceeded at y ≅ 0.5. Upon increasing Al content, the (006) and (108) peaks shift toward lower

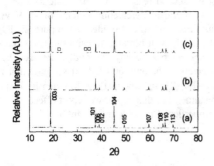

Fig. 1. Powder XRD patterns of LiAl$_y$Co$_{1-y}$O$_2$ after firing for 2 h at 800°C in air. (a) y=0, with *hkl* indexed in hexagonal setting; (b) y= ¼; (c) y=½ (□: γ-LiAlO$_2$)

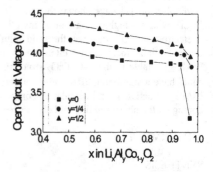

Fig. 2. Open circuit voltage as a funciton of Li and Al content in Li$_x$Al$_y$Co$_{1-x}$O$_2$.

2θ angles, resulting in a wider split of (006)/(012) and (108)/(110) peaks compared with LiCoO$_2$. The lattice parameters calculated by a least squares method from the XRD data are a = 2.816, 2.809, 2.806 Å and c = 14.049, 14.115, 14.150 Å for y = 0, 0.25, 0.5, respectively. Substitution of Al results in shorter a and larger c, increasing the c/a ratio from 4.99 for y = 0 to 5.04 for y = 0.5. No significant change can be observed in the relative intensity ratios $I_{(003)}/I_{(104)}$, which would decrease if Al or Co ion occupies the Li ions sites [15]. A strong intensity of the (003) line in Figure 1 indicates that LiAl$_y$Co$_{1-y}$O$_2$ has a well-ordered α-NaFeO$_2$ structure. These results show that the LiAl$_y$Co$_{1-y}$O$_2$ samples can be considered to be a uniform solid solution of isostructural LiCoO$_2$ and α-LiAlO$_2$ (a = 2.80 Å, c = 14.23 Å [9]).

Open circuit voltage (OCV) was measured as a function of composition by stepwise charging the cells at 0.2 mA/cm^2, and then equilibrating for 15 h. Figure 2 shows firstly that the OCV increases as the cell is charged, as expected. Note that the OCV increases with Al content, indicating that Al substitution decreases the chemical potential of Li in the compound. These results support the *ab initio* calculations by Aydinol *et al.* showing a higher voltage for LiAlO$_2$ than LiCoO$_2$ [1]. Although LiAl$_y$Co$_{1-y}$O$_2$ solid solutions have previously been studied [11], the voltage increasing effect of Al substitution was not observed. Ref. 11 reported the solid solubility limit to be y = ~0.25, which is much less than our result. The difference in the apparent solid solubility is probably due to differences in processing, and possibly homogeneity. Powders obtained in this study were in the size range 100-500 nm as observed by TEM. A homogeneous distribution of the cations was observed by STEM/EDX as shown in Figure 3.

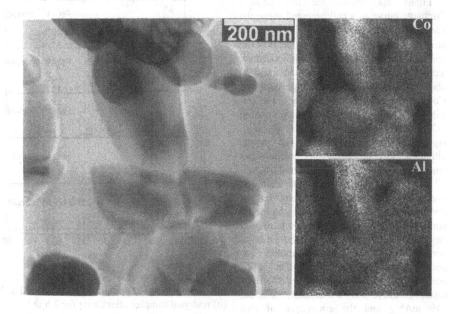

Fig. 3. STEM/EDX elemental maps for Al and Co in LiAl$_{0.5}$Co$_{0.5}$O$_2$

<u>LiAl$_y$Mn$_{1-y}$O$_2$</u>

XRD patterns of the oxide powders obtained by firing an Al-doped precursor (y = 0.25) at 945°C in various oxygen partial pressures are shown in Figure 4(a). While LiMn$_2$O$_4$, Li$_2$MnO$_3$ and γ-LiAlO$_2$ (tetragonal) are the major phases at oxygen partial pressures of 10^{-2} and 10^{-3} atm, m-LiMnO$_2$ (monoclinic) becomes the predominant phase at 10^{-5} and 10^{-6} atm oxygen partial pressures, with o-LiMnO$_2$ (orthorhombic) as a minor phase. This result shows that the predominant Mn valence is 3+ at oxygen partial pressures of 10^{-5} and 10^{-6} atm. The appearance of two diffraction lines in the 64-68° 2θ range indicates that the phase obtained in this study is m-LiMnO$_2$, not Li$_2$Mn$_2$O$_4$ [16], which has a very similar XRD pattern except for the appearance of one peak in this 2θ range [8]. Figure 4(b) shows the XRD pattern of undoped LiMnO$_2$ sample (y = 0) obtained at 945°C in 10^{-6} atm oxygen partial pressure. In the absence of Al-doping, o-LiMnO$_2$ is the predominant phase, consistent with the reported stability of this phase under these firing conditions [15]. It is therefore clear that the stabilization of m-LiMnO$_2$ phase is due to the addition of aluminum, and not to the firing conditions alone. We also note that the α-NaFeO$_2$ polymorph of LiAlO$_2$ (α-LiAlO$_2$) is not the stable phase under these conditions; it is known to irreversibly transform to γ-LiAlO$_2$ above 600°C [12]. The present LiAl$_x$Mn$_{1-x}$O$_2$ solid solution therefore crystallizes in the α-NaFeO$_2$ structure under conditions where neither endmember is stable in this structure.

Figure 5(a) shows the first charge-discharge curve of a cell prepared using the LiAl$_{0.25}$Mn$_{0.75}$O$_2$ (fired at 945°C, Po$_2$ = 10^{-6} atm) as the cathode and lithium metal as the anode. It can be seen that the cell exhibits a single voltage plateau at ~4.1 V, and has about 203 mAh/g of first-charge capacity. Compared with the result for pure m-LiMnO$_2$ obtained by ion-exchange reaction [17], there does not appear to be a significant difference in average voltage, although the comparison may not be exact since our samples and the materials in ref. 17 are prepared and tested under different conditions. In addition, our sample is not purely single phase (Figure 4(a)), and the Al concentration in the monoclinic phase may be somewhat less than the nominal composition. More detailed studies are necessary to understand the effect of Al addition on voltage in this system. The first-discharge curve shows a capacity of about 119 mAh/g and the emergence of two voltage steps. After further cycling, the voltage steps become more distinct. At the 20th cycle, shown in Figure 5(b), two

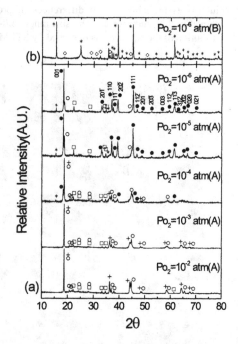

Fig. 4. Powder XRD patterns of (a) Al-doped and (b) undoped samples after firing for 2 h at 945°C in various oxygen partial pressures. (●: m-LiMnO$_2$, with hkl indicated; ✳: o-LiMnO$_2$; +: LiMn$_2$O$_4$; O: Li$_2$MnO$_3$; □: γ-LiAlO$_2$, ◇: Mn$_3$O$_4$)

plateaus at ~4 V and ~3 V are clearly seen. This is characteristic of the spinel phase $Li_xMn_2O_4$, corresponding to lithium intercalation into the tetrahedral and octahedral sites, respectively [18]. It can be concluded that the present m-$LiAl_{0.25}Mn_{0.75}O_2$, which initially has only octahedral cation occupancy, has transformed upon cycling to a material with spinel-like lithium ion occupancy. From these results and previous observations [17,19], it is clear that both m- and o-$LiMnO_2$ have a strong tendency to transform to a spinel-like structure during electrochemical cycling.

Figure 6 shows the evolution of the charge and discharge capacities during cycling between 2.0 and 4.4 V. While an initial drop in capacity is seen over the first 5 cycles, with further cycling the discharge capacity increases progressively, and saturates after about 15 cycles at ~148 mAh/g. The initial decrease in capacity is likely related to an intermediate stage of cation ordering. The fact that cycling can be conducted over both the 4 V and 3 V plateaus without capacity fade is a remarkably different behavior from that of $LiMn_2O_4$ spinel, in which the capacity decreases rapidly upon cycling into the 3 V region. Further improvement in capacity may be possible in $LiAl_yMn_{1-y}O_2$ solid solutions with additional study.

CONCLUSIONS

$LiAl_yCo_{1-y}O_2$ and $LiAl_yMn_{1-y}O_2$ solid solutions of α-$NaFeO_2$ structure have been synthesized and characterized. In $LiAl_yCo_{1-y}O_2$ ($0 < y < 0.5$), the intercalation voltage increases systematically with Al substitution. In $LiAl_yMn_{1-y}O_2$ solid solutions, Al doping stabilizes the α-$NaFeO_2$ structure under conditions where $LiMnO_2$ is not stable in this structure. The oxygen atmosphere during firing is critical in the stabilization of $LiAl_yMn_{1-y}O_2$. The composition $LiAl_{0.25}Mn_{0.75}O_2$ transforms to a spinel-like material during electrochemical cycling, but shows excellent cyclability and high reversible capacity (148 mAh/g).

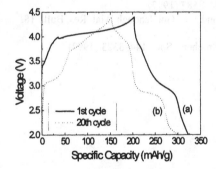

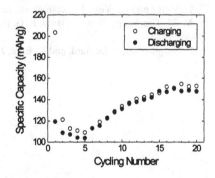

Fig. 5. (a) First charge-discharge curve for m-$LiAl_{0.25}Mn_{0.75}O_2$, tested against a lithium metal anode in a cell at 0.4 mA/cm^2 current density between 2.0 and 4.4 V. (b) Charge-discharge curve for the cell in (a) after 20 cycles at 0.4 mA/cm^2 current density between 2.0 and 4.4 V.

Fig. 6. Specific capacity vs. cycle number for m-$LiAl_{0.25}Mn_{0.75}O_2$, tested against a lithium metal anode at 0.4 mA/cm^2 current density between 2.0 and 4.4 V

ACKNOWLEDGMENTS

We thank G. Ceder and A. M. Mayes for helpful discussions. This work was supported by Furukawa Electric Company and the Idaho National Engineering and Environmental Laboratory under Grant No. C95-175002, and used instrumentation in the Shared Central Facilities at MIT, supported by NSF Grant No. 9400334-DMR.

REFERENCES

1. M. K. Aydinol, A. F. Kohan, G. Ceder, K. Cho, and J. Joannopoulos, Phys. Rev. B, 56, 1354 (1997)
2. J. N. Reimers and J. R. Dahn, J. Electrochem. Soc., 139, 2091 (1992)
3. J. R. Dahn, U. von Sacken, M. W. Jukow, and H. Al-Janaby, J. Electrochem. Soc., 138, 2207 (1991)
4. E. Rossen, C. D. W. Jones, and J. R. Dahn, Solid State Ionics, 57, 311 (1992)
5. J. N. Reimers, E. Rossen, C. D. Jones, and J. R. Dahn, Solid State Ionics, 61, 335 (1993)
6. R. Stoyanova, E. Zhecheva, and L. Zarkova, Solid State Ionics, 73, 233 (1994)
7. A. R. Armstrong and P. G. Bruce, Nature, 381, 499 (1996)
8. F. Capitaine, P. Gravereau, and C. Delmas, Solid State Ionics, 89, 197 (1996)
9. T. Ohzuku, A. Ueda, and M. Kouguchi, J. Electrochem. Soc., 142, 4033 (1995)
10. Q. Zhong and U. von Sacken, J. Power Sources, 54, 221 (1995)
11. G. A. Nazri, A. Rougier, and K. F. Kia, Mat. Res. Soc. Sym. Proc., 453, 635 (1997)
12. H. A. Lehmann and H. Hesselbarth, Z. Anorg. Allg. Chem., 313, 117 (1961)
13. Y.-M. Chiang, Y.-I. Jang, H. Wang, B. Huang, D. R. Sadoway, and P. Ye, J. Electrochem. Soc., in press
14. Y.-I. Jang, B. Huang, Y.-M. Chiang, and D. R. Sadoway, Submitted to J. Electrochem. Soc.
15. J. Morales, C. Pérez-Vicente, and J. L. Tirado, Mat. Res. Bull., 25, 623 (1990)
16. J. M. Tarascon and D. Guyomard, J. Electrochem. Soc., 144, 2587 (1991)
17. G. Vitins and K. West, J. Electrochem. Soc., 144, 2587 (1997)
18. M. M. Thackeray, W. I. F. David, P. G. Bruce, and J. B. Goodenough, Mat. Res. Bull., 18, 461 (1983)
19. L. Croguennec, P. Deniard, and R. Brec, J. Electrochem. Soc., 144, 3323 (1997)

HIGH DENSITY ACTIVE-SITE MnO$_2$ NANOFIBERS FOR ENERGY STORAGE AND CONVERSION APPLICATIONS

T.D. Xiao, D.E. Reisner, and P.R. Strutt
US Nanocorp, Inc., 20 Washington Avenue, North Haven, CT 06473

ABSTRACT

This investigation involves the synthesis of MnO$_2$ nanofibrous materials, via an aqueous chemical synthesis route. A critical step in the synthesis is the extended period of the refluxing of the reactive constituents, which enables the gradual transformation of the initial amorphous nanoparticles into a random-weave nanofibrous structure, in the form of a bird's nest superstructure. The bird's nest has a diameter of approximately 10 μm, which is an assemblage of many individual nanofibers with a diameter of about 15 nm, and a length up to 1 μm. Partial transformation of the nanostructured MnO$_2$ realizes a novel bimodal morphology, which combines a high density of chemically active sites with an enhanced percolation rate. Characterization of these nanofibers include SEM, TEM, surface area, and chemical analysis. High resolution TEM observations reveals that the as-synthesized MnO$_2$ nanofibers contain lattice defects, including molecular pores, meso- and micro-pores.

INTRODUCTION

In recent years, much interest has been focused on developing synthetic manganese oxide (MnO$_2$) [1-7]. This material is composed of MnO$_6$ octahedrons linked at their vertices and edges to form single or double chains [8-10]. These chains share corners with other chains, leading to structures with channels of empty sites. The size of these channels is related to the number of Mn-O chains on each side [11-12]. Naturally occurring MnO$_2$ coexists with many impurity oxide phases, including SiO$_2$, Fe$_2$O$_3$, Al$_2$O$_3$, and P$_2$O$_5$ [13-18]. These impurities limit possible applications and complicates the chemical and structural analyses of these materials. Approaches for the preparation of pure hollandite phases include ion-exchange reactions [19], hydrothermal synthesis [5,20], electrolytic synthesis [7], and chemical synthesis methods [5, 21-22].

Chemical methods for the preparation of phase pure MnO$_2$, in a variety of crystalline forms, were developed by McKenzie in the early 1970s [21]. Since then, several studies have been performed on the synthesis of layered MnO$_2$ by reacting manganese salts, such as MnCl$_2$ or MnSO$_4$, with strong oxidizers, such as KMnO$_4$ or ozone/oxygen mixtures [6, 23-26]. Most of this work has addressed synthetic methods and possible applications. Only a limited amount of research has been concerned with morphological aspects and lattice defect structures.

Here, we report a unique high percolation rate nanofibrous MnO$_2$ material for a wide range of applications, including rechargeable batteries and fuel cells. This nanostructured MnO$_2$ is synthesized via a patent pending wet synthesis process route that is scalable to large volume manufacturing and anticipated to be low in cost. The synthesis combines a refluxing technique with highly intense ultrasonic radiation. The morphology, microstructure, possible reaction mechanism, and possible application of these novel nanofibrous MnO$_2$ are also discussed.

Mat. Res. Soc. Symp. Proc. Vol. 496 © 1998 Materials Research Society

EXPERIMENTAL

Nanostructured MnO_2 is synthesized utilizing the chemical reaction of $KMnO_4$ and $MnSO_4$ in the presence of nitric acid (HNO_3), using a refluxing technique in combination with controlled solution mixing. The reaction may be written as follows:

$$2\ MnO_4^- + 3\ Mn^{2+} + 2\ H_2O \rightarrow 5\ MnO_2 + 4\ H^+ \qquad (1)$$

In a typical experiment, 36.8 g $KMnO_4$ (dissolved in 625 cc H_2O) was reacted with 55 g $MnSO_4$ (dissolved in 190 cc H_2O) at ambient temperature and pressure. The reaction product was then aged at 60-100 °C for 2-24 hours, ultrasonicated, filtered, washed, and dried at ~110 °C.

RESULTS AND DISCUSSIONS

The as-synthesized MnO_2 powders had a coffee-like color. Examination of the morphology by SEM revealed that the synthetic MnO_2 powders were weave-like fiber-networks (see Fig. 1). Detailed structural features could not be resolved in the SEM micrograph due to their nanostructured dimensions. At this resolution, the powders seemed to have a bird's-nest-like morphology formed by fiber networks. Each "bird's-nest" was about 10 μm in diameter, consisting of an assembly of many individual nanostructured fibers.

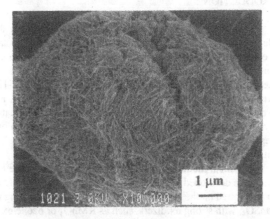

Fig. 1. SEM of the synthetic nanostructured MnO_2 material showing a bird's-nest morphology.

The XRD analysis revealed that the sample only contained the cryptomelane-type hollandite phase, with a chemical formula of KMn_8O_{16}. This phase has a monoclinic structure with the $I2/m$ space group. XRD analysis indicated that all peaks were broad, suggesting the crystallites were small. The width of the (020) peak was smaller than all other reflection peaks at lower angles, which could be attributed to an anisotropic crystallite morphology, indicating crystallites must be elongated along the <020> direction, parallel to the b-axis. Therefore, the fiber growth direction was assigned to the b-axis, or the <010> direction [22].

TEM images showed that the powders were formed by elongated nanocrystals or nanofibers which have diameters varying from a few nm to 25 nm, and lengths from several tens of nm up to 1 μm. High resolution transmission electron microscopy (HRTEM) was used to analyze individual nanofibers, in which atomic resolution was obtained. Fig. 2a is an HRTEM image of a nanofiber which reveals the (402)-plane type reflection, and Fig. 2b shows the atomic resolution of a nanofiber viewed perpendicular to the tunnel direction.

Fig. 2. HRTEM image showing (a) an individual fiber (402)-type lattice plane, (b) atomic resolution when viewed from <100> direction, or perpendicular to the fiber growth axis

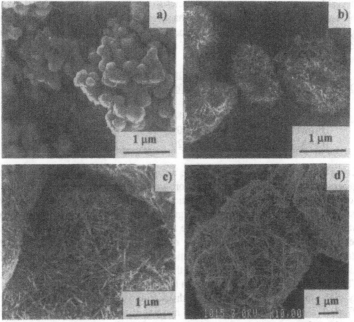

Fig. 3. SEM micrographs showing the nanofibrous bird's-nest growing after (a) precipitation, (b) refluxing for 2 hours, (c) refluxing for 8 hours, and (d) refluxing for 24 hours.

When the KMnO$_4$ solution was fed into the MnSO$_4$ solution as an aerosol, an amorphous powder was formed. A typical SEM micrograph, Fig. 3a, shows that the nanoparticles are in the form of spherical agglomerates, approximately 0.1-0.5 μm in diameter, with relatively smooth surfaces. Refluxing this solution at 60 °C for 2 h caused the size of the original nanoparticle agglomerates to increase (0.5 - 3 μm). The surfaces of the agglomerated masses began to show the development of a nanofibrous surface structure, with many nanofibers sticking out of the surfaces of the agglomerated mass, as in Fig. 3b. There were also some fiber-free agglomerates. After 4 h aging at the same temperature, the size of the agglomerated masses had grown to 1 - 5 μm, and many more of them displayed nanofibrous surface structures. There was still no change in the interior parts of the agglomerated nanoparticle masses. While the structure continued to develop more small fibers, it appeared that the largest fibers attained a maximum size of about 25 nm in cross section and approximately 0.5 μm in length. After 6 - 8 h aging, many of the larger agglomerates displayed a well-defined nanofibrous structure, or bird's-nest morphology, which consisted of a three-dimensional random weave of nanofibers (Fig. 3c). After 24 h aging, the nanofibrous structure was fully developed, and there were no visible remnants of the original nanoparticle agglomerates in their interiors. At this stage, a typical bird's-nest structure is about 3 - 10 μm in size, Fig. 3d, and is composed of many interwoven nanofibers that extend throughout the entire structure.

The as-precipitated material has the highest surface area of about 140 m^2/g. The general trend was that an extended period of refluxing resulting in a decrease in surface area, with the final surface area of about 83 m^2/g for the well developed MnO$_2$ materials. As shown in Table 1, another important characteristic of the synthetic material was that its properties could be greatly modified by changing dopant type and concentration. For example, by introducing Fe as the dopant species, the measured BET surface area can be increased to ~280 m^2/g, with a pore volume of 0.323 cc/g.

Table 1. Surface area (m^2/g) for MnO$_2$ with different dopants

Dopant Concentration	Co	Cu	Fe
0.01 M	84	85	90
0.1 M	105	95	95
0.4 M	120	115	280

Crystal imperfection, or defects, were also found in the synthetic MnO$_2$ nanofibers. Viewing perpendicular to the tunnel direction, we observed an arrangement of two edge dislocations of opposite sign, viz., a dislocation dipole [22]. These dislocations terminate along a straight line, which is normal to the growth axis, or b-axis. Specifically, the dipole creates vacancy rows (cavities) normal to the nanofiber axis. Similarly, viewing parallel to the tunnel direction (fiber cross-section), we observed the termination of a dislocation line parallel to the growth axis [27]. This dislocation line arose out of the intersection of four of the mentioned dislocation lines, which met at the nanofiber centerline.

The existence of dislocation dipoles in the nanofibers has implications with respect to the utility of the material. In ionic crystals, dislocations terminating inside the fibers would create effective charges associated with the merging dislocation sites. The sign of these charges is dependent on the atomic environment surrounding the cavities. Insertion of an element of opposite sign may invert the effective charge at the cavity by a simple sideways displacement to maintain charge balance. All these dislocation were frequently observed for the synthetic MnO_2 nanofibrous materials.

The synthetic MnO_2 is associated with many types of defects, summarized in Fig.4. Features in a nanofiber would include lattice tunnels, nano-holes, dislocation dipoles, edge dislocations, twinning, and surface steps. These features are important to electrochemical and catalytic applications, where these defects are active centers or sites for reactions and provide paths for the diffusion of mobile species including Li^+ and H^+. The cavities within a nanofiber are also centers for effective charge diffusion including electrons or ions. Other features are the high degree of the porosity, including, micro- and mesoporosity. The nature of the porosity can be altered by doping with different concentration of transition metal elements. A material with intrinsic defects is shown in Fig. 4b, and materials with dopant defect and hybrid defect features are shown in Figs. 4c and 4d.

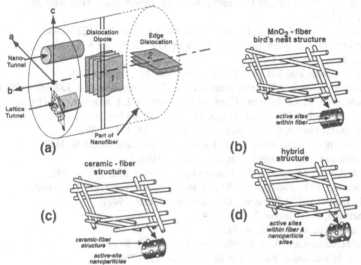

Fig. 4. Schematics of the nanostructured MnO_2 bird's-nest materials, (a) a section of the fiber, (b) intrinsic defects, (c) extrinsic defects, and (d) hybrid structure.

CONCLUSIONS

In summary, the nanostructured MnO_2 material has been synthesized via an aqueous solution technique. This new class of nanostructured active materials offers the potential for boosting the performance of rechargeable batteries and fuel cells, as well as other catalysts for environmental applications. The resultant materials have extremely high surface area, homogeneous porosity and pore size distribution. The high density of internal defects of the as-synthesized MnO_2 provide a high density of electrochemically and catalytically active sites, while the open-weave structure ensures a high percolation rate through the structure. This unique combination of high active sites density and high percolation rate has not been realized before.

ACKNOWLEDGMENTS

Portions of this work were funded under a grant from the National Science Foundation (DMI 9660798) to Inframat Corporation. Our thanks also go to Ms. H. Chen of the University of Connecticut, Dr. Mohamed Benaissa and Mr. Luis Rendon of IFUNAM, Mexico, and Alfredo Aguilar-Elguezabal of CIMAV, Mexico for technical help.

REFERENCES

[1]. T.D. Xiao, P.R. Strutt, B.H. Kear, H.Chen, and D.M. Wang, US Patent application filed 11/18/96.
[2]. K. Kordesch and M. Weissenbacher, *J. Power Source* **51**, 62 (1994).
[3]. M.H. Rossouw, D.C. Liles, M.M. Thackeray, W.I.F. David, and S. Hull, *Mat. Res. Bull.* **27**, 221 (1992).
[4]. W.H. Kao, V.J. Weibel, and M.J. Root, *J. Electrochem. Soc.* **139**, 1223 (1992).
[5]. R.N. DeGuzman, Y.F. Shen, E.J. Neth, S.L. Suib, C.L. O_Young, S. Levine, and J.M. Newsan, *Chem. Mater.* **6**, 815 (1994).
[6]. Q. Feng, H. Hanoh, and K. Ooi, *J. Electrochem. Soc.* **141**, L135 (1994).
[7]. T. Ohzuku, M. Kitagawa, K. Sawai, and T. Hirai, *J. Electrochem. Soc.* **138**, 360 (1991).
[8]. G. Parravano, in "Procs. 4th Intnl. Cong. on Catalysis," vol. 1, Moscow, 149 (1971).
[9]. J.R. Goldstein and A.C.C. Tseng, *J.Phys. Chem.* **76**, 3646 (1972).
[10]. A.I. Onuchukwn and A.B. Euru, *Mater. Chem. Phys.* **15**, 131 (1986).
[11]. J.M. Tarascon, E. Wang, and F.K. Shokoohi, *J. Electrochem. Soc.* **138**, 2859 (1991).
[12]. A. Byström and A.M. Byström, *Acta Cryst.* **3**, 146 (1950).
[13]. J.E. Post, R.B. Von Dreele, and P.R. Buseck, *Acta Cryst.* **B38**, 1056 (1982).
[14]. G.M. Faulring, W.K. Zwicker, and W.D. Forgeng, *Am. Mineral.* **45**, 946 (1960).
[15]. W.E. Richmond and M. Fleischer, *Am. Mineral.* **27**, 607 (1942).
[16]. L.S. Ramsdell, *Am. Mineral.* **27**, 611 (1942).
[17]. J.W. Gruner, *Am. Mineral.* **28**, 497 (1943).
[18]. A. McL. Mathieson and A.D. Wadsley, *Am. Mineral.* **35**, 99 (1950).
[19]. K. Ooi, Y. Miyai, and S. Katoh, *Sep. Sci. Technol.* **22**, 1779 (1987).
[20]. M. Tsuji and M. Abe, *Solvent Extrant. Ion Exch.* **2**, 253 (1984).
[21]. R.M. McKenzie, *Mineral. Magazine* **38**, 493 (1971).
[22]. T.D. Xiao, Bokimi, M. Benaissa, R. Pérez, Peter R. Strutt, and Miguel José Yacamán, *Acta Mater.* **45**, 1685 (1997).
[23]. M.M. Thackeray, *J. Electrochem. Soc.* **142**, 2558-2563 (1995).
[24]. Armstrong, A.R., & Bruce, P.G., *Nature* **381**, 499-500 (1996).
[25]. Golden, D.C., Chen, C.C., Dixon, J.B., *Science* **231**, 717-719 (1986).
[26]. Shen, Y.F., Zerger, R.P., DeGuzman, R.N., Suib, S.L., McCurdy, L., Potter, P.I., & Young, C.L., *Science* **260**, 511-515 (1993).
[27]. M. Benaissa, M.J. Yacaman, T.D. Xiao, and P.R. Strutt, *Appl. Phys. Lett.* **70(16)**, 2120 (1997)

VIBRATIONAL SPECTROSCOPIC STUDIES OF THE LOCAL ENVIRONMENT IN 4-VOLT CATHODE MATERIALS

C. JULIEN*, M. MASSOT*, C. PEREZ-VICENTE*, E. HARO-PONIATOWSKI*
G.A. NAZRI**, A. ROUGIER**
*Laboratoire des Milieux Désordonnés et Hétérogènes, UA800, Université Pierre et Marie Curie
4 place Jussieu, 75252 Paris cedex 05, France
**Physics and Physical Chemistry Department, RCEL, General Motors R&D Center
Warren, MI 48090, USA

ABSTRACT

We report the vibrational spectra of numerous 4-volt cathode materials, the transition metal oxides which are potential materials for advanced Li-ion batteries. They provide high specific energy density, high voltage, and remarkable reversibility for lithium intercalation-deintercalation process. Studied were carried out by Raman and FTIR spectroscopies. Oxides such as $LiMn_2O_4$, $LiNiVO_4$, $LiCoVO_4$ spinels, $LiMeO_2$ (Me=Co, Ni, Cr) layered compounds and their mixed compounds have been investigated. The local environment of cations against oxygen neighboring atoms has been determined by considering tetrahedral and octahedral units building the lattice. Structural modifications induced by the intercalation-deintercalation process, by the cation substitution, or by the low-temperature preparation route are also examined. The results are compared with those of end members.

INTRODUCTION

The recent interest in developing advanced rechargeable lithium batteries such as lithium-ion type cells, has stimulated investigation on high-performance positive electrodes, the so-called four-volt cathode materials [1]. Transition metal oxides are potential cathode materials providing high specific energy density, high voltage, and remarkable reversibility for lithium intercalation-deintercalation process. According to their crystallographic structure, one can distinguish two classes of materials: those with a spinel-type structure such as $LiMn_2O_4$, $LiNiVO_4$, $LiCoVO_4$, and those with a layered-type structure such as $LiMeO_2$ (Me=Co, Ni, Cr) compounds and their mixed compounds.

Since the electrochemical characteristics of the 4-volt materials depend on their crystal nature, size and shape, various kinds of preparation methods have been developed to improve the characteristics of these cathodes [2]. Also, it has been pointed out that the rechargeability can be markedly improved and the fading loss upon cycling can be minimized for mixed or for substituted transition metal oxides [3-4]. The current debate consists in the investigation of the local structure of these compounds because of the difficulty encountered in using powder x-ray and neutron diffraction for an unambiguous structural determination [5].

Vibrational spectroscopies, i.e., Raman scattering (RS) and Fourier transform infrared (FTIR), are sensitive to the short-range environment of oxygen coordination around the cations in oxide lattices. Since RS and FTIR techniques can solve the problem of phase determination when various environments are present. Spectra consist of a superposition of the components of all local entities. The frequencies and relative intensities of the bands are sensitive to coordination geometry and oxidation states. Thus, spectra are less affected by the grain size or the degree of long-range order of the lattice.

We present in this paper the vibrational spectra of numerous 4-volt cathode materials with spinel-type and layered-type structure. Oxides such as $LiMn_2O_4$, $LiMeVO_4$ (Me=Ni, Co) spinels, $LiMeO_2$ (Me=Co, Ni, Cr) layered compounds and their mixed compounds have been investigated. Lattice dynamics are studied using either a classical group theoretical analysis or a local environment model. Raman and FTIR bands are identified on the basis of vibrational modes of tetrahedral and octahedral units which are building the structure.

415

Mat. Res. Soc. Symp. Proc. Vol. 496 © 1998 Materials Research Society

EXPERIMENT

Raman scattering spectra of 4-volt cathode materials were recorded with a spectral resolution of 1 cm^{-1} using a Jobin-Yvon U1000 double-monochromator equipped with a microscope. Powdered samples were placed under a microscope objective (50x) which allows the laser beam to focus on a small selected area of the sample surface (about 10 μm^2), and the backscattered Raman signal was collected. The laser light source was the 514.5 nm line radiation excited at 10 mW from a Spectra-Physics 2020 argon-ion laser. Standard photon-counting techniques were used for detection. Each spectrum is the average of 20 scans.

Infrared absorption spectra were recorded using a Bruker IFS113 vacuum Fourier transform infrared interferometer. In the far-infrared region, this apparatus was equipped with a 3.5 μm-thick beam splitter, a globar source, and a liquid helium cooled bolometer at a spectral resolution of 2 cm^{-1}. Samples were ground to fine powders painted onto pellets of solid paraffin. The solid paraffin does not exhibit any infrared absorption peak in the studied wavenumber range.

RESULTS AND DISCUSSION

Spinel-type compounds

The cubic spinel $LiMn_2O_4$ possesses Fd3m symmetry and has a general structural formula $A[B_2]O_4$, where the manganese cations reside on the octahedral 16d sites, the oxygen anions on the 32e sites, and the lithium ions occupy the tetrahedral 8a sites. Analysis of the vibrational spectra of $LiMn_2O_4$ yields nine optic modes: five Raman-active modes ($A_{1g}+E_g+3F_{2g}$) and four infrared-active modes (F_{1u}). It is also convenient to analyse these spectra in terms of localized vibrations, considering the spinel structure built of MnO_6 octahedra and LiO_4 tetrahedra [6].

The Raman band located at ca. 625 cm^{-1} can be viewed as the symmetric Mn-O stretching vibration of MnO_6 groups. The position and the halfwidth of this band remain almost unchanged upon delithiation. This band is assigned to the A_{1g} symmetry in the O_h^7 spectroscopic space group. Its broadening may be related with the cation-anion bond lengths and polyhedral distortion occurring in $LiMn_2O_4$. The intensity of the shoulder located at 580 cm^{-1} increases upon lithium deintercalation. This may be due to the change of Mn^{3+} and Mn^{4+} proportion vs. x in the material.

It is a fact that $Li[Mn^{3+}Mn^{4+}]O_4$ is a small-polaron semiconductor. The RS peak at 304 cm^{-1} has the E_g symmetry whereas the peaks located at 154, 365 and 480 cm^{-1} have the F_{2g} symmetry. The high-frequency bands of the FTIR absorption spectrum of $LiMn_2O_4$ located at ca. 619 and 513 cm^{-1} are attributed to the asymmetric stretching modes of MnO_6 group whereas the low-frequency bands at 225, 277 and 360 cm^{-1} are assigned to the bending modes of Mn-O. Because FTIR spectroscopy is capable of probing directly the near neighbor environment of the cation, we can study the local environment of lithium ions in this material. The vibrational frequency of the LiO_4 tetrahedron has appeared out at 435 cm^{-1} in the spinel $LiFeCr_4O_8$. It has been also demonstrated that the IR resonant frequencies of alkali metal cations in their equilibrium positions in inorganic oxide glasses are cation mass dependent bands [6]. Thus, the IR resonant frequency is related to the local force constant with an effective mass of vibration which is roughly that of alkali ion. This leads to leads to the frequency at 440 cm^{-1} for oscillation of the Li^+ ion with O^{2-} neighbors in $LiMn_2O_4$.

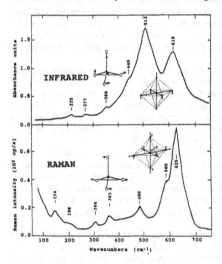

Fig. 1 Raman scattering and FTIR spectra of the spinel $LiMn_2O_4$.

It can be stated that in the ideal cubic spinel $LiMn_2O_4$, the Mn^{3+} and Mn^{4+} cations are considered as crystallographically equivalent (16d sites) in agreement with X-ray diffraction data; then, occupation probabilities of 0.5 must be affected for each cation in 16d. Hence, a loss of translation invariance certainly occurs, due to local lattice distorsions around the different Mn^{3+} and Mn^{4+} cations. As a result, a breakdown in the Raman and IR selection rules is expected, which may explain the observation of broads bands (disorder) and the fact that more modes than expected are observed in cubic $LiMn_2O_4$. Furthermore, this would be consistent with the fact that the $LiMn_2O_4$ spectra are not markedly different from those of $Li_xMn_2O_4$ where, of course, disorder necessarily takes place because of the non stoichiometry.

Vanadate compounds, $LiMeVO_4$ (Me=Ni, Co) crystallize in the inverse spinel structure with a prototypic symmetry $Fd\overline{3}m$. The pentavalent vanadium is located on the tetrahedral 8a sites, whereas Li and Ni are distributed on the octahedral 16d sites, the distribution being disordered in the simple spinel structure. Factor group theoretical analysis for inverse spinel-type compounds yields similar mode activities than for the normal spinel optic modes [7-8].

Figure 2 shows the Raman scattering spectra of single phase $LiNiVO_4$ and $LiCoVO_4$ annealed at 500°C. From these results, it can be observed that: (a) these spectra exhibit the complex pattern of broad bands in the 700-850 cm^{-1} region, (b) the high frequency vibrational bands are the dominant ones. FTIR spectra exhibit some additional peaks which are not expected from the group theory. The high-frequency vibrational bands of an inverse spinel can be assigned to a vibration between the oxygen and the highest-valency cation. Two vibrational modes, namely the stretching frequencies of the VO_4 tetrahedron are expected to be observed in this region. Considering an isolated VO_4 tetrahedron the mode frequencies of this $[VO_4]$ unit have been observed [9] thus, in $LiMeVO_4$ (Me=Ni, Co) the high-frequency band located at ca. 800 cm^{-1} corresponds to the stretching mode of VO_4 tetrahedron which has the A_1 symmetry, whereas the band situated at ca. 310-340 cm^{-1} corresponds to the bending mode of VO_4 tetrahedron with an E symmetry. These vibrations are Raman-active modes. The broadness of the high frequency band could be tentatively explained in terms of asymmetrical bonding of VO_4 tetrahedron. Two types of cations, namely Li and (Ni, Co) may be bonded with each oxygen atom of a VO_4 tetrahedron. This introduces some asymmetry in the VO_4 unit without disturbing the overall cubic symmetry of the elementary unit cell. This can occur at the origin of the appearance of additional infrared-active modes as well. The bands observed in the 700-850 cm^{-1} region is attributed to the asymmetric stretching vibrations of VO_4 tetrahedron.

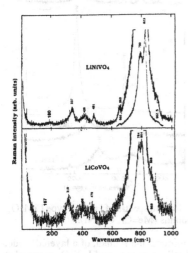

Fig. 2. Raman spectra of $LiNiVO_4$ and $LiCoVO_4$ inverse spinels.

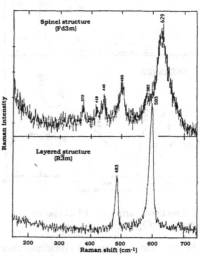

Fig. 3. Raman spectra of $LiCoO_2$ with a spinel and a layered structure.

A detailed interpretation of the medium-frequency region of Raman and FTIR spectra is difficult because different types of vibrations exist in this spectral domain, involving simultaneously significant displacement of octahedral and tetrahedral cations. Therefore, one may observe either the bending vibrations of VO_4 tetrahedron or the vibrations involving the octahedral MeO_6, LiO_6 environments. If one considers that all the Li ions are accommodated in octahedral LiO_6 environments, the F_{1u} modes are normally split into (A+2B) Raman-active and IR-active components. Therefore, IR modes having (A+2B) symmetry are intense whereas Raman modes are expected to be very weak. For $LiNiVO_4$, these two modes are observed at 420 and 435 cm^{-1}, respectively.

Figure 3 displays the Raman spectra of the spinel and layered structure of $LiCoO_2$. A spinel structure with the space group Fd3m is unambiguously assigned to $LiCoO_2$ synthesized at 350°C according to the Raman spectrum and group factor analysis. The vibrational modes involving significant motion of lithium atoms are observed at 418 and 440 cm^{-1} for low-temperature $LiCoO_2$.

<u>Layered-type compounds</u>

Materials $LiMeO_2$ (Me=Ni, Cr, Co) with a layered structure that belong to crystallographic $R\bar{3}m$ space group have a corresponding spectroscopic space group of D^5_{3d} [4]. The Bravais cell contains one molecule (Z=1) and the atoms Li, (Co, Cr, Ni), and O are located at Wyckloff sites 3b, 3a, and 6c, respectively. The corresponding symmetry for the D^5_{3d} yields four infrared-active modes ($2A_{2u}+2E_u$) and two Raman-active modes ($A_{1g}+E_g$). The Raman-active modes correspond mainly to vibrations of oxygen cages (oxygen-oxygen vibration in the c direction and in parallel to the Li and transition-metal planes).

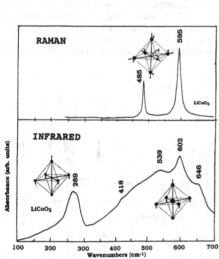

Fig. 4. Raman scattering and FTIR spectra of layered $LiCoO_2$.

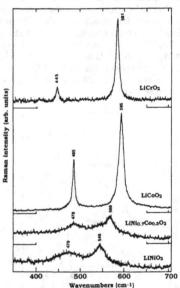

Fig. 5. Raman spectra of layered-type $LiMeO_2$ compounds (Me=Cr, Co, Ni).

Figure 4 shows the typical vibrational spectra, i.e., infrared and Raman, of a layered oxide. Figure 5 shows the Raman spectra of $LiCrO_2$, $LiCoO_2$, $LiNi_{0.7}Co_{0.3}O_2$, and $LiNiO_2$ recorded at room temperature. The two Raman-active modes for $LiMeO_2$ are located at 400-650 cm^{-1} region. One observes two strong peaks at 485 and 595 cm^{-1} for $LiCoO_2$, which are in good agreement with the factor group analysis. They are attributed to the E_g and A_{1g} Raman-active modes, respectively. The corresponding modes for $LiNiO_2$ are located at 470 and 546 cm-1, respectively. In the A_{1g} mode the two oxygen atoms vibrate in the opposite directions parallel to the c-axis of $LiMeO_2$, while they vibrate alternately in the opposite directions parallel to the lithium and transition-metal atom planes in the E_g mode.

The position of these bands is sensitive to the nature of the transition metal and their intensity is very sensitive to the long-range order in MeO_2 slabs. As pointed out by Inaba et al. [10], another reason for the decrease in intensity is an increase in electrical conductivity. It was reported that the electrical conductivity of $LiNiO_2$ is higher than that in $LiCoO_2$ because of their electronic structure. The electrical conductivity of $LiNi_{0.7}Co_{0.3}O_2$, which is considered to be similar to that of $LiNiO_2$, reduces the Raman scattering efficiency due to the weak optical skin depth of the incident laser beam from the surface of the samples.

The A_{1g} mode shift from 595 cm-1 in $LiCoO_2$ to 546 cm-1 in $LiNiO_2$, is attributed to the evolution of the hexagonal cell parameters and to the modification of the c/a ratio of the mixed $LiNi_{1-y}Co_yO_2$ phases. The increase of the inner slab bond covalency in $(Ni_{1-y}Co_yO_2)_n$ as cobalt content increases explains the decrease of the metal-metal intralayer distance which induces a frequency shift of the Raman-active modes. One possible explanation of the broadening of the vibrational bands for Ni-rich compounds is the cation mixing in the crystal layers. The partially disordered cation distribution appearing in lithium nickelate materials can also explain the observed broadening of the Raman spectrum.

In order to get a better understanding of the vibrational spectra of the layered $LiMeO_2$ compounds with $R\bar{3}m$ space group, we consider a structure which consists of compressed MeO_6 and elongated LiO_6 octahedra that yields distinct vibrations in two different frequency regions, i.e., at 400-650 cm-1 there are bands due to MeO_6 vibrations, while the LiO_6 vibrations are within the region 200-400 cm-1. The Raman peaks involve mainly the Me-O stretching and O-Me-O bending vibrations, as the Raman mode contributions originate from the motion of oxygen atoms.

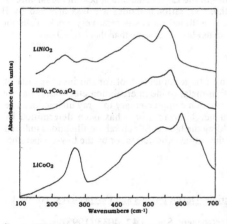 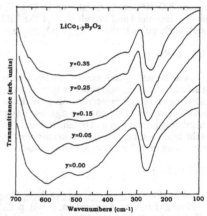

Fig. 6. FTIR absorbance spectra of $LiCoO_2$, $LiNi_{0.7}Co_{0.3}O_2$, and $LiNiO_2$.

Fig. 7. FTIR transmittance spectra of boron substituted $LiCo_{1-y}B_yO_2$ compounds.

The FTIR spectra of $LiCoO_2$, $LiNi_{0.7}Co_{0.3}O_2$, and $LiNiO_2$ recorded at room temperature are shown in Fig. 6. As expected from the factor group analysis, these spectra display four distinct infrared bands. The spectrum of $LiCoO_2$ matches well with those reported recently [11]. A closer examination of the shape of the high-wavenumber band at 602 cm-1 indicates that a shoulder exists at 646 cm-1. The infrared-active bands shown in Fig. 6 are generally broader than those observed in the Raman spectra (Fig. 5). For example, the half-width at half-maximum (HWHM) of the Raman high-wavenumber band for $LiNiO_2$ is about 30 cm-1, whereas the HWHM of the IR high-wavenumber band is 58 cm^{-1}. The pure $LiNiO_2$ phase has two IR peaks between 400-600 cm^{-1}. These two peaks are characteristic of the NiO_6 vibrations. Substitution by cobalt displaces these peaks toward higher wavenumbers. The systematic change in band frequency as a function of cobalt substitution is not surprising, as in solid solutions with various concentrations of substitution, a systematic shift has been observed [4].

The isotopic ^6Li-^7Li replacement in $LiMeO_2$ with D^5_{3d} space group has proven that the far-infrared peak between 200-300 cm^{-1} is related to an asymmetric stretching vibration of LiO_6 [11]. The IR vibrational modes of the LiO_6 octahedra appear at 269 and 237 cm^{-1} for $LiCoO_2$ and $LiNiO_2$, respectively. The frequency of the LiO_6 octahedra is sensitive to the ionic radius of the M^{III} and ionicity of the bonds in MeO_6 octahedra. For $LiNi_{0.7}Co_{0.3}O_2$ the asymmetric stretching LiO_6 vibration occurs at 250 cm^{-1}. Vibrational spectra of the MeO_6 octahedra are also affected by the way in which these octahedra are linked to each other. The broadening of the IR peaks can be mainly interpreted as an increase in MeO_6 distortion. For $LiNiO_2$, the broadening of the IR bands (Fig. 6) can be associated with the distortion of the NiO_6 octahedra due to the Jahn-teller effect of Ni^{III} in this compound. As cobalt substitutes, the shift of the IR-active modes toward higher wavenumbers confirms the formation of mixed nickel-cobalt layers in $LiNi_{0.7}Co_{0.3}O_2$, which exhibits the contraction of the elementary unit cell.

In order to improve the cycle life, several efforts have been made, such as the addition (or substitution) of aluminum, boron, titanium, or fluorine to lithium nickelate and lithium colbatate, and the effects of the addition of these species upon cyclability were described [2]. We studied the lattice vibrational properties of the $LiMeO_2$ substituted with Al and B. Figure 7 shows infrared spectra of $LiCo_{1-y}B_yO_2$ with various amounts of B substitution. The broadening of the high-frequency FTIR peak can be interpreted as an increase in CoO_6 distortion due to the addition of boron in the CoO_2 layers. It is well known that boron oxide is used as glass former, and it may reduces the long-range order in the $LiMeO_2$. This effect is also seen in x-ray diffraction of lithium nickelate doped with boron. Similar distortion of lattice is also observed when the $LiCoO_2$ is doped with Al [4]. The infrared vibrational modes of the LiO_6 octahedra appear in the far-infared region, 200-300 cm^{-1}. As shown in Fig. 7, these vibrational modes are less affected by B substitution; the band broadening of the LiO_6 vibrational modes is relatively weak. Thus the interlayer bonding in the $LiCo_{1-y}B_yO_2$ lattice remains almost the same than that $LiCoO_2$.

CONCLUSION

Vibrational spectroscopy has been shown to be a valuable tool for the investigation of structural and compositional changes in transition-metal oxide intercalation electrodes. Raman scattering and FTIR measurements provide information complementary to x-ray diffraction. The local environment of cations against oxygen neighboring atoms has been determined by considering tetrahedral and octahedral units building the lattice. Structural modifications induced by the intercalation-deintercalation process, by the cation substitution, or by the low-temperature preparation route have been also examined.

REFERENCES

1. T. Ohzuku, in Lithium Batteries, New Materials, Developments and Perspectives, ed. by G. Pistoia (Elsevier, Amsterdam, 1994), p. 239.
2. T. Ohzuku, A. Ueda, and M. Kitagawa, J. Electrochem. Soc., **142**, 4033 (1995).
3. C. Delmas, I. Saadoune, and A. Rougier, J. Power Sources **43-44**, 595 (1993).
4. G.A. Nazri, A. Rougier, and K.F. Kia, Mater. Res. Soc. Symp. Proc. **453**, 635 (1997).
5. W. Huang and R. Frech, Solid State Ionics **86-88**, 395 (1996).
6. C. Julien, A. Rougier, and G.A. Nazri, Mater. Res. Soc. Symp. Proc. **453**, 647 (1997).
7. J. Preudhomme and P. Tarte, Spectrochim. Acta **27A**, 845 (1971).
8. P. Tarte and J. Preudhomme, Spectrochim. Acta **26A**, 747 (1970).
9. N. Weinstock, H. Schulze, and A. Muller, J. Chem. Phys. 59, 5063 (1973).
10. M. Inaba, Y. Todzuka, H. Yoshida, Y. Grincourt, A. Tasaka, Y. Tomida, and Z. Ogumi, Chem. Lett. 889 (1995).
11. P. Hope and B. Schepers, Z. Anorg. Allgem. Chem. **295**, 233 (1958).

AMORPHOUS AND NANOCRYSTALLINE OXIDE ELECTRODES FOR RECHARGEABLE LITHIUM BATTERIES

A. MANTHIRAM, J. KIM AND C. TSANG
Center for Materials Science and Engineering, ETC 9.104, The University of Texas at Austin, Austin, TX 78712

ABSTRACT

Oxo ions $(MO_4)^{n-}$ (M = V, Cr, Mn and Mo) have been reduced in aqueous solutions with potassium borohydride to obtain the binary oxides $MO_{2+\delta}$. While the vanadium and manganese oxides are nanocrystalline, the chromium and molybdenum oxides are amorphous. The nanocrystalline VO_2 having a metastable structure and the amorphous CrO_2 and $MoO_{2.3}$ transform to the thermodynamically more stable phases upon heating above 300-400 °C. These metastable oxides after heating in vacuum at 200-300 °C to remove water show good electrode performance in lithium cells. VO_2, CrO_2 and $MoO_{2.3}$ show a reversible capacity of, respectively, 290 mAh/g in the range 4-1.5 V, 180 mAh/g in the range 3.3-2.3 V, and 220 mAh/g in the range 3-1 V. MnO_2 obtained by this process does not show good electrode properties.

INTRODUCTION

Lithium batteries offer higher energy density compared to other rechargeable systems and are attractive as power sources for portable electronic devices [1]. The currently available lithium-ion cells use the layered $LiCoO_2$ as the cathode and carbon as the anode. However, cobalt is expensive and toxic, which demands the development of alternate cathodes. Transition metal oxide cathodes are traditionally made by repeated grinding and firing of the reactants at elevated temperatures in order to overcome the diffusional limitations. Such high temperature procedures generally give the thermodynamically more stable phases. On the other hand, solution-based, low-temperature approaches can access metastable phases and unusual valence states that are otherwise inaccessible by conventional solid state reactions. In addition, the solution-based approach can give smaller particle size, which may be attractive to achieve faster lithium-ion diffusion and high rate capability in electrode hosts. We investigated the reduction of oxo ions $(MO_4)^{n-}$ (M = V, Cr, Mn and Mo) with potassium borohydride KBH_4 in aqueous solutions to obtain amorphous and nanocrystalline electrode hosts. The synthesis and electrode properties are presented in this paper.

EXPERIMENTAL

Borohydrides are known to be effective reducing agents in aqueous solutions [2]. The borohydride ion hydrolyzes in aqueous solutions to give hydrogen and the hydrolysis reaction is facilitated by acidic conditions. The reduction of oxo ions with borohydrides were carried out by adding a known volume of KBH_4 solution from a burette into a known volume of K_3VO_4, K_2CrO_4, K_2MoO_4, or $KMnO_4$ solution kept under constant stirring on a magnetic stirrer [3-6]. During the reduction reaction, the pH was maintained constant at a predetermined value by adding hydrochloric acid as the pH tends to increase during the reaction due to the formation of the basic KOH or KBO_2 (see reaction 1 later). The product formed was filtered, washed with water and dried in vacuum at 200-300 °C for M = V, Cr and Mo or in air at 500 °C for M = Mn before making the electrodes.

While identification of crystalline phases was carried out by X-ray diffraction, crystalline versus amorphous character of the as-prepared samples was studied by Transmission Electron Microscopy (TEM). Oxygen contents and oxidation states of the products were determined by redox titrations or thermogravimetric analysis (TGA) [4, 6, 7]. Electrochemical properties were studied with coin-type cells employing metallic lithium anodes and 1M LiClO$_4$ in propylene carbonate / 1,2-dimethoxyethane (1:1 ratio) electrolyte. The electrodes were fabricated by mixing the vacuum- or air-heated samples with 25 wt% fine carbon and 5 wt% polytetrafluoroethylene, rolling the mixture into thin sheets, and punching out circular electrodes of about 2 cm^2 area.

RESULTS AND DISCUSSION

Reduction of the oxo ions with borohydrides give both the binary oxides MO$_z$ and the ternary oxides A$_x$M$_y$O$_z$ depending upon the reaction pH and the concentration and volume of borohydride [3-6]. The chemical reaction for the formation of, for example, the binary oxide VO$_2$ can be given as

$$2K_3VO_4 + 2KBH_4 + 6H_2O \longrightarrow 2KBO_2 + 2VO_2 + 6KOH + 7H_2 \qquad (1)$$

Although the reducing power of borohydride increases with decreasing pH [2], the increasing degree of condensation of the oxo ions with decreasing pH to give poly anions [8] leads to a complex dependence of the reduction products on pH as the condensed poly anions are more difficult to reduce than the monomeric $(MO_4)^{n-}$ ions. The reaction conditions to obtain the binary oxides that were used as electrodes in this study are given in Table 1. The chemical compositions given in Table 1 refer to that of the samples used for making the electrodes after heating as described in the experimental section.

Table 1 Synthesis conditions and product compositions

Reactants	pH	Product
0.25 M K$_3$VO$_4$ (50 ml) + 0.25 M KBH$_4$ (300 ml)	4	VO$_2$
0.25 M K$_2$CrO$_4$ (50 ml) + 0.1 M KBH$_4$ (10 ml)	4	CrO$_{1.99}$
0.25 M K$_2$MoO$_4$ (50 ml) + 2.5 M KBH$_4$ (50 ml)	1	MoO$_{2.3}$
0.25 M KMnO$_4$ (100 ml) + 2.5 M KBH$_4$ (5 ml)	1	MnO$_2$

The VO$_2$ sample after heating in vacuum at 240 °C was found to be nanocrystalline as revealed by TEM and X-ray diffraction. It belong to the metastable form called VO$_2$(B) and transforms irreversibly to the more stable rutile form above 300 °C as indicated by X-ray diffraction and DSC [9]. The first discharge-charge curves and cyclability of nanocrystalline VO$_2$(B) after heating in vacuum at 240 °C are shown in Fig. 1. It shows superior performance compared to that obtained in the literature by a reduction of NH$_4$VO$_3$ [10]. While the VO$_2$(B) prepared by the literature methods insert about 0.5 lithium per vanadium, the VO$_2$(B) prepared by our solution-based approach inserts reversibly about 0.9 Li per V with a larger capacity of about 290 mAh/g and excellent cyclability. We believe, the smaller particle size achieved by the solution-based approach and the accompanying microstructure play a critical role in giving better electrode properties. In addition, VO$_2$(B) is tedious to synthesize as a single phase material by the literature procedures [11]. Our method is simple and gives readily single phase material.

The CrO$_2$ sample after annealing in vacuum at 250 °C was found to be amorphous as indicated by the absence of diffraction peaks in the X-ray pattern. The amorphous nature was also confirmed by the absence of diffraction spots in electron diffraction. However, TGA experiments followed by an examination of the products reveal that the amorphous CrO$_2$ lose oxygen and crystallize to give Cr$_2$O$_3$ around 430 °C [6]. The oxygen loss and crystallization was also indicated by a sharp exothermic peak around 430 °C in Differential Calorimeter (DSC). The first discharge-charge curves of amorphous CrO$_2$ after heating in vacuum at 250 °C are given in Fig. 2a for two current densities. With a current density of 0.1 mA/cm^2, it shows an initial capacity of 180

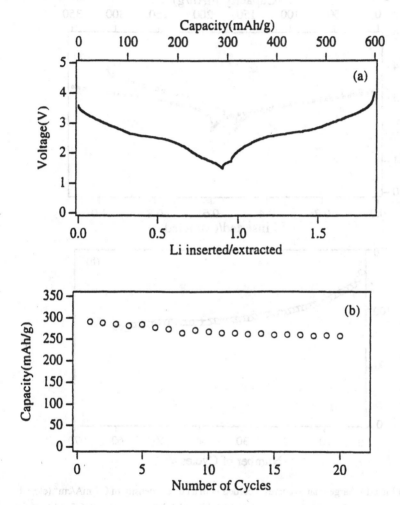

Fig. 1 (a) First discharge-charge curves, and (b) cyclability of nanocrystalline VO$_2$(B) recorded with a current density of 0.5 mA/cm^2.

mAh/g in the range 3.3 - 2.3 V, which corresponds to an insertion of about 0.6 lithium per Cr. Upon increasing the current density to 0.5 mA/cm^2, the initial capacity decreases to about 150 mAh/g. The cyclability of CrO$_2$ with a current density of 0.5 mA/cm^2 is shown in Fig. 2b. After an initial decline during the first 20 cycles, the capacity tends to stabilize around 90 mAh/g. The electrochemical behavior of amorphous CrO$_2$ differs distinctly from that of crystalline CrO$_2$. While the former shows a capacity of 150 mAh/g in the range 3.3-2.3 V, the latter shows a capacity of 50 mAh/g in the range 3-1 V [12] for the same current density.

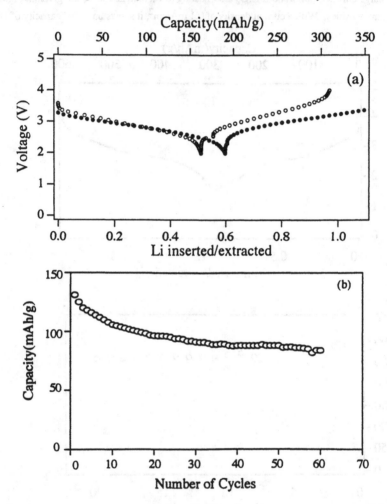

Fig.2 (a) First discharge-charge curves recorded with a current density of 0.1 mA/cm^2 (closed circles) and 0.5 mA/cm^2 (open circles), and (b) cyclability recorded with 0.5 mA/cm^2 for amorphous CrO$_2$.

MoO$_{2.3}$ obtained after heating in vacuum at 300 °C was found to be amorphous as revealed by X-ray diffraction and TEM. It transforms irreversibly to the rutile MoO$_2$ upon heating to 370 °C [7]. The first discharge-charge curves and cyclability of amorphous MoO$_{2.3}$ are shown in Fig. 3. It shows a reversible capacity of about 220 mAh/g corresponding to an insertion of 1.1 lithium per Mo with good cyclability in the range 3-1 V.

MnO$_2$ obtained after heating in air at about 300 °C was found to be nanocrystalline as revealed by TEM and contain some potassium. But X-ray diffraction did not show any discernible reflections. Thermogravimetric analysis (TGA) showed that the residual water is lost below 500 °C. The sample heated to 500 °C in air was found to have the α-MnO$_2$ structure and the electrochemical studies were carried out with this sample. The discharge-charge behavior shown in Fig. 4 reveals that the MnO$_2$ obtained by this approach shows poor electrode performance.

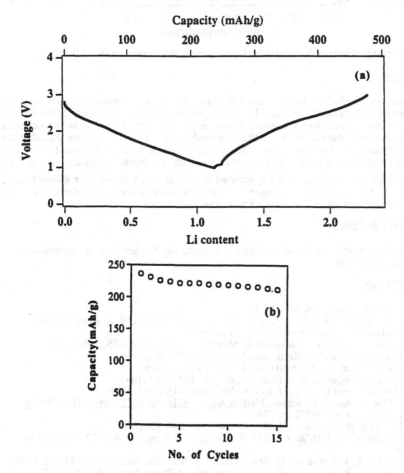

Fig. 3 (a) First discharge-charge curves, and (b) cyclability of amorphous MoO$_{2.3}$ recorded with a current density of 0.5 mA/cm^2.

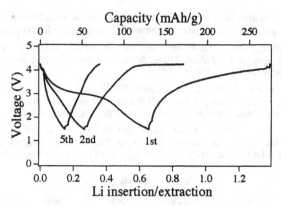

Fig. 4 Discharge-charge behavior of α-MnO$_2$ recorded with a current density of 0.5 mA/cm^2.

CONCLUSIONS

Potassium borohydride has been used to reduce oxo ions in aqueous solution to obtain amorphous or nanocrystalline MO$_{2+\delta}$ (M = V, Cr, Mn and Mo) oxides. Nanocrystalline VO$_2$(B) obtained by this solution-based approach shows about two times higher capacity and better cyclability than that obtained by conventional methods. VO$_2$(B) is an attractive candidate for polymer lithium cells. Amorphous CrO$_2$ shows electrochemical behavior different from that of crystalline CrO$_2$ reported in the literature. Amorphous MoO$_{2.3}$ exhibits a capacity of over 200 mAh/g with good cyclability. α-MnO$_2$ obtained by this approach shows poor electrochemical performance. The study shows that innovative, solution-based synthesis procedures can access metastable phases that exhibit promising electrochemical properties.

ACKNOWLEDGMENT

Financial support by the Robert A. Welch Foundation and National Science Foundation is gratefully acknowledged.

REFERENCES

1. B. Scrosati, Nature **373,** 557 (1995).
2. H. I. Schlesinger, H. C. Brown, A. E. Finholt, J. R. Gilbreath, H. R. Hoekstra and E. K. Hyde, J. Amer. Chem. Soc. **75,** 215 (1953).
3. C. Tsang, A. Dananjay, J. Kim and A. Manthiram, Inorg. Chem. **35,** 504 (1996).
4. C. Tsang, J. Kim and A. Manthiram, J. Solid State Chem. (in press).
5. C. Tsang, J. Kim and A. Manthiram, J. Mat. Chem. (in press).
6. J. Kim and A. Manthiram, J. Electrochem. Soc. **144,** 3077 (1997).
7. A. Manthiram and C. Tsang, J. Electrochem. Soc. **143,** L143 (1996).
8. H. J. Emeleus and A. G. Sharpe, Modern Aspects of Inorganic Chemistry, (John Wiley & Sons, New York, 1973) , p. 276.
9. C. Tsang and A. Manthiram, J. Electrochem. Soc. **144,** 520 (1997).
10. P. A. Christian, F. J. DiSalvo and D. W. Murphy, US Patent No. 4228226, October 14, 1980.
11. J. R. Dahn, T. V. Buurn and U. Von Sacken, US Patent No. 4965150, October 23, 1990.
12. Y. Takeda, R. Kanno, Y. Tsuji and O. Yamamoto, J. Power Sources **9,** 325 (1983).

X-ray Absorption and NMR Studies of LiNiCoO$_2$ Cathodes Prepared by a Particulate Sol-Gel Process

S. Kostov*, Y. Wang*, M. L. denBoer*, S. Greenbaum*, C.C. Chang†, and Prashant N. Kumta†
*Physics Department, Hunter College CUNY, NY
†Dept. of Materials Science and Engineering, Carnegie Mellon University, Pittsburgh, PA

Abstract

We have synthesized lithiated nickel oxide-based cathode materials containing stoichiometric and excess lithium using a new low temperature colloidal particulate sol-gel process. The process yields a xerogel precursor that transforms to the crystalline oxide at 800 °C in 2h. We studied the Li environment with nuclear magnetic resonance (NMR) and that of the transition metal ions with x-ray absorption spectroscopy. We measured samples of LiNi$_{1-x}$Co$_x$O$_2$, with x = 0 and 0.25. The effect on each composition of the incorporation of 5 mol % Li was also examined. The precursor material appears to have no Ni^{3+}, as indicated x-ray absorption measurements, and is highly disordered, showing little sign of interatomic correlations beyond the nearest neighbor in extended x-ray absorption fine structure (EXAFS) spectra. The ^{7}Li NMR line widths and spin-lattice relaxation (T$_1$) behavior are dominated by strong interactions with the paramagnetic Ni^{3+}. The presence of 5 % excess Li causes almost no change in NMR line width or T$_1$ in the mixed (Ni/Co) cathode, but does produce an almost 30% reduction in line width for the pure LiNiO$_2$, implying that Co stabilizes the structure. The near-edge x-ray absorption measurements show the local Ni environment is relatively unaffected by Co substitution, a result confirmed by EXAFS analysis. The heat-treated samples are highly ordered, and both the near-edge and extended analysis imply Co substitutes primarily for Ni^{2+}, not Ni^{3+}. The Jahn-Teller distortion is apparent in both the stoichiometric and Li-excess materials.

Introduction

LiNiO$_2$ and LiNi$_{0.75}$Co$_{0.25}$O$_2$ are technically important cathode materials for Li ion battery applications[1,2,3,4]. The conventional solid state processes normally used in generating these materials usually require prolonged heat treatments at high temperatures which are not conducive to attaining good electrochemical activity, due to the loss of lithium and the exaggerated growth of crystallites that occur during prolonged heat treatments at high temperature. The sol-gel process is extremely popular and has been widely used for generating a variety of oxide ceramics for a number of electronic applications. However, this process utilizes expensive alkoxide precursors which are moisture sensitive, and, in the case of nickel and cobalt, these alkoxides are not only expensive but are also not easily soluble in common polar solvents. Thus, the conventional sol-gel process becomes quite uneconomical for generating these materials in large quantities. There is therefore a need for an economical process to generate precursors that yield materials of good electrochemical activity, while retaining the advantages of the sol-gel process. With this in mind, we have developed a new route called the particulate sol-gel (PSG) process, and have used this to synthesize LiMO$_2$ (M = Ni, Ni $_{0.75}$ Co $_{0.25}$). This PSG process is simple and cost-effective, can be easily scaled up, and can yield good quality precursors. The homogeneous precursors the process yields have a good level of mixing at the molecular level, generating lithiated transition metal oxides with good activity, as we have previously reported.[5] We report here a detailed study of both the as-prepared powders generated by the PSG process and the heat treated powders using X-ray absorption spectroscopy and nuclear magnetic resonance spectroscopy.

Sample Preparation

The $LiNiO_2$ and $LiNi_{0.75}Co_{0.25}O_2$ materials were prepared using lithium hydroxide monohydrate (Aldrich, 99 %) and stoichiometric amounts of nickel acetate tetrahydrate (Aldrich, 98 %) and cobalt acetate tetrahydrate (Aldrich, 98%) as precursors. Lithium as well as nickel or nickel and cobalt precursors were first separately dissolved in de-ionized (DI) water to obtain clear solutions. Mixing of the individual solutions resulted in pale green (in the case of $LiNiO_2$ synthesis) or dark purple (in the case of $LiNi_{0.75}Co_{0.25}O_2$) colored suspensions. Ethyl alcohol was then added to the solutions to facilitate removal of the liquid organic by-products. The pale green or dark purple turbid solutions were stirred for 15-20 minutes prior to drying. A rotary evaporator (Buchi) was used for the subsequent drying process with an initial pressure of 500 mbar at 120 °C for 3 hours followed by a reduced pressure of 100 mbar at 140 °C for 1 hour to dry the solutions completely. The resultant precipitates were collected and then ground before being subjected to further analysis and heat treatments. Samples were heat treated at 800 °C for 2 hours, in air using alumina boats as sample carriers. Five samples were prepared as follows:

1. Precursor $LiNiO_2$ powders prepared using a starting lithium to nickel ratio of unity (Li/Ni = 1).
2. $LiNiO_2$ obtained by heating precursors with a lithium to nickel ratio of unity (Li/Ni = 1).
3. $LiNiO_2$ obtained by heating precursors with a lithium to nickel ratio of 1.05 (Li/Ni = 1.05).
4. $LiNi_{0.75}Co_{0.25}O_2$ obtained by heating precursors with a lithium to (nickel + cobalt) ratio of unity (i.e. Li/(Ni + Co) = 1).
5. $LiNi_{0.75}Co_{0.25}O_2$ obtained by heating precursors synthesized by the PSG process and having a lithium to (nickel + cobalt) ratio of 1.05 (i.e. Li/(Ni + Co) = 1.05).

Nuclear Magnetic Resonance

For NMR measurements, approximately 1 g of each sample was packed into 5 mm OD pyrex tubes. Lithium-7 NMR measurements were conducted with a Chemagnetics CMX300 NMR spectrometer operating at a ^7Li resonance frequency of 117 MHz. Wide-line NMR spectra were obtained with a quadrupole echo sequence, utilizing approximately 2 μs rf pulse widths. An inversion recovery pulse sequence was utilized for the T_1 measurements, although incomplete inversion is expected for some of the broader lines. Consequently the T_1 results are useful only when comparisons between different samples with similar line widths are made. Variable temperature control was achieved with a commercial Chemagnetics VT attachment, yielding an uncertainty of ±2°C. Aqueous LiCl was employed as a chemical shift reference.

Arrhenius plots of line width in the various materials are shown in Fig. 1. The roughly 1/T dependence observed for the line widths is expected on the basis of the magnetic susceptibility temperature dependence for strongly paramagnetic systems. Li excess appears to have little effect on the line width of the Co-substituted material, but causes a reduction in line width for the pure $LiNiO_2$. A similar trend is observed for T_1 (Table 1); Li excess causes a larger change in $LiNiO_2$ than in the Co-substituted material. These significant differences, which indicate changes in the Li ion mobility, suggest that Co substitution stabilizes the structure, possibly thereby preventing Ni on Li site disorder.

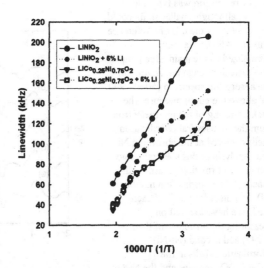

Figure 1 NMR linewidth in compounds indicated, as a function of inverse temperature.

Sample	T_1(ms)
$LiNiO_2$	3.44
$LiNiO_2$ (5% excess Li)	6.04
$LiNi_{0.75}Co_{0.25}O_2$	5.65
$LiNi_{0.75}Co_{0.25}O_2$ (5%excess Li)	5.40

Table 1.

X-ray Absorption

Each powder sample was carefully sealed between two layers of wax paper in a dry glove box to isolate it from atmospheric moisture. Near-edge x-ray absorption fine structure (NEXAFS) and extended x-ray absorption fine structure (EXAFS) measurements of the Ni K-edge were made at room temperature in transmission at beam line X-23A2 at the National Synchrotron Light Source at Brookhaven National Laboratory using a double Si(311) crystal monochromator. The energy resolution was typically ~ 0.5 eV, although smaller shifts could be detected by the use of a reference sample. All spectra were normalized to the main edge jump. The extended range modulation in the absorption spectrum (50 eV-900 eV above the edge) measures the backscattering of the photoelectron from the first several coordination shells centered at the absorbing Ni site. Analysis of this spectral range can give information regarding the spacing (R), coordination number (N), and disorder ($\Delta\sigma^2$) of each shell. We have focused on measuring the structure of the first (Ni-O) and second (Ni-Ni) coordination shells in the Li(Ni,Co)O$_2$ samples and the Ni-O shell in the precursor sample, which is the only resolved interaction in the latter material. Fig. 2 shows the background subtracted k space

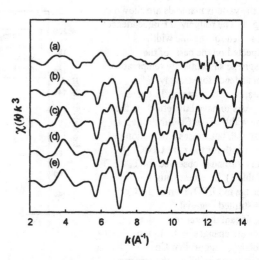

Figure 2 EXAFS k^3 weighted spectra of Ni compounds a) precursor, b) LiNiO$_2$, c) LiNiO$_2$ (5% Li), c) LiNi$_{0.75}$Co$_{0.25}$O$_2$ d) LiNi$_{0.75}$Co$_{0.25}$O$_2$ (5% Li)

EXAFS spectra for the Li(Ni,Co)O$_2$ series. All spectra are weighted by k^3 to compensate for decrease in amplitude with increase in k (Fig 2). The useful Δk range was found to be from 3 Å to 14 Å, which allows a resolution (ΔR) of $\pi /2\Delta k$ or $\Delta R \geq 0.14$Å. The precursor spectrum is significantly less structured than in that of the other samples. This indicates the Ni environment in the precursor is highly disordered, as discussed further below. In order to determine the structural parameters in the first and second shells we have fitted simulated spectra (FEFF 6.01[6]) to the experimental data. The simulations were based on the well-known LiNiO$_2$ crystallographic parameters.[7] The first shell was assumed to consist of six oxygen atoms at trial distances of 2.02 Å and the second of six Ni atoms at 2.87 Å. The first Ni-Li interaction was neglected due to the weak Li backscattering. In order to simulate the LiNi$_{0.75}$Co$_{0.25}$O$_2$ spectra, ideal Co substitution for Ni was assumed. Theoretical phase and amplitude functions for a Ni-Co distance of 2.87 Å were calculated and combined with Ni-Ni in a 1:4 amplitude ratio.

1.Near Edge

Fig. 3a shows the Ni K absorption edge for the Li(Ni, Co)O$_2$ samples and two Ni standards, Ni metal and NiO. It is apparent that the edge energy of the LiNiO$_2$ is lower than that in the LiNi$_{0.75}$Co$_{0.25}$O$_2$. The shift toward higher binding energies in Ni oxycompounds is linearly related to the increase in the mean oxidation state of the Ni ion as shown by Mansour and Melendres.[8] Using a different calibration procedure, in which the first derivative maximum was plotted against the oxidation numbers of stoichiometric standards (Fig. 4), we have extrapolated the average valency in the mixed compounds (Table 2). The value of this chemical shift ΔE per electron, while larger than that observed in Ref. 8, is similar to that reported in a study of various cobalt compounds.[9] The mean oxidation state of stoichiometric $LiNiO_2$ was taken to be +2.7 which is the value reported in Ref. 3. The small increase in average Ni valency in the cobalt substituted compounds shown in Table 2 is actually a lower bound for this effect, as it is slightly dependent on the extrapolation procedure. A different calibration scheme using the inflection point of the main edge jump yields a value closer to 2.9 for the cobalt substituted samples and no change in the excess lithium $LiNiO_2$ one. Thus it appears that Co substitutes preferentially for Ni^{2+}, thereby increasing the average Ni valence.

The small pre-edge peak (Fig. 3b) is due to $1s \rightarrow 3d$ transitions which are dipole-forbidden in a central symmetric site. Deviations from octahedral symmetry at the Ni ions give rise to this peak, the intensity of which is therefore a measure of the distortion of the NiO_6 octahedra. From Fig. 3b it is apparent that the pre-edge peak positions in $LiNiO_2$, $LiNiO_2$ +5% Li, $LiNi_{0.75}Co_{0.25}O_2$ and $LiNi_{0.25}Co_{0.25}O_2$ +5%Li are very close and are shifted ~0.6 eV above the NiO and precursor peak positions, consistent with Ref. 8. The precursor pre-edge peak is also somewhat sharper than that of the product samples.

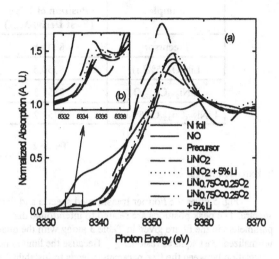

Figure 3 a) Ni $1s$ edge and b) pre-edge absorption peak in the Ni compounds indicated.

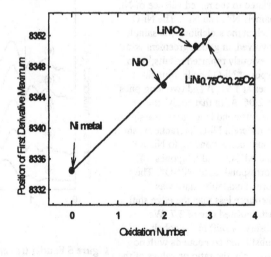

Figure 4 Position of Ni $1s$ absorption edge in $LiNi_{0.75}Co_{0.25}O_2$ and stoichiometric compounds indicated.

Sample	Position of Edge (First Deriv. Max)	Average Ni Oxidation State
Precursor	8345.8	1.98
LiNiO$_2$ (5%Li)	8350.5	2.72
LiNi$_{0.75}$Co$_{0.25}$O	8351.3	2.82
LiNi$_{0.75}$Co$_{0.25}$O (5%Li)	8351.2	2.81

Table 2.

2. Extended range

Fig. 5 gives the Fourier transformed spectra and the best fit simulation curves to these spectra. The peak positions are related to interatomic distances. The best values of the floating parameters in the fit are given in Table 3 along with the quality of the fit, expressed in terms of the normalized least squares residual χ^2_v. Because the limited range of the data and the high degree of correlation between the free parameters leads to instability in the fits, not all parameters were varied simultaneously. The distortion of the NiO$_6$ octahedra in lithiated Ni oxides has been interpreted by Pickering et al.[10] as related to the mixed valence of Ni, namely Ni^{2+} and Ni^{+3}. The Ni-O shell in the stoichiometric sample involved, in good agreement with previously reported results,[8] two types of Ni-O bonds, four short ones at 1.91 Å and two long ones at 2.06 Å. In that study[8] these various bond lengths were assigned to different Ni-O interactions, short bonds corresponding to Ni^{+2}-O$^-$ and Ni^{+3}-O^{-2} and long ones corresponding to Ni^{+2}-O^{-2}. The formal oxidation state was therefore less than the edge shift extrapolated one of 2.7. We observe a shift in the cobalt substituted compounds with no change in the ratio or values of the Ni-O separation compared to stoichiometry. Both LiNiO$_2$,

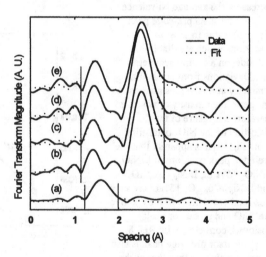

Figure 5 Fourier transform EXAFS amplitudes (not adjusted for phase shifts) for: a) precursor; b) LiNiO$_2$; c) LiNiO$_2$ + 5% Li; d) LiNi$_{0.75}$Co$_{0.25}$O$_2$; e) LiNi$_{0.75}$Co$_{0.25}$O$_2$ + 5% Li.

$LiNiO_2$ +5%Li, $LiNi_{0.75}Co_{0.25}O_2$ and $LiNi_{0.25}Co_{0.25}O_2$ +5%Li are characterized by the same distortion of the NiO_6 octahedra. If the increase in average Ni valency, as indicated by the measured edge shift, leads to a predominantly Ni^{+3} environment, a measurable decrease in the distortion of the NO_6 octahedra in all of product samples would be expected. Apart from the resolution and precision concern, the fact that no such decrease was measured could be attributed to a distortion characteristic of a purely N^{+3} environment. Such a Jahn-Teller type distortion attributed to the low $t_2^6e^1$ spin state of the N^{+3} ion has been measured by Rougier et al.[11] in a quasi 2D system of nearly pure trivalent Ni. The values obtained for the different Ni-O bond lengths and their ratio is identical to the ones measured for a presumably mixed valence $LiNiO_2$ system in which the distortion is related to the difference in ionic radii of the divalent and trivalent Ni. In the precursor a single Ni-O distance gives the best fit, implying that this first shell distortion is absent from this material. The bond lengths are consistent with the +2 Ni valency measured in the edge shift. Further coordination shells are strongly disordered and no fitting was attempted.

Sample	Shell	S_0	N	R	$\Delta\sigma^2$	E_0	χ^2_v
Precursor	Ni-O	0.46	6	2.06	0.005	-7.2	44
$LiNiO_2$	Ni-O	0.71	4	1.91	0.003	-10.8	17
	Ni-O	0.71	2	2.06	0.005	-10.8	17
	Ni-Ni	0.71	6	2.89	0.005	-10.8	17
$LiNiO_2$ (5% Li)	Ni-O	0.71	4	1.91	0.004	-8.8	17
	Ni-O	0.71	2	2.05	0.005	-8.8	17
	Ni-Ni	0.71	6	2.89	0.005	-8.8	17
$LiNi_{0.75}Co_{0.25}O_2$	Ni-O	0.82	4	1.92	0.005	-9.8	7
	Ni-O	0.82	2	2.06	0.009	-9.8	7
	Ni-Ni, Ni-Co	0.82	6	2.87	0.005	-9.8	7
$LiNi_{0.75}Co_{0.25}O_2$ (5% Li)	Ni-O	0.51	4	1.92	0.003	-7.5	10
	Ni-O	0.51	2	2.07	0.007	-7.5	10
	Ni-Ni, Ni-Co	0.51	6	2.87	0.004	-7.5	10

Table 3.

Conclusions

NMR line widths shows a 1/T dependence, consistent with the temperature dependence of magnetic susceptibility in strongly paramagnetic systems. We find that excess Li slightly reduces the line width in pure $LiNiO_2$, while it has little effect in the Co-substituted compound. In addition, in the latter the line width is considerably smaller at all temperatures and does not change with Li excess. We conclude that alloying with Co stabilizes the system by inhibiting the tendency of Ni to occupy Ni sites. The EXAFS measurements show the local structure of the Ni environment is relatively unaffected by Co substitution. The $LiNiO_2$ precursor is highly disordered and amorphous, while the final heat-treated samples are highly ordered. Both the near-edge and extended analysis imply Co substitutes preferentially for Ni^{2+}, not Ni^{3+}. The Jahn-Teller distortion is apparent in both the stoichiometric and Li-excess materials.

Acknowledgements

Support for this work was provided by the U. S. Department of Energy and the Office of Naval Research.

References

1. M. Broussely, F. Perton, P. Biensan, J.M. Bodet, J. Labat, A. Lecerf, C. Delmas, A. Rougier, and J.P. Peres, Journal of Power Sources **54**, 109 (1995).

2. Tsutomu Ohzuku, Atsushi Ueda, Masatoshi Nagayama, Yasunobu Iwakoshi and Hideki Komori; Electrochimica Acta, **38**, 1159 (1993).

3. B. Banov, J. Bourilkov, and M. Mladenov, Journal of Power Sources **54**, 268 (1995).

4. D. Caurant, N. Baffier, B. Garcia, and J.P. Pereira-Ramos, Solid State Ionics **91**, 45 (1996).

5. Chun-Chieh Chang, N. Scarr and P.N. Kumta, Proceedings of the Joint International Meeting of the ECS and ISE, Paris, France, 31 August - 5 September 1997.

6. J. J. Rehr, R. C. Albers, and S. I. Zabinsky, Phys. Rev. Lett. **69**, 3397 (1992), and *Ab initio Multiple-Scattering X-ray Absorption Fine Structure and X-ray Absorption Near Edge Structure Code*, Copyright 1992, 1993, FEFF Project, Department of Physics, FM-15 University of Washington, Seattle, WA 98195.

7. L. D. Dyer, B. S. Borie, Jr., and G. P. Smith, J. Am. Chem. Soc. **76**, 1499 (1954).

8. A. N. Mansour and C. A. Melendres, J. Phys. Chem., **102**, issue 002, 1998.

9. A. R. Chetal, P. Mahto and P.R. Sarode, J. Phys. Chem. Solids **49**, 279 (1988)

10. I. Pickering, G. George, J. Lewandowski, and A. Jacobson, J. Am. Chem. Soc., **115**, 4144 (1993).

11. A. Rougier, C. Delmas , and A. Chadwick, Sol. State Comm. **94**, 123 (1995).

ON THE ELECTROANALYTICAL CHARACTERIZATION OF Li_xCoO_2, Li_xNiO_2 AND $LiMn_2O_4$ (SPINEL) ELECTRODES IN REPEATED LITHIUM INTERCALATION-DEINTERCALATION PROCESSES

D. AURBACH*, M.D. LEVI*, E. LEVI*, B. MARKOVSKY*, G. SALITRA*, H. TELLER*, U. HEIDER**, L. HEIDER**
*Department of Chemistry, Bar-Ilan University, Ramat-Gan 52900, Israel
**Merck KGaA/Central Service, Frankfurterstrasse 250, D-64271 Darmstadt, Germany

ABSTRACT

This paper reports on electroanalytical studies of the intercalation-deintercalation of lithium into lithiated transition metal oxides which are used as cathodes for Li ion batteries. These include Li_xCoO_2 Li_xNiO_2 and $Li_xMn_2O_4$ spinel. The basic electroanalytical response of these systems in LiAsF6 1M/EC-DMC solutions was obtained from the simultaneous use of slow and fast scan cyclic voltammetry (SSCV), potentiostatic intermittent titration (PITT) (from which D vs. E was calculated), and impedance spectroscopy (EIS). Surface sensitive FTIR spectroscopy and XRD were also used for surface and 3D characterization, respectively. A large and important denominator was found in the electrochemical behavior of lithium intercalation-deintercalation into these transition metal oxides and graphite. The use of the electroanalytical response of these systems as a tool for the study of stabilization and failure mechanisms of these materials as cathodes in rechargeable Li batteries is demonstrated and discussed.

INTRODUCTION

The most popular, important and commonly used cathode materials for lithium ion batteries are lithiated cobalt, nickel and manganese oxides: $LiCoO_2$, $LiNiO_2$ and $LiMn_2O_4$ spinel. These compounds deintercalate/intercalate lithium reversibly at potentials around 4 V vs. Li/Li^+, and they are thereby very attractive for Li ion batteries with lithiated carbon anodes (which intercalate lithium in the 0 - 1 V vs. Li/Li^+ range). A tremendous amount of work has been dedicated so far to the correlation between preparation modes and performance of these materials.[1] From practical experience accumulated, it is clear that the reversibility of these cathodes is limited due to fading mechanisms which lead to deterioration of their capacity upon cycling.[2] This work was aimed at the study of the basic electroanalytical behavior of these compounds as reflected from the simultaneous application of slow scan-rate cyclic voltammetry (SSCV), potentiostatic intermittent titration (PITT), and electrochemical impedance spectroscopy (EIS), and the use of the electroanalytical response of these electrodes upon cycling as a tool for a rigorous study of their possible capacity fading mechanisms. The above is demonstrated for one type of stoichiometric $LiMn_2O_4$ spinel electrodes in which a capacity fading is observed upon prolonged cycling.

EXPERIMENTAL CONSIDERATIONS

$LiCoO_2$, $LiNiO_2$ and several types of $LiMn_2O_4$ of very high quality, as well as standard solution components including ethylene and dimethyl carbonates (EC, DMC), and LiAsF6 (Li battery grade), were obtained from Merck. Standard, thin electrodes were prepared using aluminum foil (current collector), carbon black (15% by weight for conductivity) and PVDF (5%, binder), and the active mass (85%). These were studied in standard solutions EC-DMC 1:3/LiAsF$_6$ 1M vs. Li and lithiated carbon anodes. SSCV, PITT and EIS were performed using computerized instrumentation from Arbin, EG&G and Schlumberger, and software from EG&G and Scribner. The impedance spectra were modeled by the Zsim software from Scribner. Chemical diffusion coefficients were calculated as a function of the electrode's potential from the Cottrell region in the It$^{1/2}$ vs. log t plots obtained routinely (at different potentials) from the PITT measurements).[3] In addition, X-ray diffractometry (XRD) of electrodes before and after cycling was carried out. The electrodes were also measured by FTIR spectroscopy (diffuse reflectance mode).

RESULTS AND DISCUSSION

a. The Electroanalytical Fingerprint of Li Intercalation Compounds, Including Graphite, LiNiO$_2$, LiCoO$_2$ and LiMn$_2$O$_4$ Spinel.

There is an interesting common denominator in the electroanalytical behavior of lithiated graphite[4,5] and the above transition metal oxides.
1. At sufficiently slow scan-rates sets of reversible, narrow peaks characterize the CV. In some cases, (e.g. graphite Li$_x$CoO$_2$), these peaks reflect phase transition and two phase regions. At sufficiently small scan rates and with thin and highly oriented electrodes, the SSCV may reflect an accumulation of lithium (as well as its depletion) close to equilibrium, beyond diffusion control.
2. D vs. E calculated from PITT[3] are non-monotonous functions with sharp minima at potentials at which the differential intercalation capacity is maximal. (E$_p$ in the CV).
3. The narrow CV peaks and the above shape of D vs. E may be typical of adsorption/insertion processes in which attractive interactions exist amongst intercalation sites. (A Frumkin-type adsorption/insertion isotherm).
4. The impedance spectra (Nyquist plots) reflect very clearly an overall intercalation process which occurs in a series of steps.[5]
* Li$^+$ migration through surface films (high-to-medium ω semicircle).
* Charge transfer (low ω semicircle).
* Solid state diffusion of Li$^+$ into the bulk. (A Warburg-type element).
* Accumulation consumption of Li: a capacitive behavior at the very low ω. The intercalation/deintercalation capacitance (reciprocal to the Z" values measured at the lowest ω) has a maximum at the CV peak potentials.
Consequently, the following analysis of the electrochemical behavior (developed for lithiated graphite electrodes[4,5]) is suitable for the lithiated metal oxides as well.

b. The analysis of slow scan CV data (developed in ref. 4)

The classical Frumkin isotherm was used to take into account the interactions between the intercalation sites (suitable for insertion/adsorption processes with very narrow CV peaks):

$$X / (1-X) = \exp [f (E - E_o')] \exp (-g X) \tag{1}$$

with E and E_o' standing for any insertion/deinsertion potential and its standard value, respectively, X is the intercalation level (i.e. the mole fraction of species) and g reflects the interaction between the sites in $Li_{1-x}CoO_2$.

Combination of the isotherm (1) with the Butler-Volmer equation accounting for the slow Li-ion interfacial charge transfer results in the following equation for the dimensionless current in the CV response:[4]

$I = (k_o/\delta fv) \{ (1-X) \exp (-0.5gX) \exp [0.5f(E-E_o)] -X \exp (0.5gX) \exp[-0.5f(E-E_o)]\}$,   (2)

with dimensionless rate constant $k_o/\delta fv$ (k_o = standard heterogeneous rate constant, δ = thickness of the electrode coating, f = F/RT is the conventional electrochemical constant) and symmetrical charge-transfer coefficient $\alpha = 0.5$ and the number of Li-ion transferred per unit intercalation site n=1.

c. Kinetic Models of the EIS and PITT Responses and their Link to the Model for SSCV[4,5]

Highly resolved PITT measurements require a proper choice of the potential step height and its adjustment to the shape of the corresponding slow-scan rate CV curve and the background current contribution to the overall current. Practically, 10 mV height steps around the CV peaks were sufficiently good to ensure a high resolution. The characteristic diffusion time τ and the corresponding chemical diffusion $D = l^2 / \tau$ were obtained from the simultaneous PITT and EIS measurements. l - the diffusion length is half of the average particle size.

PITT: $\tau = [Q_t\Delta X/\pi^{1/2} It^{1/2}]^2$ at $t << \tau$ (3)

EIS: $\tau = 2 [Q_tA_w dX/dE]^2$ in the Warburg region   (4)

See details in ref. (5).

Here Q_t = the overall amount of the injected charge, $It^{1/2}$ = the Cottrell slope, A_w = the Warburg slope. Both these two techniques provide with the same τ (and D) values, thus

$$A_w (It^{1/2} / \Delta E) = (2\pi)^{-1/2}$$ (5)

Low-frequency Domain

In this region, similar plots of differential intercalation capacity vs. potential are obtained from the various electroanalytical tools using the following equations:

CV: $C_{int} = I(E) / v$ (6); PITT: $C_{int} = Q_t\Delta X(E)/ \Delta E$ (7); EIS: $C_{int} = - 1/\omega Z''_{\omega\to 0}$ (8)

Medium-frequency Domain (see details in ref. 4).

The relation between the CV and the EIS data is expressed in the following equation for R_{ct} which calculated from the kinetic model and is compared to R_{ct} measured directly by EIS.

$$R_{ct} = \delta / [f Q_t k_o X^{0.5} (1-X)^{0.5}]$$ (9)

d. Surface Spectroscopic Studies

FTIR spectroscopic studies of the $LiCoO_2$, $LiNiO_2$ and $LiMn_2O_4$ electrodes show that the active mass is initially covered with a Li_2CO_3 film due to reaction between the lithiated oxides and CO_2 of the air. Upon cycling vs. Li or Li-C electrodes, surface species formed on the anodes due to solvent reduction (e.g. $ROCO_2Li$ species in the case of EC-DMC solutions)[6] saturate the electrolyte solution and re-precipitate on the cathodes. Hence, these cathodes are also SEI electrodes. Impedance spectra of these electrodes indeed show that Li ion migration through surface films is one of the necessary steps in the entire intercalation process.

e. LiCoO₂ Electrode

1. Slow scan rate cyclic voltammetry (SSCV, I *vs.* E) of these electrodes could be nicely modeled by an equation which combines a Frumkin-type adsorption process and Butler-Volmer kinetics [Eq. 2 above]. The best fitting was obtained with K=40 and g=-4.2. This reflects an intercalation process with high attractive interactions amongst intercalation sites interfered with by slow charge transfer.[4,5] D *vs.* E has minimum at the CV peak potential

2. The impedance spectra of this electrode could be modeled precisely by an equivalent circuit analog which includes a Voigt-type analog (3-4 R∥C circuits in series which reflect Li ion migration through multilayer surface films + interfacial charge transfer) in series with a finite length Warburg-type element (solid state diffusion), and with capacitor (accumulation of Li in the bulk). See details on this model as well as its relevance to lithiated graphite electrodes in refs. 4,5.

f. LiNiO₂ Electrode

*   SSCV show two major red-ox transfers (two reversible sets of peaks).
*   D *vs.* E have minima at the peak potentials.
*   This electrode is very stable upon cycling in EC-DMC/LiAsF₆ solution if the potential is lower than 4.2 V (*vs.* Li/Li⁺). This is reflected by the following:

1. D *vs.* E plots of pristine and cycled electrodes are similar. This reflects the stability of this electrode upon cycling.

2. The impedance spectra measured at constant potentials upon cycling show no pronounced changes in their structure. For example, the high-to-medium semicircle which relates to the surface film remains invariant upon repeated cycling.

g. LiMn₂O₄ Electrode

During the first cycling of this electrode, a drastic change is observed in the SSCV picture (Fig. 1), as well as in the impedance spectra (the medium frequency semicircle which reflects charge transfer increases, Fig. 2). However, the overall capacity, as well as the shape of D *vs.* E, remains unchanged (Fig. 3). These reflect changes in the kinetic behavior of this electrode which probably relates to the interfacial charge transfer, but with no significant change in the bulk of the active mass during the first cycles. During prolonged cycling, a capacity fading is observed (Fig. 4). However, after the initial changes described above, the high-to-medium part of the impedance spectra remains unchanged upon cycling, but the D *vs.* E plot becomes featureless (Figs. 3).

The XRD of the pristine and cycled electrodes show no qualitative difference, but rather a decrease in peak intensity upon cycling. (Fig. 5). Hence, we conclude that during the first cycles, surface changes slow the charge transfer of this electrode with no considerable change in the bulk (reflected by the peak-shaped D *vs.* E plot). The surface then stabilizes, and, upon prolonged cycling, there are no changes in the active mass surface, but rather in its bulk. Probably some amorphization of the active mass occurs upon prolonged cycling, although the surface films remain stable. This is reflected by the capacity fading, decrease in

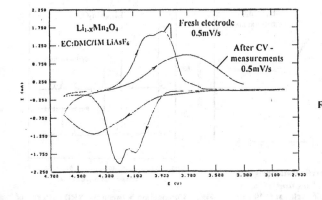

Fig.1. Comparison between cyclic voltammograms measured with fresh a $Li_{1-x}Mn_2O_4$ electrode (5 mg active mass spread on a 1.2×1.2cm Al foil and after several consecutive cycles.

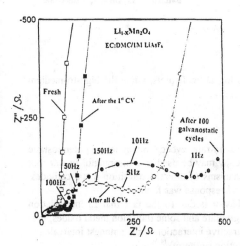

Fig.2. Variation of the Nyquist plots measured with a $Li_{1-x}Mn_2O_4$ spinel electrode as a function of cycle number referred to the OCV (E=3.0V) of the full reduced material.

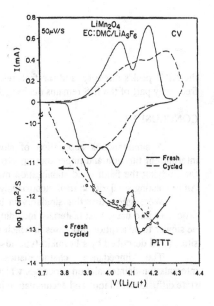

Fig.3. Slow scan rate CV and log D vs. E for $LiMn_2O_4$ electrode (1.29 mg/cm² 20 μm thick, 2.9 cm²). The CV plots relate to a fresh electrode and 6 consecutive CV cycles. Log D vs. E plots were obtained from a PITT experiment after these 6 initial cycles and the after prolonged cycling.

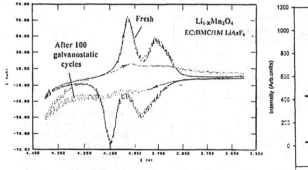

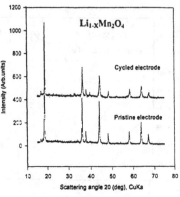

Fig.4. Comparison between the cyclic voltammograms of the fresh and cycled Li$_{1-x}$Mn$_2$O$_4$ spinel electrode (100 galvanostatic cycles at C/10 rate) measured at a low scan rate, $v = 0.01$mV/s.

Fig.5. Comparison between the XRD patterns of pristine (lower pattern) and cycled (upper pattern) Li$_{1-x}$Mn$_2$O$_4$ electrode.

the XRD peak's intensity, and the featureless log D *vs.* E plots, while the high-to-medium frequency part of the EIS remains unchanged.

CONCLUSION

A simultaneous application of slow scan-rate cyclic voltammetry, potentiostatic intermittent titration and impedance spectroscopy may be used as a meaningful *in situ* tool for studying the failure and stabilization mechanisms of lithiated transition metal cathodes. An interesting similarity in their electroanalytical response was found.

We suggest that this similarity in behavior is due to the fact that Li intercalation processes into many host materials, including graphite and some transition metal oxides, may be similar to adsorption processes in which attractive interactions exist amongst intercalation sites, (well described by a Frumkin-type adsorption isotherm).[4-5]

The impedance characteristics of these electrodes reflect multistep intercalation-deintercalation processes which include Li ion migration, charge transfer, solid state diffusion of Li ions and accumulation in series.

ACKNOWLEDGMENTS

This work is partially supported by the German Ministry of Science, BMBF, and the Israeli Ministry of Science and Technology.

REFERENCES

1.   T. Ohzuku in <u>Li Batteries. New Materials, Developments and Perspectives</u>. G. Pistoia, Ed., Elsevier, Amsterdam, London and New York, 1994, Ch. 6
2.   2.   Y. Xia, Y. Zhou and M. Yoshio, J. Electrochem. Soc. **144**, p. 2593 (1997).
3.   C.J. Wen, B.A. Boukamp, R.A. Huggins and W.J. Weppener, J. Electrochem. Soc. **126**, p. 2258 (1979).
4.   M.D. Levi and D. Aurbach, J. Phys. Chem. B **101**, p. 4630 (1997).
5.   M.D. Levi and D. Aurbach, J. Phys. Chem. B **101**, p. 4641 (1997).
6.   D. Aurbach, B. Markovsky, A. Schechter, Y. Ein-Eli and H. Cohen, J. Electrochem. Soc. **143**, pp. 3809-3820 (1996).

Part V

Battery Electrolytes, Interfaces, and Passive Films

Thermal Stability of Lithium Ion Battery Electrode Materials in Organic Electrolytes

M.N. Richard and J.R. Dahn

Departments of Physics and Chemistry, Dalhousie University, Halifax, Nova Scotia, Canada, B3H 3J5.

Abstract

The technology of lithium ion batteries has been developed so that these systems can be used to power portable electronic devices. The next step is to increase the size of the batteries so that they power larger systems such as electric vehicles. However, before such a step can be taken the thermal stability of any materials placed in such a battery must be determined. Although the thermal properties of the materials are important for the small systems, which can reach high temperatures when abused, they become even more important in large systems where heat dissipation is more difficult because of the physical size of the system. This paper will discuss our approach, which uses an Accelerated Rate Calorimeter, to the study of the thermal behaviour of lithium ion battery electrode materials.

Introduction

There are North American safety standards set out by Underwriter's Laboratories[1] concerning the behaviour of lithium ion batteries under abuse conditions. One of the abuse conditions tested is short circuit and its effects on the battery. If an 18650 cell (18mm diameter and 65mm length) were to short circuit, the heat generated could quickly raise the cell temperature to the shutdown temperature of the separator (approximately 155°C for polypropylene). At this point, the current will stop flowing, since the ions can no longer move from the anode to the cathode, and if there are no chemical reactions taking place to provide further heat, the battery will start to cool. However, if the electrode materials undergo chemical reactions at these elevated temperatures, then further heat could be generated and a battery could proceed to thermal runaway leading perhaps to vent and flame. In order to test for any chemical reactions, UL describes a test in which a cell is placed in an oven at 150°C and its temperature monitored to determine if any exotherm occurs[1]. Figure 1 shows the data obtained from one manufacturer's 18650 cell after it was placed in an oven at 150°C. The figure clearly shows that the cell experiences exothermic reactions at the electrode(s), goes through thermal runaway and did eventually vent with flame. **It should be noted that 18650 products from other manufacturers pass this test.** Also in figure 1 is a dashed line indicating the behaviour expected had the cell not experienced any chemical reactions at the electrode(s). This data clearly indicates the importance of determining the thermal behaviour of lithium ion battery electrode materials. This type of oven exposure test however, only gives qualitative data concerning the relative stability of one lithium ion battery system compared to another, it gives no quantitative data concerning the reactions that are occurring at the electrode(s) to produce the observed heat.

To determine the reactions generating heat in the system, the components (cathode + electrolyte, anode + electrolyte) must be considered separately. This is the approach that researchers at Moli Energy [2,3] have adopted and it is our method as well. The thermal behaviour of the components is measured in an Accelerated Rate Calorimeter (ARC), developed by

445

Townsend at Dow Chemical for the study of hazardous materials[4]. The advantages of the ARC are its ability to measure exotherms as small as 0.02°C/min and its tracking system which monitors any temperature increase in the sample and then increases the environmental temperature of the calorimeter, by activating heaters at the top, bottom and sides of the calorimeter, so that adiabaticity is maintained. As a result, no heat is lost from the system and the true self heating rate of the material is measured.

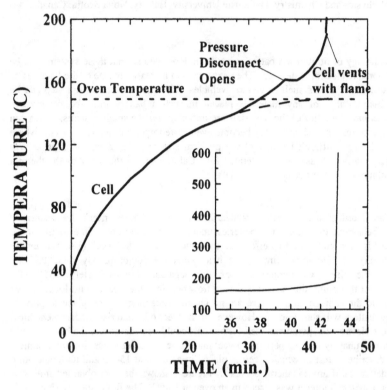

Figure 1: Data obtained from a commercially available 18650 cell placed in an oven at 150°C. Most commercially available cells are safe and follow the long dashed curve.

Experimental

Moli Energy reported the thermal behaviour of mesocarbon microbeads (MCMB) heated near 2800°C in $LiPF_6$ EC:DEC (33%:66%) electrolyte, as a function of lithium content. Their ARC samples were heated at 5°C/min to an initial temperature of 100°C after which the exotherm was monitored. They found that the greater the lithium content of the carbon material, the greater the reactivity and that the MCMB materials which have lithium metal plated on them have the greatest reactivity. This former behaviour comes as no surprise if one considers that the concentration of lithium in the carbon increases as MCMB approaches the potential of lithium metal and hence its heat of reaction increases. A model for the self heating rate was proposed. The model states that the heat observed is produced from the reaction of lithium at the surface of

446

the carbon with electrolyte, but that as the layer of reaction products on the surface of the carbon increases, the rate of the reaction slows as the inverse of the thickness.

There are four steps for the preparation of our ARC samples: 1) making pellet stock for cells, 2) making pellet cells, 3) using signature curve discharge to discharge the cells to the chosen voltage, and 4) ARC sample construction. The electrodes were made by mixing MCMB heated near 2800°C with 7% by weight poly-vinylidene fluoride(PVDF) binder and 8% by weight of Super S. Excess 1-methyl-2-pyrrolidinone (NMP) is added to the mixture in order to make it less viscous. The three components are placed in a small nalgene bottle, ceramic beads are added and the bottle is shaken so that a good mixture is obtained. The mixture is then poured into a boat made of stainless steel foil and dried for a minimum of three hours at 110°C. The powders are recovered by scraping them off the foil. They are then lightly ground in a mortar and pestle and sieved through a 300μm mesh. The resulting fine powder is pellet stock.

Electrodes are made by pressing approximately 330mg of the pellet stock material into a 0.75" diameter disk at 2500psi. The pellet is fairly robust. These pellets are brought into an argon filled glove box where cell construction proceeds. Regular 2325 coin cell hardware is used.

Cell construction proceeds as indicated in figure 2.

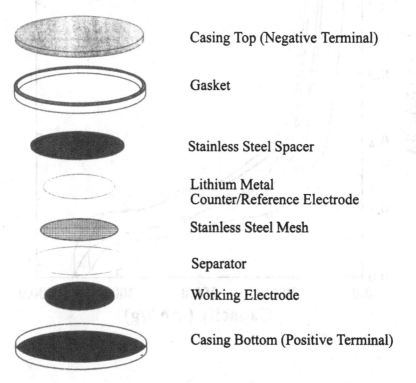

Casing Top (Negative Terminal)

Gasket

Stainless Steel Spacer

Lithium Metal
Counter/Reference Electrode

Stainless Steel Mesh

Separator

Working Electrode

Casing Bottom (Positive Terminal)

Figure 2: Exploded view of a pellet coin cell assembly.

The pellet is placed in the cell can. The LiPF$_6$ in EC:DEC (33%:67%) electrolyte is added to the surface of the pellet, then the pellet is left so that the electrolyte may be absorbed into it. While the electrolyte is being absorbed, the pellet is covered so that as little DEC as possible evaporates. If the surface of the carbon becomes dry, as the electrolyte is absorbed, more electrolyte is added. The pellet is then covered by three 2502 celgard separators. Lithium metal is pressed into a stainless steel mesh. This combination is placed above the separators, with the mesh in contact with the separator. A stainless steel spacer is added to the stack and the cell cap, complete with a polypropylene gasket, is placed on top. This ensemble is then crimped so that the edges of the bottom casing fold over the casing top, isolating the cell contents from the environment.

The MCMB cells are discharged to three voltages (0.0V, 0.0892V, 0.1265V) indicated in the voltage profile for our MCMB material shown in figure 3.

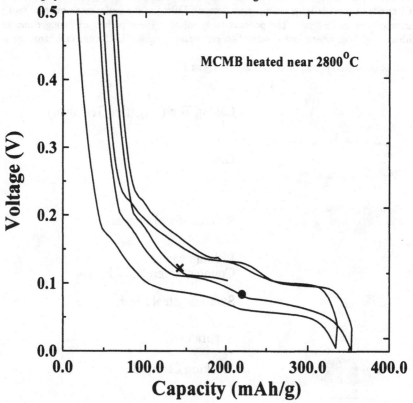

Figure 3: Voltage profile for the MCMB used here cycled versus lithium metal. The dot indicates the 0.0892V position and the cross indicates the 0.1265V position.

A signature curve discharge test is used in order to discharge the cell to the desired voltage. This test has the advantage of lowering the current at each successive discharge, thus allowing the cell to slowly equilibrate at the set voltage. An example of such a signature curve discharge for MCMB discharged to 0.0V, is shown in figure 4.

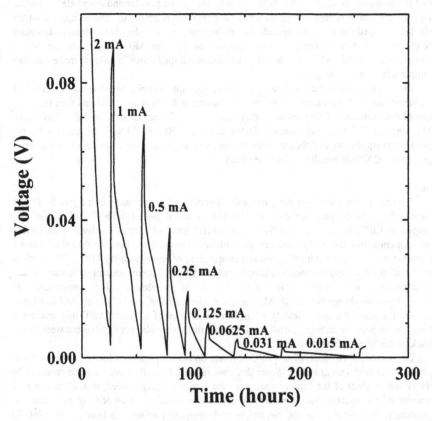

Figure 4: Signature curve discharge for MCMB in LiPF$_6$ in EC:DEC(33%:67%). The cell is discharged to 0.0V with an initial current of 2mA.

After each discharge, the cell is left to relax for 2.5 hours before the next cycle. The initial current is 2mA and the current reduction rate is 50%. The discharged cells are then brought into an argon filled glove box.

Because of the small size of our samples, approximately 300mg, a new ARC sample holder had to be designed that would not mask the self heating of the sample. For example, if the sample holder is much more massive than the sample, any heat generated by the sample will be used to heat the holder, but the resulting temperature increase in the holder would probably not be large enough to be labeled an exotherm by the ARC and hence the self heating of the sample would be masked. Our sample holder is 0.250" diameter stainless steel tubing with a 0.006" wall thickness cut into 1.54" lengths. One end of the holder is crimped flat then TIG welded shut using a Maxstar 91 TIG welder with a 0.040" tungsten tip.

The cell is carefully opened with tin snips, the pellet is recovered and ground using a mortar and pestle. The powder is added to the stainless steel tube sample holder. An equal amount of electrolyte is added to the tube with a syringe. The powder and liquid electrolyte are stirred together with a syringe needle so that a good mixture is obtained. The second end of the tube is then crimped flat and TIG welded. The entire process , including the welding, takes place in the glove box, hence the sample is never exposed to air. The ARC samples are then placed into the antechamber of a glove box and the chamber is pumped down in order to make sure that the sample tube is not leaking.

The run parameters for the ARC vary. In an attempt to duplicate the results obtained by Moli Energy, some of our samples were initially heated at 5°C/min to 100°C and then the ARC monitored the exotherm of the sample. Other samples were heated to 40°C initially, then heated in 10°C steps at 5°C/min. At each ten degree step the ARC would wait 15 minutes for the temperature to equilibrate and then monitor the temperature in an attempt to detect an exotherm greater than 0.02°C/min which it would then track.

Results

The results using both the run parameters described above are shown in figure 5. Figure 5a shows the high temperature data for MCMB material mixed with PVDF binder and discharged in $LiPF_6$ EC:DEC(33%:67%). The general trend of greater reactivity with greater lithium content demonstrated by Moli energy is observed here. However, our data also shows a very distinct maximum in the heating rate at a temperature of approximately 110°C. This peak is present in all the runs and it appears relatively independent of lithium content so it must be due to a component that is roughly equally present in all the samples. The low temperature data shown in figure 5b shows that an MCMB sample cycled in $LiPF_6$ actually starts self heating at 80°C. At this point the ARC detects a temperature change of at least 0.02°C/min and starts tracking the temperature change. Smaller self heating rates were observed earlier but were below the tracking abilities of the machine.

The distinct peak observed at 110°C in figure 5a is also present in figure 5b, but at lower temperature and with less intensity. Since this peak is present in all of our runs and seems to be relatively independent of the lithium content of the material being studied, it must be due to a component of the system that is present in all of the samples. Considering then that the temperature is too low to cause the electrolyte to decompose (we see this later at about 180°C) we have attributed this peak to the decompostion of the solid electrolyte interface (SEI) layer formed on the carbon as lithium is intercalated into it[5]. We propose that this SEI formed on discharge is quickly depleted so that we observe a decrease in the amount of heat generated, observed at T>110°C in figure 5a and at T>95°C in figure 5b. Following this decrease in the heating rate, we propose that any further heat results from the formation of new SEI from the reaction of lithium near the surface of the carbon and electrolyte solvent and the subsequent decompostion of this SEI. The rate of the decompostion of this new SEI would be limited by the rate of its formation and hence limited by the diffusion of lithium through any reaction products on the surface of the MCMB.

It is in the latter part of the exotherm, when new SEI is formed by reacting lithium at the surface of the carbon with electrolyte, that the difference between a fully lithiated material and a partially lithiated material become evident. In figures 5 a and b, we see that the exotherm for the MCMB material discharged to 0.1265V becomes weaker as time progresses. We believe this occurs because the reaction is running out of lithium whereas a fully lithiated material contains a high enough concentration of lithium near the surface of the carbon to maintain the rate of temperature increase until the electrolyte starts to decompose.

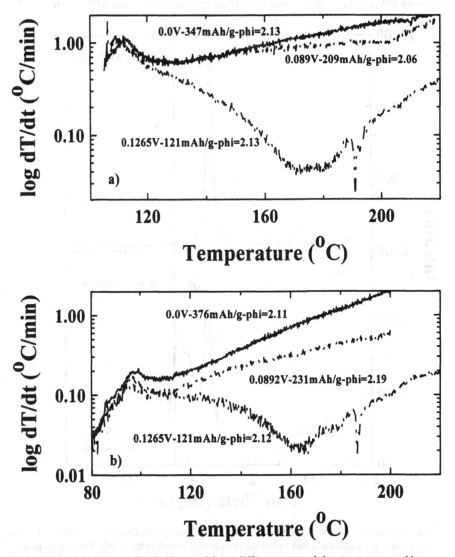

Figure 5: The heating rate of MCMB materials, at different states of charge, as measured by an Accelerated Rate Calorimeter. Figure 5a shows the data collected when the initial temperature is raised to 100°C and the bottom panel shows data collected when the exotherm was started at lower temperature.

Although we have no direct proof that the decomposition of the SEI generates the first peak, we do have proof that lithium is consumed in the overall reaction. Figure 6 compares the x-ray diffraction patterns measured on a powder sample obtained from a freshly opened cell to the x-ray diffraction pattern measured for a sample extracted from a used ARC sample.

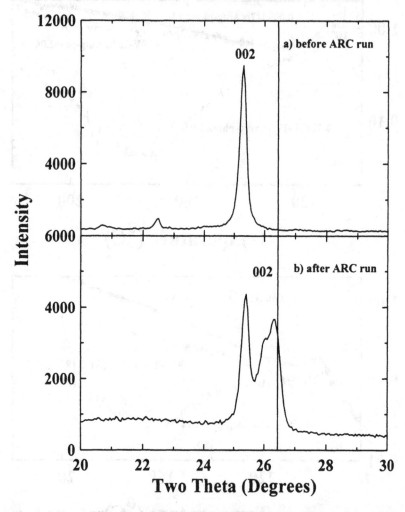

Figure 6: X-ray diffraction data collected on an MCMB sample discharged to 0.0892V. The top panel shows the data collected before the sample was heated in the ARC. The bottom panel shows the sample after the ARC run. The solid line is the measured peak position for the 002 peak in our MCMB material.

The x-ray diffraction data was measured in a D500 x-ray diffractometer equipped with a copper x-ray tube and a diffracted beam monochromator. Data were collected between 10° and 70° with a 0.05° step and a 6 second counting rate. Since the samples are air sensitive, they were placed on a piece of zero background silicon wafer (510 cut). A 0.010" thick beryllium window is placed over the sample and tightened down with four set screws. The seal is formed by tightening the screws so that the beryllium is tightly pressed into a silicone o-ring. The 002 peak position for our MCMB material is indicated as the solid vertical line. The measured 002 peak is clearly shifted to the left from its theoretical position corresponding to an increase in the lattice spacing due to lithium having intercalated into the MCMB. If this scan is compared to the x-ray diffraction measured from the same powder after it has gone through an ARC experiment, figure 6b, we find that it is still shifted to the left with respect to the theoretical 002 position but by a lesser amount. Some lithium has left the material. By comparing our measured 002 peak positions to the positions of the 002 peak versus lithium concentrations as tabulated in the literature[6], we can determine the amount of lithium in the structure for each of the cases in figure 6 a and b, and hence the amount of lithium consumed in the reaction. For the sample shown in figure 6a, comparison of the 002 peak positon to the literature gives $x \approx 0.5$ in $Li_x C_6$, while $x \approx 0.1$ in $Li_x C_6$ for the material shown in figure 6b. We can conclude then that 0.4 moles of lithium per six moles of carbon were consumed in the reaction

The heat generated in our MCMB systems is probably due to two processes occuring simultaneously: 1) the formation of new SEI and 2) the decomposition of the SEI to stable inorganic compounds. We use the generic reactions below to model these processes:

$$1) \quad Li_x C_6 + electrolyte \rightarrow Li_{x-y} C_6 + (y + y_o)SEI \qquad (1)$$

$$2) \quad 2(y + y_o)SEI \rightarrow zLi_2 CO_3 + 2(y + y_o - z)SEI \qquad (2)$$

If the amount of lithium in the MCMB is labeled x and the amount of lithium in the SEI formed on cycling is y_o, then:

$$\frac{dx}{dt} = \frac{-k_1}{p} x$$

$$\frac{dy}{dt} = \frac{k_1}{p} x - \frac{k_2}{p}(y + y_o) \qquad (3)$$

$$\frac{dz}{dt} = \frac{k_2}{2p}(y + y_0)$$

where y is the amount of new SEI formed, z is the amount of final reaction product and p is the thickness of the reaction product layer on the surface of the carbon that the lithium must diffuse through to react to form new SEI and is proportional to the amount of lithium in that layer:

$$p = y + y_0 + 2z \qquad (4)$$

The model states that the reaction proceeds as the inverse of the thickness of the reaction product layer. This is simply as a first approximation. Although we have no experimental basis for choosing it, we know that the reaction should slow as the thickness, p, increases. Other powers or functions of p could be tried.

The heat generated can be expressed as the sum of the heat generated by the formation and the decomposition of the SEI:

$$\frac{dT}{dt} = \frac{h_1}{Cp} A_1 e^{-E_1/kT} x + \frac{h_2}{Cp} A_2 e^{-E_2/kT} (y + y_o) \qquad (5)$$

where h_1 and h_2 are the heats of reaction of reactions (1) and (2) per lithium atom consumed and C is the specific heat of the entire sample (lithiated carbon + electrolyte + stainless steel tube +...). Figure 7 shows the fit obtained using a least squares approach based on equation 5.

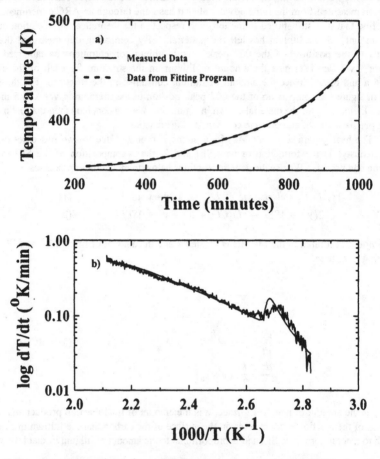

Figure 7: Compares measured data for a MCMB sample discharged to 0.0892V to the calculated data generated by the least squares fitting program based on equation 5.

The program has a lot of difficulty obtaining a good fit to the measured data. Because of the difficulties encountered in generating a good fit, it is our opinion that the model as stated above is incomplete. The actual reaction processes that generate the heat need to be experimentally identified.

We have no direct proof that the SEI is involved in the reaction to give the initial peak in the heating rate plot. However, if our assumption is correct, we should observe that the thermal behaviour of MCMB having a different SEI will be different than that presented above. Peled et al.[6] have most recently proposed that the SEI is formed of : LiF, Li_2CO_3, Li_2O and lithium semi-carbonates as well as polyolefin. Using this model, we should be able to generate a different SEI by using a different electrolyte. We used MCMB cells cycled in $LiBF_4$ in EC:DEC (50%:50%) as our samples. Cell construction, cycling and ARC sample construction proceeded as described previously. Figure 8 compares the results of $LiBF_4$ runs compared to $LiPF_6$ runs.

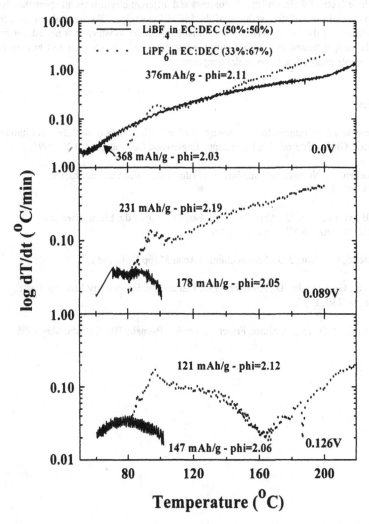

Figure 8: The heating rate of MCMB materials, measured at different states of charge, in $LiBF_4$ electrolyte.

The self heating of the LiBF$_4$ samples not only starts at lower temperature, but has a completely different behaviour supporting our idea that the SEI is involved.

Conclusions

We have developed a reliable methodology to test the thermal behaviour of small samples in a laboratory environment. The question of how our results compare to an oven exposure test is unknown for the moment, but will be determined soon. We have shown that the thermal behaviour of electrode materials can be measured easily using a small amount of material. The study of the effects of different lithium contents and different electrolytes all show that these are factors involved in the relative stability of the MCMB electrode. We have also shown that the thermal stability of the cell depends on the initial temperature to which it is heated, for example an MCMB sample heated to a higher initial temperature exhibits stronger self heating than an identical sample heated to a lower initial temperature.

References

1) Underwriters Laboratories Inc. Document number 1642, Standard for Safety in Lithium Batteries, Global Engineering Documents, Englewood Colorado, April 26, 1995.

2) von Sacken, U., Nodwell, E., Sundher, A., Dahn, J.R., Solid State Ionics, **69**, pp 284-290, 1994.

3) Way, B., von Sacken, U., Abstract # 830 , Fall Meeting of the Electrochemical Society, San Antonio Tx, pp1020, 1996.

4) Townsend, D.I., Tou, J.C., Thermochimica Acta, **37**, pp 1-30, 1980.

5) Peled, E., Golodnitsky, D., Abstract # 196, Electrochemical Society Meeting, Paris , France, pp 236, 1997.

6) Zheng, T., Ph.D. Thesis, Simon Fraser University, Burnaby BC, Canada, May 1996

EFFECT OF HF IN LiPF$_6$ BASED ELECTROLYTES ON THE PROPERTIES OF SURFACE PASSIVATION FILMS FORMED ON GRAPHITE ELECTRODES IN Li SECONDARY BATTERIES

T. SATO *, M. DESCHAMPS, H. SUZUKI, H. OTA, H. ASAHINA AND S. MORI
* Tsukuba Research Center, Mitsubishi Chemical Corporation,
8-3-1, Chuo, Ami, Ibaraki, Japan, 3709954@cc.m-kagaku.co.jp

ABSTRACT

The effect of hydrofluoric acid (HF) concentration in conventional LiPF$_6$/EC+DEC electrolyte on the properties of solid electrolyte interphase (SEI) films formed on synthetic graphite electrode and the electrochemical performance were studied using x-ray photoelectron spectroscopy (XPS), scanning electron microscopy (SEM), elemental analysis and electrochemical measurement. The morphology and composition of SEI are affected by a small amount of HF in the electrolyte. Higher F content in the SEI was observed when high HF electrolyte were used, although the irreversible capacities used for formation of SEI were nearly the same value. The cycleability of the synthetic graphite electrode was greatly improved when surface film contains larger amount of LiF.

INTRODUCTION

Interest in the formation of the solid electrolyte interphase (SEI) on graphite electrode has recently increased to improve the performance of lithium ion batteries. It is well known that surface passivating film play a key role in the cycling of lithium ion cell. The composition of SEI is extensively analyzed by some researchers concluding lithium carbonate and lithium alkylcarbonates are the main components when ethylene and diethyl carbonates (EC+DEC) solution is used [1,2]. In case of LiPF$_6$ containing electrolytes, lithium fluoride (LiF) is also analysed. Aurbach et al. proposed several possible reactions scheme to produce LiF [3]. However, it is not clearly understood the effect of HF, one of the major impurities found in LiPF$_6$, on the formation of SEI and the graphite electrode performance [1-4].

This work aims at a comparative study of graphite electrodes in electrolyte solutions based on LiPF$_6$ and EC+DEC containing different HF concentrations. HF effect on electrode performances and properties of SEI were explored.

EXPERIMENT

EC and DEC (Mitsubishi Chemical Co., battery grade) were used without further purification. LiPF$_6$ was obtained from Hashimoto (battery grade). Electrolyte solutions, 1M

LiPF$_6$ / EC+DEC (1:1), were used in this study and the different concentrations of HF were obtained by adding water to this electrolyte. The water content was measured by Karl-Fischer titration (Hiranuma, AQ7) and the HF concentration was determined by NaOH titration after 2 weeks storage. The electrolyte solutions used in this study are listed in Table I. In this paper, note that the titlated acid value was calculated as HF concentration.

Electrochemical measurements were performed using CR2016 coin-type cell. KS44 synthetic graphite (Timcal) was used as working electrode and lithium metal (Honjo Metal) was used as reference and counter electrode. All cells were assembled in glove box (VAC) filled with dry Ar. Impedance spectroscopy was performed using the same coin cell at an open-circuit potential. The amplitude of the alternating current was 10 mV and the frequency range was 10 kHz to 0.1 Hz.

The passivating film on KS44 electrode was analysed as follows. After the coin cell was cycled 1.5 V to 0 V several times to complete SEI film, the cell was carefully opened at undoping state to get KS44 electrode. The electrode was applied for XPS and SEM analysis without washing. To determine F content in SEI, the film was dissolved into pure water and analysed by ion chromatography.

Table I Concentrations of HF and H$_2$O in 1M LiPF$_6$ / EC+DEC (1:1) solutions

HF (ppm)	H$_2$O(ppm)
3	3
38	3
130	8
290	18
600	30

RESULTS

Electrochemical Study

Fig.1 shows the first reversible and irreversible capacities obtained for KS44 electrode in 1M LiPF$_6$ / EC-DEC (1:1) containing different amount of HF. Current density was set at 0.41 mA/cm^2 ($\approx$C/5). The higher HF content in the electrolyte, the lower reversible capacity obtained. However, the irreversible capacities, which were used to make SEI film, are nearly the same (around 35 mAh/g). When lower current density was applied to high HF content electolyte system, reversible capacity increased up to

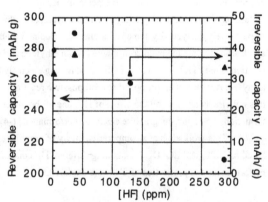

Fig. 1 Effect of HF concentration on the capacities of the cell KS44/EC+DEC, 1M LiPF$_6$/Li

the same capacity obtained for low HF content electrolyte. This means that the impedance of the cell was greatly affected by HF concentration in the electrolyte.

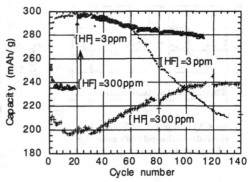

Fig.2 presents the capacity as a function of cycles number for different HF concentrations. In the case of low HF concentration (3 ppm), the reversible capacity is stable until the 50th cycle and then it rapidly decreased. In the case of high HF concentrations, the capacity gradually

Fig. 2 Effect of HF concentration on the cycleability of the cell KS44/EC+DEC, 1M LiPF6/Li

increased up to the 100th cycle and thereafter it stabilized. If we replace lithium counter electrode and change the electrolyte by one with low HF concentration in the cell after 20th cycle, then reversible capacity reach 296 mAh/g and its cycleability is improved. These results indicate that SEI films formed under high HF electrolytes are more stable.

To determine the effect of HF on electrochemical properties of SEI, impedance measurements were applied for both lithium and KS44 synthetic graphite electrodes. Fig.3 shows the 5th discharge and charge curves for KS44 / Li cells using 3 ppm and 600 ppm of HF content electrolytes. The difference in discharge curves was observed only at the end of discharge. The cell containing 150 ppm of HF was fully discharged to 0 V and measured impedance using new lithium counter electrode and low HF content electrolyte (Fig.4). It revealed that the large impedance observed at the end of the discharge was mainly come from lithium electrode and the effect of HF concentration on the impedance of KS44 electrode was negligible. Fig.5 shows the impedance change for lithium electrode stored in different HF content electrolytes. The impedance of lithium was greatly affected by HF [3,5].

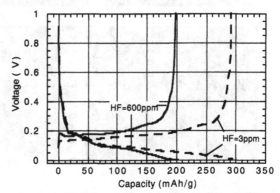

Fig.3 Effect of HF concentration on the profile of the fifth cycle discharge / charge curves.

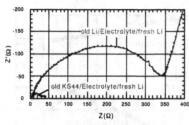

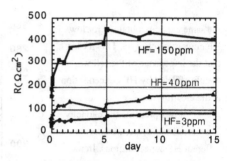

Fig.4 Nyquist plots for KS44 and Li
electrodes discharged to 0V using high
HF content electrolyte.

Fig.5 HF effect on the impedance change for
Li electrode stored at room temperature.

Analytical Study

Fig.6 shows SEM images of KS44 graphite electrode before and after the formation of passivating film. For HF=4ppm, small particles are observed on graphite surface (Fig.6b) and the morphology of SEI looks different from higher HF content. For higher HF concentrations, graphite surface was fully covered by smooth SEI (Fig.6c and 6d). From XPS anlysis (Fig.7),

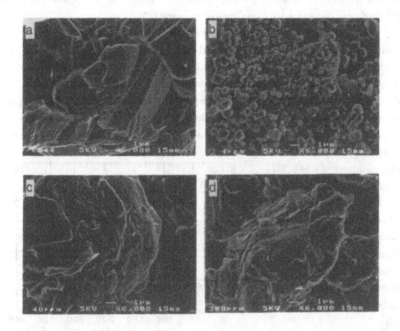

**Fig. 6 SEM images for SEI on KS44 electrodes ; (a) bare KS44
electrode, SEI formed under electrolyte containing (b) 3ppm, (c)
40ppm and (d) 300ppm of HF in EC+DEC, 1M LiPF$_6$.**

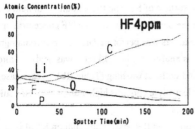

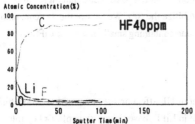

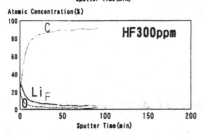

Fig.7 XPS depth profiles for SEI on KS44 formed under different HF content electrolytes, etching rate 6nm/min.

Table II Effect of HF on the F/Li ratio in SEI, analysed after washing with DEC (a) and without washing (b).

HF (ppm)	F/Li ratio (a)	F/Li ratio (b)
3	0.028	0.004
40	0.14	0.02
300	-	0.05
600	0.16	-

small particles, observed on KS44 in case of very low HF concentration, can be identified as lithium carbonate or lithium alkylcarbonate and LiF is also observed in the XPS spectrum. For higher HF electrolytes, LiF is observed on the surface of SEI and the amount of lithium carbonate and lithium alkylcarbonate looks smaller than the one formed under low HF concentration. To determine F content in the SEI, quantitative analysis was performed using ion chromatography. Table II shows the F/Li molar ratio of SEI films analysed. The F/Li ratio was analysed after washing the electrode with DEC (Table IIa) and without washing (Table IIb). In each case, higher F content in the SEI was observed when the electrolyte contains higher amount of HF, still there remains washing problems for quantitative analysis.

These results indicate that higher F content in the passivating film on KS44 is produced mainly from HF in electrolyte solution and stabilize the SEI. We believe the SEI containing LiF is much less soluble to this electrolyte than the one containing small amount of LiF.

CONCLUSIONS

We can conclude that a small amount of HF affect on the composition and stability of SEI on KS44 synthetic graphite electrode. HF in $LiPF_6$ based electrolyte play a key role to form stable surface passivating film containing LiF.

REFERENCES

1. D.Aurbach, Y.Ein-Eli, B.Markovsky, A.Zaban, S.Luski, Y.Carmeli and H.Yamin, J.Electrochem.Soc., 142, p.2882 (1995)
2. S.Mori, H.Asahina, H.Suzuki, A.Yonei and E.Yasukawa, Ext.Abs.8th Int.Meeting on Lithium Batteries, p.40 (1996)
3. D Aurbach, B.Markovsky, A.Shechter, Y.Ein-Eli and H.Cohem, J.Electrochem.Soc., 143, p.3809 (1996)
4. O.Y.Chusid, Y.Ein-Eli, D.Aurbach, M.Babei and Y.Carmeli, J.Power Sources, 43-44, p.47 (1993)
5. K.Kanamura, S.Shiraishi and Z.Takehara, J.Electrochem.Soc., 143, p.2187 (1996)

INFLUENCE OF WATER AND OTHER CONTAMINANTS IN ELECTROLYTE SOLUTIONS ON LITHIUM ELECTRODEPOSITION

T. Fujieda, S. Koike and S. Higuchi
Osaka National Research Institute Ikeda, Osaka, JAPAN

ABSTRACT

Electrochemistry of a nickel electrode in propylene carbonate [PC] containing LiClO4, LiCF3SO3, LiPF6 was studied through a micro electrode (ϕ =25 μ m) techniques in the wide potential range between +4.5 and -0.2 V vs. Li/Li$^+$. Common pronounced peaks were observed in the potential range positive to lithium electrodeposition on nickel in all electrolyte solutions examined. Thus, these peaks can be attributed to reactions related to Li$^+$ or commonly contained contaminants such as water and acids. In particular, the peak which appeared at the most negative potential seemed to be underpotential deposition (UPD) of lithium.

To prove this hypothesis a nickel electrode in highly dried PC (water content : 3 - 8 ppm) intentionally contaminated with a small amount of water and CF3COOH was examined via cyclic voltammetry. Changing the content of water and acid (and its ratio) in PC resulted in a variety of voltammograms and one of them was identical to the one observed in PC containing lithium electrolytes. These facts preclude the existence of UPD of lithium on nickel in the electrolyte solutions. Instead, the existence of NiOH on nickel and its redox reaction mechanism have been postulated. The mechanism is consistent with the experimental facts : a nickel electrode passivates in PC with a small amount of water, and a small amount of acid, CF3COOH, can prevent passivation. The vicinity of the electrode surface may be exposed to an alkaline atmosphere owing to the reduction product of water. This seems to be the cause of troubles we run into with the electrodes at cathodic potentials

INTRODUCTION

The contaminants in electrolyte solution do affect the surface chemistry of a lithium electrode [1], because all compounds will be reduced around the equilibrium potential of a lithium electrode. Thus, much attention has been paid to the purity of the chemicals used for lithium batteries. Meanwhile the electrochemistry of non-lithium electrodes (such as nickel, gold etc.) in lithium electrolyte solutions in the potential range positive to lithium deposition [2, 3, 4], where the contaminants can be reduced, has been a minor subject compared to that of lithium electrodes in electrolyte solutions. However, it could give us important information about the contaminants in solutions, which has not been obtained from experiments with a lithium electrode.

The cyclic voltammograms with a nickel electrode in PC containing 0.1 M LiAsF6 show a few peaks at potentials positive to bulk lithium deposition [2]. These were attributed to the reduction of water or oxygen-related reactions except for the pair of peaks which appear around 0.4 V(cathodic) and 0.9 V (anodic). So far, the existence of underpotential deposition (UPD) and dissolution has been presupposed for these peaks without proof. One objective of this paper is identification of these peaks on the voltammogram. Further, the role of a trace of water and acids for lithium deposition on nickel has been investigated.

EXPERIMENT

The electrolytes used for the experiments were cautiously purified in a glove box [M-Brown Ltd.] filled with recycling Ar gas (dew point <-80 °C, O2 < 1 ppm). Battery grade of LiCF3SO3 (Tomiyama chemicals Ltd.) were recystallized twice in dry dimethoxy carbonate containing less than 10 ppm of water and vacuum dried in a oven at 100 °C for 10 hours. LiClO4 [Kishida Chemical Ltd.] was recrystallized twice in dry ethanol containing less than 50 ppm of water and vacuum dried at 140 °C for 48 hours. LiPF6 was used after drying at 100 °C for 24 hours. All dry organic solvents (water content less than 10 ppm) were

463

obtained from Kishida Reagents Chemicals Ltd. (Propylene carbonate [PC] was distilled and went though a column filled with molecular sieve 3Å before being filtered by a membrane with 0.2 μ m of pores. The content of residual water in the solvents were checked via the Carl Fisher method for each bottle.)

All electrolyte solutions were degassed just before each electrochemical measurement by being treated in an ultrasonic bath to remove residual oxygen. The two electrodes cell was fabricated and sealed in a glove box and taken out for electrochemical measurements. The counter electrode was 99.9% pure lithium ribbon [Honjoh Ltd.]. The electrochemical measurements with 99.99 % pure nickel micro-disc-electrodes [Niraco Ltd.] of 25 μ m in diameter were conducted with potentio-galvanostat Model 283 [EG&G Instruments, Princeton Applied Research Ltd.] under control of the software attached to it.

RESULTS AND DISCUSSION

The typical cyclic voltammograms or a part of them for a nickel electrode in PC containing 1M of LiClO4, LiCF3SO3 and LiPF6 are shown in Fig. 1 (A), (B) and (C). In all curves, three common cathodic peaks appeared near 0.4, 0.9 and 1.4 V prior to lithium electrodeposition. Corresponding anodic peaks were also observed near 0.9, 1.8 and 2.4 V. Even in 1 mM of LiCF3SO3 / PC a similar curve was obtained as shown in Fig. 1 (D). Thus, these peaks should be related to contaminants which is unavoidably included in the electrolyte solutions or decomposition of propylene carbonate. In all solutions, water and acid (H^+) are commonly contaminated and considered to be reduced at the potentials positive to lithium deposition. Thus, PC containing water and acid of a few ppm level was cautiously investigated.

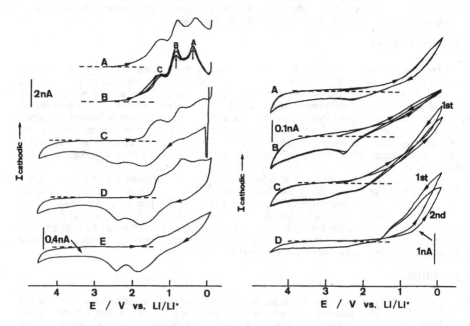

Fig.1 Typical cyclic voltammograms for a Ni electrode (ϕ = 0.025 mm) in PC containing 1 M of (A) LiClO4, (B) LiPF6 , (C) LiCF3SO3, and (D) 1mM, (E) 0.05mM of LiCF3SO3.

Fig.2 Typical cyclic voltammograms for a Ni electrode (ϕ = 0.025 mm) in PC containing (A) 0, (B) 50, (C) 100, (D) 200 volume ppm of water.

Fig. 2 shows typical cyclic voltammograms for a nickel electrode in PC with and without water. The monotonous increase in cathodic current at cathodic potential can be largely attributed to the decomposition of PC since the current continued to increase as the potential was scanned further in the negative direction. The loops observed in a dry PC were almost the same even when scanning potential was repeated up to ten times. However, a small amount of water disturbed it and passivated the nickel electrode after several cycles. The anodic peak(s) appeared around 2.4 V may be attributed to the water-related reaction and will be discussed later.

Fig. 3 shows typical cyclic voltammograms for a nickel electrode in PC with and without CF_3COOH. The addition of 50 ppm of CF_3COOH increased cathodic current probably due to reduction of H^+ or H_3O^+ formed with residual water, however, further addition of the acid did not increase the current. Meanwhile, anodic current which may be attributed to the corrosion of nickel or decomposition of PC was observed at anodic potentials more than 3.7 V even in the solution containing only 50 ppm of CF_3COOH. This fact suggests that a trace of acid can cause such trouble in dry solvents at the anodic potentials where lithium batteries are to be operated. The results in Fig. 2 and Fig. 3 indicate that water or acid individually can not be the cause for the peaks appearing in the underpotential range of lithium electrodeposition as shown in Fig. 1.

Fig. 4 shows typical cyclic voltammograms for a nickel electrode in PC with 100 ppm of CF_3COOH and 0, 50, 100, 200 ppm of water. Depending on the content of water, a variety of voltammograms which may be practically identical to those which appeared in Fig. 1 were observed. Obviously four water-acid-related reactions in addition to hydrogen evolution occur in the potential range from 0 to 4.5 V vs. Li/Li$^+$. These are not unique to CF_3COOH,

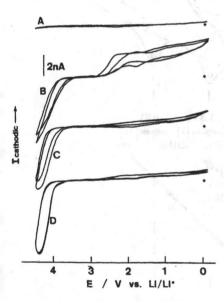

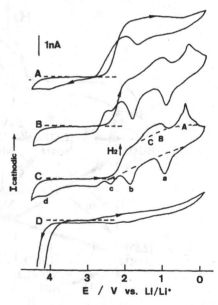

Fig.3 Typical cyclic voltammograms for a Ni electrode (ϕ = 0.025 mm) in PC containing (A) 0, (B) 50, (C) 100, (D) 200 volume ppm of CF_3COOH.

Fig.4 Typical cyclic voltammograms (2nd cycle) for a Ni electrode (ϕ = 0.025 mm) in PC containing 10 ppm of CF_3COOH and (A) 200, (B) 100, (C) 50, (D) 0 volume ppm of water.

because a similar voltammogram was obtained in PC containing 98% H_2SO_4. The existence of a pair of peaks ((A) at 0.4 V and (a) at 0.9 V) in the solution without Li ions deny the possibility of reversible UPD of lithium on nickel at those potentials.

The plausible reactions and nickel compounds formed on a nickel electrode in PC containing a small amount of water and acid are postulated in Fig. 5 in accordance with the voltammogram. Bare nickel metal can be exposed to the solution at 0 V vs. Li/Li$^+$. While hydrogen would evolve at a maximum rate on the surface.

$$2H_2O + 2e \rightarrow H_2 + 2OH^- \qquad (1)$$

When the potential is moved at 0.9 V, NiOH$_{ad}$ (metallic Ni$^+ \cdot$ adsorbed OH$^-$) would be formed according to the following reaction,

$$Ni + H_2O \rightarrow NiOH_{ad} + H^+ + e \qquad (2)$$
or
$$Ni + OH^- \rightarrow NiOH_{ad} + e \qquad (3)$$

In particular, the second reaction could occur, because the electrode surface must be abundant in water-decomposition products of OH$^-$ at such a cathodic potential.

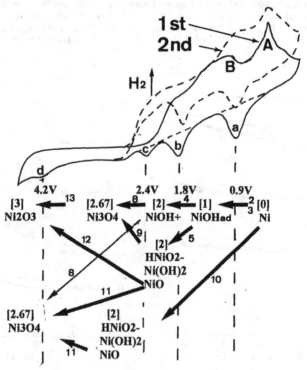

Fig.5 The postulated compounds and oxidation mechanism to explain the cyclic voltammogram obtained in PC containing water and CF3COOH.

Then, at 1.8 V, the above monovalent compound would oxidize as,

$$NiOH_{ad} \rightarrow NiOH^+ + e \qquad (4)$$

or

$$NiOH_{ad} + H_2O \rightarrow Ni(OH)_2 + H^+ + e \qquad (5)$$

A part of nickel (II) hydroxide can further chemically decompose to,

$$Ni(OH)_2 \rightarrow NiO + H_2O \qquad (6)$$
$$Ni(OH)_2 \rightarrow HNiO_2^- + H^+ \qquad (7)$$

In lithium electrolyte solutions the peaks near 1.8 are considerably broad, probably reflecting the complicated oxidation mechanism of nickel from +1 to +2 as suggested above.
At 2.4 V, those bivalent nickel compounds oxidize as,

$$3NiOH^+ + OH^- \rightarrow Ni_3O_4 + 4H^+ + e \qquad (8)$$

or

$$HNiO_2^- + 4H^+ \rightarrow Ni_3O_4 + 2H_2O + e \qquad (9)$$

In particular, the formal potential for (9) calculated with thermodynamic data, 2.312 V vs. Li/Li+ [5] corresponds well to the experimental results. At the same potential inner nickel metal may be oxidized as a minor reaction according to,

$$Ni + H_2O \rightarrow HNiO_2^- + 3H^+ + 2e \qquad (10)$$

The following are less plausible but should not be excluded.

$$3NiO + H_2O \rightarrow Ni_3O_4 + 2H^+ + 2e \qquad (11)$$
$$2NiO + H_2O \rightarrow Ni_2O_3 + 2H^+ + 2e \qquad (12)$$

This is because a mixed potential due to NO and NiOH is often observed around this potential. Thus, identification of the peak (c) could not be determined with only the facts obtained here.
At 4.2 V, in addition to (11) and (12), the reaction,

$$2Ni_3O_4 + H_2O \rightarrow Ni_2O_3 + 2H^+ + 2e \qquad (13)$$

may be plausible.
The cathodic mechanism may consist of reversed reactions of the oxidation above. However, the reverse reaction of (5) or (8) is least plausible when the vicinity of the electrode is very basic due to water decomposition. In such a case, the subsequent reduction (2) or (3) can not occur. This can be the cause for passivation of a nickel electrode in PC and explain the effect of a small amount of water on the cyclic voltammograms observed in Fig.2.
In the presence of Li $^+$, (5) may be partially replaced by

$$Ni(OH)_2 + 2Li^+ \rightarrow \leftarrow Ni + 2LiOH \qquad (14)$$

This reaction could be the reason for the shift of cathodic peak B towards the negative potential and broadening of anodic peak b in the presence of a lithium electrolyte. Furthermore, this is consistent with the fact that as the peak B becomes pronounced, the peak A which can be attributed to the reversed reaction of (2) or (3) becomes weak as shown in Fig. 1 (B).
If the above proposed mechanism is true, the addition of strong acid can inhibit the passivation of a nickel electrode during cycling and make reversible lithium electrodeposition possible. Thus, the effect of the small addition of CF₃COOH on lithium electrodeposition on nickel was investigated. Fig. 6 shows typical cyclic voltammograms for a nickel electrode in 1 mM LiCF₃SO₃ / PC with 50 and 100 ppm of water or CF₃COOH. The amount of lithium electrodeposited was larger in the solution containing acid than in others. Furthermore, it is noted that lithium is deposited and dissolves in the same way at the 5th and 20th cycles in the

acid-contaminated solution. This cyclability was confirmed up to 300th cycle as shown in Fig. 7. Therefore, pH-control in residual water is imperative for a long cycling of lithium deposition on nickel.

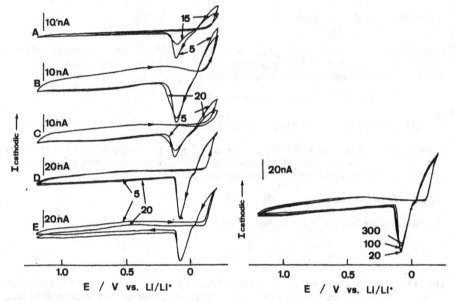

Fig.6 Typical cyclic voltammograms for a Ni electrode (ϕ = 0.025 mm) in 10 mM LiCF₃SO₃/PC containing (A) 100, (B) 50, ppm of water, (C) none, (D) 50, (E) 100 ppm of CF₃COOH. The cycle numbers are indicated in figure.

Fig.7 Typical cyclic voltammograms for a Ni electrode (ϕ = 0.025 mm) in 10 mM LiCF₃SO₃/PC and 50 ppm of CF₃COOH. The cycle numbers are indicated in the figure.

CONCLUSIONS

The possibility of the existence of UPD of lithium should be discarded to account for the peak which appeared prior to lithium bulk deposition on cyclic voltammogram. Instead the formation of $NiOH_{ad}$ is presupposed for the corresponding peak. Any electrodes would be exposed to a strong basic atmosphere in organic solvents at cathodic potentials owing to the cathodic decomposition of contaminated water and be apt to result in passivation. The addition of a small amount of CF₃COOH could prevent the passivation of a nickel electrode in LiCF₃SO₃ / PC.

REFERENCES

1. D. Aurbach and A. Zaban, J. Electroana. Chem., **348**, p. 155 (1993).
2. D. Pletcher, J. F. Rohan and A. G. Ritchie, Electrochimica Acta, **39**, p. 1369 (1994).
3. D. Wagner and H. Gerischer, Electrochimica Acta, **34**, p. 297 (1994).
4. D. Aurbach, M. Daroux, P. Faguy, and E. Yeager, J. Electroana. Chem., **297**, p. 225 (1991).
5. A. J. Arvia and D. Posadas in <u>Encyclopedia of Electrochemistry of the elements vol. 3</u> edited by A. J. Bard, Marcel Dekker, Inc., New york and Basel, 1973, p.212.

PASSIVE-FILM FORMATION ON METAL SUBSTRATES
IN 1M LiPF₄/EC-DMC SOLUTIONS

RONALD A. GUIDOTTI and GERALD C. NELSON
Sandia National Laboratories
P.O. Box 5800, Albuquerque, NM 87185-0614, raguido@sandia.gov

ABSTRACT

The initial irreversible capacity loss in lithium-ion cells has been attributed to passive-film formation on the carbon and graphite anodes during the first intercalation. We have examined the nature of these passive films on select metal substrates. We studied film formation on Cu, 304 stainless steel, and Mo cycled in a 1M LiPF₄/ethylene carbonate (EC)-dimethyl carbonate (DMC) solution (1:1 v/v) over a potential range of 3 V to below 0 V. Film formation occurs readily on each of the metals and involves both solvent and salt species. The composition, thickness, and distribution of the films depends strongly on the substrate. The redox processes that occur during film formation and the potential for Li plating are highly dependent on the substrate composition.

INTRODUCTION

The first-cycle efficiencies of lithium-ion cells are typically low during the intercalation of Li⁺ into ordered and disordered carbons in organic electrolytes, i.e., the deintercalation capacity is lower than the intercalation capacity. In the case of highly ordered graphite, this capacity loss is relatively low, typically on the order of 20%.[1-2] In disordered or turbostratic carbons, however, the capacity losses can be 50% or more for the first cycle.[3] In the case of Si-doped carbons derived from polymethacrylonitrile (PMAN)-divinyl benzene, these losses can exceed 60%.[4] These losses are the result of secondary electrochemical and chemical reactions occurring at the electrode involving the solvent and supporting electrolyte. This passive film has been referred to as a solid-electrolyte-interface (SEI) layer since it inhibits transport of solvent molecules but permits transfer of Li⁺. After formation of a stable SEI layer on carbon anodes, the coulombic efficiencies approach ~100% on subsequent cycling

Several investigators—notably Peled and co-workers[5-6] and Aurbauch et al.[7-9]—have examined the composition and characteristics of the SEI layer on various carbon anode materials. In reviewing this work, it becomes apparent that the composition of the layer is a function of the solvent/salt combination used. For example, in the case of organic-carbonate electrolytes, the SEI layer was commonly found to contain Li₂CO₃ and alkoxides, alkylcarbonates, and carboxylates of Li, along with LiF (for fluoride-based electrolytes).[10] More recently, Peled reported that the SEI layer was nonhomogeneous, i.e., there were significant compositional differences, depending on whether the SEI-electrolyte or SEI-substrate interface was examined.[6] He estimated a SEI thickness of 10-15 nm. A comparable SEI layer is observed when metallic Li is placed in contact with the same type of electrolyte solutions. Similar SEI-characterization work was also reported by Naji et al.[11]

A number of metal substrates are commonly used in testing carbons in experimental cells, with Cu and stainless steel being the most commonly used current-collector materials. It was of interest to determine the extent of passive-film formation on such materials under the normal test environments used for carbon-anode characterization. Little has been reported in the literature in this area. Some work has been done by Zaban and Aurbach measuring the time dependency of the impedance of surface films formed on Ni at low (0.2 V) potentials.[12] These workers also examined the films on Li surfaces using Fourier-transform infrared spectroscopy (FTIR), but no FTIR data were reported for the case of Ni. They modeled their impedance data to calculate film thicknesses that varied from 5 Å to over 150 Å (15 nm) in $LiPF_6$ in propylene carbonate (PC). This is comparable to the values reported by Peled.[6] However, these films were not stable over long-time storage and tended to dissolve after formation in this system. Subsequent work on film formation on a Au substrate was done using an electrochemical quartz crystal microbalance.[13] Some cyclic voltammograms (CVs) were reported by Ein-Eli *et al.* for a Pt electrode in 1M $LiAsF_6$ solutions in PC and DMC, but the intent was to study the effects of SO_2 as an additive, and not film formation.[14]

For the most part, much of the previous work has focused on the formation and nature of the film in direct contact with the highly energetic negative electrode materials, namely lithium metal and the carbon intercalation compounds. Our interest was in being able to separate the contribution and role of the solvent and salt in this film-formation process, removed from the highly reducing electrode material. In addition, we were also interested in determining to what extent the metal substrates employed as the current collectors could contribute to the initial first-cycle loss and inefficiencies.

To obtain a better understanding of the role that the passive film plays in Li-ion cells, the reduction processes were studied using a number of metal and carbon materials as substrates. In this paper, we report on metal substrates of Cu, 304 stainless steel (SS), and Mo substrates. The Cu and 304 SS were of particular interest since Cu is used in most commercial cells and both metals are used in our test cells for the characterization of carbon anodes. Film formation was examined as a function of applied potential, down to voltages where Li deposition was observed. The films were then examined *ex situ* using secondary ion mass spectroscopy (SIMS) and, in preliminary tests, by FTIR and Raman spectroscopy. Complex-impedance spectroscopy was used as a complementary analytical technique.

EXPERIMENTAL

Electrochemical Procedures

Three-electrode tee cells were used for testing the 1/2"-dia. sample discs. The cells were assembled in the dry room (RH<3%) and used Li metal counter and reference electrodes and two Celgard 2500 separator discs. The electrolyte was 1M $LiPF_6$ in 1:1: (v/v) ethylene carbonate (EC) and dimethyl carbonate (DMC) (Merck) and contained <40 ppm of H_2O.

CVs were generated using a Princeton Applied Research (PAR) Model 263 or 273A potentiostat at a scan rate of 1 mV/s. The cells were scanned from the open-circuit voltage (OCV) on the first scan and 3 V on subsequent scans to between 1.0 V and 0 V.* In several cases scans were also made to increasingly more negative potentials in order to observe Li plating and stripping. A minimum of three scans was performed over each potential range.

Complex-impedance spectroscopy was performed using a Solartron Model 1250 Frequency Response Analyzer coupled to a Solartron Model 1286 Electrochemical Interface and a 10 mV peak-to-peak sine wave over a frequency range of 65 kHz to 100 mHz. Cells were cycled between 3 V and lower potential limits of 1 V, 0.1 V, and 0.01V at 1 mV/s. Spectra were taken at both voltage limits for the first and fifth cycles.

---

* Unless otherwise noted, all voltages are referenced to Li/Li+.

RESULTS

Cyclic Voltammetry

Cu - A typical CV for a 1/2" Cu disc is shown in Figure 1 for a lower voltage limit of 0.1 V. Two reduction waves are seen on the first cycle at 0.61 V and 1.20 V, along with corresponding oxidation waves at 1.33 V and 2.05 V. The magnitude of the peaks decreased on subsequent cycles, but remained, indicating two reversible processes. The onset of Li stripping was typically observed when the Cu was taken to 0 V and sometimes 0.01 V; it showed up as a small oxidation wave (shoulder) near 0.25 V.

The coulombic efficiency on the first cycle was only 53%. This increased slightly on the following cycles and stabilized at slightly over 75% by the fourth cycle. This low number indicates that the passive film may not be stable but continues to reform when cycled in this manner. Or, its nature is such that continued access of the electrolyte to the electrode surface is possible, i.e., it may be porous. Some of the redox waves observed at the higher potential could be related to oxidation/reduction and subsequent reaction with the electrolyte.

Similar CV tests where a fresh Cu sample was only cycled between 3 V and 1 V showed a gradual increase in efficiency from 57% to almost 82.5% by the third cycle. Continued cycling between 3 V and 0.1 V further increased the efficiency from 67% to 90% after three additional cycles. Additional cycling between 3 V and 0.01 V increased the efficiency to 93% after three more cycles. Finally, cycling the same sample between 3 V and 0 V further increased the efficiency to about 95.4% after three additional cycles. The efficiency did not increase above this point with further cycling.

When the Cu was cycled at 10 and 100 mV/s, the reduction waves became less pronounced—especially at the highest rate—and the efficiencies dropped from 95.4% (at 1 mV/s) to 92% and 73%, respectively, when cycled to 0 V. These data indicate that the voltage history of the Cu, i.e., the test protocol, can have a significant effect on the nature of the film that forms.

Preliminary tests with 1M LiPF$_6$ in DMC showed a significant reduction in the degree of irreversible losses on Cu substrates during the first reduction. This indicates that EC plays a major role in the formation of the passive film, with little effect from DMC. More work is underway in this area.

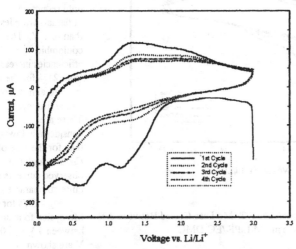

Figure 1. Cyclic Voltammograms for 1/2" Cu Disc Cycled Between 3 V and 0.1 V at 1 mV/s in 1M LiPF$_6$/EC-DMC.

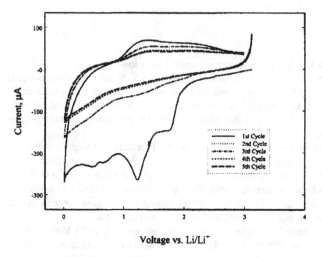

Figure 2. Cyclic Voltammograms for 1/2" 304 SS Disc Cycled Between
3 V and 0.01 V at 1 mV/s in 1M LiPF₆/EC-DMC.

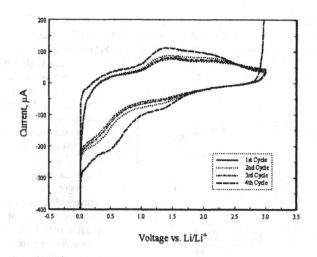

Figure 3. Cyclic Voltammograms for 1/2" Mo Disc Cycled Between 3 V
and 0.01 V at 1 mV/s in 1M LiPF₆/EC-DMC.

304 Stainless Steel - CVs for 304 stainless steel (SS) cycled five times between 3 V and 0.01 V are shown in Figure 2. Reversible reduction waves were observed at 1.75 V, 1.25 V, with two additional waves at 0.7 V and 0.5 V. The reduction waves were comparable to those observed for Cu, except Cu did not show the wave at 1.75 V (Figure 1).

A moderate oxidation wave was seen at 1.35 V along with a very broad wave near 2.25 V, similar to those for Cu. Li stripping was typically not observed until reduction potentials were less than -0.1 V. The coulombic efficiencies increased from 21% for the first cycle to almost 76% by the fifth cycle. These are considerably lower than for the case of Cu, although the observed currents are quite comparable.

Mo - CVs for Mo cycled four times between 3 V and 0.01 V are shown in

Figure 3. The shape of the CV and the potential for reduction and oxidation processes were very similar to those seen for Cu (Figure 1). The coulombic efficiencies were much higher, however.

472

SIMS

The SIMS spectra for the film deposited on Cu are shown in Figure 4. The film was not homogeneous in composition, as was noted by Peled for the SEI layer formed on carbons.[6] The films consisted of three layers. The surface of the film in contact with the solution contained a very high concentration of C_2H species. The concentration of these organic species rose again in the center of the film, tracking the concentration of P. Cu was also found as a film component, primarily as CuF_2. The CuF_2 concentration increased towards the Cu substrate but was lower than P and C_2H in the center of the film. The carbonate concentration peaked at the outside surface of the film and quickly dropped to a low value. It was not a major component of the film.

When SIMS spectra are taken at potentials starting at 1 V and stepping to increasingly more negative values, the thickness of the film rapidly increased. The data in Figure 4 are reported as sputtering time. For the conditions under which the SIMS data were taken, a sputter time of 1,000 s is equivalent to ~250 Å. The film thickness thus ranged from 25 up to 200 Å at the lowest potential examined. This is somewhat thicker than the values reported by others on carbon substrates.[6,12]

Preliminary SIMS data for the 304 SS substrate indicate a shifted elemental distribution relative to the film of Cu. There was a high concentration of P and Li at the film surface facing the solution, with the carbonate peak shifted more towards the center of the film. The oxygen spectrum tracked the carbonate, while the fluoride remained relatively constant throughout the film. The Fe profile showed high levels near the SEI-solution interface which dropped rapidly towards the substrate side. The film thickness was also much less than for the copper, being on the order of 40 to 80 Å.

FTIR/Raman

Only a cursory examination of the film deposited at 0.01 V on a Cu substrate was undertaken using FTIR and Raman spectroscopy, mainly to determine whether these techniques would be suitable for characterizing films on metal surfaces. Consequently, the data were taken without pains to protect the sample from atmospheric exposure. The preliminary results indicate that FTIR will be quite useful for characterization

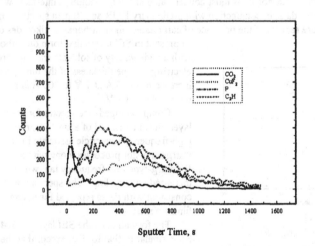

Figure 4. SIMS Spectra for Film Formed on Cu Cycling Between 3 V and 0.01V at 1 mV/s in 1M $LiPF_6$/EC-DMC.

473

of passive films on a metal substrate. Features similar to those reported by Aureate in his FTIR work with carbons were noted at 1702, 1656, 1530, 1426, 1323, 1087, and 868 cm⁻¹. Strong peaks attributed to carbonate were quite evident.

While a signal was obtained using the 457 nm laser line for Raman spectroscopy, it was not very strong. Still, the band at 1090 cm⁻¹ was consistent with the presence of carbonate as determined by FTIR, along with bands at 1572 and 2923 cm⁻¹ attributed to organic species. However, the FTIR and Raman data may have been adversely influenced by the exposure of the sample to ambient air. These tests need to be repeated in a sealed cell, to avoid possible atmospheric contamination.

Complex-Impedance Spectroscopy - Typical complex-impedance spectra for the films that form on Cu are shown in Figure 5 after one and five CVs between 3 V and 0.01 V. The spectra show a small depressed high-frequency semicircle associated with charge transfer and a low-frequency Warburg-type tail. The low-frequency tail increased dramatically after five cycles. These spectra are typical of those for polymer films on metals. Similar spectra were obtained for films on Mo.

CONCLUSIONS

The SEI layer that forms on Cu during initial cycling in 1M LiPF$_6$/EC-DMC solutions results from the inefficient redox processes involving both the salt (LiPF$_6$) and solvent (EC-DMC), with the dominant contribution from EC. After the first cycle, reversible redox processes continue to occur, with increasing coulombic efficiencies with each cycle, up to limiting values of 75%-95%, depending on the scan rate used. The actual chemical reactions that are taking place are unknown at this time but involve Cu species (i.e., CuF$_2$). SIMS analysis shows that the SEI film is actually made up of three layers, with widely differing compositions. This is consistent with the multiple reduction waves that are seen during the first cycle. SIMS confirms the presence of CuF$_2$ in the film near the Cu substrate, with C$_2$H and P present in the bulk of the SEI layer. Carbonate is most concentrated at the film-solution interface, where the C$_2$H species dominates. These data, in conjunction with preliminary FTIR and Raman spectroscopic data, are consistent with literature reports of the presence of carbonates, alkyl carbonates, alkoxides of Li present in SEI layers that form on carbons with a wide variety of solvents and supporting electrolytes. The thickness of the films on Cu increase from 25 Å at 1 V to 200 Å at a potential of 0.01 V.

Complex-impedance spectra for the SEI layer on Cu consisted of a small depressed high-frequency semicircle associated with charge-transfer processes along with a Warburg-type low-frequency tail, which increased with cycling. The spectra are consistent with those reported for polymer films on metals.

The formation of the SEI layer on 304 SS is similar to that for Cu, except that the composition of the multilayer film is shifted from what is observed for Cu. The films are much thinner in the case of 304 SS, relative to

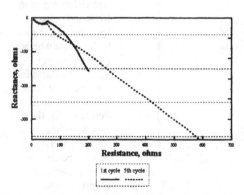

Figure 5. Complex-Impedance Spectra for Films Deposited on Cu After One and Five CVs Between 3 V and 0.01 V in 1M LiPF$_6$/EC-DMC.

Cu, and range from 40 to 80 Å. Mo behaves similarly to Cu, in terms of the persistence of reversible redox waves in the CVs.

The composition of the substrate has a major effect on the nature of the SEI layer that forms. The SEI layer forms on nonintercalating materials as well as on the surfaces of traditional carbons and elemental Li. The constituents of the various layers are similar, in that they contain many of the same species: carbonates, aklylcarbonates, alkoxides, oxides, and hydroxides of Li. The supporting electrolyte also influences the composition of the SEI layer, as it participates in the redox reactions. (We found n this work, for example, that CuF_2 is a significant component in one layer of the film.) However, the relative distribution of the various species differs for the various substrates. The exact mechanism and the pertinent chemical reactions that occur during SEI formation on metal substrates is unclear at this time but involves scission of the C-O bonds in the EC; DMC plays only a minor role in SEI formation in EC-DMC solutions.

ACKNOWLEDGMENTS

The authors wish to thank Mike Overstreet and Herb Case for cell construction and electrochemical testing, David Tallant and Regina Simpson for the Raman data, Robert Patton for the FTIR data, and David Ingersoll for his sound technical advice. This work was supported by the U.S. Department of Energy under Contract DE-AC04-94AL85000.

REFERENCES

1. Z. X. Shu, R. S. McMillan, and J. J. Murray, J. Electrochem. Soc., **140** (4), 922 (1993).
2. R. Yazami, K. Zaghib, and M Deschamps, J. Power Sources, **52**, 55 (1994).
3. T. Zheng, Y. Liu, E. W. Fuller, S. Tseng, U. von Sacken, and J. R. Dahn, J. Electrochem. Soc., **142** (8), 2581 (1995).
4. R. A. Guidotti and B. J. Johnson, Proc. 37th Power Sources Conf., 219 (1994).
5. E. Peled, D. Golodnitsky, G. Ardel, C. Machem, D. Bar Tow, and V. Eshkenazy, Mat. Res. Soc. Symp. Proc., **Vol. 393**, 209 (1995).
6. D. Bar Tow, E. Peled, and L. Burstein, Ext. Abst. 835, **Vol. 96-2**, Fall Mtg. of The Electrochen. Soc., San Antonio, TX, 1028 (1996).
7. D. Aurbach, M. Daroux, P. Faguy, and E. Yeager, J Electrochem. Soc., **134**, 1611 (1987).
8. D. Aurbach and O. Chusid, J. Electrochem. Soc., **140**, L1 (1993).
9. D. Aurbach, Y. Ein-Ely, and A. Zaban, J. Electrochem. Soc., **141**, L1 (1994).
10. D. Aurbach and A. Zaban, J. Electroanal. Chem., **365**, 41 (1994).
11. A. Naji, J. Ghanbaja, B. Humbert, P. Willmann, and D. Billaud, J. Power Sources, **63**, 33 (1996).
12. A. Zaban and D. Aurbach, J. Power Sources, **54**, 289 (1995).
13. D. Aurbach and M. Moshkovich, *op cit*. ref. 5, Extended Abst. 807, 987 (1996).
14. Y. Ein-Eli, S. R. Thomas, and V. R. Koch, J. Electrochem. Soc., **144** (4), 1159 (1997).

APPLICATION TO LITHIUM BATTERIES OF TERNARY SOLVENT ELECTROLYTES WITH ETHYLENE CARBONATE — 1,2-DIMETHOXYETHANE MIXTURE

Y. SASAKI *, N. YAMAZAKI, and M. HANDA
* Department of Industrial Chemistry, Faculty of Engineering, Tokyo Institute of Polytechnics, Atsugi, Kanagawa 243-0297 JAPAN, sasaki@chem.t-kougei.ac.jp

ABSTRACT

The electrolytic conductivity, the lithium cycling efficiency of lithium electrode and the energy density for $Li/V_2O_5(2025)$ coin-type cell were examined in ternary solvent electrolytes containing $LiPF_6$ and $LiClO_4$ with ethylene carbonate(EC) — 1,2-dimethoxyethane(DME) equimolar binary mixture at 25°C. The solvents applied to EC—DME mixture are dimethyl carbonate(DMC), ethyl methyl carbonate(EMC), diethyl carbonate(DEC), 1,3-dioxolane(DOL), 2,2-bis(trifluoromethyl)-1,3-dioxolane(TFMDOL), 1,3-dimethyl-2-imidazolidinone(DMI) and 1-methyl-2-pyrrolidinone(NMP). The order of decrease of the molar conductivities in ternary solvents electrolytes except DMI and NMP systems is agreement with that of increase in viscosities of the solvents applied to EC—DME binary mixture. The molar conductivities in ternary solvent electrolytes containing DMI and NMP are mainly affected by the dielectric constants rather than viscosities of mixed solvents. The energy density of $Li/V_2O_5(2025)$ coin-type cell in $LiPF_6$/EC—DME—DOL electrolyte with the highest molar conductivity was 500 Wh kg^{-1}, which is the highest value in every ternary electrolyte. The lithium cycling efficiency(charge - discharge coulombic cycling efficiency of lithium electrode) in EC—DME—EMC, EC—DME—DMI and EC—DME—TFMDOL electrolytes containing $LiPF_6$ is more than 75% at 40 cycle numbers. The lithium electrodeposition on the Ni(working) electrode surface in ternary solvent electrolytes by cyclic voltammetry was observed by atomic force microscopy(AFM) and scanning electron microscope(SEM).

INTRODUCTION

The search on highly conductive and electrochemically stable electrolytes for lithium batteries above 4.0 V is widely carried out by numerous workers. It is well known that the electrolytic conductivity and the charge - discharge characteristics of lithium electrode are dependent on many factors, such as a type of electrolyte, electrolyte concentration, solvent composition and a presence of impurity in electrolytes. Propylene carbonate(PC) and ethylene carbonate(EC) - based binary solvents with various ethers and esters have been examined mainly for electrolytes of lithium batteries [1-10]. In previous papers [11-13], we reported that the lithium cycling efficiencies for ternary solvent electrolytes with PC — diethyl carbonate(DEC), EC — dimethyl carbonate(DMC) and EC — 1,2-dimethoxyethane(DME) containing other ethers and esters as cosolvents became higher compared with those of PC— DEC, EC—DMC and EC—DME binary solvent electrolytes, respectively. In addition, the advantageous cyclic behavior of lithium electrode in ternary solvent electrolytes has already reported by other workers [14,15]. Accordingly, a study on the electrolytic behavior in the ternary solvent electrolytes is very important to estimate the characteristics of lithium batteries. The purpose of the present study is to examine the electrolytic behavior and the charge - discharge characteristics of lithium electrode in EC—DME—1,3-dioxolane(DOL), EC—DME —2,2-bis(trifluoro methyl)-1,3-dioxolane(TFMDOL) and other ternary solvent electrolytes with EC—DME equimolar binary mixture previously reported [13].

477

EXPERIMENT

Materials

The solvents of DMC, DEC, ethyl methyl carbonate(EMC)(Mitsubishi Petrochem. Co., Ltd.) and TFMDOL(Seimi Chem. Co., Ltd.) were used as received (battery - grade). The preparation and purification of other solvents, such as EC, DOL, DME, 1,3-dimethyl-2-imidazolidinone(DMI) and 1-methyl-2-pyrrolidinone(NMP) have been described elsewhere [16]. Each purified solvent was dehydrated by purified molecular sieves(4A) before preparation of the solutions. The lithium salts of commercial reagent grade, $LiClO_4$ and $LiPF_6$ were dried at 40 and 25°C in vacuo for 48 h, respectively. A commercial reagent grade for other electrolytes was used as received.

Apparatus and measurement

The apparatus and techniques for measurement of conductivity [11], dielectric constant [17] and viscosity [18] were similar with those previously report. The lithium cycling efficiency(charge - discharge coulombic cycling efficiency of lithium electrode) was estimated by a galvanostatic plating - stripping method which was reported by Koch and Brummer [19]. The preparation of electrolyte solutions and the assembly of the cell were carried out in an Ar filled dry box. To estimate the decomposition potentials of solvents, redox potential of ferrocene(Fc)/ferricinium ion(Fc^+) was used as a standard potential(extra thermodynamic assumption) for Li/Li^+ electrode by using two steps method of (1) and (2) as follows.

$$Ag/Ag^+ \text{ (in 0.1 mol dm}^{-3} \text{ TBAP)} | Fc/Fc^+ \text{ (in 0.1 mol dm}^{-3} \text{ TBAP), Pt} \qquad (1)$$

$$Li/Li^+ \text{ (in 0.1 mol dm}^{-3} \text{ TBAP)} | Fc/Fc^+ \text{ (in 0.1 mol dm}^{-3} \text{ TBAP), Pt} \qquad (2)$$

The $AgClO_4$ and $LiClO_4$ of 0.01 mol dm^{-3} were used as silver and lithium salts, respectively. In (1) and (2), TBAP means tetrabutylammonium perchlorate. A commercial Li/V_2O_5(2025) coin-type cell(Toshiba Battery Co., Ltd.) was used for measurement of energy density at a cutoff potential of 2.0 V vs. Li/Li^+ and current density of 1.0 mA cm^{-2}. *In situ* atomic force microscope(AFM) experiments were reported in detail in previous papers [12,13]. Direct observation on the Ni electrode after cycling was carried out with a JEOL JSM-5400 scanning electron microscope(SEM).

RESULTS and DISCUSSION

Molar conductivity

The variation of molar conductivities(λ) in ternary solvent electrolytes with EC−DME equimolar binary mixture containing 1.0 mol dm^{-3} $LiPF_6$(a) and $LiClO_4$(b) at 25°C is shown in Fig. 1. In Fig. 1, the EC−DME−DOL ternary solvent mixture seems to be a good solvent electrolyte for lithium batteries because of higher molar conductivity. In spite of increase or decrease of the viscosities(η) in ternary solvent mixtures as shown in Fig. 2(a), the molar conductivities in every ternary solvent electrolyte generally decreased with increase in the mole fraction(Xs) of the solvents applied to EC−DME mixture. In addition, the order of decrease of the molar conductivities in ternary solvent electrolytes except DMI and NMP systems was in fair agreement with that of increase in viscosity of the solvents(Table 1) applied to the EC−DME mixture. The molar conductivities in ternary solvent electrolytes containing DMI and

NMP with higher viscosities compared with those in ternary mixtures containing DEC and TFMDOL became higher as shown in Fig. 1, respectively. This means that the molar conductivities in ternary solvent electrolytes containing DMI and NMP are mainly affected by the dielectric constants(Fig. 2(b)) due to the increase of dissociation degree of electrolytes rather than viscosities of the mixed solvents, although the variation of the molar conductivities in ternary electrolytes containing DMI and NMP is very complicated in comparison with that in other ternary solvent electrolytes. On the other hand, the order of the molar conductivities in ternary solvent electrolytes containing LiPF$_6$(Fig. 1(a)) is different from that in the electrolytes containing LiClO$_4$(Fig. 1(b)). This difference seems to be dependent on the donor - acceptor properties and the structure of the mixed solvents, and the dissociation degree of electrolytes due to the radii of the solvated anions [20].

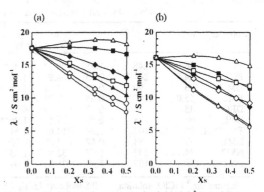

Fig. 1. Variation of molar conductivities (λ) in ternary solvent electrolytes with EC DME equimolar binary mixture containing 1.0 mol dm^{-3} LiPF$_6$(a) and LiClO$_4$(b) at 25℃. Xs means the mole fraction of DMC(■), EMC(◆), DEC(▲), DMI(◇), NMP(□), DOL(△) and TFMDOL(○) applied to EC DME eqimolar binary mixture.

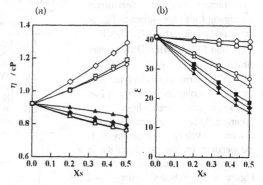

Fig. 2. Variation of viscosities(η)(a) and dielectric constants(ε)(b) in ternary solvent mixtures with EC DME equimolar binary mixture at 25℃. Xs means the mole fraction of DMC(■), EMC(◆), DEC(▲), DMI(◇), NMP(□), DOL(△) and TFMDOL(○) applied to EC DME eqimolar binary mixture.

Energy density

The energy densities for cathode of Li/V$_2$O$_5$(2025) coin-type cell in equimolar binary and ternary solvent electrolytes are given in Table 2. The energy densities in every ternary solvent electrolyte are fairly higher than those in EC − DME binary solvent electrolytes. The energy densities in the electrolytes containing LiPF$_6$ are appreciably higher than those in the electrolytes containing LiClO$_4$. Furthermore, the energy densities in ternary electrolytes containing DOL with the highest molar conductivity became higher compared with those in other ternary electrolytes. The ternary electrolytes containing TFMDOL with the lowest molar conductivity in Fig. 1 have relatively higher energy density. On the other hand, the ternary electrolytes containing NMP and DMI show relatively higher energy densities in LiPF$_6$ solutions, although these electrolytes have smaller energy densities in LiClO$_4$ solutions. These results indicate that the energy density, in other words, the intercalation of Li$^+$ ions to V$_2$O$_5$ and electrode kinetics is influenced by many fators, such as the types of solvents and electrolytes(anion size), the solvent composition and the conductivity in electrolytes.

Table 1. Physical properties of various organic solvents at 25℃.

Solvent	m.p ℃	b.p ℃	η cP	ε	ρ $10^3 kgm^{-3}$	κ [a] $mScm^{-1}$	κ [b] $mScm^{-1}$	DN	AN	M.W.
EC	36.2	248	1.9*	89.1*	1.32*	8.50*	8.58	16.4		88
PC	-54.5	241.7	2.53	64.9	1.20	4.97	5.52	15.1	18.3	102
DMC	3	90.3	0.63	3.12	1.06	0.44	2.14			90
EMC	-55.0	107.0	0.68	2.93	1.01	0.24	1.34			104
DEC	-43.0	126.8	0.75	2.82	0.97	0.11	0.92	16.0		118
DOL	-95	78	0.59	7.1	1.06**	0.75				74
TFMDOL	-30~-20	107	1.39	14.0	1.51					210
DME	-58	84.5	0.47	7.20	0.86	2.92	6.72	24		90
NMP	-24.4	202	1.67	32.0	1.03	7.43	7.08	27.3	13.3	99
DMI	-40	107	1.95	35.5	1.04	5.16	5.00	20~23		114

[a] 0.5 mol dm^{-3} LiClO$_4$ solution,   [b] 0.5 mol dm^{-3} LiPF$_6$ solution.
* 40℃, ** 20℃.

Lithium cycling efficiency

The Lithium cycling efficiencies of lithium electrode in equimolar binary and ternary solvent electrolytes containing LiPF$_6$ are shown in Fig. 3. In analogy with the EC − DME − EMC and EC − DME − DMI ternary electrolytes, the EC − DME − TFMDOL ternary electrolyte with the lowest molar conductivity in Fig. 1(a) shows considerably high lithium cycling efficiency. Accordingly, the lithium cycling efficiency is independent on the conductivity in the ternary solvent electrolytes. Figure 4 shows the i-E curves(decomposition potentials) for oxidation and reduction by use of a Pt electrode in various solvents containing 0.1 mol dm^{-3} TBAP. By comparison of the i-E curves and the lithium cycling efficiency, it was assumed that the lithium cycling efficiency is hardly dependent on the oxidation-reduction potentials of the solvents.   This means that the lithium cycling efficiency, in other words, the charge-discharge process depends mainly on the morphology and the chemical composition [21-24] of the passivating films formed on the Ni electrode arising from the

Table 2. Energy density(Wh kg^{-1}) of Li/V$_2$O$_5$(2025) coin-type cell in equimolar binary and ternary solvent electrolytes containing 1.0 mol dm^{-3} LiPF$_6$ and LiClO$_4$ at 25℃.

Mixed solvent	LiPF$_6$	LiClO$_4$
EC − DME	280	90
EC − DME − DMC	440	290
EC − DME − EMC	400	270
EC − DME − DEC	360	250
EC − DME − NMP	470	150
EC − DME − DMI	430	140
EC − DME − DOL	500	340
EC − DME − TFMDOL	450	310

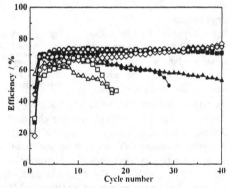

Fig. 3. Variation of lithium cycling efficiencies in EC − DME and EC − DME − S equimolar solvent electrolytes containing 1.0 mol dm^{-3} LiPF$_6$ at 25℃. S means the DMC(■), EMC(◆), DEC(▲), DMI(◇), NMP(□), DOL(△) and TFMDOL(○) applied to EC − DME eqimolar binary mixture(●). $i_D = i_a = 1.3$ mA cm^{-2}, $Q_D = 0.38$ C cm^{-2}.

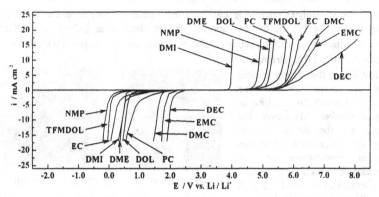

Fig. 4.   i-E curve of a Pt electrode in various solvents containing 0.1 mol dm^{-3} TBAP.   Scan rate; 2 mV s^{-1}.

reduction of reactive impurities and electrolytes itself [25,26].   On the other hand, the lithium cycling efficiencies for the binary and ternary solvent electrolytes containing LiClO$_4$ became lower than those for these electrolytes containing LiPF$_6$.   This shows that no charge-discharge process in solvent electrolytes containing LiClO$_4$ proceeds smoothly on a Ni electrode.

In situ AFM images and SEM photographs

In situ observation on the Ni(working) electrode surface with charge(deposition) - discharge(dissolution) processes in equimolar binary and ternary solvent electrolytes by cyclic voltammetry was performed with AFM.   The AFM images in EC−DME−DOL and EC − DME − TFMDOL ternary electrolytes containing LiPF$_6$ at 1st and 3rd cycle are shown in Fig. 5(a) and (b), respectively.   It is difficult to observe the AFM images at high cycle number because the formation of lithium dendrite proceeds with increasing cycle number and the films on the electrode surface become considerably heterogeneous. The formation of lithium dendrite was already observed at first cycle in every solvent electrolyte, although its

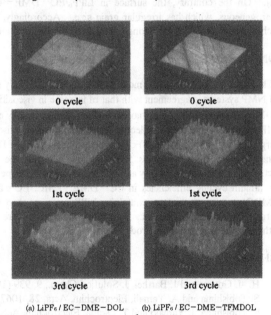

(a) LiPF$_6$ / EC−DME−DOL   (b) LiPF$_6$ / EC−DME−TFMDOL

Fig. 5.   AFM images on the Ni electrode in equimolar ternary solvent electrolytes containing 1.0 mol dm^{-3} LiPF$_6$ at open-circuit voltage after the current reached a zero value for various cycle numbers by cyclic voltammetry.

481

formation in the electrolytes containing LiClO₄ is fairly smaller than that in the electrolytes containing LiPF₆. In Fig. 5, the formation of lithium dendrite in EC−DME−DOL(a) electrolyte with the lowest lithium cycling efficiency became smaller compared with that in EC−DME−TFMDOL(b) electrolyte. This indicates that the lithium cycling efficiency is independent on the formation of the lithium dendrite at lower cycle number. However, the dependence on the lithium cycling efficiency and the growth of lithium dendrite or morphology on the electrode surface at high cycle number should be examined in detail. Fig. 6 shows the SEM photographs in EC− DME−DOL(a) and EC−DME− TFMDOL(b) ternary electrolytes containing LiPF₆ at 15th cycle. The

(a) LiPF₆ / EC−DME−DOL   (b) LiPF₆ / EC−DME−TFMDOL

Fig. 6. SEM photographs on the Ni electrode in equimolar ternary solvent electrolytes containing 1.0 mol dm⁻³ LiPF₆ at open-circuit voltage after the current reached a zero value at 15th cycle by cyclic voltammetry.

surface on the Ni electrode in LiPF₆/EC−DME−TFMDOL electrolyte with higher lithium cycling efficiency is fine and homogeneous, and consists of uniform grain size, as shown in Fig. 6(a).   On the contrary, the surface in LiPF₆/EC−DME−DOL electrolyte is coarse and heterogeneous, which has irregular grain size.   Accordingly, it was found that the charge - discharge characteristics are strongly influenced by the morphology on the electrode.

CONCLUSIONS

The order of decrease of the molar conductivities in ternary solvent electrolytes except DMI and NMP systems is agreement with that of increase in viscosities of the solvents applied to EC−DME binary mixture.   The molar conductivities in ternary electrolytes containing DMI and NMP are mainly affected by dielectric constants rather than viscosities in mixed solvents.   The energy density of Li/V₂O₅(2025) coin-type cell in LiPF₆/EC−DME−DOL electrolyte with the highest molar conductivity was 500 Wh kg⁻¹, which is the highest value in every ternary electrolyte.   The energy density as a whole, however, is independent on the molar conductivity. The lithium cycling efficiencies in EC−DME−EMC, EC−DME−DMI and EC−DME− TFMDOL ternary electrolytes containing LiPF₆ are more than 75% at 40 cycle numbers.   The charge - discharge characteristics from SEM photograph were found to be strongly influenced by the morphology on the electrode.

REFERENCES

1. H. -J. Goures and J. Barthel, J. Solution Chem., **9**, 939 (1980).
2. S. Tobishima and A. Yamaji, Electrochim. Acta, **28**, 1067 (1983).
3. S. Tobishima, J. Yamaki and T. Okada, Electrochim. Acta, **29**, 1471 (1984).
4. S. Tobishima and T. Okada, Electrochim. Acta, **30**, 1715 (1985).
5. Y. Matsuda, M. Morita, and F. Tachihara, Bull. Chem. Soc. Jpn., **59**, 1967 (1986).
6. Y. Matsuda, J. Power Sources, **20**, 19 (1987).
7. S. Tobishima, M. Arakawa, T.Hirai, and J. Yamaki, J. Power Sources, **20**, 293 (1987).

8. T. Ohzuku, M. Kitagawa, and T. Hirai, J. Electrochem. Soc., **136**, 3169 (1989).
9. Y. Matsuda, Nippon Kagaku Kaishi, 1989, 1.
10. S. Subbaro, D. H. Shen, F. Deligiannis, C. -K. Huang, and G. Halpert, J. Power Sources, **29**, 579 (1990).
11. Y. Sasaki, S. Goto, and M. Handa, Denki Kagaku, **61**, 1419 (1993).
12. Y. Sasaki, M. Hosoya, and M. Handa, J. Power Sources, (1997) in press.
13. N. Yamazaki, M. Handa, and Y. Sasaki, Denki Kagaku, **65**, 834 (1997).
14. C. Fringant and A. Tranchant, R. Messina, Electrochim. Acta, **40**, 513 (1995).
15. S. Tobishima, K. Hayashi, K. Saito, and J. Yamaki, Electrochim. Acta, **40**, 537 (1995).
16. J. A. Riddick, W. B. Bunger, and T. K. Sakano, Organic Solvents, 4th ed. (John Wiley and Sons, Inc., New York, 1986.
17. Y. Sasaki, K. Miyagawa, N. Wataru, and H. Kaido, Bull. Chem. Soc. Jpn., **66**, 1608 (1993).
18. Y. Sasaki, T. Koshiba, H. Taniguchi, and Y. Takeya, Nippon Kagaku Kaishi, 1992, 140.
19. V. R. Koch and S. B. Brummer, Electrochim. Acta, **23**, 55 (1978).
20. M. Salomon, S. Slane, E. Plichta, and M. Uchiyama, J. Solution Chem., **19**, 977 (1989).
21. D. Aurbach, M. L. Daroux, P.W. Faguy, and E.B. Yeager, J. Electrochem. Soc., **135**, 1863 (1988).
22. D. Aurbach, Y. Ein-Eli, and A. Zaban, J. Electrochem. Soc., **141**, L1 (1994).
23 K. Kanamura, H. Tamura, S. Shiraishi, and Z. Takehara, J. Electroanal. Chem., **142**, 340 (1995).
24. K. Kanamura, H. Tamura, S. Shiraishi, and Z. Takehara, J. Electroanal. Chem., **394**, 49 (1995).
25. V. R. Koch, J. Power Sources, **6**, 357 (1981).
26. M. Garreau, J. Power Sources, **20**, 9 (1987).

PYRROLE COPOLYMERS WITH ENHANCED ION DIFFUSION RATES FOR LITHIUM BATTERIES

PAUL CALVERT[*], ZACK GARDLUND[*], TREY HUNTOON[*], H.K. HALL[**] AND ANNE PADIAS[**]
[*]Department of Materials Science and Engineering, [**]Department of Chemistry, University of Arizona, Tucson, AZ, 85712.

ABSTRACT

Copolymers of pyrrole with a polyether-substituted pyrrole were tested as cathodes for lithium batteries. The charge and discharge characteristics showed that anion transport was much faster in the copolymer than in polypyrrole. As a result these electrodes store and release much more charge at higher current densities but are similar to polypyrrole at low currents. Pulse and relaxation measurements of the ion diffusion showed that this difference was due to a ten-fold increase in the anion diffusion coefficient.

INTRODUCTION

Conducting polymers originally offered the prospect of stable, flexible, light conductors and semiconductors for many applications in electrical and electronic systems. With time it has become clear that these polymers are not as versatile as had been hoped. In battery applications, there are concerns of stability against oxidation and reduction at high positive or negative potentials. The charge storage capacity of conducting polymers is also not better than comparable transition metal oxides or sulfides on a weight basis. In addition, ion diffusion through conducting polymers is slow and this limits the rate at which current can be withdrawn from a battery.

In an effort to improve this last factor, we have prepared a number of copolymers of pyrrole with pyrrole substituted with polyether side groups. The ether groups were expected to provide ionophilic channels within the polypyrrole structure and so allow rapid diffusion. We have shown that these copolymers do have better charge/discharge characteristics than polypyrrole and that these changes can be attributed to rapid diffusion [1]. A similar ether-modified polyphenylene has also been described [2].

The charging reaction for a conducting polymer cathode is:

$$Polypyrrole + ClO_4 \rightarrow Polypyrrole^+ + ClO_4 + e$$

This requires that an anion transfer from solution into the cathode and then diffuse through the cathode, down the concentration gradient from the surface, and the associated electron must be conducted out from the electrode.

Given the very slow diffusion of ions in conducting polymers, the interpretation of charge/discharge data can be difficult. Firstly, the history of the material must lead to pre-existing gradients that can influence any new measurement. It is thus hard to ensure that an electrode at some surface potential is in the same state throughout or that the starting state is reproducible from one cycle to the next. Secondly, it is known from gravimetric studies [3] that charge and discharge cycles may occur through diffusion of small cations which neutralize embedded anions in the neutral polymer state. The small cations would move much more rapidly than the large anions.

MATERIALS AND METHODS

Two ether-substituted pyrrole monomers were formed, 3-(3,6-Dioxaheptyl)pyrrole (I) and 3-(3,6,9-trioxadecanyl)pyrrole (II):

This was carried out via tosylation of pyrrole to protect the nitrogen, reaction with 2-chloroacetyl chloride to substitute the 3-position, conversion to the bromo-derivative and addition of the appropriate ether-alcohol and then reduction of the carbonyl group [1]. Considerable effort went into purifying out the di-adducts where two alcohols coupled to the chloracetlyated pyrrole. Our previous work had shown that direct substitution via an oxygen link to the pyrrole resulted in less stable material.

Pure homopolymers of these monomers were found to have low conductivity, presumably due to steric hindrance between the side groups causing twisting of the chains, but 50:50 (molar) copolymers with pyrrole formed conducting films on electropolymerization.

Polymerization was done using a Bioanalytical Systems, SP-2 potentiostat. The working and counter electrodes were platinum foil, and the reference electrode was SSCE. Using a constant potential of 0.5 volts vs. SSCE, a current of 0.5-6 milliamps resulted for the duration of the polymerization, 90 minutes. The polymerization solution was a 1:1 mixture of pyrrole and substituted monomers in 0.1 M lithium perchlorate in propylene carbonate. It was found that a complete coating of pyrrole, a black film on the substrate, was viewed after 20 minutes. Copolymer required slightly longer, approximately 40 minutes, to coat the entire substrate.

The characterization cells were formed in a three neck, ground glass, flask with a tungsten wire embedded into the stoppers. These wires protruded into the cell but were not allowed to touch the electrolyte. The ground glass joint was lubricated with high vacuum grease. These factors provided an air tight cell that could exist in ambient atmospheric conditions for a minimum of one year (judged by the metallic luster of the lithium electrodes). Each cell was constructed in a "Vac" drybox in which the water and oxygen levels were kept to less that 1 part per million, as monitored by the onboard monitors. The polymerized sample was gripped by an alligator clip soldered to the tungsten wire and placed into the cell. Thirty-five milliliters of premixed electrolyte, consisting of 0.05M lithium perchlorate in propylene carbonate, were then poured into the cell. Each lithium electrode was impaled on, and pinched around, the tungsten wire.

Measurements were made on the sealed cells in another drybox in which the atmosphere is dried by cycling over calcium sulfate and phosphorus pentoxide. Constant current was supplied by a Keithley 224 programmable current source, and a homemade potentiostat was used during constant potential experiments. The resultant passive parameter was measured on a Keithley 175A digital multimeter.

After the respective experiments were completed, the samples were removed from the cell, washed in THF, dried a vacuum oven, removed from the substrate with a razor blade and weighed in the discharged state. Thickness measurements were made with a scanning electron microscope to within +/- 2 microns. Diaz [4] gives the density of perchlorate doped polypyrrole as 1.5 g/ml, which is comparable with those given in the literature. No porosity is seen in the sample cross sections.

RESULTS

Figure 1 and 2 shows constant current charge and discharge cycles for n=2 copolymer and polypyrrole. Starting at 2.5V vs lithium, samples were charged for 1 hour, held open circuit for 1 hour and then discharged back to 2.5V. It can be seen that the copolymer charges to higher potentials and releases more charge during the discharge cycle. The copolymer can be charged at 200 microamps/cm², while polypyrrole shows overcharging and irreversible reaction at this rate. Two effects can be seen. Firstly, the copolymer shows a higher open circuit potential than polypyrrole after a given charge input. In addition, the polypyrrole collapses more rapidly to 2.5V versus lithium at a given discharge rate, implying that limited diffusion in the electrode may be causing the surface to discharge.

These and other measurements [5] confirm that the current and energy storage capacity of the copolymer is higher than that of polypyrrole except under slow charge and discharge conditions, as shown in table 1. It can be seen that polypyrrole has a greater charge storage capacity when fully charged and then discharged at low current density (30 microamps/cm²). The difference in energy density is less marked because the copolymer charges to higher potential. These energy density numbers reflect the weight of the polymer cathode and an allowance for the lithium anode and electrolyte sufficient to dissolve the full capacity of lithium perchlorate [5]. The value would be roughly 4x as high for the cathode alone. The first four data rows in table 1 are derived from figures 1 and 2 and show that the copolymer has a much higher capacity than polypyrrole at high current densities.

This difference is more significant given that about half the mass fraction of the copolymer is as ether side chains, which are presumably inactive electrochemically. Two factors are involved, the higher potential of the copolymer for a given degree of ionization increases the energy density and, more significantly, the faster ion transport, which leads to reduced internal potential gradients at any given discharge rate.

To establish this second effect, we determined ion diffusion rates in these polymers by measuring the transient response to a small potential step [6]. Typical curves for polypyrrole and copolymer are shown in figure 3 and 4. The diffusion coefficients for perchlorate anion doping were found to be $2.2 \cdot 10^{-9}$ and $2.4 \cdot 10^{-8}$ cm²/sec for polypyrrole and the copolymer respectively. The main difference arises not from the slope of the curves in figure 3 and 4 but from the greater charge passed into the copolymer. These values agree well with those derived from the constant current charge and discharge experiments outlined above [5] and the polypyrrole value agrees with those in the literature [7,8]. Thus diffusion in the copolymer is about 10 times faster than in polypyrrole. From lithium constant-current charge and discharge measurements, the diffusion of lithium is much faster than perchlorate in polypyrrole but the two are similar in the copolymer.

DISCUSSION

When initiating this work, we envisaged a phase-separating copolymer with channels of polyether allowing easy movement of ions into the polypyrrole electrode. The volumes of ether units and pyrrole units in the copolymer are roughly equal. In a block or graft copolymer, for instance the styrene-butadiene copolymers, phase separation occurs with a morphology which depends on the volume fractions of the two components [9]. At about 1:1, a layered morphology is observed, similar to the eutectic structure seen in many alloys. The scale of this structured is determined by the lengths of the blocks of either component, since the junctions between blocks predominantly lie at the interfaces between phases. In the pyrrole-ether system, kinetic limitations make it extremely difficult to copolymerize pyrrole substituted with long polyether chains. The relatively short side chains, used

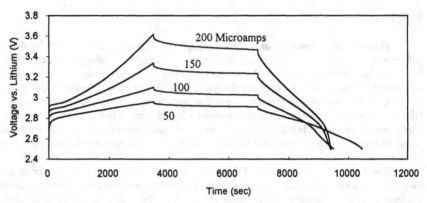

Figure 1: Copolymer charged for one hour at constant current and then discharged at the same current to 2.5V vs. Li.

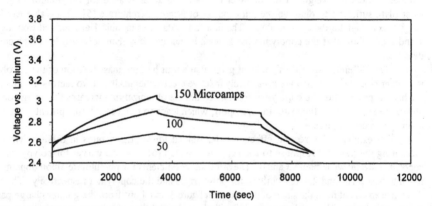

Figure 2: Polypyrrole charged for one hour at constant current and then discharged at the same current to 2.5V vs. Li.

Current, microamps	Storage Capacity Ah/kg		Energy Density Wh/kg	
	Copolymer	Polypyrrole	Copolymer	Polypyrrole
50	12.0	5.5	7.1	2.2
100	22.5	12.1	9.6	6.0
150	32.5	18.5	16.5	9.4
200	42.5	---	22.5	---
Full charge	44.4	55.5	23.9	28.3

Table I: Discharge storage capacity and energy density for copolymer and polypyrrole after charging at constant current for 1 hour and after full charging.

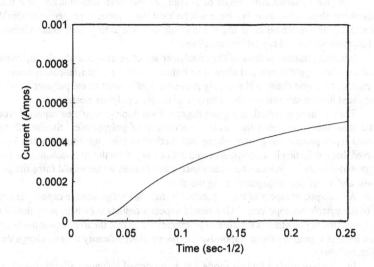

Figure 3: Response of equilibrated copolymer to a potential step from 3.3V to 3.4V vs. Li.

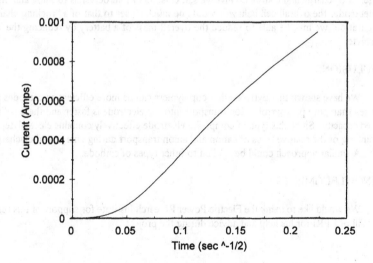

Figure 4: Response of equilibrated polypyrrole to a potential step from 3.3V to 3.4V vs. Li.

here, would give a characteristic length of 1-2 nm. At this small scale it is not clear that it is useful to talk about phase separation, but there will be local fluctuations in properties within the material. Similar ambiguous structures occur in polymers with long side chains, such as hexadecylmethacrylate and polymers with liquid crystalline side chains.

It is thus possible to think of the copolymer structure as zones of polyethyleneoxide able to solubilize lithium perchlorate and allow it to diffuse. Unlike polyethyleneoxide homopolymer, the constraints on the side chains will probably prevent crystallization of the polymer or of crystals with intercalated lithium salt that are characteristic of the bulk polymer electrolyte.

The increased potential, at a given degree of ion doping, would be expected from the effect of the substituent modifying the electronic structure of polypyrrole. Similar effects would be expected in polypyrrole and polythiophene with methyl- or ethyl-substitution at the 3-position. We observed that conducting homopolymers could not be made from the ether-substituted pyrroles. This was attributed to steric hindrance between adjacent substituents which would force the polymer chain to twist and so reduce conjugation along the chain.

An unexpected aspect of the copolymer is that the charge storage capacity is comparable to that of the pyrrole homopolymer. One would expect a reduction by half, since the 50 vol% of the copolymer that is polyether is inactive. The implication is that the irreversible oxidation reactions, which limit charging, are more responsive to volume charge density than to charge density on the conjugated chain.

In a battery using a lithium anode and anion-doped polymer cathode, the electrolyte must contain enough salt (lithium perchlorate) to charge the electrodes fully with ions and still be conducting. In the discharged state all this salt must be solubilized in the electrolyte, thus requiring a considerable volume. In contrast, in "rocking-chair" batteries, the ionic content of the electrolyte does not change during charging. In the copolymer electrode, much of the salt is probably confined within the ether regions of the electrode and the free electrolyte experiences little concentration change. If discharge and charge do involve salt clusters or microcrystals forming and dispersing in the electrode, the overall cell behavior would be much closer to that of a "rocking chair". In this configuration, we may be able to reduce the overall mass of a battery by reducing the volume of electrolyte.

CONCLUSIONS

We have shown that pyrrole-ether copolymers can be more efficient as cathodes for lithium batteries than pure polypyrrole. Ion transport into the electrode is faster and there is little cost in reduced capacity. Since this type of composite electrode effectively contains electrolyte, it requires a rethinking of the relative roles of cation and anion transport during charge and discharge.

A similar approach could be applied to other types of cathode.

ACKNOWLEDGMENTS

We would like to thank the Electric Power Research Institute for support of this research and Fritz Will of EPRI for his help and advice during the project.

REFERENCES

[1]   Moon D.-K., Padias A.B., Hall H.K. *et al.*, .Macromolecules **28** , 6205-6210 (1995)

[2]   Lauter U., Meyer W., and Wegner G., .Macromolecules **30** , 2092-2101 (1997)

[3]   Kaufman J.H., Kanazawa K.K., and Street G.B., .Phys. Rev. Lett. **53** , 2461 (1984)

[4]   Diaz A.F. and Bargon J., "Electrochemical synthesis of conducting polymers," in *Handbook of conducting polymers*, edited by T.A. Skotheim (Dekker, New York, 1986), Vol. 1, pp. 81-115.

[5]   Huntoon T.W.S., "The polymerization and electrochemical characterization of polypyrrole and ethyleneoxide-pyrrole graft copolymers," PhD., University of Arizona, 1997.

[6]   Kemp J., Instrumental Methods in Electrochemistry (Halstead Press, New York, 1985).

[7]   Rosa C. and Peres D., .J. Power Sources **40** , 299 (1992)

[8]   Naoi K. and Sakai S., .J. Power Sources **20** , 237 (1987)

[9]   Sperling L.H., Introduction to physical polymer science, 2nd ed. (Wiley, New York, 1992).

REFERENCES

LITHIUM ION DIFFUSION THROUGH GLASSY CARBON PLATE

M. INABA, S. NOHMI, A. FUNABIKI, T. ABE, Z. OGUMI
Dept. of Energy and Hydrocarbon Chem., Graduate School of Engineering, Kyoto Univ., Sakyo-ku, Kyoto 606-01, Japan

ABSTRACT

The electrochemical permeation method was applied to the determination of the diffusion coefficient of Li^+ ion (D_{Li^+}) in a glassy carbon (GC) plate. The cell was composed of two compartments, which were separated by the GC plate. Li^+ ions were inserted electrochemically from one face, and extracted from the other. The flux of the permeated Li^+ ions was monitored as an oxidation current at the latter face. The diffusion coefficient was determined by fitting the transient current curve with a theoretical one derived from Fick's law. When the potential was stepped between two potentials in the range of 0 to 0.5 V, transient curves were well fitted with the theoretical one, which gave D_{Li^+} values on the order of 10^{-8} cm^2 s^{-1}. In contrast, when the potential was stepped between two potentials across 0.5 V, significant deviation was observed. The deviation indicated the presence of trap sites as well as diffusion sites for Li^+ ions, the former of which is the origin of the irreversible capacity of GC.

INTRODUCTION

The anode reaction in lithium-ion cells is accompanied by Li^+-ion diffusion within host carbonaceous materials. Since diffusion in solid is generally a slow process, the diffusion rate usually dominates the overall reaction rate, that is, the maximum charge-discharge current. The diffusion coefficient of Li^+ ion (D_{Li^+}) within carbonaceous materials is thus a critical parameter that determines the power density of lithium-ion cells. The values of D_{Li^+} have been determined by several methods including galvanostatic intermittent titration technique (GITT),[1] current pulse relaxation (CPR),[2,3] potential step chronoamperometry (PSCA), and AC impedance spectroscopy.[4-6] These reported values of D_{Li^+} are rather scattered in a wide range of 10^{-6}-10^{-13} cm^2 s^{-1} depending on the kind of carbon and on the technique employed. To obtain D_{Li^+} using these methods, one has to know the real surface area of the sample (A), and in some cases, the variation of the open-circuit potential with lithium concentration (dV_{oc}/dx). However, precise determination of A and dV_{oc}/dx is difficult in general, which is one of the reasons for the scatterings in the data. In order to avoid these difficulties, we employed a new method for the determination of D_{Li^+} called the "electrochemical permeation method", which was originally developed for the determination of the diffusion coefficients of hydrogen in steel[7,8] and gas molecules in Nafion[9,10].

The principle of the electrochemical permeation method is briefly mentioned here. Consider a carbon plate with a thickness of L as shown in Fig. 1a. First both surfaces are kept at a potential so that Li+ concentration is uniform throughout the plate ($C = C_0$ at $0 < x < L$). At $t = 0$, the potential of one surface ($x = 0$) is stepped at a lower potential to increase Li^+ concentration at $x = 0$ ($C_{x=0} = C_1$ at $t \geq 0$). Lithium ions diffuse through the carbon plate toward the other surface ($x = L$) according to a concentration gradient formed by the potential step. After an elapse of time, Li^+ ions permeate to the other surface ($x = L$), where they are removed into the solution because Li^+ concentration at $x = L$ is electrochemically kept constant ($C_{x=L} = C_0$). The flux for permeated Li^+ ions can be monitored as a change in oxidation current at $x = L$. The current (I) increases with time, and then reaches a steady-state value (I_∞). When the diffusion obeys Fick's law, the current transient is predicted by a "build-up" curve shown in Fig. 1b.

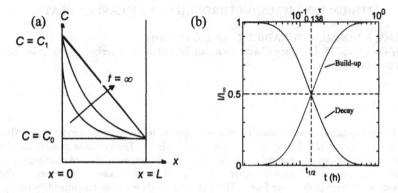

Fig. 1. (a) Principle and (b) theoretical curves for current transient of the electrochemical permeation method.

After the steady state is attained, a potential step in the opposite direction gives a current transient shown by a "decay" curve in Fig. 1b. The D_{Li^+} values can be determined from the current transients using the following relation:[7]

$$D_{Li^+} = 0.138L^2/t_{1/2} \tag{1}$$

where $t_{1/2}$ is the time when the current reaches a half of the steady-state value ($I/I_\infty = 0.5$). It should be noted that both A and dV_{OC}/dx are not necessary for the determination of D_{Li^+}. Therefore, more reliable data can be obtained.

EXPERIMENT

For the test electrode, we used a glassy carbon (GC) plate (Tokai Carbon GC-20S, HTT = 2000°C, thickness = 1 mm, density = 1.604 g cm^{-3}), which is a typical hard carbon. An electrochemical cell used in permeation tests is shown in Fig. 2. The cell was composed of two compartments, both of which were filled with 1 M LiClO$_4$/EC + DEC (1:1 by volume). The electrode (3.14 cm^2) separated the two compartments and was used as a kind of bipolar electrode.

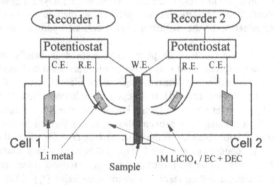

Fig. 2. Electrochemical cell for permeation tests.

Two potentiostats (Hokuto Denko, HA301) were used to control the potentials of both surfaces independently. First, both surfaces were kept at a constant potential, typically 0.75 or 1.0 V, where no Li$^+$ insertion takes place. The potential at one face was then stepped to a lower potential while that at the other face was kept constant. The current transient at the latter face was monitored. The experimental transient curve was fitted with the theoretical one to obtain D_{Li^+} using Eq.(1). The step potential was successively lowered to 0.05 V, and then returned to 1.0 V.

Charge-discharge tests of GC-20S were carried out using a conventional three-electrode cell. A small piece of GC-20S (5.5 x 4.2 x 1 mm) was used as a working electrode. The electrode was charged and discharged at a constant current (26.4 µA, C/480 rate) in 1 M LiClO$_4$/EC+DEC (1:1).

RESULTS AND DISCUSSION

Charge-Discharge Characteristics

Figure 3 shows the charge-discharge characteristics of GC20S. In the first cycle, the specific charge and discharge capacities were 130 and 109 mAhg^{-1}, respectively; that is, an irreversible capacity of 21 mAhg^{-1} was observed. In the second cycle, GC20S showed a good reversibility with a coulombic efficiency of 98%.

On the first charge curve, a potential plateau was observed at about 0.5 V, which seems to be the origin of the irreversible capacity. Figure 4 shows the derivative plots of the charge curves. A reduction peak was observed at about 0.5 V upon the first charging, which is in agreement with the presence of the plateau on the charge curve. Upon the second charging, the peak disappeared. This shows that the irreversible capacity was consumed by a reaction that takes place at ca. 0.5 V upon the first charging. In the third and fourth cycles, similar curves to the second one were observed. After the fourth discharge, the electrode was kept at 1 V for 5 days to remove Li$^+$ ions completely, and then charged again. The derivative plot of the fifth charge curve is also shown in Fig. 4. It is surprising that the peak at ca. 0.5 V appeared again. The irreversible capacity has been attributed to the solvent decomposition and surface film formation; hence, it should be totally irreversible. Therefore, there should be another origin for the irreversible capacity of GC. The origin will be discussed later.

Permeation Tests

Figure 5 shows typical fitting results for electrochemical permeation tests. Panel (a) in Fig. 5 shows a build-up curve when the potential of the Cell-1 side was stepped from 0.2 to 0.1 V while that of the Cell-2 side was kept at 0.75 V. The permeation current increased with time, and then reached a steady-state current at about 30 h. The experimental transient curve was well fitted with the theoretical one, which gave a D_{Li^+} value of 5.0 x 10^{-8} cm^{2}s^{-1}. Panel (b) shows a decay curve when the potential of the Cell-2 side was stepped from 0.5 to 0.3 V while that of the Cell-1 side was kept at 0.3 V. The current decreased with time as predicted by a theoretical decay curve with a D_{Li^+} value of 3.1 x 10^{-8} cm^{2}s^{-1}. When the potential was stepped between two potentials in the range of 0 and 0.5 V, current transients obtained were well fitted with the theoretical curves with D_{Li^+} values on the order of 10^{-8} cm^{2}s^{-1}.

In contrast, when the potential was stepped between two potentials across 0.5 V, fitting was poor as shown in Fig. 6. Panel (a) shows a build-up curve when the potential of the Cell-1 side was stepped from 1.0 to 0.05 V while that of the Cell-2 side was kept at 1.0 V. No detectable current was observed up to 10 h although the theoretical curve predicts the permeation in a shorter time. After an induction period of 10 h, the permeation current started to flow and

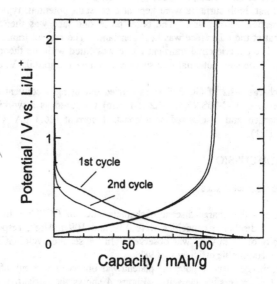

Fig. 3. Charge and discharge curves of glassy carbon. The current was 26.4 μA (C/480 rate).

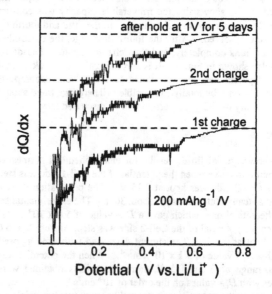

Fig. 4. Derivative plots of the first, second, and fifth charge curves of glassy carbon. After the fourth discharge step, the potential was kept at 1 V for 5 days to remove Li⁺ ion completely..

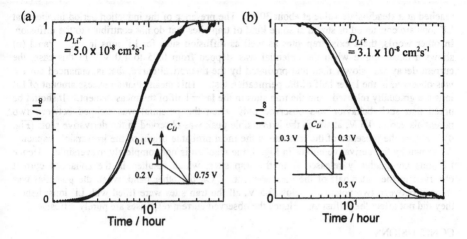

Fig. 5. Fitting results of permeation test for glassy carbon. (a) Cell 1: 0.2 ⇒ 0.1 V; Cell 2: 0.75 V. (b) Cell 1: 0.5 ⇒ 0.3 V; Cell 2: 0.3 V. Solid lines: theoretical curves; circles: experimental results.

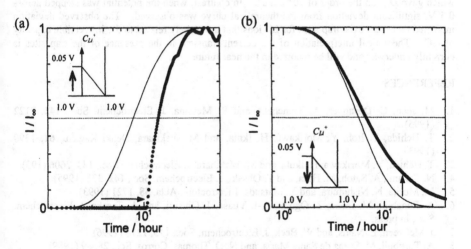

Fig. 6. Fitting results of permeation test for glassy carbon. (a) Cell 1: 1.0 ⇒ 0.05V; Cell 2: 1.0 V. (b) Cell 1: 0.05 ⇒ 1.0 V; Cell 2: 1.0 V. Solid lines: theoretical curves; circles: experimental results.

reached at a steady-state value at about 30 h. The presence of the induction period implies that Li^+ ions are consumed or stored at some kind of trap sites that do not contribute to the diffusion. In other words, there exist trap sites as well as diffusion sites for Li^+ ions in GC. Panel (b) shows a decay curve when the potential was stepped from 0.05 to 1.0 V. In this case, the current decay was slower than that predicted by the theoretical curve, that is, enhanced current was observed in the latter half of the permeation test. This means that an excess amount of Li^+ ions are gradually released from the trap sites in the latter half of the measurement. It should be noted that such behavior was observed only when the potential was stepped between two potentials across 0.5 V, at which the irreversible peak was observed in the derivative plots (Fig. 4). Hence the presence of the trap site is the most probable origin of the irreversible capacity. As shown by the derivative curves in Fig. 3, the reaction is not completely irreversible. Hence Li^+ ions are gradually released from the trap sites, which resulted in the enhanced current observed in the latter half of the measurement. On the other hand, when the potential was stepped between two potentials below 0.5 V, all the trap sites were filled with Li^+ ions; hence, they did not affect the diffusion. Hence the observed current responses are purely diffusive.

CONCLUSIONS

The results in this study showed that the electrochemical permeation method is useful for not only the determination of D_{Li^+}, but also for understanding the nature of Li^+-ion sites in carbonaceous materials. When the potential was stepped between two potentials in the range of 0 to 0.5 V, the current transient was well fitted with the theoretical one derived from Fick's law, which gave D_{Li^+} on the order of 10^{-8} cm^2s^{-1}. In contrast, when the potential was stepped across 0.5 V, significant deviation from the theoretical curve was observed. The observed deviation indicated that there exist trap sites for Li^+ ions that are closely related to the irreversible capacity of GC. Theoretical interpretation of the current transient in the presence of the trap sites is currently underway and will be reported in the near future.

REFERENCES

1. M. Jean, C. Desnoyer, A. Tranchent, and R. Messina, J. Electrochem. Soc., **142**, 2122 (1995).
2. T. Uchida, T. Itoh, Y. Morikawa, H. Ikuta, and M. Wakihara, Denki Kagaku, **61**, 1390 (1993).
3. T. Uchida, Y. Morikawa, H. Ikuta, and M. Wakihara, J. Electrochem. Soc., **143**, 2606 (199).
4. N. Takami, A. Satoh, M. Hara, and T. Ohsaki, J. Electrochem. Soc., **142**, 371 (1995).
5. M. Morita, N. Nishimura, and Y. Matsuda, Electrochim. Acta, **38**, 1721 (1993).
6. A. Funabiki, M. Inaba, Z. Ogumi, S.-i. Yuasa, J. Ohtsuji, and A. Tasaka, J. Electrochem. Soc., in press.
7. J. McBreen, L. Nanis, and W. Beck, J. Electrochem. Soc., 113, 1218 (1996).
8. A. Turnbull, M. Saenz de Santa Maria, and N. D. Thomas, Corros. Sci., 28, 89 (1989).
9. Z. Ogumi, T. Kuroe, and Z. Takehara, J. Electrochem. Soc., 132, 2601 (1985).
10. Z. Ogumi, Z. Takehara, and S. Yoshizawa, J. Electrochem. Soc., 131, 769 (1984).

POLYMER-OXIDE ANODE MATERIALS

T.A. KERR, F. LEROUX, and L.F. NAZAR
University of Waterloo, Department of Chemistry, Waterloo, Ontario Canada N2L 3G1;
lfnazar@uwaterloo.ca

ABSTRACT

The adaptable layer structure of molybdenum trioxide was exploited to insert the amino derivative form of the conductive polymer poly(para-phenylene) (PPPNH$_2$) within the van der Waals gap. Two polymer insertion routes were designed that yield novel PPPNH$_2$-MoO$_3$ materials of different composition. Characterization of these materials using powder XRD, thermal analysis, and FTIR spectroscopy shows insertion of the polymer has occurred. The properties of the nanocomposites for low potential electrochemical lithium insertion were compared to those of the sodium molybdenum bronze using the materials as cathodes in conventional lithium cells. Initial results indicate the specific charge capacity and irreversibility during the first charge are effected by polymer content whereas polarization is not.

INTRODUCTION

Nanocomposites comprised of an intercalated conductive polymer in the interlamellar gap of a transition metal oxide are of interest as cathode materials for lithium reversible batteries. Recent reports on [poly(aniline)]$_{0.24}$MoO$_3$ ([PANI]$_{0.24}$MoO$_3$), [PANI]$_{0.44}$V$_2$O$_5$, and [PANI]$_{0.34}$HMWO$_6$ (M=Ta,Nb) have each shown that this incorporation demonstrates enhanced ion and/or electronic transport on galvanostatic cycling over that of the pristine transition metal oxide.[1,2,3] In addition, the redox capacity of the intercalated PANI provides a moderate increase in cell capacity, and improves the reversibility of the Li insertion reaction. In the case of [PANI]$_{0.44}$V$_2$O$_5$ and [PANI]$_{0.34}$HMWO$_6$ the Li chemical diffusion coefficients were found to be greater for the nanocomposites compared to the transition metal oxides by over one order of magnitude. The improved kinetics appear to be the result of a lower energy pathway for Li ion migration in the nanocomposites. This may due to the polymer propping the oxide layers apart, thus modifying the Li insertion sites, and shielding the lithium ions from the polarizing effect of the oxide. The effect is most pronounced at small amounts of Li intercalation, but suppressed as more lithium is inserted.

Molybdenum trioxide has been primarily considered as a 2V material in Li batteries due to topotactic Li insertion at that potential. A recent study by Leroux and Nazar[4], however, shows that sodium molybdenum bronze ([Na$_{0.26}$MoO$_3$]) demonstrates good electrochemical performance at low potential in the range 3-0.005 V vs. Li, albeit with substantial cell polarization leading to an average charge potential of 1.3V. With this in mind, and the fact that no known investigations of polymer-oxide nanocomposites as negative electrodes have been reported, a study of the low potential electrochemical performance of the material formed from the incorporation of an electronically conducting polymer between the layers of MoO$_3$ was instituted.

Most polymers are not stable to cycling at low potential. Nevertheless, recent studies by Dubois and Billaud have reported that poly(para-phenylene) can reversibly intercalate 0.5Li between 500-0mV with low polarization (after initial irreversibility has subsided).[5] PPP is known to be a fairly crystalline polymer that is infusible and insoluble. This presents a challenge to develop a synthetic method to incorporate the

499

polymer chains within the van der Waals region. Herein we report the synthesis of the novel polymer-oxide composite, PPPNH$_2$-MoO$_3$, its characterization, and initial electrochemical and cycling behavior studies.

EXPERIMENTAL

Synthesis and characterization. The molybdenum bronze [Li$_{0.07}$Na$_{0.13}$(H$_2$0)$_{0.40}$]MoO$_3^{0.20-}$, was prepared as described previously, by the reduction of MoO$_3$ in a Li$_2$MoO$_4$ buffered aqueous solution.[6] The lithium bronze [Li^+_x(H$_2$O)$_n$]MoO$_3$ (LiMoO$_3$) was prepared by reduction of MoO$_3$ with LiBH$_4$ in diethyl ether, as reported by Kanatzidis and Marks.[7]

Poly(*para*-phenylene) was prepared by the Kovacic method of oxidative polymerization of benzene using AlCl$_3$ with a CuCl$_2$ catalyst.[8] The amino derivative form of the polymer was prepared by first nitrating the phenylene in mixed HNO$_3$/H$_2$SO$_4$ acid, followed by reduction of the nitrate group in a hydrazine-diethylene glycol mixture.[9]

The composite material [PPPNH$_2$]$_{1.37}$MoO$_3$ (I) was synthesized by mixing polymer and [Li$_{0.07}$Na$_{0.13}$(H$_2$0)$_{0.40}$]MoO$_3$ in 3 ml of H$_2$O. The pH was adjusted to 1.0 with the dropwise addition of HCl. The reagents were then placed into a 23ml Parr acid digestion bomb and heated at 130°C for 72 h with intermediate grinding. The final product was isolated by filtration and washed with methanol.

The composite material [PPPNH$_2$]$_{0.64}$MoO$_3$ (II) was prepared by mixing LiMoO$_3$, PPPNH$_2$ in 25ml of a 1M C$_{16}$H$_{33}$N(CH$_3$)$_3$OH (CTAOH) solution for 24 h. The product was then isolated by centrifugation, washed with water, then heated at 250°C for 48 h.

X-ray powder diffraction (XRD) patterns were obtained from random powder samples on a Siemens D500 X-ray diffractometer equipped with a diffracted beam monochromator, using Cu Kα radiation. Samples were scanned at a step rate of 0.05°/s. Infrared spectra were recorded on a Nicolet 520 FTIR as KBr pellets. Differential thermal analysis and thermal gravimetric analysis (DTA/TGA) were performed on a PL Thermal Sciences STA 1500 thermal analysis system. DTA and TGA curves were run simultaneously on each sample from room temperature to 600°C in a flowing atmosphere of air using a heating rate of 10°C/min.

Electrochemical Measurements. The composite materials were evaluated as cathodes prepared from active material and carbon black using PVDF as a binder in the mass ratio 80:15:5 respectively. The powders were mixed with cyclopentanone and the slurry cast onto nickel discs. The electrodes were heated at 180°C for one hour to remove trace water prior to cell assembly. Dimethyl carbonate (DMC- Aldrich) and Ethylene carbonate (EC- Aldrich) were purified by distillation under vacuum, and mixed in the ratio 2:1 followed by dissolution of LiPF$_6$ to form a 1M solution. Swagelock type cells were assembled in an argon filled glove box containing less than 2ppm of water and oxygen. Positive electrodes had a surface area of 1cm^2 and contained 1-2 mg of active material. Lithium spread onto a nickel disc was used as both the negative and the reference electrode. A multichannel galvanostatic/potentiostatic system (Mac-Pile™) was used for the electrochemical study. The specific capacities obtained are normalized per gram of composite electrode.

RESULTS AND DISCUSSION

Facile ion exchange *via* exfoliation of alkali molybdenum bronzes in aqueous solution has made them an excellent host material for use in intercalation reactions. Guests ranging in size from small alkaline earth ions to large cationic polymers have been readily inserted between the layers of MoO_3 by this route.[6,10] This property was utilized to incorporate $PPPNH_2$ into the interlamellar region of MoO_3 by means of a hydrothermal reaction in acidic media yielding the composite, $[PPPNH_2]_{1.37}MoO_3$.

The *010* reflection at 11.2Å in the XRD pattern of this material indicates that the interlayer distance is expanded by 4.3Å *vs.* that of pristine MoO_3. This is in good agreement with the width of a phenylene unit (~4.5Å), confirming the intercalation of a monolayer of polymer between the oxide layers (**Fig. 1**). The lack of subsequent harmonic reflections indicates an overall decrease in crystallite ordering compared to that of the precursor bronze.

Thermal gravimetric and differential thermal analysis curves (**Fig. 2a**) exhibit a two step exothermic mass loss from 350-450°C suggesting the presence of two separate environments in which the polymer chains reside. We assign the low temperature

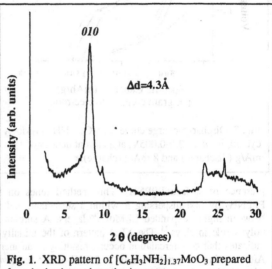

Fig. 1. XRD pattern of $[C_6H_3NH_2]_{1.37}MoO_3$ prepared from hydrothermal synthesis route

combustion to polymer on the crystallite surface and that at higher temperature to the chains stabilized within the oxide layers.

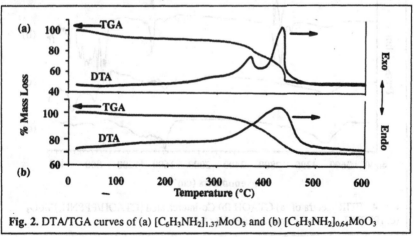

Fig. 2. DTA/TGA curves of (a) $[C_6H_3NH_2]_{1.37}MoO_3$ and (b) $[C_6H_3NH_2]_{0.64}MoO_3$

The first three cycles of the discharge-charge curves of $[C_6H_3NH_2]_{1.37}MoO_3$ in the voltage range 2.9 to 0.005V are shown in **Fig. 3**. The initial specific charge capacity of

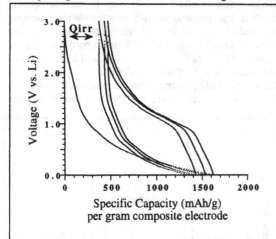

Fig. 3. Discharge-charge curve for $[[C_6H_3NH_2]_{1.37}MoO_3$ cycling between 3.0-0.005V, at a current density of 11 mA/g (discharge) and 8 mA/g (charge).

1141 mAh/g is greater than observed for the bronze (940 mAh/g), due to the added redox capacity of the polymer. The irreversible capacity (Q_{irr} = 23%) is slightly lower than observed for $Na_{0.26}MoO_3$ (31%). Incorporation of PPPNH$_2$ has therefore enhanced the electrochemical properties; however, cell hysteresis is not diminished. This may be partly due to the presence of surface polymer which is known to result in the partial entrapment of Li ions.[2] To determine if better performance could be obtained in its absence, a surfactant-mediated route was designed to reduce the presence of surface PPPNH$_2$. This method relies on facilitating introduction of the PPPNH$_2$ by first dispersing it within a surfactant, and incorporating both components between highly expanded $Li_xMoO_3^{x-}$ layers. A similar method was used to introduce polypyrrole in clays.[11] The XRD pattern of the initially obtained material (not shown) indicates that co-intercalation occurs, resulting in an increase in gallery height of 21 Å. After heating the product in air at 250°C for 48 h, however, the surfactant is removed whereas the polymer remains within the layers. **Fig. 4** shows a comparison of the IR

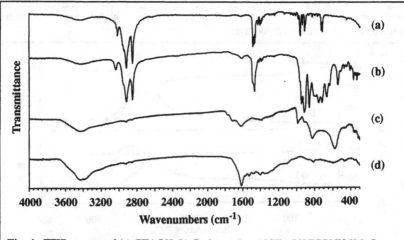

Fig. 4. FTIR spectra of (a) CTAOH (b) Co-intercalated [CTAOH/PPPNH$_2$]MoO$_3$ (c) [PPPNH$_2$]$_{0.64}$MoO$_3$, and (d) PPPNH$_2$

spectra of the two composites, before and after heating, and their components. No surfactant peaks remain in the heated composite; only absorptions due to the MoO₃ host and the PPPNH₂ are observed.

The XRD pattern for [PPPNH₂]₀.₆₄MoO₃ is shown in **Fig. 5**. An average layer expansion of 4.7Å is calculated from the three $0k0$ reflections, consistent with the interlamellar incorporation of a monolayer of PPPNH₂. This material is also somewhat better ordered than **I**, which only displayed one interlayer reflection. The slight increase in the interlayer spacing (0.3 Å) over that displayed by **I** is not significant, and probably the result of disorder in the former material.

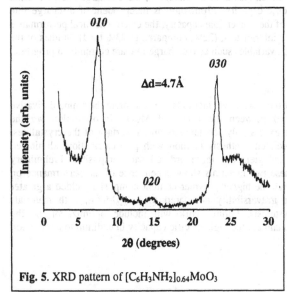

Fig. 5. XRD pattern of [C₆H₃NH₂]₀.₆₄MoO₃

The one step exothermic mass loss observed in the thermal analysis curves (**Fig. 2b**) indicates that only one environment for the polymer exists; that is, that the deposition of excess polymer on the surface of the crystallites has been minimized by this surfactant-mediated route. The lack of a discrete exothermic mass loss at 300-350°C, corresponding to surfactant combustion, also confirms that the surfactant has been removed by the thermal treatment. Calculation of the polymer content from the TGA profile gives a composition [C₆H₃NH₂]₀.₆₄MoO₃ for this material (**II**).

The discharge-charge vs. specific capacity curve for **II** is shown in **Fig. 6**. The discharge rate was approximately twice as fast as that used for Li insertion in **I** (**Fig. 3**), although the charge rate was approximately the same. Although the cycling conditions are not exactly the same, it is apparent that there are small differences in the profile of the discharge curves. An inflection between 1.4-0.7V is better developed in **II**, for example, suggesting that more well-

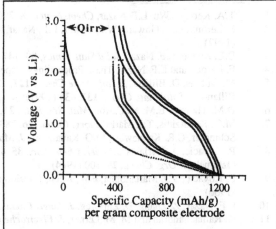

Fig. 6. Discharge-charge curve for [[C₆H₃NH₂]₀.₆₄MoO₃ cycling between 3.0-0.005V, at a current density of 22 (or 130) mA/g (discharge) and 11 mA/g (charge).

defined sites for Li insertion exist in this material. These are evident even at a more rapid discharge rate of 130 mA/g, which results in no change in the discharge-charge profile. The resultant reversible specific capacity, 770mAh/g, is independent of discharge rate within these limits. In terms of the irreversible capacity, the electrochemical performance of II is somewhat surprisingly inferior to I, (32%, compared to 23% for I). Studies of the dependence of these factors on variables such as the charge rate are currently in progress.

CONCLUSIONS

We report for the first time two novel methods to incorporate the amino derivative form of poly(*para*-phenylene) between the layers of MoO_3. Incorporation *via* the hydrothermal route results in excess polymer deposits on the surface of the crystallites. The second, a surfactant mediated synthesis method with post-calcination, eliminated polymer surface deposition and yielded a more ordered nanocomposite. Preliminary electrochemical studies of these two materials show dependence of the performance on the synthetic method. Although the higher polymer content composite yielded a greater specific capacity and lower irreversibility compared to $Na_{0.26}MoO_3$, both materials demonstrated substantial hysteresis. Future work will include optimization of the synthetic method in order to maintain the high specific capacity in addition to a reduction of the polarization.

ACKNOWLEDGEMENTS

LFN gratefully acknowledges the generous financial support of NSERC through the strategic and operating programs. Morven Duncan, Gillian Goward, and Patricia Dewar are thanked for their aid in electrolyte distillation.

REFERENCES

1.   T.A. Kerr, H. Wu, L.F. Nazar, *Chem. Mater.*, **8**, 2005 (1996)
2.   F. Leroux, G.R. Goward, W.P. Power, L.F. Nazar, *J. Electrochem. Soc.*,**144**, 3886 (1997)
3.   B.E. Koene, L.F. Nazar, *Solid State Ionics*, **89**, 147 (1996)
4   F. Leroux and L.F. Nazar, These Proceedings, Paper Y10.8.
5.   M. Dubois, D. Billaud, *J. Solid State Chem.*, **127**, 123 (1996); M. Dubois, D. Billaud, *J. Solid State Chem.*, **132**, 434 (1997);
6.   D.M. Thomas, E.M. McCarron, *Mat. Res. Bull.*, **21**, 945 (1986)
7.   M.G. Kanatzidis, T.J. Marks, *Inorg. Chem.*, **26**, 783 (1987), L. Wang, J. Schindler, C.R. Kannewurf, M.G. Kanatzidis, *J. Mater. Chem.*, **7**, 1277 (1997).
8.   P. Kovacic, A. Kyriakis, *J. Amer. Chem. Soc.*, **85**, 454 (1963); P. Kovacic, J. Oziomek, *J. Org. Chem.*, **29**, 100 (1964).
9.   P. Kovacic, V.J. Marchionna, F.W. Koch, J. Oziomek, *J. Org. Chem.*, **31**, 2467 (1966).
10. L.F. Nazar, Z. Zhang, D. Zinkweg, *J. Amer. Chem. Soc*, **114**, 6239 (1992)
11. K. Ramachandran and M. M. Lerner, *J. Electrochem. Soc.*, **144**, 3739 (1997).

IONICALLY CONDUCTING GLASSES WITH SUBAMBIENT GLASS TRANSITION TEMPERATURES

R.E. Dillon, D.F. Shriver
Department of Chemistry and Materials Research Center, Northwestern University, Evanston, Illinois, 60208-3113

ABSTRACT

Cryptands and crown ethers along with the lithium salt, $LiCF_3SO_2N(CH_2)_3OCH_3$ (LiMPSA) were employed to produce a new type of amorphous electrolyte. The key to producing an amorphous phase was the mismatch between the cavity size of the macrocycle and the diameter of the cation. The addition of poly(bis-(2(2-methoxyethoxy)ethoxy)phosphazene) (MEEP) to the amorphous complex, LiMPSA/2.2.2 Cryptand, imparts improved electrochemical and viscoelastic properties. Conversely, when poly(sodium-4-styrenesulfonate) (PS4SS) is added to the amorphous complex, LiMPSA/2.2.2 Cryptand, the product crystallizes. The ionic conductivity of the MEEP rubbery electrolyte is a full order of magnitude higher when compared to the analogous PS4SS doped electrolyte (3.8×10^{-5} S cm^{-1} (MEEP), 1.7×10^{-6} S cm^{-1} (PS4SS) both at 305°K).

INTRODUCTION

Inorganic superionic glasses and polymer electrolytes have attracted considerable attention as solid electrolytes for electrochemical devices.[1-3] Presently both electrolytes fall short of the required combination of electrochemical and mechanical properties necessary for many applications. Inorganic superionic glasses have high ionic conductivity but they are usually brittle materials. By contrast, polymer electrolytes often have lower ionic conductivity but the desired mechanical compliance. Recently Angell and co-workers reported a new type of solid electrolyte that combined superionic glasses and polymer electrolytes.[4,5] These 'rubbery electrolytes' or 'polymer-in-salt' electrolytes contain a supercooled mixture of lithium salts doped with poly(ethylene oxide) (PEO). Superionic glasses are observed to maintain their high ionic conductivity in the vitreous state. This phenomenon is best described as a decoupling of ionic motion from the structural relaxations of the glass matrix.[6,7] Ionic motion is coupled closely to local motion in the polymer matrix of polymer electrolytes and consequently ionic conductivity decreases dramatically as the glass transition temperature (T_g) is approached.

The supercooled mixture of lithium salts originally described by Angell are potentially explosive because strongly oxidizing perchlorate anions are combined with reducible organic materials.[8] The replacement of perchlorate and related oxidizing anions has recently been the subject of extensive research.[9-11] Novel ionic liquids and molten salts have also been pursued as alternatives to the supercooled mixture of lithium salts.[9-11]

In this investigation we explore the use of cation encapsulating macrocycles for the formation of amorphous electrolytes that do not contain strongly oxidizing anions. These macrocycles surround the cation and thereby reduce the coulombic interaction between the cation and anion. We found that if the cation matches the macrocycle cavity, the resulting complex is a solid with a low melting temperature (T_m), however if the macrocycle cavity is larger than the cation the complex is a glass with a subambient T_g. The ionic glass formed can be described as a viscous liquid at room temperature. Furthermore, we found that the viscoelastic properties and

505

attendant conductivity were improved by the incorporation of the polar polymer MEEP.

EXPERIMENT

Materials

All experimental manipulations were carried out under an inert atmosphere of dry nitrogen. LiMPSA was prepared as previously described.[12] Characterization included FAB⁻, FTIR, EA, ^{19}F NMR, ^{1}H NMR, and DSC. MEEP was prepared as previously described.[13] PS4SS (Aldrich) (M_w 70,000) was dried on a high vacuum line at 100°C (8x10^{-6} torr). All macrocycles were purchased from Aldrich. 18-C-6 was sublimed at 35°C (1X10^{-1} torr). 12-C-4 and 15-C-5 were dried over molecular sieves for more than 48 hours. 2.2.2 Cryptand was dried on a high vacuum line at 30°C (8x10^{-6} torr). 2.2.1 Cryptand and 2.1.1 Cryptand in 100 mg ampules were used as received. Tetrahydrofuran (THF) was distilled under nitrogen from sodium benzophenone. Methanol was distilled under nitrogen from magnesium methoxide.

Synthesis

Samples with 1:1 mole ratios of LiMPSA to macrocycle were dissolved in THF and solvent was removed under vacuum at 50°C (1x10^{-1} torr), followed by 40°C (8x10^{-6} torr). LiMPSA, 2.2.2 Cryptand and MEEP were dissolved in THF and solvent removed as described above. LiMPSA, 2.2.2 Cryptand and PS4SS were dissolved in methanol and treated as above. All samples were checked for residual solvent and water by infrared spectroscopy.

Instruments

DSC was performed at a heating rate of 40°C/min. T_g's were assigned at the middle of the transition. Melting and cold crystallization temperatures were assigned to the onset of these transitions. Crystalline samples were quenched at 200°C/min in an attempt to form glasses. Complex impedance measurements were performed with a Hewlett-Packard 4192A in the frequency range 5 Hz to 13 MHz, on samples that were annealed for 12-24 hrs at 100°C.

RESULTS

Izatt and co-workers reported extensive thermodynamic data for cation-macrocycle interactions.[14] These data indicate that the most stable complexes are formed when the cation diameter matches the size of the macrocycle cavity.[15-17] The diameter of the lithium cation matches the macrocyclic cavity of 12-C-4 and 2.1.1 Cryptand. Therefore, the most stable complexes in Table I are complexes (1) and (4). Cryptands are known to bind cations more strongly than their crown ether analogues.[15-17] As a result ion pairing is reduced in complex (4) compared with the 12-C-4 ligand in complex (1), and the reduced cation-anion coulombic interactions results in a lower T_m for complex (4). A variety of structural data indicate that complexes with macrocyclic cavities larger than the diameter of the lithium cation should exist in many conformations, resulting in an amorphous electrolyte with an attendant increase in the entropy of the sample.

The temperature dependence of amorphous polymer electrolytes follows a Vogel-Tammann-Fulcher (VTF) function (1) which indicate that ion motion is coupled to segmental motions of the polymer.[18-20]

$$\sigma = A \, T^{-0.5} \exp[B/(T-To)] \quad (1)$$

Table I. DSC data of LiMPSA in a 1:1 ratio with various cation encapsulating macrocycles. T_g(°C)-Glass transition temperature, T_g^*(°C)-Glass transition temperature after quenching, T_m(°C)-Melting temperature, T_m^*(°C)-Melting temperature after quenching, T_c(°C)-Cold crystallization temperature.

Macrocycles	T_g	T_g^*	T_m	T_m^*	T_c
(1) 12-C-4		-18	97	94	44
(2) 15-C-5	-26				
(3) 18-C-6	-31				
(4) 2.1.1 Cryptand		-35	66		
(5) 2.2.1 Cryptand	-57				
(6) 2.2.2 Cryptand	-50				

The VTF behavior is apparent in Figure 1. Complex (5) has the lowest T_g, -57°C, and consequently the highest ionic conductivity. Conversely, complex (1) a solid at room temperature with a T_g^* of -18°C has the lowest ionic conductivity.

Complexes (3) and (4) have similar ionic conductivities and not surprisingly have similar T_g's. However, complex (6) with a T_g 15°C lower than complex (4) also has a similar conductivity. VTF parameters derived from the fit of temperature dependent ionic conductivity plots indicate that a higher pseudo activation energy exists for complex (6) compared to complex (4). Conversely, a smaller number of charge carriers is determined for complex (6) compared to complex (4). The combination of a lower pseudo activation energy and a higher number of charge carriers results in complex

(4) having a similar conductivity to that of complex (6).

The effect of adding MEEP (T_g = -80°C) to complex (6) is similar to the addition of a plasticizer to polymer electrolytes. Plasticizers effectively lower the microviscosity of the electrolyte and thereby promote ion motion. In Table II mixtures (7), (8) and (9) show a decrease in the T_g of the electrolyte as the dopant levels of MEEP increase. In Figure 2 lowering the T_g of the electrolyte results in a higher ionic conductivity for these rubbery electrolytes. The number of charge carriers as derived from the VTF equation is highest for materials (9). Clearly MEEP is highly compatible with complex (6).

Crystallization of the amorphous electrolyte, complex (6) is observed (Table II) on doping with PS4SS. The resulting low T_m solid can be quenched to form an amorphous

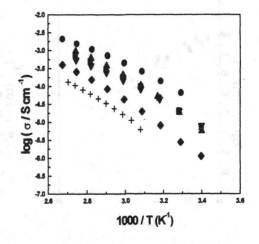

Figure 1. Temperature dependent conductivities (σ) of 1:1 ratios of LiMPSA : 2.2.2 Cryptand (■), LiMPSA : 2.2.1 Cryptand (●), LiMPSA : 2.1.1 Cryptand (▲), LiMPSA : 18-C-6 (▼), LiMPSA : 15-C-5 (◆), LIMPSA : 12-C-4 (+)

Table II. DSC data of LiMPSA : 2.2.2 Cryptand : MEEP in 10:10:1 (7), 10:10:3 (8) and 10:10:5 (9) ratios. DSC data of LiMPSA : 2.2.2 Cryptand : PS4SS in 10:9:1 (10), 10:7:3 (11) and 10:5:5 (12) ratios. T_g(°C)-Glass transition temperature, T_g^*(°C)-Glass transition temperature after quenching, T_m(°C)-Melting temperature.

Materials	T_g	T_g^*	T_m
(7)	-50		
(8)	-55		
(9)	-58		
(10)		-48	29
(11)		-42	38
(12)		-38	52

material. Ion transport is known to be favored in the amorphous phase of polymer electrolytes. Accordingly, ionic conductivity of mixtures(10), (11) and (12) (Figure 3) decrease with increasing content of PS4SS. Compatibility of PS4SS in complex (6) is not favored because the bulky, non-solvating, highly charged polystyrenesulfonate is not soluble in the amorphous electrolyte.

CONCLUSIONS

Ionic glass formation is favored by a mismatch of cation diameter and macrocycle cavity size. The ionic glasses formed have low T_g's and correspondingly good ionic conductivity. The comb polymer MEEP enhances ionic conductivity and improves mechanical compliance of the ionic glass.

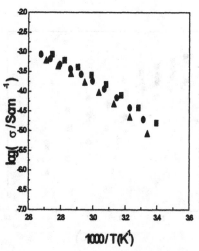

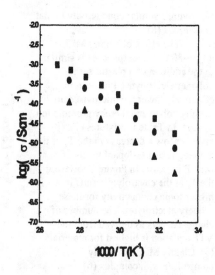

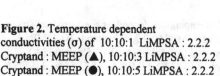

Figure 2. Temperature dependent conductivities (σ) of 10:10:1 LiMPSA : 2.2.2 Cryptand : MEEP (▲), 10:10:3 LiMPSA : 2.2.2 Cryptand : MEEP (●), 10:10:5 LiMPSA : 2.2.2 Cryptand : MEEP (■)

Figure 3. Temperature dependent conductivities (σ) of 10:9:1 LiMPSA : 2.2.2 Cryptand : PS4SS (■), 10:7:3 LiMPSA : 2.2.2 Cryptand : PS4SS (●), 10:5:5 LiMPSA : 2.2.2 Cryptand : PS4SS (▲).

ACKNOWLEDGMENTS

This work was supported by the MRSEC program of the National Science Foundation (DMR-9632472) at the Materials Research Center of Northwestern University.

REFERENCES

1. M.A. Ratner and D.F. Shriver, Chem. Rev. **88**, 109 (1988).

2. C. Julien and G. Nazri, Solid State Batteries: Materials Design and Optimization, edited by H.L. Tuller (Kluwer Academic Publishers, Boston, 1994), Chapters 3 & 5.

3. J.R. MacCallum and C.A. Vincent, Polymer Electrolyte Reviews, (Elsevier Science Publishers, London, 1987) Volumes 1 & 2.

4. C.A. Angell, C. Liu and E. Sanchez, Nature **362**, 137 (1993).

5. C.A. Angell, J. Fan, C. Liu, Q. Lu and E. Sanchez, K. Xu, Solid State Ionics **69**, 343 (1994).

6. C.A. Angell, Solid State Ionics 9&10, 3 (1983).

7. C.A. Angell, Solid State Ionics **18&19**, 72 (1986).

8. R.E Cook and P.J. Robinson, J. Chem. Research (S), 267 (1982).

9. C.A. Angell, K. Xu, S. Zhang and M. Videa, Solid State Ionics **86-88**, 17 (1996).

10. M. Watanabe, S. Yamada and N. Ogata, Electrochimica Acta **40** (13-14), 2285 (1995).

11. M. Watanabe and T. Mizumura, Solid State Ionics **86-88**, 353 (1996).

12. S. Lascaud, M. Perrier, A. Vallèe, S. Besner, J. Prud'homme and M. Armand, Macromolecules **27**, 7469 (1994).

13. H.R. Allcock, P.E Austin, T.X Neeman, J.T. Sisko, P.M. Blonsky and D.F. Shriver, Macromolecules **19**, 1508 (1986).

14. R.M Izatt, J.S. Bradshaw, S.A. Nielsen, J.D. Lamb and J.J. Christensen, Chem. Rev. **85**, 271 (1985).

15. J.M. Lehn, J.P. Sauvage, J. Am. Chem. Soc. **97** (23), 6700 (1975).

16. A.J. Smetana and A.I. Popov, J. Soln. Chem. **9** (3), 183 (1980).

17. J.S Bradshaw, R.M. Izatt, A.V. Bordunov, C.Y. Zhu, J.K. Hathaway, Comprehensive Supramolecular Chemistry, edited by G.W. Gokel (Elseveir Science Publishers, New York, 1996), Vol. 1, pp. 35; B. Dietrich, ibid., pp. 153.

18. H. Vogel, Phys. Z. **22**, 645 (1921).

19. G. Tamman and W.Z. Heese, Anorg. Allg. Chem. **254**, 165 (1926).

20. G.S. Fulcher, J. Am. Ceram. Soc. **8**, 339 (1925).

DRY COMPOSITE POLYMER ELECTROLYTES FOR LITHIUM BATTERIES

G.B. Appetecchi, F.Croce, G.Dautzenberg and B. Scrosati
Dipartimento di Chimica, Sezione di Elettrochimica
Università "La Sapienza", 00185 Rome, Italy

Abstract

The synthesis , characteristics and properties of a PEO-based ,dry composite electrolyte are presented and discussed. The major feature of this electrolyte is the high stability towards the lithium metal electrode. This unique property makes the electrolyte quite promising for the development of rechargeable polymer lithium batteries.

Introduction

Complexes between lithium salts, LiX and high molecular weight poly(ethylene) oxide, PEO, form a very well known class of polymer electrolytes [1]. These PEO-LiX electrolytes are commonly prepared by a casting procedure which uses acetonitrile as the carrier solvent. Due to their particular structure, the conductivity of the PEO-LiX complexes becomes appreciable only at temperatures above ambient, typically around 100 °C. This is of course a drawback for applications in the consumer electronic market, but it is an acceptable limitation for the use in batteries designed for electric vehicles. In fact, PEO-LiX electrolytes are presently studied for this type of application, with the use of metal lithium as preferred anode and of a suitable intercalation compound (e.g., V_6O_{13}, TiS_2 or MnO_2) as cathode [2].

Considerable progress has been achieved in these lithium polymer rechargeable batteries. However, there is still some concern on cyclability and safety due to the reactivity of the lithium metal anode with the PEO-LiX electrolytes. In fact, lithium unavoidably reacts with the polymer electrolyte with the formation of a surface passivation layer, the nature and characteristic of which vary according to the type of lithium salt and, especially, to the presence of liquid impurities [3]. Thus, it is of paramount importance to promote the most favorable type of passivation layer, namely the layer which may assure an efficient lithium deposition-stripping process. An effective approach for reaching this goal is that of dispersing low-particle size ceramic powders in the polymer electrolyte bulk [4]. Most active powders include alumina, lithium gamma aluminate and zeolites. In the attempt of further improving the lithium metal interfacial properties we have developed and characterized a new class of dry composite polymer electrolytes. The characteristics and the properties of these electrolytes are reported in this paper.

Results and discussion

Generally, the stability of the lithium interface in a given electrolyte is determined by monitoring the change upon time of the impedance response of symmetrical Li/ electrolyte/Li cells. The width of the middle frequency semicircle is associated to the interfacial resistance; thus changes in this width may be directly related to modifications of the passivation film. Figure 1 shows in comparison the impedance response of cells based on a plain PEO-LiClO$_4$ electrolyte (Figure 1A), and on two PEO-LiClO$_4$ composite electrolytes, one using lithium gamma aluminate (Figure 1B) and the other zeolite (Figure 1C), as the ceramic powder additive.

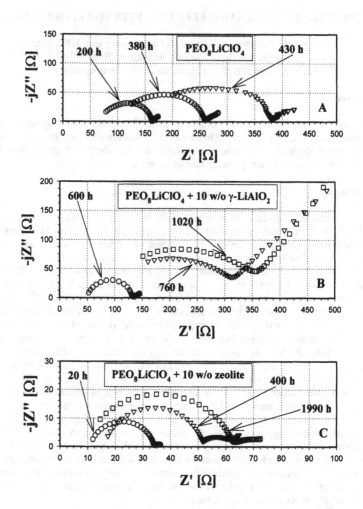

Figure 1: Time evolution of the impedance response of symmetrical Li/electrolyte/Li cells using a plain PEO-LiClO₄ (**A**), a composite PEO-LiClO₄ + 10 w/oγ-LiAlO₂ (**B**) or a composite PEO-LiClO₄ + 10 w/o zeolite (**C**) electrolyte (w/o = weight percent). The time of storage is indicated directly on the curves. Temperature: 106 °C. Frequency range: 0.1Hz - 65 kHz.

The figure clearly shows that, while in the plain electrolyte the width of the semicircle progressively increases with time (this suggesting a continuous growth of a passivation layer on the lithium electrode surface), in the two composite electrolytes the width does not vary consistently over a prolonged period of time (this suggesting the occurrence of a stable and compact passivation layer). These results confirm that the dispersed ceramic powder indeed promotes stabilization of the lithium interface in PEO-LiX electrolytes, probably by trapping water and other liquid impurities which otherwise would react with the lithium electrode with the continuous formation of not-uniform, porous passivation layers[5'].

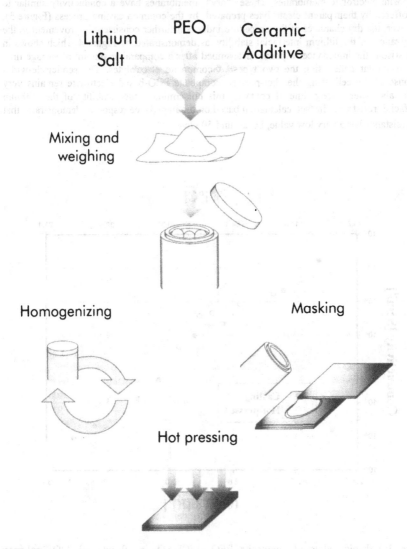

Figure 2: Scheme of the preparation procedure of dry, hot-pressed, composite electrolyte membranes.

To further enhance interfacial stabilization, we have prepared ceramic powder-added composite electrolytes using a procedure where the presence of any liquid was avoided throughout the entire process. The procedure consisted in the direct reaction of the electrolyte components obtained by hot pressing an intimate mixture of PEO, LiX and the selected ceramic additive (Figure 2).

This procedure excludes the presence of any residual solvent and gives uniform composite electrolyte membranes. These "dry" membranes have a conductivity similar to that offered by their parent electrolytes prepared by the common casting process (Figure 3). However, the dry character of the procedure induces a further consistent improvement in the stabilization of the lithium interfacial stability, as demonstrated by Figure 4 which shows in comparison the impedance response determined after a comparable amount of storage time of two similar cells using the two types of composite electrolytes. The semicircle-width response of the cell using the hot-pressed composite PEO-based electrolyte remains very small also after a long time of contact, this confirming the stability of the lithium interfacial resistance. In fact, calculation based on the impedance response demonstrates that this resistance has a very low value, i.e. around 50 Ωcm^2.

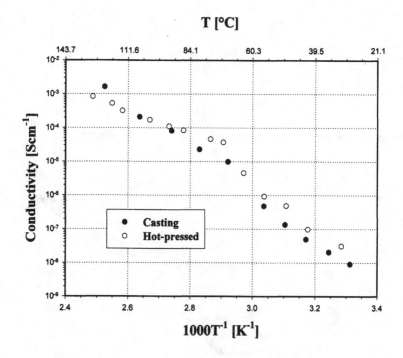

Figure 3:Arrhenius plots of composite PEO_{12}-$LiCF_3SO_3$ + 10 w/o γ-$LiAlO_2$ polymer electrolytes prepared by casting (acetonitrile carrier solvent) and by hot-pressing (dry procedure).(w/o = weight percent).

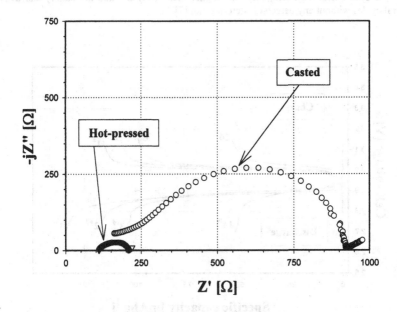

Figure 4: Impedance response of symmetrical Li / electrolyte / Li cells using a hot pressed PEO_{10}-$LiCF_3SO_3$ + 10 w/oγ-$LiAlO_2$ electrolyte and an analogous electrolyte prepared by casting (w/o = weight percent). Storage time: 24 hours Temperature: 95 °C. Frequency range: 10 Hz - 100 kHz.

These results confirm that composite polymer electrolytes where the dispersion of ceramic powders is combined with a solvent-free preparation procedure, are indeed characterized by a consistent improvement in the stability of the lithium electrode interface. It is reasonable to assume that in the absence of liquids, the lithium passivation process may only be due to the reaction with the lithium salt anion which likely gives the formation of a thin, compact , inorganic film. This type of film is considered to be the most favorable for assuring lithium interface stability [6] and this explains the favorable interfacial properties of the dry composite electrolytes.

Due to these important properties, the "dry" composite PEO-based electrolytes seem to be quite promising for assuring a good cyclability of the lithium electrode. This in turn would suggest that these electrolytes may profitably be used for the development of rechargeable lithium polymer batteries. To confirm this prevision, we have assembled and tested lithium battery prototypes using a PEO_{20}-$LiCF_3SO_3$ +20w/o $\gamma LiAlO_2$ composite electrolyte and a $LiMn_2O_4$ manganese spinel cathode. Figure 5 shows typical charge-discharge cycles of this battery: good capacity delivery and high efficiency are obtained. Test is in progress to evaluate the cycling performance and the energy content of this advanced lithium polymer battery. Preliminary results are encouraging in demonstrating that the lithium electrode cycles with high efficiency (i.e. higher than 95%) and that the battery can deliver several cycles without any detectable deterioration[7].

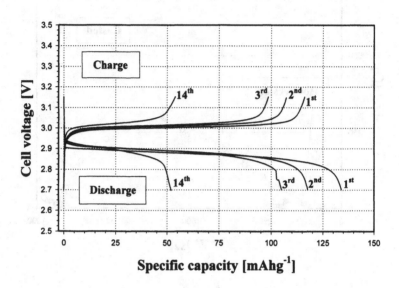

Figure 5 - Charge-discharge cycles of a Li / PEO_{20}-$LiCF_3SO_3$ + 20 w/oγ-$LiAlO_2$/ $LiMn_2O_4$ lithium polymer battery. (w/o = weight percent). Temperature: 90 °C. Rate: C/38.

Acknowledgement
The results here reported have been obtained thank to the financial support from ENEA, Italy (ALPE Project) and CNR, Italy (Progetto Strategico Batterie Leggere). Two of us (G.B.A.and G.D.) are grateful to ARCOTRONICS Italia spa for research fellowships.

References
[1]. - F.M. Gray, *Solid Polymer Electrolytes*, VCH, Weinheim, **1993**.
[2] -Neat R. and Scrosati B., in *Applications of Conductive Polymers* , Scrosati, B. Ed., **1993**, pag. 182, London, Chapman & Hall.
[3] Croce F. and Scrosati B.,*J.Power Sources,* **1993**, *43*., 43
[4]. -M.C.Borghini, M.Mastragostino, S.Passerini and B.Scrosati, *J. Electrochem.Soc.* ,**1995** , *142*, 2118
[5] -F.Capuano, F. Croce and B.Scrosati, *US Patent, N. 5,576,115*, Nov. 19, **1996**.
12.-G.B.Appetecchi, F.Croce, G.Dautzenberg and B.Scrosati, *J.Electrochem.Soc.*, submitted.
[6]. - K. Kanamura, S.Shirashi and Z. Takehara, *J.Electrochem.Soc.*,**1994**, *141*, L108 .
[7] S.Saguatti, P.Bernini, F.Alessandrini, M.Carewska, M.Conti, L.Giorgi, S.Passerini, P.Prosini, R.Vellone, M.C.Borghini, M.Mastragostino, A.Zanelli, G.B.Appetecchi, F.Croce, G.Dautzenberg and B.Scrosati, *International Conference on Applications of Conducting Polymers*, Rome, Italy, **1997**.

Part VI

Lithium Ion Rechargeable Battery – Anode Materials

COMPARISON OF CARBON AND METAL OXIDE ANODE MATERIALS FOR RECHARGEABLE LI-ION CELLS

C.-K. Huang[1], J. S. Sakamoto[2], M. C. Smart[1], S. Surampudi[1], and J. Wolfenstine[2]
[1]Jet Propulsion Lab., California Institute of Technology, 4800 Oak Grove Drive, Pasadena, CA 91109
[2]University of California, Irvine, Dept. of Chemical and Biochemical Engineering, Irvine, CA 92697-2575

ABSTRACT

The state of the art (SOA) Li-ion cells utilize carbon anodes. However, to improve specific energy, energy density, and safety of cells using carbon anodes, alternative anodes must be developed. Recently, Fuji Film Inc. has suggested the use of tin oxide based anodes in Li-ion cells. It is believed that cells containing tin oxide based anodes have the potential to meet the need for NASA's future missions. As a result, we conducted an analysis to compare the performance of cells containing carbon anodes and cells containing tin oxide anodes. The comparison between these cells involved the following: 1) reaction mechanisms between Li and carbon and reaction mechanisms between Li and tin oxide, 2) half-cell and full-cell performance characteristics, 3) interactions between the anode materials and electrolyte types and compositions, and 4) the optimization of binder composition.

INTRODUCTION

JPL is involved in the development of rechargeable lithium cells for future Mars Exploration Missions. Mars Exploration Missions can be broadly classified into four types: orbiters, landers, rovers and penetraters. These missions have some common performance requirements, such as high specific energy and energy density due to mass and volume limitations. However, each of these missions has some unique primary performance drivers, such as, long cycle capability (orbiters) and ability to operate at low temperatures (landers, rovers and penetraters). The orbiters require a cycle life greater than 30,000 cycles at 20-30% depth of discharge with a specific energy > 100 Wh/kg. The landers and rovers require batteries that can provide > 120 Wh/kg and operate at temperatures < -20 °C, whereas, the cycle life requirement is < 500 cycles (50-70% DOD). Penetraters require batteries that can operate at temperatures lower than -60 °C and withstand high shock levels. The lithium-ion system was selected for near term missions as this technology was more mature compared to the lithium metal based and lithium polymer battery systems . The specific objectives of the JPL lithium ion cell effort are : 1) Improve the low temperature performance of lithium-ion cells and demonstrate their applicability to lander, rover and penetrater missions, 2) Improve the cycle life performance of lithium-ion cells and demonstrate the ability to meet life requirements of the Mars orbiters and 3) Establish effective charge methodology and reconditioning methods for on-board battery management. To realize these objectives work is in progress in areas such as chemistry and material development, design optimization and data base development. The prime objective of the chemistry and materials subtask is to develop/select electrode materials and electrolytes that are capable of providing long cycle life and improved low temperature performance.

The SOA Ni-Cd and Ni-H$_2$ batteries are quite heavy and bulky and can not meet mass and volume requirements of future Mars missions. Furthermore, they have very poor low temperature performance capability as they use aqueous electrolytes. Rechargeable lithium-ion batteries offer significant weight, volume and cost advantages compared to SOA Ni-Cd and Ni-H$_2$ batteries and are especially attractive for future Mars Missions. The performance advantages include: higher specific energy (2 to 3 times greater than Ni-Cd and Ni-H$_2$), energy density (3-4 times greater than Ni-Cd and Ni-H$_2$), higher cell voltage, coulombic and energy efficiency, low self-discharge rate, and lower battery costs compared to the SOA Ni-Cd and Ni-H$_2$ batteries. These advantages translate into several benefits for Mars Missions including: reduced weight and

Mat. Res. Soc. Symp. Proc. Vol. 496 ©1998 Materials Research Society volume of the energy storage subsystem, improved reliability, extended mission life, and lower power system life cycle costs.

Four types of lithium cells are presently under development in the US, Europe and Japan: 1) lithium metal with liquid electrolyte, 2) lithium metal with polymer electrolyte 3) lithium-ion containing liquid electrolyte, and 4) lithium-ion containing polymer electrolyte. Among these four types of technologies, lithium-ion battery technology is the most advanced and likely candidate for future space missions (1998 and beyond). This system is also being considered by other aerospace organizations for GEO and LEO spacecraft applications. These cells employ a carbon/graphite anode (instead of metallic lithium) and liquid organic electrolytes. $LiCoO_2$, $LiNiO_2$, and $LiMn_2O_4$ are presently being evaluated by several commercial vendors as candidate cathode materials for these cells. Small capacity cylindrical cells have been introduced into the United States by Japanese manufacturers for portable electronic applications. These cells have a specific energy of about 80 - 120 Wh/kg and 200-240 Wh/l and can operate at temperatures in the range of -10 to 30 °C. These cells can deliver > 500 deep discharge cycles. For future Mars exploration applications, however, certain improvements must be made in order to meet specific mission requirements, such as improving the cycle life of the cells, extending the operating range to lower temperatures, as well as scaling-up the technology to large cell sizes (5-20 Ah).

At present, carbon materials, such as coke and graphite, are used as anode materials in rechargeable lithium-ion cells. These materials exhibit lower usable specific capacity (150-300 mAh/g) compared to metallic lithium and this results in lower specific energy. Additionally, carbon-based anodes typically have low densities, which imposes limits on the energy density of the Li-ion batteries. Moreover, carbon anodes operate at potentials very close to that of lithium and, consequently, are reactive and may exhibit safety problems especially in large size batteries.

In order to further improve the specific energy, energy density, and safety of lithium-ion cells, advanced anode materials alternative to carbon anodes with higher specific capacity and stability need to be developed. Of the alternative non-carbon anode materials, lithium alloys appear to be attractive. Lithium alloys usually have high specific capacity of lithium. The major problem which limits their use is the volumetric instability of the lithium alloy electrode during charge and discharge cycling. However, recent developments suggest that lithium alloys may work well as anode materials for Li-ion cells. Fuji Film Inc. in Japan has announced the successful development of Li-ion cells containing amorphous tin-based composite oxide anodes[1,2]. It has been suggested that the reaction of lithium and tin oxide forms tin metal and lithium oxide initially[3,4], followed by the reaction between lithium and tin metal.

At JPL, a preliminary analysis was conducted to compare and evaluate carbon and tin oxide anodes for use in Li-ion cells[5]. It is the purpose of this paper to report the results of this comparison which includes the following: i) the mechanisms that occur when Li reacts with both carbon and tin oxide, ii) cycling performance of carbon vs. tin oxide, iii) the effects of electrolyte types and compositions in cells containing carbon and tin oxide anodes, and iv) the optimization of binder composition of tin oxide anodes.

<u>EXPERIMENTAL</u>

The electrochemical evaluation of carbon, SnO, and SnO_2 anodes was conducted using half-cells with the electrodes wound in a spiral configuration. In the half-cell configuration, lithium was used as the anode and reference electrode, and carbon, SnO, and SnO_2 electrodes were used as the cathode. The carbon, SnO and SnO_2 electrodes were made by spraying a solution containing either carbon, SnO, or SnO_2 powder, Polyvinylidene fluoride (PVDF) binder, and carbon black onto a copper foil substrate. Lithium electrodes were made by cold pressing lithium foil onto nickel mesh substrates. In the full-cell configuration, $LiCoO_2$ was used as cathode material, carbon or tin oxides were used as anodes. The anodes and cathodes were separated by two layers of 1-mil thick polypropylene membranes. The cells were activated with the various mixed solvent electrolytes containing $LiPF_6$ salt dissolved in Ethylene Carbonate (EC), Dimethyl Carbonate (DMC), Diethyl Carbonate (DEC), Butylene Carbonate (BC) and Propylene Carbonate

(PC). All cell assembly operations were carried out in an oxygen- and moisture-free glove box. The experimental cells were evaluated for charge/discharge characteristics, faradaic utilization of the active material, rate capability and cycle life. Constant current was used for charging and discharging the cells. Open circuit voltages (OCV) were determined using a coulometric titration technique.

Preparation of high specific surface area crystalline SnO was achieved by using a modified precipitation technique as follows: 1) dissolving $SnCl_2.2H_2O$ (Alfa 99.99%) in distilled water, 2) the resulting solution was added dropwise into a 1N NaOH solution (Fisher Scientific), 3) additional NaOH was added to maintain a pH of 14 at all times, and lastly 4) a black aggregate formed which was then washed with distilled water, filtered, and allowed to dry in air at room temperature. The commercial tin oxides were obtained from Alfa (99.99%). X-Ray Diffraction (XRD) was used to identify crystalline phases, and determine average grain sizes, for in-house prepared SnO and commercially obtained SnO powder. A standard B.E.T. analysis determined the average specific surface areas for both SnO powders. The average particle sizes of the SnO powders were determined using a laser light scattering technique. Surface morphologies for both powders were examined using Scanning Electron Microscopy (SEM).

To determine the crystalline phases present after lithium was titrated into the SnO and SnO_2 powder the following procedure was used: 1) 5 SnO half cells and 5 SnO_2 half cells were made using the procedure mentioned above, 2) the amount of lithium titrated into each SnO or SnO_2 electrode was pre-determined, 3) after titrating lithium into each electrode the OCV were measured, 4) the cells were then opened in a glove box and the electrodes were washed in solvent and sealed in a polypropylene bag, 5) the electrodes were then analyzed using XRD to determine the crystalline phases present.

RESULTS AND DISCUSSIONS

(1) Comparison of reaction mechanism:

Graphite and Coke

The graphite structure consists of layers of carbon atoms arranged in hexagonal rings that are stacked in an ABAB.... sequence. In coke, the basic structural unit is similar to graphite, however, its carbon layers have small lateral sizes and the stacking of the layers is imperfect and characterized by a turbostratic structure. The reaction mechanism between Li and graphite is an intercalation type reaction and proceeds in stages as follows: C --> LiC_{24} -->LiC_{18} --> LiC_{12} --> LiC_6, where the formation of each of these LiC_x phases corresponds to the constant voltage plateaus in the charge/discharge curve in the Li/C system. It is important to note that during this reaction, there are no major structural reconfigurations in graphite, which results in a highly reversible reaction. However, when graphite is initially charged, the formation of a passivation layer due to electrolyte decomposition at the electrode surface is imminent, and results in an irreversible capacity loss. Additionally, the average discharge voltage for the intercalation of Li into graphite is approximately 0.1 V, which will be compared to the Li-tin oxide system, shortly. Unlike graphite, the reaction mechanism between Li and coke is a solid-solution reaction and therefore the charge/discharge curves for coke have no voltage plateaus but instead have sloped curves between 1.0 and 0 volts. In Li-ion cells containing carbon anodes, graphites are preferred to coke due to the higher reversible capacity of graphite. However, when compared to graphite, coke has higher rate capabilities which results from the open structure of coke. Additionally, compared to graphite, coke has a higher lithium potential gradient.

Crystalline SnO and SnO_2

SnO_2: The reaction that occurs when Li is titrated into tin oxide is not an intercalation reaction, instead it has been suggested that a reconstitutional reaction occurs according to the following equation for SnO_2 (rutile):

$$8.4 \text{ Li} + \text{SnO}_2 \longrightarrow 2\text{Li}_2\text{O} + \text{Li}_{4.4}\text{Sn} \qquad \text{Eqn. 1}$$

Hence, the reaction that occurs for SnO (layered type structure) is:

$$6.4 \text{ Li} + \text{SnO} \longrightarrow \text{Li}_2\text{O} + \text{Li}_{4.4}\text{Sn}. \qquad \text{Eqn. 2}$$

In Figure 1 the charge/discharge curves are shown for Li/SnO and Li/SnO$_2$. It is observed that 8.15 moles of Li reacted with SnO$_2$, which is in agreement with the 8.4 moles predicted from equation 1. From equation 1, out of the 8.4 moles of Li initially titrated into SnO$_2$, 4 moles of Li are irreversibly consumed to form 2 moles of Li$_2$O, and 4.4 moles of Li are reversibly stored in 1 mole of Li$_{4.4}$Sn. However, it is observed that only 2.9 moles of Li were extracted during charging, which is not in agreement with the 4.4 moles of Li predicted by equation 1. It is believed that the decrease in the reversible capacity can be attributed to the fact that 2 moles of Li$_2$O are formed for every mole of SnO$_2$ compared to only 1 mole of Li$_2$O formed for every mole of SnO. Although Li$_2$O is a reasonably good Li-ion conductor, its conductivity is still considerably less than that of the liquid organic electrolyte. Therefore, it is believed that the increase in Li$_2$O formation for SnO$_2$ significantly increases the polarization. By doing this, the resulting conditions are far from equilibrium. The net result is a decrease in the reversible Li capacity for SnO$_2$ for a fixed current density.

The data of Figure 1 suggests that the plateau from 0 to approximately 3 moles of Li, represents the decomposition of SnO$_2$ into Li$_2$O and Sn as indicated by the corresponding line segment spanning from $x = 0$ to $x = 4.0$ moles of Li in the proposed Li-Sn-O ternary phase diagram (Figure 2). (The ternary phase diagram is used to assist in understanding the Li-Sn-O system) However, equation 1 and the phase diagram shown in Figure 2 predict that this reaction should consume 4.0 moles of Li. It is believed that this difference is due to the fact that the voltage values indicated along the discharge/charge curve do not represent OCV (equilibrium voltages), because they are measured under an applied current. According to the OCV measurements listed in Table I, the OCV for 1.5, 2.5, and 4.0 moles of Li titrated into 1 mole of SnO$_2$ are 1.16 V, 1.10 V and 1.09 V, respectively. However, when more than 4.0 moles of Li are titrated into SnO$_2$, the voltage drops significantly to 0.76 V for 5.5 moles of Li. From this, it is apparent that the amount of Li consumed during the first plateau is approximately 4.0 moles as indicated by the significant decrease in the OCV after $x = 4.0$ as predicted by Figure 2. Altogether, by measuring the OCV for the cells listed in Table I, rather than estimating the length of the plateau using the discharge/charge curve (Figure 1), a more accurate value for the amount of Li consumed irreversibly to form Li$_2$O can be determined. Thus, the experimentally determined value of 4.0 moles of Li consumed to form 2.0 moles of Li$_2$O is in agreement with the predicted value.

Table I. OCV Measured for the SnO$_2$ Half Cells

Moles of Li	OCV vs. Li (Volts)
1.5	1.16
2.5	1.10
4.0	1.09
5.5	0.76

In Figure 3 the XRD patterns are shown for the cells listed in Table I. When Li is titrated into SnO$_2$ (from $x = 1.5$ to $x = 8.0$) the SnO$_2$ peaks are broadened and shortened when compared to pure SnO$_2$ ($x = 0$). Additionally, Sn peaks are observed in the samples containing Li. From this, it can be concluded that the initial reaction of Li with SnO$_2$ results in the formation of Sn. The absence of any Li$_2$O peaks, which should be present from the predictions made by Figure 2, most likely suggests that the Li$_2$O is amorphous. According to Figure 2 when more than 4 moles of Li react with SnO$_2$, Li-Sn alloys should be present. However, the XRD patterns in Figure 3 do not indicate the presence of any Li-Sn alloys. It is suggested that the combined effects of the increased background noise from amorphous Li$_2$O, and the fact that the Li-Sn peaks are relatively weak in intensity can explain the difficulties of identifying any Li-Sn peaks in the XRD patterns. However, to confirm this, further detailed work is required. Altogether, the confirmation that metallic Sn is formed from the reaction of Li with SnO_2, supports the validity of equation 1 and the phase diagram shown in Figure 2.

<u>SnO:</u> From Figure 1, it is observed that 6.25 moles of Li initially reacted with 1 mole of SnO, which is in good agreement with the 6.4 moles predicted by equation 2. Equation 2 predicts that out of the 6.4 moles of Li that react with SnO, 2 moles of Li are irreversibly consumed to form 1 mole of Li_2O, and 4.4 moles of Li are reversibly stored in 1 mole of $Li_{4.4}Sn$. From Figure 1, it is observed that 4.0 moles of Li are extracted from SnO, which is in close agreement with the 4.4 predicted by equation 2. The close agreement between the experimentally determined initial and reversible Li capacities with the theoretically determined capacities supports the validity of equation 2.

Table II. OCV Measured for the SnO Half Cells

Moles of Li	OCV vs. Li (Volts)
1.0	1.40
2.0	1.34
3.0	0.86
4.5	0.48
6.0	0.36

The data of Figure 1 shows that the length of the plateau, from 0 to approximately 2 moles of Li, represents the decomposition of SnO into Li_2O and Sn as indicated by the line spanning from x = 0 to x = 2.0 moles of Li shown in Figure 4. To obtain the true equilibrium voltages as function of the moles of Li, the OCV were again measured (listed in Table II), and it is observed that the OCV for 1.0 and 2.0 moles of Li reacted with 1 mole of SnO are similar (1.40 and 1.34 V, respectively). However, when more than 2.0 moles of Li react with SnO, the voltage drops significantly to 0.86 V for 3.0 moles of Li and eventually to 0.36 V for 6.0 moles of Li. The relatively constant OCV measured from 1.0 to 2.0 moles of Li further confirms that the length of the initial plateau in Figure 1 between 0 and 2.0 moles of Li. The similarities between the predicted amount of 2.0 moles of Li consumed to form 1 mole of Li_2O and the amount of Li consumed during the initial plateau in Figure 1, (approximately 2 moles Li), further supports the validity of equation 2.

In Figure 5 the XRD patterns are shown for the SnO half-cells listed in Table II. The phase diagram in Figure 4 predicts that: i) the amount of SnO should decrease from x = 0 to x = 2.0 moles of Li, and ii) at x = 2.0 moles of Li, SnO should not be present. Likewise, according to the XRD patterns for x = 0 to x= 2.0 moles of Li, the heights of the SnO diffraction peaks and therefore the amounts of SnO decrease. At x = 2.0 moles of Li, the SnO peak is absent. Furthermore, the phase diagram in Figure 4 predicts that the maximum amount of Sn present · occurs at x = 2.0 moles of Li. Correspondingly, according to the XRD pattern for x = 2.0 moles Li, the height of the Sn peak and therefore the amount of Sn is at a maximum when compared to all of the other samples. It is important to note that since the XRD peak heights change as a function of the amount of a particular phase present, the amount of material analyzed has a significant effect on the peak height. However, all of the SnO samples listed in Table II were carefully weighed to assure that the amounts of SnO analyzed were the same. Altogether, the strong agreement between the predictions made by Figure 4 and the XRD patterns shown in Figures 5 further confirms the validity of Equation 2.

Amorphous SnO_2

In Figure 6, the discharge/charge curves are shown for both the amorphous SnO_2 (SnO_2-AM) and crystalline SnO_2 (SnO_2-C). It is observed that 10.20 moles of Li were titrated into SnO_2-AM and 8.15 moles of Li were titrated into SnO_2-C. The plateau region for the SnO_2-C from x = 0 to x = 4.0 moles of Li, during discharging, represents the formation of Li_2O and Sn. Conversely, no

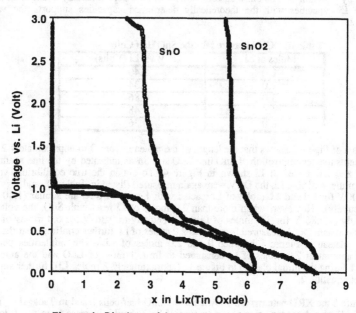

Figure 1. Discharge/charge curves for SnO2 and SnO.

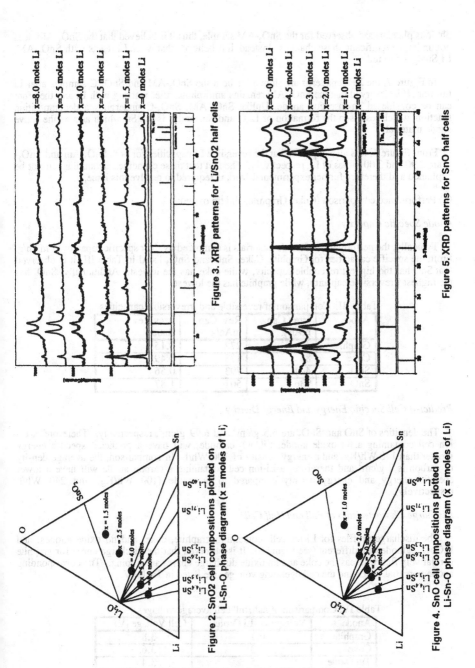

Figure 2. SnO2 cell compositions plotted on Li-Sn-O phase diagram (x = moles of Li)

Figure 3. XRD patterns for Li/SnO2 half cells

Figure 4. SnO cell compositions plotted on Li-Sn-O phase diagram (x = moles of Li)

Figure 5. XRD patterns for SnO half cells

obvious plateaus are observed for the SnO_2-AM sample, thus it is believed that the SnO_2-AM does not undergo significant phase changes. Instead, it is believed that when Li reacts with SnO_2-AM, Li_xSnO_2 is formed.

In Figure 7, the XRD patterns are shown for both the SnO_2-AM and SnO_2-C samples after Li titration. It is observed that the SnO_2-AM remains amorphous after reacting with Li and therefore can be considered an insertion anode. Unlike SnO_2-AM, SnO_2-C undergoes a decomposition reaction, which results in the formation of Li_2O and Sn, and it is the Sn which acts as the active anode material.

From Figure 6, it is also apparent that the reversible Li capacities for both SnO_2-Am and SnO_2-C are 2.90 and 3.00 moles of Li, respectively. The fact that these capacities are similar can not be explained and therefore further experimental work is required to resolve this issue.

(2) Performance comparison (Coke, Graphite, and tin oxide):

Anode Specific Capacity

Typically, the performance of anode materials is compared by their specific capacity in mAh/g. From the specific capacities for Graphite, Coke, SnO and SnO_2 listed in Table III, it is observed that SnO has the highest reversible capacity, while coke has the lowest. Additionally, SnO_2 has the highest irreversible capacity, while graphite has the lowest.

Table III. Comparison of reversible and irreversible capacities

Anode	Irrev.Cap. mAh/g	Rev. Cap mAh/g	Irrev/Rev Ratio
Graphite	42	272	0.15
Coke	85	173	0.49
SnO	444	793	0.56
SnO_2	936	501	1.87

Predicted Cell Specific Energy and Energy Density

The densities of SnO and SnO_2 are 6.5 g/cm^3 and 6.99 g/cm^3, respectively. Therefore, a Li-ion cell containing a tin oxide anode, $LiCoO_2$ cathode, will have a predicted specific energy greater than 140 Wh/Kg, and a energy density of 300 Wh/l. In comparison, the average density of carbon is 2 g/cm^3 and therefore, a Li-ion cell containing a carbon anode will have a lower specific energy and energy density compared to tin oxide (100 Wh/Kg, and 240 Wh/l, respectively).

Discharge Characteristics of Full and Half Cells

The discharge profiles for Li-ion cells containing graphite, coke, and tin oxide anodes, and $LiCoO_2$ cathodes are different (see Figure 8). It is observed that the discharge curve for graphite is relatively flat, whereas the coke and tin oxide discharge curves are sloping. The corresponding anode potential vs. Li and the cell operating voltages are listed in Table IV.

Table IV. Comparison of half/full cell average voltages

Anode	Voltage vs. Li (Volts)	Cell Voltage (V)
Graphite	0.1	3.8
Coke	0.3	3.6
Tin Oxide	0.5	3.4

Effects of Particle Size

Typically, high-specific-surface-area carbon powders are desired to maximize the accessible reaction area. This in turn, results in increased cell rate capabilities. However, for Li-ion applications, by increasing the carbon specific surface area, the relative amount of passivation film formed is correspondingly increased. Therefore, there is an optimal specific surface area (5-10 m²/g) to which a balance can be established between cell rate capability and irreversible Li capacity loss.

Unlike carbon powders, the irreversible capacity loss in tin oxide is not attributed to surface reactions with the electrolyte. Instead, the irreversible loss is considered to be due to a bulk reaction and is therefore not dependent of specific surface area. As a result, high-specific-surface-area tin oxide powders are preferred to enhance cell rate capabilities.

3) Electrolyte Analysis

An analysis was conducted to determine the effect of electrolyte types and compositions on cycle life. Different electrolytes containing $LiPF_6$ salt dissolved in EC, DMC, DEC, BC and PC were used in half cells. All cells were tested using a discharge current density of 0.344 mA/cm², and a charge density of 0.172 mA/cm². In this analysis the following effects were investigated: a) electrolyte compositions on the irreversible/reversible capacities, b) electrolyte types on cycling performance, and c) EC composition on cycling performance.

a) The irreversible, reversible, and the reversible/irreversible Li capacity ratios for the initial cycle in cells containing different EC/DMC electrolyte compositions are shown in Table V. It is observed that the 10EC/90DMC electrolyte yields the highest irreversible and reversible capacities whereas the 50EC/50DMC electrolyte yields the lowest irreversible and reversible capacities. It is important to note, however, that the reversible/irreversible capacity ratios for the 10EC/90DMC, 30EC/70DMC, and 50EC/50DMC electrolytes are the same. It is believed that the increase in the amount of EC increases electrolyte viscosity thus, resulting in increased polarization. As a result, the irreversible and reversible capacities change proportionally by varying the amount of EC in the electrolyte.

On the contrary, the irreversible and reversible Li capacities do not change proportionally for carbon anodes. For example, in cells containing carbon anodes with EC/DMC electrolyte, by increasing the amount of EC the reversible capacity remains the same whereas the irreversible capacity increases (Fig. 9). The fact that the reversible capacity for carbon is relatively insensitive to the amount of EC indicates that polarization is minimal. Therefore, since it is known that EC decomposes readily on the carbon surface to form a passivating layer, it is suggested that by increasing the amount of EC, a proportional increase in the amount of decomposition product on the carbon surface results. Hence, the irreversible Li capacity increases as the amount of EC increases.

Table V. Effects of EC/DMC composition on irreversible and reversible capacities

EC/DMC Compositions	Irreversible Cap. (mole Li /mole SnO)	Reversible Cap. (mole Li /mole SnO)	Rev./Irrev
10/90	2.330	2.408	1.033
30/70	2.059	2.232	1.084
50/50	2.001	2.101	1.050

It was determined that the irreversible Li capacity of tin oxide is insensitive to the electrolyte composition but is highly dependent on the electrolyte type.
b) The reversible Li capacity vs. cycle number plot is shown in Fig. 10 for SnO half cells containing: i) 30EC/70DMC, ii) 30BC/70DMC, and iii) 50PC/50DEC. The capacity decline is most severe for the cell containing 50PC/50DEC electrolyte, whereas the cell containing 30EC/70DMC yielded the highest cycle life performance. The reasons for this are not clear at the moment.

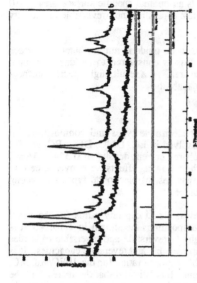

Figure 7. XRD patterns for SnO2-AM after Li was titrated in (a), and SnO2-C after Li was titrated in (b)

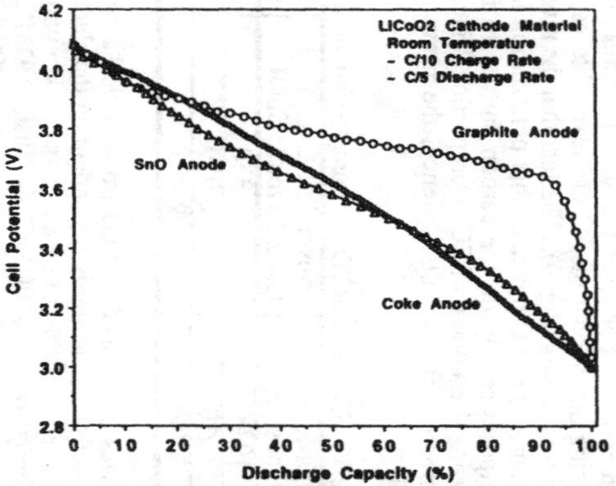

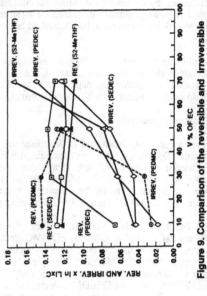

Figure 9. Comparison of the reversible and irreversible capacities in Li/C cells with various electrolytes

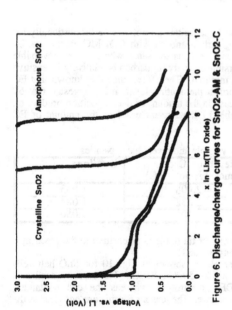

Figure 6. Discharge/charge curves for SnO2-AM & SnO2-C

Figure 8. Discharge characteristics of Li-Ion cells containing various anode materials

c) The reversible Li capacity vs. cycle number plot is shown in Fig. 11 for SnO half cells containing: i) 30BC/70DMC, ii) 10EC/30BC/60DMC, and iii) 40EC/10BC/50DMC. A trend is observed which indicates that as the amount of EC increases the cycle life performance improves. Since EC is a solid at room temperature it is believed that by increasing the amount of EC, the viscosity of the electrolyte increases, which in turn acts to reinforce the Li_2O/Sn composite electrode during cycling.

4) Binder Composition Analysis

An analysis was conducted to determine the optimum PVDF binder composition in carbon and SnO electrodes because the amount of binder affects the electrode performance. In the case of carbon electrodes, the amount of binder affects the reversible and irreversible Li capacity. The voltage vs. Li capacity is shown in Fig. 12 for SnO half cells containing 10 % carbon black and: i) 5.5% binder, ii) 11% binder, and iii) 15.4% binder. Several observations can be made. Firstly, it is observed that the initial capacity increases with decreasing amounts of binder. Secondly, the electrodes containing 5.5 and 11% binder have the same reversible capacity whereas the electrode containing 15.4% binder has a significantly lower reversible capacity.

The Li capacity vs. cycle number plot is shown in Figure 13. It is apparent that up to 7 cycles, the electrode containing 5.5% binder has the highest capacity, while the electrode containing 15.4% binder has the lowest. However, after 7 cycles, the capacity of the electrode containing 5.5% drops below that of the electrode containing 11% binder. This suggests that 5.5% binder is not sufficient to counteract the effects of severe volume changes that occurs during cycling. Therefore, to optimize cycle life performance it is suggested that the binder composition is 11%.

SUMMARY

In this study, it was determined that half and full cells containing carbon and tin oxide anodes have significantly different behavior. Firstly, the initial reaction between Li and tin oxide results in the decomposition of tin oxide into metallic Sn and Li_2O. Upon further addition of Li, Li reacts with Sn to form several Li-Sn alloys. Therefore, the reaction of Li with tin oxide is considered to be a reconstitutional reaction whereas the insertion of Li into carbonaceous materials is an intercalation-type reaction for graphite and a solid solution type reaction for coke. Secondly, the discharge profiles for cells containing graphite have a relatively constant voltage during cycling whereas the voltage of cells containing tin oxide and coke anodes decreases continuously. Additionally, it was determined that the capacity fade rate for cells containing coke anodes is the lowest. The capacity fade rate for graphite decreases steadily, but is considerably lower than that of tin oxide. Thirdly, the irreversible Li capacity of cells containing tin oxide anodes was found to be insensitive to electrolyte composition but is highly dependent on electrolyte type. In addition, it is believed that by increasing the amount of EC, the viscosity of the electrolyte increases, which in turn acts to reinforce the Li_2O/Sn composite electrode during cycling. Furthermore, for Li-ion cells containing carbon anodes, the irreversible capacity during the first cycle is due to electrolyte decomposition, which results in irreversible film formation on the carbon surface. This irreversible capacity is highly dependent on the electrolyte type and composition. In this study, electrolytes containing higher percentages of EC showed higher initial irreversible capacities. Hence, a tradeoff in the electrolyte composition may be necessary to obtain optimal rate capability and cycle life while minimizing initial irreversible loss. Lastly, it was determined that the optimum binder composition which yields the lowest capacity fade rate for SnO is 11% PVDF.

ACKNOWLEDGMENTS

The work described here was carried out at Jet Propulsion Laboratory, California Institute of Technology, under contract with the National Aeronautics and Space Administration.

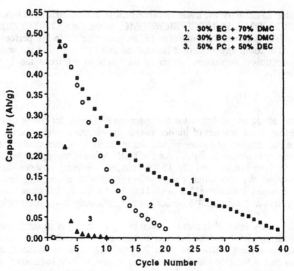

Figure 10. Effect of electrolyte type on cell cycling performance

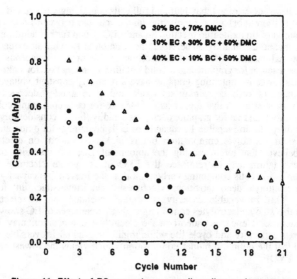

Figure 11. Effect of EC percentage on cell cycling performance

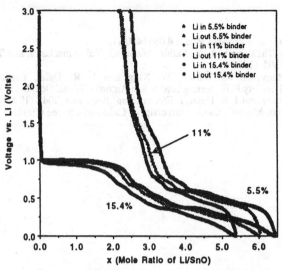

Figure 12. Effect of binder composition on initial cycle

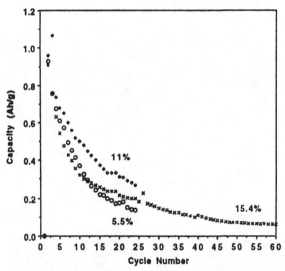

**Figure 13. Comparison of cycling performance of SnO electrodes
with various binder amounts**

REFERENCES

1. Yoshio Idota, etc., European Patent # 0651450A1.
2. Yoshio Idota, Tadahiko Kubota, Akihiro Matsufuji, Yukio maekawa, and Tsutomu Miyasaka, Science, **276**, 1395 (1997).
3. I. A. Courtney, A. M. Wilson, W. Xing and J. R. Dahn, Paper presented in the Electrochemical Society Fall Meeting held at San Antonio, Texas, Oct. 6-11, 1996.
4. Ian A. Courtney and J. R. Dahn, J. Electrochem. Soc., **144**, 2045 (1997).
5. J. S. Sakamoto, Master's thesis, University of California, Irvine (1997).

MAGNETIC AND TRANSPORT PROPERTIES OF HEAT-TREATED POLYPARAPHENYLENE-BASED CARBONS

M. J. Matthews[a], N. Kobayashi[d], M. S. Dresselhaus[a,b], M. Endo[c], T. Enoki[d], T. Karaki[c]
[a] Department of Physics, Massachusetts Institute of Technology, Cambridge, MA 02139
[b] Department of Electrical Engineering and Computer Science, Massachusetts Institute of Technology, Cambridge, MA 02139 [c] Faculty of Engineering, Shinshu University, Nagano, 380 Japan [d] Department of Chemistry, Tokyo Institute of Technology, Tokyo, 152 Japan

ABSTRACT

Electron spin resonance (ESR), magnetic susceptibility, and transport measurements were recently performed on a set of heat-treated polyparaphenylene (PPP)-based carbon samples, which are of significant interest as novel carbon-based anode electrodes in Li-ion batteries. Attention is focused on the evolution of the carbonaceous structures formed at low heat-treatment temperatures (T_{HT}) in the regime of $600°$ C $\leq T_{HT} \leq 800°$ C, where percolative transport behavior is observed. At the percolation threshold, $T_{HT}^c \approx 700°$ C the coexistence of two spin centers with peak-to-peak Lorentzian linewidths of $\Delta H_{pp}(300K) = 0.5$ and 5.0 G is observed. The relatively high ratio of hydrogen:carbon (H/C) near T_{HT}^c is believed to influence the ESR results through an unresolved hyperfine interaction. Curie-Weiss temperatures are found from measurements of $[I_{pp}(\Delta H_{pp})^2]^{-1}$, where I_{pp} is the peak-to-peak lineheight, yielding results that are in agreement with static susceptibility, $\chi(T)$, measurements. At low T_{HT}, PPP-based materials exhibit a large amount of disorder and this is evidenced by the high density of localized spins, N_C, which is obtained from a Curie-Weiss fit to $\chi(T)$, assuming a spin quantum number of $S = \frac{1}{2}$. A model explaining the microstructure and high electrochemical doping capacity of PPP samples heat-treated to 700° C can be related to Li-ion battery performance.

INTRODUCTION

In the search for the best carbon host material for use in so-called Li ion batteries, many types of carbons have been investigated and characterized.[1] Carbons heat-treated to temperatures above $\sim 2300°$C assume structures closely resembling graphite, especially for the case of soft carbons, while low-T_{HT} ($\leq 1000°$C) carbons tend to be much more disordered. The well-ordered nature of the high-T_{HT} carbons have the advantage that well-defined π bands can accommodate electronic charge in such a way so as to allow Li ions in $Li_x C_6$ systems to discharge at a voltage near that of lithium metal. The disadvantage of these systems is the limit on x to values close to 1, which is the theoretical limit of graphite intercalation compounds under ambient conditions. Because of the corresponding low discharge capacities of the well-ordered systems (≤ 372 mAh/g), a growing interest has shifted to low-T_{HT} carbon systems, which, because of the small size of their graphene-like clusters, are capable of accommodating larger amounts of lithium than their high-T_{HT} counterparts. In Fig. 1 we show the discharge capacities found in a few high-capacity/low-T_{HT} carbon systems, namely PPP[2], ribbon-like pitch-based carbon films (RCF)[3], phenol-formaldehyde

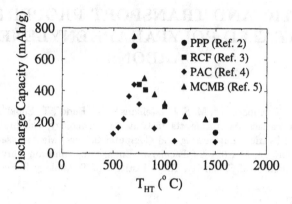

Figure 1: Discharge capacities of various carbon-based electrode materials.

resin-based "polyacenic" (PAC) carbons[4], and coal-tar pitch-based mesocarbon microbeads (MCMB)[5]. It is clear that all of the carbons shown in Fig. 1 have a maximum discharge capacity near T_{HT}=700°C. Although not shown, the discharge capacities of PPP-based carbons with T_{HT} < 700°C were all below 500 mAh/g. Due to the large amount of disorder in low-T_{HT} carbons, lithium ions are believed to bind to both sides of single graphene layers[6], resulting in a higher lithium to carbon ratio as compared to first-stage GICs. Furthermore, an increase in the amount of edge sites in low-T_{HT} disordered carbons could also increase the amount of lithium doping possible. However, the precise doping mechanism and associated Li binding sites are presently not well understood, primarily because the structure of low-T_{HT} carbons is highly disordered and quite complex. Thus, sophisticated methods for structural characterization have been employed in order to develop a clearer picture of the microstructure that allows for maximum lithium uptake (i.e., the carbon structures formed near the electrochemically critical temperature of T^c_{HT}=700°C).

In this paper we present the results of DC conductivity, magnetic susceptibility and ESR measurements covering a wide temperature range. DC conductivity measurements are used to evaluate the bulk electronic behavior of the samples, while the local electronic environment is probed through magnetic susceptibility measurements. Relevant parameters can be extracted from the data using magnetic and transport theories that are appropriate for disordered polymer/disordered carbon systems. A structural model that is consistent with the experimental observations is then suggested.

EXPERIMENTAL DETAILS

The sample preparation of heat-treated, pressed Kovacic PPP samples has been described in detail previously[7]. Electrical resistivity measurements were made using a four-point electrical resistance technique. ESR spectra were measured using a conventional X-band (~9.5 GHz) spectrometer with a rectangular TE_{102} microwave cavity within an He cryostat. Susceptibility measurements were made using a Quantum Design SQUID magnetometer over the temperature range 2.5–310 K in magnetic fields up to 5 Tesla.

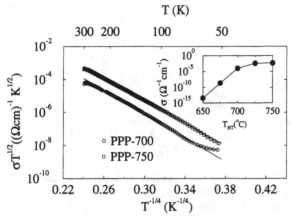

Figure 2: Temperature-dependent electrical conductivity of samples PPP-700 and PPP-750, plotted as $\sqrt{T}\sigma(T)$ vs. $T^{-1/4}$ consistent with a 3D VRH transport mechanism. The inset shows the room temperature resistivity vs. T_{HT} of samples heat-treated near 700°C.

RESULTS/DISCUSSION

Figure 2 shows the results of conductivity measurements for PPP-700 and PPP-750, plotted as $(T)^{1/2}\ln\sigma(T)$ vs. $T^{-1/4}$. Although measurements on sample PPP-650 were attempted, the high resistance of this sample (≈ 10 GΩ) prevented accurate measurements below about 250K. For samples PPP-700 and PPP-750, however, we found the temperature dependence to follow a 3D Variable Range Hopping (VRH) behavior, where the conductivity has the form[8]

$$\sigma(T) = \sigma_0(T)\exp\left[-\left(\frac{T_0}{T}\right)^{1/4}\right] \tag{1}$$

where $\sigma_0(T) = e^2 R^2 \nu_0 N(E_F)$ and $T_0 = \lambda\gamma^3/k_B N(E_F)$. Here R is the average hopping distance, ν_0 is the hopping rate prefactor, $N(E_F)$ is the density of states at the Fermi level, γ is the inverse decay length, and λ is a dimensionless constant equal to about 20[9]. For ν_0 we use $\nu_0 = \nu_{ph}$, where ν_{ph} is the frequency of strongly coupled phonons ($\approx 10^{14}$ Hz). From the plot in Fig. 2, we extract values for $N(E_F)$ and γ which are given in Table I. An increase in $N(E_F)$ is found as we go from PPP-700 to PPP-750, indicating an increase in the number of localized states near the Fermi level.

The inset of Fig. 2 shows $\sigma(300K)$ vs. T_{HT} for samples heat-treated near T_{HT}^c. As shown in the figure, $\sigma(300K)$ changes by several orders of magnitude as T_{HT} is increased from 650°C to 750°C. Essentially, PPP-650 appears to be completely insulating ($\rho \sim 10^{10}\Omega - $ cm) while PPP-700 and PPP-750 are moderately conducting. If we consider a granular conductor-insulator model, PPP-700 is considered to be roughly at the heat-treatment percolation threshold.

Magnetic susceptibility measurements for samples heat treated in the range 650°C $\leq$ $T_{HT} \leq$ 750°C are well fit by a Curie-Weiss law, $\chi = C/(T + \theta) + \chi_0$ over the measurement temperature range $2.5 < T < 20$K where the constant offset, χ_0, represents the temperature-independent contributions, while C and θ are the Curie constant and Curie-Weiss temperature, respectively. Figure 3 shows a plot of $(\chi - \chi_0)^{-1}$ vs. T which was used to accurately

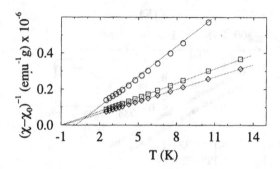

Figure 3: Inverse Curie-Weiss magnetic susceptibility vs. temperature for samples PPP-650 (○), PPP-700 (□) and PPP-750 (◇). Dotted lines extend to $(\chi - \chi_0)^{-1} \to 0$ indicating finite Curie-Weiss temperatures for all samples.

determine θ, and, from the slope of χ vs. $1/(T+\theta)$, the Curie constant C is obtained. From the measured values of C we extract the localized spin concentration (N_{loc}) for each PPP sample by assuming a spin quantum number of $S = 1/2$ and using the relation

$$C = \frac{N_{\text{loc}}S(S+1)g^2\mu_B^2}{3k_B}. \tag{2}$$

We find from the measured χ (see Table I) that the spin concentration increases as we go from PPP-650 to PPP-750. This can be explained by the thermal decomposition of PPP molecules at they progressively lose more hydrogen. Also, as we go from a disordered polymer below T_{HT}^c, to a disordered carbon above T_{HT}^c we change from a weakly ferromagnetic system (J>0) to a weakly antiferromagnetic one (J<0) where $3k_B\theta = 2JS(S+1)$.

The room temperature ESR spectra for samples PPP-650, PPP-700 and PPP-750 have been previously published[10]. Briefly, the PPP-650 ESR spectrum shows a broad peak (~5 G), that of PPP-750 shows a narrow peak (~ 1 G), while that of PPP-700 exhibits a superposition of both broad and narrow peaks (~ 4 and 0.5 G, respectively). In order to elucidate the broadening mechanism for the two types of lineshapes, we plot in Fig. 4 the peak-to-peak ESR linewidth, ΔH_{pp}, versus T for PPP both above (PPP-750) and below (PPP-650) T_{HT}^c. In all our samples, the lineshapes were Lorentzian while the isotropic g-factor was found to be T-independent (≈ 2.0027).

The temperature dependence of ΔH_{pp} of the hydrogen-rich PPP-650 sample is constant, consistent with an unresolved hyperfine broadened ESR trace.[11] However, since the lineshape was Lorentzian, we suspect that a weak motional narrowing mechanism is operative, which could explain the power saturation behavior previously reported.[10] The increase in ΔH_{pp} at low temperatures could perhaps be the result of the quenching of some slow motional narrowing process. Nonetheless, a dominant hyperfine interaction is consistent with the results of Raman studies which show the presence of coherent polymer segments existing up to the critical temperature T_{HT}^c[7]. The temperature dependence of the narrower peak of the PPP-750 ESR spectrum showed a quite different behavior which is consistent with a spin-lattice dominated relaxation process involving delocalized states. We see that at low temperatures ΔH_{pp} for PPP-750 approaches a constant value of ~ 1.2 G but rises almost

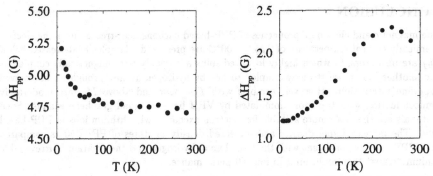

Figure 4: Temperature dependence of peak-to-peak ESR linewidths (ΔH_{pp}) of PPP-650 (left) and PPP-750 (right).

Table 1: Summary of extracted magnetic and electronic properties.

	PPP-650	PPP-700	PPP-750
$N(E_F)$ (eV^{-1} cm^{-3})	—	2.4×10^{24}	1.8×10^{25}
γ (Å^{-1})	—	1.2	0.67
θ (K)	0.12	-0.95	-0.86
C (emu K^{-1}g^{-1})	17.8	39.0	44.3
χ_0 (emu g^{-1})	1.16	1.41	1.28
$T_2/T_1(300K)^*$	0.003	0.008	0.020
N_{loc} ($\times10^{19}g^{-1}$)	3.2	4.9	5.7
H/C (molar)	0.56	0.35	0.19

* Taken from Ref. 7.

linearly with T up to ~200K. Korringa showed that the spin-lattice relaxation time $T_1 \sim 1/T$ for the case of metals, which would yield a ΔH_{pp} linear T-dependence for a T_1 dominated process.[12] Although the spin-spin relaxation time T_2 governs ESR line broadening in samples heat treated below 700°C, room temperature T_1^{-1} values appear largest in PPP-750 compared to the other heat-treated samples (see Table I) so that linewidth contributions from T_1 become noticeable.

Putting these results together, we find that the critical heat-treatment temperature, $T_{HT}^c = 700°C$, lies roughly at the electronic transport percolation threshold. This may be related to the electrochemical characteristics of low-T_{HT} carbons using the following argument. For samples below the percolation threshold, a percolative conduction path is not present, thereby preventing electrical "access" to some parts of the sample during Li intercalation. Therefore, percolation may be thought of as a minimum requirement for the maximum uptake of dopant species. At the same time, heat-treatment above $T_{HT}^c = 700°C$ causes much of the hydrogen to be expelled, leaving behind many dangling bonds which can irreversibly bind Li ions in deep traps. The creation of micropores above T_{HT}^c could also relate to a lower lithium uptake, since micropores allow electrolyte molecules to enter such pores and permanently block access to would-be Li binding sites[1].

CONCLUSION

The magnetic and electronic properties of PPP-based carbons heat-treated near the electrochemically critical temperature of $T_{HT}^c = 700°C$ are presented. Samples heat-treated below T_{HT}^c are insulators, in which highly localized spins are weakly ferromagnetically coupled to one another, but more strongly coupled to nearby hydrogen atoms through an unresolved hyperfine interaction. However, samples with T_{HT} near and above T_{HT}^c are moderately semiconducting with transport dominated by VRH between small graphene islands. Such materials are therefore more suitable for electron exchange with lithium ions in PPP-based cells. The increased spin density and increased density of states of PPP-750 as compared with PPP-700 are consistent with the large loss of hydrogen and the creation of irreversible lithium "traps" that could limit Li ion cell performance.

Acknowledgments

The MIT authors gratefully acknowledge NSF grant DMR-95-10093, and support (MJM) from an NSF Fellowship and an NSF travel grant (NSF INT-93-15165).

References

[1] W. Xing, J. S. Xue, T. Zheng, A. Gibaud, and J. R. Dahn, J. Electrochem. Soc. **143**, 3482 (1996).

[2] M. Endo, Y. Nishimura, T. Takahashi, K. Takeuchi, and M. S. Dresselhaus, J. Phys. Chem. Solids **57**, 725 (1996).

[3] Y. Matsumura, S. Wang, T. Kasuh, and T. Maeda, Synth. Met. **71**, 1755 (1995).

[4] B. Huang, Y. Huang, Z. Wang, L. Chen, R. Xue, and F. Wang, J. Power Sources **58**, 231 (1996).

[5] A. Mabuchi, K. Tokumitsu, H. Fujimoto, and T. Kasuh, J. Electrochem. Soc. **142**, 1041 (1995).

[6] Y. Liu, S. Xue, T. Zheng, and J. R. Dahn, Carbon **34**, 193 (1996).

[7] M. J. Matthews, M. S. Dresselhaus, M. Endo, Y. Sasabe, T. Takahashi, and K. Takeuchi, J. Mat. Res. **11**, 3099 (1996).

[8] N. F. Mott and E. Davis, in *Electronic processes in noncrystalline materials*, (Clarendon Press, Oxford, 1979).

[9] B. I. Shklovskii and A. L. Efros. In *Electronic Properties of Doped Semiconductors*, Springer-Verlag, Berlin, 1984. Springer series in solid state sciences; vol 45.

[10] M. J. Matthews, M. S. Dresselhaus, N. Kobayashi, T. Enoki, M. Endo, and T. Takahashi, Appl. Phys. Lett. **69**, 2042 (1996).

[11] N. R. Lerner, J. Polymer Sci. **12**, 2477 (1974).

[12] J. Korringa, Physica **16**, 601 (1950).

Electrochemical and Spectroscopic Evaluation of Lithium Intercalation in Tailored Polymethacrylonitrile Carbons

Kevin R. Zavadil*, Ronald A. Guidotti*, and William R. Even**
*Sandia National Laboratories, P.O. Box 5800, Albuquerque, NM, 87185
** Sandia National Laboratories, P.O. Box 969,Livermore, CA, 94550

Abstract

Disordered polymethacrylonitrile (PMAN) carbon monoliths have been studied as potential tailored electrodes for lithium ion batteries. A combination of electrochemical and surface spectroscopic probes have been used to investigate irreversible loss mechanisms. Voltammetric measurements show that Li intercalates readily into the carbon at potentials 1V positive of the reversible Li potential. The coulometric efficiency rises rapidly from 50% for the first potential cycle to greater than 85% for the third cycle, indicating that solvent decomposition is a self-limiting process. Surface film composition and thickness, as measured by x-ray photoelectron spectroscopy (XPS) and secondary ion mass spectrometry (SIMS), does not vary substantially when compared to more ordered carbon surfaces. Li^+ profiles are particularly useful in discriminating between the bound states of Li at the surface of solution permeable PMAN carbons.

Introduction

A primary goal in the development of Li ion batteries is to minimize the irreversible losses in the intercalating host matrix to maximize energy density. Mechanisms of irreversible loss include solvent decomposition and subsequent solid electrolyte interphase (SEI) formation which contains bound ionic lithium, surface deposition of lithium followed by solvent reactions and entrapment with the carbon matrix. Focus is currently being placed on synthetically derived carbon materials as potential anodes (1-3). The goal is to either take advantage of intrinsic properties in a given carbon or to try to tailor the properties by altering the synthesis process. The overall degree of disorder, as well as other microstructure features like open porosity, can be altered with the potential for achieving capacities in excess of that found for ordered graphite. Polymer derived carbonized polymethacrylonitrile represents one material that appears well suited for use as an optimized, tailored carbon (4).

Successful development of PMAN electrodes requires an understanding of how irreversible loss processes correlate with carbon surface activity and structure. Voltammetry, x-ray photoelectron spectroscopy (XPS) and secondary ion mass spectrometry (SIMS) have been used to study the process of solvent decomposition, Li intercalation and Li deposition on monolithic PMAN electrodes. A contiguous method has been developed to couple solution and vacuum environments to minimize the contamination and secondary chemistry that is intrinsic to ex situ analysis methods. The basal plane of highly ordered pyrolytic graphite (HOPG) and glassy carbon (GC) have been included as examples of more highly ordered, non-intercalating, non-porous carbons for comparison purposes.

Experimental

The carbon electrodes were electrochemically treated in a contiguous cell attached directly to the ultrahigh vacuum system. This cell, shown in cross-section in Figure 1, is mounted in a separate vacuum cell that is pressurized with UHP Ar during the electrochemical experiments. Carbon electrodes are transported on an electrically isolated manipulator and placed face down, immersed in the electrolyte meniscus that is formed at the top of the TFE cell. An approximate area of 0.5 cm^2 is immersed using this approach. Lithium is used as both reference and counter electrodes. The reference electrode is an annular ring that is placed on line with the top of the cell to minimize the working-reference spacing (< 1mm). The counter is a planar disk that is parallel to the working electrode surface, to ensure a uniform primary current density, and separated by 1.5 cm. The cell is alternately filled with a 1M LiPF$_6$ ethylene carbonate, dimethyl carbonate (1:1 vol., 11 mS/cm, Merck) electrolyte or pure DMC (98%, Aldrich) as a rinsing reagent. The electrolyte and rinse solutions are stored in gas tight syringes and are supplied to the cell by a TFE manifold and metering valves. The entire manifold is kept under vacuum until solution is required. A PAR 273 potentiostat was used for electrochemical control and measurement. The cell is mounted in a UHV compatible chamber added to the end of a Vacuum Generators ESCALAB 5 spectrometer. XPS measurements were made with a 300 W Mg(Kα) source using a pass energy of 20 eV. SIMS measurements were made with a Balzer QMA400 quadrupole with a three lens optic system for energy separation. Secondary ion emission was stimulated using 4 keV Xe.

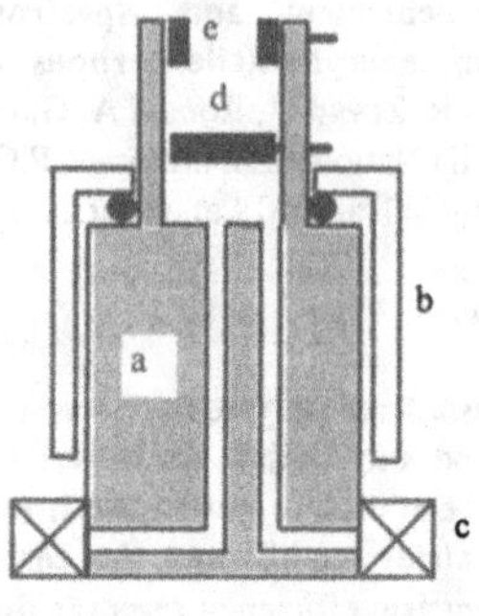

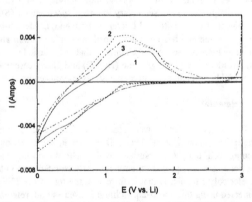

Figure 1: Cross-section View of Contiguous Electrochemical Cell - a) TFE cell body, b) polished stainless steel vacuum sealing surface, c) TFE metering valves, d) Li counter electrode, and e) Li reference electrode

The materials included in this study were glassy carbon (GC, Alpha AESAR, type II), a ZYB graphite crystal oriented to the basal plane (HOPG, Advanced Ceramics) and carbonized methacrylonitrile/divinylbenze co-polymers (PMAN, Sandia). Glassy carbon disks were spark cut to size from either sheet or rod stock, polished and cleaned by sonication in o-xylene, followed by soxhlet extraction in isopropanol and a final o-xylene rinse. The graphite crystal was cut to size using a razor blade. PMAN monoliths were pressed from 3:1 (vol.) methacrylonitrile/divinylbenzene powder mixtures and carbonized at 1100°C in an Ar atmosphere (4). The resulting monoliths are permeable to the electrolyte with geometric densities of 0.54 g/cm^3 and strut densities of 1.86 g/cm^3. The average interlayer spacing, d[002], was found to be 3.74Å compared to values of 3.35 and 3.48 Å for HOPG and GC, respectively. The glassy carbon and graphite samples were also annealed in a 5% H2/Ar atmosphere at 800°C for a period of 2 hours. After processing, samples were transferred under Ar to a glove box and stored under Ar until use to minimize contamination.

Results and Discussion

The initial carbon samples are relatively contaminant free. We find that all three types of carbon show varying degrees of surface oxygen. HOPG surfaces were the least contaminated surfaces yielding a value of 1.4 at% while PMAN monoliths gave values of 2.2 at.%. The glassy carbon surfaces show the greatest oxygen concentrations on the order of 9.8 at.%. CHNO analysis of these carbons produced O values at less than 0.8 at.%, indicating that atmospheric exposure leads to measurable O adsorption. The PMAN surfaces show residual traces of nitrogen in the 1.0 at.% range, consistent with bulk chemical analysis results that yielded a value of 1.7 at.%.

Figure 2: Cyclic Voltammetric Response of a PMAN Electrode in 1M LiPF6 EC/DMC at 1 mV/s - cycle numbers are designated

The similarity of these values argues that the surface is not appreciably enriched with N. SIMS measurements show that all of the carbons contain some quantity of surface oxygen with the glassy carbon yielding the highest levels. These three surfaces can also be distinguished by the width of their respective C(1s) transitions with the HOPG yielding the most narrow value of 1.04 eV while PMAN yields a value of 1.45 and GC of 1.24 eV.

The PMAN carbon shows clear evidence of lithium intercalation at potentials above 0V. Figure 2 shows a series of voltammetric scans from initial open circuit (ca. 3.1 V) to 0V at 1 mV/s. The first cycle voltammogram shows a cathodic current threshold at approximately 1.5 V, followed by a rapidly increasing current as 0V is approached. The return first cycle shows significant recovery of charge with de-lithiation (efficiency = 49%). De-intercalation is characterized by a broad band of current with several inflection points, indicating that the process of Li release is occurring in stages. Subsequent cycles show a decrease in the potential of cathodic current threshold to 1.2 V and an overall decrease in cathodic current levels to 0V. The higher currents of cycle 2 are an artifact of a meniscus fluctuation. The shift in threshold and lower currents indicate that intercalation and solvent decomposition are occurring simultaneously. The efficiency rapidly climbs during the first several cycles reaching a value of 85% after three cycles. These electrochemical results are very similar to what we observe for PMAN fully immersed in the same electrolyte using an alternate electrochemical cell design. Comparison of the meniscus and fully immersed techniques indicate that capacities of 29 mAhr/g are achieved in these 1 mV/s cycles and that only 26 % of the total monolith mass, or 9.4 mg, is being accessed using this meniscus method. Full immersion of the PMAN electrode also appears to produce a more rapid rise in efficiency with cycle number (> 95% for second cycle).

Table 1: Variation in Surface Composition for Electrochemically Pretreated Carbon Electrodes

Concentration (at. %)

Sample	Li	F	O	C	P
HOPG cycled to 0V	38.3	47.2	5.0	7.5	2.0
GC cycled to 0V	52.8	33.2	8.4	2.1	3.5
GC cycled and held at -0.3V	54.5	33.9	5.6	3.8	2.3
PMAN cycled to 0V	43.7	28.2	6.9	19.4	1.8
PMAN cycled and held at 0V	43.2	26.0	13.1	15.3	1.9

The structure of the solid electrolyte interphase (SEI) layer appears to be comprised of two components. XPS measurements indicate that the outer layer of the SEI is predominantly comprised of Li and F, as shown in the data of Table 1. Considerably lower levels of both C and O are measured along with trace levels of P. The position of the Li(1s) is subject to considerable surface potential induced shifting. We find a consistent energy difference of 26.1 eV between the Li(1s) and the F(2s) indicating a similar LiF outer layer composition for the films formed on all carbon types explored. The position of the F(1s) is consistent with a surface LiF species. This predominance of surface F indicates that residual moisture is present and may be responsible for HF-based modification of the SEI layer (6). The outer surface also shows a broad, asymmetric O(1s) with a

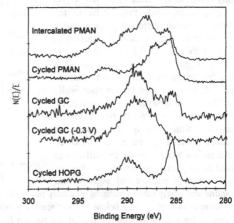

Figure 3: Variation in C(1s) Lineshape with Carbon Type and Electrochemical Pretreatment

halfwidth typically of 3 eV indicative of multiple states of surface O. C(1s) spectra show a variety of forms of carbon. Representative spectra are shown in Figure 3. We find evidence for aliphatic carbon in the range of 285 to 286 eV, carbon bound to oxygen in the range of 286 to 289 eV, and carbonate and alkylcarbonate at 290 eV (5). The very high binding energy species observed on PMAN at 293 eV most likely results from a CF complex and is not always observed. All of the carbons are seen to possess some level of near-surface carbonate. HOPG shows the lowest overall degree of product carbon build-up with its substrate transition (full width at half maximum - 1.05 eV) still easily detectable. The fact that a product layer is observed in SIMS analysis suggests that the surface film on HOPG may be discontinuous, resulting in electrolyte decomposition and product nucleation and growth at select sites on the surface.

SIMS data are consistent with the XPS results. Representative positive and negative ion spectra for PMAN cycled to 0V are shown in Figure 4. We find a series of positive ions separated by 26 amu starting at 33 amu and extending up to 137 amu. We attribute these ions to a series of $Li_{x+1}F_x$ species. Negative ion spectra show the predominance of F^-. Additional positive ions are detected at 37, 47 and 56 amu in substantially lower yields than the $Li_{x+1}F_x$ ions. We attribute these ions as being due to $LiCH_2O^+$, $CH_3O_2^+$ or $C_2H_4F^+$, and $C_3H_4O^+$, respectively. Additional negative ions are detected at 35, 38, 45 and 71 amu. Possible assignments for these ions are $LiCO^-$, F_2^-, CHO_2^- and $C_4H_7O^-$, respectively. These observed $LiCH_nO^{+/-}$ ions may result from carbonate or they may be derived from alkylcarbonates. We do not detect CO_3^{2-} directly. We find that an Li_2CO_3 standard (powder pressed into In foil) yields predominantly Li^+, O^-, LiO^+, Li_2O^+, $LiCO^-$ and $LiCH_2O^{+/-}$ ions, indicating that the inert gases are not effective at generating appreciable carbonate ion yields. We do observe higher order (C_2 or greater) alkoxy ions on PMAN, but do not detect analogs complexed with Li. We also observe low intensity emission consistent with C_2H_n, C_3H_n and C_4H_n ions, that may indicate the possibility for some degree of polymerization of ethylene carbonate. In addition, emission of a series of PO_n^- ions (63 and 79 amu) is detected in negative ion spectra. These same ions are detected on a reference PMAN surface that is kept in the vacuum system specifically for tracking the adsorption of volatile species. P, along with F, appear to be ubiquitous contaminants within our system. Large scale differences in SIMS spectra are not observed between carbon types, despite the differences observed in the photoemission spectra. We see the same series of positive and negative ions

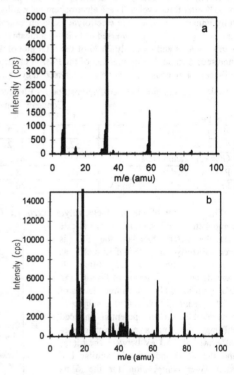

Figure 4: Positive (a) and Negative Ion (b) Spectra of PMAN Cycled to 0V

for graphite and glassy carbon surfaces with only minor differences in abundance.

The composition of the SEI below this surface layer can be evaluated by ion sputtering. SIMS has been used to track both the composition of the surface film as well as the Li profile into the carbon bulk. The spectra displayed in Figure 5 show the relative secondary ion intensities as a function of

542

sputtering time for glassy carbon. The glassy carbon possesses a clear interface between substrate and surface film because of it's lack of porosity. The positive ion profile shows an immediate decay for $CH_3O_2^+$ ions, an initial increase, maximum and eventual decrease for Li_2F^+ and $LiCHO^+$, and a similar but extended response for Li. The negative ion profile shows that P is rapidly removed from the surface, followed by a maximum signal for both oxygenated and fluorinated forms of carbon, and the eventual rise of the substrate signals. This data argues that F^-, in various forms, is distributed throughout the film along with $LiCO^-$ ions that may be derived from either carbonate or alkylcarbonates. The mismatch in the $LiCO^-$ and $LiCH_2O^+$ profiles suggests two different molecular sources of these ions or a significant change in local chemical environment resulting in varying ion production probability. The cause of the extension in time of the Li signal is unknown. Sequential 0 to 200 amu spectra were acquired at increments of 200 seconds for the first 1000 seconds of sputtering. No unique positive or negative ions were detected. Possible interpretations of this data could include either an

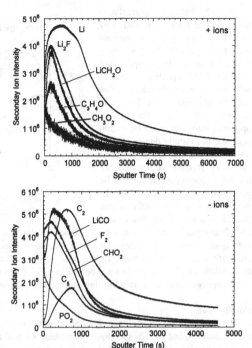

Figure 5: SIMS Profiles for Glassy Carbon Cycled to 0V - intensities have been adjusted for display purposes

aliphatic layer adjacent to the glassy carbon substrate or preferential ion mixing of Li with the substrate and a resulting ion yield change. Support for the latter explanation can be found in a more narrow profile for Li^- when compared to Li^+. We have calibrated our sputter rate using polydimethylsiloxane films ranging in thickness from 20 nm to 100 nm, spin cast onto Si wafers. An average sputter rate of 0.08 nm/s was calculated for this thickness regime. Based on the inflection point for the product ion curves in these sputter profiles, we estimate a mean film thickness of approximately 70 nm (±10%) for glassy carbon.

The profiles generated on PMAN substrates are more difficult to interpret due to the permeable nature of this material and the lack of a reference plane to define the substrate/surface film interface. However, the same general trends are observed. We find that Li is preserved well into the bulk of the film, after the decay of surface fluoride and alkylcarbonate derived species. Similar mean film thicknesses of 70 nm are measured using ion emission from the film. The primary difference in the profiles is the continued presence of surface product long after the appearance of ions derived from the substrate. We find no significant change in the ion distribution, indicating that what forms on the outer surface of the monolith is very similar to what forms on the internal surface.

The SIMS profiles can be used to detect Li intercalation into PMAN monolith electrodes. Figure 6 shows a series of Li^+ distribution curves for both PMAN and glassy carbon subjected to different electrochemical pretreatment. Comparison of these profiles shows the retention of a relatively high Li signal for PMAN due to the porous nature of the monolith, as previously discussed. Comparison of Li profiles from PMAN cycled from open circuit to 0V and back to open circuit versus PMAN cycled to 0V and emersed, show a greater relative signal, indicating Li has intercalated into the monolith. We can distinguish between intercalation and substrate surface deposition by comparing the previous profiles with that generated from a glassy carbon surface cycled to -0.3 V and emersed. This profile shows a discrete region of additional Li that extends the maximum ion yield portion of the curve by approximately an additional 500 seconds, but rapidly decays. Based on previous metal and semi-metal sputtering studies, we estimate the mean Li underlayer thickness to be approximately 34 nm. The Li profiles for PMAN show that where the Li intensity from the cycled

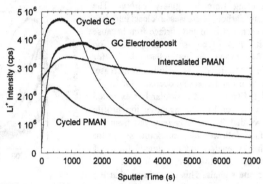

Figure 6: Variation in Li$^+$ Secondary Ion Sputter Profiles with Carbon Type and Electrochemical Treatment

PMAN eventually stabilizes at a nearly constant value, the intercalated PMAN signal gradually decays with sputter time. This behavior would be expected for the cell design used because the region of the monolith immediately adjacent to the bulk solution should be in contact with the highest local concentration of Li and would be expected to yield a higher extent of intercalation, provided solution diffusion constraints are as rate limiting as intercalation kinetics.

Conclusions

The response of monolithic PMAN electrodes has been studied using a combination of electrochemical and surface spectroscopic probes. We find that Li intercalation occurs at 1.2 V positive of the reversible Li potential. Solvent decomposition occurs with initial intercalation below this threshold potential. The initial efficiency of the monolith can be as low as 50% and rapidly rises to 85% with the third full cycle (higher values are seen for full electrode immersion). XPS analysis demonstrates that the near-surface of the SEI that forms is predominately LiF, indicating moisture contamination in the solution phase treatment of these electrodes. Evidence for hydrocarbon, carbonate and alkylcarbonate formation in the surface film are found with combined XPS and SIMS analysis. The chemical identity of the film is not strongly dependent on the carbon type. We find that these films are approximately 60 to 80 nm in thickness and show some compositional variation. Li distribution profiles generated by dynamic SIMS can be used to discriminate between surface film formation, Li surface deposition and intercalation, even for the electrolyte permeable PMAN carbons.

Acknowledgments

Sandia is a multiprogram laboratory operated by Sandia Corporation, a Lockheed Martin Company, for the United States Department of Energy under Contract No. DE-AC04-94AL85000.

References

1. S. Wang, Y. Zhang, L. Yang and Q Liu, Solid State Ionics, **86-88**, 919 (1996).
2. A. Naji, J. Ghanbaja, B. Humbert, P. Willmann and D. Billaud, J. Power Sources, **63**, 33 (1996).
3. Y. Matsumura, S. Wang and J. Mondori, J. Electrochem. Soc., **142** (9), 2914 (1995).
4. W. Even and D. Gregory, MRS Bulletin, **19**, 29 (1994).
5. D. Aurbach, I. Weissman, A. Schechter and H. Cohen, Langmuir, **12**, 3991 (1996).

NEW 7Li-NMR EVIDENCE FOR LITHIUM INSERTION IN DISORDERED CARBON

S. WANG *, H. MATSUI *, Y. MATSUMURA* and T. YAMABE**
*Fundamental Research Laboratories, Osaka Gas Co., Ltd., Osaka, Japan,
**Division of Molecular Engineering, Faculty of Engineering, Kyoto University, Kyoto, Japan

ABSTRACT

In fully-charged carbon materials, we tried to determine the nature of the interaction between Li species and carbon using variable temperature measurements of the Li Knight shift. Only one interaction between the Li species and graphite carbon was observed from room temperature to -100℃. However, while only one average interaction between the Li species and the disordered carbon was observed at room temperature, three kinds of interactions were observed at low temperature. These results provide direct evidence for the model which explains why the discharge capacity of the carbon electrode with a disordered structure as an anode can exceed the theoretical capacity of a graphite anode in lithium ion rechargeable batteries.

INTRODUCTION

Lithium ion rechargeable batteries(LIB) using carbon as an anode material have attracted a great deal of worldwide attention because of their high cyclability and high energy density. A key issue of the lithium ion rechargeable batteries is its high discharge capacity. Recently, it was reported that when disordered carbon materials were used as an anode, these materials could store much more lithium with discharge capacities surpassing the theoretical capacity of the graphite anode[1-3]. These results showed that the disordered carbon materials may store Li via a mechanism completely different from that in graphite intercalated compounds since it has a special structure (small crystallite size and random crystallite orientation). In order to explain the high discharge capacity, several models have been proposed, such as a model of a single layer[1], a model having two Li layers between the graphitic layers[2] and a model of Li_2[3]. Recently, although the Li-NMR measurements at various temperatures have carried out [4], the interactions between Li species and carbon in fully-charged disordered carbon are not clear yet. Here, we report that the Li-NMR spectrum of a fully-charged disordered carbon clearly shows three different signals at -100 ℃. These results support a model with three kinds of interactions between Li species and carbon in a fully-charged disordered carbon.

EXPERIMENT

The disordered carbon material employed in the present study was prepared from isotropic coal-tar pitch heat-treated at 1100 ℃ under a nitrogen atmosphere. The graphite carbon material was prepared from anisotropic pitch heat-treated at 2800 ℃ under a nitrogen atmosphere. The disordered carbon material has a small crystallite size (Lc = 11 Å) and a large space distance (d_{002} = 3.7 Å).

For electrochemical Li doping and undoping, a mixture of disordered carbon material and polymer binder was pressed into a sheet-shaped electrode. The size of the sheet-shaped electrode was 2.0 cm x 10.0 cm with a thickness of 110 μm. The electrode was dried at 180 °C under vacuum for 4h. It was used as the anode and a Li sheet and a Li chip were used as the counter and reference electrodes, respectively. A solution of 1M propylene carbonate/LiClO4 was used as the electrolyte. A solution of 1M EC / DEC / LiClO4 was used as the electrolyte for the graphite carbon.

In order to obtain a sample of the discharged carbon electrode, the cell is charged from its rest voltage (usually in the range of 2.5 - 3.0 V vs. Li / Li$^+$) to 1mV vs Li / Li$^+$ at a constant current, and kept at 1mV until a steady-state current was reached (10^{-5} - 10^{-6} A), and then discharged to 2.0 V (Li / Li$^+$) under a constant current and kept at 2.0 V(Li / Li$^+$) until a steady-state current was reached (10^{-5} - 10^{-6} A) by employing a charge/discharge unit (Hokuto HJ-201B). Charge and discharge current densities were both 1.0 mA/cm^2. The charged and discharged carbon electrodes were rinsed with propylene carbonate to remove electrolyte, dried in a vacuum and then put into a solid NMR sample tube under an argon atmosphere.

The NMR measurements were made using a Bruker ASX 200 spectrometer with a multinuclear MAS solid-state NMR probe. ^{7}Li solid-state NMR spectra were recorded at a frequency of 77.7 MHz. The actual measurements were carried out by adopting a magic angle spinning of 5000 rpm. The ^{7}Li-NMR shift was calibrated in ppm relative to 1M LiCl (-1.19 ppm) as the external reference standard.

RESULTS AND DISCUSSION

Knight shift at room temperature

The ^{7}Li-NMR spectra at room temperature for the fully-charged disordered carbon and discharged disordered carbon are shown in Fig.1(a) and Fig.1(b), respectively. From the ^{7}Li-NMR spectrum shown in Fig.1(a), one can see that there are two types of broad NMR lines. One broad resonance is shifted to low field at 81 ppm with respect to that of the LiCl reference. Another resonance shift appears at -1.40 ppm. The former is plausibly explained in terms of the Fermi contact interaction between a Li nucleus and conduction electrons, that is, the Knight shift of the Li nucleus. The latter is observed at the almost same position with LiCl (-1.19 ppm) reference without the Knight shift. These results indicate that two different electronic states surrounding the Li nucleus exist in the fully-charged disordered carbon. In comparison with Fig.1 (a) and Fig.1(b), we suggest that the broad resonance shift at 81 ppm

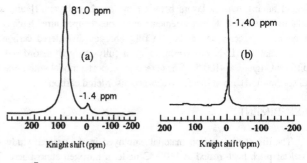

Fig. 1 ^{7}Li-NMR spectra of (a) the fully-charged disordered carbon and (b) discharged disordered carbon

corresponds to reversible capacity, because it disappears after the discharge process, and the resonance shift at -1.40 ppm corresponds to the irreversible capacity loss because it is still observed after the discharge process.

Our results of the Fourier Transform Infrared and X-ray Photoelectron Spectroscopy have confirmed that the irreversible capacity loss arose from Li_2CO_3 produced by solvent decomposition on the surface of the carbon electrode and Li species produced by some side reactions in the bulk of the carbon electrode during the first cycle [5]. In order to identify the nature of the residual Li species which cause irreversible capacity loss, the discharged carbon sample was kept in air for 5 minutes, then its NMR spectrum was measured. As shown in Fig. 2, a new band with a large line width at -0.41 ppm and some spinning side bands are observed. These results reveal that part of the residual Li species is unstable in air. After it is exposed in air, some chemical reactions occur. Therefore, it is considered that the residual Li species which causes irreversible capacity loss not only contributes to Li_2CO_3 which is stable in air, but also to some Li compounds which are unstable in air. This speculation is in agreement with the ESCA result. The Li_{1s} binding energy of Li species in the bulk of the discharged carbon is lower by 0.5 eV than that of Li_2CO_3 on the surface of the discharged carbon [5].

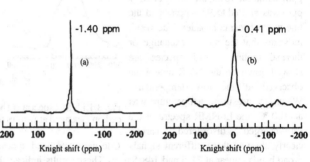

Fig. 2 ^{7}Li-NMR spectra of (a) discharged disordered carbon and (b) discharged disordered carbon in air for 5 minuts

It is well known that both methyllithium and phenyllithiun with a Li-C bond are unstable in air and can be easily decomposed by water or oxidized by oxygen. Their resonance shifts are at -1.0 ppm and -1.3 ppm, respectively[6]. The properties of the Li-C valence bonds are different from that of the Li valence bond which is intercalated in graphite carbon. Therefore, it can be considered that if similar compounds with Li-C exist in the fully-charged carbon electrode, they can not be discharged from the carbon electrode. Therefore, the resonance shift at -1.4 ppm in a discharged carbon electrode can be explained by the formation of Li_2CO_3 on the surface of the carbon electrode and formation of Li compounds with Li-C in the bulk of the discharged carbon electrode.

Knight shift at various temperatures

^{7}Li-NMR spectroscopy can provide meaningful information on the nature of lithium species in a fully-charged carbon anode and especially for the equilibrium arrangements of lithium species at low temperature. We tried to measure the Li Knight shift using the solid-state magic-angle spinning ^{7}Li-NMR at various temperatures and to determine the equilibrium arrangements of lithium species in the fully-charged disordered carbon. The results are shown in Fig. 3. The fully-charged disordered carbon shows a resonance shift at 81 ppm which corresponds to reversible capacity at room temperature. The lithium species at 81 ppm is quite different from that in graphite carbon (46.6 ppm see Fig. 4). It may depend on the different crystal structures. It is well known that a resonance shift not only depends on the electronic density in a Li nuclei, but also strongly depends on the starting material and the heat-treatment of the carbon material[7]. When the measured temperature was down to -25 ℃, the resonance shift at 81 ppm is shifted to 82.7 ppm, and the line width become broad. With decreasing measurement temperature, a new resonance shift appears at about 20 ppm while the resonance shift at 82.7 ppm was shifted to 92.0 ppm, and the line width became broader. The result indicates that the rate of exchange or thermal motion of Li species are slowed down on the NMR time scale observed at the low temperature. When the measuring temperature was at -100 ℃, the Li-NMR spectra of the fully-charged disordered carbon

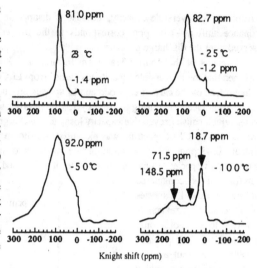

Fig. 3 Li-NMR spectra of the fully-charged disordered carbon measured at various temperatures.

clearly showed three different signals. One narrower band appears at 18.7 ppm and two very broad bands appear at 71.5 and 148.5 ppm. These results indicate that an equilibrium arrangement of Li species in the fully-charged disordered carbon can be observed at low temperature. It is well known that the Knight shift of the Li species depends on the electronic density at the Li nucleus.

The three bands with different shifts indicate that three states with different electronic densities exist in the fully-charged disordered carbon. It should be emphasized that the average value of the resonance shift for three bands is 79.6 ppm. This value is similar to the 81 ppm observed at room temperature. This result further indicates that the Li electronic state observed at room temperature is a mixture of the three different electronic states.

On the other hand, the fully-charged graphite carbon shows a resonance shift at 46.6 ppm and many spinning side bands at room temperature as shown in Fig. 4. With decreasing measurement

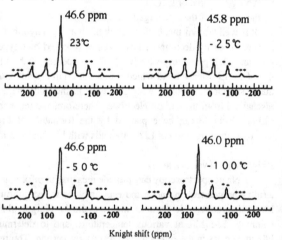

Fig. 4 Li-NMR spectra of the fully-charged graphite carbon measured at various temperatures.

548

temperature, the resonance shift and line width could not be changed, even the measurement temperature was at - 100 °C the Li-NMR spectra of the fully-charged graphite carbon shown the almost same shift and line width comparison with the room temperature. These results indicate that only one interaction between lithium species and graphite carbon exists in the fully-charged graphite carbon. This may depend its crystal structure. Since the graphite carbon has a very large crystallite size comparison with the disordered carbon, lithium is only located between the graphitic layers.

In a previous paper [8], we presented a model which explains why the discharge capacity of the carbon electrode with a disordered structure as an anode can exceed the theoretical capacity (C_6Li, 372Ah/kg) of the graphite anode in lithium ion rechargeable batteries. We predicted that because disordered carbon has a small crystallite size, there are three kinds of interactions between the lithium species and carbon in a fully-charged disordered carbon material. In general, lithium is not only located between the graphitic layers, like graphite intercalation compounds(GICs), but also located at the edge of the graphitic layers and on the surface of the crystallite as shown in Fig. 5. In other words, due to the unique structure of the disordered carbon, more kinds of interactions with lithium are possible. These interactions allow disordered carbon to store much more lithium compared with graphite. Our Li-NMR result with three kinds of Li species at low temperature strongly support our model.

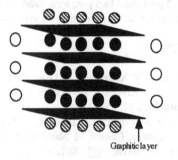

Graphitic layer

● lithium species between graphitic layers

◈ lithium species on the surface of the crystallites

○ lithium species at the edge of graphitic layers

Fig. 5 A model of the interactions between disordered carbon and lithium species

Hence, we suggest that for the fully-charged disordered carbon electrode there are three types of bindings between carbon and the lithium species. The band at 18.7 ppm is assigned to lithium species located between the graphitic layers. The bands at 71.5 and 148.5 ppm may be assigned to the lithium species located on the surface of a crystallite and at the edge of the graphitic layers, respectively.

The results of theoretical calculations[9] show that lithium species located between graphitic layers are the most stable and fully ionized, and lithium species located at the edge of the graphitic layers are less positive than those located on the surface of a crystallite. Detailed results will be reported in another paper.

According to the above discussions, it can be concluded that the mechanism of the interactions between carbon and Li species in disordered carbon is different from that of carbon and lithium in graphite. Since graphite has a large crystallite size, the interactions mainly occur between the carbon atoms of the graphitic layers and lithium. In the case of disordered carbon with a small crystallite size, Li is not only intercalated between the graphitic layers, but is also doped at the edge of the graphitic layers, and even more Li species are doped onto the surface of the crystallite. These interactions allow the disordered carbon to store much more lithium than graphite. In other words, it is the unique structure of the   disordered carbon that leads to its higher discharge capacity than graphite.   However, due to the thermal motion of the Li species, these interactions can only be observed at low temperature.

Conclusions

We have performed the Li-NMR measurements for the discharged disordered carbon at room temperature and the fully-charged disordered carbon at various temperatures. Li-NMR results conform that part the residual Li species in the discharged disordered carbon is unstable in air. These Li species are explained by formation of Li compounds with Li-C bond in the bulk of the electrode during the first cycle. The equilibrium arrangement of Li species in the fully-charged disordered carbon was found by Li-NMR at low temperature. The three different signals at 18.7 ppm, 71.5 ppm and 148.5 ppm were clearly observed at -100 °C. These results are assigned to the Li species located at the different positions of the crystallite.

REFERENCES

1. T. Zheng, W. Xing and J. R. Dahn, Carbon 34, 1501 (1996).
2. S. Yata, H. Kinoshita, M. Komori, N. Endo, T. Kashiwamura, T. Harada, and T. Yamabe, Synth. Met. 62, 153 (1994).
3. K. Sato, M. Noguchi, A. Demachi, N. Oki, and M. Endo, Science 264, 556 (1994).
4. K. Tatsumi, J. Conard, M. Nakahara, S. Menu, P. Lauginie, Y. Sawada and Z. Ogumi, Chem. Commun. 1997 , 687.
5. Y. Matsumura, S. Wang and J. Mondori, J. Electrochem. Soc. 142, 2914 (1995).
6. P. A. Scherr, R. J. Hogan and J. P. Oliver, J. Am. Chem. Soc. 96, 6050 (1974).
7. Y. Mori,T. Iriyama, T. Hashimoto, S. Yamazaki, F. Kawakami, H. Shiroki, and T. Yamabe, J. Power, Sources 56, 205 (1995).
8. Y. Matsumura, S. Wang and J. Mondori, Carbon 33, 1457 (1995).
9. H. Ago, K. Nagata, K. Yoshizawa, K. Tanaka and T. Yamabe, Bull. Chem. Soc. Jpn. 70, 1717 (1997).

VOLTAGE HYSTERESIS AND HEAT GENERATION BEHAVIOR IN LITHIUM-ION BATTERIES

Y. SAITO, K. TAKANO, K. KANARI and K. NOZAKI
Electrotechnical Laboratory, Tsukuba, Ibaraki, 305 JAPAN, ysaito@etl.go.jp

ABSTRACT

In order to characterize the heat generation behavior of a lithium-ion cell during charge and discharge, calorimetry has been carried out. The influence of past treatments applied to the cell is examined based on heat generation. The test cell shows a voltage hysteresis during charge and discharge, and the heat generation behavior is related to this hysteresis. A hard carbon is used in the cell as the negative electrode material, and the hysteresis mechanism and heat generation are discussed based on the reactions in the hard carbon electrode.

INTRODUCTION

Since lithium-ion secondary batteries have excellent characteristics such as high energy density and long cycle life, they are candidate batteries for an electric energy storage system. Lithium-ion batteries for portable electronic applications are currently very common. However, the scaling-up of the batteries for dispersed energy storage is still difficult because there are many complicated problems associated with thermal safety and reliability. In order to evaluate the scale-up of these batteries, simulation studies on the electrical and thermal behaviors of the batteries have been carried out in our group [1, 2]. It is important to clarify the heat generation mechanism during charge and discharge, therefore, a calorimetric study has been undertaken using commercially available lithium-ion cells [3-5]. It was found that the main heat generation factors were the entropy change of the cell reaction and the electrochemical polarization, however, it was also suggested there were other heat generation factors along with the voltage hysteresis. In this work, the additional mechanism of heat generation is discussed based on calorimetry under various initial cell conditions focusing on the hysteresis.

EXPERIMENT

A cylindrical type lithium-ion cell (US14500, Sony Energytec) was used as the test sample. The active electrode materials are $LiCoO_2$ in the positive electrode and hard carbon in the negative. The nominal cell voltage is 3.6 V and the nominal electric capacity is 500 mA·h.

A twin-type heat conduction calorimeter (C-80, Setaram) was used for calorimetry that was placed in a room in which the temperature was controlled to 278 K. The cell was introduced into a specially designed sample holder made of stainless steel that was connected to external instruments, i.e., a nanovoltmeter (Model 181, Keithley) to measure the cell voltage and a DC source (Model 7651, Yokogawa) for charge or discharge through two pairs of lead wires [3]. The heat loss through the wires was at most 1.5% of the generated heat. The time constant of the delay due to heat transfer from the sample to the thermopile of the calorimeter was 341 s. The influence of this delay was corrected before analysis.

In the following discussion, the cell charged at constant voltage of 4.30 V at a constant temperature of 303 K is defined as being in the fully charged state, and the state of the battery is described as the quantity of electricity, Q, discharged from the fully charged state.

RESULTS

Energy balance during one cycle

Figure 1 shows typical heat generation and voltage curves during one cycle of the test cell. The cell was discharged at a constant current of 100 mA to 2.75 V. After 5 hours of rest, it was charged at the same current to 4.30 V followed by constant voltage charging. The total charging time was 10 hours. The temperature in the sample holder was held at 303 K. Heat absorption is observed during charge while heat generation is noted during discharge, and these are the same results observed by Kobayashi [6] and Hong [7]. Several exothermic and endothermic peaks observed in both the discharging and charging curves are reversible and are caused by crystal phase transitions of the positive electrode material [4]. Even during rest, heat flow is observed, but gradually decreasing. The delay in the response of the experimental system is not the cause of this heat flow because its influence has been already corrected in the curve. It is suggested that side-reactions or the diffusion of lithium ions in the active electrode materials cause the heat. Figure 2 shows the input and output energy during each cycle. The total electrical input is the difference in energy input during the charge and output during the discharge. This is due to the difference in voltage during the charge and discharge. The total thermal outputs are the integration of the measured heat flow corresponding to the energy dissipated in the cell during one cycle. The input is compensated by the output if the heat generation during rest time is also included in the output. The same cycling procedure was repeated several times and the good cyclability of the cell is shown in Fig. 2, suggesting there was little decline in the cell performance. Thus, side-reactions accompanying the decline is not the main factor of heat generation during the rest time.

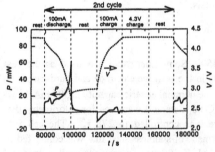

Fig. 1 Heat generation (P) and voltage (V) curves of a lithium-ion cell during one cycle at 303 K.

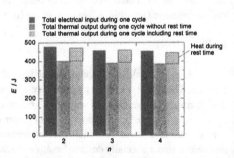

Fig. 2 Energy balance of each cycle. Total cycle number is described as n.

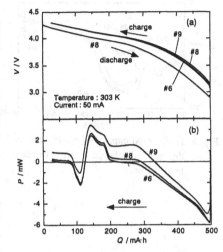

Fig. 3 Charging behaviors depending on temperature during the preceding discharge, that is 283 K (#6 ; 6th cycle), 303 K (#8) or 323 K (#9).

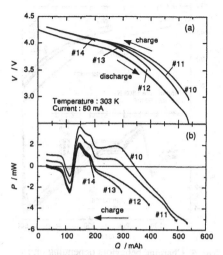

Fig. 4 Charging behaviors depending on initial DOD that is 537 mA·h (#10), 498 mA·h (#11), 399 mA·h (#12), 299 mA·h (#13) or 199 mA·h (#14),.

Heat generation during charge

For the calorimetric study, the cell was charged and discharged under various conditions. As a result, it was found that the past treatment applied to the cell affected the heat generation behavior during the next procedure, especially the charging process. In Fig. 3, Fig. 4, and Fig. 5, many curves of the cell voltage and the heat generation during charge are drawn. In all the measurement, both the charging and discharging currents were constant at 50 mA, and the cell temperature was held at 303 K during charge.

In Fig. 3, the cell was charged from the 498 mA·h discharged state and the rest time before the charge was 10 hours in all cases. It is the temperature during the preceding discharge that affects the charging behavior. After discharge at a higher temperature, the cell voltage is slightly higher, and more heat is generated during the following charge. Figure 6 shows differential capacity curves, dQ/dV, during discharge as a function of Q. Peaks observed before 200 mA·h correspond to the crystal phase transition of the positive electrode material [4]. Note that variation depending on the temperature occurs after about 450 mA·h. This suggests that a reaction which is rate dependent on the temperature occurs in the region.

The depth of discharge (DOD) also affects the charging behavior as shown in Fig. 4. The temperature was held at 303 K not only during the charge but also during the discharge and rest time. The rest time is 5 hours in all cases. The cell voltage curves of the discharges are almost same. On the other hand, a variety of charging curves are shown depending on the initial DOD. The curves indicate hysteresis if the cell is deeply discharged. It should be noted that a shoulder peak at about 300 mA·h in the heat generation and a large hysteresis in the voltage are observed after discharges deeper than 450 mA·h. It is also noteworthy that the hysteresis disappears

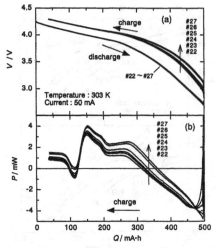

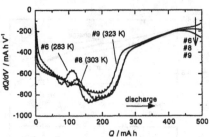

Fig. 6 Differential capacity (dQ/dV) curves during discharge at various temperatures.

Fig. 5 Charging behaviors depending on rest time of 0 h (#21), 0.5 h (#22), 1.0 h (#23), 3.5 h (#24), 24.0 h (#25) or 48.0 h (#26).

after the peak for every case in Figs. 3, 4 and 5. In addition, the excess heat generation continues after the peak while the charging voltage curves agree with each other.

Figure 5 shows the influence of the rest time before charge. The temperature was again held at 303 K during the discharge, charge, and rest time. The initial DOD before charge is 498 mA·h discharged. A larger amount of heat is observed if the cell is allowed to rest longer before charging. The cell voltage is also higher after a longer rest time. It is suggested that a process which generates heat with an increase in the open circuit voltage (OCV) during rest time as shown in Fig. 1 has a relation to the voltage hysteresis and the heat generation during charge. It seems to be that the process is also occurring during discharge in a deeply discharged state, especially after approximately 450 mA·h. In the final period of discharge, the heat flow rapidly increases as shown in Fig. 1. At first glance, it is thought to be an increase in the polarization. However, the polarization is not large enough to explain such a large heat flow [5].

<u>The reaction mechanism</u>

During the electrode reaction of LiCoO$_2$, any obvious voltage hysteresis was not reported in previous studies. On the other hand, it is remarkable in disordered carbon materials. There are many kinds of carbon materials in which lithium can be reversibly inserted, and those characteristics are dependent on the structure [8]. In the case of hard carbon, which is used in the test cell, the voltage profile generally consists of two parts: a steeply sloping region and a low voltage plateau. In addition, some hard carbon materials show a voltage hysteresis in the former region. Thus, it is postulated that the hysteresis of the test cell originated during the reaction of the hard carbon electrode. Ishikawa *et al.* reported that there were at least two kinds

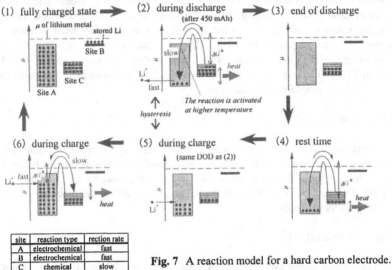

(1) fully charged state ➡ (2) during discharge ➡ (3) end of discharge
                              (after 450 mAh)

site	reaction type	rection rate
A	electrochemical	fast
B	electrochemical	fast
C	chemical	slow

Fig. 7 A reaction model for a hard carbon electrode.

of sites in hard carbon materials where the electronic environment around inserted lithium is different from each other based on ^{7}Li-NMR measurement [9]. Some carbon materials that are pyrolyzed at temperatures near 700 ℃ show a large voltage hysteresis, of which the voltage profiles are not similar with those of hard carbons. Dahn's group explained that the cause of the hysteresis in these materials is the formation of Li-C binding [10]. There are few reports on the mechanism to explain the hysteresis of hard carbon materials.

Considering the heat generation behavior of the test cell, we propose a reaction model in the hard carbon electrode during one cycle as shown in Fig. 7. Lithium electrochemically reacts with site A in the sloping region and with B in the plateau. The chemical potential, μ, of lithium in site A increases with the number of lithiums in the site. On the other hand, μ of lithium in site B is almost constant and close to that of lithium metal. In Fig. 7, there is another site, C, where lithium can be stored and released hardly by direct electrochemical reaction but mainly by chemical transfer between site C and the other sites. The chemical transfer rate is considerably slower than the electrochemical reaction rate at sites A or B. During discharge, lithium of a higher chemical potential is first extracted from B and then from A. However, in the deeply discharged state, after about 450 mA·h, the potential of lithium in site A becomes lower than that in site C. In this situation, lithium in site C is transferred to A with heat generation as shown in Fig. 7 (2). This lithium transfer with heat generation is also occurring during the rest time after the discharge that is shown in Fig. 7 (4). In the following charging, a large amount of heat due to the lithium transfer from site A of higher potential to site C is dissipated if there are many empty sites in C as shown in Fig. 7 (6). The potential of lithium in site C does not relate to the OCV, and thus hysteresis is observed in the voltage profile. In Fig. 7 for (2) and (5), the OCV is different for each other even if same number of lithiums is stored. When site A is filled with lithium, a shoulder peak appears in the heat generation, and then storage into site B becomes the main reaction resulting in disappearance of voltage hysteresis. In conclusion, the distribution of lithium between sites A and C has a relation to the voltage hysteresis and the variation in heat generation during charge. In our earlier work of staircase voltage step coulometry, it was actually recognized that there was an extremely slow rate reaction with a fast main reaction in the sloping region [11]. Recently, Inaba *et al.* characterized the diffusion behavior of lithium in a glassy carbon electrode and reported there were trap sites for lithium in this material [12]. The trap site might correspond to site C in our model. The identification of each site will be accomplished in future work.

CONCLUSIONS

A calorimetric method was used to characterize the performance of a lithium-ion cell. The influence of past treatments has been examined based on the heat generation behavior during charge. The heat generation behavior has a relation to the charging and discharging voltages hysteresis of the cell. A new reaction model has been proposed for the hard carbon electrode in order to explain these results.

ACKNOWLEDGEMENT

The authors would like to thank Mr. Kunio Ishii, who had been a managing director of Sony Energytec Inc., for providing the lithium-ion cells.

REFERENCES

1. K. Takano, T. Hirayama and T. Nakano, Bull. Electrotechnical Laboratory, **60**, 57 (1996).
2. K. Kanari, K. Takano and Y. Saito, Bull. Electrotechnical Laboratory, **60**, 65 (1996); Extended Abstracts of 8th Intern. Meeting on Lithium Batteries, Nagoya, 1996, pp. 360.
3. Y. Saito, K. Takano, K. Kanari and T. Masuda, Proc. Intern. Workshop on Advanced Batteries, Osaka, 1995, pp. 283; Bull. Electrotechnical Laboratory, **60**, 11 (1996).
4. Y. Saito, K. Kanari and K. Takano, Abstracts of 37th Battery Symposium in Japan, Tokyo, 1996, pp. 97.
5. Y. Saito, K. Takano and K. Kanari and T. Masuda, Abstracts of 36th Battery Symposium in Japan, Kyoto, 1995, pp. 209.
6. Y. Kobayashi, H. Miyashiro, K. Takei, K. Kumai and N. Terada, Extended Abstracts of 190th ECS Fall Meeting, **96-2**, 172 (1996).
7. J.-S. Hong, H. Maleki, A. Al Hallaj, J. R. Selman and L. Redey, Meeting Abstracts of 192nd ECS Meeting and 48th ICE Annual Meeting, **97-2**, 145 (1997).
8. J. R. Dahn, T. Zheng, Y. Liu and J. S. Xue, Science, **270**, 590 (1995).
9. M. Ishikawa, N. Sonobe, H. Chuman and T. Iwasaki, Abstracts of 35th Battery Symposium in Japan, Nagoya, 1994, pp. 49.
10. T. Zheng, W. R. McKinnon and J. R. Dahn, J. Electrochem. Soc., **143**, 2137 (1996).
11. K. Nozaki, A. Negishi, K. Kato and K. Takano, Abstracts of 36th Battery Symposium in Japan, Kyoto, 1995, pp. 207.
12. M. Inaba, S. Nohmi, A. Funabiki, T. Abe and Z. Ogumi, Abstracts of 64th Electrochemical Society in Japan, Yokohama, 1997. pp. 20; This Meeting, Y11.6.

A NOVEL METHOD FOR OBTAINING A HIGH PERFORMANCE
CARBON ANODE FOR LI-ION SECONDARY BATTERIES

T. TAKAMURA*, K. SUMIYA**, Y.NISHIJIMA**, J.SUZUKI** AND K.SEKINE**

*Advanced Material Division, Petoca, Ltd, 3-6 Kioicho, Chiyodaku, Tokyo 102-0094, Japan
** Department of Chemistry, Rikkyo University, 3-34-1 Nishiikebukuro, Toshimaku, Tokyo 171-0021, Japan

ABSTRACT

Entire surfaces of the pitch-based well graphitized and poorly graphitized carbon fibers were covered with an evaporated metallic film of Au, Ag, Sn, or Zn, whose thickness was varied from 100 to 600 Å. Li dope and undope characteristics as the anode of Li-ion secondary battery were compared for the samples with and without evaporated film. Not only high rate charge/discharge characteristics but cycleability have been realized to be improved remarkably for the sample whose surface was covered with a thin metal film.

INTRODUCTION

Recent explosive increase in the production amount of Li-ion secondary batteries is due to the dramatically rapid production expansion of cellular phones, handy personal computers and handy camcorders. These electronic appliances require of the secondary battery to be improved further in the battery characteristics of high rate charge/discharge and cycleablity as well.

For the purpose of obtaining much better characteristics we have pointed out that improvement of the surface condition of active materials are one of the most important key factors in addition to the improvement of the internal structure of carbon[1]-[5]. Mild oxidation and heating *in vacuo* are the effective treatment, both of which have been proposed by us at first[5]-[7]. The proposed methods, however, still allow the carbon surface to be in contact with electrolyte directly, so that the so called SEI(solid electrolyte interphase) found on the treated carbon may resemble the SEI formed on the untreated carbon surface, causing a deterioraton although it is not so serious. It is, therefore, worthwhile to explore the possibility of finding an entirely different SEI which is effective to keep the Li dope/undope reactions quite stable.

In the present study the authors would like to provide a novel method to obtain very stable SEI, *i.e.*, the entire carbon surface is coated with a film of stable soft metal which has a good affinity with lithium. In this case the SEI is formed on the surface of metal and not on the carbon surface, hence, it is expected to form a new type of SEI having a very good electrode performance.

EXPERIMENTAL

Materials

Carbon fibers used were mesophase pitch-based carbon fiber felts(average fiber diameter of 10 μm) prepared at different temperatures(Melblon, provided by Petoca, Ltd). Examples of SEM photographs are shown in Fig. 1. The fiber felt was heated *in vacuo* at 250° C for 2 hours before use. The felt sample was sliced 3mm thick and placed in a vacuum chamber for vacuum evaporation. The vacuum evaporation of metal was performed by heating a pure metal rod kept in a tungsten crucible at a temperature slightly higher than the melting point in the evaporation chamber(10^{-8} Torr). The sample was mounted on a rotating holder so that both sides of the felt specimen could be covered with the evaporated film. The metal film thickness was monitored with a quartz vibrating micro-balance mounted near the specimen in the vacuum chamber.

Test Electrode Preparation

About 3 mg of the fiber felt having a size of 1x1 cm square was sandwiched between two 50 mesh Ni grids whose rims were spot-welded at many points. After spot-welding a Ni wire to the grid the sadwiched specimen was offered for the electrochemical evaluation.

Electrochemical Evaluation

A cylindrical 25 ml glass cell having three electrodes was used for the evaluation. Metallic lithium foils were used as counter and reference electrodes, respectively. Ethylene carbonate(EC)+ dimethyl carbonate(DMC) 1:1 mixture containing 1 M LiClO4 and propylene carbonate(PC) containing 1 M LiClO4 were used as electrolytic solutions. The measurement was done mainly with a cyclic voltammetry(cv) and a constant current charge/ discharge cycle test was also done if necessary. All the measurements were carried out at room temperature in a glove box in which dried Ar gas was filled.

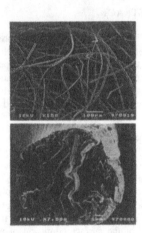

Fig. 1 SEM photographs of Melblon fiber mat prepared at 3100°C. Top: low magnification; Bottom: high magnification cross section.

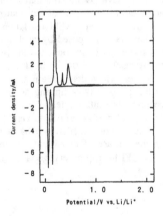

Fig. 2 Cyclic voltammograms of evaporated gold film (300Å thick) on a Ni plate measured in EC+ DEC containing 1 M LiClO4; sweep rate: 1 mV/s.

Fig. 3 SEM photographs of graphitized carbon fiber (Melblon 3100) evaporated with gold film(500Å thick), a part of the front being removed

RESULTS AND DISCUSSION

Effect of gold film

Cyclic voltammograms of a gold thin film (300 Å thick) evaporated on a Ni plate are shown in Fig. 2 where several sharp cathodic and anodic peaks are recognized to be very reversible. This means that lithium can be doped and undoped in the gold film reversibly. Gold was successfully evaporated on 3100°C carbon fiber as a smooth film covering over the entire surface of carbon as shown in Fig. 3. Yellow golden color could be recognized by eye when the thickness of the gold film was over 100 Å. Evaporation of gold was found to enhance the Li dope/undope reaction rate as shown in Fig. 4, where cv's are compared for the cases with and without gold film evaporated (100 Å hick). Large anodic peak at around 250 mV due to the deintercalation of Li from the carbon electrode is enhanced in peak height for the gold evaporated sample while somewhat irreversible. Upon increasing the film thickness the peak height and reversibility were more pronounced, but tend to saturate, then began to decrease when the thickness was over 600 Å. This decreasing tendency is explained by assuming that the overall rate is controlled by the lithium mobility which is slower in the gold film.

The results shown above indicate that the SEI formed on gold surface in contact with the present electrolyte is more effective to enhance the rate than that on carbon. It is surprizing that lithium can move easily across the interface between carbon and gold which may not stick together tightly on atomic level all over the interface.

In spite of much cheaper price of PC, graphitized carbon cannot be utilized in this solution since PC decomposes continuously on a graphite surface. There is a possibility, however, PC can be utilized when graphite is covered with gold. The results were not satisfactory but far better as compared with those obtained with the bare graphite surface, indicating that further examination is worthwhile.

Effect of silver film

Silver has been examined by Hitachi group to be quite effective for obtaining high performance characteristics[8]. They, however, have used a method of chemical reduction

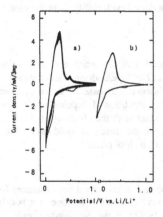

Fig. 4 Cyclic voltammograms of graphitized carbon fiber(Melblon 3100) with (a) and without(b) gold film (100 Å thick) in EC+DMC containing 1 M LiClO₄; sweep rate: 1 mV/s.

Fig. 5 SEM photographs of graphitized carbon fiber (Melblon 3100) evaporated with silver film(450Å thick). A small part of the film was removed at the front.

559

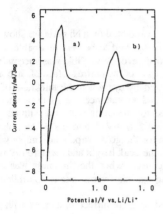
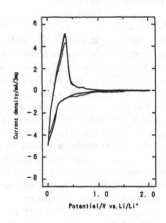

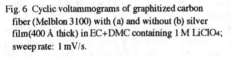

Fig. 6 Cyclic voltammograms of graphitized carbon fiber (Melblon 3100) with (a) and without (b) silver film(400 Å thick) in EC+DMC containing 1 M LiClO4; sweep rate: 1 mV/s.

Fig. 7 Cyclic voltammograms of graphitized carbon fiber (Melblon 3100) having a Zn film(300 Å thick) in EC+DMC containing 1 M LiClO4 ; the peak height being elevated during the cycles. Sweep rate; 1 mV/s.

of Ag^+ in solution, where metallic silver was deposited randomly as fine particles on graphite crystallites[8]. In our experiment the silver film was found to cover very uniformly all over the graphite surface as shown in Fig. 5. The cyclic voltammograms with and without silver film are compared in Fig. 6, where we see that silver is quite effective to provide an improved Li dope/undope rate with a very good cycleability. The performance was dependent on the silver film thickness and the film thicker than 600 Å gave rather reduced cycleability as in the case of gold film.

Effect of zinc film

Zinc was found also effective for the improvement of the anode characteristics as shown in Fig. 7. The reason why zinc film gives rise to a very good reversible dope/undope characteristics is explained by assuming that at first lithium is favor to dissolve and ease to move in zinc, secondly, zinc crystal keeps its original structure even after the repetition of dope/undope cycles, and thirdly, good SEI is formed on the surface of zinc in contact with the electrolyte. Such an argument may be applicable to any good metal like as in the cases of gold and silver. Application of zinc is more realistic for practical use because of its low price.

Effect of tin film

Since tin has similar physical properties to zinc it is expected to give good performances for carbon when evaporated. Two different types of carbon were examined; the one is a poorly graphitized carbon, and the other is well graphitized one. In Fig. 8 the dope/undope cv's are compared for with and without evaporated tin on the 980°C prepared carbon in PC containing 1 M LiClO4. As seen in the figure reversibility as well as the capacity in the negative voltage region are improved for tin deposited carbon. Small spikes appearing on the cv's of tin covered

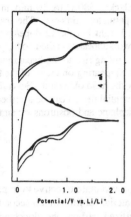

Fig. 8 Cyclic voltammograms of poorly graphitized carbon fiber (Melblon 950) with(bottom) and without (top) tin film (300 Å thick) in PC containing 1 M LiClO4; sweep rate: 1 mV/s.

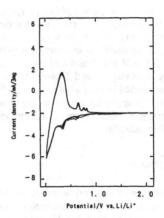

Fig. 9 Cyclic voltammorgrams of graphitized carbon fiber (Melblon 3100) having tin film(300 Å thick) in EC+DMC containing 1 M LiClO4; sweep rate: 1 mV/s.

carbon are due to the dope and undope of lithium into and from the tin film respectively. It is very interesting to note on the anodic branch that the lithium doped in carbon begins to be undoped at more negative potential than that of the lithium doped in tin film which covers all over the surface of carbon. The film is expected to have a pinning action to prevent the undoping of lithium from carbon in the potential region negative than the spike potential where lithium begins to be undoped. This phenomenon is left to be explained in the future studies.

For the well graphitized carbon the effect of evaporated tin film is shown in Fig. 9. As seen in the figure the anodic peak height is enhanced for the tin evaporated carbon which, however, shows rather poor cycleability. We attributed the poor cycleability to a gradual collapse of tin crystal in the evaporated film caused by the change in the crystal form during the lithium dope/undope cycling process. According to the recent studies by Idota *et al.* tin oxides give rise to a very stable dope/undope properties for lithium[9], so we examined oxidized tin film. The tin covered sample was offered to the mild oxidation

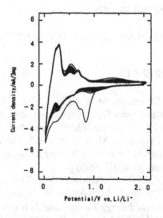

Fig. 10 Cyclic voltammograms of graphitized carbon fiber (Melblon 3100) having tin film (300 Å thick) in EC+DMC containing 1 M LiClO4. The fiber evaporated with tin was at first exposed to the mild oxidation by heating in air covered with acetylene-black powder at 580°C for 10 min. and afterwards offered to the electrochemical measurement; sweep rate: 1 mV/s.

, process by heating the sample in a large amount of acetylene-black at 580°C for 10 min. in air. The oxidized sample was examined for the electrochemical dope/undope process. The results are shown in Fig. 10. As seen in the figure not only the peak height but the Li dope/undope cycleability for carbon is quite much improved. Splitting into two subpeaks on the anodic peak at about 250 mV is indicative of the increase in undoping rate, giving rise to the separation of undoping from stages 1 and 2 respectively. Large cathodic peaks appearing on the cv of the first cycle is attributed to the partial reduction of tin oxides formed by the mild oxidation. Several ill-defined spikes appearing on the anodic branch in the positive potential region can be ascribed to the undoping of lithium from the amorphous tin film whose peak shape and positions are similar to that of metallic tin.

CONCLUSIONS

Evaporation of appropriate metals on carbon fiber has been proved quite effective to improve the lithium dope/undope characteristics of carbon as the anode material for lithium-ion secondary battery. Not only well graphitized carbon but poorly graphitized carbon, the dope/undope characteristics could be improved markedly. Lithium was found to move with ease across the carbon/metal interface once doped in the metal film from the electrolyte.

All the metal examined (Au, Zn, Ag, Sn) exhibited their excellent capability to improve the anode characteristics when properly treated, implying that there is a possibility for practical application.

ACKNOWLEDGEMENT

Authors are thankful to PEC of MITI of Japan whose fund suppored a part of this work.

REFERENCES

1. T. Takamura, M. Kikuchi, J. Ebana, M. Nagashima, Y. Ikezawa, in *New Sealed Rechargeable Batteries ans Super Capacitors*, edited by B. M. Barnett, E. Dowgiallo, G. Halpert, Y. Matuda and Z-i Takehara (Electrochem. Soc. Proc., **93-23**, Pennington, NJ 1993) p. 228.
2. T. Takamura, M. Kikuchi and Y. Ikezawa, in *Rechargeable Lithium and Lithium-ion Batteries*, edited by S. Megahead, B. M. Barnett and L. Xie (Electrochem. Soc. Proc. **94-28** Pennington, NJ 1995) p. 213.
3. M. Kikuchi, Y. Ikezawa and T. Takamura, *J. Electroanal. Chem.*, **396**(1995) 451.
4. T. Takamura, M. Kikuchi, H. Awano, T. Ura and Y. Ikezawa, in *Materials for Electrochemical Energy Storage and Conversion*, edited by D. H. Doughty, B. Vyas, T. Takamura and J. R. Huff (Mater. Res. Soc. Proc. **393**, Pittsburgh, NJ 1995) p. 345.
5. T. Takamura, H. Awano, T. Ura and Y. Ikezawa, *J. Korean Soc. Anal. Sci.*, **8**(1995)583.
6. E. Peled, G. Menachem, D. Bar-Tow and A. Melman, *J. Electrochem. Soc.*, **143**(1996)L4.
7. T. Takamura, H. Awano, T. Ura and K. Sumiya, *J. Power Sources*, **68**(1997)114.
8. H. Honbo, *et al.*, Extended Abstr. (*38th Japanese Battery Symp.*, in Osaka, Japan 1997) 225.
9. Y. Idota, *et al*, *Science*, **276**(1997) 1395.

LOCAL STRUCTURE OF BALL-MILLED CARBONS FOR LITHIUM ION BATTERIES: A PAIR DISTRIBUTION FUNCTION ANALYSIS

A. CLAYE*, P. ZHOU*, J. E. FISCHER*, F. DISMA**, J-M. TARASCON**
* LRSM, University of Pennsylvania, Philadelphia PA 19104, USA
** LRCS, Université de Picardie Jules Verne, 80039 Amiens, France

ABSTRACT

The local atomic structure of ball-milled carbons was investigated by radial distribution function (RDF) analysis using pulsed time-of-flight neutron diffraction. The results exhibit a gradual loss of long-range order as a function of milling time. Modeling of the elastic structure factors and of the differential correlation functions identified the structure of ball-milled carbons as finite-size graphene fragments whose size decreases continuously with milling time. The large increase in lithium reversible capacity after 20 hours of milling was correlated with the loss of interlayer correlation between graphite flakes, similar to the structure of hard carbons in the "House of Cards" model.

INTRODUCTION

Disordered carbons have recently received tremendous interest because of their exceptional lithium intercalation capacities which make them potential anode materials in rechargeable lithium ion batteries. Many structural studies of disordered carbons obtained through pyrolysis of organic precursors [1,2] or through ball milling [3] have been reported. Although widely used for these studies, X-ray diffraction can only provide limited information on these highly disordered systems. Diffraction patterns of disordered carbons usually consist of a few broad and weak maxima corresponding to the (002), (100) and (101) reflections of graphite [4]. In disordered materials, peak positions and peak widths can be affected by Hendricks Teller disorder or paracrystallinity [5], and therefore Bragg's law and Scherrer's formula cannot be used as readily as in crystalline materials to estimate lattice parameters and crystallite size. Hence, local coordination, lattice spacings and coherence lengths in disordered carbons are difficult to estimate using standard X-ray diffraction.

We recently reported a novel approach using pulsed time-of-flight neutron diffraction and radial distribution function (RDF) analysis for the structural study of amorphous carbons [5]. Extensive details on this method were reported earlier [4,5]. As a summary, the method relies on the following relationship between the experimentally determined structure factor S(Q) and the differential correlation function D(r)

$$D(r) = \frac{2}{\pi} \int_0^\infty Q[S(Q) - 1]\sin(Qr)dQ = 4\pi r[\rho(r) - \rho_0] \quad \ldots\ldots\ldots(1)$$

in which r is the real space distance from a reference origin, Q is the momentum transfer, $\rho(r)$ is the number of atoms per unit volume at r, and ρ_0 is the average density. The differential correlation function provides very valuable information such as bond lengths, bond angles, and local coordination. When combined with real space modeling, it can also help estimating coherence lengths in disordered carbons. This novel approach was taken to study the structure of ball-milled carbons. Ball milling was reported to increase reversible lithium intercalation capacities up to 708 mAh/g [3]. The goal of this work was to understand the structural changes occurring during ball-milling, so as to explain the enhanced electrochemical properties of these disordered carbons.

563

EXPERIMENT

Carbon samples were obtained by ball-milling commercial graphite using a Spex 8000 mixer which generates normal mechanical strains. Graphite was ground in a stainless steel vessel sealed under argon atmosphere to avoid air contamination. Detailed sample preparation and chemical analysis of the ground carbons were reported earlier [3,6]. The milled materials were kept and handled in an argon-filled glove box to avoid moisture adsorption (the presence of hydrogen in the sample could corrupt the quality of neutron diffraction data because of the high incoherent scattering of hydrogen).

Five samples ground respectively for 0, 5, 20, 40 and 80 hours were sealed in air-tight vanadium cans and analyzed by pulsed time-of-flight neutron powder diffraction. Measurements were carried out using the glass-liquid-amorphous material diffractometer (GLAD) at the intense pulsed neutron source (IPNS) of Argonne National Laboratory. This diffractometer probes a wide domain of reciprocal space ($Q_{max} = 45 Å^{-1}$), which is necessary to avoid truncation effects when Fourier transforming the data directly for RDF analysis [7]. Details of GLAD instrumentation are presented elsewhere [8]. Data were corrected for absorption, multiple scattering and inelastic scattering, then were normalized to a vanadium standard using the GLAD user software. Differential correlation functions were obtained through simple Fourier transform of the normalized structure factors using equation (1).

RESULTS AND DISCUSSION

The normalized structure factors obtained for all five samples are presented in figure 1. These spectra illustrate the gradual loss of long range order resulting from the milling treatment, as already observed by X-ray diffraction [4]. One of the advantages of neutron over X-ray diffraction is the large peak intensities observed even at high angles. This allows for the resolution of in-plane reflections of graphite more easily than in an X-ray pattern. In figure 1, peaks corresponding to both in-plane ((100) (110)) and out-of-plane ((220) (400)) reflections of graphite broaden and weaken continuously upon milling. This indicates that exfoliation and reduction of the in-plane coherence length occur simultaneously from the very beginning of the process. After 20 hours of milling, the spectrum only presents very broad peaks from which coherence lengths cannot be estimated. It is also interesting to notice that the very small angle scattering ($Q \leq 1 Å^{-1}$) intensities increase with milling time, suggesting an increase in porosity.

The differential correlation functions obtained by Fourier transform of the respective structure factors are presented in figure 2. The weakening of the intermediate-r features and the disappearance of the high-r features at long milling times confirm the progressive loss of long-range order due to grinding. As long-range order disappears, distances between individual atoms loose their well-defined character, and peaks in D(r) become broader and weaker at high r. While crystalline order is lost, it is clear that short-range order is retained even for the longest milling times. The first three peaks in D(r) corresponding to the first, second and third nearest neighbors in graphite do not shift position with milling time. The first peak, which corresponds to the smallest C-C distance (bond length), occurs at 1.42Å for all samples. The second peak is located at 2.46Å in all spectra, indicating that the 120°-bond angle of graphite hexagons is conserved in milled carbons. Finally the third peak, corresponding to the intrahexagon distance, occurs at 2.84Å (twice the bond length) in all cases, showing that the hexagons remain planar. These observations confirm that the short-range structure of graphite is conserved in the milled carbons. The main effect of the milling process is therefore a severe reduction in coherence

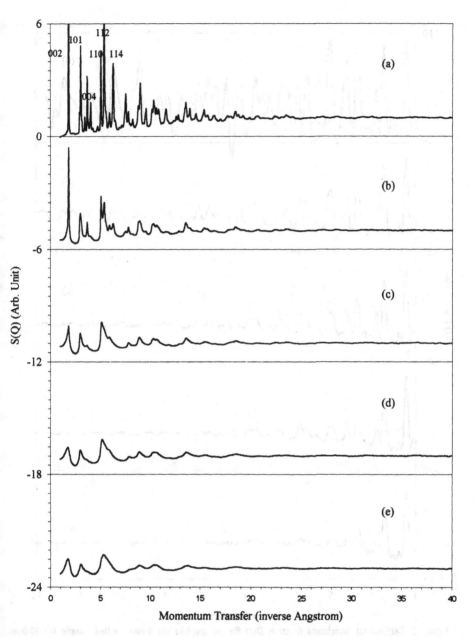

Figure 1: Neutron powder diffraction structure factors S(Q) for (a) graphite (spectrum is cropped for scaling), (b) 5-hour milled sample (c) 20-hour milled sample, (d) 40-hour milled sample and (e) 80-hour milled sample. Spectra are offseted for clarity.

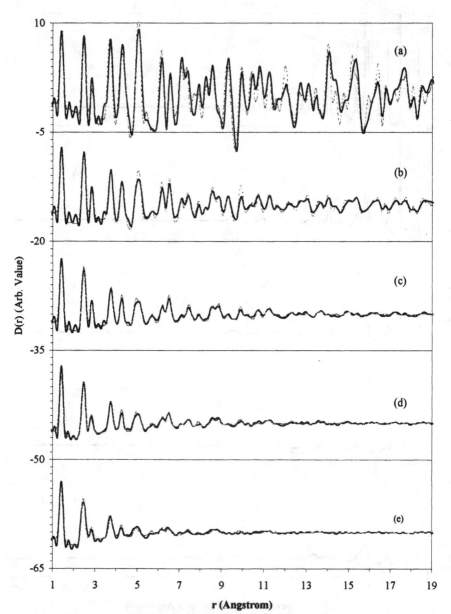

Figure 2: Differential correlation function D(r) for (a) graphite (b) 5-hour milled sample (c) 20-hour milled sample (d) 40-hour milled sample (e)80-hour milled sample. Spectra are offseted for clarity. Continuous dark line represents experimental data; light dotted line represents the calculated models, made of (a) 20 layers of 25x25 repeat units, (b) 3 layers of 20x20 repeat unit and 1 layer of 10x12 repeat units, (c) 1 layer of (8x6) repeat units, (d) 1 layer of 8x3 repeat units, and (e) 1 layer of 6x2 repeat units.

length rather than a disruption of the local structure.

The coherence length in the milled carbons was investigated by real-space and reciprocal-space modeling. Since the short-range structure of graphite is conserved during milling, structural building blocks were constructed using the unit cell of graphite (a=b=2.42Å, c=6.7Å, $\alpha=\beta=90°$, $\gamma=120°$), and by imposing a finite number of repeating units in the x, y and z directions. Finite-size graphite fragments were thus obtained, as illustrated in figure 3.

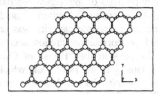

Figure 3: Example of structural building block used for modeling S(Q) and D(r). Here a single-layer graphene fragment of (5x5) repeat units in the (xy) plane is represented.

Coordinates were generated for all atoms in the model, and hypothetical structure factors were calculated using Debye's scattering equation:

$$S(Q) = \sum_{i,j} \frac{\sin(Qr_{ij})}{Qr_{ij}} \quad \dots\dots(2)$$

in which r_{ij} is the distance between the ith and the jth atom in the model. The Debye equation provides a spherical average for S(Q) by assuming a random orientation of the building blocks with respect to one other. A Debye-Waller factor $<u^2>$ of 0.0045 was incorporated in the calculations to account for the temperature effects. Hypothetical D(r) were obtained through direct Fourier transform of the corresponding hypothetical S(Q). Modeling was completed by fitting the experimentally obtained S(Q) and D(r) by those calculated for the structural models. Fits were optimized by adjusting the dimensions of the models along x, y and z until the best match of experimental data was obtained, both in reciprocal space for S(Q) and in real space for D(r).

The results of this modeling exercise are presented in figure 2, where the experimental and calculated D(r) are compared for each milled carbon. Graphite was reasonably well modeled by a structure composed of 20 graphene layers arranged with A-B stacking, each layer containing (25x25) unit cells in the (xy) plane. Although this building block is much smaller than in real graphite, it is large enough to fit D(r) up to 20 Å. This confirmed the validity of the approach. For the 5-hour milled sample, the experimental data could not be fitted by a unique system with finite dimensions along x, y and z. The best fit was obtained for a mixture of 3-layer building blocks containing (20x20) unit cells in the (xy) plane, together with single-layer graphene fragments made of (10x12) unit cells. This indicates that after 5 hours of milling, the material has significantly exfoliated, and is now composed of two phases: regions of well stacked graphene sheets (although the coherence lengths in all directions are reduced compared with those of the starting material), separated by smaller isolated graphene flakes. Beyond 20 hours of milling, data could only be fitted by sets of single-layer graphene flakes. Any interlayer correlation in the models worsened the fits significantly. Samples milled for 20, 40 and 80 hours were best fitted by single-layer flakes respectively composed of (8x6), (8x3) and (6x2) repeat units, corresponding to coherence lengths L_a of 25Å, 20Å and 15Å. This indicates that after 20 hours, the milled carbons are essentially composed of small, uncorrelated graphene planes whose size decreases upon further milling. We believe that these flakes are not totally isolated, but rather interconnected with one another. Isolated flakes would involve a large density of broken bonds, and therefore of unpaired electrons. Electron spin measurements on these systems show no signal for unpaired electrons, indicating that the dangling bonds are saturated by some chemical species such as hydrogen or oxygen, or that the fragments are interconnected. Considering the large number of hydrogen or oxygen atoms required to saturate all dangling bonds in these small structures, and based on chemical analysis reported earlier [3], we believe the latter option to hold. The hypothesis of interconnected flakes has already been reported for pyrolized resins [4] and explained in terms of a "butterfly" model.

A large increase in lithium reversible capacity was reported after 20 hours of milling [3,6], time after which all interlayer correlations are lost according to our results presented here. This seems similar to what happens in the "House of Cards" model [9] in which lithium intercalates on both sides of isolated graphite flakes, and which proposes a relationship between lithium capacity and fraction of isolated flakes. A similar mechanism could be involved in the increased lithium capacity of ball-milled carbons.

CONCLUSIONS

RDF analysis is a very useful technique to investigate the local structure of disordered carbons. It allowed us to show that ball milled carbons consist of graphite fragments whose size decrease as a function of milling time. Beyond 20 hours of milling, the material has lost essentially all stacking order along the c-axis. The loss of interlayer correlation coincides with a large increase in lithium reversible capacity. The reasons for enhanced lithium capacity are not obvious. Small angle neutron scattering experiments are currently in progress in order to see whether porosity changes upon milling, as in the "House of Cards" model for hard carbons. A very useful experiment would be to repeat the RDF analysis on lithium-doped samples. Lithium-carbon and lithium-lithium correlations would allow us to understand where the lithium ions go in the structure and thus understand the capacity enhancement.

AKNOWLEDGEMENTS

This work was supported by the MRSEC Program of the National Science Foundation under award Number DMR96-32598 and by Hughes Electronics Corp, and has benefited from the use of the IPNS at ANL which is funded by the U.S. Dept. of Energy, BES-Materials Science, under contract W-31-109-ENG-38.

REFERENCES

[1]- T. Zheng, Y. Liu, E. W. Fuller, U. Von Sacken, J. R. Dahn, *Journal of the Electrochemical Society*, **142**, 2581 (1995).

[2]- K. Sato, M. Noguchi, A. Demachi, N. Oki, M. Endo, *Science*, **264**, 556 (1994).

[3]- F. Disma, L. Aymard, L. Dupont, J-M. Tarascon, *Journal of the Electrochemical Society*, **143**, 3959 (1996).

[4]- P. Zhou, R. Lee, A. Claye, J. E. Fischer, submitted to *Carbon*.

[5]- P. Zhou, P. Papanek, R. Lee, J. E. Fischer, W. A. Kamitakahara, *Journal of the Electrochemical Society*, **144**, 1744 (1997).

[6]- F. Disma, C. Lenain, B. Beaudoin, L. Aymard, J-M. Tarascon, *Solid State Ionics*, **98**, 145 (1997).

[7]- H. Toby, T. Egami, *Acta Crystallographica*, **A48**, 336 (1992).

[8]- R. K. Crawford, D. L. Price, J. R. Haumann, R. Kleb, D. G. Montague, J. M. Carpenter, S. Suman, and R. J. Dejus, in *Proceedings of the 10th Meeting on Advanced Neutron Sources*, Vol.97, p.419, Institute of Physics, IOP Publishing, Limited, New York, 1989.

[9]- J. R. Dahn, T. Zheng, Y. Liu and J. S. Xue, *Science*, **270**, 5236 (1995).

CHARACTERISTICS OF BORON DOPED MESOPHASE PITCH-BASED CARBON FIBERS AS ANODE MATERIALS FOR LITHIUM SECONDARY CELLS

Toshio Tamaki, Toshifumi Kawamura, Yoshinori Yamazaki ,
PETOCA, LTD. Researcher, Research Dept. Advanced Material Division
4 Touwada Kamisumachi, Ibaraki PREF, JAPAN

Abstract

Mesophase pitch-based Carbon Fibers(MCF) have been investigated as anode materials for lithium secondary cells by examining their physical and electrochemical properties. Discharge capacity and initial charge-discharge efficiency of the materials were studied in relation to the heat treatment temperatures of MCF. MCF heat treated at about $3,000°$ C gave high discharge capacity over 310mAh/g , good efficiency (93 %) and superior current capability of 600mA/g (6mA/cm2).   On the other hand, to improve the battery capacity, Boron was doped to the fiber about several % by adding B_4C to the pre-carbonized milled fibers and then heat-treated up to 3000°C in Ar. Then heat treated at 2,500°C under vacuum condition to remove remained B_4C .The structure of Boron-doped fibers was characterized and compared with that of non-doped standard fibers, and also Li ion battery performances are evaluated. The Boron-doped MCF indicated improvement in graphitization and increased discharge capacity as high as 360mAh/g. The voltammograms of both fibers are different from each other. The cell mechanism is discussed based on the unique structure of Boron-doping to the MCF is very effective for the battery performance.

Experimental

Sample

Mesophase Pitch-based Carbon Fibers(MCF), which we used were "MELBLON$^{T·}$ ^{M}" short carbon fibers manufactured by PETOCA, LTD. MCF were carbonized at 65 0° C, then milled. The milled MCF's average particle diameter is 18μm. Boron was doped to the fiber 0~10 wt % by adding $B_4C(10\mu$m)   to the pre-carbonized milled fibers and then heat-treated at several temperatures up to 3000° C in Ar, to obtain various grades of crystallized MCF.

Measurement

The characteristics of boron doped MCF were examined by XRD and electrochemical methods. Structural parameters of the boron doped MCF were measured by XRD with Cu-Kα radiation.

Electrochemical measurements

1. Anode sheet

Milled MCF is mixed with 7 wt% of PVDF binder and NMP for solvent, and then the mixed paste is spread on copper foil. The sheet of MCF is pressed at 1.5 ton/cm^2, and dried at 150°C under vacuum for 3 hours.

2. Three-electrodes cell

A sheet of electrode (20mm in square and about 80μm in thickness) is set in a three-electrodes cell with metallic lithium foil electrodes as the counter and the reference electrodes in excess electrolyte.

3. Electrolyte

1 Mol/l solution of LiClO$_4$ in mixtures of ethylene carbonate (EC) and dimethyle carbonate (DMC) (volume ratio = 1/1) is used as the electrolyte .

4. Charge and discharge condition

1~5th cycle

Charge condition : 1 mA/cm^2 constant current density to 0.01V vs Li/Li$^+$ potential and keep constant voltage at 0.01V total charge time 8 hours.

Discharge condition : 1 mA/cm^2 constant current density discharge to 1.5V.

6~10th cycle

Charge condition : 2 mA/cm^2 constant current density to 0.01V and keep constant voltage at 0.01V total charge time 5 hours.

Discharge condition : 2 mA/cm^2 constant current density discharge to 1.5V.

11~15th cycle

Charge condition : 6 mA/cm^2 constant current density to 0.01V and keep constant voltage at 0.01V total charge time 3 hours.

Discharge condition : 6 mA/cm^2 constant current density discharge to 1.5V.

5. Cyclic voltammograms (CV)

The potential sweep rate: 5μV/sec,

The potential sweep range: 350~0 mV(vs Li/Li$^+$ potential)

RESULTS

Fig. 1 shows the relationship between various B$_4$C addition MCF's d002 value and heat treated temperature. It shows that B$_4$C addition group b-MCF are highly graphitized over 2,200°C by boron doping. Boron is the graphitization catalyst. Boron is the only metal that can replace a carbon atom in the crystal structure [1]. When the boron doping , it cut the carbon sp^3 bonding and accelerate the graphitization of MCF.

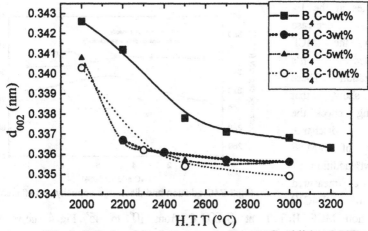

Fig.1 Relationships between various B_4C addition MCF

d_{002} value and heat treated temperature

Fig. 2 shows the relationship between B_4C addition and MCF's peak intensity of I 101/1100 , d_{002} value heat treated at 3,000° C. I 101/1100 peak intensity shows graphitization of b-doped MCF[2]. It indicate that boron remarkably accelerated the graphitization of MCF.

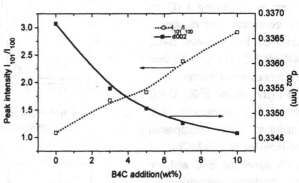

Fig.2 Relationships between B4C addition and peak intensity, d002 value at 3000°C

Fig. 3 shows the B_4C addition and discharge capacities , charge-discharge efficiencies at the 1st cycle for b-MCF. Discharge capacity of b-MCF increase with B4C addition to 5wt%, 5wt%B 4C addition MCF shows340mAh/g discharge capacity, and 93%efficiency at the 1st cycle[3]. But more than 7wt%addition of B4C, discharge capacities ,

charge-discharge efficiencies at the 1st cycle decrease although the MCF accelerated it's graphitization . These facts show that something check the charge discharge property of b-MCF over 5wt%addition .
Then we measured XRD of B4C10

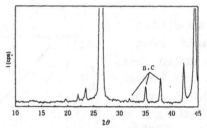

Fig.3 B4C addition and discharge capacity ,1st efficiency.

wt%addition MCF H.T.T at 3,000°C from 10° to 45°. Fig.4 shows B4C10 wt%MCF-3,000°C in Ar XRD pattern. It indicates clearly B₄C remaining peak at 37.8°. B₄C is non-electroconductive material and we confirmed that B₄C remained on the

surface of b-MCF by SEM observation. Thereupon we try to remove remaining B₄C by heat treatment under vacuum at 2,500°C. Fig.5 shows B4C1 0wt%MCF-3,000°C in Ar then heat treated under vacuum at 2, 500°C(for short 25R)XRD pattern. It indicates B₄C remaining peak disappeared and removed from b-MCF.

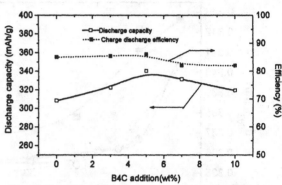

Fig.4 B4C-10wt%MCF-3,000°C in Ar Xray diffraction pattern

Fig.6 shows the B4C addition and discharge capacities , charge-discharge efficiencies at the 1st cycle for b-MCF and heat treated under vacuum at 2, 500°C. B₄C10wt%-MCF-3, 000°C - 25R gives superior discharge capacity of 368 mAh/g and high efficiency(93 %). Therefore it is clear that remaining B₄C check the

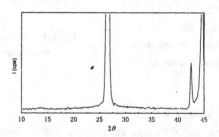

Fig.5 B4C-10wt%MCF-3,000°C in Ar-2500°C under vacuum Xray diffraction pattern

charge discharge property of b-MCF over 5 wt%addition .

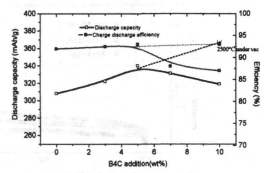

Fig. 6 B4C addition and discharge capacity ,1st efficiency.

Fig. 7 shows the cyclic voltammogram (CV) of B_4C10wt%-MCF-3,000° C - 25R and non doped MCF-3,000° C. At such a low sweep rate , two large cathodic and anodic peaks were clearly observed for both samples. These peaks are labeled with arabic numerals. Peak 1 in Fig. 7 correspond to the equilibrium reaction of the dilute stage 1(C_6Li) and stage 2 (C_{12}Li) two phase coexistence. Peak 2 correspond to the stage 2(C_{12}Li) and stage 2 (C_{18}Li) two phase coexistence[4].

Both peaks 1, 2 of B_4C10wt%-MCF-3,000° C - 25R are larger than that of non doped MCF-3,000° C.

Therefore it is clear that B_4C10wt%-MCF-3,000° C - 25R gives superior discharge capacity of 368 mAh/g .

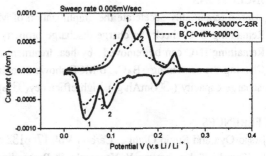

Fig.7 CV curves of B_4C-10wt%-3000°C-25R and non doped MCF-3000°C

Fig. 8 shows high rate discharge curves of B_4C 10wt%-MCF-3,000° C - 25R. At current densities of 1mA/cm² , 3mA/cm² and even at 6mA/cm² it shows superior discharge capacity more than 350mAh/g . Because in the negative electrode composed of the highly graphitized b-MCF keep the radial-like texture [5], the intercalation of lithium ions into the graphite layers is simultaneously carried out through the lithium ion inlet opening widely spread all over the surface of the b- MCF. So we can use this b-MCF for high rate charge-discharge batteries.

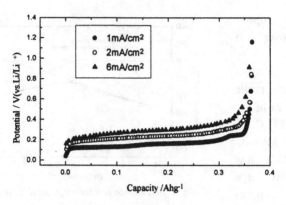

Fig.8 Discharge curves of B4C-10wt%-3000 C-25R
at 1,2,6 mA/cm^2 current density.

CONCLUSIONS

1. Boron remarkably accelerate the graphitization of MCF.
2. Remaining B₄C check the charge discharge property of b-MCF over 5wt%addition .
3. Remaining B₄C can be removed by heat treatment at 2,500° C under vacuum.
4. Clearing the remaining B₄C, b-MCF shows essential performance of high
   discharge capacity (368mAh/g), high efficiency(93%) and superior current capability.

REFERENCES

[1]Asao Oya and Sugio Otani , Carbon Vol. 17, p132 1978
[2]T. Tamaki, T. Kawamura, Y. Yamazaki, in Proceedings of the 38th Battery
    Symposium in Japan , Osaka p257 1997
[3]T. Tamaki, T. Kawamura, T. Maeda, Y. Yamazaki, in Proceedings of the 37th
        Battery Symposium in Japan , Tokyo p55 1996
[4]N. Takami, A. Satoh, M. Hara, T. Ohsaki . J. Electrochem. Soc. , Vol. 142, No. 8,
    p2566 August 1995
[5]T. Tamaki, M. Tamaki, in Proceedings of the 38th Battery Symposium in Japan ,
    Hirosima p89 1993

COMMERCIAL COKES AND GRAPHITES AS ANODE
MATERIALS
FOR LITHIUM - ION CELLS

David J. Derwin *, Kim Kinoshita **, Tri D. Tran +, Peter Zaleski *

*Superior Graphite Co., 6540 S. Laramie Ave. Chicago, Ill. 60638
** Energy and Environmental Division
Lawrence Berkeley National Laboratory, Berkeley, Ca. 94720
+Chemistry & Materials Science Department,
Lawrence Livermore National Laboratory, Livermore, Ca. 94550

Abstract

Several types of carbonaceous materials from Superior Graphite Co. were investigated for lithium ion intercalation. These commercially available cokes, graphitized cokes and graphites have a wide range of physical and chemical properties. The coke materials were investigated in propylene carbonate based electrolytes and the graphitic materials were studied in ethylene carbonate / dimethyl solutions to prevent exfoliation. The reversible capacities of disordered cokes are below 230 mAh / g and those for many highly ordered synthetic (artificial) and natural graphites approached 372 mAh / g (LiC_6). The irreversible capacity losses vary between 15 to as much as 200 % of reversible capacities for various types of carbon. Heat treated cokes with the average particle size of 10 microns showed marked improvements in reversible capacity for lithium intercalation. The electrochemical characteristics are correlated with data obtained from scanning electron microscopy (SEM), high resolution transmission electron microscopy (TEM), X - ray diffraction (XRD) and BET surface area analysis. The electrochemical performance, availability, cost and manufacturability of these commercial carbons will be discussed.

INTRODUCTION

The carbonaceous materials (both graphite and coke) that will be discussed were supplied by Superior Graphite Co., a leading producer of quality graphite, carbon and related materials for 80 years. Currently producing graphite related products at three different plants, Superior Graphite Co. has been able to be a leader in the graphite industry by expanding their operations from graphite mining to graphite processing and finally to the development of engineered graphitic materials. Utilizing Superior Graphite Co.'s unique high temperature furnaces (at the Hopkinsville KY and Russelville, Ark. plants) and advanced process technologies with raw material supplies (at the Chicago plant), several different types of carbon were examined and processed under a variety of conditions. Working in conjunction with Lawrence Berkeley National Laboratory and Lawrence Livermore National Laboratory, this study focused on how different carbon based materials, by varying processing steps, could produce low cost materials suitable for application in lithium - ion cells.

EXPERIMENTAL

The cokes and graphites used in this study have been divided into two classes (graphitized and non graphitized) based on the highest heat treatment temperature to which the carbon was exposed to. This distinction was made because different electrolytes were used to evaluate these two classes of materials to minimize the irreversible capacity loss. Briefly, the non graphitized carbons were tested in propylene carbonate based electrolytes with 0.5 M lithium trifluoromethanesulfonimide. The graphitized carbons were studied in 0.5 M (50/50) ethylene carbonate / dimethyl carbonate.

The powdered materials of known size were used "as received" directly in the electrode fabrication procedure using carbon based binder (1). The larger, more granular carbons were first ground in a mortar and pestle, then sieved to 30 - 40 microns.

Briefly, the electrodes were prepared using a commercial carbon fiber sheet as the support matrix and carbonized phenolic resin binder (10-15%). A slurry containing the carbon particles, the phenolic resin precursor and furfuryl alcohol was spread on the carbon fiber support and allowed to dry. The composites were pyrolyzed in N_2 at 1050° C. Cycling experiments for each carbon electrode were carried out in a 15 ml three electrode cylindrical cell. Lithium foils were used as the counter and reference electrodes. Whatman fiberglass filters were used as the separator between the working and counter electrodes. The cells were constructed and tested at 16 +/- 2° C in a dry argon atmosphere glove box (<10 ppm water). The electrode was charged (intercalated) at a constant rate corresponding to 1 mole of lithium per 6 moles of carbon in 24 hours (ca. C/24 rate) to a cut off voltage of 0.005V (Li^+ / Li) and held at this potential for 4 hours. It was then discharged (deintercalated) at the same rate to insure complete lithium intercalation / deintercalation. Electrochemical studies were performed using a 64 channel Maccor battery tester.

Various equipment, High Resolution Transmission Electron Microscopy (HRTEM) X - ray Diffraction (XRD), Scanning Electron Microscope (SEM), BET Surface Area machine and Laser Particle analyzers were used to study and characterize these carbonaceous materials.

RESULTS AND DISCUSSION

Table 1 lists materials used in this study. This information was obtained from specification sheets or measured in our laboratories.

COKES

This group of carbons consisted of two grades of calcined petroleum cokes that vary in sulfur content. The two cokes tested, SGC Coke # 1 (< 1 % sulfur) and SGC Coke #3 (< 3% sulfur) in general showed different results. The curves for these non graphitic materials have a sloping shape with varying lithium capacity, typical for amorphous carbons. The low sulfur coke (SGC Coke #1) have x - values near 0.70 with an irreversible capacity around 83 mAh / g. This performance data is comparable to other cokes that have been tested in the laboratory. High sulfur cokes tend to have lower capacities (x approx. 0.51) and larger irreversible capacities (110 mA / g). The capacity of SGC Coke #1 was stable for more than 20 cycles whereas the capacity of SGC Coke #3 decreases significantly after 4 cycles. The higher sulfur content of SGC Coke #3 may affect its performance.

Note : further studies are underway to determine which physical and microstructure properties of calcined petroleum cokes are important for lithium battery usage. These studies include, but are not limited to, different heat treatment temperatures, grinding techniques (before and after heat treatment), tighter particle distributions and surface characteristics.

GRAPHITIZED COKES

The graphitized petroleum cokes discussed in this section are derived from Superior Graphite Co. patented thermal purification process based in Hopkinsville, KY. This continuous thermal purification process converts calcined petroleum cokes to a purified / partially graphitized product (DESULCO). The temperatures reached during this process, near 2700° C, eliminate sulfur and other volatile components from the calcined petroleum cokes. DESULCO carbons (SGC 9035) have capacities near 230 mAh / g (x = 0.60). They exhibit graphite like potential curves albeit at lower capacities than those of more crystalline graphites (LiC_6).

Note : as with the cokes, further studies are underway to determine what physical and microstructure properties of DESULCO are important for lithium battery usage.

GRAPHITE

Through the same patented thermal purification process used to produce DESULCO, it is possible to take an impure (high ash content) natural graphite and thermally upgrade the carbon value to levels of 99.9 +% pure LOI (ash content less than 0.1 %). Once these impurities have been thermally removed, the resulting graphite can be a candidate for lithium battery usage. These thermally purified graphites can be highly crystalline materials that vary in microstructure and morphology, depending on the source. Two of graphites that will be discussed in this section are the thermally purified natural crystalline flake graphite and the thermally purified natural crystalline vein graphite. The thermally purified natural crystalline flake graphite is highly anisotropic in nature. The graphite morphology resembles flat platelets (flat lamellae) with well defined basal planes. The thermally purified natural crystalline vein graphite is more microcrystalline in nature i.e. smaller crystallites). The graphite morphology is more needle / grain like in shape.

The reversible capacities observed with these materials are about 320 - 360 mAh / g, which correspond to the x values between 0.85 - 0.95 (Table 1). Thermally purified natural flake graphite (SGC 2933 & SGC BG 39) tend to have a considerably lower capacity than the LiC_6 composition. All types of thermally purified natural graphites tested in this study exhibited the graphite like potential curves (i.e., deintercalation / intercalation takes place at potentials below 0.3 V). However, there are distinct differences between the profiles of the flake and vein materials. Flake graphites showed plateaus below 0.3 V that could be identified with the formation of staged phases. For the vein graphites, the curve appears smooth in this region with no apparent plateaus (probably attributed to the microcrystalline nature of the vein graphites).

Note : further studies are underway to determine which characteristics of the thermally purified natural graphites affect lithium battery performance. These future studies on the thermally purified natural graphites do include, but are not limited to, the lowering of surface area (below 3.0 sq.m./g.) and the classifying of the particle size to optimize performance. Also, another source (thermally purified Amorphous graphite) is under investigation.

AVAILABILITY AND COST

One of the goals of this project was to identify suitable carbon / graphite materials that are inexpensive and commercially available in large quantities. The materials covered in this report are commercially available in large quantities. Superior Graphite CO., an ISO 9001 certified company, has the technology and capability to custom process these materials, and guarantee their sourcing for consistent products. In general, the cost of the thermally purified graphites covered in this report is below $5.00 / lb, with the cokes priced significantly lower.

SUMMARY

The results show the performance of Superior Graphite CO. carbonaceous materials for lithium ion intercalation. The material shows a wide range of performance characteristics. The electrochemical behavior is related to their physical and structural properties. Many types of carbon appear suitable for application in lithium - ion cells. Work is currently underway to produce and test lower surface area thermally purified natural graphite which have a more controlled particle distribution for lithium battery usage.

TABLE 1.

SUPERIOR GRAPHITE CO.

GRAPHITIZED AND NON GRAPHITIZED CARBONS

PROPERTIES AND PERFORMANCE.

Sample	Type	Part.size (microns)	x in (Li_xC_6)	Irr. Cap. Loss (mAhr / g)
Coke #1	Calcined Pet. Coke (low sulfur < 1 %)	30 - 40	0.70	83
Coke #3	Calcined Pet. Coke (sulfur > 2 %)	30 - 40	0.51	110
SGC 9035	DESULCO (grap. Pet. Coke 2700 deg C)	20 - 30	0.59	129
SGC 2933	Purified Flake (natural graphite)	30 - 40 *	0.86	76
SGC BG-39	Battery Grade (purified natural flake graphite)	10 - 20	0.87	98
SGC 4941A	Purified Vein (natural graphite)	7 - 10	0.98	130

* Sieved to between indicated range.

ACKNOWLEDMENT

The author would like to Tri D. Tran and Kim Kinoshita for their contributions to this paper. This work was supported by the Assistant Secretary for Energy Efficiency and Renewable Energy, Office of Transportation Technologies, Electric & Hybrid Propulsion Division of the U.S. Department of Energy under Contract No. W-7405-ENG-48 (Lawrence Livermore National Laboratory) and Contract No. DE-AC0376SF00098 (Lawrence Berkeley Laboratory).

REFERENCES

1. T. D. Tran, J. H. Feikert, X. Song, and K. Kinoshita, *J. Electrochem. Soc.* **142,** 3297 (1995)

ELECTROCHEMICAL PROPERTIES OF NITROGEN-SUBSTITUTED CARBON AND ORGANOFLUORINE COMPOUNDS

T. NAKAJIMA, M. KOH, K. DAN
Division of Polymer Chemistry, Graduate School of Engineering, Kyoto University, Sakyo-ku, Kyoto, 606-01, Japan

ABSTRACT

The C_xN samples($C_{14}N$-$C_{62}N$) prepared with a nickel catalyst had the higher crystallinity and less pyridine-type nitrogens existing at the edge of graphene layers than C_xN prepared in the absence of a catalyst. With increase in the deposition temperature of C_xN, the cycleability for electrochemical intercalation-deintercalation of lithium ions was improved and the profile of the charge-discharge curve approached that of graphite due to increase in the crystallinity and decrease in the incorporated nitrogens. C_xN-coated graphites demonstrated gradual increase in the potential at the last stage of lithium ion deintercalation process. The effect of fluoroester-mixing in 1 M $LiClO_4$-EC/DEC was also investigated at a low temperature. It was found that CHF_2COOCH_3 with a low molecular weight and a small number of fluorine atoms was effective as a mixing agent.

INTRODUCTION

Recently boron- and/or nitrogen-substituted carbons, BC_2N, BC_x and C_xN were synthesized by chemical vapor deposition (CVD), and their electrochemical behavior as negative electrodes of lithium secondary batteries were investigated [1-4]. These compounds would consitute a new class of candidates for negative electrodes with modified electronic structures, having different chemical interaction with lithium ions. Well crystallized C_xN filaments and particles have been synthesized by CVD of acetonitrile or pyridine using a nickel or cobalt catalyst [5]. In this paper, we report the electrochemical behavior of these C_xN compounds in 1 M $LiClO_4$-EC/DEC. In addition, we have examined the effect of fluoroester-mixing into 1 M $LiClO_4$-EC/DEC at a low temperature. The result is also reported.

EXPERIMENT

The C_xN samples were synthesized by thermal decomposition of acetonitrile at 800-1100°C using a nickel catalyst. Flow rate of a mixture of acetonitrile and nitrogen was 55 mlmin^{-1} (partial pressure of acetonitrile:9.2×10^3 Pa). C_xN coating was performed in a similar manner at 950°C using natural graphite powder($\approx 7\mu$m) oxidized by 94% HNO_3. The obtained samples were analyzed by elemental analysis, X-ray diffractometry, SEM, TEM, XPS and Raman spectroscopy.

Electrochemical intercalation and deintercalation of lithium ion into and from C_xN samples were performed in 1 M $LiClO_4$-EC/DEC at a current density of 30 or 60 mAg^{-1} at 25°C. Counter and reference electrodes were metallic lithium. The effect of fluoroester-mixing was examined at 25°C and 0°C by adding a reagent grade fluoroester (2.5ml) : CHF_2COOCH_3 1, $CF_3CF_2COOCH_2CH_3$ 2, $(CF_3)_2CHCOOCH_3$ 3, $F(CF_2)_3COOCH_3$ 4, $H(CF_2)_4COOCH_2CH_3$ 5, $F(CF_2)_7COOCH_3$ 6 or $F(CF_2)_7COOCH_2CH_3$ 7, to 1 M $LiClO_4$-EC/DEC (50ml). Cyclic voltammetry at 0.1 mVsec^{-1} and charge-discharge cycling at 80mAg^{-1} were conducted for natural graphite($\approx 7\mu$m) electrode in fluoroester-mixed 1 M LiClO$_4$-EC/DEC solutions.

RESULTS

Compositions, morphologies and crystallinity of C$_x$N

Table 1 shows the typical examples of compositions, morphologies and X-ray diffraction data of C$_x$N samples prepared at 800-1100°C with nickel catalyst in comparison with those of C$_x$N prepared without catalyst and carbon prepared from benzene with nickel. C$_x$N samples with compositions C$_{14}$N-C$_{62}$N were obtained in the temperature range of 800-1100°C. No hydrogen was detected by elemental analysis for all the samples prepared with nickel catalyst whereas less crystallized C$_{14}$NH$_{0.6}$ having hydrogen was obtained at 1000°C in the absence of the catalyst. The products prepared 900°C and 1000°C had the highest nitrogen contents, as C$_{14}$N-C$_{21}$N. Nitrogen concentrations were much less in the products deposited at the lower and higher temperatures, 800°C, 1050°C and 1100°C.

The C$_x$N samples prepared were apparently black powder. However, SEM and TEM observations revealed that the products obtained at 800°C were fibrous C$_x$N with diameters of ca. 3 μm and 0.3 μm, those at 900°C were mixtures of fibrous C$_x$N with diameters of ca. 0.5 μm and fine particles with diameters of 2-3 μm, and those at 1000°C were particles with diameters of 2-3 μm.

The C$_x$N samples exhibited the same diffraction lines as usual carbon materials. Table 1 shows that both d(002) values and half widths were smaller than those for C$_{14}$NH$_{0.6}$ prepared without nickel, indicating that the C$_x$N samples synthesized using nickel have much higher crystallinity than that prepared without nickel.

XPS analysis shows the existence of pyridine type nitrogen at the edge of graphene layer (398.8eV), nitrogen bonded to three carbon atoms (400.9eV), and pyridine-N-oxide (402.3eV) [6]. The peak at 400.9eV has higher intensity than other two peaks corresponding to nitrogen atoms existing at the edge of graphene layer when C$_x$N is prepared with nickel catalyst, whereas C$_x$N prepared without nickel has a large amount of pyridine type nitrogens.

Table 1. Compositions,morphologies and XRD data of C$_x$N and carbon samples.

React. temp. (°C)	Composition	Morphology	Diameter (μm)	d(002) (nm)	FWHM (°)	L_c(002) (nm)
1100	C$_{62}$N[a]	Particle	—	0.335	0.575	17
1050	C$_{33}$N[a]	Particle	—	0.335	0.663	15
1000	C$_{20}$N[a]	Particle	2~3	0.336	0.775	13
900	C$_{21}$N[a]	Particle	2~3	0.335	0.900	11
		Fiber	$\approx$0.5			
800	C$_{40}$N[a]	Fiber	$\approx$3,0.3	0.337	1.250	8
1000	C$_{14}$NH$_{0.6}$[b]	—	—	0.340	2.850	3
900	C[c]	—	—	0.336	0.938	10

a Prepared from acetonitrile (partial pressure: 9.2$\times$10^{3}Pa) with Ni catalyst,

b Prepared from acetonitrile (partial pressure: 9.2$\times$10^{3}Pa) without catalyst,

c Prepared from benzene (partial pressure: 9.2$\times$10^{3}Pa) with Ni catalyst.

Electrochemical behavior of C_xN

The potential of C_xN electrodes gradually decreased and increased with lithium ion intercalation and deintercalation. Nitrogen incorporation in carbon induced the larger polarization for lithium ion deintercalation process and the slightly smaller one for the intercalation process as shown in Fig. 1. The observed polarization suggests that lithium ion interacts with a lone pair of nitrogen atom in C_xN. Table 1 indicates that nitrogen content decreased and crystallinity increased with increase in the deposition temperature from 900°C to 1100°C. According to this trend, the profile of charge-discharge curves of C_xN samples approached those of benzene-derived carbon and natural graphite powder. The cycleability was improved with increasing deposition temperature for C_xN probably due to the increase in the crystallinity while the capacity was slightly decreased. Surface oxidation of C_xN samples by nitric acid solution improved the cycleability, however, induced the larger polarization at the same time. The oxidation with nitric acid increased the amount of surface oxygen by about four times and the surface areas by about two times. XPS also showed that the surface oxidation changed the pyridine-N-oxide to amine ($-NH_2$).

$C_{20}N$ prepared at 1000°C with nickel exhibited lower charge capacities than $C_{14}NH_{0.6}$ prepared without nickel at a first cycle, however, their capacities approached each other after 7th cycle. The charge capacity of $C_{20}N$ was larger than that of $C_{21}N$ prepared at 900°C probably because of slightly higher crystallinity of $C_{20}N$.

The coulombic efficiencies of $C_{14}NH_{0.6}$ were inferior to those of other C_xN samples prepared with nickel, which were in the range of 68% to 75% lower than that of natural graphite powder. Surface oxidation of C_xN further reduced the first coulombic efficiencies by 9%.

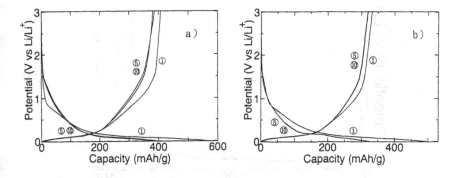

Fig. 1 Charge-discharge curves for $C_{20}N$(a) and $C_{33}N$(b) prepared from acetonitrile at
    1000°C and 1050°C with nickel catalyst.
    1:1st, 5:5th, and 10:10th cycles at $30mAg^{-1}$.

Electrochemical behavior of C_xN-coated graphite

There was a large difference between the Raman spectra of a mixture of NHO_3-treated natural graphite (HNO_3-TNG) with C_xN and C_xN-coated HNO_3-TNG, while X-ray diffraction profile of the C_xN-coated sample was almost the same as those of HNO_3-TNG itself and C_xN-mixed sample. The Raman spectrum of C_xN-coated sample gave two broad peaks at $1580cm^{-1}$ and $1360cm^{-1}$, however, the Raman spectrum of the mixture of HNO_3-TNG and C_xN was similar to that of natural graphite powder. Since detection depth by Raman spectroscopy is much shallower than that by X-ray diffractometry, the result concides with that C_xN is coated on HNO_3-TNG.

Fig. 2 shows the charge curves at 10th cycle for natural graphite (NG) and C_xN-coated HNO_3-TNG samples prepared by 0.5, 1.5 and 3 hrs deposition. Table 2 summarizes the electrochemical results. NG shows a sharp increase in the potential at the last stage of lithium ion deintercalation with the first charge capacity of 348 mAhg^{-1} between 0 V and 1 V vs Li/Li$^+$. On the other hand, C_xN-coated HNO_3-TNG sample prepared by 0.5 and 1.5 hrs reactions demonstrated gradual increase in the potential at the last stage of charging with the first charge capacities of 351 and 353 mAhg^{-1}, respectively. No decrease in the charge capacity was observed for these two C_xN-coated HNO_3-TNG samples. The potential of C_xN-coated HNO_3-TNG prepared by 3 hrs deposition started to increase at the earlier stage than those for other two C_xN-coated HNO_3-TNG samples. It is due to the larger amount of C_xN deposited by longer duration of the reaction. Thus the thin C_xN coating to HNO_3-TNG would have several advantages without decrease in the cycleability. Because the C_xN deposition to the HNO_3-TNG proceeds in a reductive atmosphere, it may be possible that carbon-carbon bond formation occurs in part between HNO_3-TNG and deposited C_xN by eliminating the oxygen atoms as carbon oxide gases.

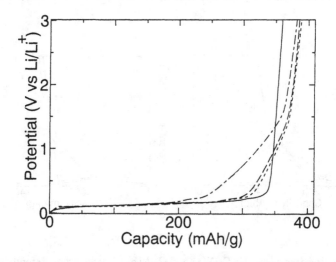

Fig. 2 Charge curves at 10th cycle for natural graphite and C_xN-coated natural graphite.
———— :natural graphite, - - - - - -:C_xN-coated natural graphite after oxidized with 94% HNO_3 (C_xN deposition:0.5hr), — — — :C_xN-coated natural graphite after oxidized with 94% HNO_3 (C_xN deposition:1.5hr),
— - — :C_xN-coated natural graphite after oxidized with 94% HNO_3 (C_xN deposition:3hrs).

Table 2. Charge and discharge capacities between 0 and 1V vs. Li/Li$^+$ and coulombic efficiencies of natural graphite, surface-treated graphite and C$_x$N- or carbon-coated graphites.

Sample name	1st cycle			10th cycle		
	Discharge (mAhg^{-1})	Charge (mAhg^{-1})	Q (%)	Discharge (mAhg^{-1})	Charge (mAhg^{-1})	Q (%)
NG	444	348	78.4	364	354	97.3
HT-94NG^a	451	320	71.0	350	332	94.9
CN-NG^b	497	349	70.2	354	328	92.7
CN-94NG1^c	542	351	64.8	384	355	92.6
CN-94NG2^c	529	353	66.7	380	351	92.4
CN-94NG3^c	588	354	60.3	362	323	89.3
C-94NG^d	452	314	69.5	334	308	92.3

NG : Natural graphite (average diameter:ca.7μm).

a : Heat-treated natural graphite after oxidized with 94% HNO$_3$.

b : C$_x$N-coated natural graphite.

c : C$_x$N-coated natural graphite after oxidized with 94% HNO$_3$.

d : Carbon-coated natural graphite after oxidized with 94% HNO$_3$.

Current : 60mAg^{-1}

Effect of fluoroesters on the low temperature characteristics of graphite electrode

The cyclic voltammetry revealed that the fluoroesters used were electrochemically reduced at more positive potentials between 0.87 V and 1.41 V vs Li/Li$^+$, mainly at 0.99-1.10 V, than EC reduced at 0.60 V, and with decreasing reduction potential of the fluoroesters, the larger redox currents reversibly flowed in the fluoroester-mixed solvents than in EC/DEC at 0℃. Among fluoroesters examined, CHF$_2$COOCH$_3$ gave the lowest reduction potential of 0.87 V. The reversible redox currents increased with decreasing reduction potential of the fluoroester.

The lower charge capacities were observed as a whole at 0℃ compared with those at 25℃. However, the capacity in CHF$_2$COOCH$_3$-mixed solvent was relatively much higher than those in EC/DEC (Fig. 3) and other fluoroester-mixed solvents. Not only the capacities but also the coulombic efficiencies were higher in CHF$_2$COOCH$_3$-mixed EC/DEC than in other solvents. The first charge capacities and coulombic efficiencies in the fluoroester-mixed solvents decreased with increasing reduction potentials of the fluoroesters. The results indicate that CHF$_2$COOCH$_3$ with a low molecular weight and a small number of fluorine atoms would have several advantages of the lower reduction potential, and higher chemical interaction and miscibility with lithium salt and EC/DEC than other fluoroesters.

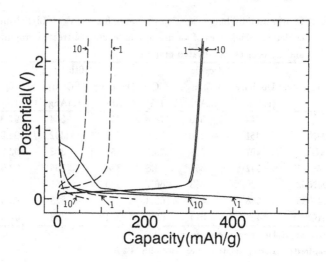

Fig. 3 Charge-discharge curves for natural graphite in 1 M LiClO$_4$-EC/DEC and CHF$_2$COOCH$_3$-mixed EC/DEC at 0℃.
- - - - -:EC/DEC, ——————:CHF$_2$COOCH$_3$-mixed EC/DEC, 1:1st cycle, 10:10th cycle.

CONCLUSIONS

Well crystallized nitrogen-substituted carbons, C$_{14}$N-C$_{62}$N were synthesized by CVD of acetonitrile using a nickel catalyst. Polarization due to interaction of lithium ion with nitrogen atom was observed. C$_x$N-coated graphite showed gradual increase in the potential at the end of lithium ion deintercalation. The larger charge capacities were achieved in CHF$_2$COOCH$_3$-mixed 1 M LiClO$_4$-EC/DEC than in EC/DEC itself and other fluoroester-mixed ones at 0℃ and -4℃.

REFERENCES

1. M. Morita, T. Hanada, H. Tsutsumi, Y. Matsuda and M. Kawaguchi, J. Electrochem. Soc. **139**, 1,227 (1992).
2. W. J. Weydanz, B. M. Way, T. van Buuren and J. R. Dahn, J. Electrochem. Soc. **141**, 900 (1994).
3. M. Way and J. R. Dahn, J. Electrochem. Soc. **141**, 907 (1994).
4. M. Ishikawa, T. Nakajima, M. Morita, Y. Matsuda, S. Tsujioka and T. Kawashima, J. Power Sources **55**, 127 (1995).
5. T. Nakajima and M. Koh, Carbon **35**, 203 (1997).
6. J. R. Pels, F. Kapteijn, J. A. Moulijn, Q. Zhu and K. M. Thomas, Carbon **33**, 1,641 (1995).

RECENT STUDIES OF INTERFACIAL PHENOMENA WHICH DETERMINE THE ELECTROCHEMICAL BEHAVIOR OF LITHIUM AND LITHIATED CARBON ANODES WITH THE EMPHASIS ON *IN SITU* TECHNIQUES

D. AURBACH, A. SCHECHTER, B. MARKOVSKY, Y. COHEN, I. WEISSMAN AND M. MOSHKOVICH
Department of Chemistry, Bar-Ilan University, Ramat-Gan 52900, Israel

ABSTRACT

This paper reports on some new results on the application of surface sensitive techniques for the study of the correlation of surface chemistry, morphology and electrochemical behavior of lithium and lithiated graphite as anodes for rechargeable batteries. Surface sensitive FTIR spectroscopy, atomic force microscopy (AFM), electrochemical quartz crystal microbalance (EQCM) were applied to Li and Li-graphite electrodes in a variety of electrolyte solutions of interest, in conjunction with standard electrochemical techniques. The similarity in the surface chemistry developed on Li and lithiated graphite in solutions is demonstrated and discussed. We demonstrate the strong impact of the surface chemistry on the morphology of Li deposition-dissolution processes, and the use of *in situ* EQCM measurements for the choice of optimal electrolyte solutions for rechargeable batteries with Li metal anodes.

INTRODUCTION

It is generally accepted that the behavior of both lithium and lithiated carbon anodes for rechargeable, high energy nonaqueous batteries depends on their surface chemistry. This surface chemistry develops spontaneously as lithium or lithiated carbon are exposed to any nonaqueous solution based on polar aprotic solvents.

Insoluble Li organic or inorganic salts precipitate on the active electrode surfaces as surface films. Their growth stops as they reach a certain thickness which blocks electron tunneling from the active electrode through the surface films to solution species. While these surface films become electronically insulating at a certain thickness, they remain Li$^+$ ionically conducting, due to the special properties of the bonds between Li and elements such as oxygen, carbon, halides, nitrogen, sulfur, etc., which allow some lability of the Li ion in the solid state (especially when the relevant compounds are thin surface films with defects and imperfections). The structure of these surface films is usually complicated because they are mosaic-like, containing different regions due to the large variety of surface reactions possible for each Li-solution or Li-C-solution system. In addition, these films have a multilayer structure which includes a bi- or three-layer compact part and an out-porous part in the solution side. [1,2]

As these surface films determine the electrochemical behavior of both Li and Li-C electrodes, identification of their chemical structure and studying their electrical properties and morphology is very important for systematic R&D of rechargeable batteries based on Li and Li-C anodes. In this paper we review our recent work on several channels of information, including *in situ* and *ex situ* surface sensitive FTIR spectroscopy, *in situ* electrochemical quartz crystal microbalance (EQCM) and *in situ* electrochemical atomic force microscopy (AFM).

Mat. Res. Soc. Symp. Proc. Vol. 496 © 1998 Materials Research Society

EXPERIMENTAL CONSIDERATIONS

All the considerations for the various *in situ* techniques and the experimental set-up (with the emphasis on the application to Li and Li-C electrodes) were already described. The relevant references for EIS, XPS, *in situ* FTIR spectroscopy, AFM and EQCM are 1-5, respectively. Typical studies on Li-C electrodes are described in ref. 6, and experiments with practical rechargeable Li-Li$_x$MnO$_2$ batteries are described in ref. 7.

All the solvents mentioned in this work were highly pure battery grade, obtained from Tomiyama or Merck (Selectipure series). LiPF$_6$ was obtained from Hashimoto or Tomiyama, and LiAsF$_6$ was obtained from FMC (USA).

RESULTS AND DISCUSSION

a. Examples for Studies of the Surface Chemistry of Li and Li-C Electrodes

As already found, non-active electrodes such as noble metals or metals which are inactive at low temperature, develop surface chemistry similar to that of Li electrodes when polarized to low potentials in nonaqueous Li-salt solutions. This could be further confirmed by FTIR spectroscopy of Ni, Pt and Au electrodes polarized to low potentials, or Ni electrodes on which Li was deposited electrochemically.

Fig. 1 presents FTIR spectra obtained *in situ* from Ni or Pt electrodes (indicated) polarized to low potentials (indicated) in ethylene carbonate (EC) based solutions with dimethyl carbonate (DMC), or tetrahydrofurane (THF) as cosolvents, and LiAsF$_6$ or LiPF$_6$ as the electrolytes (indicated near each spectrum). Each spectrum in Fig. 1 was obtained by subtracting the solution spectrum (measured at OCV) from the spectrum obtained *in situ* at the low potential which includes absorption of both solution and surface species.

It should be noted that subtraction of an OCV spectrum from spectra measured from an electrode covered with surface films is very difficult, and thus masking of part of the surface species spectra by solution absorption is a problem. Thus, not all the spectra obtained are of high quality. Nevertheless, these studies were found to be quite useful.

These FTIR spectra reflect some differences in the surface chemistry of both Li and Ni (low potentials) in LiAsF$_6$ and LiPF$_6$ solutions.

In LiAsF$_6$ solutions, EC reduction to (CH$_2$OCO$_2$Li)$_2$ is dominant (major peaks at 1700-1600 cm^{-1} $\nu_{C=O, as}$; 1450-1400 cm^{-1} δ_{CH_2}; 1350-1300 cm^{-1} $\nu_{C=O, a}$; 1100-1000 cm^{-1} $\nu_{C=O}$ and 850-800 $\delta_{CO_3^-}$).

In LiPF$_6$ solutions, HF contamination is always present, and thus partial solubilization at the ROCO$_2$Li species by their reaction with HF takes place, allow salt reduction to be pronounced.

Thus, species of the Li$_x$PF$_y$ and Li$_x$PO$_y$F$_z$ types are formed (ν_{P-F} at 850 cm^{-2} and ν_{P-O} at 1100-1000 cm^{-1}).

The application of *in situ* FTIR spectroscopy for the study of surface films formed on lithium electrodes is very important because both the Li surfaces and the surface species formed on Li in many polar aprotic solvents of interest are highly reactive with all major atmospheric components.

However, as shown above, these measurements may have a serious drawback: a possible masking effect of the solution species.

Therefore, the surface chemistry of Li electrodes has also to be studied by *ex situ* FTIR measurements whose reliability is confirmed and complemented by the above described *in situ* measurements. It is well known that the reversibility and cyclability of lithiated carbon anodes in Li ion batteries depends very strongly on their surface chemistry. The stability of lithiated carbon electrodes, especially those of graphitic structure, depends on the properties of the surface films formed on them at low potentials. [6]

Analysis of these surface films on carbonaceous material by FTIR spectroscopy is difficult due to the poor reflectivity of the carbon particles. However, since similar surface films are formed on Li and Li-C electrodes in the same solutions, a comparative study which links surface film identification on lithium surfaces (by FTIR spectroscopy using grazing angle reflectance mode) and diffuse reflectance FTIR spectroscopy of carbon electrodes after being cycled in the same solution, can be very useful. This is demonstrated below in Figs. 2-4.

Fig. 2 compares a FTIR spectrum measured from *ex situ* Li surface prepared fresh in ethyl methyl carbonate (EMC)/LiAsF$_6$ 1M solution and stored for 3 days in solution (external reflectance mode) with a spectrum measured from composite graphite electrodes (comprised of KS-6 Lonza graphite powder and 5% PVDF by weight) after one intercalation-deintercalation cycle (within 2-0 V *vs.* Li/Li$^+$) with Li in the same solution (diffuse reflectance mode). Figs. 3 and 4 show similar spectra from similar electrodes and experimental set-up when the solutions were propylmethyl carbonate (PMC)/LiAsF$_6$ and EMC/LiPF$_6$ 1M, respectively. Both solvents are important for Li ion batteries, as the intercalation of lithium into graphite in their solutions is highly reversible and their anodic stability is high. The partial spectral analysis and the derived identification of the various surface species formed are also shown in Figs. 2-4.

These FTIR spectra reflect the formation of surface films comprised of ROCO$_2$Li (major) and ROLi (minor) formed by solvent reduction. Li$_2$CO$_3$ is also formed, probably in a secondary reaction between the ROCO$_2$Li and trace water.

In the case of LiPF$_6$ solutions, the spectral studies reflect the reaction of trace HF unavoidably present in solutions with the ROCO$_2$Li initially formed. LiF, Li$_x$PF$_y$ and Li$_x$PO$_y$F$_z$ surface apecies were identified. This contamination effect is pronounced for Li electrodes and less pronounced for the carbon electrodes because the active surface area of the latter electrode is high, and thus the effect of trace HF is less important.

The interfacial resistance of both Li and graphite electrodes in alkyl carbonate solutions reflects their surface chemistry. In LiAsF$_6$ solutions where the surface chemistry is dominated by the carbonates' formation, the resistance is low (<100 Ω cm^2). In LiPF$_6$ solutions, it may be one or two orders of magnitude higher due to LiF formation, which replaces the ROCO$_2$Li species formed initially. Adding basic additives such as Li$_2$CO$_3$ or tributylamine, which neutralize the HF, decreases the interfacial chemistry due to the shift in the surface chemistry.

The relative amount of F on Li surfaces treated in the alkyl carbonate solutions measured by XPS is a very good probe for the processes dominating the surface chemistry. The F concentration is relatively low for LiAsF$_6$ solutions, and high, and increasing considerably in time, for LiPF$_6$ solutions. In the presence of TBA (HF neutralization), the F concentration is lower.

Hence, there is a good correlation among the results obtained by FTIR, impedance and photoelectron spectroscopies. The role of trace impurities such as HF or H$_2$O in determining the surface chemistry, and thus the interfacial electrical properties of Li and lithiated carbon electrodes, may be important. However, it should be noted that in practical batteries in which the ratio between the electrodes' area and the total amount of the electrolyte solution is high, the influence of impurities may be less important than in laboratory tests.

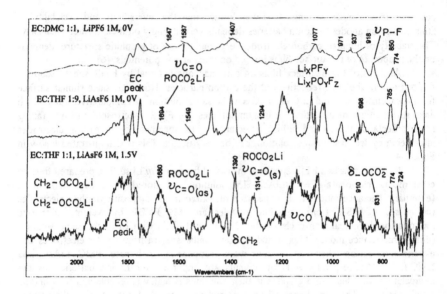

Fig. 1. FTIR spectra measured *in situ* from nickel electrodes polarized to low potentials (*vs.* Li/Li$^+$) as indicated in three different EC-based solutions (as indicated). Partial peak identification is also presented.

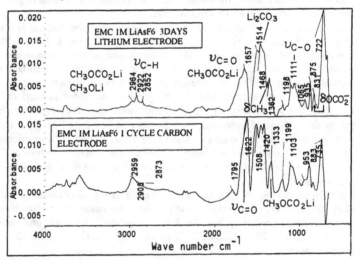

Fig. 2. FTIR spectra measured *ex situ* from lithium electrodes prepared fresh in EMC- LiAsF$_6$ 1M solution (external reflectance mode) and stored for 3 days, and from a graphite electrode intercalated-deintercalated with lithium in the same solutions (diffuse reflectance mode). Partial peak assignment is also presented.

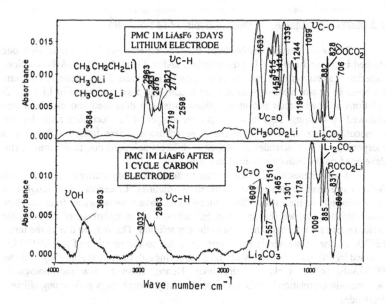

Fig. 3. Same as Fig. 2, PMC-1M LiAsF₆ solution.

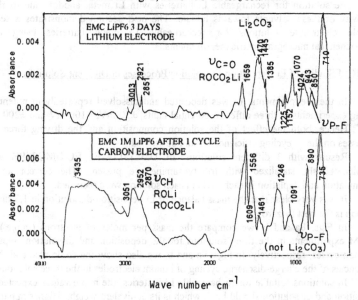

Fig. 4. Same as Figs. 2 and 3, EMC LiPF₆ 1M solutions.

b. Examples for *in situ* morphological studies of Li electrodes.

Below we present results from different experiments in which Li electrodes were studied in dry and wet PC/ LiAsF$_6$ solutions and in EC:DMC 1:1 LiAsF$_6$ solution. . Each experiment starts with the imaging of the pristine electrode. The solution is then introduced, followed by *in situ* measurement at open circuit. Fig. 5 shows images of Li under the three solutions at open circuit potential. Lithium was then deposited and dissolved repeatedly, followed by imaging. Figs. 6 and 7 present images of Li electrodes after deposition and consecutive dissolution of Li in the three solutions. These images reflect the different morphology of Li electrodes developed in different solutions due to different surface films developed on the active metal in each solution.

It was interesting to discover that Li dissolution may increase the surface roughness factor more than Li deposition. In all three solutions, Li deposition is rough due to the non-uniformity of the surface films formed which have a mosaic-type structure. However, it was interesting to see that in a wet PC solution the morphology of Li deposition on a lithium substrate was smoother than in the dry solution. This may be due to the formation of Li$_2$CO$_3$ on the Li surface in the former case due to the reaction of the ROCO$_2$Li species (formed by solvent reduction) with trace water. Indeed, Li cycling efficiency is higher in wet PC/LiAsF$_6$ solutions than in dry ones. Hence, these *in situ* morphological studies demonstrate correlation between surface chemistry, morphology and cycling efficiency of Li electrodes.

None of the above solutions based on the alkyl carbonate solvents are suitable as an electrolyte solution for rechargeable Li batteries with Li metal anodes, due to the high reactivity of these solvents towards lithium. The next section demonstrates a search of suitable electrolyte solutions for rechargeable Li metal batteries using *in situ* electrochemical microgravimetric measurements.

c. EQCM Studies of Li Deposition-Dissolution Processes in Different Solutions

In these experiments, Li was deposited and dissolved repeatedly in potentiostatic cycling experiments at three different potential limits, ±50 mV, ±100 mV and ±200 mV *vs.* Li/Li$^+$. Hence, both the effect of the solution composition and the driving force for the processes on the Li cycling efficiencies could be studied.

Results with 3 different solutions: EC:DMC/ LiAsF$_6$, EC:DMC/LiPF$_6$ and 1,3 dioxolane/LiAsF$_6$ stabilized with tributyl amine are present. The current and mass accumulation and depletion in each cycle in each solution were recorded.

It was very clear that the mass balance of Li deposition-dissolution in 1,3 dioxolane solution is the best ($\approx$ zero).

In Figs. 8a and b, we compare the mass per moles of electrons (m.p.e.) in the EQCM experiments in the three solutions for the deposition and dissolution steps in the repeated cycles, respectively. In Figs. 9a and b, we also show the coulombic and the mass efficiencies of the charge-discharge cycling of lithium electrodes in the three solutions.

In solutions suitable for rechargeable Li batteries, the m.p.e values expected for Li deposition and dissolution should be 7, which is its equivalent weight. Higher m.p.e. values reflect the reaction of the lithium deposited with solutions species, which form surface species. For EC:DMC solutions, Li cycling efficiency is low due to pronounced reactions of lithium with the solutions species (m.p.e. > 7).

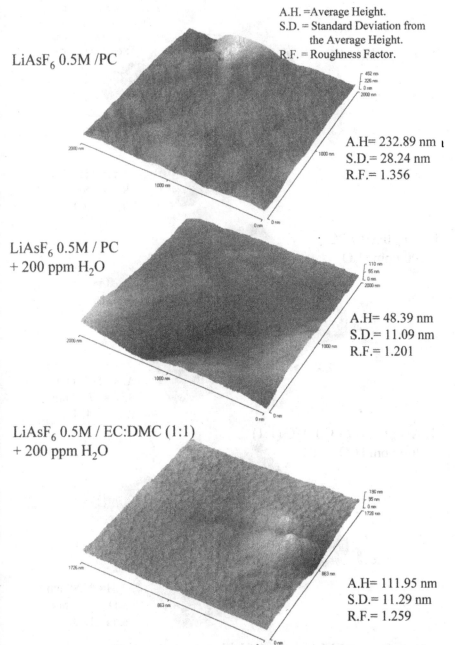

A.H. = Average Height.
S.D. = Standard Deviation from
      the Average Height.
R.F. = Roughness Factor.

LiAsF$_6$ 0.5M /PC

A.H= 232.89 nm
S.D.= 28.24 nm
R.F.= 1.356

LiAsF$_6$ 0.5M / PC
+ 200 ppm H$_2$O

A.H= 48.39 nm
S.D.= 11.09 nm
R.F.= 1.201

LiAsF$_6$ 0.5M / EC:DMC (1:1)
+ 200 ppm H$_2$O

A.H= 111.95 nm
S.D.= 11.29 nm
R.F.= 1.259

Fig. 5   AFM Images of Lithium at OCV (under solution)

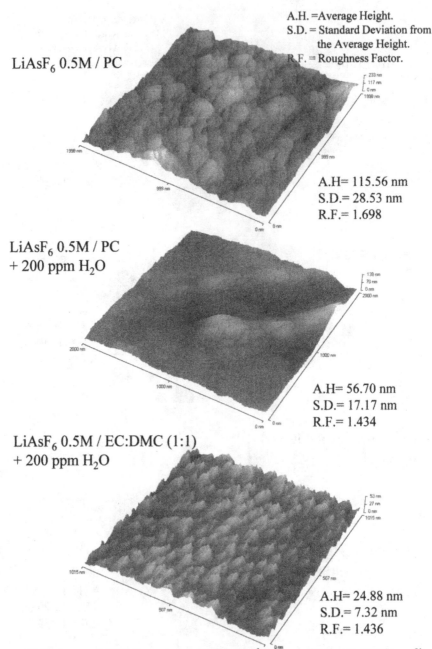

LiAsF$_6$ 0.5M / PC

A.H. =Average Height.
S.D. = Standard Deviation from
the Average Height.
R.F. = Roughness Factor.

A.H= 115.56 nm
S.D.= 28.53 nm
R.F.= 1.698

LiAsF$_6$ 0.5M / PC
+ 200 ppm H$_2$O

A.H= 56.70 nm
S.D.= 17.17 nm
R.F.= 1.434

LiAsF$_6$ 0.5M / EC:DMC (1:1)
+ 200 ppm H$_2$O

A.H= 24.88 nm
S.D.= 7.32 nm
R.F.= 1.436

Fig. 6   AFM Images of Lithium Deposition (1.75C/cm^{2}).

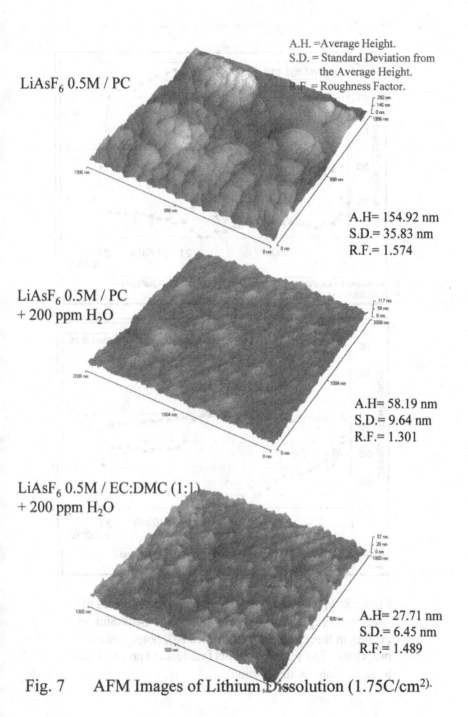

A.H. = Average Height.
S.D. = Standard Deviation from the Average Height.
R.F. = Roughness Factor.

LiAsF$_6$ 0.5M / PC

A.H= 154.92 nm
S.D.= 35.83 nm
R.F.= 1.574

LiAsF$_6$ 0.5M / PC
+ 200 ppm H$_2$O

A.H= 58.19 nm
S.D.= 9.64 nm
R.F.= 1.301

LiAsF$_6$ 0.5M / EC:DMC (1:1)
+ 200 ppm H$_2$O

A.H= 27.71 nm
S.D.= 6.45 nm
R.F.= 1.489

Fig. 7   AFM Images of Lithium Dissolution (1.75C/cm^2).

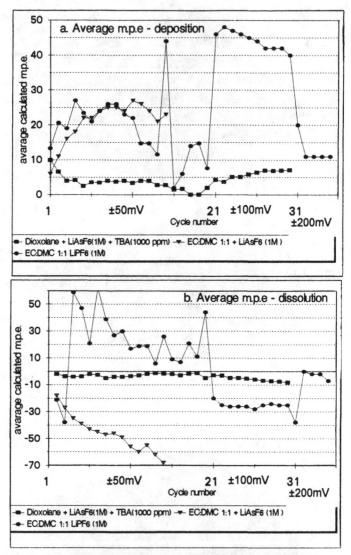

Figure 8: Mass accumulated and depleted per moles of
electrons (m.p.e) during lithium deposition-dissolution
cycling in three solutions (as indicated), potentiostatic
processes. The potential limits (vs. Li/Li+) are indicated:
a. deposition, b. dissolution

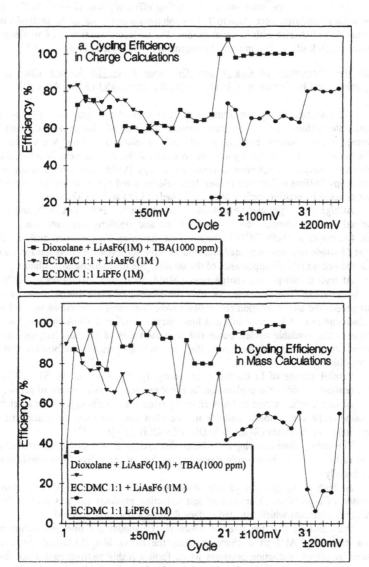

Figure 9: Cycling efficiencies of lithium during deposition-dissolution cycles calculated during EQCM experiments in three solutions (as indicated), potentiostatic processes.
The potential limits (vs. Li/Li+) are indicated: a. coulombic efficiencies, b. mass efficiencies

For 1,3 dioxolane solutions, the Li cycling efficiency was close to 100%, and the m.p.e. values calculated were close to 7. This solutions was chosen as the preferred one for rechargeable Li batteries with Li metal anodes. The surface chemistry of Li anodes in these solutions which leads to their improved performance is described in refs. 2 and 7.

d. On the Correlation of Charge and Discharge Rates, Li Surface Chemistry and Morphology, and Performance of Rechargeable Batteries with Li Metal Anodes.

We investigated the performance of lithium anodes in practical AA Li-Li$_x$MnO$_2$ rechargeable batteries [7] at different current densities of charge and discharge, and performed a post-mortem analysis on cells after prolonged cycling. We correlated the results to parallel Li anode testing in laboratory cells at the same current densities, followed by electrode analysis by scanning electron microscopy (SEM) and surface sensitive FTIR spectroscopy (diffuse reflectance mode). It should be noted that practical Li electrodes are always covered by pristine surface films comprised of Li$_2$O and Li$_2$CO$_3$.

At high discharge rates (>1mA/cm^2), low charge rate (<0.5 mA/cm^2), and selecting optimized solution composition in terms of salt and stabilizer concentration, we could obtain 400 cycles at 100% DOD in the practical batteries. Post-mortem analysis revealed that the Li anode remains integrated and smooth after 400 cycles. The reason for the end of life for the cell was the disappearance of the solution due to reactions with lithium.

At high discharge and charge rates (>1mA/cm^2), only 150 cycles at 100% DOD could be obtained. The Li electrodes after the life-end of the batteries were fragile and disintegrated, and all the solution components were consumed by reactions with lithium. At low discharge rates < 0.5 mA/cm^2) and low charging rates (<0.5 mA/cm^2), a pronounced decrease in the available cycles, compared with the cases of high discharging rates, was observed. Post-mortem analysis showed that the anode appears non-uniform (dendritic) and partially disintegrated.

Parallel studies of Li electrodes in laboratory cells followed by SEM, FTIR and EDAX analysis provided the explanation for the above results, as summarized below:

1. At high discharge rate and low charging rates, the morphology of the Li electrodes was found to be very smooth, and the surface films covering them are comprised of the solvent reduction products (ROLi, poly DN, HCOOLi). [2,7]).

2. The surface films covering the lithium after cycling at slow discharge rates contained lithium carbonate as a major surface species, residual of the native films which cover the anode initially.

3. At high charging rates Li deposition was dendritic, and the surface films were comprised of a pronounced amount of salt reduction products (as was evident from the EDAX measurements which detected surface fluorine and arsenic species).

We conclude that the rates at which the Li anodes are operated influence their surface chemistry. At high discharging rates, the native film (Li$_2$O-Li$_2$CO$_3$) is rapidly replaced by solvent reduction products which form a highly passivating surface layer. Li deposition occurs beneath this film in a flake-like formation with no dendrite formation. Thus, the lithium cycling efficiency (and thus, the practical battery performance) is high. At too slow a discharge rate, the replacement of the native surface films by the new ones is incomplete, leading to a non-uniform structure of the surface layers which induces non-uniform Li deposition and dissolution (dendrite formation). At too high a charging rate, Li deposition is too fast, leading to the exposure of fresh Li deposits to the solution. The reaction of these deposits with solution species seems to be selective, and the salt anion

(AsF$_6^-$) is predominantly reduced. This situation also induces non-uniform Li deposition and dissolution, and enhanced reactions between reactive lithium and the solution components.

CONCLUSION

This work demonstrates again the correlation of surface chemistry, morphology and the electrochemical behavior of lithium (and lithiated graphite) electrodes. The surface chemistry of Li and lithiated graphite are very similar (in the same electrolyte solutions). On both electrodes, surface films precipitate and may or may not passivate these electrodes. Their chemical and physical structure is the result of a delicate balance between competing reduction processes of various solution species (solvent, salt anion, contaminants) by the active surfaces.

For instance, when the solvents are alkyl carbonates and the salt is LiAsF$_6$, the major surface species are ROCO$_2$Li and ROLi compounds where R = alkyl. We are able to identify both types of compounds quite rigorously by FTIR spectroscopy. Li$_2$CO$_3$ is also formed in a secondary reaction of the ROCO$_2$Li with trace H$_2$O. In LiPF$_6$ solutions, the ROCO$_2$Li are partially solubilized by reaction with trace HF. We identify Li$_x$PF$_y$ and Li$_x$PO$_y$F$_z$ surface species. *In situ* AFM studies reflect clearly the impact of differences in the Li surface chemistry on the morphology of Li deposition-dissolution processes. *In situ* EQCM studies were able to characterize and identify solutions which are suitable as electrolyte systems for rechargeable Li batteries with Li-metal anodes. 1,3 Dioxolane LiAsF$_6$ solution containing tributyl amine as a Lewis acid neutralizer was found to be the best for Li-metal rechargeable Li batteries using liquid electrolyte solutions.

ACKNOWLEDGMENTS

This work was partially supported by the Israeli Academy of Science (National Science Foundation), and the Battery Division of Tadiran (Israel).

REFERENCES

1.   D. Aurbach, E. Zinigrad and A. Zaban, J. Phys. Chem. **100**, p. 3098 (1996).

2.   D. Aurbach, I. Weissman, A. Schechter and H. Cohen, Langmuir **12**, pp. 3991-4007 (1996).

3.   O. Chusid (Youngman) and D. Aurbach, J. Electrochem. Soc. **140**, pp. L1, L155 (1993).

4.   D. Aurbach and Y. Cohen, J. Electrochem. Soc. **143**, p. 2352 (1996).

5.   D. Aurbach and A. Zaban, J. Electroanal. Chem. **393**, p. 43 (1995).

6.   D. Aurbach, Y. Ein-Eli, O. Chusid, M. Babai, Y. Carmeli and H. Yamin, J. Electrochem. Soc. **141**, p. 603 (1994).

7.   D. Aurbach, I. Weissman, A. Zaban, Y. Ein-Eli, E. Mengeristky and P. Dan, J. Electrochem. Soc. **143**, p. 2110 (1996).

LOW POTENTIAL Li INSERTION IN TRANSITION METAL OXIDES

F. LEROUX, and L.F. NAZAR*
University of Waterloo, Department of Chemistry, Waterloo, Ontario Canada N2L 3G1;
lfnazar@uwaterloo.ca

ABSTRACT

Low-potential Li insertion materials comprised of molybdenum oxides (A_xMoO_3) have been prepared by a "chimie douce" route. Li insertion below 200 mV is associated with dramatic transformation of the structure, leading to a material which displays good cyclability with a high reversible specific capacity of 940 mA/g in the voltage window 3.0-0.005V (volumetric capacity of 4000 mAh/cc), albeit with notable polarization on charge. The structural and compositional changes on discharge to 200 mV have been studied by a combination of XRD, and XAS. The interlayer ions have also been exchanged for Sn, and the electrochemical characteristics of these materials are compared with the alkali derivatives.

INTRODUCTION

The ability of graphitic materials to intercalate Li at low potential forms the basis of current rocking-chair batteries. Despite many advantages, these anodes suffer from a few drawbacks; namely they are limited to an insertion of approximately $Li_{1+x}C_6$ (practical capacity of ~400 mA/g) and display poor volumetric energy density. The latter is also a problem of disordered carbonaceous type-materials which suffer from somewhat higher cell polarization. Renewed interest in negative electrode materials has resulted in the discovery of promising Li insertion properties of various new materials, including amorphous nanocomposites,[1] and crystalline materials such as metal oxides. The latter were first proposed 10 years ago by Auborn and Barberio, based on the MoO_2 or WO_2 /$LiCoO_2$ based "rocking-chair" battery.[2] Unfortunately, their studies were limited by the characteristics of these materials and poor stability of the electrolyte at low potential. Recently, however, oxides have experienced a renaissance, with the report of the manufacture of a Li ion battery that utilizes an amorphous tin-based oxide as a negative electrode.[3] Studies have also detailed low-potential Li insertion in $LiMVO_4$ (M=Zn, Cd, Ni),[4] Li_xMO_z and $Li_xM_yV_{1-y}O_z$ (M=Ti, V, Mn, Co, Fe, Ni, Cr, Nb and Mo)[5] and in amorphous compounds such as $InVO_4$ and MV_2O_{6+d} (M=Fe, Mn, Co).[6]

Our study describes the electrochemical and cycling behavior of simple molybdenum bronzes, A_xMoO_3 (x ~ 0.25; A= Na, Li, Sn), which show promising behavior as Li hosts at low potential. The performance is described as a function of heat treatment and the voltage window, and the electrochemical characteristics are compared to other systems. X-ray absorption spectroscopy studies on these materials as a function of the discharge potential have provided insight on the local structural modification that occurs during the Li "insertion" process.

EXPERIMENTAL

Synthesis and characterization. The molybdenum bronze, $[Na^+(H_2O)_n]_y[MoO_3]^{y^-}$, was prepared as described previously, by reduction of MoO_3 in aqueous media.[7] The final product analyzed (TGA and ICP) as $([Na(H_2O)_2]_{0.26}MoO_3$, in agreement with previous reports. This material is referred to as "$NaMoO_3$".

The lithium bronze, $[Li_{0.07}Na_{0.13}(H_2O)_{0.40}]MoO_3$, or "$LiNaMoO_3$" was prepared by the same method, substituting Li_2MoO_4 as the buffer in place of Na_2MoO_4. To obtain more highly lithiated bronzes, a cation exchange reaction was also carried out by ion-exchange of fresh $NaMoO_3$ (2g) with LiI in 50ml of butanol, as previously reported.[8] Dehydration *in vacuo* for 24

601

hours at 120°C yielded a material of stoichiometry $[Li_{0.20}Na_{0.06}(H_2O)_{0.52}]MoO_3$, "$LiMoO_3$". The Sn-exchanged material ("$Sn_{Na}MoO_3$", was prepared similarly, by exchange with $SnCl_2 \cdot H_2O$.

X-ray powder diffraction (XRD) patterns were obtained on a Siemens D500 diffractometer equipped with a diffracted beam monochromator, using Cu-K$_\alpha$ radiation. A Varian Liberty 100-ICP-AES was used to determine the ratio of alkali (Li, Na) to molybdenum. XAFS studies were undertaken at LURE (Orsay, France) using X-ray synchrotron radiation emitted by the DCI storage ring (1.85 GeV positrons, average intensity of 250 mA). Data were collected in transmission mode at the Mo K edge (20,000 eV), using a a two crystal Si (311) monochromator. The study was not conducted using the L-edge due to lack of sufficient beam time. For each run, three EXAFS spectra were recorded from 19,900 to 21,000 eV with a 3eV step and two seconds of accumulation time per point. For each XANES run, two spectra were recorded from 19,950 to 20,150 eV with a 1 eV and one second of accumulation time per point. The analysis was performed using standard XAS software.[9]

Electrochemical measurements.

The materials were evaluated (*vs* Li) as positive electrodes prepared from active material, and carbon black (Super S, Chemetals Inc.) using PVDF as a binder. The powders in the weight proportion 85, 12, 3, respectively, were mixed in cyclopentanone and spread onto a nickel disc. The electrodes were heated at 170°C in a vacuum for 2 hours prior to cell assembly in order to remove the free water. Dimethyl carbonate (DMC- Aldich) and ethylene carbonate (EC- Aldrich) were distilled under vacuum, mixed in the ratio 2:1 and $LiPF_6$ was added to prepare the electrolyte as a 1.0M solution. Swagelock type cells were assembled in an argon filled glove box containing less than 2ppm of water and oxygen. Positive electrodes had a surface area of $1cm^2$ and contained 1 - 2 mg of active material. Li metal, spread on a Ni disk, was used as the anode and reference electrode. A multichannel galvanostatic/potentiostatic system (Mac-Pile™) was used for the electrochemical study. The specific capacities obtained are normalized per gram of composite electrode (if not indicated), giving rise to an overestimation of the carbon contribution (*vide infra*). The capacity retention of at the nth cycle is defined as the charge capacity at the nth cycle divided by the first charge capacity.

RESULTS AND DISCUSSION

Numerous papers have demonstrated that the insertion of Li ions into MoO_3 based materials in the potential range 4-2V *vs* Li occurs via a topotactic Li insertion process of the lamellar transition metal oxide in this regime, along with poor ionic and electronic conductivity of MoO_3.[10] If the lower voltage limit is kept above >500 mV *vs* Li, the features characteristic of a single phase transformation are maintained. At very *low potential*, however, these materials exhibited unanticipated results. To examine the response, we first estimated the contribution of the carbon black in this low voltage window, by examining a "blank" cell containing only carbon black (223 mAh/g irreversible capacity, attributed to the formation of a passivation layer, and a reversible capacity of 242 mAh/g). This is small, but not insignificant compared to the total capacity of the composite electrode.

Deep discharge. The discharge-charge curves for $NaMoO_3$ were evaluated as a function of the discharge cut-off voltage. The latter, V_d, strongly influences the subsequent charge process. After discharge to V_d=500 or 300mV, for example, the charge curve retains an additional feature at 1.3V in addition to the 2.2V plateau (corresponding to a reduction process at 2.4V). A lower discharge limit (V_d=200 *vs* Li, **Fig. 1a**) results in a reversible capacity of 652 mA/g, *after* the contribution of the carbon (12%) is removed. We observe complete disappearance of oxidation features in the high potential region; *i.e.* no well-defined Li de-insertion is observed above a potential of 1.5V. The material therefore undergoes a transformation during the first cycle below 500 mV. Ex-situ X-ray diffraction experiments on the electrode materials show a complete amorphization of the starting compounds for V_d<500mV. Discharge to 0.005V (**Fig. 1b**)

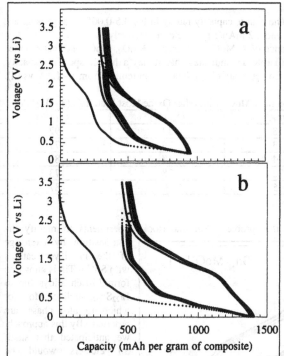

Fig 1. Voltage profiles of the first 5 cycles of NaMoO$_3$ after a heat treatment at 170°C, in the window a) 3.5-0.2V and b) 3.5-0.005V, at a current density of 71 and 44 mA/g in discharge and charge respectively. Capacity is expressed per g of composite electrode.

increases the reversible capacity Q_C to 938 mAh/g per gram of composite, although the irreversible component also increases. The latter is the result of the intrinsic nature of the material (irreversible reaction of Li, and/or decomposition of the electrolyte catalysed on the surface of the active material).

The behavior on the charge capacity, $R_{20/1}$ on cycling was studied in two different voltage windows (3.5-0.2 and 3.5-0.005V vs Li), at a constant current density of 62 and 35 mA/g in discharge and charge, respectively (**Fig. 2**). This corresponds to C/11 and C/22, respectively, for V_d= 200 mV, and to C/19 and C/30, respectively, for V_d= 5 mV. Discharge to V_d = 200mV gives rise to a stabilization of the charge capacity on cycling, such that the charge capacity retention is 88% after the 20th cycle and 75% at the 50th cycle. Over a wider voltage range a greater specific capacity is recovered. The partial loss of the cyclability is due to an increase of

the polarization, which is observed to a lesser degree in the 3.5-0.2V voltage window. The electronic density vs voltage curves (not shown), also indicate the sensitivity of the capacity retention with V_d. The electronic density profile is not maintained on cycling between 3.5-0.005V. For V_d=200mV, on the other hand, there is only a small change in the differential capacity between the 2nd and 100th cycle. The material undergoes a reversible transformation under these conditions. These observations emphasize the fact that different Li insertion mechanisms occur as function

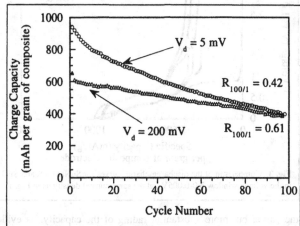

Fig. 2. Specific capacity (per g of composite electrode) vs cycle number for NaMoO$_3$ after treatment at 170°C, in the voltage windows 3.5 - 0.2V (triangles) and 3.5 - 0.005 V vs Li (circles).

of the discharge cut-off potential. The charge capacity fading in the 3.5-0.005 V range is almost twice than in 3.5-0.2 V portion (4.3 and 2.2 mAh/g per cycle, respectively).

Comparison of the performance of A_xMoO_3 with that of MoO_2, MoO_3 and that reported for other transition metal oxides (**Table 1**), indicates that it has a higher specific capacity, although all of these systems suffer from a relatively high charge potential (from 1 - 1.5 V vs Li).

Table 1. Comparison of Li insertion in A_xMoO_3 with other Oxide Host Materials (3 - 0.005V)

	Q_{1C} (mAh/g)	Q_{irr} (%)	$R_{5/1}$ (%)
$NaMoO_3$	938	33	92
MoO_3	773	29	77
MoO_2	674	17	83
SnO	722	28	73
$Sn_{Na}MoO_3$	965	30	87

Sn-exchanged molybdenum bronze.

In an effort to overcome this problem, we undertook experiments to modify the composition by exchange of the interlayer cations with Sn^{2+}. Tin is known to form Li-rich alloys (up to $Li_{22}Sn_5$), unlike Mo, for which no alloy phases are reported. By this approach, we anticipated that small Sn clusters would be formed on reduction that would be sequestered by the Li-Mo-O matrix, similar to the mechanism proposed for Li reaction with tin oxides.[11]

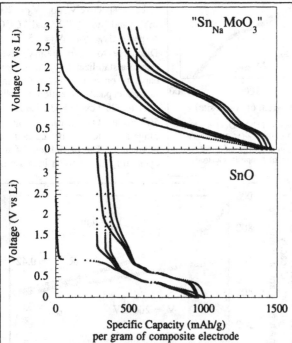

The charge-discharge curves for the $SnNaMoO_3$ (Sn-exchanged bronze) are compared with SnO in **Fig. 3**. The shape of the corresponding curve is what one would expect for a mixture of SnO and $NaMoO_3$. The addition of Sn in the structure results in the desired lowering of polarization during the charge process, evident in the low potential region of the curve; but more substantial fading of the capacity, as evident within the first few cycles. This results in a capacity loss of 60% on the 25th cycle on discharge to 5mV (slightly higher than SnO itself), compared to 25% for $NaMoO_3$ under the same conditions.

Fig. 3. Comparison of the discharge-charge curves for $SnNaMoO_3$ and SnO in the voltage window 3.0-0.005V, at the same current density as in Fig. 1.

XAFS studies.

To examine how the local environment of the Mo is affected by reaction with Li, we carried out ex-situ XAFS experiments on the molybdenum oxide as a function of Li insertion potential. The absorption features (XANES) of the lithiated materials were first analyzed at different discharge cut-off potential, by consideration of the main absorption edge energy shift. The features in such a curve are attributed to dipole-allowed electronic transitions from 1s to 4p states. The so-called chemical shifts vary linearly with the electronic valence state of the absorbing atom. We found the 1s-> 4p transition energy systematically decreased with increased Li content in the molybdenum bronze. The electrode material discharged to 1.0V has a main edge at 31.9eV (energy normalized to Mo metal); the main edge for the corresponding 5mV electrode is at 27.8eV. The former corresponds to a formal valency of ~Mo^{IV}, the latter, Mo^{II}. Some discrepancies appear in the analysis of the first derivative XANES plots (**Fig. 4**) For 1V >

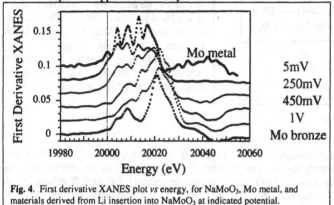

Vd > 250mV, the derivative dA/dE maintains the same features as the pristine NaMoO$_3$ bronze. whereas the structure of the curve for the 5mV electrode material resembles that of Mo metal. The question is whether the Mo is indeed reduced down to its metallic state, and where the the lithium and oxygen are located in the material.

Fig. 4. First derivative XANES plot *vs* energy, for NaMoO$_3$, Mo metal, and materials derived from Li insertion into NaMoO$_3$ at indicated potential.

To further explore this issue, we examined the radial distribution functions (RDF) derived from the EXAFS spectra, for two reference molybdenum oxide materials, α-MoO$_3$ and

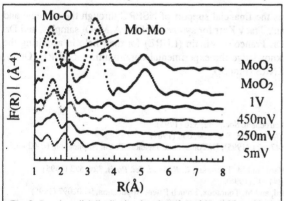

MoO$_2$ along with the lineshapes for the lithiated electrode molybdenum bronzes (**Fig. 5**). It is evident from this data that the local enviroment of the 1V-electrode material differs somewhat from the structural arrangement of the starting material α-MoO$_3$ or NaMoO$_3$, as expected. The most striking difference corresponds to the loss of the longer Mo-Mo contact in the second metal shell. Progressive reduction corresponding to the lineshapes at 450mV, 250mV and 5mV suggests that the material becomes more disordered as a

Fig. 5. Pseudo-radial distribution functions for reference Mo-oxides, and materials derived from Li insertion into NaMoO$_3$ at potential indicated; distances are not corrected for phase shift

function of reduction potential. Diminution of the first shell Mo-O interaction is also indicated, as shown by the decay of the peak corresponding to this contribution. This implies a loss of oxygen within the structure at low potential, consistent with the electrochemical data.

Successful fitting of the Fourier transform of the first two shell contributions (Mo-O and Mo-Mo) were obtained for all of these materials, except for that at 5mV due to the high degree of disorder present in that sample. A representative example at 250mV is shown in **Fig. 6**. The data shows that with increasing reduction level, the average Mo-O distance increases slightly as a result of a wider bond length distribution. The Mo-Mo distance decreases from 3.46Å in α-MoO$_3$ (corresponding to the edge-shared octahedra in the structure), to 2.58Å in the 1V electrode material. Below 1V, no additional significant changes in the Mo-Mo distance occur. This span is similar to the <Mo-Mo> distance of 2.51Å between edge-sharing octahedra within chains of MoO$_2$ in the rutile structure, but shorter than the Mo-Mo bond length in Mo metal (8 Mo at 2.725Å, and 6 at 3.147Å).

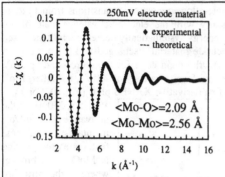

Fig. 6. Fourier transform (exptl and fit) for material from Li insertion in NaMoO$_3$ discharged to 250mV.

CONCLUSIONS

The combined electrochemical and XAS are consistent with a mechanism in which low potential Li insertion in molybdenum bronzes occurs to give rise to an irreversible capacity corresponding to the formation of Li$_2$O, and a reversible capacity (up to 950 mAh/g in the voltage window 3.0-005V), that corresponds to the formation of a disordered, oxygen- deficient molybdenum material, viz:

$$Na_{0.25}MoO_3 + 7.8 (Li^+ + e^-) \rightarrow 1.3 Li_2O + Li_{5.2}MoO_{1.7}$$

ACKNOWLEDGMENTS

LFN gratefully acknowledges the financial support of NSERC through the strategic and operating programs. The authors thank Tracy Kerr for synthesis of the A$_x$MoO$_3$ samples, and Dr. Guy Ouvrard (IMN, Nantes), and Dr. Francoise Villain (LURE) for their help in acquiring the XANES and EXAFS data. We acknowledge the experimental opportunities at LURE (Orsay, France), which is supported by the CNRS (France).

REFERENCES

1. A.M. Wilson and J.R. Dahn, J. Electrochem. Soc., 142, 326 (1995).
2. J.J Auborn and Y.L. Barberio, *J. Electrochem. Soc.*, **134**, 638 (1987)
3. Y. Idota, US Patent Application, 5478671 (1995); H. Tomayama, Japanese Patent Application, 07-029608 (1995).
4. C. Sigala, D. Guyomard, Yves Piffard, and M. Tournoux, *C. R. Acad. Sci. Paris*, **320**, 523 (1995)
5. N. Kumagai, Japanese Patent, 08-241707 (1996).
6. F. Leroux, D. Guyomard, Y. Piffard, and M. Tournoux, French Patent Application, 95.02097 (1995)
7. D.M. Thomas and E.M. McCarron, *Mat. Res. Bull.*, **21**, 945 (1986)
8. T. A. Kerr and L.F. Nazar, *J. Mater. Chem.*, 8, 2005 (1996).
9. A. Michalowicz, in *Logiciels pour la chimie*, SFC, Paris (1991).
10. N. Margalit, *J. Electrochem. Soc.*, **121**, 1460 (1974); J.O Besenhard, J. Heydecke and H.P. Fritz, *Solid State Ionics*, **6**, 215 (1982)
11. I.A. Courtney, and J.R. Dahn, *J. Electrochem. Soc.*, **144**, 2943 (1997).

EFFECTS OF PROCESSING CONDITIONS ON THE PHYSICAL AND ELECTROCHEMICAL PROPERTIES OF CARBON AEROGEL COMPOSITES

T. D. TRAN*, D. LENZ*, K. KINOSHITA**, AND M. DROEGE***

*Chemistry & Materials Science Department,
Lawrence Livermore National Laboratory, Livermore, CA 94550
**Energy and Environmental Division,
Lawrence Berkeley National Laboratory, Berkeley, CA 94720
***Ocellus, Inc., 887 A Industrial Road, San Carlos, CA 94070

ABSTRACT

The carbon aerogel/carbon paper composites have physical properties similar to those of monolithic carbon aerogels but do not require supercritical extraction during fabrication. The resorcinol-formaldehyde based carbon aerogel phase is intertwined between the fibers of a commercial carbon paper. The resulting composites have variable densities (0.4-0.6 g/cc), high surface areas (300-600 m^2/g), and controllable pore sizes and pore distribution. The effects of the resorcinol-formaldehyde concentrations (50-70% w/v) and the pyrolysis temperature (600-1050°C) were studied in an attempt to tailor the aerogel microstructure and properties. The composite physical properties and structure were analyzed by transmission electron microscopy and multipoint-BET analyses and related to electrochemical capacitive data in 5M KOH. These thin carbon aerogel/carbon paper composite electrodes are used in experiments with electrochemical double-layer capacitors and capacitive deionization.

INTRODUCTION

Thin carbon aerogel/carbon paper composites are attractive materials for many applications, most notably, as electrodes in electrochemical double-layer capacitors [1] and capacitive deionization [2, 3]. These high-density composites (>0.4 g/cc) have high electrical conductivity (20-100 S/cm), high surface areas (200-500 m^2/g) and favorable microporous structure (average pore diameter, 40-200Å). The carbon aerogel phase in the composite is derived from an organic aerogel precursor made by the aqueous polycondensation of resorcinol (1,3-dihydroxylbenzene) and formaldehyde. In contrast to monolithic microporous aerogels which are supercritically dried (with CO$_2$), the composite organic aerogel is generally air dried. These high-density cross-linked gels have sufficient strength to retain a large fraction of the porous structure from collapsing due to capillary forces associated with solvent evaporation. The simpler preparation procedure makes the composite aerogel materials more economically attractive than monolithic samples.

While extensive work has been done with monolithic resorcinol-formaldehyde (RF) aerogels and their carbon derivatives [4-8], studies on carbon composite materials have been limited. Only results with high-density materials (> 0.5 g/cc) have been reported [3,9]. With the current development of commercial electrochemical devices using thin carbon aerogel composites, the availability of these types of electrode materials with a wide range of properties are desirable.

This collaborative effort investigates two changes in the processing conditions of the composite materials. The effects of the resorcinol-formaldehyde concentration and the pyrolysis temperature (600-1050°C) are studied to tailor the composite structure and properties. Results of these effects on the materials properties and their electrochemical capacitive behavior are reported.

EXPERIMENTAL

Carbon aerogel/carbon paper composites - The preparation of composite electrode materials containing RF aerogels integrated into a commercial carbon paper has been described before [9]. In general, resorcinol (1,3 dihydroxybenzene) and formaldehyde were mixed in a 1:2 molar ratio, respectively. Deionized/distilled water was added as the diluent and sodium carbonate as the base catalyst. A porous carbon paper (0.127-mm thick, Lydall Technical Papers, Rochester, NH) was soaked with the catalyzed-RF solution and placed in between two glass plates. The impregnated sample was cured at room temperature, 50°C and 85°C successively for one day each at the various temperatures. The cured composite was then soaked in a copious amount of acetone and subsequently dried at room temperature. Care had to be taken to assure that the pore water was adequately exchanged with acetone. Finally, the RF/carbon paper composite was heat treated at 1050°C in a nitrogen atmosphere for two hours to pyrolyze the RF component. Table 1 lists several monolithic and composite samples and their characteristics. Conventional high-density monolithic and composite RF materials (samples M1, M2 and C1) were prepared with RF concentrations between 50-70% weight per volume (w/v) using a molar ratio of resorcinol to catalyst (sodium carbonate) of 200.

A number of the carbon aerogel composite materials were commercial products obtained from Ocellus, Inc. (San Carlos, CA). These materials were prepared in a similar procedure as described above except with modified formulation and drying steps. The solvent (acetone) exchange step was eliminated in this case and the pore fluid (water) was dried directly at ambient conditions. The concentration of resorcinol-formaldehyde was varied between 50 to 70% weight per volume (w/v) at a constant resorcinol to catalyst molar ratio. After curing and drying, the resulting RF composites were then pyrolyzed under an inert atmosphere using a temperature ramp of 8 to 13 hours duration, depending on the final pyrolysis temperature (between 600 and 1050°C.) Electrode materials were then tested directly without any additional treatment.

Electrochemical capacitance measurements - Electrochemical studies were performed using a 64-channel Maccor battery tester. Experiments were done using circular (4.5-cm diameter) Teflon cells. The two identical 1.5-cm diameter electrodes were separated by 2 pieces of Whatman fiberglass filter papers (934-AH). Nickel foils were used as current collectors in sodium hydroxide (5M KOH) electrolytes. The whole assembly is sandwiched between 2 Teflon plates and held together by 0.6-cm diameter Teflon screws. Electrolyte filling was accomplished via three successive evacuation (2 psi) and pressurization (24.5 psi) stages in approximately 30 minutes. The assembly and testing of aqueous cells were carried out under ambient conditions. The charge/discharge experiments were carried out at constant currents (25 mA) to 0.85 and 1 V upper limits for 40 cycles. The capacitance, C, was determined by dividing the discharge capacity, Q, by the charging voltage, V. The total dry weight and volume of the two carbon electrodes were used in the capacitance density calculations.

Physical characterization - The BET surface area and pore size distribution were obtained using a five-point N_2 gas adsorption technique (Micromeritics ASAP 2000). The average pore size and pore size distribution were determined from the desorption branch according to a theory developed by Barrett, Joyner, and Halenda (the BJH method) [10]. The composite morphology and microstructure was also examined by scanning (Hitachi S570) and transmission electron microscopy (JEM-200CX).

RESULTS AND DISCUSSION

Carbon resorcinol-formaldehyde aerogel composite - Table 1 compares the difference in properties and specific capacitances between monolithic and composite RF carbon aerogels. Selected information on monolithic materials with formulation comparable to the range studied in this work are summarized in Table 1 as the basis for comparison with those for composite materials. More extensive data on monolithic carbon aerogels can be found in the literature [4-8]. Supercritically-dried monolithic carbon aerogels showed the most desirable properties. Surface areas typically range from 400-1000 m^2/g and thermally treated samples can have BET surface areas as high as 3000 m^2/g. Specific capacitance in 5M KOH electrolytes was as high as 35 F/g [9]. On the other hand, thin, air-dried carbonized RF composites tend to have lower surface areas. The samples from this study as well as several similar materials from other work [9, 11, 12] show BET surface areas between 200-600 m^2/g, a reduction of about 40% from supercritically-dried monolithic samples. Nevertheless, the thin composite and the supporting carbon paper matrix apparently prevent a total collapse of the pore network during ambient temperature drying. At sufficiently high-density (>0.4 g/cc) a fraction of the bulk is rigid and should retain a microporous structure, even in subsequent pyrolysis. The pore size and distribution are adequate (see Table 1) for double-layer formation. The electrochemical capacitive measurements in 5M KOH show specific capacitance values in the range of 17-19 F/g (see table 1). These values are also consistent with those obtained from other samples.

Effects of resorcinol-formaldehyde sol concentration - A series of samples with different concentration of resorcinol-formaldehyde in the gel was prepared to study the effect of density on the properties and microstructure. The concentration of the dissolved solids (resorcinol and formaldehyde) was varied between 50 to 70% (weight/volume) at a constant resorcinol/catalyst molar ratio. The results are shown in Table 1. Samples C52 and C62 (at 50 to 60 %w/v, respectively) show high BET surface areas with reasonable large average pore sizes. High-resolution transmission electron microscopy show porous structure with average pore diameters in the same range with values reported in Table 1. Our initial investigations with Raman spectroscopy reveal a small up-shift in the Raman frequencies in the aerogel/paper composites compared to the monolithic materials. Some broadening of the graphite peak

Table 1. List of carbon aerogels and their characteristics

Samples	Aerogel type	R-F content, % wt./vol	Density, g/cc	BET area, m^2/g	Average pore size, Å	Specific capacitance, F/g
M1[a]	monolithic	40	0.58	666	65	34
M2[a]	monolithic	50	0.83	580	50	28
C1	composite	70	0.6	360	80	17
C52	composite	50	0.41	389	170	17.8
C62	"	60	0.47	408	56	19.0
C72	"	70	0.60	297	37	0.9
C600	"	60	0.40	623	77	14.9
C700	"	60	0.43			23.0
C800	"	60	0.43	589	77	28.4
C900	"	60	0.45			25.8
C1050	"	60	0.44	399	61	17.8

[a] data from references 8 and 9.

(1603 cm^{-1}) and the disordered-induced peak (1350 cm^{-1}) was evident. The in-plane crystallite sizes (L$_a$) for composite aerogels are 20-30% larger than the value of 25 Å reported for monolithic aerogels [9]. Additional discussion on the microstructure of these materials will be reported in future communications. The specific capacitances for C52 and C62 samples (see Table 1) were 17.8 and 19 F/g, respectively.

The sample with an RF content of 70% w/v, C72, was apparently affected by a significant structure change (including pore collapse). While the standard procedure (sample C1 in Table 1) with acetone exchange prior to drying apparently was successful in preserving the porous structure, the direct drying of this similar high-density sample was not effective here. A smaller average pore size and an abundance of micropores (< 20 Å) were observed. Such a structure apparently could not form a double-layer and thus yield a low capacitance. A comparable result was observed with two batches of similarly-prepared samples. Transmission electron microscopy shows evidence of highly interconnected particles with limited pore volume. The simple air drying of the pore water, which did not appear critical for the two lower-density samples, was detrimental at this RF content. The reason for this variation in the effects of the drying conditions is not obvious at this point. Additional study is being considered.

Effects of pyrolysis temperature between 600 to 1050°C - Another important means for controlling the structure and properties of carbon aerogels is by changing the pyrolysis temperature. Figure 1 shows the specific and volumetric capacitance densities of a series of composite materials (C600-C1050) as a function of pyrolysis temperature. Both curves show a peak capacitance near 800°C. The trend here is in agreement with an earlier study with monolithic carbon aerogels [9]. The peak specific capacitance of 29 F/g was the highest ever observed for composite electrode materials. This value approaches those obtained from monolithic samples as seen in Table 1. Considering the drastic change in processing conditions (air dried vs. supercritical extraction), the performance improvement is significant. Higher energy densities can be expected from capacitors using these modified electrodes.

Several characterization experiments including gas adsorption analysis, transmission electron microscopy and Raman spectroscopy are being conducted to obtain additional information on these materials. Surface areas may be the critical factor for this increased

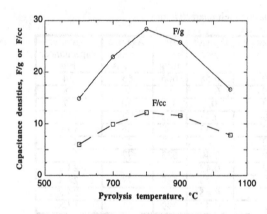

Figure 1.
Specific and volumetric capacitance densities (F/g and F/cc, respectively) of RF carbon aerogel composites as a function of pyrolysis temperature.

capacitance as observed for the sample pyrolyzed at 800°C (*e.g.*, 589 m^2/g). A question remains on the contribution of the pseudo-capacitive components to the electrochemical activities of these electrodes. Continued cycling of C800 (Table 1) electrodes in 5 M KOH was carried out to study the reversibility of the capacitive behavior and to elucidate irreversibility arising from any pseudo-capacitive redox reactions. Stable capacities (over 650 cycles) at constant current charge/discharge conditions at two voltage limits of 0.85 V and 1 V suggest that the carbon surface remains unchanged under these cycling conditions.

CONCLUSIONS

The results demonstrate that sufficiently high-surface-area carbon aerogel composites can be prepared with a simple air drying procedure. Electrode materials with different properties can be prepared by changing the processing conditions. Composites pyrolyzed at 800°C showed capacitances approaching those of monolithic materials. The availability of these materials with high electrochemical capacitance is expected to improve the energy densities of supercapacitors and electrosorptive capacities of electrochemical deionization cells.

ACKNOWLEDGMENTS

The authors thank Ms. S. Hulsey (LLNL) for N$_2$ gas adsorption measurements, Mr. X. Song (LBNL) for electron microscopy analysis and Ms. S. Brown (Department of Physics, Massachusetts Institute of Technology) for Raman study. This work was performed under the auspices of the U.S. Department of Energy by Lawrence Livermore National Laboratory under Contract No. W-7405-ENG-48 with financial support from the Strategic Environmental Research and Development Program (SERDP) and the Office of Basic Energy Sciences - Division of Advanced Energy Projects.

REFERENCES

1. S. T. Mayer, R. W. Pekala, and J. L. Kaschimitter, J. Electrochem. Soc., 140, 446 (1993).
2. J. C. Farmer, D. Fix, G. Mack, R. Pekala, and J. Poco, J. Appl. Electrochem., 26, 1007 (1996).
3. R. W. Pekala, J. C. Farmer, C. T. Alviso, T. D. Tran, S. T. Mayer, J. M. Miller, and B. Dunn, Proceeding volume of the Fifth International Society of Aerogels, Montellier, France, September 8-10, (1997).
4. R. W. Pekala, *J. Mat. Sci.*, 24, 3221 (1989).
5. R. W. Pekala, C. T. Alviso, X. Lu, J. Gross and J. Fricke, J. of Non-crystalline Solids, 188, 34 (1995).
6. H. Tamon, H. Ishizaka, M. Mikami, and M. Okazaki, Carbon, 35, 791 (1997).
7. J. H. Song, H. J. Lee, and J. H. Kim, Han'guk Chaelyo Hakhoechi, 6, 1082 (1996).
8. R. W. Pekala, and C. T. Alviso in Novel Forms of Carbon, edited by C. L. Renschler, J. J. Pouch, and D. M. Cox, p. 3, (1992).
9. R. W. Pekala, S. T. Mayer, J. F. Poco and J. L. Kaschmitter in Novel Forms of Carbon II, edited by C. L. Renschler, D. M. Cox, and J. J. Pouch, p. 79, (1994).
10. E. P. Barret, L. G. Joyner and P. Halenda, J. Amer. Chem. Soc., 73, 373 (1951).
11. GenCorp./Aerojet Company (Sacramento, CA)
12. T. D. Tran, L. Murguia, J. Wills and R. W. Pekala, Abstract Y8.30, Materials Research Society fall meeting, Boston, MA, December 1-5, (1997).

IMPROVED CARBON ANODE MATERIALS FOR LITHIUM-ION CELLS

DR. J. FLYNN, DR. C. MARSH
Yardney Technical Products, Inc. 82 Mechanic Street, Pawcatuck, CT 06379

ABSTRACT

Several carbon materials have been studied for suitability as anode materials in lithium-ion cells. Carbons that have been included in this evaluation are three grades of commercially available mesophase carbon microbeads (MCMB) 6-28, 10-28 and 25-28, two specially prepared mesophase fibers (Amoco), a foreign mesophase fiber and KS-15 graphite (Lonza). Differences in cycling behavior between the three types of MCMB material are shown. Data of full lithium-ion cells demonstrate the effect that the choice of carbon material has on the cell discharge voltage and capacity. Lithium reference electrode experiments in full cells (3.0-4.0Ah capacity), elucidate the dynamics under several charge/discharge regimes and provide a comparison between the performance of carbon fiber and graphite anode materials. These test results indicate that the fibers can be charged at significantly higher rates than graphite without showing polarization at the anode. Full and half cell data also demonstrates the high coulombic efficiencies of the mesophase materials and first cycle efficiencies as compared to graphite. A comparison of two mesophase materials with different textures in full cells under strenuous cycling conditions shows significant differences in capacity retention. SEM photos of fibers showing the different textures are also presented.

INTRODUCTION

Carbon materials play a unique and important role in lithium-ion cells. Pure lithium metal anodes are plagued by dendritic formation resulting in poor safety of such cells. Under a SBIR Phase II Program, Yardney Technical Products has been researching and developing a 20Ah lithium-ion cell for satellite applications. For this effort, various carbon materials have been evaluated for suitability in Lithium-ion cells. Important characteristics for the carbons include, the ability to rapidly intercalate/deintercalate the lithium-ion, the stability of the passivation layer on the carbon surface and a flat voltage curve versus lithium metal. A carbon used as an anode material with the above attributes should result in a cell with long cycle life, increased cell voltage and rapid charging ability. This paper presents some results of the evaluation process.

In Lithium-ion systems, a key component for high rate charge/discharge capability is the carbon's ability to rapidly intercalate/deintercalate the lithium ion into its matrix. Some recently introduced carbon materials, which have found use as lithium-ion anode materials, include the mesophase pitch carbons, microbeads and fibers. The unique feature shared by the two mesophase carbon materials are the radial orientation of the graphite layers in the carbon structure. Mesophase microbeads (MCMB) have been studied extensively in the literature and have been shown to be attractive host materials for lithium-ion cells because the surfaces of MCMB [1,2] have graphite crystallite edges exposed. This structural characteristic has the advantage of supporting high current densities because lithium intercalation proceeds from the edges of the crystallite; thus it has found use in commercial batteries. [1,3] A major supplier of this material is Osaka Gas, which has three commercially available varieties designated 6-28, 10-28 and 25-28. The first number indicates the particle size while the second designates the temperature of graphitization, 2800°C.

Recently, a foreign manufacturer of mesophase fibers developed an optimized pitch fiber for Lithium-ion applications and reported some very exciting results using this type of mesophase carbon fiber material [4] as an anode. The authors of this study attribute the performance attributes of cells to the radial structure of the mesophase fiber material. [4]

This paper presents lithium reference electrical experiments in full lithium-ion cells (3.0-4.0Ah capacity) using mesophase pitch fiber material as the anode and compares them with graphite anodes. These experiments indicate the fiber containing cell significantly resists polarization at the anode during high rates of charge compared to the cells using graphite as an anode. These tests indicate that mesophase fibers are the preferred anode material for high rate applications instead of graphite. In addition, the other dynamics of the cell elucidated from the reference electrode measurements are presented. Results from a cycling experiment on the three commercially available mesophase bead materials are shown and possible explanations of the different behavior presented. Other important parameters effecting cell performance, such as the effect of the carbon materials on cell discharge voltage, coulombic and first cycle efficiencies are also discussed.

EXPERIMENT

Cell Construction

Test cells were made with lithiated nickel oxide cathodes and carbon anodes. The cathodes were made from a slurry of lithiated nickel oxide, graphite, carbon black and PVDF binder dissolved in DMF. After mixing, this slurry was coated onto thin aluminum foil, dried and then compressed. The anodes were prepared in a similar manner where the carbon material (graphite, fiber, bead or petroleum coke) was added to PVDF/DMF solution and coated onto thin copper foil. The anodes were then bagged with Celgard separator material. The prismatic cells have the following parallel configuration anode/separator/cathode....etc. The lithium reference electrode was prepared by rolling lithium onto a narrow strip of copper foil (<1/16" wide). The reference electrode was then bagged with Celgard separator and placed between a cathode/anode pair in the middle of the cell pack. The reference electrode lead was then fed through a hole in the cell cover which was later sealed with epoxy. The cell was leak tested, vacuum filled with electrolyte and under went five formation cycles prior to cycles involving ref/anode and ref/cathode measurements.

Data Acquisition

Data was acquired for this experiment using the Arbin battery cycler and the Helios Data Acquisition System (Fluke, Inc.) operating in parallel. The cycling conditions for the cell were controlled by the Arbin while the positive versus lithium reference and the negative versus lithium reference were measured and stored by the Helios. The cell voltage was monitored by both the Arbin and the Helios. As a check to ensure that all the leads were properly attached, the difference between the cathode versus lithium reference and the anode versus lithium reference voltages equaled the cell voltage.

RESULTS

Preliminary work done at Yardney with several samples of pitch carbon fiber material provided by Amoco have yielded some very encouraging results. When comparing cycling results of cells made with graphite and those with mesophase pitch fiber anodes, the fibers showed more resistance to polarization than graphite. Figure 1 is a plot of the 4 Ah test data of a cell using lithiated nickel oxide cathodes with graphite anodes. The cell was cycled under a variety of conditions where both the lithium references to cathode voltage and anode voltage were measured. Inspection of the curve of the negative electrode to reference voltage, during constant current charging at 0.5 amps (0.5 mA/cm^2) shows the negative electrode to reference voltage dropping below zero volts versus lithium. At voltages below zero, lithium metal can plate onto the surface of the graphite anode and shorten cycle life or cause safety problems. At an even higher charge current, 1.0 amps, (1.0 mA/cm^2) the negative electrode to reference voltage is even more polarized, increasing the likelihood of lithium metal plating at the anode. The kinetics of intercalation are slowed by larger particle size and the increased degree of lithiation of the graphite. [5] Polarization has been suggested as a major source of cycle life loss in Lithium-ion cells. [6] A carbon material which can rapidly intercalate lithium will resist polarization and eliminate metallic lithium plating at the carbon surface.

Figure 1: 4Ah Test Data Graphite Containing Cell

Figure 2 is a plot of a similar test cell using a mesophase pitch fiber carbon anode. In this case, the negative to reference voltage curve, even at the higher constant current charge of 2.0 amps, (2.0 mA/cm^2) does not drop below zero volts versus lithium. Only at constant voltage charge, where the current cut-off can be adjusted, does this voltage approach zero versus lithium. This data indicates the mesophase fibers with their unique radial structure demonstrate some remarkable improvements in intercalation/deintercalation kinetics that allow for increased utilization of the carbon material during high rate charge/discharge cycling of lithium-ion cells.

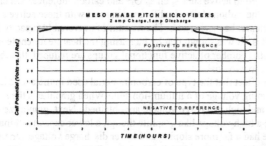

Figure 2: Plot of Mesophase Pitch Fiber Carbon Anode

The effect of a strenuous cycling regime on two different fiber textures has been explored with samples from Amoco. As shown in Figure 3, the texture and other physical characteristics of the fiber have dramatic effect on the cycling behavior of the materials. With such preliminary results, the direction of future fiber optimization will be determined. SEM photos of the cross section of two fibers studied are shown in Figure 4. One fiber, Sample A, has more radial texture than the other, Sample B, allowing for better cycling. These findings are consistent with what has been reported in the literature where mesophase pitch fibers with varying textures performed differently under high rate cycling conditions. [7] From this study, two important parameters emerge as necessary for good cycling behavior of mesophase fibers, high mechanical strength derived from fiber texture and specific crystallographic orientation. To maintain fiber integrity strength during intercalation and deintercalation, a texture that retains mechanical strength in the carbon host is needed. In addition, the fiber must have a crystal orientation that allows for rapid diffusion of lithium to the inner part of the fiber to allow for high utilization of the carbon host during high rate charge/discharge cycling. [4, 7, 8]

Sample A   Sample B

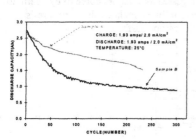

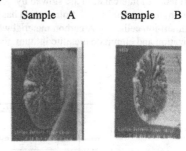

Figure 3: Cycle Life of Two Mesophase Fibers

Figure 4: Cross Sectional SEM of Fibers A and B

Figure 5 is a comparison of first cycle intercalation/deintercalation process of three types of carbon materials, a Japanese supplied mesophase fiber, mesophase microbead 6-28 and KS-15 graphite. Both the fiber and bead materials gave high first cycle coulombic efficiencies >93% compared to 75% for the KS-15 material. Some findings for the first cycle irreversible capacity loss associated with mesophase pitch fibers have been reported in the literature. [9] In this study, the amount of lithium was studied on different sections of mesophase pitch carbon ribbons after intercalation/deintercalation cycle. In this system, the amount of lithium was found to be higher at the surface of the deintercalated carbon but significant amounts were also found in the carbon bulk. This has been attributed to active sites such as OH and carbon moieties. On charge, when lithium is intercalated into the carbon electrode, lithium can react with these active sites to form stable compounds which cannot be discharged. The amount of irreversible capacity is dependent on the number of active sites. It was shown that by increasing the heat treatment temperature for mesophase pitch systems, the number of active functional groups are reduced, thereby decreasing the first cycle irreversible capacity loss.

Figure 6 shows the effect that the type of carbon material used in the anode has on the specific capacity and operating voltage of a lithium-ion cell. When graphite is the anode material, highest capacity and high discharge voltage are obtained. Next, graphitized mesophase fibers and beads also have a high voltage like graphite though a lower capacity. Finally, cells made with petroleum coke had a far more sloping and lower discharge voltage versus time curve and half the discharge capacity as compared to the graphite cell.

A comparison was conducted on the three types of mesophase microbeads commercially available 6-28, 10-28 and 25-28. From Figure 7, the 25-28 shows the greatest rate of capacity fade of the three materials tested. SEM photos of the pure material indicated that the 25-28 material is the most agglomerated of the three. [10] The increased agglomeration could reduce surface area of the carbon host and slow the kinetics of intercalation, decreasing cycle life.

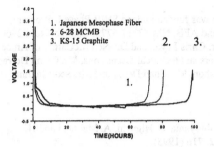

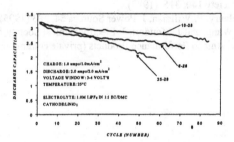

Figure 5: Comparison of Anode Materials During 1st Cycle. 0.1 mA/cm² CCD to 0.005V (Li⁺/Li) Cutoff. Followed by 0.1 mA/cm² CCC to 1.5V (Li⁺/Li).

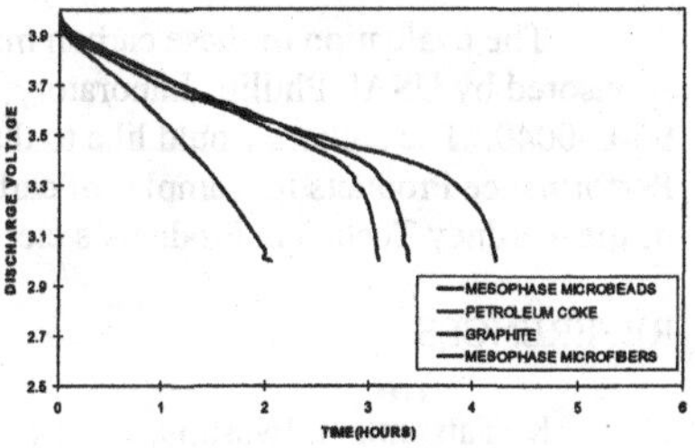

Figure 6: Effect of Carbon Material on Cell Capacity and Voltage

CONCLUSIONS

An important property for a carbon lithium ion anode material is radial texture that allows for greater access for lithium ion to intercalate into the carbon host matrix. Reference electrode measurements in full lithium-ion cells demonstrated that the cell made with a mesophase

Figure 7: Cycle Life of Three Mesophase Microbeads

material with a radial texture could be charged at a higher rate without anode polarization than the cell made with graphite anode. In addition, the mesophase fiber with a small diameter will minimize the distance of diffusion of lithium-ion into the bulk carbon host structure. With samples supplied from Amoco Performance Products, the mesophase pitch carbon fiber with the greater radial texture has better cycle life presumably due to improved transport of lithium ion into the carbon host. An important study would be to measure the diffusion coefficients of lithium ion into these carbon materials. High temperatures of heat treatment are desirable to eliminate active species in the carbon host that undergo irreversible processes, in return this would increase the first cycle coulombic efficiencies, increase the graphitic nature of the carbon and increase the cell voltage.

ACKNOWLEDGMENTS

The evaluation of these carbon materials was funded under a SBIR Phase II Program, sponsored by USAF Phillips Laboratory, Kirkland AFB, NM 87117-5772, Contract No. F29601-96-C-0040. The authors would like to thank Mr. Chris Levan and Dr. Neil McCarthy of Amoco Performance Products for samples of carbon fibers and technical discussions, Mr. Glenn Derrah of the Yardney Technical Products's electronic shop, Mr. Chad Deroy and Mr. Scott Ingram.

REFERENCES

1. K. Tatsumi, N. Iwashita, H. Sakebe, H. Shioyama, S. Higuchi, A. Mabuchi and H. Fujimoto, J. Electrochemical Society **142**, 716 (1995).

2. A. Mabuchi, H. Fujimoto, K. Tokumitsu and T. Kasuh, J. Electrochemical Society **142**, 3049 (1995).

3. M. Kitagawa, H. Koshina, A. Ohta, The 14th International Seminar on Primary and Secondary Battery Technology and Application, March, 1997.

4. N. Takami, A Satoh, M. Haraand, T. Ohsaki, J. Electrochemical Society **142**, 2564 (1995).

5. T.D. Tran, J. H. Feikert, R.W. Pekala and K. Kinoshita, J. Applied Electrochemistry **26**, 1161 (1996).

6. Proceedings for the Electrochemical Society, Montreal Conference, Spring Meeting, 1997.

7. N. Imanishi, H. Kashiwagi, T. Ichikawa, Y. Takeda, O. Yamamoto and M. Inagaki, J. Electrochemical Society **140**, 315, (1993).

8. D. Billaud, F. X. Henry, P. Willman, J. Power Sources **54**, 383 (1995).

9. Y. Matsumura, S. Wang, and J. Mondori, J. Electrochemical Society **142**, 2914 (1995).

10. Dr. Neil McCarthy, Amoco Performance Products (private communication).

THE EFFECTS OF NEAR SURFACE AND BULK MICROSTRUCTURE ON LITHIUM INTERCALATION OF DISORDERED, "HARD" CARBONS

WILLIAM R. EVEN*, LAWRENCE W. PENG*, NANCY YANG*, RONALD GUIDOTTI, ** THOMAS HEADLEY**
Sandia National Laboratories
*P.O. Box 969, Livermore, CA 94551, wreven@sandia.gov
**P.O. Box 5800, Albuquerque, NM 87185-0614

ABSTRACT

Disordered carbons were synthesized at 700°C from methacrylonitrile-divinylbenzene precursors. The disorder, even at the free surface, was confirmed with TEM. These powdered carbons were subjected to rapid surface heating by a pulsed infrared laser. While the bulk structure remained essentially unchanged, there was substantial surface reconstruction to a depth of 0.25μm after heating (5.9 W average power at 10Hz, 10 ns pulse width, 1064nm wavelength). The surface ordering appears similar to the bulk microstructure of carbons isothermally annealed at 2,200°C (i.e., turbostratic). Improvements were observed in first cycle irreversible loss, rate capability, and coulombic efficiencies of the reconstructed carbons, relative to the untreated carbon.

INTRODUCTION

Ordered carbons and true graphites dominate production, but multi-phase and disordered carbons are being pursued in R&D because of their capacity advantages. Ordered-carbon systems imposes their structural order onto the intercalated lithium, thereby limiting its overall capacity to 372 mAhr/g at standard temperatures and pressures (i.e. LiC_6 formation). Both the phenomenology and energetics of this interaction, as well as, surface losses are reasonably well understood.[1-3] Disordered carbons, on the other hand, lack the obvious "templating" which occurs in graphites and achieve reversible intercalation capacities in excess of 650 mAhr/g (e.g., on the 20th cycle). On the surface, the disordered carbon seems to be behaving more like a sponge, picking up lithium irrespective of the influences or lack of local microstructural order. This disorder, however, profoundly affect lithium losses at surface defects or functionalities. In these carbons loss can vary from 35% to 100% of the first-cycle reversible capacity as a result of irreversible electrolyte reduction to form a solid-electrolyte interface (SEI). Similarly, microstructure influences both intercalation and deintercalation rate capability. [4-6]

We have been studying disordered ("hard") carbons derived from polymerized methacrylonitrile (PMAN)-divinylbenzene, in order to better understand lithium insertion, transport, storage and trapping, i.e. the overall lithium inventory. As part of this overall broad program, we are examining laser-induced surface reconstruction to transform the PMAN disordered carbon surface into a more-ordered, turbostratic state. Irreversible lithium losses associated with solvent decomposition and SEI formation are normally much lower for ordered carbons, whether turbostatic or fully graphitized. The intent of this work was to examine the potential for combining the better properties of disordered carbons, such as high capacity, with the low irreversible losses and good rate behavior of more ordered systems. To date, we have not attempted to maximize the entire performance space, rather initially intending to demonstrate that

619

synergistic performance can be realized by the combination of two microstructural polymorphs obtained via surface reconstruction.

EXPERIMENTAL
Synthesis

The carbon precursor was synthesized as an inverse emulsion containing a methacrylonitrile / divinylbenzene monomer molar ratio of 1.6:1. The polymerization was carried out using a persulfate catalyst and a 65°C cure for 16 h. The resulting low-density foam was forced air dried, removing the water pore former, and attrited to a fine powder, <400 mesh. The precursor powder was oxidatively stabilized in air at 240°C for 6 hours.[7] Pyrolysis took place in a 6"-dia. tube furnace under high-purity argon flowing at 2 L/min. The sample was heated to 700°C at a rate of 2°C/min and was held at at this temperature for 5 h. The low pyrolysis temperature was chosen for the preparation of the PMAN because it offered the potential for the greatest performance differential.

Laser reconstruction was accomplished under argon in a sealed quartz cuvette containing about 4 g of powder. The cuvette was attached to a vibratory table via stretched "O"-rings such that on activation the carbon bed was fluidized. A 9-mm-dia. infrared laser beam (1064 nm wavelength) was directed orthogonally to the cuvette. The power of the laser was measured at 5.9 W, time averaged at a 10-Hz repetition rate. Normalizing for pulse length, the peak energy deposition was 59 MW per laser pulse. The cuvette was translated beneath the beam to assure more uniform carbon exposure. (For the experimental configuration, see the schematic of Figure 1.)

Physical and Chemical Characterization

The resulting carbon was a free flowing powder which is typical of a low-temperature pyrolysis and a nitrogen-containing precursor..The final composition was 1.3% H, 3.5% N, 5.7% O, with the balance C.[1] The surface area of the carbon was measured via nitrogen-BET and found to be >180 m^2/g. This material is nearly completely disordered; exhibiting a very, very broad x-ray diffraction (XRD) maximum at a two-theta angle equivalent to a d_{002} lattice spacing of 3.75Å; no other maxima were observed

Microstructural imaging was carried out on a Phillips CM-30 transmission electron microscope (TEM). Because of the emulsion origin of the carbon powder, no TEM sample preparation was necessary. The natural meniscus edges provided both sufficiently thin specimen locations for examination and original-surface identification. Specimen powders were supported on "holey" carbon films.

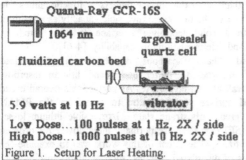
Figure 1. Setup for Laser Heating.

Electrochemical Characterization

The three-electrode cells used for testing the carbons were made from 1/2" Swagelok® perfluoroalkoxy (PFA) tees.

---

[1] Unless otherwise noted, all compositions are in weight percent.

The electrodes cells were prepared by "doctor blading" a paste onto Cu-foil substrates. The paste consisted of a mixture of 80% carbon, 5% Super 'S' carbon black (as a conductive additive), and 15% polyvinylidene difluoride (PVDF). Discs 1.27 cm (0.5" in) in diameter were then punched from the dried, pasted Cu sheets for used in the cells. The mass of carbon in each electrode ranged from 5 to 6 mg.

The tee cells were assembled in the dry room (RH<3%) and used a Li counter and reference electrodes and two Celgard 2500 separator discs. The cells were transported into a glovebox (<10 ppm each H_2O and O_2) where they were evacuated and backfilled with 1M $LiPF_6$/ethylene carbonate (EC)-dimethyl carbonate (DMC) (1:1 v/v). The electrolyte typically contained <40 ppm of H_2O. Generally, the cells were allowed to stand overnight before being placed on test.

The cells were tested galvanostatically between voltage limits of 2 V and 0.01 V vs. Li/Li$^+$ on an Arbin Corp. Battery Test System. Two cells were run for each of the test conditions. The galvanostatic cycling involved two test regimes: a 32-cycle protocol (Table I) and an 18-cycle protocol (Table II). The 18-cycle test immediately followed the 32-cycle test, for a total of 50 cycles.

Table I. Testing Protocol for 32-Cycle Galvanostatic Cycling Between 2 V and 0.01 V.

Number of Cycles	Intercalation Rate	Deintercalation Rate
20	C/5	C/5
3	C/10	C/10
3	C/2.5	C/2.5
3	C/1.25	C/1.25
3	C/0.625	C/0.625

Table II. Testing Protocol for 18-Cycle Galvanostatic Cycling Between 2 V and 0.01 V.

Number of Cycles	Intercalation Rate	Deintercalation Rate
3	C/10	C/10
3	C/10	C/5
3	C/10	C/2.5
3	C/10	C/1.25
3	C/10	C/0.625
3	C/10	C/0.313

RESULTS

The laser pulses produced carbon surface heating along with visible light, carbon material volatilization, and a visible deposit on the inside of the cuvette walls. Measurement of the emitted light yielded an estimated black-body temperature of 1,250°, which was lower than originally expected. There were signs of carbon ionization in the spectrum, indicating that some of the carbon—either the immediate surface or material expelled from the surface—experienced significantly higher temperatures. Considering thermal conduction, convection, and volatilization, 1,250°C probably represents a lower bound for the "average" particle surface temperature.

There is no way of assuring that every particle or all surfaces of every particle received laser surface reconstruction. The small areal exposure cross-section, the number of repeated pulses, and the randomization associated with the fluidized bed reasonable guarantee a very large fraction of "affected" particles. TEM examination of random samples from the treated powder confirm the large population of treatment particles. Figure 2 illustrates the surface changes associated

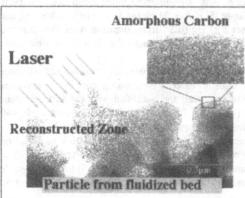

Figure 2. Typical Area of Localized Ordering Caused by Laser Heating

with laser contact for a typical carbon particle, ranging from completely disordered bulk structure to a turbostratic surface layer.

Detailed examination of particles, including specimen tilting to maximize alignment with the incident TEM insident beam, revealed substantial reconstructive ordering involving the immediate surface [material potentially associated with SEI layer formation] and the subsurface microstructure [involving intercalation capacity]. Figure 3 illustrates that the laser-reconstructed surface is quite similar to the 2,200°C isothermal case.

Surface area measurements of the reconstructed carbons showed a decrease in surface area of about 25% relative to the unheated material. This is consistent with the expected loss of fine structure due to volatilization during pulse reconstruction. The aggregate effect of laser surface reconstruction on cycling performance is shown in Table III. Once again, the 700°C pyrolysis carbon was specifically chosen because of its greater potential for microstructural reordering thereby reducing its large passivation. These data were extracted for the first 20 cycles of the 32-cycle testing where the carbons were intercalated and deintercalated at the same rate. This test was then followed by18 more cycles: three cycles each at deintercalation rates of C/10, C/2.5, C/1.25, C/0.625,, and C/0.313, while maintaining the same C/10 intercalation rate. (The latter approach enhances the rate capability during deintercalation.) The data are averages of two cells.

There areno significant differences in the first cycle reversible capacities of all the carbons. Laser heating did increase the first-cycle efficiency for the high-power sample which resulted in a significant reduction in the first-cycle irreversible loss. This observation is consistent with surface reconstruction invoking a more "graphine" and less "disordered" presentation to the electrolyte. By the 20th cycle, however, the high-power sample actually showed a slight reduction in efficiency and had a reversible capacity similar to the control sample. The material heated with low-power-density laser pulses had a slightly higher efficiency and reversible capacity by the 20th cycle. There was no difference in fade for these three samples.

The effects of surface reconstruction at the C/10 rate and the higher

Figure 3. Typical Localized Surface Ordering as Compared to Anneals.

Table III. Summary of Galvanostatic Cycling Tests with Laser-Heated Samples.

| Treatment | First Cycle | | | | 20th Cycle | | | |
	Load mAh/g	Unload mAh/g	Eff. %	Qirr mAh/g	Load mAh/g	Unload mAh/g	Eff. %	Fade mAh/g-cyc[#]
Neat [control]	746.3 [20.5]*	337.3 [11.2]	45.2	409.0 [9.3]	233.6 [2.5]	225.2 [2.5]	96.4	-1.94 [0.31]
Low Dose	779.9 [12.2]	354.9 [11.6]	45.5	425.0 [0.6]	252.9 [4.7]	245.6 [4.2]	97.1	-1.97 [0.06]
High Dose	641.6 [35.7]	326.5 [22.2]	50.9	315.1 [0.6]	242.2 [13.5]	228.4 [15.3]	94.3	-1.94 [0.12]

[#] Fade = (Capacity for Cycle 20 - Capacity for Cycle 11)/10 Cycles.
* Values in parentheses are one standard deviation about the mean.

deintercalation rates on the reversible capacities are summarized in Figure 4. The low-power-treated sample showed a greater capacity than the control over the entire range of rates. The high-power-treated sample was superior to the control above 0.15 C and outperformed the low-power-treated sample at rates >0.4C. This sample also showed a lower loss in capacity with rate than the other samples

Figure 5 shows the reversible capacities as a function of rate when the intercalation rate was reduced by 50% to C/10. The deintercalation rates were the same as for Figure 4, except that an addition rate of C/0.313 (3.19C)was incorporated in the test. Similar trends were obtained as for Figure 4, in that laser heat treatment improved the high-rate capability of the carbon. The overall reversible capacities were similar at low rates (below 0.4C or C/2.5) but were substantially improved at the higher rates. Under these test conditions, the low-power laser heat treatment was more beneficial than the high-power heat treatment. The conclusions remain the same, however: laser-induced reconstruction improves the high-rate capabilities of this PMAN-derived carbon.

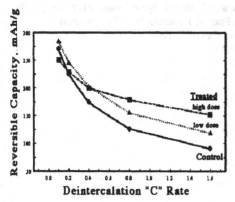

Figure 4. Reversible Capacities as a Function of Dintercalation Rate for the Same Rate of Intercalation and Deintercalation.

CONCLUSIONS

In summary, laser-induced surface reconstruction of PMAN-derived carbons has overall beneficial effects on the performance. Microstructural changes can be observed and linked to performance variations. The power level used has a significant effect on the observed results. A reduction in the first-cycle irreversible loss in capacity is observed only under high-power conditions. This condition, however, results in lower coulombic efficiencies at rates above

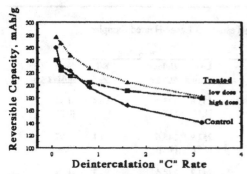

Figure 5. Reversible Capacities as Function of Deintercalation Rate for Same Intercalation Rate of C/10 (0.1C).

0.2C (C/5) when the same rate of intercalation and deintercalation of C/5 is used. Intercalation at a C/10 rate greatly improved the overall performance of the carbon—especially for the heat-treated carbons. There will be some tradeoff between coulombic efficiency and the reversible capacity for the various carbons, depending on the charging regime and the power level used during reconstruction. It should be emphasized that these data are preliminary and the treatment conditions have not optimized. These data suggest that this is an area for additional fruitful research. It would also be interesting to determine if this technique has applicability to other types of disordered or polymorphic (e.g., PAN-derived and MCMB) carbons as other authors have examined. [8]

ACKNOWLEDGMENTS

The authors wish to thank Marion Hunter for her assistance in preparing and characterizing the carbons. Mike Overstreet and Herb Case assisted in the electrochemical testing. This work was supported by the U.S. Department of Energy under Contract DE-AC04-94AL85000.

REFERENCES
1. D. Aurbach, et. al., JECS, 141(3), p. 603, (1994).
2. M. Masayuki, T. Ichimura, M. Ishikawa, and Y. Matsuda, JECS, 143(2), p.L26, (1996).
3. F. Disma, L. Aymard, L. Dupont, and J-M. Tarascon, JECS, 143(12), p.3959, (1996).
4. S. Wang, Y. Matsumura, and T. Maeda, Synth. Mater. 71, p. 1759 (1995).
5. Y. Matsumura, W. Wang, , and K Shinohara, and T. Maeda, Synth. Mater. 71, p1757 (1995).
6. C. Bindra, V. Nalimova, D. Sklovsky, and J. Fischer, J. Electrochem. Soc., in press.
7. W. R. Even and D. P. Gregory, MRS Bulletin, 19(4), 29, 1994.
8. M. Inagaki, Solis State Ionics, 86, p.833, (1996).

Part VII

Supercapacitors

ELECTRIC DOUBLE- LAYER CAPACITOR
USING ORGANIC ELECTROLYTE

T.Morimoto, M.Tsushima, M.Suhara, K.Hiratsuka, Y.Sanada and T.Kawasato,
Asahi Glass Co. Ltd., Research Center Hazawa-cho, Kanagawa-ku Yokohama
221, Japan

ABSTRACT

Electric double-layer capacitors based on charge storage at the interface between a high surface area activated carbon electrode and a propylene carbonate solution are widely used as maintenance-free power sources for IC memories and microcomputers. New applications for electric double-layer capacitors have been proposed in recent years. The popularity of these devices is derived from their high energy density compared with conventional capacitors and their long cycle life and high power density relative to batteries.

The performance of the capacitor depends not only on the materials used in the cells but also on the construction of the cells. The performance of power capacitors for large power sources as well as materials and construction of these capacitors are described.

INTRODUCTION

Electric double-layer capacitors based on the charge storage at the interface between a high surface area activated carbon electrode and an organic electrolyte solution have been developed, and are widely used as maintenance-free power source for IC memories and microcomputers [1]. The achievement of excellent performances of capacitors requires an electrode fabricated with high surface area activated carbon of suitable surface property and pore geometry, an electrolyte solution which provides high conductivity over a wide temperature range and good electrochemical stability to allow the capacitor to be operated at higher voltage, and a cell construction material which does not corrode electrochemically during anodic polarization.

New applications for electric double-layer capacitors have been proposed in recent years. The popularity of these devices is due to their high energy density relative to conventional capacitors, and their long cycle life and high power density relative to batteries. Electric double-layer capacitors exhibit in principle provide unlimited cycle life and maintenance free operation as an alternative to batteries in power circuits. A new promising application for electric double-layer capacitor is a pulse power source in electric and hybrid vehicle application. The pulse power source provides the peak power during acceleration and recovers energy during braking using a generator.

In this paper the performance of the electric double-layer capacitor for power sources as well as the material and the construction of the capacitor are described.

MATERIAL

Electrolyte

As the breakdown voltage of the capacitor is strongly related to the decomposition voltage of the electrolyte, it is important to use organic electrolytes having a high decomposition voltage to obtain a capacitor of both stable performance and high energy density [2],[3]. It is also important to use an electrolyte of high conductivity to obtain a capacitor of high power density.

Therefore electrolyte solutions obtained by dissolving a 1.1M tetraethyl phosphonium tetrafluoroborate of the highest grade available in propylene carbonate(PC) were used.

The specific conductivity at room temperature and the decomposition voltage of the solution are 0.010S/cm and 3.0V, respectively.

Electrode

Activated carbon electrode should have a large volumetric capacitance and high electrochemical stability, to obtain a capacitor of high energy density and stable performance.

The volumetric capacitance is a function of the apparent density of the electrode and the specific surface area of the activated carbon. As the apparent density of the electrode decreases as the specific surface area of activated carbon increases, the maximum volumetric capacitance of the electrode is obtained using the activated carbon of the specific surface area of $2000m^2/g$ by a compromise between the increase in the specific surface area of activated carbon and the decrease in the apparent density of the electrode.

Stability of the cell against voltage application is dependent on the electrochemical decomposition of PC, which is strongly related to the existence of oxygen-containing functional groups such as carboxylic and quinone groups on the surface of activated carbons [4]. This fact suggests that the activated carbon of low oxygen content is effective to obtain a capacitor of stable performance during voltage application.

Therefore activated carbon of low oxygen content having the specific surface area of $2000m^2/g$ is used as the electrode material.

CELL CONSTRUCTION

Power capacitors having the capacitance of 1300F shown in Fig.1, were developed using the material mentioned before. The power capacitors consist of activated carbon sheet electrodes attached to aluminum foil current collectors and separators. Rectangular polarizable anodes and cathodes separated by separators are stacked alternatively and electrically connected in parallel using a current lead [5].

The whole assembly immersed in an organic electrolyte solution was housed in a aluminum boxes of the size of $70 \times 150 \times 22 \text{mm}^3$. Internal resistance of the capacitors was $4 \text{m} \Omega$. The specification of the capacitors is shown in Table1.

PERFORMANCE

The performance of the electric double-layer capacitor is defined in terms of capacitance,C in Farad, cell internal resistance, R in Ohm.

Charge and discharge curve of the electric double-layer capacitor is schematically shown in Fig. 2. In Fig. 2, V_0, V_1 and V_2 in volt are the charged voltage, cell voltage at the beginning of discharge and cell voltage at the end of discharge, respectively. In the discharge curve, a constant discharge current of I Amperes flows for T seconds from the beginning of the discharge to the end of the discharge. Initial cell voltage drop ΔV is expressed by the equation (1).

$$\Delta V = V_0 - V_1 = IR \tag{1}$$

Capacitance of the cell is expressed by the equation (2).

$$C = IT / (V_1 - V_2) \tag{2}$$

Generating power in watts (W), during discharge is expressed by the equation (3).

$$W = (V_1 + V_2)/2 \times I = V_0 I - (R + T/2C)I^2 \tag{3}$$

Discharged energy in watt hours (E), is expressed by the equation (4).

$$E = (V_1 + V_2)/2 \times I \times T/3600 = 1/2C(V_1^2 - V_2^2)/3600 \tag{4}$$

Performance of 1300F capacitor

Fig.3 shows the charge characteristics of the 1300F capacitor under constant charge current of 100A and 200A to the cell voltage of 2.5V. The capacitor can be fully charged within 30 seconds for 100A and 15 seconds for 200A, respectively. Fig.4 shows

629

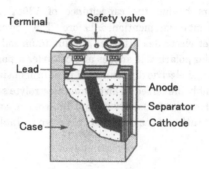

Fig.1 Structure of 1300F cell

Table 1 Specification of 1300F cell

Voltage (V)	Capacitance (F)	Int.Resistance (mΩ)	Dimension (mm)	Volume (cm^3)	Weight (kg)
2.5	1300	4±1	150(H)X70(W)X22(D)	231	0.38

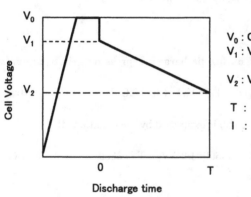

V_0 : Charged voltage (V)
V_1 : Voltage at the beginning of discharge (V)
V_2 : Voltage at the end of discharge (V)
T : Discharge time (sec.)
I : Constant discharge current(A)

Fig.2 Schematic constant current charge and discharge curve

the discharge characteristics of the 1300F capacitor under various constant discharge currents. The cell voltage of the capacitor decreases linearly under constant discharge current from 2.5V to 1.0V in 40 seconds for discharge current of 50A and 20 seconds for discharge current of 100A, respectively. As shown in Fig.4, the discharge characteristics of the capacitor appear to be ideal in that the curves are linear with discharge time for constant discharge current indicating that the capacitance and resistance are constant, essentially independent of the cell voltage. Fig.5 shows constant power discharge curves of the capacitor. The capacitor can discharge for 7 seconds from the cell voltage of 2.5V to 1.25V under the constant power density of 600W/kg. Constant power discharge curves are not linear as the current changes during the discharge due to the decreasing cell voltage. It is clear from these charge and discharge characteristics that the electric double-layer capacitor can pass a large current and as the result can be quickly charged and discharged at a high rate.

Fig.6 and Fig.7 show the cycle life performance of the 1300F capacitor under accelerated deterioration condition of 45°C. As shown in Fig.6, the deterioration in capacitance is less than 15% after 400,000 times charge and discharge cycles under constant charge and discharge current of 50A between the cell voltage of 2.5V and 1.25V.

It is important to keep the increase in internal resistance of the cell as small as possible to prevent the energy density of the capacitor from deteriorating with respect to the number of charge and discharge cycles, because the discharged energy decreases as the internal resistance of the cell increases, as shown by equation (1) and (4). The internal resistance of the cell increases from the initial 4mΩ to 5mΩ and changes little after 400,000 times charge and discharge cycles. As a result, the change in the energy density of the capacitor with respect to the number of charge and discharge cycles shows a similar tendency with the change in the capacitance of the cell and becomes about 17% after 400,000 times charge and discharge cycles test, as shown in Fig.7.

Fig.8 shows the temperature performance of the capacitor. As shown in Fig. 8, the discharged energy under constant discharge power density of 600W/kg decreases with the lowering of temperature and the simultaneous increase in resistance of the electrolyte and internal resistance of the cell.

When the voltage is applied to the capacitor, three different kind of current flow through the capacitor and decrease with time. First, the charging current flows to store the charge electrostatically. After a time, the absorption current flows to supply charge that is not recovered upon discharge of the capacitor. Finally only the leakage current flows through the cell. Due to ultra high capacitance of the electric double-layer capacitor, it takes about 30 hours to reach a steady current: i.e. the leakage current. The magnitude of this current will determine the rate of self-discharge. A leakage current, aside from mechanical fault such as a separator rupture, is the result of impurities in the electrolyte and the electrode and surface species of the electrode participating in charge transfer processes. Fig.9 shows the current that flows through

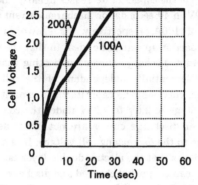

Fig. 3 Charge characteristics of 1300F cell

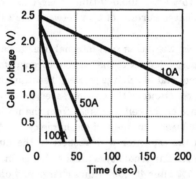

Fig. 4 Discharge characteristics of 1300F cell

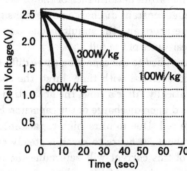

Fig.5 Constant power discharge curve of 1300F cell

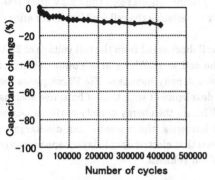

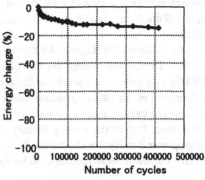

Fig.6 Capacitance change in cycle life test
of 1300F cell under constant charge-
discharge current of 50A at 45℃.

Fig.7 Energy change in cycle life test
of 1300F cell under constant charge-
discharge current of 50A at 45℃.

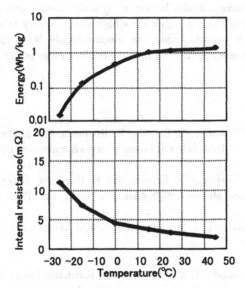

Fig.8 Temperature dependence of discharged energy
and internal resistance of 1300F cell under constant
discharge power density of 600W/kg.

the 1300F capacitor during constant voltage application of 2.5 V. It is seen that the leakage current of the capacitor is below 3mA. Fig.10 shows the retained voltage after constant voltage charging at 2.5V. As the capacitor retains more than 2.2V after 1000 hours self discharge time, it is not necessary to recharge the cell for 1000 hours after it is charged.

Fig.11 shows the Ragone plot for the cell discharged from the cell voltage of 2.5V to 1.25V. Discharged energy density of the cell is 1.0Wh/kg at a power density of 600W/kg and decreases linearly as the power density increases. Fig.12 compares the performance of the 1300F capacitor under development with that of batteries such as Ni-hydrogen battery and Li-ion battery. Although the energy density of the capacitor is less than 1/10 of the energy density of batteries, the capacitor can discharge at higher power density than batteries. It is seen that electric double layer capacitors are suitable for use as rechargeable cells for hybrid vehicles.

CONCLUSIONS

Electric double-layer capacitors based on the charge storage at the interface between a high surface area activated carbon electrode and an electrolyte solution is a new type of energy storage device having many outstanding features compared with conventional batteries. The capacitors show high power density as well as extremely long cycle life. It is expected that the electric-double layer capacitor of large capacitance will find new applications in many fields, such as power sources in hybrid cars.

REFERENCES

1. T. Morimoto, K. Hiratsuka, Y. Sanada, K. Kurihara, S. Ohkubo and Y. Kimura, Proceedings of the 33rd International Power Sources Symposium, 618 (1988).

2. K. Hiratsuka, T. Morimoto, Y. Sanada and K. Kurihara, Extended Abstracts of 178th Electrochem. Soc. Meeting, 90-2, Abstracts No. 81, 129 (1990).

3. K. Hiratsuka, Y. Sanada, T. Morimoto and K. Kurihara, Denki Kagaku, 59, 209 (1991).

4. K. Hiratsuka, Y. Sanada, T. Morimoto and K. Kurihara, Denki Kagaku, 59, 607 (1991).

5. T. Morimoto, K. Hiratsuka, Y. Sanada and K. Kurihara in Materials for Electrochemical Energy Storage and Conversion-Batteries,Capacitors and Fuel Cells, edited by D. H. Doughty, B. Vyas, T. Takamura and J. R. Huff (Mat. Res. Soc. Proc., 393, San Francisco, CA, 397(1995).

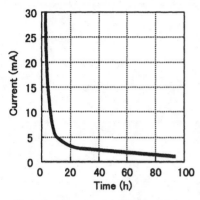

Fig.9 Current through 1300F cell during voltage application of 2.5V at 25℃.

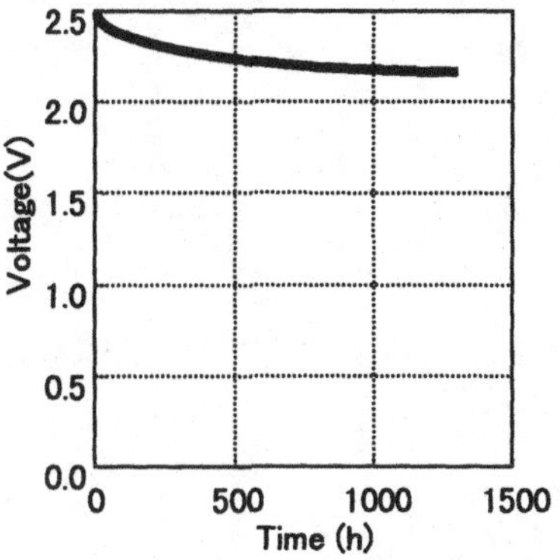

Fig.10 Retained voltage of 1300F cell at 25℃ after charging at 2.5V for 100hrs.

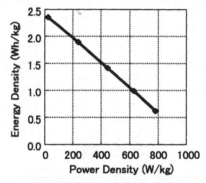

Fig.11 Ragone plot for 1300F cell discharged from 2.5V to 1.25V.

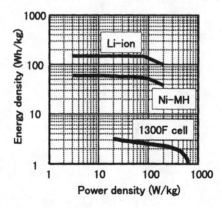

Fig.12 Comparison of performance of 1300F cell with batteries.

STRONTIUM RUTHENATE PEROVSKITES WITH HIGH SPECIFIC CAPACITANCE FOR USE IN ELECTROCHEMICAL CAPACITORS

P. M. WILDE, T. J. GUTHER, R. OESTEN*, J. GARCHE

ZSW   Center for Solar Energy and Hydrogen Research Baden–Wuerttemberg,
      Division 3: Energy Storage and Conversion, Helmholtzstrasse 8, 89081 Ulm, Germany
* present address: E. Merck KGaA, Frankfurter Straße 250, 64293 Darmstadt, Germany

ABSTRACT

Strontium ruthenates with the perovskite type structure ABO_3 have been shown to exhibit attractive capacitive properties. Doping on the A site with La lead to typical capacitance values of 21 F/g. These materials were synthesized by coprecipitating metal hydroxides from a stoichiometric salt solution and subsequent firing at 800 °C in air. In this paper we present a new procedure to synthesize the materials which are crystalline and nevertheless show appreciable capacitances in contrast to ruthenium dioxide material, which only works in a hydrated amorphous structure. The process basically consists in a pyrolysis of concentrated metal salt solutions of the respective chlorides and nitrates at 500 °C for several minutes. Excess soluble phases are removed by washing out with water. X-ray diffraction experiments revealed similar phase purity and crystallinity as known from the coprecipitated materials. However the measured capacitances of undoped perovskites reached high values of 200 F/g exceeding twenty times the value of respective coprecipitated materials. First experiments on doping the materials promise further progress. The new synthesis route introduces a higher surface area by leaving cavities from leached soluble phases and bulk defects into the crystal structure. The first effect increases the number of active sites in contact with the electrolyte while the latter enhances the protonic conduction which is necessary to keep the charge balance within the material during cycling.

INTRODUCTION

Under the materials that exhibit pseudocapacitive behavior in an electrochemical capacitor the family of metallic conducting oxides has gained very much attention. The amorphous form of hydrous ruthenium oxide is considered to be the one of the materials with the largest specific capacitances. Values up to 768 F/g are reported [1]. Devices for use in electric vehicles as peak power sources using this kind of material are under development [2]. A major drawback of this material is its high cost.

It was our intention to screen the metallic conducting perovskites for pseudocapacitive behavior. Perovskites are widely used in electrochemical energy storage and conversion, but mainly at higher temperatures either as electrodes or as proton conducting electrolytes in fuel cells. [3, 4]. Under the metallic conducting perovskites are ruthenates [5] and we started with strontium ruthenate $SrRuO_3$ [6, 7]. Modifications in the chemical composition by doping on either the A or B site of the perovskite ABO_3 were investigated on their influence of the capacitance [7, 8]. Because the results are fairly topical, some important facts are elucidated in this paper again. Moreover here a proposal on the charge storage mechanism is given.

It is the aim of this paper to illustrate our approach to increase the capacitive properties of $SrRuO_3$ by doping and by introducing a new preparation technique which multiplied the capacitance value by a factor of 25.

EXPERIMENTAL

The sample perovskites were prepared by a coprecipitation process: A stoichiometric aqueous solution of strontium nitrate and ruthenium chloride $RuCl_3$ was slowly added to 3 M KOH under vigorous stirring. After 30 minutes the solution was filtered and the precipitate washed several times with distilled water until chloride was removed completely. The precursor was dried and calcined at 500 to 800 °C for several hours in air. Raney type samples were prepared by a pyrolysis process. A solution of $RuCl_3$ and $Sr(NO_3)_2$ with the stoichiometric ratio of Ru:Sr = 1:5/2 was given dropwise in an alumina crucible kept at 500 °C in an upright standing tube furace. The powder was kept at the desired temperature for 10 to 20 minutes, was then removed quickly from the crucible and ground by a mortar and pestle. Excess soluble phases were removed by washing the sample several times with water.

Phase composition was determined using a Siemens D 5000 X-ray powder diffractometer, equipped with a graphite secondary monochromator, Bragg-Brentano focusing and Cu K_α radiation. BET surface area by nitrogen adsorption was determined with a Carlo–Erba–Instruments Sorptomatic model 1990.

Electrode preparation was done by painting a slurry of the sample with Triton–X–100 on one side of a nickel foil, previously etched in concentrated HCl to clean and to roughen the surface. The foil was 15 mm x 15 mm and spot welded to a nickel wire as a current lead before. To adhere the coating and to burn out the binder the electrodes were annealed at 400 °C for 2 hours in air. Typical mass loading was 3 mg/cm^2, which was determined by difference weighing the electrode before and after the coating process. There was no detectable weight change for an uncoated nickel foil during such an annealing procedure. In some cases the pyrolysis was done directly on the nickel foil. Three drops of the precursor solution were given on the substrate, the water was evaporated and the electrode given in the preheated furnace at 500 °C. After 10 to 20 minutes the electrode was removed and quenched in water. Excess soluble phases were washed out by dipping the electrode in distilled water several times until it was apparently chloride free. The oxide coating withstood gentle rinsing but not sonication.

Electrochemical properties were examined using cyclic voltammetry (CV) experiments under various voltage sweep rates. These were done in a beaker type glass cell with a large platinum sheet as counter electrode and Hg/HgO as reference electrode in the same electrolyte, which was 6 M KOH. All potentials are given with respect to this reference potential. CV driving and recording was done using an EG&G model 273 potentiostat controlled by a PC. Specific capacitances were calculated by integration of a CV half cycle taken at a voltage sweep rate of 20 mV/s.

RESULTS AND DISCUSSION

A cyclic voltammogram of coprecipitated $SrRuO_3$ is shown in fig. 1a. There is a nearly rectangular shape of the curve which indicates almost pure double layer capacitance. At potentials of around +100 mV there is a redox couple noticeable. The integration of the CV yields a specific capacitance of 8 F/g [7].

The charge storage process for the strontium ruthenate perovskites is mainly due to the double layer capacitance. Plotting the mean currents at different voltage sweep rates within the potential limits of the waist region of the CV in fig. 1a yields a straight line (fig. 2). From the slope the true double layer capacitance can be calculated.

For the case of coprecipitated $SrRuO_3$, annealed at 800 °C, a value of 5.55 mF/cm^2 is obtained. Relating this value to the published capacitance of smooth oxidic surfaces of 60 $\mu F/cm^2$ [9, 10] one gets a roughness factor of about 90. The roughness is made by pores, cracks and inter particle spacings which result from the electrode coating process.

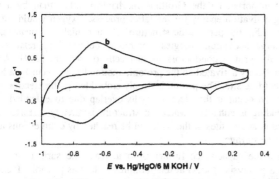

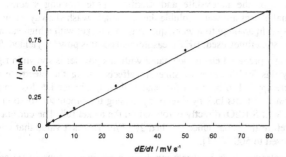

fig. 1: **a**: CV (20 mV/s) of coprecipitated SrRuO$_3$, annealed at 800 °C, 8 F/g, **b**: CV (20 mV/s) of coprecipitated Sr$_{0.8}$La$_{0.2}$RuO$_3$, annealed at 800 °C, 21 F/g

fig. 2: Mean currents caused by different voltage sweep rates, coprecipitated SrRuO$_3$, annealed at 800 °C

The redox couple at +100 mV can be attributed to the transition of Ru(IV) to Ru(V) and/or to Ru(VI). The initial open circuit potential of freshly prepared samples is always in the region of 0 to +20 mV. This is the state where the perovskite B cation Ru is tetravalent. Neither the bivalent A cation Sr nor the oxygen framework is thought to change upon cycling, so Ru is the redox active species. The redox transitions requires counter ions coming from the electrolyte to keep the charge balance. The transition of Ru from IV to V/VI can be described by surface take up of hydroxyl ions from the electrolyte:

$$SrRuO_3 + xOH^- \longrightarrow SrRuO_3(OH)_x + e^-, \qquad x = 1, 2 \tag{1}$$

The reduction to Ru(III) can be formulated as follows:

$$SrRuO_3 + H^+ \text{ (from electrolyte: } H_2O = H^+ + OH^-) + e^- \longrightarrow SrRuO_2OH \tag{2}$$

There is almost no activity noticeable in negative potential regions, so this process must be inhibited. One reason can be that protonic motion within the perovskite structure is inhibited for the undoped case. If 20 % of Sr are replaced by La, which is stable trivalently but of a different ionic radius, one finds well developed redox waves in the mentioned negative potential region (fig. 1b). This can be explained by facilitated protonic motion due to lattice distortion caused by the different ionic radius of the dopant.

The protons must be transported by the Grotthus mechanism rather than by a vehicle mechanism because of lacking activation energy for the latter process. Oxygen would act as the vehicle and can itself migrate within the perovskitic structure via interstitials or vacancies. Both processes require relatively large activation energies, commonly delivered by relatively high temperatures, as is the case for proton conductors in fuel cells or sensors [2, 4]. At room temperature such processes are hardly activated so the protonic motion has to occur by a hopping from an occupied to an adjacent free site. The proton is deeply embedded in the electron shell of the OH⁻ ion, a place change can occur if the electron shells overlap due to extended O–B–O binding bending [4]. This bending is enhanced when the structure is already distorted due to mismatching ionic radii, so redox active sites in the bulk can be reached by counter ions and can so contibute to the overall charge storage.

Another way to increase the capacitance further is to increase the surface area of the powder. This can be done by a pyrolysis process at temperatures as low as possible. Preliminary experiments yielded that a minimum temperature of 500 °C is necessary to form the perovskite phase. At this temperature high surface area powders tend to sinter to quite an extent and therefore this process seems not suitable to obtain large capacitances. One can avoid the sintering by a simultaneous formation of the perovskite and sacrificial pore forming species at high temperatures. If the additional phases are water soluble the can simply washed away leaving pores and cavities within the powder. In analogy to a technique to obtain nickel with a high surface area the term Raney type oxide is sometimes used. BET measurements of this powder yielded 70 m²/g.

A CV of a conventionally prepared electrode coated with this powder is shown in fig. 3a. It is almost featureless and yields 70 F/g. Due to sintering effects during the electrode coating process, wehere 400 °C were applied to burn out the binder, the capacitance has lost an amount of its original value. One can avoid this loss by directy pyrolysing the precursor solution ion the substrate. Fig. 3b shows a CV of SrRuO₃ directly pyrolysed on the nickel foil. The current density is again magnified and therefore the capacitance reached 200 F/g. This is a value that cannot be reached by RuO₂ when exposed to 500 °C [1].

The capacitance of electrodes made by direct pyrolysis is sensitve to subsequent annealing procedures, as shown in fig. 4. The capacitance drops begins at an annealing temperature of 400 °C. Therefore it can be concluded that the charge storage for undoped pyrolysed SrRuO₃ is mainly due to surface effects. Surface area is reduced upon annealing and sintering and therefore the capacitance decreases.

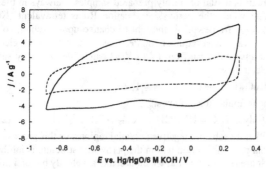

fig. 3: CV (20 mV/s) of pyrolysed and washed SrRuO₃, a: electrode prepared by painting and annealing at 400 °C, 70 F/g, b: electrode prepared by direct pyrolysis on the substrate without subsequent annealing, 200 F/g

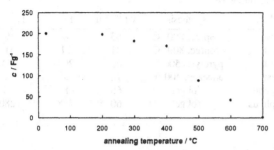

fig. 4: Influence of sintering on the capacitance (CV at 20 mV/s) of pyrolysed SrRuO₃, electrode sequentially annealed at increased temperatures for 1 hour each time.

There is evidence for bulk utilisation though. Powder with a BET surface area of 70 m²/g with a considerable amount of micropores yielded 200 F/g. We assume the same BET surface area for the differently treated samples. The preparation parameters were the same for the BET powder and the direct electrode respectively, 500 C for 15 min with subsequent washing. Due to the fact that micropores with radii of <2 nm do not contribute to double layer capacitance [12] the active surface area must be lower. The value of the measured capacitance is too high as could be explained by 60 µF/cm² surface area of the oxide. Therefore additional redox active sites in the bulk participate in the charge storage. They can be reached by the protons to crystal imperfections due to the pyrolysis process. These defects are thought to be healed during the sintering procedure discussed the paragraph above as well and therefore reducing the measured capacitance.

CONCLUSIONS

Crystalline SrRuO₃ for use in electrochemical capacitors were prepared in two ways, a coprecipitation and a pyrolysis process. The former yielded capacitance values of 8 F/g and the latter, if done directly on the electrode substrate, 200 F/g. A preparation temperature of 500 °C is necessary to obtain the perovskite phase. The obtained capacitances are far larger than for crystalline RuO₂ if treated at 500 °C also. The surface area is increased and maintained due to simultaneous formation of the active and sacrificial pore forming phases, which can easily be removed by washing in water. A comparison of results is summarized in table 1. The effect of reduction of the Ru content while maintaining the capacitive properties can be seen by relating the obtained capacitance to the Ru mass. Further optimisation of this process including introduction of dopants or additives to the precursor solution promise further increase of the capacitive behavior of the perovskite powders. The developed pyrolysis technique is time and energy saving and has the capability to produce electrodes for supercapacitors very economically.

Table 1: Specific capacitances for differently prepared strontium ruthenate perovskites and ruthenium dioxide as reference.

material	crystallinity	synthesis	wt % Ru	F/g	F/g Ru
SrRuO$_3$	cryst.	coprec. 800 °C	43	8	19
La$_{0,2}$Sr$_{0,8}$RuO$_3$	cryst.	coprec. 800 °C	41	21	51
SrRuO$_3$	cryst.	pyrolysis 500 °C	43	200	454
RuO$_2$	cryst.	annealed 400 °C	76	20	26
RuO$_2$ x 2 H$_2$O [2]	amorphous	sol gel	60	485	811
RuO$_2$ x 2 H$_2$O [1]	amorphous	sol gel	60	768	1280

ACKNOWLEDGEMENTS

This work was supported by the German Ministry of Education and Science (BMBF) under contract no. 03N 3007 E7 and was carried out in cooperation with Dornier GmbH, Friedrichshafen, Germany.

REFERENCES

[1]   J.P. Zheng, P.J. Cygan, T.R. Jow, *J. Electrochem. Soc.,* **142** (1995) 2699.

[2]   P. Kurzweil, O. Schmid, *Proceedings of the 6th International Seminar on Double Layer Capacitors and Ssimilar Energy Storage Devices*, Deerfield Beach, Fl, (1996)

[3]   RT.L. Cook, A.F. Sammells, *Solid State Ionics*, **45** (1991) 311.

[4]   K.-D. Kreuer, A. Fuchs, J. Maier, *Solid State Ionics,* **77** (1995) 157.

[5]   F.S. Galasso: *Perovskites and High T$_c$ Superconductors*, Gordon and Breach, New York (1990).

[6]   R. Oesten, T. Guther, J. Garche, *Proceedings of the 6th International Seminar on Double Layer Capacitors and Similar Energy Storage Devices*, Deerfield Beach (1996).

[7]   T. J. Guther, R. Oesten, J. Garche, in *Electrochemical Capacitors II*, F.M. Delnick, D. Ingersoll, X. Andrieu and K. Naoi, editors, PV 96-25, p. 16, the Electrochemical Society Proceeding Series, Pennington, NJ (1997).

[8]   P.M. Wilde, T.J. Guther, R. Oesten, J. Garche, paper # 253 presented at the Joint International Meeting of the Electrochemical Society and the International Society of Electrochemistry, Paris, (1997).

[9]   S. Trasatti: "Physical Electrochemistry of Ceramic Oxides", *Electrochimica Acta*, **36-2** (1991) 225.

[10]   S.P. Singh, R.N. Singh, G. Poillearat, P. Chartier, *Int. J. Hydr. Energy* , **20-3** (1995) 203.

[11]   R. Oesten, R.A. Huggins, *Ionics* **1**, 427 (1995).

[12]   N. Marinzic, *Proceedings of the 6th International Seminar on Double Layer Capacitors and Similar Energy Storage Devices*, Deerfield Beach (1996).

HIGH SURFACE AREA METAL CARBIDE AND METAL NITRIDE ELECTRODES

M. R. WIXOM*, D. J. TARNOWSKI*, J. M. PARKER*, J. Q. LEE*, P.-L. CHEN*, I. SONG*
AND L.T. THOMPSON**
*T/J Technologies, Inc., P.O. Box 2150, Ann Arbor, MI 48106;
**University of Michigan, Department of Chemical Engineering, Ann Arbor, MI 48109-2136

ABSTRACT

Processes for fabricating new high surface area ceramic electrode materials have been developed. These electrode materials have been applied in electrochemical capacitors and related energy storage and conversion devices. Several synthetic approaches have been developed for producing high surface area carbide or nitride active materials. The fabrication methods provide the capability to vary the composition and microstructure of the electrode material. A number of new candidate high surface area electrode materials have been synthesized. Compositional and microstructural information is presented. Electrodes have been evaluated by cyclic voltammetry, chronopotentiometry, and impedance spectroscopy in acidic and basic aqueous electrolyte systems. Single electrode and single cell performance data are presented. Intrinsic properties such as open circuit potential, electrochemical stability and specific capacitance are discussed with respect to electrode composition. The influence on performance of extrinsic factors such as electrode thickness, particle size, and pore structure is also discussed. The performance of these new materials is compared to carbon, with emphasis on advantages with respect to volumetric energy and power density.

INTRODUCTION

Ultracapacitors store charge at the interface between a polarized electrode and an electrolyte. Since the energy storage mechanism is an interfacial phenomenon, high power density ultracapacitors generally require the incorporation of high surface area electrode materials. In addition to possessing high surface areas, ultracapacitor electrode materials should also be low cost, electrically conductive, and electrochemically stable.

Present commercially available ultracapacitors are based on high surface area carbon or ruthenium oxide electrode materials. Ruthenium oxide-based electrodes offer good performance, but ruthenium is expensive and reserves are limited. Ruthenium is probably not practical for high volume applications. Carbon electrodes are based on low cost raw materials, but improvements will be needed to increase volumetric power density and to minimize the need for costly processing steps.

T/J Technologies is developing a new class of high surface area (HSA) electrode materials based on early transition metal nitrides, carbides, and related compounds [1]. These are interstitial compounds with nitrogen or carbon atoms generally occupying a substantial fraction of the octahedral sites in the metal host lattice. Early transition metal nitrides and carbides are metallic conductors possessing significantly higher bulk electronic conductivities than carbon (see Table I). The electrochemical properties of selected metal nitride and carbide electrode materials will be presented and discussed below.

Table I. Densities and Conductivities of Selected Nitrides and Carbides

Benchmark electrodes	Cond.[a]	Nitrides		Cond.[a]	Carbides	Cond.[a]
			ρ (g/cm^3)		ρ (g/cm^3)	
Aluminum	37,000	VN	6.1	5.0	VC	5.8	6.7
RuO$_2$ (xtal)	20.0	ZrN	7.1	73.0	ZrC	6.7	14.0
MMO (film)[b]	0.1	MoN	8.4	140.0	MoC	8.2	10.0
Carbon	<0.4	TaN	16.3	7.4	TaC	13.9	33.0

[a] Conductivities are in (ohm * m)$^{-1}$ x 10^3, measured at room temperature

[b] MMO refers to ruthenium-based mixed metal oxide film electrodes.

Four distinct processes have been developed for producing HSA metal nitride and carbide electrodes. Each process provides a particular combination of advantages and challenges with respect to substrate compatibility, precursor cost, manufacturing scale-up, microstructural control, and compatibility with down-stream device fabrication steps. Proprietary considerations limit the amount of detail provided with regard to each process.

One process is referred to as a temperature programmed reaction (TPR). In the TPR process, a high density precursor is converted into a lower density electrode material. Through careful control of the TPR conditions, the density change is accommodated by crack or pore formation, and we are able to recover a high surface area electrode material. A key advantage of the TPR process is the low cost of the precursors and reactants involved.

A second process uses sol-gel methods to produce high surface area precursors. These precursors are then converted to HSA nitride or carbide electrode materials by controlled thermal processing in the presence of gas phase reactants. As in the TPR process, reaction conditions are carefully controlled to minimize competing densification processes that can reduce surface areas. The sol-gel process affords a high degree of control over microstructure and composition.

A third process has been developed to produce HSA carbide electrode materials. This process may be broadly described as a two-step template process, in which a mesoporous template is created and then converted through solid state reactions to the desired carbide product. Among the advantages of this process are that it is easily scaleable and based on low-cost precursors. The precursors are amenable to casting or extrusion of electrode films.

In the fourth process electrochemical synthesis and deposition methods are used to fabricate HSA nitride electrodes. This process is very flexible with respect to electrode composition, and avoids potentially undesirable compounds as precursors and intermediates.

These processes have been applied to produce over 30 nitrides and carbides based on Group IV-VI metals. Presented here are results for four electrode materials selected from this extensive matrix of processes and constituents.

EXPERIMENT

The methods used to prepare electrode materials are proprietary and described in issued and pending patent applications [1].

Crystalline phase constituents of electrodes were analyzed by x-ray diffraction using a Rigaku DMAX-B x-ray diffractometer with a Digital VAXstation II/GPX for data acquisition and peak identification. Quantititative analysis of metals was determined by neutron activation analysis (NAA) performed at the Ford Nuclear Reactor, Phoenix Memorial Laboratory, at the University of Michigan. Nitrogen contents were determined by CHN analysis services by the University of Michigan Department of Materials Science.

Electrode surface areas and pore sizes were evaluated by sorption analysis. Single point BET measurements of the surface area were performed using a Quantasorb Quantachrome Jr. Pore size distribution measurements were made using a Micromeritics ASAP 2000M sorption analyzer. Density Functional Theory (DFT) models were used to derive pore size distributions and average pore sizes from isotherms. Prior to porosimetry samples were degassed for two hours at 250°C under flowing nitrogen.

Cyclic Voltammetry (CV) and Chronopotentiometric (CP) constant current charge/discharge data were recorded in open cells using EG&G Versastat and EG&G 263 potentiostats controlled by EG&G Model 270 Research Electrochemistry software. Electrochemical Impedance Spectroscopy (EIS) was recorded using the EG&G 263 potentiostat with EG&G 5210 lock-in amplifier and EG&G Model 398 Electrochemical Impedance Software. Capacitances were evaluated at a sweep rate of 20 mV/s, unless another rate is specified, using the relationship C=I/(dV/dt), where C is the capacitance, I is the current, and dV/dt is the sweep rate. Electrochemical stability was evidenced by the absence of irreversible charging or loss of capacitance with repeated cycling (> 50 cycles).

Unless otherwise noted, sulfuric acid and potassium hydroxide electrolytes were prepared using reagent grade materials and 18 MΩ deionized water. Saturated calomel (SCE) reference electrodes were used in aqueous electrolytes. Various current collecting substrates were chosen according to process and electrolyte compatibilities, generally consisting of 99.9+% purity (metals basis) nickel, or titanium or tungsten foils of 25 to 100 μm thickness.

Several prototype devices have been assembled from selected electrode materials. Prototype configurations consisted of single cell devices employing powdered electrode materials consolidated onto current collecting substrates, wetted with electrolyte, and sealed with polymer gaskets. Each cell consisted of a pair of electrodes containing 20 to 100 mg of electrode material. The active electrode material was a disk with a diameter of 0.75". Cells were compression loaded to reduce contributions of interparticle contact resistance to the equivalent series resistance. Prototypes were subjected to several two terminal tests described in DOD-C-29501, including ESR, constant current charge/discharge, self-discharge, and leakage current measurements. The prototype devices were also evaluated by EIS. Prototype cells were fabricated by JME, Inc. (Shaker Heights, OH).

RESULTS AND DISCUSSION

Composition and Microstructure

Figure 1 illustrates the monometallic nitrides and carbides that we have produced to date using each of the four mentioned processes. The figure also includes the highest BET surface area that we have recorded for each material. XRD was used to confirm the identities of the nitride or carbide phases noted in figure 1. In some cases, more than one possible phase of the nitride or carbide were produced, e.g. γ–Mo_2N and δ–MoN. In general, we have been successful in controlling the reaction conditions to recover a single phase product.

The XRD data show peak broadening indicative of small crystallite sizes. Average crystallite sizes were calculated using the Scherrer equation. Crystallite sizes correlated well with particle sizes estimated using the following equation

$$d = 6/S\rho$$

where S is the BET surface area and ρ is the density of the electrode material. For several of the monometallic compounds, slight shifts in the XRD peak positions suggested possible deviations from stoichiometry. This is quite possible for these early transition metal nitrides and carbides as they tend to have phases that are stable over wide compositional ranges.

In addition to single point BET measurements, we have collected multipoint adsorption isotherms for the calculation of pore size distributions using models based on Density Functional Theory. Pore size distributions are influential on the frequency response of porous electrodes. Pore sizes must be large enough to enable mobility of the ionic electrolyte species in response to changes in electrode polarization. In the case of aqueous electrodes, pore sizes of 3 - 10 nm are preferred. Organic electrolyte ions require larger pores, and for these pore sizes of 10 - 30 nm are preferred. Larger pore sizes are thus preferred for electrodes employed in high power devices and/or devices operated with bulkier organic electrolytes. Figure 2 illustrates pore size distributions for two samples of vanadium carbide electrode materials. Markedly different pore size distributions were produced by varying the processing conditions.

carbides

Ti	V	Cr
	AC 253	
Zr	Nb	Mo
B 112	AC 152	
Hf	Ta	W
		ACD 136

nitrides

Ti	V	Cr
BD 243	ABD 90	D 67
Zr	Nb	Mo
BD 127	BD 26	ABD 110
Hf	Ta	W
	BD 54	ABD 76

Figure 1.   High surface area transition metal carbides and nitrides produced by T/J Technologies. Letters A, B, C, and D indicate materials synthesized by TPR, sol-gel, templating, or electrochemical processes, respectively. Highest measured surface areas (m^2/g) are reported in the lower right corner for each product.

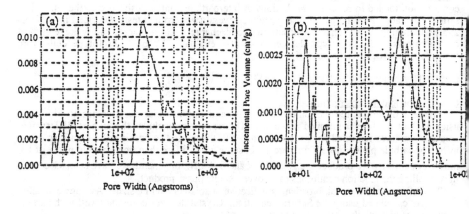

Figure 2.   Pore size distribution analyses of vanadium carbide electrode materials. The PSD for material (a) peaks at 20 nm, and for material (b) at 30 nm. Material (a) also shows a narrower distribution and fewer micropores.

The compositional flexibility of our sol-gel process has enabled us to prepare and evaluate bimetallic carbide and nitride electrode materials. Two key questions are addressed in characterizing these new electrode materials: does the metal ratio in the product reflect the ratios used in the precursor, and are the two species homogeneously distributed, i.e. forming a alloy or solid solution rather than a composite? The ratios of metal constituents in the products were evaluated by NAA. The nitrogen content was determined by CHN analysis. Figure 3 shows XRD data for TiTaN as a representative bimetallic electrode material.

646

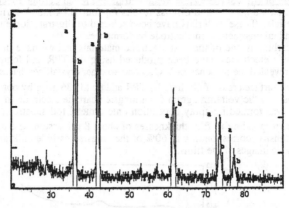

Figure 3. XRD patterns for TiTaN superimposed with peak positions for phase pure TaN (a) and TiN (b).

The XRD data for TiTaN indicates a homogeneous, single phase crystalline product with the absence of peaks associated with either elementary nitride compound. The positions of the reference peaks for the two monometallic species, TiN and TaN are overlaid on the spectrum. The XRD pattern for TiTaN shows the same structure as the monometallic species. Scattering angles for the TiTaN peaks appear between the peaks for TiN and TaN. If a composite crystalline material were formed, two sets of peaks would be expected. Table II presents the data for compositional analysis of several new bimetallic nitrides we have prepared. The Balance wt% indicates the per cent not accounted for as metal atoms or as nitrogen. The balance may indicate presence of atoms such as oxygen, carbon, or hydrogen in the products. NAA data is not yet available for W or Ta, hence their fractions are indicated by "x".

Table II - Precursor and Product Compositions for Bimetallic Electrode Materials

Metal atomic ratio in precursor	Metal atomic ratio in product	Nitrogen wt%	Balance wt%	Composition
V:Ti = 1:1	1.01:1	17.7	11	$VTiN_{1.7}$
V:Ta = 1:1	1:x	8.7	25	$VTa_xN_{2.2}$
Ti:Ta = 1:1	1:x	10.5	(5)	$TiTa_xN_{1.8}$
Ti:Nb = 1:1	0.93:1	14.4	.11	$TiNbN_2$
Ta:W = 1:1	x:x'	4.7	26	$Ta_xW_{x'}N_{1.8}$
Ti:W = 9:1	11.7:1	13.6	19	$Ti_{11.8}WN_{10.8}$
V:Ti = 9:1	4.3:1	19.7	2.3	$V_{4.3}TiN_{4.9}$
V:Ta = 9:1	1:x	15.6	(0.8)	$V_9Ta_xN_{10.7}$

Electrochemistry of High Surface Area Metal Nitride and Metal Carbide Electrodes

In this section we present and discuss results for four HSA metal nitride or metal carbide electrode materials. These materials have been selected to illustrate some of the influences of composition and microstructure on electrode performance.

Tungsten carbide is one of the first electrode materials that we have prepared and studied. Tungsten carbide electrodes have been produced using the TPR and template processes [2]. XRD analysis revealed the presence of WC_{1-x} and no other crystalline phases. Single point BET analysis showed surface areas of 60 m^2/g by TPR and up to 136 m^2/g by our templating method. Figure 4 shows a cyclic voltammogram for a tungsten carbide electrode in 1.0 N sulfuric acid. This electrode was formed by spray deposition onto a metal foil substrate. Film loading was 8.1 mg/cm^2, corresponding to film thicknesses of about 8 μm assuming a density of 10 g/cm^3. The assumed density, corresponding to 60% of the tungsten carbide bulk density, is consistent with micrographic images of the films.

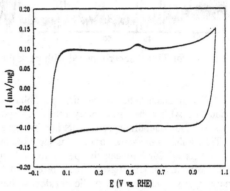

Figure 4.   CV of 9.1 mg tungsten carbide electrode recorded at 20 mV/s in 1.0 M H_2SO_4.

The symmetric rectangular voltammogram for tungsten carbide is indicative of a nearly ideal double layer capacitive response. The voltammetry indicates that the tungsten carbide is electrochemically stable over a > 1V window. There is a small anodic wave at ~500 mV_{NHE} that may be attributable to a surface oxidation process. The process giving rise to this wave appears to be highly reversible.

Vanadium nitride has also been considered as a high energy density capacitor electrode material. We have produced high surface area vanadium nitride by TPR, sol-gel methods, and electrochemical synthesis methods. Electrodes have been formed from these materials by several processes, including spray deposition, electrodeposition, compression molding (with binders), and wet powder casting. Processing methods and film thickness have been shown to influence the energy density and frequency response of these electrode materials.

Figure 5 shows a cyclic voltammogram for a vanadium nitride electrode in 1.0 M KOH electrolyte. The material used in this electrode was evaluated by XRD and VN was the only detected crystalline phase. Single point BET analysis indicated surface area of 60 m^2/g. The most noteworthy feature of the vanadium nitride is the exceptional charge storage capacity (> 200 F/g). Electrodes produced with this material have been cycled 500 times at 20 mV/sec over a 1.0 V window with minimal changes in voltammetry. In contrast to the tungsten carbide results shown in figure 4, the VN voltammogram deviates significantly from a purely double layer response. The voltammogram displays at least two waves at -350 mV_{NHE} and at -800 mV on the anodic sweep. These peaks appear to be shifted by about 250 mV from corresponding waves in the cathodic sweep. We speculate that the process associated with these peaks is a surface confined redox (pseudocapacitive) reaction giving rise to the high charge storage capacity for this material.

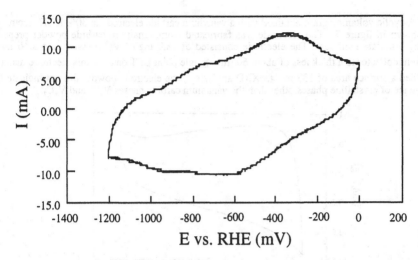

Figure 5. Cyclic voltammogram recorded for a 1.4 mg VN electrode in 1.0 M KOH at a scan rate of 20 mV/s at 25°C.

The kinetic limitations associated with pseudocapacitive processes raise concerns regarding reversibility and possible device power performance. Figure 6 displays the charge/discharge behavior for a VN electrode at various charge/discharge rates. Greater than 2/3 of the capacitance displayed at a charge/discharge rate of 4 mA/mg VN is still available at a charge/discharge rate of 70 mA/mg VN. These results indicate that high energy densities can be achieved under rapid charge/discharge conditions. Much of the loss in capacitance at higher charge/discharge rates may due to ionic diffusion limitations within the porous VN electrode structure. Higher power capacitor electrodes may be produced by utilizing thin film VN electrodes.

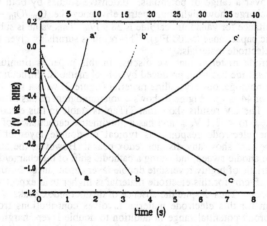

Figure 6. Charge/discharge curves for VN (0.9 mg VN on Ti substrate) in 30% KOH electrolyte over a 1.1 V potential window. Voltages were recorded during galvanostatic applications of 64 mA (a and a'), 32 mA (b and b'), and 4 mA charge/discharge currents. Time scale for 4 mA experiments (c & c') displayed on the upper axis.

A cyclic voltammogram obtained with a vanadium carbide electrode in 30% KOH electrolyte is shown in figure 7. The electrode was fabricated from vanadium carbide powder prepared using template methods. The electrode consisted of 18.7 mg of VC pressed onto a Ni mesh current collector to a thickness of about 60 μm. Single point BET data for this electrode material yielded a surface area of 133 m²/g. XRD analysis of the electrode powder did not indicate the presence of crystalline phases other than the vanadium carbide phases V_4C_3 and V_8C_7.

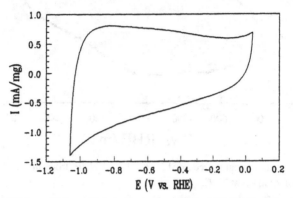

Figure 7. CV for a 18.7 mg VC electrode recorded in 30% KOH at 20 mV/s.

The noteworthy feature of the vanadium carbide electrode voltammogram is the asymmetry. The asymmetry indicates that charge stored during a cathodic sweep is not released at the corresponding potential during an anodic sweep. This is suggestive of a Faradaic charge storage mechanism. Whereas the vanadium nitride electrode CV features suggested the participation of oxidation/reduction processes at discrete potentials, in vanadium carbide these processes appear to be distributed over a range of potentials. Extensive studies by both CV and by constant current charge/discharge show high coulometric efficiency (i.e. $Q_{out}/Q_{in} \sim 1$) for this material. Gravimetric capacitance for vanadium carbide is 38.9 F/g. This value is substantially lower than the VN gravimetric capacitance (>200 F/g). This result is surprising given the otherwise similar nature of the two electrode materials.

The last electrode material that we discuss in this paper is titanium tantalum nitride. Titanium tantalum nitride has been produced by sol-gel methods. Structural and compositional analysis showed a homogenous crystalline material (figure 3) with a stoichiometry of $TiTaN_{1.9}$ and a surface area of 44 m²/g. Figure 8 shows a voltammogram of this material cycled in 1.0 M KOH electrolyte. The CV results show that TiTaN is stable in this electrolyte over potential limits from 0.1 V_{SCE} to < -1.1 V_{SCE} and has a gravimetric capacitance of 21 F/g. During the anodic sweep, the electrode response is typical of double layer charge storage. The voltammogram does not show any distinct redox peaks. However, the cathodic sweep is not symmetric with the anodic sweep, indicating a cathodic shift of the charge/discharge mechanism that is not characteristic of highly reversible double layer capacitance. Furthermore, the specific capacitance of 47 μF/cm² for this electrode material is higher than expected for a double layer charge storage process. Thus despite the absence of distinct redox peaks, we believe the charge storage mechanism for this electrode material involves contributions from a redox process occurring over a broad potential range in addition to double layer charging. As an electrode material, TiTaN appears to be stable in this electrolyte, but a higher capacitance is desired.

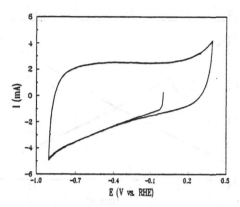

Figure 8.   CV for an 4.6 mg TiTaN electrode recorded at 20 mV/s in 1.0 M KOH.

In comparing the voltammetry of the numerous new nitride and carbide electrode materials that we have prepared, we have observed a high degree of variation in the capacitance and electrochemical stability of these materials. In Table III several of the materials that we have prepared are compared to an optimized carbon electrode material [3]. The column labeled Specific Capacitance gives values for each electrode material calculated by dividing the gravimetric capacitance by the BET surface area. Specific capacitances attributable to double layer charge storage mechanisms in aqueous electrolyte systems are typically 20 $\mu F/cm^2$. Specific capacitances >30 $\mu F/cm^2$ generally indicate Faradaic charge storage mechanisms. The observation of Faradaic charge storage mechanisms is a key attractive feature for these new capacitor electrode materials. Faradaic mechanisms suggest the possibility of substantially higher energy densities than can be attained by exploiting only double layer charge storage.

Table III. Derived Values of Specific Capacitance for Selected Electrode Materials.

	Surface Area (m^2/g)	Electrolyte	Gravimetric Capacitance (F/g)	Specific Cap. ($\mu F/cm^2$)
Mo_2N	98	H_2SO_4	70	71
WC_{1-x}	65	H_2SO_4	17	26
VN	80	KOH	226	283
VC	130	KOH	44	34
TiN	150	H_2SO_4	124	83
TiN	150	KOH	77	51
V_oTaN	62	KOH	89	143
TiTaN	44	H_2SO_4	21	47
C	551	KOH	232	42

<u>Prototype Device Evaluation</u>
Results are presented for a prototype device comprising a VN cathode (34 mg) prepared by the TPR process and a carbon anode (25 mg). Device performance was evaluated according to DOE-INEL protocols [4], including constant current charge-discharge at various rates, leakage current, self-discharge, and EIS. Figure 9 depicts constant current charge-discharge curves between 0 and 1.0 V for the prototype capacitor resulting in discharge times of approximately 3, 30, and 120 seconds. Each cycle showed greater than 95% efficiency (charge in/charge out). This device was cycled more than 1000 times over a 1 V window with no loss in capacitance.

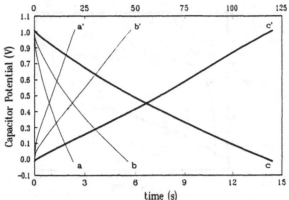

Figure 9. Charge/discharge curves for a prototype capacitor composed of a VN cathode (34 mg) and a C anode (25 mg) with KOH electrolyte. Device was charged and at constant currents of 10 mA (a & a') , 160 mA (b and b'), and 320 mA (c and c'). Time scale for 10 mA experiments (c & c') displayed on the upper axis.

Leakage current (< 15 µA at 1.0 V) and self-discharge tests indicated good performance for the device biased at 1.0 V. Similarly, EIS analysis revealed no evidence of a parallel leakage path at a frequencies down to 1 mHz. The device frequency response is being improved by reducing electrode thickness through the use of thinner, higher capacitance electrode materials.

Figure 10 is a Ragone plot showing the performance of a prototype device using T/J's electrode materials compared to identically designed and fabricated prototype devices using carbon electrodes. The Ragone plot is based on *volumetric* energy and power density. Plotting the volumetric performance illustrates the advantages of metal nitride and carbide electrode

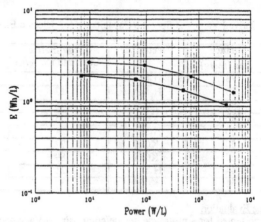

Figure 10. Ragone plots for carbon (25 mg cathode)/carbon (25 mg anode) device (lower curve) and vanadium nitride (34 mg cathode)/carbon (25 mg cathode) devices (upper curve). Data are calculated from EIS spectra obtained at 1 V_{dc}, 10 mV_{ac} between 1 Hz and 1 mHz. Volumetric densities are calculated for active materials only, assuming the density of carbon and vanadium nitride was 0.7 g/cm^3 and 2.5 g/cm^3, respectively.

materials relative to carbon. On a volumetric perspective, the metal nitride and carbide electrodes are favored by their higher bulk density relative to carbon. On the other hand, the volumetric performance of carbon electrode materials is generally limited by subtractive processes such as activation or by the use of fibers or felts that tend to result in macroporous materials. For most customers, improved volumetric power density appears to be the highest weighted selection criterion, hence the volumetric Ragone plot may be more indicative of commercial potential.

By using identical fabrication processes and designs for the devices, and plotting the performance on the basis of active electrode material only, we have attempted to portray a direct comparison of one of our electrode materials with carbon. The logarithmic scale of the Ragone plot diminishes the apparent impact of incorporating VN into a device as just the cathode. Careful inspection shows that vanadium nitride improves the active material energy density by greater than 50% over a similar carbon-carbon device. Further improvements in volumetric energy density will result from employing optimized VN in a similar configuration.

CONCLUSIONS

T/J Technologies has developed a new class of early transition metal nitride and carbide materials are being developed for application as electrochemical capacitor electrodes. Over 30 materials have been prepared as porous electrodes and evaluated to determine surface area, composition, electrochemical stability, and gravimetric capacitance. Several materials showing high capacitance appear to store charge by Faradaic or pseudocapacitive mechanisms. To date, single electrode and prototype device tests have demonstrated volumetric power densities exceeding that available from carbon electrodes. Our expectation is that further development will result in additional new electrode materials and advanced understanding of the charge storage mechanisms. Performance improvements will continue as the rate and reversibility of the charge storage process are increased.

ACKNOWLEDGMENTS

T/J Technologies gratefully acknowledges the financial support provided for this work by the United States Air Force, NASA, the Ballistic Missile Defense Organization, and the Department of Energy. We also thank Charlie Peiter for collecting and formatting data, and Lynne Owens for contributions in developing sol-gel methods used to prepare electrode materials. Researchers at the University of Michigan also acknowledge financial support from the NSF Presidential Young Investigator Program.

REFERENCES

1. Levi T. Thompson, Jr., Michael R. Wixom, and Jeffery M. Parker, U.S. Patent No. 5,680,292 (21 October 1997).
2. M. Wixom, L. Owens, J. Parker, J. Lee, and L.T. Thompson, in Electrochemical Capacitors II, edited by F. M. Delnick, D. Ingersoll, X. Andrieu, and K. Naoi, PV96-025, The Electrochemical Society Proceedings Series, Pennington, NJ (1997).
3. V. Horowitz in addendum to The Fifth International Seminar on Double Layer Capacitors and Similar Energy Storage Devices, Florida Educational Seminars, Inc., Boca Raton, FL (December 4-6, 1995).
4. INEL Report DOE/ID 10491, "Electric Vehicle Capacitor Test Procedure Manual," Rev 0, Department of Energy, October, 1994.

MODIFICATION OF A CARBON ELECTRODE SURFACE BY COLD PLASMA TREATMENT FOR ELECTRIC DOUBLE LAYER CAPACITORS

MASASHI ISHIKAWA*, ATSUSHI SAKAMOTO*, MASAYUKI MORITA*,
YOSHIHARU MATSUDA** AND KOICHI ISHIDA***
*Department of Applied Chemistry and Chemical Engineering, Faculty of Engineering,
Yamaguchi University, 2557 Tokiwadai, Ube 755, Japan
**Department of Applied Chemistry, Faculty of Engineering, Kansai University,
3-3-35 Yamate-cho, Suita 564, Japan
***Industrial Technology Institute, Yamaguchi Prefectural Government,
585-1 Asada-Yugaki, Yamaguchi 753, Japan

ABSTRACT

Activated carbon fiber cloth (ACFC) electrodes whose surface was modified by "pulsed cold plasma", i.e., low-temperature plasma generated by a pulsed electric power in argon-oxygen mixed gas at a reduced pressure, were applied to electric double layer (EDL) capacitors with an organic electrolyte composed of propylene carbonate and tetraethylammonium tetrafluoroborate (TEABF$_4$). The treatment of the ACFC electrodes with the pulsed cold plasma increased total capacitance in the EDL capacitors. The observed increase in the total capacitance was ascribed mainly to an ascertained increase in capacitance of a negative ACFC electrode involving TEA$^+$ cation adsorption/desorption with the cold plasma treatment. No obvious increase in capacitance of a positive ACFC electrode involving BF$_4^-$ anion adsorption/desorption was observed with the plasma treatment. The chemical and electrochemical characteristics of a treated ACFC interface were found to be favorable for TEA$^+$ cation adsorption/desorption.

INTRODUCTION

Electric double layer (EDL) capacitors are typically composed of an organic or aqueous electrolyte and polarizable electrodes represented by activated carbon electrodes whose specific surface area is large.[1-15] Electric charge is stored in an EDL at an interface between an activated carbon electrode and an electrolyte. It has been widely known that EDL capacitor performance is governed by physical properties of the carbon electrodes such as specific surface area and pore size distribution.[1-5,12] Chemical properties of the electrodes, e.g., atomic and functional group distribution at a carbon electrode surface, also affect the capacitor performance.[4,13] To modify such physical and chemical properties of the electrodes, heat, chemical, and electrochemical treatment methods have been proposed.[3,12,14-18]

Recently, treatment with "cold plasma", i.e., low-temperature plasma generated at reduced gas pressure, has been used for modifying various carbon materials.[19-23] This treatment as a dry process can modify surface properties of carbon materials without changing chemical and physical properties of material bulk. We previously reported that total capacitance in EDL capacitors composed of activated carbon fiber cloth (ACFC) electrodes, propylene carbonate (PC) as an organic solvent, and tetraethylammonium tetrafluoroborate (TEABF$_4$) as an electrolytic salt increased with the treatment of the ACFC electrodes with argon-oxygen (Ar-O$_2$) cold plasma.[13] Especially, "pulsed cold plasma", cold plasma generated by a pulsed electric power, was found to improve effectively the total capacitance without a great loss of ACFC electrode weight.[13] The present study focuses on the chemical and electrochemical characterization of a treated ACFC electrode interface to elucidate the origin of an increase in the capacitance with the pulsed cold plasma treatment.

EXPERIMENTAL

ACFC (BW552, Toyobo; weight per geometric area: 180 g m^{-2}, thickness: 0.62 mm, specific surface area: ca. 1370 m^2 g^{-1}) made from a phenolic resin was used for a EDL capacitor electrode with and without the pulsed cold plasma treatment. This plasma treatment of the ACFC was performed in a plasma apparatus (Sinko Seiki, APP-330) as shown in Fig. 1. This apparatus has

655

high-frequency power supply (13.56 MHz) generating a pulsed electric power (10 kW). After the ACFC was introduced into a reaction chamber filled with Ar, this chamber was degassed until 0.1 Torr. The ACFC was then exposed to plasma generated by the pulsed power supply for 30 min.; the pulse on-time and off-time were 0.15 and 34.85 ms, respectively. The supplying gas for the pulsed plasma generation was Ar-O_2 (2 vol.%), Ar-O_2 (6 vol.%), Ar-O_2 (10 vol.%), or Ar-O_2 (20 vol.%); vol.% stands for volumetric percentage of O_2 in the supplying Ar-O_2 gas. Throughout this treatment the Ar-O_2 mixed gas was supplied intermittently into the chamber; the flow rate was 500 cm^3 min.$^{-1}$, the on-time and off-time of the gas supply were 7 ms.

X-ray photoelectron spectroscopy (XPS) was performed to analyze the distribution of atoms and functional groups at the surface of the treated and untreated ACFC. BET method was applied to determine the specific surface area and the pore size at the ACFC surface.

High purity propylene carbonate (PC; Mitsubishi Chemical, Battery Grade) was used as a solvent without further purification. An electrolytic salt, tetraethylammonium tetrafluoroborate (TEABF$_4$), was used after drying under a vacuum at 80 °C for 24 h. Two types of model EDL capacitor were fabricated with Teflon cell cases as shown in Fig. 2a and b; one has two ACFC electrodes with an identical area (diameter: 10 mm) and the other contains two ACFC electrodes with different areas (diameter: 6 mm for a working electrode, 35 mm for a counter electrode). The former, "a symmetric capacitor", as a conventional configuration was used for the total capacitance measurement. On the other hand, the latter, "an asymmetric capacitor", was applied to the determination of single electrode capacitance, i.e., negative or positive electrode capacitance, and to ac impedance analysis as described below. The current collector of the electrode was nickel mesh. A polypropylene non-woven cloth was used as a separator, which was impregnated with the electrolyte solution. The charge-discharge characteristics of the model capacitors were measured under a constant current cycling. The cutoff voltages of the capacitors were 2 V for charging and 1 V for discharging. The ac impedance at a ACFC working electrode interface in the asymmetric model capacitor shown in Fig. 2b was measured by a frequency analyzer (Solartron, 1250) and a potentiostat (Solartron, 1186) at polarization of 1 or -1 V vs. the counter electrode. The ac potential was 10 mVp-p, and the frequency was scanned from 65 kHz to 10 mHz. Capacitor assembly and all capacitor tests were carried out under a dry Ar atmosphere at room temperature (20-25 °C).

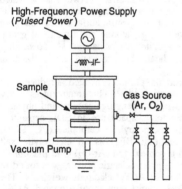

Fig. 1 Apparatus for treatment of ACFCs with pulsed cold plasma.

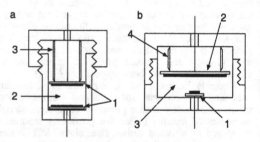

Fig.2 Configurations of model EDL capacitors. a: symmetric capacitor; 1: ACFC electrodes, 2: electrolyte + separator, 3: Teflon spacer. b: asymmetric capacitor; 1: working ACFC electrode, 2: counter ACFC electrode, 3: electrolyte + separator, 4: Teflon spacer.

RESULTS AND DISCUSSION

The ACFCs were treated with the pulsed cold plasma generated under various gas conditions (see Experimental). No obvious change in both the specific surface area and the pore size at the surface of the ACFCs was observed under any plasma conditions; specific surface area: 1370 m^2 g^{-1} (untreated ACFC), 1340-1390 m^2 g^{-1} (treated ACFCs). The radii of most pores were < 50 Å with and without the treatment. A little weight loss of the ACFCs with any plasma treatment was observed (typically ca. 5 %).

To examine the performance of EDL capacitors with the treated ACFC electrodes, charge-discharge cycling test of the symmetric model capacitors including two treated ACFC electrodes with an identical area (Fig. 2a) was curried out at a constant current. An applied electrolyte was PC containing 0.8 M TEABF$_4$. A charge-discharge current density and an operation voltage range were 1.27 mA cm^{-2} of the electrode and between 1 and 2 V, respectively.

Figure 3 shows the discharge capacitance defined as observed discharge capacitance of the model capacitor per total ACFC weight (F g^{-1}) with charge-discharge cycling. The discharge capacitance increased with increasing oxygen content in the Ar-O$_2$ gas for the plasma generation; a highest capacitance was observed at Ar-O$_2$ (20 vol.%) among applied gas conditions.

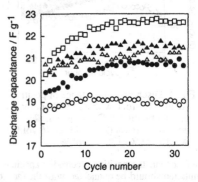

Fig. 3 Discharge capacitance of symmetric capacitor with ACFC electrodes with and without the plasma treatment; ○:untreated; ●: treated in Ar-O$_2$ (2 vol.%), △: (6 vol.%), ▲: (10 vol.%), □: (20 vol.%).

In general, observed total discharge capacitance in EDL capacitors should be affected by both a positive electrode, an ACFC electrode involving anion adsorption/desorption, and a negative electrode, alternative ACFC electrode involving cation adsorption/desorption. Therefore, we investigated the effect of the plasma treatment on each individual capacitance of the negative and positive electrodes to understand which electrode (the positive or negative electrode) is more improved by the plasma treatment. For this investigation, we used the asymmetric model capacitor with a small electrode as a working electrode and a large electrode as a counter electrode (see Fig. 2b). The total capacitance of this model capacitor is to be limited by the working electrode with relatively small area; the total capacitance observed in the capacitor can be regarded as working electrode capacitance. The discharge capacitance in the asymmetric capacitor is, therefore, indicated as observed discharge capacitance per ACFC weight of the working electrode (F g^{-1}).

Figure 4 shows the discharge capacitance of the working electrode as a positive electrode involving BF$_4^-$ anion adsorption/desorption in the asymmetric model capacitor; a charge-discharge current density was 1.27 mA cm^{-2} (of the working electrode), and an operation voltage range was between 1 and 2 V vs. the counter electrode, i.e., positive polarization of the working electrode. The capacitance of the positive ACFC electrodes treated under various plasma conditions was higher than that of the ACFC electrode without the treatment during initial cycles, and declined toward a value of the untreated ACFC electrode with the following cycles. The increase in the capacitance during the initial cycles may be due to removal of contaminants at the ACFC surface by the plasma treatment. However, the ACFC electrodes would lose this activation effect in the following cycles probably owing to the alternative contaminants from the electrolyte. It is, therefore, unlikely that the plasma treatment can essentially improve the efficiency of BF$_4^-$ anion adsorption/desorption at the positive electrode.

On the other hand, when the working electrode was operated as a negative electrode involving TEA$^+$ cation absorption/desorption; an operation voltage range of the working electrode was between -1 and -2 V vs. the counter electrode, the observed discharge capacitance, i.e., the negative electrode capacitance, obviously increased with the plasma treatment as shown in Fig. 5. This suggests that the plasma treatment should improve TEA$^+$ cation adsorption/desorption phenomena at the negative electrode. The order of observed capacitance of the treated negative electrodes corresponds to that of O$_2$ content in the Ar-O$_2$ gas for the plasma generation, and also correlates with that of the total capacitance observed in the symmetric capacitor (Fig. 3). The total cell capacitance in the symmetric capacitor is to be regulated considerably by the negative electrode capacitance, judging from the result that the negative electrode capacitances given in Fig. 5 were

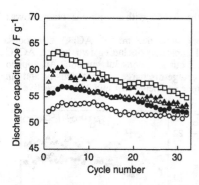

Fig. 4 Discharge capacitance of positive ACFC electrode with and without the plasma treatment; ○: untreated; ●: treated in Ar-O_2 (2 vol.%), △: (6 vol.%), ▲: (10 vol.%), □: (20 vol.%).

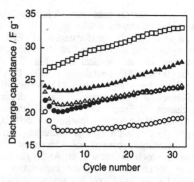

Fig. 5 Discharge capacitance of negative ACFC electrode with and without the plasma treatment; ○: untreated; ●: treated in Ar-O_2 (2 vol.%), △: (6 vol.%), ▲: (10 vol.%), □: (20 vol.%).

smaller than the positive electrode capacitances in Fig. 4 even with the plasma treatment. This limitation should be connected with the fact that Stokes' radius of TEA+ cation as an adsorbate at the negative electrode is larger than that of BF_4^- as an adsorbate at the positive electrode in the applied organic solvent, PC.[24] It is thus reasonable that the order of the symmetric cell capacitance correlated with that of the negative electrode capacitance with and without the plasma treatment. No obvious change in typical physical properties, the specific surface area and the pore size, at the surface of the ACFCs was observed under any plasma conditions as described above. However, it is likely that the chemical properties at the ACFC surface may be changed by the plasma treatment, because the capacitance of the treated ACFC electrode was affected by the O_2 content in the applied gas.

In this context, we focused on the chemical properties at the surface of the ACFC electrodes, i.e., the surface atomic concentration and the functional groups at the carbon fiber surface. The surface atomic concentration of ACFCs treated with the plasma is listed in Table I. Major atoms existing at the surface of ACFC were carbon and oxygen. The amount of the other atoms (P, N etc.) was negligible; these atoms may be derived mainly from contaminations in the plasma chamber. The ratio of oxygen at the ACFC surface increased with an increase in O_2 content in the mixed gas for the plasma generation. On the basis of the XPS analysis, functional groups containing oxygen were composed of carbon-oxygen single and double bond (C-O, C=O), and carboxyl moieties (-COO, -CO_3). As described above, the plasma treatment improved the capacitance of the negative electrode where TEA+ cation adsorption/desorption occurs. It is, therefore, supposed that oxygen in the functional groups can stabilize a negative charge at the negative electrode surface during cell charging, and can increase the amount of the adsorbed cation via the electrostatic interaction between the stabilized negative charge and the cation, resulting in an increase in the negative electrode capacitance and the total cell capacitance. On the other hand, oxygen in the functional groups would be unable to improve the anion adsorption at the positive electrode.

Table I Surface atomic concentration of ACFCs untreated and treated with the cold plasma.

Condition	C (%)	O (%)	P (%)	N (%)	Si (%)
Untreated	94.1	4.02	0.25	-	0.59
Ar-O_2 (2 vol.%)	73.9	23.1	0.44	0.56	0.71
Ar-O_2 (6 vol.%)	72.1	24.8	0.61	0.37	0.82
Ar-O_2 (10 vol.%)	71.8	25.2	0.59	0.30	0.76
Ar-O_2 (20 vol.%)	68.9	26.8	0.54	0.41	1.31

To characterize electrochemically the plasma treatment effect on the ion adsorption process at the electrode interface, ac impedance measurements were performed. Figure 6 shows Cole-Cole plots obtained by the impedance analysis at the positive electrode interface with and without the plasma treatment when the working electrode was positively polarized at 1 V vs. the counter electrode in the asymmetric capacitor. All the positive electrodes with the BF_4^- anion adsorption showed a typical capacitive response. The difference among them is hardly recognized regardless of the plasma treatment. This result is consistent essentially with the result of the positive electrode capacitance tests; no remarkable improvement of the positive electrode capacitance with the plasma treatment was observed (Fig. 4).

On the other hand, the impedance response at the negative electrode polarized at -1 V vs. the counter electrode was changed by the plasma treatment as given in Fig. 7 where a new arc-shaped response appeared on the typical response curve with the plasma treatment. In addition, this arc response disappeared with the following polarization change from -1 to 0 V, suggesting that this response should correspond to a reversible electrochemical process at the negative electrode interface. Thus, this arc response would represent reversible electrostatic interaction between the functional groups containing oxygen at the negative electrode interface and TEA^+ cation. On the basis of the titration tests for the treated ACFCs, the functional groups with oxygen formed by the cold plasma were found to be hardly acidic, whereas acid functional groups with oxygen formed by conventional treatment methods have been known to induce the negative effects, e.g., the degradation of electrolytes and self-discharge.[4] The interaction between the present functional groups and the TEA^+ cation may contribute to the enhancement in the capacitance of the treated negative electrode.

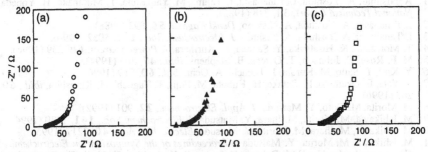

Fig. 6 Impedance response of positive ACFC electrode interface with and without the plasma treatment at 1 V polarization vs. counter electrode; O: untreated; ▲: treated in Ar-O_2 (10 vol.%), □: (20 vol.%).

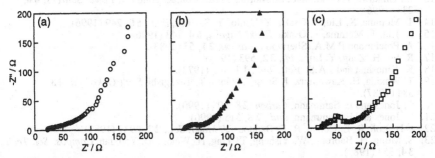

Fig. 7 Impedance response of negative ACFC electrode interface with and without the plasma treatment at -1 V polarization vs. counter electrode; O: untreated; ▲: treated in Ar-O_2 (10 vol.%), □: (20 vol.%).

CONCLUSIONS

The treatment with the pulsed cold plasma generated in the Ar-O$_2$ mixed gas improved the negative electrode capacitance, i.e., the capacity of the TEA$^+$ cation adsorption/desorption at the negative electrode surface, and not the positive electrode capacitance involving BF$_4^-$ anion. The treated negative electrode should regulate the total capacitance in the symmetric capacitor as a conventional configuration, because the negative electrode capacitance was smaller than the positive electrode capacitance even with the plasma treatment. The enhanced capacitance of the negative electrode should be the origin of the observed improvement of the total capacitance. No obvious change in both the specific surface area and the pore size at the surface of the ACFCs was observed under any plasma conditions. On the other hand, the chemical and electrochemical properties at the ACFC interface changed with the treatment; the oxygen-including functional groups formed by the plasma treatment at the electrode surface would affect the efficiency of the TEA$^+$ cation adsorption/desorption process.

ACKNOWLEDGMENT

We are grateful to Toyobo Co., Ltd. for supplying the ACFCs and the measurement of the surface properties of the ACFCs with BET method.

REFERENCES

1. A. Nishino, A. Yoshida, I. Tanahashi, I. Tajima, M. Yamashita, T. Muranaka, H. Yoneda, *National Technical Report*, **31**, 318 (1985).
2. I. Tanahashi, A. Yoshida, A. Nishino, *Denki Kagaku*, **56**, 892 (1988).
3. I. Tanahashi, A. Yoshida, A. Nishino, *J. Electrochem. Soc.*, **137**, 3052 (1990).
4. T. Morimoto, K. Hiratsuka, Y. Sanada, K. Kurihara, *J. Power Sources*, **60**, 239 (1996).
5. M. F. Rose, C. Johnson, T. Owens, B. Stephens, *ibid.*, **47**, 303 (1994).
6. Y. Kibi, T. Saito, M. Kurata, J. Tabuchi, A. Ochi, *ibid.*, **60**, 219 (1996).
7. S. Yata, E. Okamoto, H. Satake, H. Kubota, M. Fujii, T. Taguchi, H. Kinoshita, *ibid.*, **60**, 207 (1996).
8. M. Morita, M. Goto, Y. Matsuda, *J. Appl. Electrochem.*, **22**, 901 (1992).
9. M. Ishikawa, M. Morita, M. Ihara, Y. Matsuda, *J. Electrochem. Soc.*, **141**, 1730 (1994).
10. M. Ishikawa, M. Ihara, M. Morita, Y. Matsuda, *Electrochim. Acta*, **40**, 2217 (1995).
11. M. Ishikawa, M. Morita, Y. Matsuda in *Proceedings of the Symposium on Electrochemical Capacitors II*, edited by F.M. Delnick, D. Ingersoll, X. Andrieu, K. Naoi (The Electrochem. Soc. Inc., Pennington, 1997) Vol. 96-25, p. 325.
12. M. Ishikawa, M. Morita, S. Yamashita ,Y. Matsuda, *Progress in Batteries & Battery Materials*, **13**, 404 (1994).
13. M. Ishikawa, A. Sakamoto, M. Morita, Y. Matsuda, K. Ishida, *J. Power Sources*, **60**, 233 (1996).
14. T. Momma, X. Liu, T. Osaka, Y. Ushio. Y. Sawada, *ibid.*, **60**, 249 (1996).
15. X. Liu, T. Momma, T. Osaka, *Denki Kagaku*, **64**, 143 (1996).
16. A. Proctor and P.M.A. Sherwood, *Carbon*, **21**, 53 (1983).
17. R. Fu, H. Zeng, Y. Lu, *ibid.*, **32**, 593 (1994).
18. K. Kinoshita and J.A.S. Bett, *ibid.*, **11**, 403 (1973).
19. T. Wakida, H. Kawamura, J. Song, T. Goto, T. Takagishi, *Sen'i Gakkaishi*, **43**, 384 (1987).
20. C. Jones and E. Sammann, *Carbon*, **28**, 509 (1990).
21. C. Jones and E. Sammann, *ibid.*, **28**, 515 (1990).
22. I. H. Loh, R. E. Cohen, R. F. Baddour, *J. Mater. Sci.*, **22**, 2937 (1987).
23. S. Mujin, H. Baorong, W. Yisheng, T. Ying, H. Weiqin, D. Youxian, *Compos. Sci. Tech.*, **34**, 353 (1989).
24. Y. Matsuda, H. Nakashima, M. Morita, Y. Takasu, *J. Electrochem. Soc.*, **128**, 2552 (1981).

Experimental Electrochemical Capacitor Test Results

R. B. Wright*, T. C. Murphy*, Susan A. Rogers**, Raymond A. Sutula**
* Idaho National Engineering & Environmental Laboratory, P.O. Box 1625, Idaho Falls, ID
83415-3830, rbw2@inel.gov
**U. S. Department of Energy, Washington, D. C. 20585

ABSTRACT

Various electrochemical capacitors (ultracapacitors) are being developed for hybrid vehicles as candidate power assist devices for the Partnership for a New Generation of Vehicles (PNGV) fast-response engine. The envisioned primary functions of the ultracapacitor are to level the dynamic power loads on the primary propulsion device and recover available energy from regenerative breaking during off-peak power periods. This paper will present test data from selected U.S. Department of Energy (DOE) supported ultracapacitor projects designed to meet the fast response engine requirements.

This paper will address the temperature dependence of test data obtained from a set of three devices provided from Maxwell Energy Products, Inc. These devices are rated at 2300 F at 2.3 V. Constant-current, constant-power, and self-discharge testing as a function of temperature have been conducted. From these tests were determined the capacitance, equivalent series resistance, specific energy and power, and the self-discharge energy loss factor as a function of the device operating temperature.

INTRODUCTION

A number of highly promising materials have been studied for use as electrochemical capacitor electrodes that utilize both Faradaic and non-Faradaic processes for charge separation and storage. These include carbon, conducting doped and un-doped polymers, and metal oxides in conjunction with aqueous and non-aqueous electrolytes. In 1991 the U.S. Department of Energy (DOE) established a program to development and evaluate electrochemical capacitors (ultracapacitors) as an enabling electric/hybrid vehicle technology [1]. Minimum and desired goals for specific energy (6 to 18 Wh/kg) and specific power (800 to 1800 W/kg), and volumetric energy density (9 to 24 Wh/L) were established to meet advanced hybrid vehicle requirements. Identified elsewhere are the major contributors to the program that include national laboratories, universities, and private industry [2]. Coordination of the development, testing and evaluation of prototype capacitors is the responsibility of the Idaho National Engineering and Environmental Laboratory (INEEL). This paper describes the temperature dependence of the test results obtained on prototype capacitors prepared at Maxwell Energy Products, Inc. These devices were tested according to test procedures developed at the INEEL [3].

DISCUSSION OF TEST RESULTS

Maxwell Energy Products, Inc. produced and supplied to the INEEL three devices (designated as #1021, #1022 and #1023) rated at 2300 F at 2.3 V with an average weight of 0.63 kg. These capacitors [4] consist of two porous electrodes isolated from electrical contact by a porous separator. Both the separator and the electrodes were impregnated with a proprietary organic based electrolytic solution. The electrodes used in these capacitors were made from a composite of activated carbon and aluminum. The carbon has an extremely high volumetric surface area that is responsible for the high capacitance of the electrodes. The aluminum matrix is added as a means of reducing the number of carbon/carbon contact points the current must flow through before it reaches the current collecting phase. This decrease in the number of contact points lowers the internal resistance of the capacitor. The aluminum composite structure also minimizes packaging pressure required to achieve the lowest internal resistance in the device. The improved packaging also increases the energy density of the capacitor by reducing the weight of the device. The capacitors were constructed by connecting many single cells in parallel in one individual housing that was fabricated out of aluminum and had overall dimensions of 13.5 cm x 6 cm x 5.7 cm (462 cm^3; 0.462 liters) [4]. The capacitor case consists of two cells, each side acting as a terminal for current collection.

Constant-current tests [3] were performed with discharges ranging from 10 A to 300 A over a voltage range of 3 V to 0.05 V. The tests were conducted at -20°C, 23°C and 55°C using temperature controlled ovens. From these data the capacitance and ESR values were calculated [3] for each of the three devices and the results are plotted in Figures 1 and 2 respectively. At 23°C the capacitance values generally decreased as a function of discharge current and varied from 2686 F to 2792 F at 10 A decreasing to 2312 F to 2440 F at 100 A; capacitor #1023 being the "best" device. At 55°C the capacitance values ranged from 2487 F to 2768 F at 10 A; decreasing to 2307 F to 2337 F at a 100 A discharge. The extent of decrease in the capacitance values with increasing discharge current was not as great as for the 23°C data. Only one of the capacitors, #1022, was tested at -20°C. Its capacitance values decreased from 2337 F at 10 A to 1899 F at 100 A; a greater degree of decrease with increasing current than at the higher temperatures. The ESR values presented in Figure 2 show that the lowest values were generally observed at the higher temperature (55°C) and their dependence on the discharge current increased to a maximum and then decreased again as the discharge current was increased. There was a more pronounced increase and then decrease in the values as the test temperature was lowered.

The constant-power tests were conducted over the range of 10 W to 330 W and also as a function of device temperature. A 10 A constant-current charge was used to charge the capacitors to 3.0 V; the constant-power discharge was varied to discharge the capacitors to a voltage of 1.5 V. Measured values for the specific energy as a function of specific power are shown in Figure 3. The specific energy decreased in a fairly linear manner with increasing specific power. The highest specific energies were observed at 23°C. Increasing the test temperature decreased the values; lowering the temperature to -20°C lowered the values even further and also caused the specific energy to show a greater degree of decrease with increasing specific power than observed for the 23°C and 55°C test data. The rather dramatic effect on the capacitance when decreasing the temperature to -20°C as discussed above is the principal cause for this lowering of the specific energy. These values of specific energy and specific power are below the minimum goals defined for the DOE PNGV Program.

Leakage-current tests which measure the current required to maintain, in this case, a voltage of 3 V on the capacitor were conducted on the three capacitors only at 23°C. The collected test data are shown in Figure 4. Capacitor #1021 is not very good as its leakage current stabilized at 85 milliamperes after ~8 hours while devices #1022 and #1023 have leakage currents of approximately 10 milliamperes after ~21 hours. The corresponding equivalent parallel resistance (EPR) values can be calculated from these data [4]. Devices #1022 and #1023

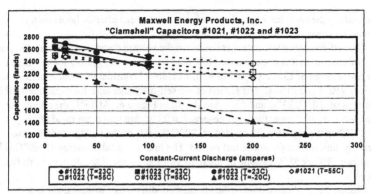

Figure 1. Capacitance values.

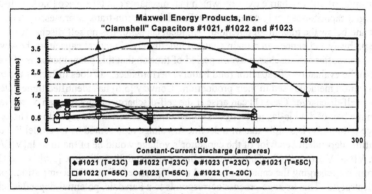

Figure 2. Equivalent Series Resistance (ESR) values.

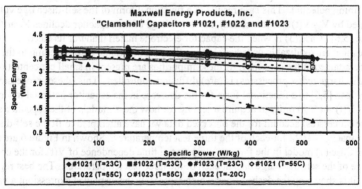

Figure 3. Specific energy and power values from constant-power tests (3 to 1.5 discharge range).

were essentially the same having an EPR value of ~300 ohms after 24 hours on test while that of #1021 was 3.5 ohms.

The self-discharge test data that measures the voltage decrease of the capacitor as a function of time when it is not connected to a load are shown in Figure 5 for the three capacitors at a test temperature of 23 °C. Capacitor #1021 exhibited considerably worse performance than did devices #1022 and #1023. The results of the self-discharge tests for capacitor #1022 at test temperatures of -20 °C, 23 °C and 55 °C are shown in Figure 6. At 55 °C the rate of the voltage decrease was initially slower than that observed at 23 °C, but was more rapid after approximately 19 hours. When the capacitor was tested at -20 °C the rate of the voltage decrease was considerably slower over the entire test period. The total extractable energy at -20 °C is higher than at either 23 °C or 55 °C in spite of the reduced capacitance of the device at this temperature (Figure 3).

Electrochemical capacitors in the charged condition, like batteries, are in a state of high energy relative to that of the system in the discharged state. There is, therefore, a driving force corresponding to the free energy of discharge that tends to diminish the charge if some mechanism(s) of self-discharge exist by which that discharge can take place [5,6]. Self-discharge of a charged capacitor can only occur if some Faradaic electron-transfer process(es) can take place at and below the maximum potential. An ideal capacitor has no self-discharge or current-leakage pathway and hence can remain charged indefinitely. Practical capacitors, however, like batteries, can suffer self-discharge over a period of time so that this phenomenon is of major interest in the evaluation of capacitor (and battery) performance and does influence the materials and fabrication methods used in their production. Conway [5,6] has identified three potential causes of self-discharge. Case (1) can arise if the self-discharge occurs from an ohmic leakage between the pair of double-layer electrodes constituting one cell of the capacitor. The self-discharge in this case is simply that of a decline in voltage represented by a parallel RC circuit so that the time dependent behavior of the capacitor's voltage would be of the form $\ln[V(t)/V_i] = -t/RC$ or, $V(t) = V_i \exp(-t/RC)$. Thus, V(t) declines exponentially with time [5,6]. Case (2) can result from overcharging the capacitor beyond the respective "+" or "-" decomposition potential of the electrolyte. This leakage process corresponds to a Faradaic, potential-dependent charge-transfer reaction. An equivalent circuit representation of the behavior of the electrode interphase or double-layer would be a charge leakage process that corresponds to a potential-dependent Faradaic resistance operating in parallel across the double-layer capacitance; its value increasing with declining potential. This case results in a functional form of the self-discharge voltage represented by $V_i - V(t) \propto \ln(t + \text{const.})$, i.e. the voltage of the capacitor declines logarithmically with time [5,6]. Case (3) is when the capacitor electrode material and/or its electrolyte contain impurities that are oxidizable or reducible over the potential range corresponding to the potential difference attained in the capacitor. If these impurities are present in small concentrations then the self-discharge redox process(es) is (are) diffusion-controlled. This (these) process(es) generally lead [5,6] to a functional form for the decrease in the capacitor potential having the form $V_i - V(t) \propto t^{1/2}$. In order to acquire a feeling for how these models of the capacitor self-discharge process(es) account for the observed data we have attempted to fit the self-discharge data for capacitor #1022 for each of the three test temperatures. Shown in Figure 6 are the best fits to the Case (1) model in the form an exponential time dependence of V(t) for the entire test time range of the self-discharge data at each of the three test temperatures. The best regression analysis fit is shown by the dashed lines which represents the displayed expression; a value for the quality of fit is given by the R^2 value. The shapes of the trend lines do not really correspond to the entire range of the data curves; perhaps if this form of the time dependence was applied

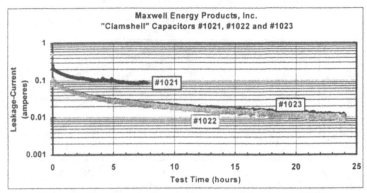

Figure 4. Data from leakage-current test (temperature = 23°C).

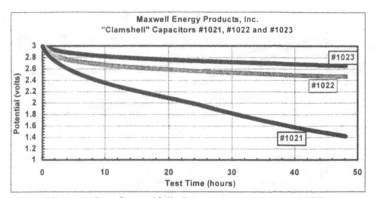

Figure 5. Data from self-discharge test (temperature = 23°C).

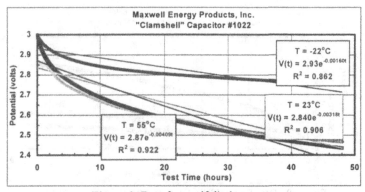

Figure 6. Data from self-discharge test.

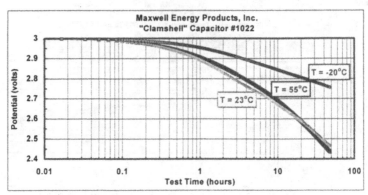

Figure 7. Data from self-discharge test.

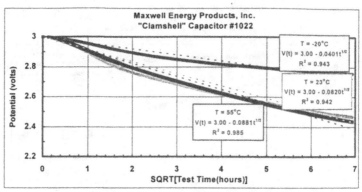

Figure 8. Data from self-discharge test.

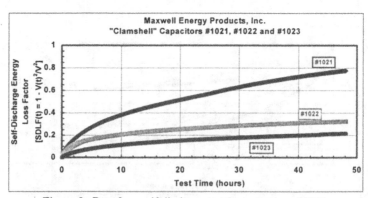

Figure 9. Data from self-discharge test (temperature = 23 °C).

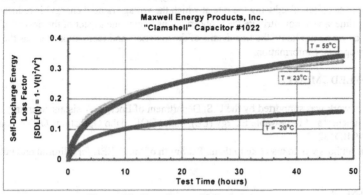

Figure 10. Data from self-discharge test.

only over limited ranges of the data better fits with multiple time constants could be obtained. Similarly, in Figure 7 the Case (2) model is applied to the data in the form of a logarithmic time dependence of the self-discharge data. There perhaps may be certain time regions of the voltage vs. time data, in particular at the longer test times, over which a linear fit could be obtained, but certainly the major portions of the data do not correspond to this model. Figure 8 plots the self-discharge data in form of the Case (3) model that predicts an $\sim t^{\frac{1}{2}}$ time dependence for the capacitor self-discharge voltage. Once again the regression fits to the entire self-discharge data set is not very good, but some correlation may exist at the longer test times. Given the rather poor fits to the data we do not feel it appropriate at this time to discuss possible temperature dependent effects when discussing these results. Perhaps as more capacitor self-discharge data is acquired and analyzed by the these and additional model predictions concerning the time (and temperature) dependence of the self-discharge voltage then a better understanding of this important phenomenon will be obtained.

The self-discharge data can also be used to calculate the self-discharge energy loss factor [4], SDLF(t), that measures how much energy the capacitor looses as a function of time after it has been charged to its working voltage and then left in an open-circuit state. Since the total potential energy stored in a capacitor is equal to $E(t) = \frac{1}{2}CV(t)^2$ where $E(t)$ is the energy in joules after a time t, C is the capacitance of the capacitor in farads, and $V(t)$ is the voltage on the capacitor at a time t after it has been charged to its initial voltage and left in an open-circuit condition. A decrease in the voltage on the capacitor will then have a dramatic affect on its total stored energy. The self-discharge energy loss factor is calculated using: $SDLF(t) = 1 - V(t)^2/V_i^2$ where $V(t)$ is the time dependence of the voltage and V_i (equal to 3 V in this case) is the initial potential of the capacitor. The SDLF(t) functions were calculated from the data displayed in Figures 5 and 6 and are presented in Figures 9 and 10 respectively. The data shown in Figure 9 are for the three capacitors measured at 23°C. Capacitor #1023 only looses $\sim22\%$ of its initial stored energy after 48 hours; capacitor #1022 looses $\sim34\%$ of its total stored energy while capacitor #1021 looses $\sim78\%$ of its stored energy after this same period of time The better the capacitor the slower will be the decrease in its potential as function of time in the open-circuit state. Obviously capacitor #1023 was the "best" of the three tested capacitors. The temperature dependence of the SDLF(t) shows that increasing the temperature causes a slight increase in the rate at which the capacitor's voltage (i.e. stored energy) is reduced. Lowering the temperature has the opposite effect of reducing the rate at which the voltage (stored energy) is reduced. At the present time we cannot attribute these observations to any one aspect of the design of these capacitors as we do not know the fine details of the materials and fabrication steps as they remain Maxwell proprietary information.

ACKNOWLEDGMENT

This work was supported by the U.S. Department of Energy, Assistant Secretary for Energy Efficiency and Renewable Energy (EE), under DOE Idaho Operations Office Contract DE-AC07-94ID13223.

The authors would also like to thank Pat Smith of the INEEL for editorial and publication assistance.

REFERENCES

1.   Office of Transportation Technologies, Energy Efficiency and Renewable Energy, U.S. Department of Energy, *Ultracapacitor Program Plan*, (1994).

2.   A. F. Burke, J. E. Hardin, and E. J. Dowgiallo, "*Applications of Ultracapacitors in Electric Vehicle Propulsion Systems*," Proceedings of the 34th Power Sources Symposium, Cherry Hill, NJ (1990).

3.   J. R. Miller and A. F. Burke, *Electric Vehicle Capacitor Test Procedures Manual, Revision 0*, DOE/ID-10491 (October, 1994).

4.   C. J. Farahmandi, J. Dispennette, and E. Blank, "*High Power 2,300 Farad Ultracapacitors Based on Aluminum/Carbon Electrode Technology*,"Proceedings of the Symposium on Electrochemical Capacitors, F. M. Delnick and M. Tomkiewcz, eds., Proceedings Volume 95-29 (The Electrochemical Society, Inc., Pennington, NJ), p. 187.

5.   B. E. Conway and J. R. Miller, "*Fundamentals and Applications of Electrochemical Capacitors*," manual for an Electrochemical Society Short Course, May 4, 1997, Chapter VIII.

6.   B. E. Conway, T-C Liu, and W. G. Pell, "*Experimental Evaluation and Interpretation of Self-Discharge and Recovery Behavior of RuO_2 and Carbon Electrodes*," Proceedings of the 6th International Seminal on Double Layer Capacitors and Similar Energy Storage Devices, December 9-11, 1996, Volume 6.

GROWTH AND ELECTROCHEMICAL CHARACTERIZATION OF SINGLE PHASE Mo$_x$N FILMS FOR THE FABRICATION OF HYBRID DOUBLE LAYER CAPACITORS

S.L. Roberson*, D. Evans**, D. Finello[†], R.F. Davis
Department of Materials Science & Engineering, North Carolina State University, Raleigh, NC 27695

* U.S. Air Force Palace Knight student attending North Carolina State University
** President, Evans Capacitor Corporation
[†] U.S. Air Force Research Labs, Armament Directorate, Eglin AFB, FL 32542

Corresponding Author: roberson@eglin.af.mil

ABSTRACT

The electrochemical stability and capacitance in H$_2$SO$_4$, KOH, and propylene carbonate of single phase polycrystalline γ-Mo$_2$N and δ-MoN thin film electrodes deposited at 350 and 700° C, respectively, at a rate of 0.5 μm/min on Ti substrates have been determined. The films prepared by chemical vapor deposition at 100 torr using Mo(CO)$_6$ and NH$_3$ flowing at 1.5 standard liters per minute (slm). Cyclic voltammetry referenced to a standard Ag/AgCl electrode indicated that both phases possessed a more positive voltage stability limit in H$_2$SO$_4$ than KOH. Films of γ-Mo$_2$N had voltage stability ranges of -0.3 to 0.6 V and -1.3 to - 0.3 V in 4.4 M H$_2$SO$_4$ and 7.6 M KOH, respectively. Films of δ-MoN possessed voltage stability ranges of -0.3 to 0.7 V and -1.3 to -0.3 V in the same respective electrolytes. Both phases had a voltage stability of approximately one volt in the propylene carbonate electrolyte. These results were used to design and fabricate a hybrid capacitor composed of a Ta/Ta$_2$O$_5$ anode and a δ-MoN cathode contained in an electrolyte of 4.4M H$_2$SO$_4$. The hybrid device had an operating voltage range between 0 and 50 V, a temperature range of -55 to + 90°C, a capacitance of ≈ 5.0 mF and an energy density of ≈ 1.32 J/cm^3. This device is now in pre-production at Evans Capacitor Corporation.

INTRODUCTION

In recent years, the use of high surface area carbon powders [1-5] and RuO$_2$ films [6-11] as electrode materials in high energy density capacitors (supercapacitors) has resulted in the fabrication of low voltage devices with energy densities on the order of 1 J/cm^3 [12-13]. However, the high electrical resistance of carbon and the expense of RuO$_2$ has resulted in the investigation of alternative electrode materials. Much of this research has focused on transition metal carbides and nitrides which are more conductive than carbon and less expensive than RuO$_2$. In these efforts, the surface area necessary for high capacitance was created by conversions of less dense phases of transition metal oxides to more dense phases of M$_x$C$_y$ and M$_x$N$_y$ using temperature programmed reactions with various carbon gases for the former and NH$_3$ or N$_2$+ H$_2$ mixtures for the latter [14-18]. These processes have resulted in films and powders with surface areas to 44 and 200 m^2/g [14, 18], respectively; however, the conversion process can be lengthy.

In the present research, chemical vapor deposition (CVD) of the precursors Mo(CO)$_6$ and NH$_3$ has been employed to grow single-phase γ-Mo$_2$N at 350° C and δ-MoN at 700° C. The electrochemical stability and additional chemical and physical characteristics of each Mo$_x$N phase were determined. Hybrid capacitors containing a Ta/Ta$_2$O$_5$ anode and a δ-MoN cathode in 4.4M H$_2$SO$_4$ were fabricated and characterized.

The functional operation of a hybrid capacitor [13, 19] is described by the equation

$$q = CV \tag{1}$$

where q is the charge in coulombs, C is the capacitance in farads and V is the voltage in volts. When a positive voltage is supplied to the Ta/Ta$_2$O$_5$ anode, a charge develops on the surface of the dielectric due to the movement of ions in the solution. An equal but opposite charge, q$_c$, must be developed on the surface of the cathode for charge conservation. Thus,

Mat. Res. Soc. Symp. Proc. Vol. 496 © 1998 Materials Research Society

$$q_a = q_c \qquad (2)$$

and many combinations of C and V on the cathode will satisfy q_c. As the capacitance of the cathode increases, the voltage required to satisfy both equations 1 and 2 for the cathode decreases and vice versa [19]. However, the overall capacitance of the device is determined by both the Ta/Ta_2O_5 anode and the Mo_xN cathode. These two electrodes add in series as shown in Eq. 3.

$$1/C_T = 1/C_A + 1/C_C \qquad (3)$$

The anode capacitance (C_A) is usually the limiting factor [19].

EXPERIMENTAL

As-received 50 μm thick polycrystalline Ti substrates were cut into 4 cm squares and etched in 1.0 M HCl at 90° C for ten minutes. The substrates were sequentially cleaned in trichloroethylene, acetone, and methanol, rinsed in deionized water, and dried in N_2 before being loaded onto a molybdenum substrate holder contained in a cold-wall, vertical, pancake-style CVD system. The system was evacuated to 10^{-3} torr and backfilled with ultra high purity (UHP) N_2 to 760 torr. This last procedure was repeated five times to reduce the oxygen content in the system. The substrates were heated to the desired deposition temperature in two slm of UHP N_2 by a resistive SiC coated graphite heater. Nitrogen was also used as the carrier gas for the $Mo(CO)_6$ precursor which was contained in a stainless steel metalorganic bubbler and heated to 55° C. Films were deposited at 0.5 μm/minute with 1.5 slm of NH_3 at 350 and 700° C.

The electrochemical evaluations of the Mo_xN electrodes and the hybrid capacitor were conducted using cyclic voltammetry and AC impedance spectroscopy in 4.4M H_2SO_4, 7.6M KOH and propylene carbonate (PC). The former measurements were taken with a Ag/AgCl reference electrode, a platinum counter electrode and a CH Instruments model 800 Electrochemical Detector. Ultra high purity Ar was bubbled through each electrolyte to remove any dissolved oxygen. A Solotron 1250 frequency analyzer and an EEG Model 273 potentiostat were used for the latter studies.

Bulk chemical analyses of the electrode materials were determined from integrated phase ratios and lattice parameters via X-ray diffraction (XRD) using a Rigaku Model A operating at 27.5 kV and 15 A. Surface area measurements were determined by nitrogen adsorption and desorption isotherms in tandem with the BET equation.

RESULTS AND DISCUSSION

ELECTRODE EVALUATION

The successful deposition of single phase γ-Mo_2N and δ-MoN films has allowed the physical and chemical characterization and the determination of the electrochemical stabilities of each phase. The XRD patterns shown in Figure 1 reveal that single-phase randomly oriented γ-Mo_2N and δ-MoN films were deposited at 350 and 700° C, respectively. The broadened peaks were due to the small average grain sizes. The surface areas of these respective films were 27 and 34 m^2/g.

AC impedance spectroscopy was conducted on the γ-Mo_2N films to determine the electrochemical stability in each of the three electrolytes. The plot of the magnitude of the impedance, |Z|, as a function of the frequency presented in Figure 2 shows that the γ-Mo_2N films biased at - 0.5 V and + 0.5 V in 7.6M KOH and 4.4M H_2SO_4, respectively, had a constant ideal capacitance from $\approx$ 1,000 to 0.01 Hz. This is indicated by the slope, which is $\approx$ -1 within this frequency range. The magnitude of the impedance, |Z|, is related to the capacitance and the frequency by the relation

$$|Z| \propto 1/\omega C \qquad (4).$$

In an ideal capacitor |Z| increases as ω decreases. A decrease in |Z| at the lowest frequencies in the 4.4M H₂SO₄ indicated an increase in capacitance and, most likely, an electrochemical reaction on the film. A similar plot for films biased at 1.0 V in PC did not have a slope of -1 and thus did not have ideal capacitance. Although not shown here, similar results were achieved with single phase δ-MoN films.

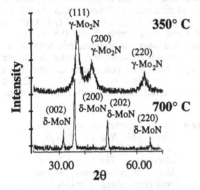

Figure 1. X-ray diffraction patterns of single phase γ-Mo₂N and δ-MoN films deposited at 350 and 700 °C. The peak broadening is due to the small average grain sizes.

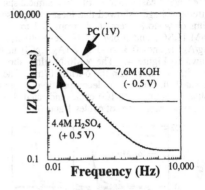

Figure 2. AC impedance spectroscopy of the impedance |Z| as a function of frequency. Films of γ-Mo₂N in 7.6M KOH show ideal capacitor behavior. The γ-Mo₂N films in 4.4 M H₂SO₄ and PC do not show ideal capacitance.

A plot of the current and voltage phase angle as a function of frequency provides an alternative method of evaluating the ideality of the capacitance behavior of an electrode in a given electrolyte. This is calculated using

$$\theta = -\tan^{-1}(Z''/Z') \qquad (5)$$

where Z'' and Z' are the imaginary and real components, respectively, of the impedance. An ideal capacitor will have a current-voltage phase angle shift of -90°. This is shown in Figure 3 for a γ-Mo₂N film biased at + 0.5 V in 4.4M H₂SO₄, - 0.5 V in 7.6M KOH and 1.0 V in PC. The phase angle of the γ-Mo₂N in the 4.4M H₂SO₄ electrolyte approaches ideal behavior to 100 Hz. Below this frequency a decrease in the phase angle indicates that an electrochemical reaction is occurring on the film. The γ-Mo₂N film in 7.6M KOH, approaches ideal capacitance behavior at all frequencies and reaches a maximum of - 87° out of phase at 1 Hz. The capacitance of the γ-Mo₂N film immersed in PC was further from ideal behavior at all frequencies and reached a maximum of -73°

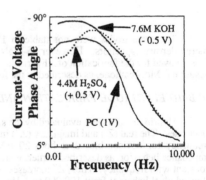

Figure 3. Current-voltage phase angle shifts for γ-Mo₂N films biased at -0.5, +0.5 and +1.0 V in 7.6M KOH, 4.4M H₂SO₄, and PC. Ideal capacitance values were obtained at -90°. Films in KOH approach more closely ideal capacitance behavior than films in either H₂SO₄ or PC.

out of phase at 1 Hz. Similar results were obtained for single-phase δ-MoN films.

Cyclic voltammetry was conducted on both γ-Mo$_2$N and δ-MoN films in 7.6M KOH and 4.4M H$_2$SO$_4$ to determine the individual stabilities of each phase in the two electrolytes. Evaluation in PC was not conducted due to the low capacitance values obtained in the AC impedance measurements. Electrochemical stabilities, indicated by peaks in the voltammograms at extreme voltages, showed that these two phases possessed different stabilities in each electrolyte. Films of γ-Mo$_2$N were stable from -0.3 to + 0.6 V (Ag/AgCl) and -0.3 to -1.3 V (Ag/AgCl) in 4.4M H$_2$SO$_4$ and 7.6M KOH, respectively; δ-MoN films were stable from -0.3 to 0.7 V (Ag/AgCl) and -0.3 to -1.3 V (Ag/AgCl) in the former and latter electrolytes, respectively, as shown in Figure 4. The γ-Mo$_2$N films showed more pronounced oxidation peaks and lower capacitance values than the δ-MoN films. The increasing current with potential in the latter films indicated that either a decomposition or passivation of the films was occurring. Cyclic voltammograms after 100 cycles indicated that the latter occurred, due to an overlapping of the voltammogram traces of cycles 1 and 100.

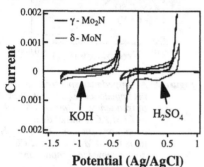

Figure 4. Cyclic voltammograms for single phase γ-Mo$_2$N and δ-MoN films in 4.4M H$_2$SO$_4$ and 7.6M KOH electrolytes. The former had a larger oxidation current peak in both electrolytes. The latter showed an increasing current with potential due to the oxide passivation of the films. Higher capacitance values were obtained for δ-MoN films due to higher surface areas than films of γ-Mo$_2$N.

These electrodes were not stable to 1V, the voltage which is desired to fabricate high energy density capacitors. However, δ-MoN films showed excellent negative voltage stability. This allowed for the fabrication of a hybrid capacitor incorporating a positive Ta/Ta$_2$O$_5$ anode and a negative δ-MoN cathode, as described in the following section.

HYBRID ELECTRODE FABRICATION AND EVALUATION

Hybrid devices were evaluated by the same AC impedance measurements as the individual electrodes. The real (Z′) and imaginary (Z″) impedances are plotted in Figure 5 as a function of the frequency for devices biased at both 20 mV and 1V. These devices show good imaginary impedance behavior, as described earlier, with a slope of -1. This indicates that the capacitance is constant with frequency. Thus, Z″ increases as the frequency decreases. The real impedance, Z′, showed ideal behavior from 100,000 to 1 Hz. The change in the slope at 100 Hz, indicates that the devices were responding to the decreasing frequency and were capacitive. The change in slope at 1 Hz to a less positive value indicates the occurrence of an electrochemical reaction involving the passivation of the surface with an oxide layer. The device behaved more ideally when biased at 1 V and was affected less at low frequencies than the device biased at 20 mV. This was most likely due to the initial evaluation of the device at 20 mV and the subsequent passivation of the surface. When evaluated at 1V, less passivation occurred and the device more closely approached ideality.

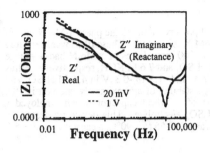

Figure 5. Plot of the Real (Z') and Imaginary (Z") impedance versus frequency for devices biased at 20 mV and 1 V. The latter approaches ideal behavior at low frequencies.

The graph of the real (Z') versus imaginary (Z") impedance also indicates the equivalent series resistance (ESR) of a device [19] and is determined by the intersection of the frequency values with the x or the real axis. The ESR for devices biased at both 20 mV and 1V was 0.042 Ω, as shown in Figure 6. This value is similar to that for commercial devices. The slope of the line at higher frequencies near the real axis is ≈ 45° which is an ideal value for a porous electrode [19]. At lower frequencies, the slope should be and was more vertical. The plot for the device biased at 1 V at high frequencies overlapped the graph for the device biased at 20 mV.

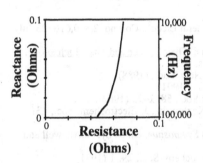

Figure 6. Plot of the Real (resistance) versus Imaginary (reactance) impedance at each frequency value from 100,000 to 10,000 Hz for devices biased at 20 mV and 1 V. The slope of the line at high frequencies (45°) is indicative of a porous electrode and the vertical slope at high frequencies indicates ideal capacitor behavior. The values for 20 mV and 1V overlap.

The plot of the current-voltage phase shift angle as a function of frequency represented in Figure 7 for a device biased at 20 mV and 1V shows that the phase shift angle approaches -90 at ≈ 40 Hz. At lower frequencies the phase angle increased to - 70°; it then decreased to - 85° at lower frequencies due to an electrochemical passivation of the surface. It may be argued that decomposition of the film was occurring; however, the phase angle would have increased to 0°. The device showed a more ideal (approaching -90°) behavior when biased at 1 V. The slight dip at ≈ 1 Hz was due to the continuing passivation of the surface.

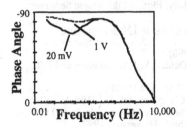

Figure 7. Plot of the voltage-current phase angle versus frequency for devices biased at 20 mV and 1 V. The latter approaches more closely ideal behavior because it approaches -90° over a larger frequency range than does the former. The dip at 1 Hz is due to the passivation of the surface with an oxide.

CONCLUSIONS

The electrochemical stability and capacitance of polycrystalline single phase γ-Mo$_2$N and δ-MoN films have been determined in 4.4M H$_2$SO$_4$, 7.6M KOH and PC. These films were prepared by chemical vapor deposition of Mo(CO)$_6$ and NH$_3$ at 100 torr. Films of γ-Mo$_2$N had a voltage stability of -0.3 to 0.6 V and -1.3 to - 0.3 V in 4.4 M H$_2$SO$_4$ and 7.6 M KOH, respectively. Films of δ-MoN possessed a voltage stability of -0.3 to 0.7 V and -1.3 to -0.3 V in 4.4 M H$_2$SO$_4$ and 7.6 M KOH, respectively. Both phases had a voltage stability of approximately one volt in PC. These results were used to facilitate the fabrication of a hybrid capacitor which employed a Ta/Ta$_2$O$_5$ anode and a δ-MoN cathode in a 4.4M H$_2$SO$_4$ electrolyte. This device possessed an operating voltage between 0 and 50 V, a temperature range of -55 to + 90°C, a capacitance of $\approx$ 5.0 mF and an energy density of $\approx$ 1.32 J/cm^3.

ACKNOWLEDGMENTS

The authors express their appreciation to Dave Evans, President Evans Capacitor Corporation for his diligent efforts and timely fabrication of the hybrid capacitors. One of the authors (SLR) would like to thank the U.S. Air Force Palace Knight Program for their support. R. Davis was partially supported by the Kobe Steel Ltd. Professorship. This work was sponsored by the U.S. Air Force through the Office of Naval Research under contract N00014-97-1-0744.

REFERENCES

1. H.E. Becker and D. Ferry, U.S. Patent (to General Electric Co.) no. 2,800,616 (23 July 1957).
2. B.E. Conway, Electrochemical Supercapacitors: Their Science and Their Technology, Plenum Publishing Corp., New York, in press, 1998.
3. K. Kinoshita, Carbon, John Wiley and Sons, New York, (1988).
4. B.E. Conway, J. Electrochem. Soc., 138, 1539 (1991).
5. B.E. Conway and E. Gileadi, Trans. Faraday Soc., 58, 2493 (1962).
6. S. Hadzi-Jordanov, B.E. Conway and H.A. Kozlowska, J. Electrochem. Soc., 131, 1502 (1984).
7. B.E. Conway, in International Power Sources Symposium, edited by. A. Attewell and T. Keily, Power Sources, 15, 65 (1995).
8. J. Birss, R. Myers and B.E. Conway, J. Electrochem. Soc., 29, 1 (1971).
9. S. Ardizoni, G. Fregonara and S. Trasatti, Electrochim, Acta, 35, 263 (1990).
10. S. Trasatti and G. Buzzanca, Electroanal. Chem., 29, 1 (1971).
11. B.E. Conway, in Proceedings of the Symposium on Electrochemical Capacitors, The Electrochemical Society, Chicago, October 1995, The Electrochemical Society Inc., Pennignton NJ, (1996).
12. C.J. Farahmandi and E. Blank, in Proceedings of the 4th International Seminar on Double Layer Capacitors and Similar Energy Storage Devices, edited by S. Wolsky, Florida Educational Seminars, 1994.
13. D. Evans, in Proceedings of the 6th International Seminar on Double Layer Capacitors and Similar Energy Storage Devices, edited by S. Wolsky, Florida Educational Seminars, 1997.
14. S.L. Roberson, D. Finello and R.F. Davis in Nanophase and Nanocomposite Materials II, edited by S. Komarneni, J.C. Parker, and H.J. Wollenberger (Mater. Res. Soc. Symp. Proc., 457, Pittsburgh, PA, 1996).
15. J. G. Choi, J. R. Brenner, C. W. Colling, B. G. Demczyk, J. L. Dunning and L.T. Thompson, Catal. Today, 15, 201 (1992).
16. R.S. Wise and E.J. Markel, Journal of Catalysis, 145, 344 (1994).
17. L. Volpe and M. Boudart, Journal of Solid State Chemistry, 59, 332 (1985).
18. C.H. Jaggers, J. M. Michaels and A. M. Stacy, Chemistry of Materials, 2, 150 (1990).
19. S.L. Roberson, PhD dissertation, North Carolina State University, 1997.

AUTHOR INDEX

677

SUBJECT INDEX

α-NaFeO$_2$ structure, 403

ab initio, 121
   methods, 65
activated carbon, 655
   electrode, 627
AFM, 587
air electrode, 205
amorphous and nanocrystalline electrode
   hosts, 421
anion diffusion, 485
application growth areas, 3
aqueous chemical synthesis, 409
auxiliary, 359

Ba$_2$In$_2$O$_5$, 193
ball-milled carbons, 563
batteries, 533
battery voltages, 77
beta alumina, 367
bifunctional air electrodes, 43
bipolar
   batteries, 303
   plate, 243
bird's nest, 409
birnessite, 385
breakdown in the conductive network, 25

calorimetry, 551
capacitance, equivalent series, 661
capacitor(s), 15, 637, 669
capacity, 397
carbide, 643
carbon(s), 139, 303, 539, 619
   aerogel, 607
   anode(s), 557
     materials, 613
   corrosion, 43
   fiber, 243
   structure and physical properties, 43
catalyst, 139
cathode(s), 109, 275, 367, 397
cation ordering, 77
ceramic membranes, 167
ceria, 185
charge and discharge efficiency, 25
charged, 613
chemical vapor
   deposition (CVD), 581
   infiltration, 243
coin lithium battery, 57
cold plasma, 655
commercial cokes and graphites, 575
composite(s), 139, 607
   electrolyte, 185
composition-graded electrolyte-electrode
   interface, 155
conducting polymers, 485

conduction, 129
conductivity, 101
constant
   current, 661
   power, 661
contaminants, 463
continuous deposition, 155
copolymers, 485
copper rhodizonate, 263
corrosion, 139, 231
   of the carbon matrix, 43
   phenomena, 57
   process, 57
   rate, 43
crossover, 223
crystal structure stability, 373
cyclic voltammetry, 463

density functional theory, 115
diffusion(-)
   coefficient, 493
   deformation FEM analyses, 51
discharging cycles, 51
disordered, 539, 619
   carbon, 95, 545
double layer, 669

EIS, 435
electric double-layer capacitors, 627, 665
electrical conductivity, 167
electroanalytical response, 435
electrochemical
   behavior, 367
   capacitors, 643
   cell, 275
   characterization, 669
   Li insertion, 385
   potential, 115
electrode, 139, 519
   materials, 257
electrolyte(s), 101
   plate, 211
electrolytic conductivity, 477
endohedral lithium complexes, 95
energy density, 477
EQCM, 587
ESCA, 139
ethylene carbonate, 469
EXAFS, 427
extraction, 391

first-principles
   pseudopotential supercell calculations,
     199
   total energy theory, 77
Fourier transform IR, 415
fuel cell, 139, 231
function analysis, 563

Printed in the United States
By Bookmasters